AF746819

Students, please read this page carefully. Sign it and, for campus students, submit it to your professor or teaching assistant the first week of classes. For distance learning students, please sign it, make a PDF, and send it as an email attachment to your professor the first week of class.

FMS 110: Introduction to New Media
Code of Academic Integrity

You are expected to turn in original work for this course. Quotations or ideas paraphrased from other work must be properly cited. Taking credit for another's ideas or writing is plagiarism, which is a serious violation of the University's Code of Academic Integrity (http://www.asu.edu/studentaffairs/studentlife/srr/index.htm)

In the "Student Academic Integrity Manual," ASU define "plagiarism [as] using another's words, ideas, materials, or work without properly acknowledging and documenting the source. Students are responsible for knowing the rules governing the use of another's work or materials and for acknowledging and documenting the source appropriately."

Academic dishonesty, including inappropriate collaboration, will not be tolerated. There are severe sanctions for cheating, plagiarizing, and any other form of dishonesty.

Your signature indicates your understanding of ASU's Code of Academic Integrity and definition regarding plagiarism.

______________________________ ____________

Name Date

COLLEGE of LIBERAL ARTS & SCIENCES
ARIZONA STATE UNIVERSITY

Dear Student,

Welcome to your custom textbook! This book represents a unique opportunity for the faculty in Film and Media Studies (FMS) and Human Communication (HDSHC) at Arizona State University (ASU) to tailor material to your specific course and to offer it to you at an affordable price. Although designed to support FMS 110: Introduction to New Media, at ASU, we have designed this book to appeal to courses that cover this topic at other universities across the country.

Film and Media Studies and Human Communication professors have taken great care and attention in selecting the most relevant and contemporary readings for each book. Rather than trying to fit a given course around a general text, we have instead planned each textbook so that the chosen readings pertain to specific course lessons. This way, all the learning material for your course, including screenings, lectures, clips, and interactive features, dovetails precisely with your readings.

To flexibly serve all of our students, we have designed the textbook to be used both in-class and online. This is particularly important to our students as the program accrues an ever-larger online profile. We have dozens of online classes either existing or in development to support an online undergraduate major and online graduate programs. We also serve students in other film and media programs across the country and around the world.

Film and Media Studies and Human Communication at ASU have devised this textbook as a way to keep the cost manageable for students. We have been able to assemble a wealth of material in each book; material that acquired independently might have necessitated that the student buy a number of books. The comprehensive nature of our selected readings allows students to buy one reasonably priced book only. Film and Media Studies and Human Communication also give students the option of purchasing the eBook version of this textbook compared to the paperback, further reducing costs and protecting the environment at the same time.

The faculty hopes that you are as excited about the material as we are. Because we have chosen articles and book chapters that we find most interesting and applicable, these books are a joy for us to teach. We hope that you take the same pleasure in reading and studying them as we did in putting them together.

Feel free to contact us at filminfo@asu.edu. To learn more about our program and online offerings, please visit our websites at www.film.asu.edu and http://humancommunication.clas.asu.edu.

Sincerely,

Daniel Bernardi, PhD
Director

Pauline Hope Cheong, PhD
Associate Professor

INTRODUCTION TO NEW MEDIA

Daniel Bernardi and Pauline Hope Cheong, Editors
Arizona State University

Custom Publishing

New York Boston San Francisco
London Toronto Sydney Tokyo Singapore Madrid
Mexico City Munich Paris Cape Town Hong Kong Montreal

Cover photographs courtesy of Getty Images USA, Inc.

Printed in the United States of America

10 9 8 7 6 5 4 3 2 1

2009420020

LR/LP

Pearson Custom Publishing is a division of

PEARSON

www.pearsonhighered.com

ISBN 10: 0-558-24630-3
ISBN 13: 978-0-558-24630-3

Acknowledgments

Grateful acknowledgment is made to the following sources for permission to reprint material copyrighted or controlled by them:

"Social Implications of the Internet," by P. DiMaggio et al., reprinted from *Annual Review of Sociology* 27 (2001), by permission of Annual Reviews, Inc.

"Hypes, Hopes and Actualities: New Digital Cartesianism and Bodies in Cyberspace," by Megan Boler, reprinted from *New Media & Society* 9, no. 1 (2007), by permission of Sage Publications, Inc.

"Why I Am Not Going to Buy a Computer," by Wendell Berry, reprinted from *New England Review and Bread Loaf Quarterly* 10, no. 1 (Autumn 1987).

"Old Rituals for New Space: Rites de Passage and William Gibson's Cultural Model of Cyberspace," by David Tomas, reprinted from *Cyberspace: First Steps*, edited by Michael Benedikt (1992), copyright © Massachusetts Institute of Technology, by permission of The MIT Press.

"The Mother of All Networks," by Tom Standage, reprinted from *The Victorian Internet: The Remarkable Story of the Telegraph and the Nineteenth Century's On-line Pioneers* (1998), Walker & Co.

"The Ecological Impact of Telecommunications," by Itheil de Sola Pool, reprinted from *Technologies without Boundaries: On Telecommunications in a Global Age*, edited by Eli M. Noam (1990), by permission of Harvard University Press.

"The Blurring of Private and Public Behaviors," by J. Meyrowitz, reprinted from *The Impact of Electronic Media on Social Behavior* (1985), Oxford University Press.

"Remembrance of Films Past: Cable Television and Classic Hollywood Cinema," by Barbara Klinger, reprinted from *Beyond Multiplex: Cinema, New Technologies, and the Home* (2006), by permission of University of California Press.

"Personal Networks and the Personal Communication System," by Jeffrey Boase, reprinted from *Information, Communication and Society* 11, no. 4 (2008), Taylor and Francis.

"Localizing the Internet Beyond Communities and Networks," by John Postill, reprinted from *New Media & Society* 10 (2008), by permission of Sage Publications, Inc.

"Multimodal Computer-Mediated Communication and Social Support among older Chinese Internet Users," by Bo Xie, reprinted from *Journal of Computer-Mediated Communications* 13 (2008), International Communication Association.

"Buying into American Idol: How We Are Being Sold on Reality TV," by Henry Jenkins, reprinted from *Convergence Culture: Where Old and New Media Collide* (2006), New York University Press.

"Digital Film," by Stephen Keane, reprinted from *CineTech: Film, Convergence, and New Media* (2007), Palgrave Macmillan.

"Call and Response: Storytelling in the Neverwinter Vault," by Robin Woods, reprinted from *Digital Gameplay: Essays on the Nexus of Game and Gamer*, edited by Nate Garrelts (2005), McFarland & Company, Inc.

"From Barbie to Mortal Kombat: Further Reflections," by Henry Jenkins, reprinted from *New Media: Theories and Practices of Digitextuality*, edited by Anna Everett and John T. Caldwell (2003), Taylor and Francis.

"Image Future," by Lev Manovich.

"Revealing the 'Real' Me, Searching for the 'Actual' You: Presentations of Self on an Internet Dating Site," by M.T. Whitty, reprinted from *Computers in Human Behavior* 24, no. 4 (2007), Elsevier Science, Ltd.

"The Chronicles of Me: Understanding Blogging as a Religious Practice," by Pauline Hope Cheong, Alexander Halavais, and Kyounghee Kwon, reprinted from *Journal of Media and Religion* 7, no. 3 (2008), by permission of Taylor and Francis.

"Feminist Sexualities, Race, and the Internet: An Investigation of Suicidegirls.com," by Shoshana Magnet, reprinted from *New Media & Society* 9 (2007), by permission of Sage Publications, Inc.

"'I'll Take My Stand in Dixie-Net': White Guys, the South, and Cyberspace," by Tara McPherson, reprinted from *Race in Cyberspace*, edited by Beth E. Kolko, Lisa Nakamura, and Gilbert B. Rodman (2000), Taylor and Francis.

"A Digital Inequality: Differences in Young Adults' Use of the Internet," by Eszter Hargittai and Amanda Hinnant, reprinted from *Communication Research* 35 (2008), by permission of Sage Publications, Inc.

"The Young and the Techless? Internet Use and Problem Solving Behaviors among Young Adults in Singapore," by Pauline Hope Cheong, reprinted from *New Media and Society* 10, no. 5 (2008), by permission of Sage Publications, Inc.

"Puerto Rican Language Use on MySpace.com," by Kevin S. Carroll, reprinted by permission from *Centro Journal* 20, no. 1 (2008).

"'Access Denied': Arguments about Equality and Access to New Media in the Information Age," by Stephen Lax, reprinted from *New Media Worlds: Challengers for Convergence*, edited by Virginia Nightingale and Tim Dwyer (2007), Oxford University Press.

"Grooming, Gossip, Facebook and MySpace: What Can We Learn about These Sites from Those Who Won't Assimilate?" by Zeynep Tufekci, reprinted from *Information, Communication, Society* 11, no. 4 (2008), Taylor and Francis.

"Weaving Webs of Faith: Examining Internet Use and Religious Communication among Chinse Protestant Transmigrants," by P.H. Cheong and J.P.H. Poon, reprinted from *Journal of International and Intercultural Communication*, by permission of Taylor and Francis.

"Community and Social Interaction in the Wireless City: Wi-Fi Use in Public and Semi-Public Spaces," by Keith N. Hampton and Neeti Gupta, reprinted from *New Media & Society* 10 (2008), by permission of Sage Publications, Inc.

"The Rise and Fall of a Facebook Hate Group," by Andre Oboler, reprinted from *First Monday* 11, no. 3 (November 2008), by permission of the author.

"Smart Mobs: The Power of the Mobile Many," by H. Rheingold, reprinted from *Smart Mobs: The Next Social Revolution* (2003), by permission of Perseus Books Group.

"Interactive Audiences? The 'Collective Intelligence' of Media Fans," by Henry Jenkins, reprinted from *Fans, Bloggers, and Gamers: Exploring Participatory Culture* (2006), New York University Press.

"The Practices of Internet Networking—A Resource for Alternative Political Movements," by Tobias Olsson, reprinted from *Information, Communication & Society* 11, no. 5 (2008), by permission of Taylor and Francis.

"Net Gains in Political Participation: Secondary Effects of Internet on Community," by Andrea Kavanaugh et al., reprinted from *Information, Communication, & Society* 11, no. 7, by permission of Taylor and Francis.

"ECash," by David D. Friedman, reprinted from *Technology and Freedom in an Uncertain World* (2008), Cambridge University Press.

"Consuming Code: Use-Value, Exchange-Value, and the Role of Virtual Goods in Second Life," by Jennifer Martin, reprinted by permission from *Journal of Virtual Worlds Research* 1, no. 2 (November 2008).

"My Obsession: I Thought I was Immune to the Net. Then I Got Bitten by Ebay," by William Gibson, reprinted from *Wired*, (2006), Conde Nast Publications.

"Fortune Telling on Ebay: Early African American Textual Artifacts and the Marketplace," by Eric Gardner, reprinted from *Everyday Ebay* (2006), Routledge Publishing, Inc.

"Situating Privacy Online: Complex Perceptions and Everyday Practices," by Ana Viseu, Andrew Clement, and Jane Aspinall, reprinted from *Information, Communication & Society* 7, no. 1 (2004), by permission of Taylor and Francis.

"Publicly Private and Privately Public: Social Networks," by Patricia G. Lange, reprinted from *Journal of Computer-Mediated Communication* (2008), International Communication Association.

"Imagined Communities Awareness, Information Sharing, and Privacy on the Facebook," by Alessandro Acquisti and Ralph Gross, reprinted from *Proceedings of the Fifth Workshop on the Economics of Information Security (WEIS)*, June 2006.

"Discussion: MySpace and Deleting Online Predators Act (DOPA)," by danah boyd and Henry Jenkins, May 26, 2006, by permission of danah boyd.

Contents

PART I

Features, Uses, and Cultural Implications of New Media

MODULE ONE

Thinking about New(er)

Social Implications of the Internet

Paul DiMaggio[1], Eszter Hargittai[1], W. Russell Neuman[2], and John P. Robinson[3]

The Internet is a critically important research site for sociologists testing theories of technology diffusion and media effects, particularly because it is a medium uniquely capable of integrating modes of communication and forms of content. Current research tends to focus on the Internet's implications in five domains: 1) inequality (the "digital divide"); 2) community and social capital; 3) political participation; 4) organizations and other economic institutions; and 5) cultural participation and cultural diversity. A recurrent theme across domains is that the Internet tends to complement rather than displace existing media and patterns of behavior. Thus in each domain, utopian claims and dystopic warnings based on extrapolations from technical possibilities have given way to more nuanced and circumscribed understandings of how Internet use adapts to existing patterns, permits certain innovations, and reinforces particular kinds of change. Moreover, in each domain the ultimate social implications of this new technology depend on economic, legal, and policy decisions that are shaping the Internet as it becomes institutionalized. Sociologists need to study the Internet more actively and, particularly, to synthesize research findings on individual user behavior with macroscopic analyses of institutional and political-economic factors that constrain that behavior.

Keywords: World Wide Web, communications, media, technology

Introduction

By "Internet" we refer to the electronic network of networks that links people and information through computers and other digital devices allowing person-to-person communication and information retrieval. Although the late 1960s saw the inception of an ancestral network dedicated to scientific (and, after 1975, military) communication, the Internet did not emerge until 1982; it began its rapid ascent only in the early 1990s, when graphical interfaces became widely available and commercial interests were allowed to participate (Abbate 1999, Castells 2001). Access to and use of the medium diffused widely and swiftly. The number of Americans online grew from 25 million in 1995 (when only 3% of Americans had ever used the Internet) (Pew Research Center for People and the Press 1995) to 83 million in 1999 (Intelli-Quest 1999), with 55 million Americans going online on a typical day in mid-2000 (Howard et al, forthcoming). The amount of information available on the World Wide Web has also risen exponentially, from fewer than 20,000 Web sites in 1995 (Prettejohn 1996) to over 10 million in 2000 (Netcraft 2000), representing over two billion Web pages, with as many as two million pages added daily (Lake 2000).

Our focus in this chapter is on the Internet's implications for social change. The Internet presents researchers with a moving target: Agre (1998a) describes it as "a meta-medium: a set of layered services that make it easy to construct new media with almost any properties one likes." We use Internet to refer both to technical infrastructure (public TCP/IP networks, other large-scale networks like AOL, and foundational protocols), and to uses to which this infrastructure is put (World Wide Web, electronic mail, online multiperson interactive spaces). We focus primarily on general, public uses. Among the topics we do not address systematically are the use of digital technologies for communication *within* formal organizations, the technology's potential contribution to the conduct of social-science research and scholarly communication, or the much broader topic of social antecedents and consequences of computerization.

Many observers allege that the Internet is changing society. Perhaps not surprisingly, given the novelty of the new digital media, there is little agreement about what those changes are. Our purpose here is to summarize research by social scientists about the Internet and to encourage more sociologists to contribute actively to such research. We believe that it is important for sociologists to address these issues for three reasons. First, the medium's rapid growth offers a once-in-a-lifetime opportunity for scholars to test theories of technology diffusion and media effects *during the early stages of a new medium's diffusion and institutionalization.* Second, the Internet is unique because it integrates both different modalities of communication (reciprocal interaction, broadcasting, individual reference-searching, group discussion, person/machine interaction) and different kinds of content (text, video, visual images, audio) in a single medium. This versatility renders plausible claims that the technology will be implicated in many kinds of social change, perhaps more deeply than television or radio. Finally, choices are being made—systems developed, money invested, laws passed, regulations promulgated—that will shape the system's technical and normative structure for decades to come. Many of these choices are based on behavioral assumptions about how people and the Internet interact. We believe such assumptions should represent more than guesswork.

Theoretical Context

Sociology's major theoretical traditions emphasize different aspects of electronic media. For Durkheimians, point-to-point communications media like telephones reinforce organic solidarity, while broadcast media like radio or television yield powerful collective representations (Alexander 1988). Marxists focus upon exploitation of communications media to enhance elite control of both politics and production through cultural hegemony and enhanced surveillance (Schiller 1996, Davis et al 1997). Weberians attend to the ways in which point-to-point media advance rationalization by reducing limits of time and space, and broadcast media provide the elements of distinctive status cultures (Collins 1979).

Other traditions also offer perspectives on the digital media. Technological determinists suggest that structural features of new media induce social change by enabling new forms of communication and cultivating distinctive skills and sensibilities (McLuhan 1967, Eisenstein 1979). In the 1960s, students of social change suggested that in the face of new developments in communications technology, industrial society would yield to the "information society," with consequences in every institutional realm (Machlup 1962, Bell 1973). Critical theorists problematize the effects of technological change on political deliberation and the integrity of civil society (Habermas 1989, Calhoun 1998).

Daniel Bell (1977) appears to have been the first sociologist to write about the social impact of digital communications media themselves. Bell predicted that major social consequences would derive from two related developments: the invention of miniature electronic and optical circuits capable of speeding the flow of information through networks; and the impending integration of computer processing and telecommunications into what Harvard's Anthony Oettinger dubbed "compunications" technology. Anticipating the democratization of electronic mail and telefaxing, as well as digital transmission of newspapers and magazines, Bell explored the policy dilemmas these changes would raise, calling "the social organization of the new 'compunications' technology" the most central issue "for the postindustrial society" (1977:38).

More recently, Manuel Castells has argued that the world is entering an "information age" in which digital information technology "provides the material basis" for the "pervasive expansion" of what he calls "the networking form of organization" in every realm of social structure (1996:468). According to Castells, the Internet's integration of print, oral, and audiovisual modalities into a single system promises an impact on society comparable to that of the alphabet (p. 328), creating new forms of identity and inequality, submerging power in decentered flows, and establishing new forms of social organization.

The comprehensive visions of Bell and Castells, like the other theoretical traditions we have described, suggest a range of empirical questions one must answer to understand the Internet's influence upon society. From the Marxian and Weberian traditions come concerns about power and inequality in the access to the new technology. The Durkheimian perspective sensitizes us to the new media's impact on community and social capital. The work of Habermas and Calhoun leads us to ask how the Internet may alter the practice of politics. The Weberian tradition raises the question of the effect of Internet technology on bureaucracy and economic institutions. Critical theory raises important questions of how the Internet may affect the arts and entertainment media.

We address each of these five topics in turn, summarizing the results of research undertaken by social scientists and other investigators. In most of these areas, the research literature is limited, and many questions remain. But there is a pattern: Early writings projected utopian hopes onto the new technology, eliciting a dystopian response. Research on each topic yields two conclusions. First, the Internet's impact is more limited than either the utopian or dystopian visions suggest. Second, the nature of that impact will vary depending upon how economic actors, government regulation, and users collectively organize the evolving Internet technology.

Major Research Questions

The Internet and Inequality: Opportunity or Reproduction?

Enthusiasts predicted that the Internet would reduce inequality by lowering the cost of information and thus enhancing the ability of low-income men and women to gain human capital, find and compete for good jobs, and otherwise enhance their life chances (Anderson et al 1995). By contrast, cyber-skeptics suggest that the greatest benefits will accrue to high-SES persons, who may use their resources to employ the Internet sooner and more productively than their less privileged peers, and that this tendency would be reinforced by better Internet connections and easier access to social support (DiMaggio & Hargittai 2001).

As in other areas, early research results suggest that the outcome is more complex than either of these predictions, and that the Internet's effects on inequality will depend on the social organization of its use. In this section, we examine research on individual-level inequality among users, as well as cross-national differences in Internet penetration and inequality in effective Internet access for content producers.

The "Digital Divide" in the United States

Anderson et al (1995) were among the first to highlight the potential of inequality in Internet access to limit people's opportunities to find jobs, obtain education, access government information, participate in political dialog, and build networks of social support. By "digital divide," we refer to inequalities in access to the Internet, extent of use, knowledge of search strategies, quality of technical connections and social support, ability to evaluate the quality of information, and diversity of uses. Although some speculate that current intergroup differences will evaporate as the Internet diffuses (Compaine 2000), Schement (1999) points out that inequalities in access to information services (e.g., telephone, cable) tend to persist in contrast to the rapid diffusion of information goods (e.g., radio, television, VCRs) that reach near saturation relatively quickly. This is because the former require ongoing expenditures, whereas the latter are based on one-time purchases. For example, although 94% of all American households have telephones, this figure drops below 80% for the low-income elderly and female-headed households below the poverty level (Schement 1996).

Because sociologists have conducted so little research on the digital divide, to chart the dimensions of inequality we must rely primarily on studies reporting bivariate statistics. Reports of the National Telecommunications and Information Administration (NTIA 1995, 1998, 1999, 2000) documented differences in Internet access favoring the college educated, the wealthy, whites, people under the age of 55 and, especially in earlier years, men and urban dwellers. (Moreover, less affluent and less well-educated users are more likely to become nonusers after trying it out [Katz & Aspden 1997].) Interestingly, despite the focus of early reports on income differences, the impact of educational attainment on Internet use is twice that of income after multivariate controls (Robinson et al 2000b). Research has also found that Internet non-users report as reasons for not going online that they are not computer users, they do not want their children to have Internet access, they lack time or interest, or they cannot afford it (Strover & Straubhaar 2000). There is some evidence that measures of access reflect resource control, whereas measures of intensity of use are driven more by demand. Thus teenagers are less likely to report Internet access than adults between the ages of 25 and 54 (NTIA 1998); but when homes have Internet access, teenagers are online much more than adults (Kraut et al 1996).

Patterns of inequality are likely to reflect such changing factors as public connection availability, private subscription price, services available, and the technology necessary to access them effectively, as well as the diffusion of knowledge and the evolution of informal technical-support networks. Therefore, it is crucial to examine change in inequality over time. Three surveys conducted between 1996 and 1998 found that the gap in access between whites and African Americans had increased over time (Hoffman et al 2000), but NTIA surveys (1998, 2000) found that divide diminishing between 1998 and 2000. Wilhelm (2000) reports that significant differences persist in Internet use among racial and ethnic groups, with socioeconomic status held constant, and he argues that access to telecommunications tools and lack of easy access to Spanish-language content explain lower usage rates among Hispanics. By contrast, broad evidence suggests that two gaps, the advantage of men over women and of the young over the old, have declined as the technology has diffused and become more user-friendly (Roper Starch 1998, Clemente 1998, Bimber 2000, NTIA 2000, Howard et al forthcoming). Other evidence suggests that late adopters have less formal education and lower incomes than earlier cohorts (Howard et al, forthcoming, Katz et al, forthcoming).

Several exemplary studies go beyond description to analysis. In a study notable for its use of multivariate analysis and multiple outcome measures, Bimber (2000) found that the gap between men and women in *access* to the Internet reflected male/female differences in income and other resources; but that women with access used the Internet less frequently than did otherwise similar men, a result he attributed to the fact that full-time employment had a significant effect on frequency of use for men, but not for women. In a study exemplary for tying individual-level inequality to institutional arrangements, Strover (1999) compared dial-up Internet connectivity in four rural US counties, concluding that low levels of commercial investment in telecommunications infrastructure in sparsely populated areas limits use by generating less choice among service providers and higher connection fees.

Other research has focused on public settings that provide Internet access for persons unable to reach the Internet at home or work. A national survey of public libraries reported that urban libraries are almost three times as likely as rural libraries to offer high-speed Internet connections; and that because many urban libraries serve high-poverty areas, access to high-speed connections is relatively available to the urban poor (Bertot & McClure 1998). An evaluation of Internet access programs at two public libraries and two community centers indicated that effectiveness was a function of the extent to which staff were trained to assist Internet users and potential users found the atmosphere welcoming and nonthreatening (Lentz et al 2000). Research on schools, another key site for public access, indicates that the proportion of US public schools offering Internet access rose from

3% in 1994 to 63% in 1999 (US Department of Education 2000), but that training and support staffing necessary for teachers to incorporate the technology effectively in instructional plans has lagged behind (Bolt & Crawford 2000).

Much research and policy assumes that people can convert Internet access into other valued goods, services, and life outcomes. Researchers have not yet tested this premise for Internet access, but research on general computer use sustains its plausibility, while leaving much to be done. Krueger (1993) reported a substantial wage premium accruing to workers who use computers. Attewell & Battle (1999) found that home computer use was significantly related to students' test scores in mathematics and reading, with higher returns for boys, whites, and the well-to-do.

The Global Digital Divide

The number of Internet users globally skyrocketed from 16 million in 1995 to almost 360 million by mid-2000 (NUA 2000a). Despite this rapid diffusion, this number represents just 5% of the world's population. As is the case with other communications devices, access across countries is very uneven, with 97% of Internet host computers located in developed countries (ITU 1998). With respect to content, US producers dominate the Web, creating and hosting a large percentage of the most visited Web sites (OECD 1997) and so establishing English as the Internet's dominant language.

Studies of cross-national variation in levels of Internet connectivity and use are few. Most reports on global Internet diffusion present little more than descriptive statistics, emphasizing correlations with national wealth and education (ITU 1997, 1999, Paltridge & Ypsilanti 1997). Cross-national differences reflect differences in the availability of local-language programming, but not that alone. Hargittai (1996) called attention to institutional factors, reporting that in 1995 three quarters of highly developed countries, but only 10% of LDCs, had commercial access providers (an indicator of private-sector involvement and thus additional impetus for diffusion). Although data quality constrains generalization, the divide between developed and less developed nations appears not to have lessened as the Internet has diffused.

Better data make it possible to analyze Internet diffusion in OECD countries in more detail. Using multivariate analyses of OECD nations, Hargittai (1999) demonstrated that national wealth and competition in the telecommunications sector (and regulatory environments fostering competition) were the strongest predictors of connectivity (see also Guillén & Suarez 2001).

Wilson (2000) distinguishes between "formal access" (physical availability) and "effective access" (affordable connectivity and diffusion of skills people need to benefit from the technology). In-depth case studies help develop this distinction. Rao et al (1999) suggest that lack of local content in native languages in South Asia discourages use. Based on a detailed review of statistics and case reports, Norris (2001) concludes that the Internet is reproducing cross-national inequalities in use of newspapers, telephones, radio, and television because diffusion largely depends on economic development and research and development investments that are unequally distributed across societies.

Yet a case study of Trinidad reports that by 1999 penetration was deep (approximately 30% of households had at least one regular user) and, while stratified by income, relatively broad. The authors attribute this both to Trinidad's comparatively strong communications infrastructure and healthy economy, and equally important, to the premium placed on email by residents of an island nation that exports its most successful young people abroad (Miller & Slater 2000). Technologies shape themselves to the contours of local priorities and ways of life: Just as some less developed countries were vanguard adopters of sound cassettes and cell phones, some may embrace the Internet relatively quickly, especially as wireless transmission creates convergence between Internet and cell phone technologies.

Inequality in Content Providers' Access to Attention

Sociologists should be concerned not only with inequality in access to the Internet, but with inequality in access to the attention of those who use the Internet. By dramatically reducing the cost of the replication and distribution of information, the Internet has the potential to create arenas for more voices than any other previous communication medium by putting product dissemination within the reach of the individual.

Information abundance creates a new problem, however: attention scarcity (Goldhaber 1997). Content creators can only reach large audiences if online gatekeepers—Web services that categorize online information and provide links and search facilities to other sites—channel users to them (Hargittai 2000b). Yet Internet traffic is highly concentrated: 80% of site visits are to just .5% of Web sites (Waxman 2000a). As was the case with broadcast media, the growth and commercialization of the Internet has been accompanied by a commodification of attention. A rapidly evolving mosaic of search engines and point-of-entry sites compete for dominance (NUA 2000a), playing a pivotal role in channeling users' attention toward some contents and away from others (Hargittai 2000b).

During the late 1990s, entrepreneurs developed comprehensive and strongly branded "portals"—Web sites containing search engines, category guides, and various shopping and information services—to match users and content. Such sites now account for one in four of the most visited destinations of the Web (Waxman 2000b). The search engines they feature are often biased in their identification and, especially, ranking of sites in response to user queries (Introna & Nissenbaum 2000). The effects of bias are compounded by the tendency of engine users to employ simple search terms and to satisfice by terminating searches at the first acceptable site. [A 1998 analysis of almost one billion queries on the Altavista search engine revealed that 77% of sessions included but one query and 85% of users viewed only the first screen of search results (Silverstein et al 1998)]. Thus, Web destinations that are displayed prominently on portal sites or ranked high by search engines are likely to monopolize the attention of all but the most sophisticated and committed Internet users. Understanding the processes by which such display opportunities and ranks are awarded is an important research tack.

Research on inequality in access to and use of the Internet—among individual users, groups, organizations, countries, and content creators—should be an important priority for sociologists. At the individual level, the priority should be on using multivariate methods to explore the determinants of different measures of inequality: not just whether or not one has "access," but inequality in location of access (home, work, public facilities); the quality of hardware, software, and connections; skill in using the technology; and access to social support networks. Because inequality reflects the technology's organization, not inherent qualities, special priority should be placed on studies of how inequality is affected by such factors as government programs, industry structure and pricing policies, and approaches to the provision and organization of content.

Impact on Time Use and Community: Social Isolation or Social Capital Formation

Initial enthusiasts anticipated that the Internet would boost efficiency, making people more productive and enabling them to avoid unnecessary transportation by accomplishing online tasks like banking, shopping, library research, even socializing online. The results (less stress, more time, new online contacts) would make individuals more fulfilled and build social capital for society at large. More recently, two studies have suggested that the Internet may induce anomie and erode social capital by enabling users to retreat into an artificial world (Kraut et al 1998, Nie & Erbring 2000). In this section, we explore research on what Internet users do with their time, how the Internet affects their well-being, and how the Internet influences communities, both real and virtual.

Time Displacement

Much of the debate over social capital is about whether the Internet attenuates users' human relationships, or whether it serves to reinforce them. Experience with earlier communications technologies suggests that Internet users may substitute time online for attention to functionally equivalent social and media activities (Weiss 1970). Thus, when television appeared in the United States, it had rapid impact on use of other media: Audiences abandoned their radio sets, movie theaters closed, and general-interest magazines stopped publishing fiction and eventually folded. Early studies documented reductions in time spent going to the movies, listening to radio, and reading fiction as television viewing time increased (Coffin 1954, Bogart 1956). Subsequent research replicated these results cross-nationally and also documented significant declines in out-of-home socializing, in-home conversation, housework, personal care activities, and even sleep (Robinson & Godby 1999).

If television, a unidirectional mass medium, displaced so many activities, then it stands to reason that the Internet, which permits interactive as well as one-way communication, might substitute for even more. Observers have expressed particular concern that Internet users may reduce the time devoted to off-line social interaction and spend less time with print media, as well as with television and other media (Nie & Erbring 2000).

The functional-equivalence model that described the effects of television thus far appears *not* to fit the experience of Internet users. Analyses of 1995 and 1998 national surveys by the Pew Center for the People and the Press, which asked respondents about activities "yesterday," have found Internet use to be unrelated or positively associated with social interaction (Robinson et al 1997, 2000a). Moreover, analysis of 1997 data from the federal Survey of Public Participation in the Arts indicates that Internet users (with appropriate controls) read more literature, attended more arts events, went to more movies, and watched *and* played more sports than comparable nonusers (Robinson & Kestnbaum 1999). A more recent study based on 1998 Pew Center data indicates intriguing changes associated with the Internet's diffusion: Among users who had been early adopters, Internet use was associated with *greater* use of print media. Among new Internet users, however, this relationship had disappeared (Robinson et al 2000b). No significant decline in TV viewing was found after demographic controls. Overall, then, these analyses provide scant support for time displacement due to functional equivalence with respect to other media. (See also Cole 2000, who found lower TV use among Internet users but slightly higher use of other media).

The situation with respect to social interaction is more complicated. Two well-publicized studies reported indications that Internet use substituted for other interactions. Kraut et al (1998), who used a rare longitudinal design to study 169 Pittsburgh-area families who were given computers and Internet connections over a two-year period, reported that higher levels of Internet use were "associated with declines in communication with family members, declines in social circles, and increased loneliness and depression." The authors inferred that heavy users substituted interactions with weak ties on the Internet for time spent with close friends and relatives. Yet as the researchers followed their sample they discovered that, except for increased stress, negative psychological effects decayed to statistical insignificance and some positive outcomes emerged. They attribute these changes to increases in experience and competence and, more speculatively, to the Internet's greater utility in the later period and to a change in sign of network externalities from negative to positive as more of these users' friends and family went online (Kraut et al forthcoming).

An innovative study that used special use-logging software to compare the online behavior of experienced and novice Web users reinforces the notion that the effect of Internet use may vary with user competence. Compared to experienced Internet users, the novices engaged in more aimless surfing, were less successful in finding information, and were more likely to report feeling a souring of affect over the course of their sessions. Their negative reactions reflected not the Internet experience per se but the frustration and sense of impotence of the inexperienced user without immediate access to social support (Neuman et al 1996).

Nie & Erbring (2000) surveyed four thousand Internet users online and asked how the Internet had changed their lives. Most reported no change, but heavier users reported declines in socializing, media use, shopping, and other activities. By contrast, analyses of national (off-line) sample surveys (from both 1995 and 1998) using more fine-grained activity measures indicate that Internet users are no less likely (with controls) to engage in social visiting or to call friends on the telephone. More recent surveys (online *and* off) have revealed that Internet users have *higher* levels of generalized trust and larger social networks than nonusers (Uslaner 1999, Robinson et al 2000b, Hampton & Wellman 2000, Cole 2000). Results from survey analyses also suggest that Internet use serves to complement rather than substitute for print media and offline socialization. Indeed, a detailed time diary study also found Internet users to be no less active media users or offline socializers than nonusers, though they did do less housework, devote less time to family care, and sleep less (Robinson et al 2000b).

COMMUNITY

Wellman (2001) argues that the Internet has contributed to a shift from a group-based to a network-based society that is decoupling community and geographic propinquity, and thus requiring new understandings and operationalizations of the former. Consistent with this insight, Katz et al (forthcoming) report that Internet users visit friends more and talk with them by telephone more frequently, but that they also travel more and have fewer friends in their immediate neighborhoods.

To some extent, whether one views the Internet as corrosive to or supportive of community depends in part on how one evaluates the things people do with it. For example, Nie & Erbring (2000:4) view moderate to heavy-users' self-reported substitution of email for telephone contact as part of their loss of "contact with their social environment." By contrast, Lin (2001) regards online communication, including email, as markedly expanding the stock of social capital.

Indeed, an increasing body of literature suggests that the Internet enhances social ties defined in many ways, often by reinforcing existing behavior patterns. A report on a national survey of users (Howard et al forthcoming) revealed that the Internet puts users in more frequent contact with families and friends, with email being an important avenue of communication. This study also suggests that research on Internet use and social capital should distinguish among different types of Internet use: The Internet seems particularly unlikely to corrode the social capital of women, more of whom than men employ the medium as a complement to other channels of social interaction. Similarly, a longitudinal study by Kraut et al (forthcoming) found that Internet use increased interaction with family members and reported closeness to friends, especially for users whose perceived social support networks were strong *before* they began using the Internet.

The Internet is unique among media in making it easy for people to assemble (at a distance) and communicate with many others at the same time in such settings as chat rooms or online discussion forums. "Online communities" come in very different shapes and sizes, ranging from virtual communities that connect geographically distant people with no prior acquaintance who share similar interests, to settings that facilitate interactions among friendship networks or family members, to community networks that focus on issues relevant to a geographically defined neighborhood (Smith & Kollock 1999, Wellman & Gulia 1999, Preece 2000). Research on "online community" should distinguish among these forms, lest results appear contradictory and confusing.

Early studies tended to focus on online role-playing games [e.g., multi-user dungeons or MUDs (Turkle 1995)] and newsgroups (Hauben & Hauben 1997). These were among the first online communities and are still popular research sites, in part because researchers can obtain full transcripts of discussions and events. Such "online ethnography" has provided useful insights into issues of identity formation (Paccagnella 1997) and the status and concerns of particular groups (e.g., Kolko et al 2000 on race in cyberspace). But as the technology matures, ever smaller percentages of Internet users participate in online games and newsgroups. Increasingly, researchers must follow users into newer kinds of online communities based on shared interests or (physical) community networks.

The number of case studies of online communities is large and growing. Participants value such online settings for making it easy (and inexpensive) to communicate across large distances, providing opportunities for participation by the homebound aged or infirm, and enabling people with minority interests or lifestyles to find companionship and counsel unavailable in their communities of residence (Etzioni & Etzioni 1997). Rheingold's (1993) classic study of an online community emphasized the capacity of online networks to provide their members with social support. And other researchers have noted that, compared to real-life social networks, online communities are more often based on participants' shared interests rather than shared demographic characteristics or mere propinquity (Wellman & Gulia 1999). Nonetheless, issues related to racial, gender, and sexual dynamics do permeate and complicate online interactions [e.g., requiring communities to establish norms for dealing with intimidating or offensive language (Lessig 1999, Silver 2000)].

Whereas some studies focus on "virtual" communities, others explore the impact of the Internet on geographic communities. An exemplary study of a highly wired residential community underscores the importance of examining online interactions in the context of offline everyday life

(Hampton & Wellman 2000). It revealed that Internet users maintain community ties through both computer-mediated communication and face-to-face interaction. Although they maintain more long-distance relationships than do non-Internet users, they communicate even more with their neighbors—and are acquainted with three times as many of their neighbors as are their unwired peers. A study of a similar community revealed that residents make much use of the Internet for "social-capital building activities," but that individual-level community involvement and attachment increased only for residents who were already very active at the experiment's inception (Kavanaugh & Patterson forthcoming). Similarly, a study of scholarly networks found that although the Internet helps maintain contact over long distances, most email contacts are between people who also interact face-to-face (Koku et al 2001). In other words, research suggests that the Internet sustains the bonds of community by complementing, not replacing, other channels of interaction.

Social Capital

Many scholars believe that the Internet facilitates the creation of social capital and other public goods by making information flow more efficiently through residential or professional communities (Lin 2001, Wellman 2001). Yet Putnam (2000) reports that, after demographic controls, Internet users are no different than non-users on measures of civic engagement. He notes, however, that it is premature to project this result onto future user cohorts, and he is agnostic about the Internet's contribution to social capital at the community level. Putnam calls attention to the need to understand qualitative differences between mediated and face-to-face interaction and to explore a tension between the technology's potential and the dangers of unequal access and "cyberbalkanization" (Putnam 2000:177; for an operationalization, see Van Alstyne & Brynjolfsson 1997).

Other studies indicate that, under some circumstances at least, Internet use may enhance social capital. In a longitudinal study of Pittsburgh residents, Kraut et al (forthcoming) found Internet use associated with greater participation in community activities and more trust (though less commitment to remaining in their community), with the positive effects greater for more extroverted participants. An analysis of online survey respondents from the United States, United Kingdom, Canada, and Australia found that increased Internet use tended to have a direct positive effect on social capital (operationalized as participation in community networks and activities) and a positive indirect effect (through social capital) on political participation (Gibson et al 2000).

There is much anecdotal evidence that the Internet provides significant benefits to people with unusual identities or concerns (e.g., rare medical conditions). But there is some evidence that "social capital" produced by less focused networks is rather thin. For example, a survey of users of Amsterdam's "Digital City," a multiuse space created to encourage Internet access and public-spirited interaction, found that, despite soaring membership figures, most users participated relatively infrequently and for recreational purposes (Van den Besselaar & Beckers 1998).

It has also been argued that the Internet builds social capital by enhancing the effectiveness of community-level voluntary associations, but little research evaluates this claim. The Internet has also been described as an inexpensive and effective means of organizing oppositional social movement. Lin (2001) describes the fascinating case of China's Falun Gong organization, which used the Internet to establish a powerful, hierarchical religious movement under the noses of an authoritarian regime. Whether similar movements will follow suit will depend on the success of states in monitoring and controlling such activities.

We draw five morals from the research to date. First, the Internet has no intrinsic effect on social interaction and civic participation. This nonfinding should challenge scholars to understand the circumstances under which different effects are produced, which will doubtless lead them to distinguish different profiles of Web use and different orientations of users. Second, Internet use tends to intensify already existing inclinations toward sociability or community involvement, rather than creating them ab initio. Third, we need to know more than we do about the qualitative character of online relationships. Fourth, we know that virtual communities exist in large number, but we know relatively little about their performance. Research on how virtual communities address problems of commitment and trust [like Kollock's (1999) innovative study of institutionalized reputation on E-Bay and Usenet barter sites] is necessary to understand the limits and possibilities of community online. Fifth, we need more systematic studies of how civic associations and social movements use the Internet, so that we can move beyond single cases to understanding the institutional conditions that encourage or discourage successful exploitation of this technology for collective ends.

Impact on Politics: Renewed Public Sphere or Electronic Battleground?

In the political domain we again find utopians and doomsayers at odds. Enthusiasts find early evidence of a re-engaged, more deliberative, more equitable political community (Browning 1996, Hill & Hughes 1998, Negroponte 1995). Skeptics foresee the re-emergence of an unresponsive commercial sphere dominated by the usual corporate players—but with an increased capacity to invade the privacy of individual citizens (Beniger 1996, Lessig 1999). Most research suggests that effects thus far have been mixed and modest.

Drawing conclusions at such early stages of technology diffusion before the emergence of stable norms is risky because it is difficult to disentangle: 1) the unique characteristics of early adopters from the characteristics of the medium in question; 2) the primitive limitations of the early Web from the technology's mature characteristics; and 3) the Web's explosive growth from other political trends (Rogers 1995, Bimber 1999). As with other topics, the literature about politics on the Internet has progressed through three stages: unjustifiable euphoria, abrupt and equally unjustifiable skepticism, and gradual realization that Web-based human interaction really does have unique and politically significant properties.

An Informed Public

Empirical research on mass political knowledge in industrial democracies, and particularly in the United States, has drawn heavily on the 'information cost' perspective of Downs (1957) and Schumpeter (1947) to explain why the public is so poorly informed. Because it takes time and energy to seek out, interpret, and remember political information, it may be rational to free-ride on the civic attentiveness of others. The political promise of the Internet is that it significantly lowers the behavioral costs of finding, storing, and communicating specific and personally relevant political information at convenient, timely intervals.

The literature reveals, however, that after controlling for education and political interest, there is little evidence of an effect of Internet use on political knowledge. Those who seek political information online are generally well informed to begin with, politically oriented, and heavy users of other media (Bimber 2000, Johnson & Kaye 1998). At present, the Internet supplements and complements rather than replaces traditional sources of political information (Pew 1998, 1999, Robinson et al 2000b). A June 2000 survey revealed that 33% of US adults (and 46% of those under thirty) go online for news at least once a week, compared to 20% in 1998, and 15% they say do so every day. About half say they seek out political news, fewer than report that they look for weather, technology, business, and sports news (Howard et al forthcoming). In some cases they access news not readily available through print or broadcast media, but often the Web is a supplementary medium through which conventional news organizations distribute information available through other means.

An Engaged Public

The economic and psychological dynamics of Web-based human communication, however, are potentially distinct enough from those of traditional print and broadcast news media that in time we may see evidence of an Internet effect. For example, news sites often provide interactive links that encourage users to "send a copy of this article to a friend or colleague." The capacity for horizontal interpersonal communication, to rebroadcast a news article with personal commentary, enhances the capacity for discussion, engagement, and the two-step flow that serves as the critical antidote to anomic mass communication (Kornhauser 1968). Evolving third-voice technologies would permit users to unilaterally convert every mass-medium Web site into an open public discussion (Dibbell 1999). Discussion groups on the Web at present lack the selective, highly edited character of letters to the editor and citizen op-eds. But though they may not achieve the ideal of deliberative discourse envisioned by Habermas (1981, Elster 1998), they would appear to be a step in that direction.

There is great concern about the political malaise and disengagement presumably reflected in low voter turnouts in US national elections. Will reduced costs of gathering political information produce higher voting rates? Probably not, due to the complex and tangled influences of multiple historical, cultural and economic trends, which render bivariate analyses of relationships between media use and electoral participation ill advised. Schudson (1998) points out that US electoral participation rates were highest in the second half of the nineteenth century, when citizens were generally uninformed and uneducated, the media were limited and sensationalistic, and quality of public debate was largely undistinguished. Bimber (2000) argues that political impact derives less from the character of the medium than from the character of information and the day-to-day culture of its use. The successful Jesse Ventura candidacy in Minnesota is widely cited as an example of grass-roots Internet populism; but in that case the Net was primarily used to organize the already engaged, not to mobilize disaffected or uninterested voters (Stromer-Galley 2000). Online financial contributions and voting online by the already politically active may prove more significant in the long run (Mintz 2000).

Political Polarization

Perhaps the most central question for sociological analysis of changing technical structures of interpersonal and mass communication is the tension between forces of social integration and polarization (Neuman 2000). Many fear that the Internet will weaken the cultural center and "political commons" that network television and metropolitan newspapers provided (Neuman 1991, Hirsch 1978). Negroponte, for example, predicts that an artificially intelligent Web-based Daily Me will select news and information based on the predilections and prejudices of the individual cybercitizen and further displace the cultural commons (Negroponte 1995).

Research on earlier media, however, indicates that individuals tend to be aware of the most popular cultural artifacts and to monitor the latest hot programs and motion pictures (Neuman 1991). Ideologically inclined individuals do choose to attend to media that reinforce their prejudices (e.g., conservatives listen to conservatively oriented radio talk shows), but expose themselves as well to opposing views (Freedman & Sears 1965, Frey 1986). The Net's capacity for anonymous communication may heighten the level of extremist and hate speech in the early stages of diffusion. But institutions of self-regulation may emerge to constrain such expression in cyberspace, as they have in nonelectronic public forums (Lessig 1999).

Deliberative Democracy

Web proponents may concede that historically apolitical social strata are unlikely to be mobilized overnight by Internet political content, and agree that there are few signs thus far that the Internet has increased political fragmentation and polarization. But they insist that the Internet will enhance the quality of political discussion and the viability, meaningfulness, and diversity of the public sphere by lowering the access barrier to meaningful public speech. No longer is it necessary to own a newspaper or television station to participate: The Web is a two-way medium, and every Internet receiver can be a publisher as well (Compaine & Gomery 2000, Todreas 1999). Such claims provide critics of commercial (and especially American) dominance of the mass media and the international flow of news and culture with a new focal point for inquiry (Bennett 1995, Bourdieu 1999, Gamham 1990, McChesney 1996, Schiller 1989).

Can the Web make a real difference? It is clear that the Internet significantly lowers entry barriers and other Downsian cost factors for participation in the electronic public sphere. Bimber finds that many of the distortions of group discussion resulting from dominant personalities and group dynamics are reproduced in cyberspace, but he concludes that virtual political space (notably Usenet-style threaded discussion groups) has its place as a significant supplement to, if not replacement for, the face-to-face discussions of Habermas' idealized nineteenth-century salon (Bimber 2000, Hill & Hughes 1998, Schneider 1996). Lowering the economic costs to initiate and sustain an accessible political voice—compare a teenager's bedroom-based Web site to the cost of sustaining a printed magazine or broadcasting facility—can lower access barriers for minority voices, as well.

Skeptics argue that the commercial incentives of advertising-based media may lead ultimately to an Internet culturally indistinguishable from modern commercial television (Davis 1998, Margolis & Resnick 1999, Rheingold 1993). This debate is particularly interesting in the case of Web-based political campaigning in the United States, where by 2000, most candidates had their own Web sites, many with detailed issue and policy information unavailable through traditional media (Schneider 2000b). Will such diverse sites attract sufficient traffic to sustain themselves? Or will dominant commercial portals like AOL or specialized startups like *voter.com* dominate attention, paying for access to the public sphere through political advertising? As of this writing the jury is out, but researchers are actively studying elite and mass behavior (Schneider 2000a).

The Politics of the Internet

A final note: It may be that the battle for control of the Net and for dominance in the electronic marketplace of ideas will prove to be the most fruitful arena for sociological inquiry. The tension between political ideals of openness and the strong economic incentives to sustain and protect scarcity and its corresponding economic return should sustain significant scholarship in this domain for years to come (Lessig 1999, Neuman et al 1998, Shapiro 1999).

Impact on Organizations: Flexible Networks or Panopticons?

Some management writers depict information technology as transforming organizations: replacing hierarchical bureaucracy with flat, networked structures in which local initiative supplants authoritative command; and replacing formal organizations themselves with "network organizations" in which agency is interstitial and strategy constantly renegotiated (Tapscott 1999). Others suggest that digital telecommunications may increase management control by permitting unprecedented degrees of surveillance (Zuboff 1989). In this section, we focus primarily on organizations' use of the public Internet, rather than on communications networks internal to the firm (the use of which is reviewed in Sproull & Kiesler 1991, Wellman et al 1996, and O'Mahony & Barley 1999).

Limits on Internet Impact

Little research bears directly on these claims, and what there is finds limited effects for three reasons. First, authors who make the strongest claims often conflate different types of digital technology, including workplace applications, local area networks, and the Internet. The Internet is less central to some notable organizational trends (e.g., the shrinkage of middle management) than computerization of internal functions (Board on Science, Technology & Economic Policy 1999).

Second, many structural changes associated with the "networked firm" predate the rise of information technologies alleged to have caused them (Powell 2001, Castells 1996). Although some argue that the Internet causes large firms to devolve into loosely integrated production networks by reducing information and transaction costs (Brynjolfsson et al 1994), the move toward network organizations was under way before the Internet became popular. (The Internet, though not determinant, is important. Although network forms emerged in response to competitive environments, new information technologies contributed to their rapid development [Castells 1996].)

Third, technology's effects reflect *not* its inherent potential, as futurists assume, but active choices that are shaped by technology owners' perceived interests, existing organizational structures and routines, and by cultural norms (O'Mahony & Barley 1999, Orlikowski & Iacono 2000). Many traditional firms heavily constrain use of email and the Internet, especially by clerical and service employees, and such firms often implement systems that facilitate surveillance rather than enabling flexible, decentralized interaction (Zuboff 1989, Wellman et al 1996, Frenkel et al 1999).

Telecommuting, once predicted to rise exponentially, is a good example. Of a national sample of 1050 workers interviewed in late 1999, 41% believed they could work effectively from home, but only 10% reported their employers provided that option (and 9% reported doing so at least once a week) (Heldritch Center 2000). Other evidence suggests that most employees use home Internet connections to supplement hours at the workplace, not to substitute for them (O'Mahony & Barley 1999:131).

Work Group Effects

Research on work groups (much of it in laboratory settings) suggests that electronic communications influence interaction style and work flow. Use of electronic mail compared to telephones, for example, enables workers to control the pace of their response and thus facilitates multitasking. Digital conferencing may make employees less risk-averse and render group decision-making less predictable, more time-consuming, and more egalitarian (Sproull & Kiesler 1991, Wellman et al 1996). Whether such effects enhance organizational performance or will persist as the technologies evolve is uncertain, in part because they depend on details of system design and implementation (Sproull & Kiesler 1991, O'Mahony & Barley 1999). In vivo research suggests that formally egalitarian "network" structures may coexist with substantial hierarchy and centralization in patterns of communication (Ahuja & Carley 1998).

Firm Structure Effects

There is little evidence that the Internet is reshaping organizational structures. O'Mahony & Barley report that "whether information technologies further centralization or decentralization" varies depending on how managements uses them. The few empirical studies of the relationships between digital technology and organizational size, buy-or-sell decisions, and organizational boundaries are anecdotal or inconclusive (O'Mahony & Barley 1999:143–45). The Internet may induce change, but we will not know until researchers undertake large-sample studies that specify changes precisely, treat separately different kinds of information technology, and distinguish effects on different kinds of workers and different business functions.

Take, for example, technology's impact on interfirm networks. Most electronic networks complement, rather than substitute for, more intimate media. For many "network organizations" propinquity is crucial in breeding trust and rapport among participants, for example enabling companies in small-firm networks to share information and exchange specialized assets (Harrison 1994). Spatial agglomeration is also central to the success of biotechnology firms (and to venture capitalists who sustain them) (Powell 2001). The most thorough review of technology-transfer research emphasizes the role of "the mobility and activity of technically trained people" over that of impersonal networks (Board on Science, Technology & Economic Policy 1999). Digital telecommunications seem most important for routine transactions (e.g., inventory systems in which multiple firms share a data base) and for communication among knowledge workers accustomed to scientific norms of exchange (e.g., R&D), and least sufficient when interactions entail risk and require interpersonal judgment.

Industry Specific Effects

Studies of specific industries indicate that digital telecommunications can facilitate transformative change when market pressures require it and organizational resources and structures render it possible. For example, when fierce competition and deregulation provoked change and rising demand made it profitable, trucking firms used the Internet (with such other technologies as global positioning) to develop logistics capacity and reposition themselves as transportation-services companies (Nagarajan Bander & White 2000). Hospitals and medical practices have used the Internet to pool information across entities, enabling the emergence of the "integrated healthcare systems" that are transforming many regional healthcare markets (Scott et al 2000, Starr 1997). Other industries that have used Internet technology to effect significant change are banking and financial services (Rochlin 1997) and, with distance learning, higher education (Brown & Duguid 2000:25). In each case, firms adapted the technology to specific strategies, rather than yielding to general technological imperatives.

Thus, the Internet is implicated in profound changes in organizational structures, practices, and strategies. But the extent and nature of these changes—which business functions they restructure, which employees they affect—vary markedly by industry. And rather than causing change, digital technologies are ordinarily pressed into the service of developments to which managers are already committed. The area is ripe for both organizational case studies that focus on the Internet's use in particular industries and organizational surveys that permit confident generalization.

The Internet is also implicated in organizational change in the public sector, where enthusiasts have hailed its potential for saving tax dollars, reducing red tape, and making government more responsive. In an empirically detailed and theoretically sophisticated study, Fountain (2001a,b) has demonstrated both the potential and impediments to its realization. Implementing digital technology saves governments money, but how much depends on network externalities. It reduces some aspects of bureaucratic rigidity but strengthens others by embedding them in code. It enhances the flow of information to citizens and enables government workers to cut through red tape, but in expanding the latter's discretion, it risks imposing new forms of inequality among citizens in their relations with the state (Fountain 2001a,b).

Impact on Culture: Bountiful Diversity, Hypersegmentation, or Massification?

Many sociologists feared that the original mass media (general-interest magazines, radio, and television) would inexorably "massify" taste, as profit-seeking firms produced only those homogeneous and banal programs or texts with the greatest audience appeal (Shils 1963). Since 1980, changes in consumer demand have combined with new media technologies to segment markets and differentiate cultural goods, enabling individuals and groups to individualize their media habits. As an "interconnected network of audio, video, and electronic text communication that will blur the distinction between interpersonal and mass communications and between public and private communications" (Neuman 1991:12), the Internet seems designed to take these trends to their logical conclusion.

Not surprisingly, early observers viewed the new technology as profoundly liberating, opening up outlets for the creative energies of people of every taste and persuasion (Barlow 1996). Because posting information on the Web is so inexpensive, the technology's enthusiasts believed it would virtually eliminate barriers to entry in fields like music recording, book publishing, and even filmmaking. In this view, the Internet would democratize the flow of information, supplanting top-down dependence on traditional news and media organizations with bottom-up sharing among consumers themselves.

Such optimistic scenarios assume that the Internet's only impact is a direct one on costs (of cultural goods to consumers and of publication to producers). But a second, perhaps more important, effect of the Internet may be to induce the restructuring of the culture industries themselves. When goods are distributed on the Internet, they can be repackaged in many ways: newspapers, for example, can be disassembled, their parts distributed separately; recorded music can come with more textual documentation than will fit in a CD jewel box. New distribution systems may also alter the size distribution of firms within industries, the relative power of gatekeepers and artists, and the nature of competitive strategies. The Web's earliest cultural impact has been in the music industry, where it has reinforced existing trends toward deconcentration, product differentiation, and the multiplication of market channels (Dowd 2000, Caves 2000).

Some observers suggest that economic imperatives will keep the Internet from realizing its technical potential as a font of cultural abundance (Neuman 1991, Castells 1996). True, barriers to entry are formally lower; but savage competition for users' limited attention may erect new barriers based on investments in marketing and production. The major media producers are developing the Internet commercially after the model of earlier media (albeit with more interactivity), with the expectation that Internet content, broadcast entertainment, and news will soon enter homes through a single system (Castells 1996, but see Owen 1999).

A third position holds that corporate power will overwhelm the Web's liberating potential by radically accelerating long-term trends toward narrower market segmentation and more fine-grained product customization. In this view, Web sites' ability to use "cookies" to track users' browsing habits provides an unprecedented opportunity for targeting appeals. Marketers will divide the public into countless market segments and bombard them with messages that reinforce dispositions and tastes their previous browsing-and-buying patterns have revealed, engendering isolation and myopia (Turow 1997).

We have little purchase on which perspective is right for two reasons. First, aside from industry reports that many users have robust appetites for free music and sexual images, we know little about cultural practice on the Web. To be sure, the Web offers a remarkable smorgasbord of free cultural products and services. But we know little about who uses them, due to the lack of scholarly research on the extent to which, and ways in which, Internet users listen to music, visit museum sites, or read literature online. Nor do social scientists know to what extent culture consumers use the Web to cultivate existing tastes or, instead, to explore unfamiliar genres.

Second, we cannot yet tell to what extent (and how) media firms will be able to wring profits from Web-based entertainment. They will develop the Web as mass medium only if consumer demand for entertainment suffices to justify large investments (Castells 1996:365). Tendencies toward cultural fragmentation may be repelled by cultural omnivores: well-educated consumers with eclectic tastes for many (finely differentiated) genres (Peterson & Kern 1996). Government will influence the outcome through legislation and court rulings (e.g., the Napster injunction) that define intellectual property rights.

The Internet's cultural effects may vary among user groups. Because marketers are most interested in reaching people who consume the most, their "fragmenting" efforts may focus on the well-to-do; but such users, especially when they are highly schooled, are the ones most likely to use sophisticated search strategies, so their online behavior may be less easily affected. Castells (1996:371) predicts a Web "populated by two essentially distinct populations, the *interacting and the interacted,*" the first using the medium's full capacity, the latter limited to a "restricted number of prepackaged choices."

The Evolving Internet

Research on technological change teaches us that the relationship between technology and society is never unidirectional. Rather technologies are often developed in response to the agendas of powerful social actors. Initially, they shape themselves to the contours of custom; ultimately, they follow paths selected through struggles among groups seeking to turn technologies to their own interests (McGuire & Granovetter 1998).

We see this malleability in the history of the telephone, which was created as a business tool (and even a broadcasting device), but which became an instrument of sociable interaction (Fischer 1992). We see it especially in the history of radio, which emerged as an interactive medium tailored to the needs of military communication, grew into a point-to-point communications device linking amateur enthusiasts, developed into a commercial broadcast system beaming a standardized mass culture across national societies, and finally, under the impact of television's competition, transformed itself into a finely differentiated medium specializing in broadcasting musical genres to narrowly defined subcultural market segments (Douglas 1988, Hargittai 2000a, Owen 1999).

If anything, the Internet is even more pliant because it combines point-to-point and broadcast capability within a single network (Robinson et al 2000b, Wellman 2001). It can be a telephone: literally, or through email, chat rooms, and other forms of real-time communication between individuals. It can serve as a library: specialized Web sites "narrowcast" information to users interested enough to use search engines to find them. It can act as a soapbox for individuals expressing themselves to e-lists and discussion forums. Or it can operate as a conventional mass medium: Internet Service Providers like AOL and services like RealMedia let providers broadcast information to huge user publics simultaneously. Precisely because *it can be all of these things at once*—because it affords users choices among multiple *modes of appropriation* that coexist at any given time—the Internet is unprecedentedly malleable. This malleability raises the stakes for actors who wish to shape its evolution (Hargittai 2000b).

The findings of individual-level research on Internet use reflect the technology as it has emerged, not patterns intrinsic to the medium itself. Economic competition and public policies will shape the extent to which the Internet develops as point-to-point communicator, library, or mass medium; and this, in turn, will alter the incentives and opportunities for different kinds of individuals to use it. Thus, the social impact of the Internet depends on the impact of society on what the Internet becomes. It follows that sociologists should be studying carefully the organization of the Internet field, as well as the manner in which different ways of organizing content shape patterns of use, because such research holds the key to anticipating and understanding the Internet's effects.

Sociologists have been largely remiss in meeting this challenge. A useful exception is Aldrich's (1999:312) analysis of the Web from the standpoint of community, in which he distinguishes among governance structures (regulators and informal consortia), commercial users, service providers to those users, browser developers, and other "infrastructural populations" (hardware and software firms, ISPs, search engines, and portal sites) that occupy niches in the Web's ecology. Owen (1999: ch. 11) presents a useful overview of the Internet industry from an economics perspective, with a particularly thorough treatment of the underlying technology and of the firms that maintain the network and offer connection services. A small but interesting literature explores the Internet's emergent structure through analysis of the network created by the hyperlinks that Web sites send to one another (Zook forthcoming). Early studies using huge data sets were able to map sites onto coherent topical clusters (Larson 1996) and also reported high levels of integration, with most sites reachable from most others at a path distance of four or less (Jackson 1997).

The Internet's future, and thus its social impact, will be influenced by the resolution of three crucial policy issues. The first, establishing equality in Internet access, is necessary to ensure that less well-to-do or technically sophisticated citizens are not excluded from the political, economic, and social opportunities that the Internet increasingly provides. As our discussion of empirical work on this issue demonstrates, a sociological perspective calls attention to the need to go beyond the conventional focus on access per se to explore inequality in the combination of technical and social resources required for effective participation (DiMaggio & Hargittai 2001).

The second, establishing meaningful and enforceable norms of privacy for Internet users, involves the quest for balance between the functionality that people and businesses seek from the Internet and the sacrifice of access to personal information that the technology currently requires people to offer in exchange (Lessig 1999). Sociological research on the beliefs and practices of Internet users, online vendors, and service providers is necessary to inform policy deliberations in this area.

The third, defining rules governing intellectual property for a world in which copying and transmitting cultural works is essentially costless, entails the search for a balance between incentives necessary to motivate creative workers and the interest of society in maximizing access to works of the mind (Computer Science and Telecommunications Board 2000). Currently the pendulum has swung far in the direction of the companies that control rights to intellectual property, with implications not just for music-loving teenagers but for social scientists eager to access formerly public data bases as well (Lessig 1999). Sociologists can contribute to this debate by testing behavioral assumptions about motivations of creative workers and the requirements of markets sufficient to sustain production of intellectual goods.

Lessig (1999) makes a valuable distinction among three ways that states and private interests can regulate communications media: law, norms, and code. The Internet, he argues, is distinctive in that code—the details of the programs that facilitate the exchange of messages and information—is a particularly powerful source of social control, with direct regulation relatively less effective. His work calls attention to the importance of studying aspects of the technology that remain invisible to most observers (and of the need for sociologists studying the Web to acquire sufficient technical expertise to address these questions).

Conclusion

Sociology has been slow to take advantage of the unique opportunity to study the emergence of a potentially transformative technology in situ. Too much of the basic research has been undertaken by nonacademic survey organizations, yielding theoretically unmotivated description at best, and technically flawed and/or proprietarily-held data at worst. (Fortunately, this is changing with such new data-collection efforts as the 2000 General Social Survey's topical module on Internet use, and with increased accessibility of data, much of which is now available on the statistically interactive web site *www.webuse.umd.edu.*) The relatively few sociologists who study the Internet have focused disproportionately on virtual communities, a worthy topic, but not the only one. And in that area, as well as in research on the Internet's impact on inequality, politics, organizations, and culture, we need to develop explanatory models that distinguish between different modes of Internet use and that tie behavior directly to social and institutional context.

Research has suffered, as well, from a disproportionate emphasis on individuals, implicitly treating the nature of the Internet itself as fixed. This is regrettable because this protean technology's character and effects will reflect the outcome of ongoing struggles among powerful economic and political actors. Yet few sociologists have examined the Internet's institutional structure, industrial organization, or political economy. Some sociologists *are* doing important work; but unless their numbers grow, a magnificent opportunity to build and test theories of social and technical change may go unexploited.

If sociology needs the Internet as a laboratory, policy makers need sociology to illuminate the collective choices that will shape the Internet's future. As Philip Agre (1998b:19) has written, discussions of the Internet are often informed less by positive knowledge than by "the cultural system of myths and ideas that our society projects onto the technology." Social science remains the best hope for substituting knowledge for myth and informing public discourse about current conditions and policy alternatives.

Acknowledgments

We are grateful to Phil Agre, Philip Howard, and Barry Wellman for wise and helpful comments on earlier drafts, and we take full responsibility for persistent defects and limitations. Research support to the authors from the National Science Foundation (grants SBR9710662, SES9819907, and IIS0086143), the Russell Sage Foundation, the Markle Foundation, and the Pew Charitable Trusts is gratefully acknowledged.

Notes

1. Department of Sociology, Princeton University, Princeton, New Jersey 08540; e-mail: dimaggio@princeton.edu, eszter@princeton.edu
2. Annenberg School for Communication, University of Pennsylvania, Philadelphia, Pennsylvania 19104; e-mail: rneuman@asc.upenn.edu
3. Department of Sociology, University of Maryland, College Park, Maryland; e-mail: robinson@bss1.umd.edu

Literature Cited

Abbate J. 1999. *Inventing the Internet.* Cambridge, MA: MIT Press.

Agre P. 1998a. *The Internet and public discourse. First Monday 3.* http://www.firstmondaydk/issues/issue3_3/agre/index.html.

Agre P. 1998b. Presentation in *Proceeding of a Congressional Breakfast Seminar on Information Technology and Social Change,* pp. 14–19. Washington, DC: Consortium of Soc. Sci. Assoc.

Ahuja MK, Carley KM. 1998. Network structure in virtual organizations. *J. Computer Mediated Commun.* http://www.ascusc.org/jcmc/vol3/issue4/ahuja.html.

Aldrich H. 1999. *Organizations Evolving.* Beverly Hills, CA: Sage.

Alexander JC, ed. 1988. *Durkheimian Sociology: Cultural Studies.* New York: Columbia Univ. Press.

Anderson RH, Bikson TK, Law SA, Mitchell BM. 1995. *Universal Access to E-Mail—Feasability and Societal Implications.* Santa Monica, CA: RAND.

Attewell P, Battle J. 1999. Home computers and school performance. *Info. Soc.* 15:1–10.

Barlow JP. 1996. *A Declaration of the Independence of Cyberspace.* http://www.eff.org/~barlow/Declaration-Final.html.

Bell D. 1973. *The Coming of Post-Industrial Society: A Venture in Social Forecasting.* New York: Basic.

Bell D. 1977 [1980]. Teletext and technology: new networks of knowledge and information in postindustrial society. In *The Winding Passage: Essays and Sociological Journeys, 1960–1980,* ed. D Bell, pp. 34–65. New York: Basic.

Beniger JR. 1996. Who shall control cyberspace? In *Communication and Cyberspace: Social Interaction in an Electronic Environment,* ed. L Srate, R Jacobson, SB Gibson, pp. 49–58. Cresskill, NJ: Hampton.

Bennett WL. 1995. *News: The Politics of Illusion.* New York: Longman. 3rd ed.

Bertot JC, McClure CR. 1998. *The 1998 National Survey of U.S. Public Library Outlet Internet Connectivity: Final Report.* Washington, DC: Am. Library Assoc. Natl. Com. on Libraries & Info. Sci.

Bimber B. 1998. The Internet and political transformation: populism, community and accelerated pluralism. *Polity* 31:133–60.

Bimber B. 1999. The Internet and citizen communication with government: Does the medium matter. *Polit. Commun.* 16:409–28.

Bimber B. 2000a. The gender gap on the Internet. *Soc. Sci. Q.* 81:868–76.

Bimber B. 2000b. The study of information technology and civic engagement. *Polit. Commun.* 17: In press.

Bimber B. 2001. Information and civic engagement in America: The search for political effects of the Internet. *Polit. Res. Q.*

Board on Science, Technology and Economic Policy. National Research Council. 1999. *Securing America's Industrial Strength.* Washington, DC: Natl. Acad. Press.

Bogart L. 1956. *The Age of Television: A Study of Viewing Habits and the Impact of Television on American Life.* New York: Ungar.

Bolt D, Crawford R. 2000. *Digital Divide: Computers and Our Children's Future.* New York: TV Books.

Bourdieu P. 1999. *On Television.* New York: New Press.

Brown JS, Duguid P. 2000. *The Social Life of Information.* Boston: Harvard Bus. School Press.

Browning G. 1996. *Electronic Democracy: Using the Internet to Influence American Politics.* Wilton, CT: Pemberton.

Brynjolfsson E, Malone T, Gurbaxani V, Kambil A. 1994. Does information technology lead to smaller firms? *Mgmt. Sci.* 40:1628–44.

Calhoun C. 1998. Community without propinquity revisited: communication technology and the transformation of the urban public sphere. *Soc. Inquiry* 68:373–97.

Castells M. 1996. *The Rise of the Network Society.* Vol. 1 of *The Information Age: Economy, Society and Culture.* Oxford, UK: Blackwell's.

Castells M. 2001. *Internet Galaxy: Reflections on the Internet, Business and Society.* New York: Oxford Univ. Press. Forthcoming.

Caves R. 2000. *Creative Industries: Contracts Between Art and Commerce.* Cambridge: Harvard Univ. Press.

Clemente PC. 1998. *State of the Net: The New Frontier.* New York: McGraw Hill.

Coffin T. 1955. Television's impact on society. *Am. Psychol.* 10:630–41.

Cole J. 2000. *Surveying the Digital Future.* Los Angeles: UCLA Ctr. Telecommun. Policy (www.ccp.ucla.edu).

Collins R. 1979. *The Credential Society.* New York: Academic.

Compaine B. 2000. *Re-examining the digital divide.* Pap. pres. 28th Annual Telecommun. Policy Res. Conf., Arlington, VA.

Compaine B, Gomery D, eds. 2000. *Who Owns the Media? Competition and Concentration in the Mass Media Industry.* Mahwah, NJ: Erlbaum.

Computer Science and Telecommunications Board, National Research Council. 2000. *The Digital Dilemma: Intellectual Property in the Information Age.* Washington, DC: Natl. Acad. Press.

Davis J, Hirschl T, Stack M, eds. 1997. *Cutting Edge: Technology, Information Capitalism, and Social Revolution.* New York: Verso.

Davis R. 1998. *The Web of Politics: The Internet's Impact on the American Political System.* New York: Oxford Univ. Press.

Dibbell J. 1999. Let third voice be heard. *Intellectual Capital* (August 19). http://www.intellectualcapital.com/issues/issue282/item6125.asp.

DiMaggio P, Hargittai E. 2001. From the 'Digital Divide' to digital inequality: studying Internet use as penetration increases. Work. Pap, Ctr. for Arts Cult. Policy Stud., Princeton Univ.

Douglas S. 1987. *Inventing American Broadcasting, 1899–1922.* Baltimore: Johns Hopkins Univ. Press.

Dowd T. 2001. Musical diversity and the mainstream recording market, 1955–1990. *Rassegna Italiana di Sociol.* Forthcoming.

Downs A. 1957. *An Economic Theory of Democracy.* New York: Harper & Row.

Eisenstein EL. 1979. *The Printing Press as an Agent of Change.* Cambridge Univ. Press.

Elster J, ed. 1998. *Deliberative Democracy.* New York: Cambridge Univ. Press.

Etzioni A, Etzioni O. 1997. Communities: virtual vs. real. *Science* 277:295.

Fischer C. 1992. *America Calling: A Social History of the Telephone to 1940.* Berkeley: Univ. Calif. Press.

Fountain JE. 2001a. The economic impact of the Internet on the government sector. In *The Economic Payoff from the Internet Revolution,* Rep. Brookings Task Force on the Internet. Washington, DC: Brookings Inst.

Fountain JE. 2001b. *Building the Virtual State: Information Technology and Institutional Change.* Washington, DC: Brookings Inst.

Freedman JL, Sears D. 1965. Selective exposure. In *Advances in Experimental Social Psychology,* ed. L Berkowitz, 2:58–98. Orlando: Academic Press.

Frenkel SJ, Korczynski M, Shire KA, Tam M. 1999. *On the Front Line: Organization of Work in the Information Economy.* Ithaca, NY: Cornell Univ. Press.

Frey D. 1986. Recent research on selective exposure to information. *Adv. Exp. Soc. Psychol.* 19:41–80.

Garnham N. 1990. *Capitalism and Communication: Global Culture and the Economics of Information.* Newbury Park, CA: Sage.

Gibson RK, Howard PEN, Ward S. 2000. Social capital, Internet connectedness and political participation: A four-country study. Pap. pres. 2000 Int. Polit. Sci. Assoc. Meet., Quebecé, Canada.

Guillén M, Suarez S. 2001. Developing the Internet: entrepreneurship and public policy in Ireland, Singapore, Argentina and Spain. *Telecommun. Policy* 25.

Goldhaber MH. 1997. The attention economy and the Net. *First Monday.*

Habermas J. 1981. *The Theory of Communicative Action,* Vol. 1. *Reason and the Rationalization of Society.* Boston: Beacon.

Habermas J. 1989. *The Structural Transformation of the Public Sphere.* Cambridge, MA: MIT Press.

Hampton K, Wellman B. 2000. Examining community in the digital neighborhood: early results from Canada's wired suburb. In *Digital Cities: Experiences, Technologies and Future Perspectives,* ed. T Ishida, K Isbister, pp. 475–92. Heidelberg, Germany: Springer-Verlag.

Hargittai E. 1996. *Holes in the Net: The Internet and International Stratification.* Senior Honors Thesis. Smith College. (http://cs.smith.edu/~hargitta/Thesis).

Hargittai E. 1999. Weaving the Western Web: Explaining difference in Internet connectivity among OECD countries. *Telecommun. Policy* 23:701–18.

Hargittai E. 2000a. Radio's lessons for the Internet. *Commun. ACM* 43:50–57.

Hargittai E. 2000b. Open portals or closed gates? Channeling content on the World Wide Web. *Poetics.* 27:233–53.

Harrison B. 1994. *Lean and Mean: The Changing Landscape of Corporate Power in the Age of Flexibility.* New York: Basic.

Hauben M, Hauben R. 1997. *Netizens: On the History and Impact of Usenet and the Internet.* Los Alamitos, CA: IEEE Computer Soc. Press.

Heldritch Center for Workforce Development (Rutgers Univ.) and Center for Survey Research and Analysis (Univ. Conn.). 2000. *Nothing but Net: American Workers and the Information Economy.* New Brunswick NJ: Heldritch Ctr.

Hill KA, Hughes JE. 1998. *Cyberpolitics: Citizen Activism in the Age of the Internet.* Lanham, MD: Rowman & Littlefield.

Hirsch PM. 1978. Television as a national medium: Its cultural and political role in American society. In *Handbook of Urban Life,* ed. David Street, pp. 389–427. San Francisco: Jossey-Bass.

Hoffman DL, Novak TP, Schlosser A. 2000. The evolution of the Digital Divide: How gaps in Internet access may impact electronic commerce. *J. Computer-Mediated Commun.* 5 March.

Howard PEN, Rainie L, Jones S. 2001. Days and nights on the Internet: the impact of a diffusing technology. Special issue of *Am. Behav. Sci.,* ed. B Wellman, C Haythornthwaite. Forthcoming.

Intelli-Quest. 1999. Intelliquest study shows 83 million U.S. Internet users and 56 million online shoppers. Press release, April 19. http://www.intelliquest.com/press/release78.asp.

ITU (International Telecommunications Union). 1997. *Challenges to the Network: Telecoms and the Internet.* Geneva: ITU Press.

ITU. 1998. *World Telecommunication Development Report.* Geneva: ITU Press.

ITU. 1999. *Challenges to the Network: Internet for Development 1999.* Geneva: ITU Press.

Introna L, Nissenbaum H. 2000. Shaping the Web: Why the politics of search engines matters. *Info. Soc.* 16.

Jackson M. 1997. Assessing the communication structure of the World Wide Web. *J. Computer-Mediated Commun.* 3. http://www.ascusc.org/jcmc/vol3/issuel/jackson.html.

Johnson TJ, Kaye BK. 1998. A vehicle for engagement or a haven for the disaffected? Internet use, political alienation and voter participation. In *Engaging the Public: How Government and the Media Can Reinvigorate American Democracy,* ed. TJ Johnson, CE Hays, SP Hays. New York: Rowman & Littlefield.

Katz JE, Aspden P. 1997. Motives, hurdles and dropouts. *Commun. ACM* 40:97–102.

Katz JE, Rice R, Aspden P. 2001. The Internet, 1995–2000: Access, civic involvement and social interaction. Special issue of *Am. Behav. Sci.,* ed. B Wellman, C Haythornthwaite. Forthcoming.

Kavanaugh AL, Patterson SJ. 2001. The impact of community computer networks on social capital and community. involvement. Special issue of *Am. Behav. Sci.,* ed. B Wellman, C Haythornthwaite. Forthcoming.

Koku E, Nazer N, Wellman B. 2001. International scholarly networks. *Am. Behav. Sci.* Forthcoming.

Kolko BE, Nakamura L, Rodman GB. 2000. *Race in Cyberspace.* New York: Routledge.

Kollock P. 1999. The production of trust in online markets. *Adv. Group Processes* 16:99–123.

Kornhauser W. 1968. Mass society. In *The Encyclopedia of the Social Sciences,* ed. D. Sills. New York: Free Press/Macmillan.

Kraut R, Scherlis W, Mukhopadhyay T, Manning J, Kiesler S. 1996. The HomeNet field trial of residential Internet services. *Commun. ACM* 39:55–63.

Kraut R, Patterson M, Lundmark V, Kiesler S, Mukophadhyay T, Scherlis W. 1998. Internet paradox: A social technology that reduces social involvement and psychological well-being? *Am. Psychol.* 53:1011–31.

Kraut R, Kiesler S, Boneva B, Cummings J, Helgeson V. 2001. Internet paradox revisited. *J. Soc. Issues.* Forthcoming.

Krueger AB. 1993. How computers have changed the wage structure: evidence from micro data. *Q. J. Econ.* 108:33–60.

Lake D. 2000. The Web: growing by 2 million pages a day. *Indust. Standard* Feb 28.

Larson R. 1996. Bibliometrics of the World Wide Web: an exploratory analysis of the intellectual structure of Cyberspace. Ray R. Larson. In *ASIS '96 Proceedings of the 59th ASIS Annu. Mtg.,* ed. S Hardin. Baltimore, MD, Oct 21–24, 1996. Medford, NJ: Info. Today. http://sherlock.berkeley.edu/asis96/asis96.html.

Lentz B, Straubhaar J, LaPastina A, Main S, Taylor J. 2000. *Structuring access: the role of public access centers in the "Digital Divide."* Pap. pres. Annu. Meet. Int. Commun. Assoc., Acapulco, June.

Lessig L. 1999. *Code and Other Laws of Cyberspace.* New York: Basic.

Lin N. 2001. *Social Capital: A Theory of Social Structure and Action.* New York: Cambridge Univ. Press.

Machlup F. 1962. *The Production and Distribution of Knowledge in the United States.* Princeton, NJ: Princeton Univ. Press.

Margolis M, Resnick D. 1999. *Taming the Cyber-Revolution: How Money and Politics Domesticate the Web.* Thousand Oaks, CA: Sage.

McChesney RW. 1996. The Internet and U.S communication policy-making in historical and critical perspective. *J. Commun.* 46:98–124.

McGuire P, Granovetter M. 1998. *Business and bias in public policy formation: the National Civic Federation and the social construction of electric utility regulation, 1905–1907.* Pap. pres. meet. Am. Sociol. Assoc., San Francisco.

McLuhan M. 1967. *Understanding Media: The Extensions of Man.* New York: McGraw Hill.

Miller D, Slater D. 2000. *The Internet: An Ethnographic Approach.* New York: Berg.

Mintz J. 2000. McCain camp enjoys a big Net advantage. *Washington Post.* Feb 9.

Nagarajan A, Bander JL, White CC. 2000. Trucking. In *U.S. Industry in 2000: Studies in Competitive Performance,* ed. Board on Sci., Technol. Econ. Policy, Natl. Res. Coun., pp. 123–53. Washington, DC: Natl. Acad. Press.

NTIA (National Telecommunications and Information Administration). 1995. *Falling Through the Net: A Survey of the 'Have Nots' in Rural and Urban America.* Washington, DC: US Dep. Commerce.

NTIA. 1998. *Falling Through the Net II: New Data on the Digital Divide.* Washington, DC: US Dep. Commerce.

NTIA. 1999. *Falling Through the Net III: Defining the Digital Divide.* Washington, DC: US Dep. Commerce.

NTIA. 2000. *Falling through the Net: Toward Digital Inclusion.* Washington, DC: US Dep. Commerce.

Negroponte N. 1995. *Being Digital.* New York: Knopf.

Netcraft. 2000. *The Netcraft Web Server Survey.* Online document available at http://www.netcraft.com/survey/ (last accessed Aug. 25, 2000).

Neuman WR. 1991. *The Future of the Mass Audience.* New York: Cambridge Univ. Press.

Neuman WR. 2000. The impact of the new media: fragmentation, stratification and political evolution. In *Mediated Politics: Communication in the Future of Democracy,* ed. WL Bennett, RM Entman. New York: Cambridge Univ. Press.

Neuman WR, McKnight LW, Solomon RJ. 1998. *The Gordian Knot: Political Gridlock on the Information Highway.* Cambridge: MIT Press.

Neuman WR, O'Donnell SR, Schneider SM. 1996. The Web's next wave: a field study of Internet diffusion and use patterns. Ms., MIT Media Lab.

Nie NH, Ebring L. 2000. *Internet and Society: A Preliminary Report.* Stanford, CA: Inst. for Quant. Stud. Soc.

Norris P. 2001. *Digital Divide? Civic Engagement, Information Poverty and the Internet in Democratic Societies.* New York: Cambridge Univ. Press.

NUA. 2000a. How many online? *NUA Internet Surveys.* Online document available at http://www.nua.ie/surveys/how_many_online/world.html.

NUA. 2000b. February 14. Portals draw lion's share of audiences. *NUA Internet Surveys.* http://www.nua.ie/surveys/?f= VS&art_id=905355592&rel=true.

O'Mahony S, Barley SR. 1999. Do digital telecommunications affect work and organization? The state of our knowledge. *Res. Org. Behav.* 21:125–61.

OECD. 1997. *Webcasting and Convergence: Policy Implications.* Paris: OECD. (http://www.oecd.org/dsti/sti/it/cm/prod/e_97-221.htm)

Orlikowski WJ, Iacono CS. 2000. The truth is not out there: an enacted view of the 'digital economy.' In *Understanding the Digital Economy: Data, Tools, and Research,* ed. E Brynjolfsson, B Kahin, pp. 352–80. Cambridge, MA: MIT Press.

Owen BM. 1999. *The Internet Challenge to Television.* Cambridge: Harvard Univ. Press.

Paccagnella L. 1997. Getting the seats of your pants dirty: strategies for ethnographic research on virtual communities. *J. Computer-Mediated Commun.* http://www.ascusc.org/jcmc/vol3/issue1/paccagnella.html.

Paltridge S, Ypsilanti D. 1997. A bright outlook for communications. *OECD Observer* 205:19–22.

Peterson RA, Kern RM. 1996. Changing highbrow taste: from snob to omnivore. *Am. Sociol. Rev.* 61:900–7.

Pew Center for the People and the Press. 1995. *Technology in the American household.* Washington, DC. http://www.people-press.org/tech.htm.

Pew Center for the People and the Press. 1998. *Internet news takes off.* Washington, DC. http://www.people-press.org/med98rpt.htm.

Pew Center for the People and the Press. 1999. *The Internet news audience goes ordinary.* January. Washington, DC.

Powell WW. 2001. The capitalist firm in the 21st century: emerging patterns. In *The 21st Century Firm: Changing Economic Organization in International Perspective,* ed. P DiMaggio. Princeton, NJ: Princeton Univ. Press.

Preece J. 2000. *On-line Communities: Designing Usability and Supporting Sociability.* New York: Wiley.

Prettejohn M. 1996. *The first year: August 1995–August 1996.* Netcraft. http://www.net-craft.com/survey/year1.html.

Putnam RD. 2000. *Bowling Alone: The Collapse and Revival of American Community.* New York: Simon & Schuster.

Rao M, Rashid I, Rizvi H, Subba R. 2000. *Online content in South Asia.* South Asia Networks Organisation. Online document http://www.sasianet.org/onlinecont.html last accessed August 26.

Rheingold H. 1993. *The Virtual Community: Homesteading on the Electronic Frontier.* Reading, MA: Addison-Wesley.

Robinson JP, Barth K, Kohut A. 1997. Personal computers, mass media, and use of time. *Soc. Sci. Computer Rev.* 15:65–82.

Robinson JP, Kestnbaum M. 1999. The personal computer, culture and other uses of free time. *Soc. Sci. Computer Rev.* Summer: 209–16.

Robinson JP, Godbey G. 1999. *Time for Life.* State College, PA: Penn State Univ. Press. 2nd ed.

Robinson JP, Kestnbaum M, Neustadtl A, Alvarez A. 2000. *IT, the Internet, and time displacement.* Pap. pres. Annu. Meet. Am. Assoc. Pub. Opin. Res, Portland, OR, May 2000.

Rochlin GI. 1997. *Trapped in the Net: The Unanticipated Consequences of Computerization.* Princeton: Princeton Univ. Press.

Rogers EM. 1995. *Diffusion of Innovations,* New York: Free Press. 4th ed.

Roper Starch Worldwide Inc. 1998. *America Online Roper Starch Cyberstudy 1998.* New York.

Sandvig C. 2000. *The information apologue: play and Internet access in the children's library.* Pap. pres. Int. Commun. Assoc. Annu. Meet., Acapulco, Mexico. June 1–5.

Schement J. 1996. *Beyond Universal Service: Characteristics of Americans without Telephones, 1980–1993.* Commun. Policy Work. Pap. No. 1. Washington, DC: Benton Found.

Schement J. 1999. Of gaps by which democracy we measure. *Info. Impacts.* Dec.

Schiller HI. 1989. *Culture, Inc.: The Corporate Takeover of Public Expression.* New York: Oxford Univ. Press.

Schiller H. 1996. *Information Inequality: The Deepening Social Crisis in America.* New York: Routledge.

Schneider SM. 1996. Creating a democratic public sphere through political discussion: a case study of abortion conversation on the Internet. *Soc. Sci. Computer Rev.* 14:373–93.

Schneider SM. 2000a. *Political portals and democracy: threats and promises.* May. netelection. org/commentary/2000015.php3.

Schneider SM. 2000b. *The dot-not candidates.* July. netelection.org/commentary/2000023.php3.

Schudson M. 1998. *The Good Citizen: A History of American Civic Life.* New York: Free Press.

Schumpeter JA. 1947. *Capitalism, Socialism and Democracy.* New York: Harper & Row.

Scott WR, Ruef M, Mendel P, Caronna CA. *2000. Institutional Change and Organizations: Transformation of a Healthcare Field.* Chicago: Univ. Chicago Press.

Shapiro AL. 1999. *The Control Revolution: How the Internet is Putting Individuals in Charge and Changing the World We Know.* New York: Century Found.

Shils E. 1963. The theory of mass society. In *American as a Mass Society,* ed. Polson, pp. 30–50. Glencoe, IL: Free Press.

Silver D. 2000. Margins in the wires: looking for race, gender and sexuality in the Blacksburg Electronic Village. In *Race in Cyberspace,* ed. BE Kolko, L Nakamura, GB Rodman. New York: Routledge.

Silverstein C, Henzinger M, Marais H, Moricz M. 1998. Analysis of a very large AltaVista query log. *SRC Tech. Note* 1998-014. Oct 26.

Smith M, Kollock P, eds. 1999. *Communities in Cyberspace.* London: Routledge.

Sproull LS, Kiesler SB. 1991. *Connections: New Ways of Working in the Networked Organization.* Boston: MIT Press.

Starr P. 1997. Smart technology, stunted policy: developing health information networks. *Health Affairs* 15:91–105.

Stoll C. 1995. *Silicon Snake Oil: Second Thoughts on the Information Highway.* New York: Doubleday.

Stromer-Galley J. 2000. Online interaction and why candidates avoid it. *J. Commun.* 50. In press.

Strover S. 1999. *Rural Internet Connectivity.* Rural Policy Res. Inst. Rep. P99-13. http://www.rupri.org/pubs/archive/reports/P99-13/.

Strover S, Straubhaar J. 2000. *E-Government Services and Computer and Internet Use in Texas. A Report from the Telecommunications and Information Policy Institute.* Austin, TX. http://www.utexas.edu/research/tipi/reports/dir_final2.htm.

Tapscott D. 1999. Introduction. *Creating Value in the Network Economy,* ed. D. Tapscott, pp. vii–xxvi. Boston: Harvard Bus. School Press.

Todreas TM. 1999. *Value Creation and Branding in Television's Digital Age.* Westport, CT: Quorum.

Turkle S. 1995. *Life on the Screen: Identity in the Age of the Internet.* New York: Simon & Schuster.

Turow J. 1997. *Breaking Up America: Advertisers and the New Media World.* Chicago: Univ. Chicago Press.

US Dep. Education. 2000. Internet access in U.S. public schools and classrooms: 1994–1999. Stats in Brief. Nat. Ctr. Educ. Stat. February.

Uslaner E. 2001. The Internet and social capital. *Proc. ACM.* Forthcoming.

Van Alstyne M, Brynjolfsson E. 1997. *Global village or cyberbalkans.* http://web.mit.edu/marshall/www/papers/CyberBalkans.pdf.

Van den Besselaar P, Beckers D. 1998. Demographics and sociographics of the "Digital City." In *Community Computing and Support Systems Social Interaction in Networked Communities,* ed. T Ishida. Heidelberg: Springer. http://www.swi.psy.uva.nl/usr/beckers/publications/kyoto.html.

Waxman J. 2000a. *The Old 80/20 Rule Take One on the Jaw. Internet Trends Report 1999 Review.* San Francisco: Alexa Res.

Waxman J. 2000b. *Leading the Pack . . . Internet Trends Report 1999 Review.* San Francisco: Alexa Res.

Weber M. 1968 [1924]. *Economy and Society,* ed. G Roth, C Wittich. New York: Bedminster.

Weiss R. 1970. Effects of mass media of communication. In *Handbook of Social Psychology,* ed. G Lindzey, E Aronson, 5:77–195. Reading, MA: Addison-Wesley.

Wellman B. 2001. Physical place and cyber-place: Changing portals and the rise of networked individualism. *Int. J. Urban Regional Res.* Forthcoming.

Wellman B, Salaff J, Dimitrova D, Garton L, Gulia M, Haythornthwaite C. 1996. Computer networks as social networks: collaborative work, telework, and virtual community. *Annu. Rev. Sociol.* 22:213–38.

Wellman B, Gulia M. 1999. Net surfers don't ride alone: virtual community as community. In *Networks in the Global Village,* ed. Barry Wellman, pp. 331–67. Boulder, CO: Westview.

Wilhelm A. 2000. *Democracy in the Digital Age.* New York: Routledge.

Wilson EJ. 2000. *Closing the Digital Divide: An Initial Review. Briefing the President.* Washington, DC: Internet Policy Inst. May. http://www.internetpolicy.org/briefing/ErnestWilson0700.html.

Zook M. 2001. Old hierarchies or new networks of centrality? The global geography of the Internet content market. *Am. Behav. Sci.* 44: Forthcoming.

Zuboff S. 1988. *In the Age of the Smart Machine: The Future of Work and Power.* New York: Basic.

Hypes, Hopes and Actualities: New Digital Cartesianism and Bodies in Cyberspace

Megan Boler

'New Digital Cartesianism' investigates the socio-material power inequities embedded in text-based, computer-mediated communication (CMC). Is the body really transcended in text-based computer-mediated communication? This article summarizes software and hardware advertising 'hypes', cyber-enthusiast 'hopes', and the 'actualities' of CMC which contradict this virtual dream of pure minds communicating. Marketing hypes and cyberhopes mythologize disembodied CMC with promises of anonymity and fluid identities. However, the actualities of how users interpret and derive meaning from text-based communication often involve reductive bodily markers that re-invoke stereotypes of racialized, sexualized and gendered bodies. Ironically, despite claims that CMC achieves Descartes' dream of 'pure minds' and the transcendence of body, users frequently rely on stereotyped images and descriptions of bodies in order to confer authenticity and signification to textual utterances. In digital Cartesianism, the body actually functions as a necessary arbiter of meaning and final signifier of what is accepted as 'real' and 'true'.

Keywords: advertising, body, Cartesian, computer-mediated communication, cyberculture, dualism, feminism, gender, identity, race

'New Digital Cartesianism' investigates the virtual vapor trails of existing socio-material power inequities in text-based computer-mediated communication (CMC). While marketing hypes and cyber-enthusiast hopes actively mythologize the potentials of disembodied CMC with promises of anonymity and fluid identities, the actualities of the way in which users interpret and derive meaning from text-based communication often involve reductive bodily markers that re-invoke stereotypical notions of racialized, sexualized and gendered bodies. Ironically, this new digital Cartesianism, initiated by a rhetorical cheerleading of the mind/body split as a desirable aim of CMC, ultimately results in the invocation of stereotyped bodies in order to confer authenticity and signification to textual utterances.

At stake is not just whether we can achieve a new adaptation of Descartes' disembodied dream or not, but how the imagined construction of online spaces functions as a material productive force that reinscribes existing social relations.[1] Evoking Henri Lefebvre, David Harvey writes, 'spaces of representation . . . have the potential not only to affect representation of space but also to act as a material productive force with respect to spatial practices' (1990: 196; see also Lefebvre, 1981). This article makes the case that in digital Cartesianism, although the body is allegedly 'transcended' in virtual environments according to the hypes and hopes, it actually functions as a necessary arbiter of meaning and final signifier of what is accepted as 'real' and 'true'. To illustrate this paradox, it summarizes the 'hypes' of the technology advertisers and the 'hopes' of cyber-enthusiasts, then outlines the 'actualities' of CMC which contradict these very hypes and hopes.

The purpose of this article is to demonstrate the potency of these cultural stereotypes which appear requisite to meaning-making within communication and dialogue, and how these reductive conceptions of gender, race and bodies invade even the supposedly 'transcendent' space of online CMC which lacks visible markers. The intention is neither to disparage social use of online environments such as MUDs (multi-user domains) and MOOs (MUD object-oriented), nor to castigate users, nor even to dismiss those users who do experience 'freedom' from traditional identities in online environments. Rather, the aim is to substantiate that, despite the hypes and hopes of the freedom offered by transcending usual images of the other, there comes a point at which users crave information about traditional markers of the body—to know the 'meat' that accompanies the textual bones (see Gibson, 1994).

While a fair amount has been written about the representations of bodies in social online environments such as MOOS and MUDs, little has been written about the implications of new digital Cartesianism for online educational environments. Thus, this inquiry is situated at the intersection of several nascent disciplines. First, feminist poststructuralist and postcolonial analyses of the importance of material and embodied experience (Fanon, 1967[1952]; Grosz, 1993; Probyn, 1993) and how this has been prematurely dismissed (Adam, 1998; Balsamo, 2000; Boler, 2002).[2] Second, cultural geographies that analyse the 'time–space compression' and 'death of distance' arguments within the context of globalization, which also signal the necessary intersections of theorizations of 'space and place' alongside analyses of discourses and experiences of cyberculture (Harvey, 2000; McDowell, 1999; Massey, 1994; Robins and Webster, 1999). Third, the implications of these socio-theoretical shifts in relation to educational theory and practice (Boler, 2002; De Castell and Jenson, 2004; Zembylas, 2005). The central question underpinning this critique of digital Cartesianism is: whose goal is it to transcend the body and what may be lost in this migration to new spatial imaginaries? These interdisciplinary approaches share a common concern with how bodies, space and social relations are understood best within materialist terms, and how the shift from proximity to CMC redefines the way in which we conceptualize social networks and relationships of self and other, thus placing radical, critical and transformative pedagogies at risk.

This analysis focuses on text-based CMC. As broader bandwidth communication such as multidirectional webcams become more common, we will need to revisit how users do or do not use markers of the body to interpret communication in ways that are similar to or different from face-to-face (F2F) communication. Online users who know nothing about one another's 'bodies' sometimes do experience fulfillment of the Cartesian dream of 'pure minds'. Although it is argued that, at present, the spaces created in online, text-based CMC frequently fail to deliver on this promise, the eventual accomplishment of communication entirely divorced from stereotyped significance of the uttering body is not precluded.

Important distinctions need to be made between social MUDs and MOOs and other environments such as web-enhanced educational environments, for example where threaded discussions or instant messaging (IM) is used to complement F2F education. Some may challenge that critiques of social CMC are not applicable to educational uses of CMC (or to newsgroups, listservs or other chat environments) because adopting a 'fictional' identity in these latter spaces is unlikely. However, in web-enhanced education (i.e. where part of the course is experienced in F2F and the rest is conducted online through threaded discussions), there is a greater likelihood, especially in large class sizes, that users will 'fill in' an image of the other. Further, in educational environments that are entirely online, users do 'fill in' identity markers, often with stereotyped conceptions of others' age, sex and location, just as in other environments (Boler and Vellenga, 2000).[3]

This article is organized in five sections. First, the relation of the body to CMC and education will be discussed, followed by the discourse of hypes. The discourse of hopes is then addressed, followed by the actualities. Finally, the implications of new digital Cartesianism for progressive pedagogies are considered.

What's the Problem? CMC, Bodies and Social Change

Some educators place a premium on the pedagogical significance of the body. Text-based, computer-mediated digital culture repackages the Cartesian desire to transcend the 'truth-polluting' body. As such, this apparent disembodiment created in cyberculture poses a genuine dilemma for critical, feminist and progressive educators who have invested decades in ensuring that 'the body' be recognized as essential to knowledge production (Grosz, 1993; Probyn, 1993; Senft, 1996).

In an article published in the *Chronicle of Higher Education,* Ingrid Banks argues that her body is a central aspect of the curriculum and pedagogy:

> *As a black female teaching African-American Studies, I am troubled by the prospect of being asked to teach Online or televised courses. I fear that they would obscure or even erase my presence in the classroom. When I walk into class on the first day of the term, I know that my presence there sends a political message. By standing in front of my students, I challenge not only their conception of the typical—that is, white and male—professor, but also their images of black people. (1998: B6)*

Yet other educators, equally committed to progressive pedagogies, argue that the relative anonymity of online interaction encourages students to participate more freely in dialogue and conversation.[4] Some studies suggest that students who are typically silent in the F2F classroom participate more frequently in online environments. The relative anonymity of disembodiment may encourage freedom of expression, emboldening students to express politically incorrect views that warrant discussion. As Banks further expresses,

> *some colleagues have told me that certain students are willing to say things online that they would not mention in the classroom. Given that I teach sensitive material that challenges students to think critically, the Internet could be a great way of pushing students to discuss why they think the way they do. (1998: B6)*

There are also strong counter-arguments in favour of having these difficult conversations in person. Banks states clearly:

> *I don't like the idea of students hiding behind a computer monitor. I want to engage them and I want them to engage each other, face to face. Over the course of the semester, I want to see their uncomfortable facial expressions and body language change. Of course, some expressions and gestures won't change, which is also information that I want to have. (1998: B6)*

This value of an embodied pedagogy has been developed by a range of progressive, radical and feminist theorists, points which are taken up in the concluding section of this article (Boler, 1997a, 1999; Ellsworth, 1989). Does disembodiment allow for growth and transformation? No doubt, to express difference in the absence of F2F may be more comfortable. But is comfort a goal of education? Given the aims of pedagogies of discomfort (see Boler, 1999), of learning to *witness* rather than *spectate* in order to understand the social responsibilities inherent to seeing differently, what is the cost of replacing discomfort with the convenience of anonymity? At present, there exists a significant gulf between the research undertaken about instructional technologies and how CMC can be used towards liberatory pedagogical goals.[5] A scrutiny of digital Cartesianism has implications for how we develop a critical digital pedagogy.

The Hypes

'Hypes' refers to the marketing discourses employed to promote and sell software, hardware and other CMC technologies. Surveying the images and texts of advertisements used to target consumers from around 2000 reveals several, frequently recurring themes: transcending bodies, transcending differences and transcending space and place. To discern the rationale and narrative thrust behind these hypes, let us begin by illustrating how the socio-historical frame of neo-liberalism functions as a cornerstone for advertising hypes.[6]

Neo-Liberal Cartesianism

A neo-liberal version of the Cartesian binary of mind and body sells differences by simultaneously highlighting difference of race and nation while erasing them; differences or bodies are recognized, situated within the discourse as an obstacle of some sort and 'displaced' through a re-emphasis of mind or transcendence of difference or body.[7]

The neo-liberal discourse of simultaneously marking difference in order to erase it is well exemplified in a series of World.com television advertisements for Generation Digital.[8] This series depicts a series of seemingly multicultural, ambiguously categorized bodies against high-speed backdrops of urban landscapes. In rapid succession seven diverse persons repeat, in different accents, the phrase 'I was born into a new generation', or in the case of the one older white man, 'I was *reborn* into a new generation'. Then one person states, 'It isn't about country', another, 'It isn't about culture', and then a white (Australian) girl states emphatically, 'It's about attitude'. Finally, a distinctly 'other' voice or body speaking in a heavy accent utters, 'But we speak the same language'. We're told, 'You've come to the right generation. Generation Digital'. World.com depicts a utopian, globalized world where 'you can be whomever you want to be'. Country and culture do not matter.

What defines Generation Digital is a 'shared language'. In 'Gen D', whatever our color or nationality, we all speak 'digital'. The implication is that digital is cleansed of any cultural determinants. In reality, digital is English, thus revealing an imperialism which governs much of the internet and software (Adam, 1998). In addition, Gen D bodies do not merely transcend; they transcend through a mediated language. Language as a technology and technologies themselves are not neutral (Castells, 2000; King, 2003), and in this instance we are able to see the dialectic of difference concurrently produced and erased.

The World.com commercial reveals the hypocrisy in the claim that on the internet there is neither 'race', nor class, nor gender. Rather, these racialized and gendered identities are necessary placeholders of difference which enable the vision of the idealized community of neo-liberal sameness. Difference is actually required for a world to exist in which differences do not matter. In short, these selves are variations of a sovereign liberal individual.[9] Jodi O'Brien words this succinctly: '[On the internet,] can I really expect to be treated just like everyone else? Does "just like everyone else" mean "just like one of the [white] guys?"' (1996: 56). A 'white global consumer' who transcends the body, local space and time in their fundamentally interchangeable nature has replaced Descartes' 'man of reason'.

Hype 1: Transcending Bodies

This advertisement for 'ThinkAd' explicitly represents one form of digital Cartesian hype. 'Free the body', states the text of the original advertisement, 'and the mind will happily follow'. Shown in the foreground is a white male professional. A white woman is in the background, seated on a desk, next to a computer, her bare legs seductively crossed, and her gaze turning 'demurely' downward (at what? her ankle?), and away from us. Both represent workers as consumers/producers, in an ambiguous space—part office, part yuppie loft—where technology allows people to be 'free to roam' from their desks. (ThinkAd) gives people (read 'bodies') and ideas (read 'minds') extra space to roam (read 'transcend') . . . beyond the realm of their desk (read 'space and place').

Here again, stereotyped bodies are crucial for meaning-making (see Hall, 1997, ch.4). The woman's body is classically eroticized, even in her 'free' relation to the machine and her implied ability to 'roam' from the desk. However, her position is all-too familiar in its fixed and static position as appendage, subservient and subordinate, while the 'older, wiser, white man' roams in the foreground. Arguably, she is a bonus prize; a fringe benefit of the freedom afforded to him through technology.

Hype 2: Transcending Differences

Closely related the hype of transcending the *body* is the hype of transcending *difference*. Each requires the other. However, within advertising discourses, often the two hypes are represented as distinct from one another. The classic MCI commercial proclaims: 'There is no race. There is no gender. There are no infirmities. There are only minds'.[10] Pure and unsullied communication can take place regardless of embodied differences—in this case, an African-American man in formal attire and a bright-faced, corn-fed, freckled, ambiguously gendered youth can communicate across differences, thanks to new technologies. Picture Tel asks, 'Who will your students meet today?' and shows a colorful line up of what appears to be a Thai princess, an astronaut, a ballerina and a Japanese geisha. These advertising hypes express a world in which difference is transcended by technology; difference, first seen and experienced, is then bridged or erased. The neo-liberal, concurrent erasure or dependence on difference is foundational for each of these hypes. The differently marked bodies are indispensable as a backdrop against which to sell technologies that bridge or transcend.

What is wrong with this picture? Some might reply, 'Nothing! Why not allow for users to engage difference through these computer-mediated forms?' However, 'drive-by difference' presents difference and the other as something that can be 'safely' met or experienced—at a distance. A voyeuristic 'experience' of otherness is replicated, working against the explicit aims of radical pedagogies, postcolonial critiques and feminist poststructuralism that have worked for decades to establish the importance of embodied difference encountered not through reductive notions of otherness, but through careful recognition of precisely how difference is constructed and perceived (Boler, 1999; Lugones, 1989; Mohanty, 1989–90). In educational practice, drive-by difference may result in superficial CMC about difference that fails to challenge deeply the kinds of prejudice and stereotypes that people hold about others (Riel, 1996). For example, in one account, two groups of young children had been interacting online between Louisiana and New York for some months. When the Louisiana group received a photograph of their pen pals, they expressed shock and asked 'Where are the brothers?' (Riel, 1996). They had assumed their pen pals to be 'black like them'. In this case, the gaze brought to bear appears not necessarily to be the dominant cultural gaze but the gaze of the self-seeking itself. In short, while there is a promise of freedom in the anonymity of text-based virtual communities and a promise that we transcend assumed differences in our online interactions, in fact, users may tend to reproduce themselves in imagined others.

There is no guarantee that F2F classrooms will ensure that beliefs and assumptions about others are profoundly challenged, but certainly one must question whether packaging multicultural encounters through technology achieves the kinds of interactions, communities and education envisioned as progressive, feminist or liberatory. The 'actualities' of these educational uses of technology will be discussed further in the final section of the article.

Hype 3: Transcending Space and Place

A persistent illustration of the new digital Cartesianism involves hype about transcending space and place.[11] In this advertisement for iSell (a parodic recreation of the actual print ad image), we are told that, 'Redesigning a campus is easy. You just knock down a few walls'. Brick and mortar universities and traditional walls are replaceable with the connectivity of computer wireless systems. The original image depicts a male and female couple happily seated outdoors on a campus looking at a laptop screen together. The text of the original ad describes creating a classroom from the entire world, following the instructed wishes of a particular university, with the promise that staff as well as students will be connected via the internet at all places and times.

In another ad from this series (not depicted visually here due to interesting copyright issues faced by this author as might now be clear), the same software company depicts parents and teachers seated in children's school chairs in a classroom, apparently at a meeting with one frustrated parent raising their hand.

This advertisement builds on the discourse of 'time–space compression'[12]—that technology closes the gap of both space and time through its capacity to transcend the limitations of material communication and bodies. It also exemplifies the hype about the efficiency of CMC in relation to education. Not only do we find the discourse that F2F communication in traditional space and place is failing us in matters of education, but that schools need to model themselves after business management to achieve success. In an even more dramatic fashion, freedom from constraining space is thrust upon us by Academic Systems, with the image of a teacher's body roped to the blackboard, sideways and partially upside down, chalk dangling in hand, with the caption: 'Not as accessible to your students as you'd like to be?' The smaller text reads:

> *If routine interaction is tying you down, break free with Academic Systems. With our interactive English and Mathematics learning courses, much of the basic learning is shifted to the software, enabling you to interact, engage and connect with your students on an individual basis.*

As in the previous advertisement, the assumption is that educational software promises efficiency and frees up the 'economy of attention' (De Castell and Jenson, 2004). Teachers can be 'more accessible' to their students. Left out of the equation is the question of the quality of interaction or educational experience, which is not in any way to dismiss or critique appropriate use of web-enhanced education, but rather to highlight the particular hypes about the limitations of bodies and space within educational technology advertising. Moreover, this implication begs the question of access to technology, whether on the part of teachers or the parents.

The hypes represented in educational technology advertisements reflect the new digital Cartesianism. Promises are made that by transcending the body and difference, the mind will be accentuated, freed and enhanced. We can escape the trappings and limitations of bodies and place through the promises of technology. Efficiency and time are rewarded in exchange for bodies and space.

The Hopes

Social theorists of cyberculture recognize the potential for challenging notions of fixed and static identities, for fluidity of identity though gender play and an escape from binaries. Such analyses have tended to focus on the social interactions that take place in MOOs and MUDs.[13]

Whereas the primary impetus here is to show the overlap of the new digital Cartesianism common to both the advertising hypes and these sociotheoretical hopes, there is a key difference between the hypes and the hopes: for the most part, cybertheorists do not want to ignore differences of ethnicity, class, gender and sexual orientation. In fact, many are painfully aware of these realities. Rather, their expressed desire to transcend the body appears to reflect a yearning to change how we think about, experience and envision social identities. The hopes reflect a desire to challenge static notions of identity, engage fluidity and break down traditional binaries.

Although cyber-enthusiast motives are quite different from the neo-liberal hype discussed in the previous section, the outcome remains the same. In advocating transcendence of the limiting aspects of embodiment, we nonetheless carry the 'virtual vapor trails' of stereotyped notions of difference embedded in our language, social imaginaries and practices. We reinscribe the stereotypes that we sought to undo and lose the embodied differences that feminists and poststructuralists see as precious aspects of subjectivity and epistemology.

Hope 1: Transcending Bodies and Difference

Can we challenge stereotypic experiences or perceptions of identity in CMC environments? In 1993, Amy Bruckman wrote:

> *MUDs are an identity workshop . . . Gender swapping is an extreme example of a fundamental fact: the network is in the process of changing not just how we work, but how we think of ourselves—and ultimately, who we are. (1996: 5)*

Over a decade later, we find similar hopes echoed: 'On the Net, race and gender distinctions dissolve' (Gray, 2001: 188) The hopes invested in cyborgs that transcend traditional binaries suggest, somewhat optimistically, that 'bodies and persons are things that can be made' (Gray, 2001: 191). These hopes are reiterated sometimes within cyberwritings in language not very different from the MCI commercial hype. To seek spaces in which we can engage dialogue without having our utterances dismissed and foreclosed by presuppositions is an admirable and worthy hope.

Progressive thinkers hold out the utopian hopes of transcendence, a hope that in digital environments identities may be fluid and 'queer', neither fixed nor in static form and may avoid the traps of binary and oppressive assumptions about identities. The transcendence of bodies and difference is achieved through two primary avenues: fluidity of online identities and transcending binaries (e.g. of human/machine).

Fluidity of Online Identities

According to the hopes, one promise of online spaces is that users have freedom to construct virtual, 'fictional' identities that allow them to move beyond the constraints of the usual social markers of ethnicity, class, gender, age and ability. The 'queer' hope, as it might be called, beholds cyberspace as potential for emergent identities. In this idealized view, cyberspace promises fluid subjectivities; uninhibited gender play where 'real life' men, for example, can 'try on' the experience of being a woman and vice versa, or where anyone can experiment with homosexual experiences without fear of ostracization.

As one cybertheorist describes the hope for moving beyond static identities:

> *The greatest freedom cyberspace promises is that of recasting the self: from static beings, bound by the body and betrayed by appearances, Net surfers may reconstruct themselves in a multiplicity of dazzling roles, changing from moment to moment according to whim. (Stallabrass, quoted in Wakeford, 1999: 179)*

Sherry Turkle's work was one of the early studies to demonstrate users' desire to experience a fluidity of social class identity. The hopes for fluidity often specifically address gender, particularly within social use of MUDs and MOOs, advocating the possibilities of gender fluidity usually in terms of users adopting gender-neutral avatars or a name or persona of the opposite sex (Bruckman, 1996; Stone, 1995; Turkle, 1995). Lori Kendall's recent book *Hanging Out in the Virtual Pub* (2002) offers an excellent summary of the discourse of hopes offered by a variety of cybertheorists:

> *As quasi-physical 'hangouts' that allow for speachline interactions among groups of people, muds in particular seem to inspire these kinds of hopes. For instance, Turkle states that MUDs allow people 'the chance to discover . . . that for both sexes, gender is constructed' (1995: 223). Similarly, Bruckman suggests that 'MUDding throws issues of the impact of gender on human relations into high relief . . . it allows people to experience rather than merely observe what it feels like to be the opposite gender or have no gender at all' (1993: 4) . . . Dickel also suggests that, rather than 'reifying conventional, hegemonic gender bias' (1995: 105), which he acknowledges as a possibility existing on other online forums, MUDs allow a 'play of gender within the imagination [which] opens up . . . a destabilization of gender positions which might spread beyond the internet into the larger culture' (106). (Kendall, 2002: 11)*

The hope of such 'queered' spaces is not merely that such play occurs online, but that somehow this online gender-play may affect offline social relations:

> *Gender switching appears to be a relatively infrequent behavior practiced by a minority of users . . . Approximately 60% of social MOO and 40% of role-play MOO users had never engaged in this behavior. An additional 15–23% had tried gender switching but stopped. Even among those currently gender switching, the majority did so infrequently. Less than 6% of the social MOO participants spent over half their time as a gender switched character. (Roberts and Parks, 2002: 280)*

Transcending Binaries

Closely related to the desires for fluid identities are the hopes of transcending binaries. To make sense of our conceptual and linguistic world, we rely on the differences created through these binaries—i.e. we understand male in relation to female, etc. However, binaries are not only a problem in terms of their apparent intractability in our conceptual grasp of the world, but they have very real material effects on how people experience themselves in the world. The trouble with binaries is that the two named terms of binary pairs—e.g. male/female, good/bad, black/white—are not equal; one becomes privileged over the other. The privileged term of the binary pair translates into social values, norms and power that marginalize those marked by the 'lesser' of the binary term. Binaries not only torment feminist and poststructural theorists but also represent a seemingly intractable foundation of traditional western thinking.

Cyber-hopefuls extol the possibility of transcending binaries and dualisms through cybermorphing. Poster, in even stronger terms, suggests that the ability of MUDders to adopt a 'fictional role that may be different from their actual gender . . . drastically calls into question the gender system of the dominant culture as a fixed binary' (quoted in Kendall, 2002: 11). For example, a cyborg is itself both human and machine, thus defying that binary opposite. In Stone's vision, virtual interactions allow us to enact our cyborg identities. Cyberspace represents the potential for 'emergent behavior, for new social forms that arise in a circumstance in which body, meet, place and even space mean something quite different from our accustomed understanding' (1995: 37). Offering a strong counter to the hopes, another cybertheorist retorts:

> *If I had a dime for every paper I received from enthusiasts announcing that participating in a MOO 'breaks down gender barriers' because of its 'performance elements', I would be the next Bill Gates. This line of thought, generally described by my friends on the Net as 'gender fucking', has the following levels of naiveté: first, it carries a wrong assumption that an online textual body is performative, whereas a biological body at the end of the terminal is stable. Second, it presents gender fucking primarily as an issue of choice, thus reinforcing an idea that you put on gender, like a change of clothing and that gender doesn't wear* you. *In short, online or off it, identity and gender are complicated performances, particularly immune to Utopias. (Senft, 1996: 17)*

Hope 2: Transcending Space and Place

Echoing the advertising hype of transcending the limitations of material space and F2F communication, cyber-discourse invokes notions of a 'postgeographical world'. Such utopian hopes envision digital worlds where we connect across differences of national and geographic boundaries: 'the new technologically-mediated world will be a post-geographical world, "profoundly anti-spatial" in its nature' (Mitchell, quoted in Robins and Webster, 1999: 240–1).

The hope discourse offers a transspatial, transcendental vision; a world less defined by violent boundaries of nations or segregated communities. But it also may harken a 'postmodern' version of the classical wish to go beyond this world and arrive into God's:

> *The postmodern will to virtuality parallels the medieval religious will to transcendence . . . Cyberspace is socially constructed as the postmodern paradise and all our hopes for virtuality express our desire to escape the limitations of our bodies and the ills of our society. (Fisher, 1997: 122).*

'Virtual paradise' discourses bear more similarity to Cartesian desires to transcend 'polluting' bodies and the impermanence of space and time.[14] This discourse is illustrated in popularized form in *Wired* magazine, in which a cover story in 2002 advertised computers as theology: 'computation seems almost a theological process. It takes as its fodder the primeval choice between yes and no, the fundamental state of 1 or 0' (Kelly, 2002: 180).

To reiterate, this argument is neither meant to be a critique of users, nor of engagement in online social spaces where, ostensibly, users can experiment with fluidity and transcending the body or difference. Rather, it interrogates how the discourses of hypes and hopes overlap and in turn how these are collectively contradicted by the actualities, to which we now turn.

The Actualities

Cartesianism with a Twist

While the hypes and hopes suggest that we can inhabit a communicative world where anonymity reigns and freedom of expression rules regardless of one's bodily identity, the actualities evidence that in fact, users rely on markers such as age, sex and location to make sense of online communication. The mythologies of anonymity and fluidity appear to have their limits: users require that the other offers 'essential' data about their 'real life' identity so that sense can be made of textual utterances. In short, we seem to be able to go only so long without asking or imagining these familiar social markers and cues.

Hence 'digital Cartesianism with a twist': the online context promising pure, disembodied communication and the 'twist' of the body required to make sense of textual communications. This 'twist' is well illustrated by the frequency with which users inquire about others' 'age/sex/location' ('asl') in order to interpret communication and/or to confirm one's projection of the other's identity.[15] In many environments users are obsessively concerned with 'asl'. As O'Brien in her study 'Writing in the Body: Gender (Re)production in Online Interaction' succinctly puts it,

> *even when the body is anchored elsewhere and unavailable as the source of symbolic cueing, central distinctions that reference the body as connected to self will still be evoked as the basis of meaningful communication. (O'Brien, 1999: 85)*

Consider this random selection from a bulletin board in which users debate the practice of demanding 'asl':

soby18	Wots with everyone's first question alway asking asl (age, sex, location) i just want have a chat with someone with out people asking this as the first question.
bensam59	Oh! i know excately how u feel
mandy1978	ITS JUST SOMETHING TO SAY TO BREAK THE ICE AND GET THE CONVERSATION GOING, INTRESTING 2 SEE WHERE PEOPLE LIVE AND WHO THEY ARE THATS ALL
en1pappy	I agree with you one 100 % every chat room i go to alway ask asl . . . just about every time i try to get a friend the big asl? comes up . . . I have my age on my profile so why invite me to be a friend then turn around and ask asl . . . ok age does matter, kids must be very careful who they talk to, but what does the sex and location have to do with it . . .
nanasangel	it burns me up when the first question is always asl? what ever happen to hello? how r u doing? thank you very much . . . soby 18 you got a smiley from me . . .
winterborn_44	i feel just like every body i don't want a date i just want to chat
Foxxie73	Hey, I understand where you are coming from, but also try and look at it this way . . . When you are chatting with someone wouldn't you want to kinda know what age group there in so then you have some idea of what to talk about? I mean if you were talking to someone who is 40 you wouldn't be talking about celebs and gossip because they wouldn't know about it as well as the teens do or location, It is a great conversation starter. Gender, well that doesnt really matter but u might as well ask all in once when u type asl and its just a quick answer. (http://www.grab.com/community/boards/viewtopic.php?t=873)

Curiously, the users themselves interrogate the significance of the 'asl' practice, revealing their own perception of the tension between the promise of disembodied exchange and this reinvocation of stereotyped bodies. These debates about the practice of requiring that users declare their age, sex and location illustrate the concerns of critics:

> *[T]here remains persistent and apparent cultural resistance to gender model change . . . Whatever social construction is, it is not something that appears to be easily made malleable or ambiguous even in the rapidly changing world of technological civilization. (Ihde, 2002: 31)*

This *reinscription* of stereotypes of the body occurs not only in purely social online environments but also in other discussion groups such as Usenet. In an article titled 'Reading Race Online: Discovering Racial Identity in Usenet Discussions', Burkhalter demonstrates that in his studies of newsgroup users:

> *Discussion between groups, which progressive people might hope would alleviate racial stereotypes, instead is a site where previously held stereotypes are made into self-fulfilling prophecies . . . In online discussions, readers treat racial identities as entailing particular perspectives . . . A discrepancy arises when a person identified as a member of a particular racial group by his or her physical characteristics offers a perspective that is inconsistent with the stereotype of that group. (1999: 72–3)*

Readers make an essentialized link between a user's racial identity and the views that this person is expected to hold. For example, the following post responds to an author who was troubled by blacks she had seen on a talk show and finds a discrepancy between the author's identity as black and the perspective that she offers:

> *It is a shame that you even have to ask these questions because I would hope that you see more blacks on Ricki Lake, whether you are black or white. But being black, I am truly amazed at what you have asked. I will just guess that you are still a teen (as opposed to a hick that has never seen a black person) and haven't been out in the world and exposed to much. (1999: 72–3)*

The Usenet member is confused by how a 'black' person could express a particular view that does not conform to a stereotyped notion of 'the perspective that 'should be' held by a black person. The user reconciles the dissonance by 'filling in' and attributing the discrepancy to the other's age and ignorance:

> *[R]esolving these puzzles by modifying the author's identity allows readers to maintain the connection between racial identities and perspectives . . . perspective and race are made to conform online. Far from being a site where race, racism, ethnocentrism or stereotyping are banished, these phenomena flourish in newsgroups. (Burkhalter, 1999: 73)*

Despite the hypes and hopes that online interaction might open up new spaces of communication, there is ample evidence that users invoke habitual assumptions and stereotypes about bodies in order to make sense of the other.[16] As Stone ascribes,

> *for symbolic exchange originating at and relating to the surface of the body, narrowing the bandwidth has startling effects. A deep need is revealed to create extremely detailed images of the absent and invisible body . . . Frequently in narrow-bandwidth communication the interpretative faculties of one participant or another are powerfully, even obsessively, engaged. (1995: 93)*

Following Lacan, Stone argues that desire

> *theorized as a response to perceived lack, arises as a product of the tension between embodied reality and the emptiness of the token, in the forces that maintain the preexisting codes . . . for body . . . that are absent from the token. (1995: 95)*

And Baym reiterates:

> *Because computer-mediated interactants are unable to see, hear and feel one another they cannot use the usual contextualization cues conveyed by appearance, nonverbal signals and features of the physical context. With these cues to social context removed, the discourse is left in a social vacuum quite different from face to face interaction. (1995: 139–40)*

The body's role as final arbiter of authentic identity is evident across studies of online interaction. One of the most revealing crises of interaction in which the body plays a most decisive role is one user's experiences of 'deception' at the hands of another user.[17]

The anger expressed towards the deception turns on the body as being the final arbiter of authenticity. As O'Brien argues:

> *Ultimately, one has either a vagina or a penis and the presence of one or the other of these physical attributes marks an 'authentic' immutable presence in time and space. Or so we will continue to believe. (1999: 95–6).*

In the case of 'deception', the physical attribute of one's 'real', biological identity functions to determine the 'real' person. Stone notes that

> *the societal imperative with which we have been raised is that there is one primary persona or 'true identity' and that in the off-line world—the 'real' world—this persona is firmly attached to a single physical body. (1995: 73)*[18]

Even so, computer-mediated relationships still may have the potential to transcend these limited and stereotyped conceptions of the self and its relationship to an essentialized body. But as O'Brien (1996, 1999) emphasizes, for online relationships to represent and inhabit truly 'queer' or 'transgendered space', the social meanings shared by those inhabitants both on and offline must be social meanings unconstrained by omnipresent and dominant assumptions about such categories as gender and ethnicity. Unfortunately, there is not yet evidence that the majority of users inhabiting online spaces represent a demographics of democracy, much less a population who share 'queered' understandings of the relationship of sex to gender. And the question remains: 'How might these narrow bandwidth educational interactions disrupt fixed assumptions?' One must examine the complexity of people's social networks and cannot examine solely 'whether and how an individual transforms her ideas in an online educational environment'. Any person's experience is occurring in myriad spaces, places and interactions. It is not at all clear that the disembodied, anonymous space of online communication can ensure that users will challenge fixed notions of gender and racial identities in any way:

> *[T]he metaphor of the internet as space masks the disassociation of Netters from their bodies, masks the fact that bodies are elsewhere, real, material—invested with a responsible subjectivity. Here it is clearest that gender is information, discourse rather than nature—and sexuality, offline or on it, functions linguistically. But the deployment of this potentially dematerialized information has material effects and historically, they have been to the detriment of women. We appropriate technology to our own physiques and resolutely refuse to change our perceptions to fit the parameters of our inventions. We may indeed be becoming cyborgs, as Haraway suggests, but if so, we are cyborgs who perform traditional biography,* Blade Runner *androids passing as standard human bodies and the standard body, the subject's body, is male. (Gilbert, 1996: 137)*

Instances in which users 'deceive' others through misleading online self-representations illustrate precisely how the body functions as the final arbiter of truth, authenticity and meaning. You can be whomever you want to be online, but quite often you will be asked to reveal your 'true' identity—i.e. a shorthand reference to your gender or ethnicity. And once you have uttered 'male' or 'female', 'black' or 'white', there is little fluidity or ambiguity about what this nomenclature means.

Implications for Digital Pedagogy

Whose desire is it to transcend the body? What sense do we make of the coincidence that, just as feminist poststructuralists, postcolonial theorists and neo-Marxist accounts had begun to establish the significance of the body to epistemology and signification, we encounter a new version of Descartes' disembodied fantasy? In this concluding section, new digital Cartesianism is read against some of the educational aims outlined by radical, progressive and feminist educators. There persists a gulf within educational theory between the discourses of cognitive psychology and social constructionism that frame the development of instructional technologies and the critical theorists who are beginning to articulate what would count as 'creative' and 'critical' digital pedagogies. While there is no space to outline this vast discrepancy here, the discourses of digital Cartesianism align with the learning models of the emphasis of instructional technology on 'scaffolded learning' and 'banking education' long eschewed within progressive and radical educational practices.

The feminist philosopher Susan Bordo's critique of Descartes and her suggested goal of 'dynamic objectivity' offers an initial answer to the question: 'Whose desire is it to transcend?' She elaborates:

> *If the key terms in the Cartesian hierarchy of epistemological values are clarity and distinctness—qualities which marked each object off from the other and from the nowhere—the key term in this alternative scheme of values might be designated as sympathy . . . [Sympathetic understanding] means granting personal or intuitive response a positive epistemological value, even (perhaps especially) when such response is contradictory or fragmented. 'Sympathetic' thinking, Marcuse suggests, is the only mode which truly respects the object, that is, which allows the variety of its meanings to unfold without coercion were too focused interrogation. (1987: 102–3)*

Bordo outlines Evelyn Fox Keller's notion of 'dynamic objectivity' and argues that:

> *[I]n contrast to the conception of dynamic objectivity, Descartes' program for the purification of the understanding . . . has as its ideal the rendering impossible of any such continuity between subject and object. The scientific mind must be cleansed of all its 'sympathies' toward the objects it tries to understand. It must cultivate absolute detachment. (1987: 103–4)*

Sandra Harding calls this a supreme characterization of modern science and a 'super masculinization of rational thought' (quoted in Bordo, 1987: 104). Worthy of note is that, in striving to transcend binaries, some of the discourses of hope outlined earlier actually reinscribe potential binary divisions between subject and object.

Feminists and others have extended these worries about the dismissal of embodied subjectivity directly to their analyses of digital identities. As Anne Balsamo reflects:

> *Upon analyzing the 'lived' experience of virtual reality, I discovered that this conceptual denial of the body is accomplished through the material repression of the physical body. The phenomenological experience of cyberspace depends upon and in fact requires the willful repression of the material body . . . From a feminist perspective is clear that the repression of the material body belies a gender bias in the supposedly disembodied (gender free) world of virtual reality. (2000: 493)*

Balsamo's argument confirms the longstanding feminist critique of Cartesian rationality. Namely, that the ideal of a reality 'free from bodies' reflects the masculinist ideal which entails a repression of materiality and the body. In a moving story ('Come in CQ') that echoes this concern, Ullman writes:

> *Ironically those of us who most believe in physical, operational eloquence are the very ones most cut off from the body. To build the working thing that is a program, we perform 'labor' that is sedentary to the point of near immobility and*

> *we must give ourselves up almost entirely to language . . . Software engineering is an oxymoron: we are engineers, but we don't build anything in the physical sense of the word. We think. We type. It's all grammar.*
>
> *And, cut off from the real body, we construct a substitute body: ourselves online. We treat it as if it were our actual self, our real life. Over time, it does indeed become our life. (Ullman, 1996: 12)*

These notions of dynamic objectivity and insistence on the importance of embodiment as part of epistemology and education are in strong tension with the constrained imaginations of bodies and differences outlined in new digital Cartesianism.

What educational experiences lead to growth and transformation? How do we comprehend self and other as situated in a lived, embodied and material environment? In *Experience and Education,* Dewey cautions:

> *A primary responsibility of educators is that they not only be aware of the general principle of the shaping of actual experience by environing conditions, but that they also recognize in the concrete what surroundings are conducive to having experiences that lead to growth. Above all, they should know how to utilize the surroundings, physical and social, that exist so as to extract from them all that they have to contribute to building up experiences that are worthwhile. (1997[1938]: 40)*

Dewey emphasizes the educator's responsibility to consider the concrete surroundings that shape educational experience. Educational growth and transformation cannot be divorced from the material environment:

> *[W]e live from birth to death in a world of persons and things which in large measure is what it is because of what has been done and transmitted from previous human activities. When this fact is ignored, experience is treated as if it were something which goes on exclusively inside an individual's body and mind. It ought not to be necessary to say that experience does not occur in a vacuum. (1997[1938]: 39–40)*

Dewey's urging to consider the distributed experience of the self suggests a fundamental epistemological and ontological problem for online education. Why discourage users from engaging difference through computer-mediated forms? As the problem of 'drive-by difference' was described earlier, the concern is that in CMC we encounter the other only through voyeuristic perception as opposed to a radical encounter with embodied difference that leads to self-reflexive change.

Foreshadowing the work of thinkers such as Foucault as well as contemporary analyses of how space is defined not merely in terms of its absolute, physical stasis, but more importantly, how spaces dynamically construct social experience and vice versa, Dewey's critique of traditional education parallels contemporary concerns about online education:

> *Traditional education . . . could systematically dodge this responsibility. The school environment of desks, blackboards, a small school yard, was supposed to suffice. There was no demand that the teacher should become intimately acquainted with the conditions of the local community, physical, historical, economic, occupational, etc., in order to utilize them as educational resources. (1997[1938]: 40)*

Dewey's insistence on the local, physical, historical conditions as fundamental to a progressive pedagogy suggests that, at minimum, education that is conducted entirely online needs to revisit what it might look like to invoke this material historicity as part of the educational exchange.[19] In contemporary terms, Alison Adam asks in *Artificial Knowing: Gender and the Thinking Machine*:

> *[H]ow far is the body or embodiment necessary for having knowledge[?] . . . [A]t least two aspects of situatedness are of interest—being physically situated in an*

> *environment (which relates to the embodiment problem) and being socially situated in a culture. (1998: 129)*

For many educators—as illustrated by Ingrid Banks' comments earlier—the body represents a kind of shared 'text' within a physically proximate environment and is recognized as part of the subject who experiences growth or transformation. The risk of losing the body as part of transformative pedagogy in web-enhanced education is a real one.

More systematic studies are needed to investigate precisely what is gained when users feel more 'freedom' to speak in online discussions and what this means about our ability to engage in productive F2F dialogue (Boler, 2004). We need to know how and when students feel recognized or heard during online discussions about volatile questions regarding gender, ethnicity, sexual orientation and disabilities. For example, in an article titled 'Type Normal Like the Rest of Us', Alison Regan analyzes the pedagogical and ethical dilemmas that she faces when, in an online chat in her composition course, a student expresses death threats towards homosexuals and in the ensuing discussion the students use the pronoun 'we', revealing the assumption that all the participants were heterosexual. Regan analyzes how such homophobic expressions easily result in the silencing and 'exclusion of lesbian and gay participants from networked conversation' (1994: 118). As Gilbert puts it:

> *The Net is not 'just words' . . . but a space of social action, in which subjects are responsible for their utterances and performances and in which discursive actions can mobilize material effects. Like other social spaces, it is not safe . . . That these spaces are discursive rather than material does not lessen their phenomenological reality. (1996: 137)*

Finally, given that one of the primary arguments about digital participation is that users feel freer to participate because of the safety offered by relative anonymity, how can we measure dialogue in its transformative sense as engaged in CMC? Paulo Freire contends that dialogue

> *cannot exist unless the dialoguers engage in critical thinking—thinking which discerns an indivisible solidarity between the world and the people and admits of no dichotomy between them—thinking which perceives reality as process, as transformation, rather than as a static entity-thinking which does not separate itself from action, but constantly immerses itself in temporality without fear of the risks involved. (1970: 92)*

Freire's vision seems in part antithetical to the distanced anonymity of interaction in digital environments. This sense of safety is not the kind of risk-taking Freire calls for: rather, he envisions a risk-taking based in action, self-reflexive praxis and dialogue that recognizes the body, the word and world as material constituents of identity and change.

Conclusion

In conclusion, it is possible to imagine engaging CMC in ways that do not promote new digital Cartesianism. It is possible to imagine web-enhanced education that engages bodies in F2F environments in ways that can be taken up with complexity in CMC environments. Nevertheless, the risks are great. We live in an era of time–space compression where attention is at a premium and the economy is fueled by oversaturated attention. We have invested decades of theory, practice and politics to create physical and theoretical spaces in which bodies and differences are recognized—in all of their complexities—as fundamental to the production of meaning. Bodies are messy and differences fraught. What would it take to preserve messiness and conflict as foundational values, rather than being tempted by neat and tidy drive-by difference, as we shape the digital future in all its material significance?

About the Author

Megan Boler is an associate professor at the University of Toronto, where she teaches in philosophy, feminist theory and media studies. Her publications include *Feeling Power: Emotions and Education* (Routledge, 1999) and as editor, *Democratic Dialogue in Education: Troubling Speech, Disturbing Silences* (Peter Lang, 2004). She is currently working on a three-year SSHRC-funded research project, 'Rethinking Media, Democracy and Citizenship: New Media Practices And Online Digital Dissent After September 11' (see: www.oise.utoronto.ca/research/sshrc)

Address: OISE/University of Toronto, 252 Bloor St West, Toronto, Ontario M5S 1V6, Canada. [email: mboler@oise.utoronto.ca]

Notes

1. This notion of materiality is referenced in Sofoulis, in which she writes that the 'objects and bodies studied/produced by technoscience . . . are . . . "material/semiotic actors", whose boundaries are not predefined but 'materialize in social interaction' (2003: 6). To some extent this portrait drawn from Latour and Haraway characterizes the needed understanding of communication (even in so-called 'disembodied' spaces) as materially productive. Similarly, in *Technologies of Gender* (1987), De Lauretis addresses how technology shapes and mediates our relationships in the physical and material world.
2. Here I signal two distinct fields: feminist poststructuralism and postcolonialism that consider the centrality of the body to epistemology and perception, and more recent critiques of cyberculture that raise worries about the 'disembodiment' discourses that threaten the inroads made by earlier decades of feminist and postcolonial work. For example, Fanon's considerations of how the body is perceived through the gaze of the other and how that look is then internalized into the self, is not a theoretical issue widely considered by cyber-enthusiasts that I discuss in the 'hopes' section. There are a handful of feminist analyses of the representation of bodies in cyberculture that have begun to develop the urgently needed 'materialist' analyses of the body in relation to technology—a concern certainly initiated in Haraway's (1991[1985]) *A Cyborg Manifesto*. King (2003) references Haraway's notion of technology as 'frozen social relations', which may be another way of talking about the intractability of stereotyped notions of bodies in relation to lived practices. The representation of the body has been taken up by some cultural and certainly literary studies (Foster, 1999; Joyce, 1999; Nakamura, 2000; O'Brien, 1999); but there is far less to be found in the way of a neo-Marxist or feminist analysis of the relationship of the body to technology in everyday practices.
3. For example, I co-directed a documentary about a women's studies class that had been conducted entirely online during a summer session (Boler and Vellenga, 2000). This project developed when one of the students came to me at the end of the class and described the 'wow' experience that she and other students had during the entirely online course. Many of the students had experienced some measure of transformative growth. As we interviewed students on camera, it was revealed that it was the frequency and 'intimacy' of IM exchange with the professor, who was conducting the course from hundreds of miles away, that established the basis for this sense of transformative 'wow'. The student who originally approached me told me that students were prepared to rent a car and take a road trip to meet the professor who was in Kansas: in fact, they were obsessed with meeting this fabulous professor 'face to face'.
4. Yet others argue, to the contrary, that degrading and harmful norms and stereotypes are reinscribed within the social spaces of digital relations (Burkhalter, 1999; O'Brien, 1996; Regan, 1994).
5. In, 2002, I conducted a study with James Dwight of 10 top tier journals in education and educational technology (see also Dwight et al., 2006). Striking in these findings of journals between 1997–2002 was that in journals devoted to educational theory, only about 5 percent of the articles addressed critical theories of technology or new media in relation to education. A

corresponding absence was found in top information technology journals: only a very small percentage addressed questions of social context, ethnicity, class, gender or critical perspectives on technology. This signals a serious gap in the way that the field of education is conversing about the progressive uses of new media.

6. For discussions of globalization and neo-liberalism as relevant to an analysis of information technology, see for example, Castells (2000), Harvey (2000) and Sassen (1998).
7. Particularly fascinating within cyberculture hypes as well as hopes, is the association of body–space–femininity. The philosophical roots of the masculinist repudiations of the body as polluting knowledge and truth are discussed extensively in numerous feminist critiques (Benjamin, 1988; Bordo, 1987; Kaufman-Osborn, 1997, to signal a few). In a forthcoming work I draw on feminist cultural geography to analyze how space is represented and recreated in 'nostalgic' forms within online discourses.
8. This advertisement can be seen as Quick time online at http://it.stlawu.edu/uglobal/pagessemiotics/montagewcom.html
9. Discussing avatars, Nicola Green states that the 'digitized bodies' represent the

    historically constructed western individualist subject whom Grosz suggests is a historical abstraction because it evokes sameness, similarity and continuity. As Robins notes in the case of virtual systems . . . all identities can be rendered as one of these 'universal' (masculine) digital identities, thus suppressing the material effects of difference in digital interaction. (2001: 154)

10. The MCI ad can also be seen at http://www.brillomag.net/NO3/erasism.htm
11. Frequently, in traditional philosophy 'space' is considered something to be transcended and has been associated with femininity (McDowell, 1999). Correlated with 'being', space is understood traditionally as a fixed and static materiality. Thus space is often opposed to time or the masculine progress associated with 'becoming'. Feminist geographer Doreen Massey (1994) makes a powerful case that we need to understand space not as fixed and static, but as dynamic and changing. She defines 'place' in part as 'localities' which are shared social spaces. Places are defined as a locality of shared space, dynamic and changing because they are defined not by 'boundaries and containment' but by social interactions.
12. This refers in part to the effect of information technologies and the creation of the so-called 'global village'. Massey (1994) points out that the recent emphasis and anxiety expressed by theorists regarding time–space compression reflects two aspects of patriarchal ethnocentrism. First, the ethnocentric or colonizer's view that this compression of time–space is 'new'. In fact, many ethnic groups and communities have been exiled, forced to move or colonized for centuries and so have experienced time–space compression for generations. The anxieties associated with time-space compression are new only for the colonizer. Second, the depiction of time–space compression as creating much greater 'global movement' ignores the fact that mobility—say, for women and many poor persons—is still greatly constricted and highly dependent on 'power geometry' or access to social and economic class and power. In short, while time–space compression affects many persons across socioeconomic classes, it creates 'increased mobility' only for the few 'jetsetters' who fly business class from one country to another calmed by the presence of Starbucks in every international airport and city.
13. There is, by 2005, a well-established trajectory of sociological and ethnographic literature studying the experience of identities online. I conceptualize the early work in this field as having established its own unique perimeters (Rheingold, 1994; Stone, 1995; Turkle, 1995). There was then an influx of feminist analyses that both embraced the potential for gender-shifting as well as critiquing the 'hopes' for reasons similar to those I pose in this article (Baym, 1995; Bruckman, 1996; Herring, 1994; Kendall, 2002).
14. The theme of transcendence as connected to religious discourses occurs frequently both in reflective and unreflective forms in cyberculture literature. The range of writings that discuss

explicitly the religious elements of the discourses include Heim (1993) and Wertheim (1999). In another forthcoming article I analyze the significance of the invocation of transcendence from space and place, reading cultural geography against these 'religious' cyberculture discourses to examine the gendered nature of the repudiation of space and place.

15. In another example, the purpose of the online chat is advertised precisely to search for people on the basis of their 'asl':

    Communities.com is an international online community where people from all over the world (184 countries) meet, chat, flirt and interact every day. There are loads of things to do and it's FREE to join!

    Make new friends

    Whether you're looking for new friends in a different part of the world or in your local area you should be able to find them here. If you're looking for someone in particular you can search for members by age, sex, location, etc. When you find someone that seems interesting you simply send them a message and start chatting. And who knows, maybe you'll even find the love of your life inside!

16. To the extent that this is accurate, psychoanalytic analyses of self–other relations constructed in online communities is promising (Benjamin, 1988). Unfortunately, there is no evidence yet that the majority of users inhabiting online spaces represent a demographics of democracy, much less a population who share 'queered' understandings of the relationship of sex to gender. How are educators, in particular, to challenge cultural habits, values and norms within the narrowed bandwidth on text-based interactions?

17. One of the more famous is the incident of Joan, the 'wheelchair therapist'. Despite that he was 'in fact' a man, the user Alex presented himself online (or was misperceived once and then came to assume this continuing identity) as a disabled, wheelchair-bound female therapist. As an online persona, over the course of several months 'Joan' developed numerous highly intimate relations with other women. When eventually 'some of the online friends wanted to meet her in person', real-life Alex freaked out and decided that online Joan needed to die. Joan's fictitious husband then gets online and tells the friends that Joan is deathly ill in a hospital. When the online friends offer an outpouring of financial and emotional assistance, real-life Alex finally decides Joan needs to recover. However, when 'real' cards and flowers are sent to the supposed Manhattan 'hospital', the fiction unraveled and Joan is revealed to be Alex.

    There are numerous accounts of this particular event. Joan's friends were furious with the deception. As Sherry Turkle summarizes, some of the anger is simply anger at being lured into intimacy by a man who posed as a woman to win their secret confidence. Some of the anger

    centers on the fact that Joan had introduced some of her online women friends to lesbian netsex and the women involved felt violated by Joan's virtual actions [when in fact he was a man] . . . In other accounts, Joan introduced online friends to Alex, a Manhattan psychiatrist, who had real-life affairs with several of them. (1995: 229)

18. In social theories of online identities and communities, there is debate about what constitutes the 'real' persona and disagreement about what such ethics must be based upon; for example, when this crisis occurred in the CompuServe community,

    the hackers . . . just smiled tiredly . . . All of them had understood from the beginning that the net presaged radical changes in social conventions, some of which would go unnoticed. That is, until an event like the disabled woman who is revealed to be 'only' a persona—not a true name at all—along with the violated confidences that resulted from the different sense in which various actors understood the term person, all acted together to push these changes to the foreground. (Stone, 1995: 80–1)

At what point did the transgression of acceptable ethical norms occur? Where does one draw the line regarding which self counts as real in online interactions? However, these questions are beyond the scope of this article. My focus is simply to recount how the body functions as an emergent 'metaphysics of presence'.

19. The materiality of the body is ignored also in the hype of the 'accessibility' of online education for certain populations: for example, offering homebound women with children access to online education reinscribes women's isolation from the public sphere. Alternatives include providing women with the option of adequate childcare so they can engage in face-to-face education. To isolate certain bodies geographically may deliver less on the promise of 'connection' across borders and instead reinscribe women's exclusion from the public sphere. Further, a concern expressed by many is that engaging with others via digital representations of language and images atomizes and isolates social experience. Whether or not one sees time spent online as socially isolating or not, relatively few studies exist that take as a central question the body's material relationship to computing practices through a materialist or phenomenological frame.

References

Adam, Alison (1998) *Artificial Knowing. Gender and the Thinking Machine*. London and New York: Routledge.

Balsamo, Anne (2000) 'The Virtual Body in Cyberspace', in David Bell and Barbara Kennedy (eds) *The Cybercultures Reader*. London: Routledge.

Banks, Ingrid (1998) 'Reliance on Technology Threatens the Essence of Teaching', *Chronicle of Higher Education,* 16 October, p. A24.

Baym, Nancy (1995) 'The Emergence of Community in Computer-Mediated Communication', in Steven G. Jones (ed.) *Cybersociety*. Thousand Oaks, CA: Sage.

Benjamin, Jessica (1988) *The Bonds of Love*. NY: Pantheon.

Boler, Megan (2004) (ed.) *Democratic Dialogue in Education: Troubling Speech, Disturbing Silence*. New York: Peter Lang.

Boler, Megan (2003) 'The New Digital Cartesianism: Bodies and Spaces in Online Education', *Philosophy of Education Society 2002* (ed. Scott Fletcher). Champaign, IL: Philosophy of Education Society.

Boler, Megan (1999) *Feeling Power: Emotions and Education*. NY: Routledge.

Boler, Megan (1997a) 'Disciplined Emotions: Philosophies of Educated Feelings', *Educational Theory* 47(3): 203–27.

Boler, Megan (1997b) 'The Risks of Empathy: Interrogating Multiculturalism's Gaze', *Cultural Studies* 11(2): 253–73.

Bordo, Susan (1987) *The Flight to Objectivity*. New York: SUNY Press.

Bruckman, Amy S.(1996) 'Gender Swapping on the Internet', in Peter Ludlow (ed.) *High Noon on the Electronic Frontier. Conceptual Issues in Cyberspace,* pp. 317–26. Cambridge, MA: MIT Press.

Burkhalter, Byron (1999) 'Reading Race Online: Discovering Racial Identity in Usenet Discussions', in Mark A. Smith and Peter Kollock (eds) *Communities in Cyberspace,* pp. 60–75. London: Routledge.

Castells, Manuel (2000) *The Rise of the Network Society*. London: Blackwell Science.

De Castell, S. and J. Jenson (2004) 'Paying Attention to Attention: New Economies for Learning', *Educational Theory* 54(4): 381–98.

De Lauretis, Teresa (1987) *Technologies of Gender*. Bloomington, IN: Indiana University Press.

Dewey, John (1997[1938]) *Experience and Education*. New York: Macmillan.

Dwight, James, Megan Boler and Pris Sears (2006) 'Reconstructing the Fables: Women on the Educational Cyberfrontier', in Joel Weiss, Jason Nolan, Jeremy Hunsinger and Peter Trifonas (eds) *International Handbook of Virtual Learning Environments,* pp. 1107–29. New York: Springer.

Fanon, Frantz (1967[1952]) *Black Skin, White Masks*. New York: Grove Press.

Fisher, Jeffrey (1997) 'The Postmodern Paradiso: Dante, Cyberpunk and the Technosophy of Cyberspace', in David Porter (ed.) *Internet Culture,* pp. 111–32. London: Routledge.

Foster, Thomas (1999) '"The Souls of Cyber Folk": Perfomativity, Virtual Embodiment and Racial Histories', in Marie-Laure Ryan (ed.) *Cyberspace Textuality, Computer Technology and Literary Theory,* pp. 137–63. Bloomington, IN: Indiana University Press.

Freire, Paulo (1970) *Pedagogy of the Oppressed*. New York: Continuum.

Gibson, William (1984) *Neuromancer*. New York: Ace Books.

Gilbert, Pamela (1996) 'On Space, Sex and being Stalked', in Theresa M. Senft and Stacy Horn (eds) *Women & Performance: A Journal of Feminist Theory,* pp.125–50. New York: Women and Performance Project Inc.

Grosz, Elizabeth (1993) 'Bodies and Knowledges; Feminism and the Crisis of Reason', in Linda Alcoff and Elizabeth Potter (eds) *Feminist Epistemologies,* pp. 187–216. London: Routledge.

Gray, Chris Hables (2001) *Cyborg Citizen*. New York: Routledge.

Hall, Stuart (1997) *Representation: Cultural Representations and Signifying Practices*. London: Sage.

Haraway, Donna (1991[1985]) 'A Cyborg Manifesto', in *Simians, Cyborgs and Women: the Reinvention of Nature*. New York: Routledge.

Harvey, David (1990) *The Condition of Postmodernity*. London: Blackwell.

Harvey, David (2000) *Spaces of Hope*. Berkeley and Los Angeles, CA: University of California Press.

Heim, Michael (1993) *The Metaphysics of Virtual Reality*. Oxford: Oxford University Press.

Herring, Susan (1994) 'Gender Differences in Computer-Mediated Communication: Bringing 'Familiar Baggage to the New Frontier', paper presented at 'Making the Net *Work*: Is There a Z39.50 in Gender Communication?', American Library Association Annual Convention, Miami, 27 June (available at: http://cpsr.org/cpsr/ gender/herring.txt).

Ihde, Don (2002. *Bodies in Technology*. Minneapolis, MN: University of Minnesota Press.

Joyce, Michael (1999) 'On Boundfulness: the Space of Hypertext Bodies', in Mike Crang, Phil Crang and Jon May (eds) *Virtual Geographies,* pp. 222–43. London and New York: Routledge.

Kaufman-Osborn, Timothy V. (1997) *Creatures of Prometheus*. Lanham, MD: Rowman & Littlefield.

Kelly, Kevin (2002) 'God Is the Machine', *Wired Magazine* 10(12), available at: http://www.wired.com/wired/archive/10.12/holytech.html

Kendall, Lori (2002) *Hanging Out in the Virtual Pub*. Berkeley, CA: University of California Press.

King, Katie (2003) 'Women in the Web', in Marc Bousquet and Katherine Wills (eds) *The Politics of Information,* pp. 303–14, URL (consulted 23 May 2006): http://www.altx.com/ebooks/infopol.html

Lefebvre, Henri (1981) *The Production of Space*. Paris: Anthropos.

Lugones, Maria (1989) 'Playfulness', "World" Traveling and Loving Perception', in Ann Garry and Marilyn Pearsall (eds) *Women, Knowledge and Reality,* pp. 275–90. Boston, MA: Unwin Hyman.

Massey, Doreen (1994) *Space, Place and Gender*. Minneapolis, MN: University of Minnesota Press.

McDowell, Linda (1999) *Gender, Identity, and Place: Understanding Feminist Geographies*. Minneapolis, MN: University of Minnesota Press.

Mohanty, Chandra Talpade (1989–90) 'On Race and Voice: Challenges for Liberal Education in the 1990s', *Cultural Critique* 14(winter): 179–208.

Nakamura, Lisa (2000) *Race in Cyberspace*. New York: Routledge.

O'Brian, Jodi (1996) 'Changing the Subject', in Theresa M. Senft and Stacy Horn (eds) *Women & Performance: A Journal of Feminist Theory,* pp. 55–68. New York: Women and Performance Project Inc.

O'Brien, Jodi (1999) 'Writing in the Body: Gender (re)production in Online Interaction', in Mark A. Smith and Peter Kollock (eds) *Communities in Cyberspace,* pp. 76–106. London: Routledge.

Probyn, Elspeth (1993) *Sexing the Self: Gendered Positions in Cultural Studies*. London: Routledge.

Regan, Alison (1994) 'Type Normal Like the Rest of Us: Writing, Power and Homophobia in the Networked Composition Classroom', in Linda Garber (ed.) *Tilting the Tower,* pp. 154–67. New York: Routledge.

Rheingold, Howard (1994) *The Virtual Community*. Amherst, MA: MIT Press.

Riel, Margaret (1996) 'Cross-Classroom Collaboration: Communication and Education', in Timothy Koschmann (ed.) *CSCL: Theory and Practice of an Emerging Paradigm*. Mahwah, NJ: Lawrence Erlbaum Associates.

Roberts, Lynn D. and Malcolm R. Parks (2002) 'The Social Geography of Gender Switching', in Eileen Green and Alison Adam (eds) *Virtual Gender,* pp. 265–87. New York: Routledge.

Robins, Kevin and Frank Webster (1999) *Times of the Technoculture*. London: Routledge.

Sassen, Saskia (1998) *Globalization and its Discontents*. New York: New Press.

Senft, Theresa M. (1996) 'Introduction: Performing the Digital Body—A Ghost Story', in Theresa M. Senft and Stacy Horn (eds) *Women & Performance: A Journal of Feminist Theory*, pp. 9–34. New York: Women and Performance Project Inc.

Sofoulis, Zoé (2003) 'Cyberquake: Haraway's Manifesto', in Darren Tofts, Annemarie Jonson and Alessio Cavallaro (eds) *Prefiguring Cyberculture: an Intellectual History,* pp. 9–34. Cambridge, MA: MIT Press.

Stone, Allucquère Rosanne (1995) *The War of Desire and Technology at the Close of the Mechanical Age*. Cambridge, MA: MIT Press.

Turkle, Sherry (1995) *Life on the Screen: Identity on the Age of the Internet*. New York: Simon and Schuster.

Ullman, Ellen (1996) 'Come In, CQ: the Body on the Wire', in Lynn Cherny and Elisabeth Weise (eds) *Wired_Women,* pp. 3–23. Seattle, WA: Seal Press.

Wakeford, Nina (1999) 'Gender and the Landscapes of Computing in an Internet Café', in Mike Crang, Phil Crang and Jon May (eds) *Virtual Geographies,* pp. 178–202. London: Routledge.

Wertheim, Margaret (1999) *The Pearly Gates of Cyberspace: a History of Space from Dante to the Internet.* New York: W.W. Norton.

Zembylas, M. and C. Vrasidas (2005) 'Levinas and the "Inter-face": the Ethical Challenge of Online Education', *Educational Theory* 55(1): 61–78.

Why I Am Not Going to Buy a Computer

Wendell Berry

Like almost everybody else, I am hooked to the energy corporations, which I do not admire. I hope to become less hooked to them. In my work, I try to be as little hooked to them as possible. As a farmer, I do almost all of my work with horses. As a writer, I work with a pencil or a pen and a piece of paper.

My wife types my work on a Royal standard typewriter bought new in 1956, and as good now as it was then. As she types, she sees things that are wrong, and marks them with small checks in the margins. She is my best critic because she is the one most familiar with my habitual errors and weaknesses. She also understands, sometimes better than I do, what *ought* to be said. We have, I think, a literary cottage industry that works well and pleasantly. I do not see anything wrong with it.

A number of people, by now, have told me that I could greatly improve things by buying a computer. My answer is that I am not going to do it. I have several reasons, and they are good ones.

The first is the one I mentioned at the beginning. I would hate to think that my work as a writer could not be done without a direct dependence on strip-mined coal. How could I write conscientiously against the rape of nature if I were, in the act of writing, implicated in the rape? For the same reason, it matters to me that my writing is done in the daytime, without electric light.

I do not admire the computer manufacturers a great deal more than I admire the energy industries. I have seen their advertisements, attempting to seduce struggling or failing farmers into the belief that they can solve their problems by buying yet another piece of expensive equipment. I am familiar with their propaganda campaigns that have put computers into public schools in need of books. That computers are expected to become as common as TV sets in "the future" does not impress me or matter to me. I do not own a TV set. I do not see that computers are bringing us one step nearer to anything that does matter to me: peace, economic justice, ecological health, political honesty, family and community stability, good work.

What would a computer cost me? More money, for one thing, than I can afford, and more than I wish to pay to people whom I do not admire. But the cost would not be just monetary. It is well understood that technological innovation always requires the discarding of the "old model"—the "old model" in this case being not just our old Royal standard, but my wife, my critic, my closest reader, my fellow worker. Thus (and I think this is typical of present day technological innovation), what would be superseded would be not only some thing, but some body. In order to be technologically up-to-date as a writer, I would have to sacrifice an association that I am dependent upon and that I treasure.

My final and perhaps my best reason for not owning a computer is that I do not wish to fool myself. I disbelieve, and therefore strongly resent, the assertion that I or anybody else could write better or more easily with a computer than with a pencil. I do not see why I should not be as scientific about this as the next fellow: When somebody has used a computer to write work that is demonstrably better than Dante's, and when this better is demonstrably attributable to the use of a computer, then I will speak of computers with a more respectful tone of voice, though I still will not buy one.

To make myself as plain as I can, I should give my standards for technological innovation in my own work. They are as follows:

1. The new tool should be cheaper than the one it replaces.
2. It should be at least as small in scale as the one it replaces.
3. It should do work that is clearly and demonstrably better than the one it replaces.
4. It should use less energy than the one it replaces.
5. If possible, it should use some form of solar energy, such as that of the body.
6. It should be repairable by a person of ordinary intelligence, provided that he or she has the necessary tools.
7. It should be purchasable and repairable as near to home as possible.
8. It should come from a small, privately-owned shop or store that will take it back for maintenance and repair.
9. It should not replace or disrupt anything good that already exists, and this includes family and community relationships.

Old Rituals for New Space: Rites de Passage *and William Gibson's Cultural Model of Cyberspace*

David Tomas

Across the communications landscape move the specters of sinister technologies and the dreams that money can buy.

—*J. G. Ballard,* Crash

The experience of subjective and intersubjective flow in ritual performance, whatever its sociobiological or personalogical concomitants may be, often convinces performers that the ritual situation is *indeed informed with powers both transcendental and immanent.*

—*Victor Turner, "Social Dramas and Stories about Them"*

The computer-generated interactive virtual environment of cyberspace has recently engaged the creative imaginations of a narrow spectrum of government, corporate, and academic researchers from various disciplines, as well as an assortment of other individuals including artists and science fiction writers. It has especially gained notoriety among cultural literati through the writings of William Gibson (Gibson 1984, 1987a, 1987b, 1988; cf. Greenfield 1988; Hamburg 1989), whose work has been most closely identified with a new science fiction genre and subculture of popular music known collectively under the rubric *cyberpunk* (cf. Sterling 1988, Dery 1989). Popular culture is in fact a principal ingredient of the synergetic technovisionary world of "cyberpunk literature" to the extent that one finds that *Neuromancer* (1984), "surely the quintessential cyberpunk novel" (Sterling 1988: xiv), was not considered by its author to be a book but rather "a pop artifact" (Gibson quoted in Dery 1989: 77).

Although cyberspace has been popularized by Gibson's books, it is neither a pure "pop" phenomenon nor a simple technological artifact, but rather a powerful, collective, mnemonic technology that promises to have an important, if not revolutionary, impact on the future compositions of human identities and cultures (cf. Tomas 1989). Gibson, in particular, has devoted considerable attention to the chilling socioeconomic implications of this space and its postindustrial context. His depiction of an information society whose governing economy is transnational and cyberspatial is sobering and worthy of consideration, especially since he has chosen to depict it, as Sterling has pointed out, "from the belly up, as it is lived, not merely as dry speculation" (1986: xi). In fact, Gibson has presented us with the most sophisticated and detailed "anthropological" vision of cyberspace to date: its social and economic facets, and the outlines of its advanced postindustrial form. Although this is not the place to engage the totality of Gibson's dystopic vision, it is important to note that his novels and short stories are noteworthy social documents because they "show," as Sterling notes, "with exaggerated clarity, the hidden bulk of an iceberg of social change" that "now glides with sinister majesty across the surface of the late twentieth century" (1986: xi). In the following pages, I examine the anthropological implications of Gibsonian cyberspace and, in particular, its connections to a panhuman ritual process known as *rites de passage.*

Why choose to treat cyberspace from an anthropological, indeed ritual, perspective? First, as Sterling (cf. 1988: xi) has pointed out, science fiction is an important tool that allows us to make sense of a rapidly emerging postindustrial culture. It is a *spatial operator* (Serres 1983a: 42–43) connecting pasts and futures by way of the present (a function that it shares, as we shall see, with a cyberspace *rite de passage*). Second, it allows us to make sense of an advanced information technology that has the potential to not only change the economic structure of human societies but also overthrow the sensorial and organic architecture of the human body, this by disembodying and reformatting its sensorium in powerful, computer-generated, digitalized spaces. This latter consequence of cyberspace will force anthropologists to take into account not only the jaded and contested question of the organic Other, which in Michel Serres's opinion (1983b: 67), "is only a variety or variation—of the Same," but also the postorganic, the classical (hardware-interfaced) *cyborg* and the postclassical, (software-interfaced) transorganic data-based cyborg or personality construct (Tomas 1989). This latter point is, as we shall see, intimately connected to the social and symbolic functions of traditional *rites of passage* rituals. Third, advanced digital technologies, such as those that generate cyberspace, hold considerable promise as a testing ground for postritual theories and practices, in particular as conceptualized by a critical postindustrial, postorganic anthropology.

Can cyberspace be considered to be a new social space? According to Gibson's description and contemporary work on virtual worlds technologies,[1] it does indeed hold the promise of new spatial configurations and related postorganic life forms. There is little doubt, therefore, that the composition of cyberspace must become the focus of intense speculation and the site of contested engagement in the immediate future, especially since corporations as well as research and development companies, university departments, and individual researchers are currently engaged in discussing and engineering its specifications. Anthropologists also have a vested interest in engaging virtual worlds technologies during this early stage of speculation and development, especially those who are interested in engaging advanced forms of Western technology from the points of view of their modes of social production. For there is reason to believe that these technologies might constitute the central phase in a postindustrial "rite of passage" between organically human and cyberpsychically digital life-forms as reconfigured through computer software systems. Should this prove to be correct, existing theories of ritual processes can provide important insights into the socially engineered cultural dimensions of cyberspace, and its social function as presently conceived. On the other hand, one can imagine that anthropological theories will not remain unaffected by their contact with advanced information systems of the caliber of cyberspace. One might envisage, for example, the outlines of another postorganic form of anthropology developing in the context of cyberspace, an anthropology specifically engaged in addressing the problems of engineering cyberspatial forms of intelligence as opposed to the more conventional humanistic, more or less reflexive, *study* of premodernist, modernist, or postmodernist humankind.

From Euclidean Space to Cyberspace

Walls of shadow, walls of ice.

—*William Gibson,* Burning Chrome

The identity of a historical culture is, according to Serres, the product of "very precise and particular connections" between "spatial varieties." "This construction," this "original intersection" is "that culture's very history." Cultures are therefore to be "differentiated by the form of the set of junctions, its appearance, its place, as well as by its changes of state, its fluctuations." "But what they have in common," Serres points out, "and what constitutes them as such is the operation itself of joining, of connecting." As a consequence, "the identity of a culture is to be read on a map, its identification card: this is the map of its homeomorphisms" (Serres 1983a: 45).

Serres argues, moreover, that an individual is also a continuously constructed product of a similar intersection or junction of social spaces and cannot therefore be considered to exist in a unified homogeneous space. The body, in fact, "works in Euclidean space, but it only works there. It sees in a projective space; it touches, caresses, and feels in a topological space; it suffers in another; hears and communicates in a third; and so forth, as far as one wishes to go" (Serres 1983a: 44). However, Serres cautions that

> *this [intersection] is not simply given or is not* always already *there, as the saying goes. This intersection, these junctions, always need to be constructed. And in general whoever is unsuccessful in this undertaking is considered sick. His body explodes from the disconnection of spaces. (1983a: 44)*

What Serres is addressing is the way both cultures and individuals are constituted in terms of junctures or multiplicities of more or less fluid social spaces that must always be understood as particular historical, indeed cultural, constructs. One such space is Euclidean space, the master space of Western "work-oriented cultures." It is a master space because it governs communication—"the space of measure and transport," "linked homogeneity," and "congruent identity" (Serres 1983a: 44, 52)—and also "because it is the space of work—of the mason, the surveyor, or the architect" (Serres 1983a: 44). It is the space of Western geometry: the geometry of vision, the road, the building, and the machine. On the other hand, this master space is a binary construct, consisting of the everyday social or profane spaces that Serres has initially described in Euclidean terms *and* sacred or "liminal" spaces. As Victor Turner notes, "for every major social formation there is a dominant mode of public liminality, the subjunctive space/time that is the counterstroke to its pragmatic indicative texture" (1977c: 34). Film, in his opinion, is the dominant form of public liminality in electronically advanced societies.

In recent years, there has emerged a new form of electronic space that holds revolutionary promise as *the* fin de siècle metasocial postindustrial work space. Variously described as a "space that wasn't space," a "nonplace," and a space in which "there are no shadows" (Gibson 1987a: 38, 166; Gibson 1988: 219), cyberspace is a postindustrial work environment predicated on a new hardwired communications interface that provides a direct and total sensorial *access* to a parallel world of potential work spaces. This interface, which is a world removed from the indirect and limited access provided by older print-based paradigms of visual literacy (Gibson 1984: 170), mediates between the sensorial world of the organically human and a parallel virtual world of pure digitalized information. Thus, in the words of one of Gibson's characters (1988: 13):

> *People jacked in so they could hustle. Put the trodes on and they were out there, all the data in the world stacked up like one big neon city, so you could cruise around and have a kind of grip on it, visually anyway, because if you didn't, it was too complicated, trying to find your way to a particular piece of data you needed.*

At its most extreme, the corporeality of "laboring" human bodies is replaced by pure information whose configurations "signify" disembodied human sensoria, personality constructs, and artificial intelligences.

The "abstract representation of the relationships between data systems" (Gibson 1987b: 169) in cyberspace is, however, highly plastic, and can take any form ranging from pure geometric color-coded copyrighted shapes or architectural representations (cf. Gibson 1988: 64, Gibson 1987b: 178, Gibson 1984: 256–257) signifying corporate ownership to "photo-realistic" illusions (cf. Gibson 1988: 148–149, 174, 221–224). *Such sites are the essence of a postindustrial society*—pure information *duplicated in metasocial form*: a global information economy articulated as a metropolis of bright data constructs, whose plasticity is governed by a Euclidean model based on the given problematic of visualizing data, a problematic subordinated, in Gibsonian cyberspace, to the dictates of a transnational computer-based economy. Although predominantly conceived as a virtual Euclidean work space for legitimate computer programmers to visualize their employer's data (cf. Gibson 1987b: 169–170, Gibson 1988: 13), cyberspace does, in principal, also allow a *direct* hardwired experience of other non-Euclidean spaces and spatialized consciousnesses to occur. The latter includes virus programs that have "replicated" and become smart (Gibson 1987a: 169) and artificial intelligences that have permeated its digitalized expanse and have finally attained "cosmic sentience" (Gibson 1988: 259). Gibsonian cyberspace is, as a consequence, digitally and socially Durkheimian in the sense that it is both profane (a metropolis of data) and sacred (a cybernetic godhead) (Gibson 1988: 192).

Gibsonian cyberspace can be distinguished according to three, dominant, "Euclidean" characteristics. First, it is conceived as a common transnational work environment. Second, it is a transportation space designed to accomplish work related tasks—both a space in which one can travel in real-time or by way of "bodiless, instantaneous shifts" (Gibson 1988: 220) *and* a space *through* which human memory and identity are transported globally. Third, it redefines and restructures *what it means to be human* in technoeconomic terms through a data-based collectivization of the human sensorium (cf. Tomas 1989) or, in the more radical customized and therefore "individualistic" terms, of "personality" or synthetic data constructs. This potential for a radical, postcorporeal, economic transubstantiation of the human body's traditional organic and sensorial architecture forcefully remind one of Serres's comments on the plasticity of interconnections between particular cultures and related human bodies. Cyberspace's dominant Euclidean form confirms Serres's observations on the Euclidean articulation of work and space in general in Western cultures. It raises, moreover, the critical question of the governing "logic" and social, cultural, and economic control of this new sensorial and mnemonic space.

In order to further clarify these issues and to highlight their importance for future cyberspace research one must attend to other, *post*-Euclidean, social and cultural alternatives that are germane to cyberspace. These are best addressed from the point of view of parallels that can be drawn between the postindustrial, cyberspatial transubstantiation of the human body and related social and symbolic transformations of the human body in traditional *rites de passage* rituals. First described in the cases of so-called, small-scale, preindustrial, pre–Industrial Revolution societies, these rituals draw attention to the role of "the social" as a mediative force in the many fundamental biological, cultural, and technological changes that have been experienced by humankind.

Rites de Passage

> *She slid the trodes on over the orange silk headscarf and smoothed the contacts against her forehead.*
>
> *"Let's go," she said.*
>
> *Now and ever was, fast forward, Jammer's deck jacked up so high above the neon hotcores, a topography of data he didn't know. Big stuff, mountain-high, sharp and corporate in the nonplace that was cyberspace.*
>
> —*William Gibson,* Count Zero

In 1909, Arnold Van Gennep published *Les Rites de Passage,* an examination of a class of rituals that mark major social stages in an individual's life or in the collective existence of a group. Van Gennep's study was the first to focus on this important class of rituals and note its tripartite structure.

Rites of passage are ideally distinguished by three successive phases in ritual or nonsecular space and time that function to engineer the transposition and transformation of an initiand (or initiands in the case of a social group) from one socioenvironmental, sociobiological, social position or stage to another. Notable events that are mediated by this ritual process include birth, puberty, marriage, and death. However, other more general transpositions from one significant social category, spatial area, cosmic, or seasonal phase to another are also subject to ritual, rites of passage, mediation. The sequence of rites, in such cases, is designed to acknowledge the importance and abnormality of the events while negotiating the movement between stages by integrating them into a succession of socially sanctioned activities while ensuring a certain spatial, temporal, and symbolic continuity.

The first phase of rites of passage, or rites of separation, consists of symbolic behavior that initiates the transition from the secular and profane world of the social group to a second phase known as a *liminal* period. This phase is marked by a time of ritual metamorphosis that mediates between the initiand's or group's previous and subsequent state. It is an especially dangerous and unclean period, since the initiand (or group of initiands) is considered to possess little of her/his (their) former or later attributes. This period is not only symbolically abnormal, but the location(s) and time in which the transformation takes place is considered outside of society. Turner (1977a) has described it as "betwixt and between" neither one nor the other—a state of nonbeing, death, or nothingness. He has argued, moreover, that "the liminal phase is the essential, *anti*secular component in true ritual, whether it be labeled 'religious' or 'magical'" (Turner 1980: 161). When the desired sociobiological or symbolic transformation has taken place, a procedure of renormalization is needed to reinstate the individual or group into society. The third phase, or rite of aggregation, functions as a "decontamination chamber," metaphorically speaking, in order that initiands can enter into a new social or sociobiological status in society without polluting everyday space and time with vestiges of former attributes or residues of the ritual process itself.

If rites of passage clearly *articulate* human bodies and identities from one social position to another, they do so because they function as structured conduits between normally distinct social realities. However, as Van Gennep originally pointed out,

> *Although a complete scheme of rites of passage theoretically includes preliminal rites (rites of separation), liminal rites (rites of transition), and postliminal rites (rites of incorporation), in specific instances these three types are not always equally important or equally elaborated. (1960: 11)*

This observation was later confirmed by Turner's brilliant examination and reconceptualization of the social and symbolic functions of the central, or liminal, phase of rites of passage (1974, 1977a, 1977b, 1977c).

Reflexive transformation is the governing function of the liminal, interstitial, and antistructural phase of the rites of passage in "small-scale, relatively stable and cyclical" tribal societies. In these societies, the liminal functions in relation to a "structure of positions" (Turner 1977a: 93). Liminality is, within this patchwork of stability, an interstructural space and time that always functions in relation to the structured, between which the "passenger," to use Turner's term (1977 a: 94), is symbolically as well as spatially and temporally articulated.

Transformation is signified and rendered effective by the unusual and "bizarre" symbolism found in the liminal phase (Turner 1977a: 96). This symbolism, associated with bodily and biological processes, marks the "ambiguous" and "paradoxical" condition that the subject or initiand is in. It is an extraclassifactory stage in which the initiand is "no longer classified and not yet classified" (Turner 1977a: 96). From the former point of view, liminality is defined by symbols of death, invisibility, structural dissolution, and other negative physical and organic processes. From the latter point of view, the condition is defined by symbols of gestation and parturition. Insofar as the liminal phase encompasses the destructuring and restructuring of the subject, it is also marked by the denial of structure and the origin of structure per se. Turner has pointed out, in connection with its creative aspect, that liminality is "a realm of pure possibility whence novel configurations of ideas and relations may arise" (1977a: 97). This creativity is well illustrated in his discussion of the communication of "sacra" to the initiand in the form of "exhibitions," "actions," and "instructions":

> *The communication of* sacra *and other forms of esoteric instruction really involves three processes, though these should not be regarded as in series but as in parallel. The first is the reduction of culture into recognized components or factors; the second is their recombination in fantastic or monstrous patterns and shapes; and the third is their recombination in ways that make sense with regard to the new state and status that the neophytes will enter. (Turner 1977a: 106)*

The result is "a promiscuous intermingling and juxtaposing of the categories of event, experience, and knowledge, with a pedagogic intention" (Turner 1977a: 106).

One begins to have a sense, in the case of small-scale societies, of the creative possibilities—of social and individual regeneration—inhering in this ambiguous phase of the rites of passage, a phase in which "the central cluster of nonlogical *sacra* is then the symbolic template of the whole system of beliefs and values in a given culture, its archetypal paradigm and ultimate measure" (Turner 1977a: 108). Turner describes this as a process in which "to look at itself a society must cut out a piece of itself for inspection." He continues,

> *To do this it must set up a frame within which images and symbols of what has been sectioned off can be scrutinized, assessed, and, if need be, remodeled and rearranged. (Turner 1977c: 35)*

This phase operates like a mythic cultural weaver whose function is, according to Serres,

> *to link, to tie, to open bridges, pathways, wells, or relays among radically different spaces; to say (*dire*) what takes place between them; to inter-dict (*interdire*). (1983a: 45)*

Moreover, as Serres points out in an observation that could have been made in regard to liminality in general, "the category of *between* is fundamental in topology and for our purposes here: to interdict in the rupture and cracks between varieties completely enclosed upon themselves" (1983a: 45). Thus, if the social is never, according to both Serres and Turner, a completely stable construct but must always be considered to be in a process of creative regeneration, then rites of passage offer exemplary instances of a social process that achieves unstable stability.

There are a number of similarities between the overall structure of rites of passages and cyberspace that suggest that the latter might be closely related to the former in a functional sense. First, the acts of "jacking in" and out of cyberspace by way of cyberdecks and matrix simulators suggest radically truncated versions of separation and aggregation "rites" in which the hardware serves as portal to, and exit from, a parallel virtual reality. The full emphasis of this "access" technology is on cyberspace itself, or on a central liminal—virtual—condition. Moreover, access and exit are *sequential,*[2] but *not* necessarily unidirectional (cf. Turner 1980: 160–161) in the sense that programmers and console cowboys can jack in and out of cyberspace at will. At its most extreme, however, as in the case of Bobby Newmark in *Mona Lisa Overdrive,* the technology can function as a permanent prosthesis or hardwired interface (Turner 1988: 240). Second, jacking into cyberspace involves a passage from the everyday space and finite time of the organically human or postorganic hardware-based cyborg to a digital—as opposed to an analogical—space and time that is both transorganic and cyberpsychically collective. It is collective in the sense that cyberspace exists, in the words of Continuity (an artificial intelligence in *Mona Lisa Overdrive*), "insofar as it can be said to exist, by virtue of human agency" (Turner 1988: 107). But the nature and form of the "human" portion of "human agency" is now in question. Finally, the nature of human identity at both an individual and collective level is rendered problematic in terms of agency, for cyberspace itself has the potential to become consensually posthuman as well as "post-present," as when its matrix is inhabited by what appear to human protagonists to be transcendent panhuman "intelligences" whose power and creative inventiveness is signaled by intracyberspatial influence or by way of the extracyberspatial production of cultural sacra (cf. Gibson 1988; Gibson 1987a: 217, 225–227).[3] The "mytho-logic" of these "presences" is explained by Continuity in an exchange with a human, Angela Mitchell (Gibson 1988: 107):

> "The folklore of console jockeys, Continuity. What do you know about that?"
>
> "What would you like to know, Angie?" . . .
>
> "'When It changed' . . ."
>
> "The mythform is usually encountered in one of two modes. One mode assumes that the cyberspace matrix is inhabited, or perhaps visited, by entities whose characteristics correspond with the primary mythform of a 'hidden people.' The other involves assumptions of omniscience, omnipotence, and incomprehensibility on the part of the matrix itself."

"That the matrix is God?"

"In a manner of speaking, although it would be more accurate, in terms of the mythform, to say that the matrix *has* a God, since this being's omniscience and omnipotence are assumed to be limited to the matrix."

"If it has limits, it isn't omnipotent."

"Exactly. Notice that the mythform doesn't credit the being with immortality, as would ordinarily be the case in belief systems positing a supreme being, at least in terms of your particular culture. Cyberspace exists, insofar as it can be said to exist, by virtue of human agency."

This mytho-logic suggests that one of cyberspace's more fundamental social functions is to serve as a medium to communicate a form of "*gnosis,* mystical knowledge about the nature of things and how they came to be what they are" (Turner 1977a: 107). In cyberspace, the classical hardware interfaced cyborg and the postclassical data-based cyborg or personality construct meet with new posthuman intelligences that engage in revelatory and pedagogic activities reminiscent of the activities of shaministic figures who mediate between traditional sacred and profane worlds. Such mediations between human and posthuman, analogue and digital spaces, suggest that cyberspace must be understood not only in narrowly socioeconomic terms, or in terms of a conventional parallel culture, but also and more importantly as an inherently original and inventive metasocial operator *and* potential creative cybernetic godhead.

On the Metasocial in Postindustrial Society

> *"Yeah," the Finn said, turning out onto the long straight empty highway, "but nobody's talkin'* human, *see?"*
>
> —*William Gibson,* Mona Lisa Overdrive

Turner has proposed a distinction between "liminal" and "liminoid," or "ergic-ludic ritual liminality and anergic-ludic liminoid genres of action" (1974: 83) that acknowledges an important, indeed decisive, difference between types of society called "tribal" and "post-tribal" or pre- and post–Industrial Revolution societies. This distinction is especially expressed in the opposition between work and leisure.

> *It would seem that with industrialization, urbanization, spreading literacy, labor migration, specialization, professionalization, bureaucracy, the division of the leisure sphere from the work sphere, the former integrity of the orchestrated religious gestalt that once constituted ritual has burst open and many specialized performative genres have been born from the death of that mighty* opus deorum hominumque. . . . *Disintegration has been accompanied by secularization. Traditional religions, their rituals denuded of much of their former symbolic wealth and meaning, hence their transformative capacity, persist in the leisure sphere but have not adapted well to modernity. Modernity means the exaltation of the indicative mood; but in what Ihab Hassan has called the "postmodern turn" we may be seeing a return to subjunctivity and a rediscovery of cultural transformative modes. . . . Dismembering may be a prelude to remembering, which is not merely restoring some past intact but setting it in living relationship to the present. (1980: 166–167)*

The transition from the liminal to the liminoid is broadly marked by a transition from the ritual collective to the secular individual, individualistic, or individualizing. While the ritual collective is "centrally integrated into the total social process, forming with all its other aspects a complete whole," the secular individual develops "most characteristically *outside* the central economic and political processes, along their *margins,* on their *interfaces,* in their 'tacit dimensions'" (Turner

1977b: 44). Liminal behavior is constrained by obligations associated with status, while liminoid behavior is characterized by relative contractual freedom (Turner 1974: 84; Turner 1977c: 50). The "liminal" is also distinguished by wide social and cultural affect in contrast to the more idiosyncratic orientations of the "liminoid." Finally, the liminal is more prone to "the inversion or reversal of secular, mundane reality and social structure" because of its overtly socially integrative functions, while liminoid phenomena, on the other hand, "are not merely reversive, they are often subversive, representing radical critiques of the central structures and proposing utopian alternative models" (Turner 1977b: 45). In conclusion, Turner notes that

> *liminal genres put much stress on social frames, plural reflexivity, and mass flow, shared flow, while liminoid genres emphasize idiosyncratic framing, individual reflexivity, subjective flow, and see the social as problem not datum. (1977c: 52)*

There are, however, connections between the two that suggest that they are not exclusive products of each type of society, that "in complex societies today's liminoid is yesterday's liminal" (1977b: 46). In this case, it might be true that the dismembering aspect of the liminal is continued in its fragmentation into different genres of social activity. As Turner points out:

> *In complex modern societies both types coexist in a sort of cultural pluralism. . . . The liminoid is more like a commodity indeed, often* is *a commodity, which one selects and pays for—than the liminal, which elicits loyalty and is bound up with one's membership or desired membership in some highly corporate group. One* works at *the liminal, one* plays with *the liminoid. (1974: 86)*

While this is not the place to enter into details concerning Turner's important distinction between liminal and liminoid, one might nevertheless query the uncritical celebration of a return to subjunctivity, with its implied displacement of a politics of the present in terms of utopian potentialities. For example, our experiences of "tribal" rites of passages have taught us that "voluntary *sparagmos* or self-dismemberment of order, in the subjunctive depths of liminality" (Turner 1980: 164), is achieved in relation to a set of actual positional states. This is also the case with "post-tribal" cyberspace. Not only does it have, in its Gibsonian version, a determinate Euclidean form and particular economic functions that are linked to actual socioeconomic formations (the military/industrial complex), but its principal mode of access and exit is by way of sophisticated communications equipment (cyberdecks or matrix simulators), the use of which is predicated on privileged or differential institutional access and specialized hardware and software expertise. In the case of legitimate programmers this access is clearly defined in terms of corporate filiation. Thus, a dominant Western spatial paradigm, corporate capitalism, differential access to technology, and specialized knowledge and expertise imply, at the very least, a more complex set of relations between so-called postmodern cultural phenomena—for cyberspace can also become, as we have seen, a culturally creative arena—and a given postindustrial economic context.

However, the distinction between liminal and liminoid, the collective and individualistic, does sharpen the focus on who will control cyberspace's governing socioeconomic logic—governments, corporations, or free-lance entrepreneurs? It does so because the issue of collective vs. individual control, which is not as clearly demarcated in the context of traditional rites-of-passage rituals, is effectively foregrounded in the contractual, marginal, subversive, and individualistic realms of the liminoid. This question can be extended to the design and operation of alternative hardware/software artifacts (the latter extending to the design of personality constructs and artificial intelligences) that might be used to mask or foreground cyberspace's governing logic and minimize or maximize social and cultural diversity. Thus, although problematic, Turner's attempt to account for pre- and posttribal ritual forms is valuable in that it serves as a corrective to uncritical trans- or ahistorical accounts of ritual processes; and the distinctions that he makes between liminal and liminoid allow for a more refined understanding of cyberspace's potential postindustrial forms and functions. At least, with this refinement and corrective in mind, we have an initial framework for addressing the complex issue of control in posttribal rituals of the caliber of cyberspace.

Gibsonian cyberspace exhibits liminoid characteristics connected to its economic functions as articulated in a complex open-ended postindustrial society. First, it exhibits a specific logic that structures the matrix in the name of a transnational information economy. This logic, replicating the binary (0-1) logic of computer languages, is visibly articulated in the matrix, sculpted in the copyrighted forms of data that structure, in turn, social activity. Walls of data, rather than walls of brick and glass, divide a hardwired, or postorganic, humanity into economic protagonists—those included and those excluded from, say, the dominant military/industrial complex. Those, in other words, that do and those that do not have direct access to information, hardware technology and software expertise. The latter include the "industrial-espionage artists and hustlers," computer or console cowboys like Bobby Quine in Gibson's short story "Burning Chrome":

> *Bobby was a cracksman, a burglar, casing mankind's extended electronic nervous system, rustling data and credit in the crowded matrix, monochrome nonspace where the only stars are dense concentrations of information, and high above it all burn corporate galaxies and the cold spiral arms of military systems. (1987b: 170)*

Cyberspace's binary structure governs the activities of anonymous corporate programmers, and individualistic console cowboys or jockeys who are always trying to gain an edge on liminal corporate artificial intelligences who are in the business of continually upgrading corporate ICE (Intrusion Countermeasures Electronics), but who periodically mutate into independent creative entities. This structure, as expressed in the walls of data, is therefore, on the one hand, literally considered as datum, while on the other hand it is considered as problem. Second, although primarily a work space, and thus not a space of "play" per se, cyberspace has the potential to become inventively and playfully antistructural in the sense proposed by Turner:

> *Anti-structure . . . can generate and store a plurality of alternative models of living, from utopias to programs, which are capable of influencing the behavior of those in mainstream social and political roles (whether authoritative or dependent, in control or rebelling against it) in the direction of radical change. (Turner 1974: 65)*

Bearing in mind the often complex relationship between the structuring functions of hardware and software systems and their potential for creative flexibility, one can nevertheless note that cyberspace's antistructural potential is best exemplified in the creative activities of console cowboys, in particular Bobby Newmark, who have transcended the economic dualism of the matrix in order to contemplate its creative logic, a logic rendered intelligible through the machinations of artificial intelligences of the caliber of Wintermute and Neuromancer. The activities of individuals such as Newmark and panhuman intelligences highlight cyberspace's inventive potential and its ability to sustain collective *and* individual activity, work *and* play, normative *and* spontaneous behavior in a manner that might directly and strategically challenge its economic logic. Thus, there is evidence to suggest that cyberspace might not only be a paradigmatic, postindustrial liminal/liminoid space—an "eye and eyestalk which society bends round upon its own condition, whether healthy or unsound" (Turner 1977c: 40)—but also a possible generative site for other creative logics and sensorial regimes.

Conclusion: On a Postorganic Anthropology

> *The world before television equates with the world before the Net—the mass culture and mechanisms of Information. And we are of the Net; to recall another mode of being is to admit to having once been something other than human.*
>
> —*William Gibson, "Rocket Radio"*

Serres's conception of the social is radically modernist in the sense that it takes into account the fracture and fluidity of cultural and individual identities whose junctions are organic human bodies, while positing the social realm as a topological construct governed by a Euclidean master space and inhabited by the mythic figure of the weaver. The weaver is, however, a *holographic* mytho-logical form to be found wherever information is being produced, transmitted, received. Its mytho-logic is, therefore, eminently cyberspatial, in the sense that its communicative function is digitally replicated throughout cyberspace. Serres's weaver is concentrated in particular in the prescriptive roles of artificial intelligences such as Wintermute, Neuromancer, Continuity, and Colin—entities that ensure that cyberspace retains its potential as a generative locale for other, more unusual, spatial configurations and cyborg intelligences.

Gibson's powerful vision is now beginning to influence the way virtual reality and cyberspace researchers are structuring their research agendas and problematics.[4] But if cyberspace represents, at the very least, the birth of a new postindustrial, metasocial spatial operator, it will remain for most part stillborn if its parameters are engineered primarily to function, following Gibson's dystopic vision, as a virtual world of contestatory economic activity. In order to counter this vision, one must actively and strategically seek alternative spatial and creative logics, social and cultural configurations. If such creative flexibility is critically foregrounded in current research agendas, cyberspace will indeed become a site of considerable cultural promise, and a locale for a new postorganic anthropology.

Acknowledgments

I would like to thank Jody Berland for comments that have contributed to the clarity of the argument presented in this chapter.

Notes

1. Cf. The First Conference on Cyberspace held May 4 and 5, 1990, at the University of Texas at Austin.
2. Turner draws attention to the definitive importance of "performative *sequencing*" for any understanding of ritual because of its role in ensuring a ritual's unidirectional transformative power (1980: 160–161).
3. In Turner's words, ritual is "prescribed formal behavior for occasions not given over to technological routine, having reference to beliefs in invisible beings or powers regarded as first and final causes of all effects" (1980: 159).
4. Cf. Tim McFadden's "The Structure of Cyberspace and the Ballistic Actors Model—an Extended Abstract" presented at The First Conference on Cyberspace, University of Texas at Austin.

References

Ballard, J. G. 1984 [1974]. Introduction to the French Edition. *Crash.* New York: Vintage Books.

Dery, Mark. 1989. Cyberpunk: Riding the Shockwave with the Toxic Underground. *Keyboard* 15-5: 74–89.

Gibson, William. 1984. *Neuromancer.* New York: Ace Books.

———. 1987a. *Count Zero.* New York: Ace Books.

———. 1987b. *Burning Chrome.* New York: Ace Books.

———. 1988. *Mona Lisa Overdrive.* Toronto, New York: Bantam Books.

———. 1989. Rocket Radio. *Rolling Stone* 554, 85–87.

Greenfield, Adam. 1988. New romancer. *Spin* 4-9, 96–99, 119. (Interview with William Gibson).

Hamburg, Victoria. 1989. The King of Cyberpunk Talks to Victoria Hamburg. *Interview* 19-1, 84–86, 91.

Serres, Michel. 1983a [1977]. Language and Space: From Oedipus to Zola. In *Hermes: Literature, Science, Philosophy.* Josue V. Harai and David F. Bell (eds.), 39–53. Baltimore: The Johns Hopkins University Press.

———. 1983b [1968]. Platonic Dialogue. In *Hermes: Literature, Science, Philosophy.* Josue V. Harai and David F. Bell (eds.), 65–70.

Sterling, Bruce. 1986. Preface. In William Gibson, *Burning Chrome,* ix–xii. New York: Ace Books.

———. 1988. Preface. *Mirrorshades,* Bruce Sterling (ed.), New York: Ace Books, ix–xvi.

Tomas, David. 1989. The Technophilic Body: On Technicity in William Gibson's Cyborg Culture. *New Formations* 8, 113–129.

Turner, Victor. 1974. Liminal to Liminoid, in Play, Flow, and Ritual: An Essay in Comparative Symbology. *Rice University Studies* 60-4, 53–92.

———. 1977a [1964]. Betwixt and Between: the Liminal Period in *Rites de Passage.* In V. Turner, *The Forest of Symbols,* 93–111. Ithaca: Cornell University Press.

———. 1977b. Variations on a Theme of Liminality. In *Secular Ritual.* Sally F. Moore and Barbara G. Myerhoff (eds.), 36–52. Amsterdam: Van Gortum, Assen.

———. 1977c. Frame, Flow and Reflection: Ritual and Drama as Public Liminality. In *Performance in Postmodern Culture.* Michel Benamou and Charles Caramello (eds.), 33–55. Madison, Wisconsin: Coda Press.

———. 1980. Social Dramas and Stories about Them. *Critical Inquiry* 7-1, 141–168.

Van Gennep, Arnold. 1960 [1909]. *The Rites of Passage.* Trans. M. B. Vizedom and G. L. Caffee. Chicago: The University of Chicago Press.

MODULE TWO

Looking at the Internet's Rear View Mirror

The Mother of All Networks

Tom Standage

telegraph. n.—a system of or instrument for sending messages or information to a distant place; v.—to signal (from French télégraphe*)*

On an April day in 1746 at the grand convent of the Carthusians in Paris, about two hundred monks arranged themselves in a long, snaking line. Each monk held one end of a twenty-five-foot iron wire in each hand, connecting him to his neighbor on either side. Together, the monks and their connecting wires formed a line over a mile long.

Once the line was complete, the abbé Jean-Antoine Nollet, a noted French scientist, took a primitive electrical battery and, without warning, connected it to the line of monks—giving all of them a powerful electric shock.

Nollet did not go around zapping monks with static electricity for fun; his experiment had a serious scientific objective. Like many scientists of the time, he was measuring the properties of electricity to find out how far it could be transmitted along wires and how fast it traveled. The simultaneous exclamations and contortions of a mile-long line of monks revealed that electricity could be transmitted over a great distance; and as far as Nollet could tell, it covered that distance instantly.

This was a big deal.

It suggested that in theory, it ought to be possible to harness electricity to build a signaling device capable of sending messages over great distances incomparably faster than a human messenger could carry them.

At the time, sending a message to someone a hundred miles away took the best part of a day—the time it took a messenger traveling on horseback to cover the distance. This unavoidable delay had remained constant for thousands of years; it was as much a fact of life for George Washington as it was for Henry VIII, Charlemagne, and Julius Caesar.

As a result, the pace of life was slow. Rulers dispatched armies to distant lands and waited months for news of victory or defeat; ships sailed over the horizon on epic voyages, and those on board were not seen or heard from again for years. News of an event spread outward in a slowly growing circle, like a ripple in a pond, whose edge moved no faster than a galloping horse or a swift-sailing ship.

To transmit information any more quickly, something that moved faster than a horse or a ship was clearly required. Sound, which travels at a speed of about twelve miles per minute, is one means of speedier communication. If a church bell strikes one o'clock, a monk standing in a field half a mile away knows what time it is about two seconds later. A horse-borne messenger, in contrast, setting out from the church precisely on the hour to deliver the message "It is one o'clock," would take a couple of minutes to cover the same distance.

Light also offers an expeditious way to communicate. If the monk has keen eyesight and the air is clear, he may be able to make out the hands of the church clock. And since light (which travels at nearly 200,000 miles per second) covers short distances almost instantly, the information that it is a particular time of day effectively travels from the clock face to the monk in what seems to be no time at all.

Now experiments by Nollet and others showed that electricity also seemed capable of traveling great distances instantaneously. Unlike light, electricity could be transmitted along wires and around corners; a line of sight from one place to another was not needed. This meant that if an electric shock was administered at one o'clock via a half-mile-long wire running from the church to a distant monk, he would know exactly what time it was, even if he was underground or indoors or otherwise out of sight of the clock tower. Electricity held out the promise of high-speed signaling from one place to another, at any time of day.

But the advantage of a horse-borne message was that it could say anything at all; instead of saying "It is one o'clock," it could just as easily say "Come to lunch" or "Happy birthday." An electrical pulse, on the other hand, was like the strike of a church bell, the simplest of all possible signals. What was needed was a way to transmit a complicated message using simple signals. But how could it be done?

Since the late sixteenth century there had been persistent rumors across Europe of a magical device that allowed people many miles apart to spell out messages to each other letter by letter. There was no truth in these tales, but by Nollet's time the stories had acquired the status of what might today be called an urban myth. Nobody had actually seen one of these devices, which relied on magical "sympathetic" needles that could somehow influence each other over great distances, but they were widely believed to exist. Cardinal Richelieu, for example, the ruthless and widely feared first minister of France, was thought to have a set because he always seemed so well informed about goings-on in distant places. (Then again, he was also thought to be the owner of a magical all-seeing eye.)

Perhaps the best-known description of the sympathetic needles was published by Famianus Strada, a learned Italian who provided a detailed explanation in his book *Prolusiones Academicae,* published in 1617. He wrote of "a species of lodestone which possesses such virtue, that if two needles be touched with it, and then balanced on separate pivots, and the one turned in a particular direction, the other will sympathetically move parallel to it." Each needle, he explained, was to be mounted in the center of a dial, with the letters of the alphabet written around its edge. Turning one of the needles to point to the letter "A" on its dial would then supposedly cause the other sympathetic needle to indicate the same letter. And this was all said to work no matter how far apart the two needles were. By indicating several letters in succession, a message could then be sent from one place to another.

"Hither and thither turn the style and touch the letters, now this one, and now that," wrote Strada. "Wonderful to relate, the far-distant friend sees the voluble iron tremble without the touch of any person, and run now hither, now thither: he bends over it, and marks the teaching of the rod. When he sees the rod stand still, he, in his turn, if he thinks there is anything to be answered, in like manner, by touching the various letters, writes it back to his friend."

The story of the needles was based on a germ of truth: There are indeed naturally occurring minerals, known as lodestones, which can be used to magnetize needles and other metallic objects. And if two magnets are placed on pivots very close together, moving one will indeed cause the other to move in response, as a result of the interaction of their magnetic fields. But it is not the case that the two magnets will always remain parallel, and the effect is only noticeable when they are right next to each other. The kind of needles described by Strada, which could interact over great distances, simply did not exist.

But that didn't stop people from talking about them. One wily salesman is even said to have tried to sell a set of needles to Galileo Galilei, the Italian astronomer and physicist. A firm and early believer in experimental evidence and direct observation, Galileo demanded a demonstration of the needles on the spot. The salesman refused, claiming that they worked only over very great distances. Galileo laughed him out of town.

Yet the talk of magic needles continued, along with the research into the properties of electricity. But no progress toward a practical signaling device was made until 1790. When the breakthrough finally came, it didn't involve needles or lodestones or electric wires; in fact, it was surprising that nobody had thought of it sooner.

Clocks and cooking pans hardly seem the stuff of which communications revolutions are made. But that was what Claude Chappe ended up using for his first working signaling system.

Chappe was one of many researchers who had tried and failed to harness electricity for the purpose of sending messages from one place to another. Born into a well-to-do French family, he planned on a career as a member of the clergy but was derailed by the French Revolution in 1789. He took up scientific research instead, concentrating on physics and, in particular, the problems associated with building an electrical signaling system. Having made no more progress than anyone else, he decided to try a simpler approach. Before long he had figured out a way to send messages using the deafening "clang" made by striking a casserole dish—a sound that could be heard a quarter of a mile away—in conjunction with two specially modified clocks. They had no hour or minute hands, just a second hand that went twice as fast as usual, completing two revolutions per minute, and a clock face with ten instead of the usual twelve numbers around its edge.

To send a message, Claude Chappe and his brother René, stationed a few hundred yards apart behind their parents' house, would begin by synchronizing their clocks. Claude would make a "clang" as the second hand on his clock reached the twelve o'clock position, so that his brother could synchronize his clock accordingly. Claude could then transmit numbers by going "clang" as the second hand passed over the number on the clock face that he wished to send. Using a numbered dictionary as a codebook, the Chappe brothers translated these numbers into letters, words, and phrases, and thus sent simple messages. It is uncertain how their original code worked, but the brothers probably transmitted digits in twos or threes and looked up the resulting two- or three-digit number in the codebook to see what word or phrase it corresponded to.

In other words, a complicated message could be sent using simple signals. However, the problem with this design (apart from the incessant clanging noise) was that the person receiving the message had to be within earshot of the sender and, depending on the direction of the wind, this was a few hundred yards away at most. Rather than replacing the copper pans with something louder, Chappe realized that it would be simpler to replace the audible signal with a visible one.

So out went the casserole dishes, and in their place was substituted a pivoting wooden panel, five feet tall, painted black on one side and white on the other. By flipping it from one color to the other as the second hand passed over a particular number, Chappe could transmit that number. This improved design had the obvious advantage that it allowed messages to be sent over very great distances very quickly—particularly if a telescope was used to observe the panel.

At 11 A.M. on March 2, 1791, Chappe and his brother used their black-and-white panels, clocks, telescopes, and codebooks to send a message between a castle in their hometown of Brûlon, in northern France, and a house in Parcé, ten miles away. In the presence of local officials, it took them about four minutes to transmit a phrase chosen by the local doctor— "si vous réussissez, vous serez bientôt couvert de gloire" (if you succeed, you will soon bask in glory)—from one location to the other.

Chappe wanted to call his invention the *tachygraphe*—from the Greek for "fast writer"—to signify the unprecedented speed with which it transmitted information. However, he was talked out of it by his friend Miot de Mélito, a government official and classical scholar, who suggested the name *télégraphe,* or "far writer," instead.

Et voilà: The telegraph was born.

Having shown that his invention worked, Chappe started to galvanize support for it in Paris with the help of another of his brothers, Ignace, who had been elected to the ruling Legislative Assembly. But the turmoil of revolutionary France was a difficult environment in which to start promoting a new invention, and Ignace didn't get very far. When the Chappe brothers staged another test in the town of Belleville, near Paris, in 1792, their apparatus was destroyed by a mob who suspected that they were trying to communicate with royalist prisoners being held in Temple Prison. The Chappes were lucky to escape with their lives.

By this time Claude Chappe had found a way to do without the synchronized clocks: He devised a completely new design that consisted of two small rotating arms on the end of a longer rotating bar. This bar, called the regulator, could be aligned horizontally or vertically, and each of the small arms, called indicators, could be rotated into one of seven positions in forty-five-degree increments. The design allowed for a total of 98 different combinations, 6 of which were reserved for "special use," leaving 92 codes to represent numbers, letters, and common syllables. A special codebook with 92 numbered pages, each of which listed 92 numbered meanings, meant that an additional 92 times 92, or 8,464, words and phrases could be represented by transmitting two codes in succession. The first indicated the page number in the codebook, and the second indicated the intended word or phrase on that page.

Abraham-Louis Bréguet, the noted clock maker, built Chappe a clever control mechanism for his new design: Through a system of pulleys, a scaled-down model of the rotating arms could be used to control the positions of a much bigger set of arms. The big arms could then be mounted on the roof of a tower and controlled from the inside by an operator. Chappe believed it would be possible to send messages quickly over great distances by constructing several such towers a few miles apart in a long line, each within sight of the next.

In 1793, Chappe sent details of this new design to the National Convention, the ruling body that had replaced the Legislative Assembly. His proposal was picked up by Charles-Gilbert Romme, president of the Committee of Public Instruction, who grasped its potential and suggested that the convention fund an experiment to evaluate its military applications.

A committee was duly appointed, consisting of Joseph Lakanal, a respected scientist; Louis Arbogast, a professor of mathematics; and Pierre Claude François Daunou, a legislator and historian. Money was allocated for the construction of a line of three telegraph stations in Belleville, Écouen, and Saint-Martin-du-Tertre, spanning a distance of about twenty miles. If a message could be passed along a network of three towers successfully, the system would obviously work with a larger number of towers over greater distances. Following the Chappe brothers' run-in with the Paris mob, the mayors of each of the three towns were made responsible for the safety of the telegraphs and their operators.

Within a few weeks the towers had been constructed, and the committee was invited to a demonstration on July 12, 1793. Transmission of the first message began at 4:26 P.M., with two operators at each station, one operating the little whirling arms, the other watching the next station through a telescope. The position of the arms at the sending station was reported by the observer in the middle station, where the operator would then move the arms to form the same signal; each signal was held in place for a few seconds, and the message rippled down the line to the receiving station. The three telegraph towers took eleven minutes to send a rather boring message ("Daunou has arrived here. He announced that the national convention has just authorized his committee of general security to put seals on the papers of the deputies") along the line one way, and nine minutes to send an equally nondescript reply back again. But the experiment was a success; the committee members, and Lakanal in particular, were highly impressed.

Two weeks later Lakanal addressed the convention in glowing terms on the potential of this great new invention, and how wonderful it was that it had been invented by a Frenchman. "What brilliant destiny do science and the arts not reserve for a republic which, by the genius of its inhabitants, is called to become the nation to instruct Europe," he gushed. Much of his enthusiasm seems to have stemmed from the potential application of the telegraph in holding the newly founded French Republic together, by ensuring that the central government in Paris could keep a firm grip on the provinces. At any rate, following his speech, the construction of a fifteen-station line from Paris to Lille, about 130 miles to the north, was proposed. Chappe was put on a government salary, complete with the use of a horse.

The Paris-Lille line, the first arm of the French State Telegraph, started operation in May 1794, and on August 15 it was used to report the recapture of a town from the Austrians and Prussians within an hour of the battle's end. As the French army advanced north into Holland, further victories were reported via the telegraph, and the government's appreciation for Chappe's invention grew. By 1798, a second line had been built to the east of Paris as far as Strasbourg, and the Lille line had been extended to Dunkirk.

Napoleon Bonaparte, who seized power in 1799, was a firm believer in the telegraph; he ordered further extension of the network, including the construction of a line to Boulogne in preparation for an invasion of England. He also asked Abraham Chappe, Claude's younger brother, to design a telegraph capable of signaling across the English Channel. (A successful prototype was built and tested between Belleville and Saint-Martin-du-Tertre, the two stations on the original experimental line, the distance between which was roughly equivalent to the shortest distance across the Channel. The station on the French side was later installed in Boulogne, but Napoleon's plans for the invasion never materialized, so neither did the British station.) In 1804, Napoleon ordered the construction of a line from Paris to Milan, via Dijon, Lyons, and Turin. This was to expand the network farther than ever before.

Lakanal's prediction had by this time come true, and France had indeed become "the nation to instruct Europe." Recognizing the military value of the telegraph, the governments of other European countries, notably Sweden and Britain, had quickly copied Chappe' s design or adopted variations upon it. In Britain, the Admiralty ordered the construction of a line of telegraph towers in 1795 to facilitate communication between London and the ports of the south coast during the war with France. The British telegraph was designed by George Murray, a clergyman and amateur scientist, and it consisted of six wooden shutters, each of which could be opened or closed to give sixty-four possible combinations (since $64 = 2 \times 2 \times 2 \times 2 \times 2 \times 2 = 2^6$). Soon, telegraph towers were springing up all over Europe.

The telegraph system was rightly regarded as a technological wonder of its time. An entry in the 1797 edition of the *Encyclopaedia Britannica* strikes a note of technological optimism that sounds familiar today: "The capitals of distant nations might be united by chains of posts, and the settling of those disputes which at present take up months or years might then be accomplished in as many hours." The author of the encyclopedia entry also suggested that it might be worth opening up the network to paying customers: "An establishment of telegraphs might then be made like that of the post; and instead of being an expense, it would produce a revenue."

Chappe also had all sorts of ambitious plans for his invention; he hadn't intended its use to be so predominantly military in nature, and wanted to promote its employment in business. He suggested a European network relaying commodity prices between Paris and Amsterdam, Cádiz, and even London via a cross-Channel telegraph. He also advocated a state-sanctioned daily national news bulletin. But Napoleon rejected both ideas, though he did agree to allow the weekly transmission of winning national lottery numbers. This meant the numbers were known throughout the country on the day of the draw, dramatically reducing the level of cheating.

But despite the success of his invention, Claude Chappe was not a happy man. He faced increasing criticism from rival inventors, who claimed either to have invented superior forms of telegraph or to have had the idea for the telegraph before him. Even his former associate, the clock maker Bréguet, turned against him and claimed to have contributed far more to Chappe's design than simply the control mechanism. Chappe sank into a deep depression and became increasingly paranoid, even going as far as to accuse his rivals of having slipped something into his food when he suffered from a bout of food poisoning. Finally, on January 23, 1805, he killed himself by jumping into the well outside the Telegraph Administration building in Paris. He was buried under a tombstone decorated with a telegraph tower showing the sign for "at rest."

Even so, his invention continued to flourish; by the mid-1830s, lines of telegraph towers stretched across much of western Europe, forming a sort of mechanical Internet of whirling arms and blinking shutters, and passing news and official messages from one place to another. The continental network eventually reached from Paris to Perpignan and Toulon in the south, Amsterdam in the north, and from Brest in the west to Venice in the east, with other networks in Finland, Denmark, Sweden, Russia, and Britain bringing the total number of telegraph towers in Europe to almost a thousand.

As the network grew, telegraph fever took hold in Britain, where amateur scientists' part-time inventors, and quacks were soon hard at work devising improvements to the nation's telegraphs. The Admiralty, which had spent much of the eighteenth century fending off idiotic suggestions about the best way to determine longitude at sea, now found itself on the receiving end of dozens of equally well meant but often crackpot schemes suggesting ways in which the telegraphs could be made faster or cheaper, or both. Some inventors advocated modifications to the six-panel shutter design that had been adopted in Britain; others proposed new and improved codebooks; and some called for the whole system to be scrapped and replaced with a completely new approach. One or two correspondents even claimed to have constructed telegraphs that used electricity.

The earliest suggestion of a scheme for using electricity to send messages had appeared in the *Scots' Magazine* of February 17, 1753. A letter from an unknown author, who signed himself simply "C. M.," was published under the heading "An Expeditious Method of Conveying Intelligence." The letter described a simple signaling system consisting of a wire for each letter of the alphabet, and a frictional generator that sent shocks down the wires. However, there is no evidence that C. M. ever actually built such a telegraph, and his identity remains a mystery.

But between the publication of C. M.'s letter and Queen Victoria's accession in 1837, at least sixty experimental electric telegraphs based on various electric and electrochemical processes are known to have been constructed by a number of researchers. Different designs used bubbling chemicals, sparks, or the twitching of pith balls to detect tiny electric shocks sent along wires. Some, like the telegraph described by C. M., involved twenty-six wires (one for each letter of the alphabet), while others used combinations of a smaller number of wires. But the scientists who constructed them worked independently, and each had to start from scratch—and, crucially, none of them ever managed to stage a decisive demonstration like the one Chappe had used to prove the value of his optical system.

In fact, as far as most people were concerned, so little progress appeared to have been made toward the goal of a practical electric telegraph compared to the highly successful optical design that anyone who expressed an interest in electric telegraphy was regarded as something of an eccentric. As one satirical verse of 1813 put it:

Our telegraphs, just as they are, let us keep,
They forward good news from afar;
And still may send better—that Boney's asleep
And ended oppression and war.
Electrical telegraphs all must deplore,
Their service would merely be mocking;
Unfit to afford us intelligence more
Than such as would really be shocking.

One example of a working electric telegraph was constructed in 1816 by a twenty-eight-year-old Englishman, Francis Ronalds. Similar to Chappe's original design, it involved synchronized clocks whose dials were marked with letters, rather like the dials supposedly used with the sympathetic needles. Instead of hands, each clock had a rotating disk with a notch cut into it so that only one of the letters on the dial was visible at any time. And instead of the clash of copper pans or the turning of a black-and-white shutter to signal each letter, Ronalds used electricity. When an electric shock, generated by a frictional generator, was sent down a wire from the transmitting station, it caused a pair of pith balls suspended from the wire at the receiving station to become electrically charged; the balls twitched as they briefly repelled one another, and the operator would write down the letter indicated on the dial at that moment.

Ronalds set up an experimental system in his garden and wrote to the government, requesting an interview with Lord Melville, the first lord of the Admiralty, to arrange a demonstration. "Why has no serious trial yet been made of the qualifications of so diligent a courier?" he asked. "And if he should be proved competent to the task, why should not our kings hold councils at Brighton with their ministers in London? Why should not our government govern at Portsmouth almost as promptly as at Downing Street? Let us have electric conversazione offices, communicating with each other all over the kingdom, if we can."

However, along with all the other suggestions for ways to improve the telegraphs, Ronalds's far-sighted ideas were politely but firmly rejected. John Barrow, secretary of the Admiralty, wrote back saying that since the war with France had ended, the telegraph system was in no need of improvement. "Telegraphs of any kind are now wholly unnecessary," he wrote, adding that "no other than the one now in use will be adopted."

The Admiralty's position was understandable; it could hardly waste its time investigating every quack's claim to have done the impossible and built a practical electric telegraph. Ronalds never got the chance to demonstrate his invention, but, surprisingly, he seems to have taken his rejection very well. "Everyone knows," he noted, "that telegraphs have long been great bores at the Admiralty." He gave up telegraphy and took up weather forecasting instead.

Ultimately, the success of the optical telegraph designs inspired by Chappe was limited because they were so expensive to run. They required shifts of skilled operators at each station and involved building towers all over the place, so that only governments could afford to run them; and their limited information-carrying capacity meant they were just used for official business. Optical telegraphs had shown that complex messages could be sent using combinations of simple signs; but other than noticing the appearance of a tower on top of the nearest hill, most people's lives were not directly affected. (Today, all that is left of the original telegraph network is a few place-names; several hills are still known as Telegraph Hill.)

As well as being expensive, optical telegraphs suffered from the drawback of not working in the dark, despite various experimental schemes that involved the use of colored lanterns on the end of the indicator arms. But at least the fall of darkness could be predicted; fog and mist, on the other hand, could arise at any time. When choosing the locations for new telegraph towers, one had to ensure that there were no marshes, rivers, or lakes along the line of sight between adjacent towers; local inhabitants were often consulted to determine the likelihood of mists arising.

If it could ever be constructed, however, a practical electric telegraph would work over any terrain, in any kind of weather, at any time of the day or night. It would be able to send messages around corners and over mountains. Yet for all its supposed advantages, and despite the work of Ronalds and others, the electric telegraph was still widely believed to be nothing more than an impossible dream.

The Ecological Impact of Telecommunications

Ithiel de Sola Pool

In a world of scarce resources, thought is pleasingly abundant; like air, it is a free good. But the means for communicating it are not equally abundant. Travel is a wanton drain of energy. The post and telephone are much cheaper, though not always cheap enough for the poor. Mass media are inexpensive when their audience is large—the secret of their success. Individualized communication and group communication are falling in cost toward near the low level of the mass media. Communication, in short, is one of the good things in life that can be had without straining the world's scarce resources. In communication we are very far from the limits of growth.

That fact underlies most of the key developments whose social implications we examine in this book. The falling cost of electronic logic supports the trend toward individualization. The growing abundance of bandwidth in transmission and better management of the electromagnetic spectrum creates the technical opportunities for small-group communication. Satellites and fiber links are making costs more distance-insensitive and underlie the internationalization of communication. Telecommunications are also likely to promote urban decentralization—the movement of business and cultural life from overcrowded cities to exurbia.

The relationship of computers and telecommunications to regional planning and decentralization has been extensively discussed in Europe. In France, Japan, Sweden, Great Britain, and many other countries there is a widespread desire to check the growth of the capital city and to disperse more of business, industry, and cultural activities to smaller urban centers and new towns. A number of authorities, for example Peter Goldmark, argue that with adequate communication facilities, rural life can be made attractive enough to check migration to urban metropolises, and indeed attractive enough to draw people out of the metropolises. Such decentralization, it is argued, could produce vast savings, for the capital required to enable millions of people to cohabit the same few square miles of ground is enormous. One must dig deep under the ground and build steel and cement towers above it. Transporting food and other consumables into cities and removing waste are complex and expensive operations. On the other hand, there are costs of decentralization, in roads that must be built, for example. But certainly if improved communication permitted life to be attractive either close-in or far-out from cities, as economic considerations dictated, and permitted business to be conducted efficiently from either place, then activities could be located wherever economically optimal in terms of transportation and other resource costs. Removal of one constraint on location cannot help but bring costs down as locations are picked to satisfy the remaining constraints.

Communication and the Pattern of Urban Settlement

The available means of communication and transportation affect the pattern of human settlement. If people can deal with one another only by walking, they will settle much more closely than if they can transmit messages or phone one another. Earlier we saw what a profound impact a change in communications technology can have on spatial patterns. Before the telephone, doctors, for example, had to live near their offices to be readily available when needed; the office was, in fact, typically in the doctor's home. The telephone allowed many doctors to separate home and office and put the office where it was convenient for patients to visit.[1] Once the telephone was available, business firms could move to cheaper quarters and still keep in touch. A firm could move outward, as many businesses did, or move up to the tenth or twentieth story of one of the new tall buildings. Instead of a checkerboard of different specialized neighborhoods, the new urban pattern became one of a downtown containing a miscellany of those commercial and marketing activities that needed to be accessible to clients and customers, as well as a growing set of satellite downtowns, for more convenient shopping and services, and the exiling of those activities that needed little outside contact (like manufacturing) to peripheral locations.

The telephone favored the growth of skyscrapers in several ways. As mentioned above, without telephones human messengers would have required too many elevators at the core of the building to make it economical. Furthermore, telephones were useful in construction; the superintendent at the bottom had to keep in touch with the workers on the scaffolding, and phones were used for that. As the building went up, a line was dropped from the upper girders to the ground.

Thus, contrary to the simple notion that the telephone led to the growth of suburbs and to urban sprawl, we find that it initially led to concentration downtown. Moreover, the movement out to residential suburbs actually began in America in the decade before the telephone and long before the automobile; it was based on the street car.[2] The conventional assumption is that the automobile and the telephone—between them—were responsible for the vast growth of American suburbia and the phenomenon of urban sprawl is challenged by the reverse proposition that the telephone also made possible the skyscraper and an increased downtown concentration.

In these contradictory ways, the telephone helped create regional agglomerations. As Jean Gottmann stresses, the telephone favored megalopolis, not antipolis.[3] A megalopolis (Gottmann's term) such as the Boston-to-Washington corridor is not an undifferentiated sprawl of medium-density settlement. It is a highly differentiated structure with numerous centers and subcenters having complex interrelations. Other commentators have seen this development as the destroyer of great urban cultural centers; Gottmann disagrees.

One can find quite early predictions of the process that we now call the formation of megalopolis. In 1902 H. G. Wells forecast centrifugal forces on cities that may lead "to the complete reduction of all our present congestions."[4] A pedestrian city, he said, "is inexorably limited by a radius of about four miles, and a horse-using city may grow out to seven or eight." With street railways the modern city thrust "out arms along every available railway line."

> *It follows that the available area of a city which can offer a cheap suburban journey of thirty miles an hour is a circle with a radius of thirty miles . . . But thirty miles is only a very moderate estimate of speed . . . I think, that the available area . . . will have a radius of over one hundred miles . . . Indeed, it is not too much to say . . . that the vast stretch of country from Washington to Albany will be all of it "available" to the active citizen of New York and Philadelphia.*[5]

Wells saw the telephone as one factor fostering this development; there was no reason "why a telephone call from any point in such a small country as England to any other should cost more than a post-card."[6] Yet Wells, like Gottmann later, emphasized that urban sprawl did not mean uniform density. Shopping and entertainment centers would continue to make downtowns, while people in some occupations would prefer to move out to the country and work by phone from home.[7]

A *Scientific American* article in 1914 played on similar themes, but with special stress on the picture-phone as likely to make dispersion possible. "It is evident," it began, "that something will soon have to be done to check the congestion" of the city. "The fundamental difficulty . . . seems to be that it is necessary for individuals to come into close proximity to each other if they are to transact business." The telephone and picture-phone, the article argued, would change all that.[8]

The telephone not only contributed to the break-up of single-trade neighborhoods but also tended, at an early stage of its development, to encourage the stabilization of "good" neighborhoods and business districts and their separation from areas of decay. Phone entrepreneurs, when they were getting started, like cable television entrepreneurs today, could not afford to wire the whole city at once. They preferred to lay their lines where they could expect to recruit many subscribers, that is, in affluent neighborhoods and business districts. Having a concentration of potential subscribers in a confined area was economically advantageous to the utility. Shifting and deteriorating neighborhoods were not good for business as capital-intensive infrastructure. Zoning of a city helped in planning for future services, so the phone companies (along with other utilities) became supporters of the zoning movement. The Department of Commerce's zoning primer of 1923 states: "Expensive public services are maintained at great waste in order to get through the blighted districts to the more distant and fashionable locations."

Zoning, along with other efforts at urban planning, became popular around the turn of the century; it actually began in New York in 1916. In the crucial second decade of the century, phone companies were one of the main sources of information for new urban plans. The Bell System collected large amounts of neighborhood data on the population trends in the city, its businesses and neighborhoods.[9]

In the nineteenth century, well before the invention of the telephone, directories of cities were already being published. From the beginning of switched telephone systems it was recognized as important to give subscribers a list of other subscribers. This list was particularly important if subscribers were to be called by number. These listings were originally printed on a card to hang by the phone. By 1897 the national phone directory had become too big a book and gave way to local directories. As telephones became universal among businesses and later the public, the phone book became the most widely used city directory. It had to be issued often to keep up with growth and change; it was available everywhere; and it was free. The inclusion of yellow pages made it useful for many business purposes, and it became the basis for much canvassing and sampling. Other city directories, sold for profit, were forced to provide more information to survive. For the ordinary citizen the telephone book became the standard way not only to find phone numbers, but also mailing addresses and business locations.[10]

In a variety of ways, then, the telephone has facilitated the coordination of a complex urban system. When the telephone was fifty years old, Arthur Pound in an anniversary volume noted how the life of a city could be brought to a crawl if the phones all suddenly stopped working.[11] The trains, the produce dealers, the hospitals would all grind down. It is hard to conceive of a metropolis running its myriad functions well without extensive use of the telephone. Some cities function with much less use of the phone than we take for granted. Beijing, Calcutta, and Moscow are large cities with very limited telephone service by American standards. Yet they function. But it hardly need be doubted that the way they operate suffers severe restrictions from their limited use of the telephone and would suffer worse were there none at all.

The telephone also shaped the countryside, together with the mail and the automobile. Together, they led farmers to do their banking, buying, and selling in larger, more remote centers rather than at the nearest rural village, and led villagers to take their trade to better stocked, more developed but more distant centers. Rural free delivery, in particular, favored the growth of mail-order business and reduced the need for farmers to come to the local post office. Parcel post was established by Congress in 1912 over the vigorous opposition of the small towns—the backbone of American democracy—which feared their commercial role would suffer by the increased competition of mail-order vendors.[12]

Most rural phone systems were at first tiny local networks providing village communication only. By the second decade of this century these systems were being rapidly interconnected with the national network. They no longer just linked farmers to each other or to the nearest small town but connected both to the national urban system. It was no longer necessary to take a trip to deal with an urban center. One could substitute a quick phone call.

The Communications/Transportation Tradeoff

The notion that improving communications facilities may reduce the need to travel is not a new one. Within three years after the invention of the telephone in 1876, the London *Spectator* predicted that the new device would substitute for personal meetings.[13]

However unrealistic the early prediction of reduced urban congestion may seem to us today, there was good empirical evidence behind these speculations. It was noted that the pattern of activity of traveling salesmen was changing. They continued to call on customers, but less often in person; they could phone between visits, and for some purposes could prepare for a visit by a phone call.[14] Some railroads claimed to be able to feel a resulting decline in their traffic.

With hindsight we know that travel, transportation, and traffic congestion has grown in this period of improved communication as never before. Indeed, even those who in the early years of the century were anticipating the replacement of travel by communication could not help but notice how delayed the process was. Herbert N. Casson, an early chronicler of the wonders of the telephone, claimed in 1910 that "slowly and with much effort the public was taught to substitute the telephone for travel."[15] Yet in this early literature we do not find the counterproposition that is often put forward today, namely, that the telephone increased relationships with people at a distance, thus leading to an increase in travel.[16]

A careful review of the relation of the telephone to transportation would show a number of complex directions of causality. The telephone, as well as the telegraph, facilitated better coordination of trains, traffic, and deliveries, making service prompter and more reliable. One notable result was an increase in the remote marketing of perishable goods.[17]

We cannot draw any single-valued conclusions about the relationship of the telephone to transportation. Often the indirect effects through such mediating factors as the ecology of the city were much more important than the direct relations themselves. In the short run one can identify many trips that do not get made because a letter or phone call takes care of the matter with greater ease. Just as the traveling salesman at the turn of the century could point to the trips he did not take, so any mother calling a doctor or manager calling the factory can easily identify trips saved. What one cannot report from common sense is how one's overall social world has been changed over time by the existence of a given communications system—and thus how the amount of transportation in society has been affected. These indirect effects are beyond our commonsense grasp.

The problem is paradigmatic of many in policy analysis; clear short-run effects of a policy can often be identified, but without being sure whether the long-run consequences are the same or indeed the opposite. Arms build-ups increase one's security in the short run, but critics argue that they make a more dangerous world. Wage increases help those who get them, but what if the resulting inflation damages the economy? What strategies are available for looking at such problems with something better than hopeless agnosticism?

One approach, which does not solve the problem, is to disregard the long run. But optimizing with respect to short-run considerations, of which we can be fairly sure, is to treat our ability to guess beyond that as zero, which is not accurate.

A better approach is to push the edges of short-run analysis toward some insights about middle-range consequences; the long run, after all, is but a series of short runs. Most serious research on the communications/transportation tradeoff is of this character. Let us briefly review three such types of studies.

Office Traffic Studies

In a number of countries, office activities have been studied to try to identify those that can be handled well by communications and those that require travel. Such studies are often done as a prelude to possible decentralization of offices—private or governmental—out of metropolitan centers. The question posed is how far the provision of good communications facilities can allow a move away from the center without causing a vast increase in travel.

Important work has been done by Bertil Thorngren in Sweden on the spatial distribution of contracts,[18] and by Chapanis and others[19] on the differences between communication with or without visual contact.

Alex Reid, who is responsible for some of the best studies, concludes that telephonic communication appears to be as effective or more so than face-to-face "for exchange of information among people who already know and trust each other, but that travel for face-to-face contact is important for creating relationships or for persuasion."[20] Thorngren found that most telephonic communication is with persons who are close enough for easy face-to-face contact and with whom there occasionally is such contact. In the absence of occasional face-to-face conversations the association attenuates; thus telecommunication and personal contact reinforce rather than substitute for one another.[21] From all this it follows that in organizations in which good working relations already exist and are being periodically reinforced by contact, improved telecommunications can save time and money. However, some meetings must continue if the organization is to remain viable and solve problems. If the sole goal were to reduce the amount of traveling, it would be better done by cutting all phone lines and letting organized interaction atrophy.

Econometric Studies

Economic studies can measure the cross elasticities of demand between communications and transportation. The argument is sometimes made that the cost of communication is so low that decisions to communicate are made without much reference to their cost. In business, for example, where large deals are being handled, communication charges do not make one hesitate to facilitate matters by a telegram or long-distance call, which costs only a tiny fraction of the money at stake.

For the communications/transportation trade-off, the implication would seem to be that the prospects of substitution are not very good. If demand is inelastic, consumer use of communications in place of transportation will not be encouraged by price changes. However, econometric analyses of the elasticity of demand may, for two reasons, be too static for our purposes, making the conclusion dubious. In the first place, such analyses usually overlook the time lag that takes place in learning to use a new service. In the second place, econometric analyses of time series measure incremental changes and, unless specified with adequate structural equations, are likely to miss the prospect of very important tip points. There are two quite different mechanisms by which demand for any product grows as its price falls. One of these is the gradual greater attractiveness of the good purchased as its cost falls, which may be considered the trade-off between the purchased object and money itself. The other is the sometimes very sudden shift that may occur when there are two ways of doing something and one can choose the cheaper one. A study of the price elasticity of telegrams done before low-cost long-distance telephone calls came onto the scene would have shown some sensitivity of usage to price, but would have given no clue to the drastic change that occurred once it became cheaper to send long-distance messages by phone than by telegram. Under the changed circumstances the only continued major use of telegrams was for those of its characteristics for which a phone call was no substitute, such as having a written text for record.

My suspicion that there is likely to be a very elastic demand for telecommunications in the coming decades is predicated on the expectation that it may become possible to do certain things cheaply by telecommunication that are now being done in other ways. The most persuasive such case is the prospect that electronic mail and facsimile will displace manually carried first-class mail. If a cut in the cost of data processing, facsimile equipment, and transmissions brings the expense of an electronic message below that of a first-class letter, then no time series regression analysis of the years in which electronic mail was more expensive than first-class mail will give a good prediction of what will happen after that tip point is reached.

The prospects of a tip point might also be explored regarding information retrieval and library use, remote and local computing, telemarketing and shopping, or tele-education and schools. At least for large portions of some of these services it seems possible that the cost of teledelivery will fall below that of conventional delivery. These possibilities point to the importance of not taking the elasticity figure during a "normal" period as being predictive of these tip points.

There is good reason to believe that over the coming two or three decades a variety of different services will be delivered over a largely shared telecommunications plant. We have already discussed this convergence of modes. If information services, text, mail, and conversations increasingly use the same electronic network, then, conclusions reached by looking at the elasticity of demand for any one service may be misleading.

By now it should be clear that any simple statement that communications will either increase or decrease travel is not only empirically wrong but conceptually confused. As we can see from a simple model, the relationship can go either way, depending on the parameters of the situation, and indeed it can go both ways successively over a period of time.

The agenda for policy making will not wait for the social science research to be completed. Policy decisions rarely await ideal intellectual solutions. In general, it seems fair to anticipate that achieving an overall reduction in travel is not a very likely outcome of social policy, even in the face of severe constraints on energy resources. The goal of reducing total travel runs athwart of too many other social goals which depend upon increasing the range and cosmopolitan nature of human interaction.

On the other hand, reducing the incidence of unnecessary trips is a reasonable social goal, both for its own sake and for the sake of immediate energy conservation. If we find later that measures taken to produce such an immediate reduction in travel have second-order consequences that ultimately increase travel, these results would demonstrate marketplace evidence of the high value that people put on face-to-face interaction.

Regardless of ultimate trends, it would seem to be rational policy to provide improved communication facilities that offer a low-cost alternative to trips people would rather avoid. That can be socially useful regardless of whether the overall trend in society is toward increasing or decreasing travel. Marginal savings, and the associated, even if limited, direct impact of those savings on the size of the aggregate trend are reasonable goals for social policy. In short, making efficient use of available resources is a good idea regardless of how human beings in the marketplace choose to use the resulting bonus.

Megalopolis

What kind of settlements can we expect to see evolve in an age of two-way electronic terminals in every home? Some futurists have described twenty-first-century life in terms of a return to the countryside. It is a rather odd return to the countryside. The picture usually given is of a family indoors, looking out on a beautiful vista of mountains or sea, with no other human beings in view, and rarely emerging from the comfortable air-conditioned cocoon of their house. There they sit in front of a big screen two-way videophone and engage, by communication, in all the activities for which people now commute to work, to school, to shop, or to a doctor's office.

There is no need to censor such fictional fantasies for practicality or consistency. The rural fantasy of turning all the rest of mankind into shadows on the screen, manipulated like genies of the lamp, sits side by side in world's fair models with 100-story urban constellations with monorails like flying buttresses threading silently among the towers, and helicopters available for an occasional quick escape to the unspoiled countryside nearby.

As with all fantasies, there is much to be learned from them about human desires. People do crave an escape from overcrowded urban concrete deserts. They are fond of having their private piece of turf, but only if they can keep the comforts and human contacts for which they, or their parents or grandparents, exchanged rural for urban life. They tolerate commuting as a compromise, but resent the discipline of the relentless time clock.

Fantasies give us a clue of directions in which people may choose to use the devices coming to hand. There is reason to believe that the twenty-first-century fantasy of rural life reflects at least two uses that people will make of the increased powers of communications technologies. People are likely to conduct a good many activities from their homes that they now go to other places to do, and the megalopolitan pattern of settlement is likely to be even more dispersed than in the twentieth century, allowing more people to live in some contact with green things.

An increase in work at home would represent a reversal of a 300-year trend. Karl Marx, Max Weber, and virtually all analysts of modern industrial society have pointed to the separation of place of work from place of residence as a key feature. It began when the factory system in the sixteenth and seventeenth centuries replaced work by home craftsmen. To Marx, factories were part of the process of alienation of the worker from his means of production; the proletarian went to work at a plant that was none of his own. To Weber, it was a trend that applied not just to industrial production but also to intellectual production. He defined modern bureaucracy by a number of features, one of which was that the official owned none of his means of administration and did not work from his home, but went from home to office to work there with the organization's files. The same point could be made for education. The near universality of the modern-day school, to which the student comes, for a few hours, as if to an office, can be contrasted with many other patterns in the history of education before the nineteenth century: tutors came to the homes of the affluent; children went to the church, or parish house, or monastery, where men of the cloth who taught them lived; if they went to school it was probably to a boarding school where they went to live as well as to study.

The separation of place of work from place of residence occurred because it had many advantages. Some of these remain just as valid as ever, but some are sharply reduced when modern communications are available. The factory or office allows tight discipline by supervisors observing workers on the shop floor. To have the same degree of control of people working at home would require a video monitor over each worker, something that is technically possible but expensive to supervise, and not economical except for such special applications as security operations. But intermediate levels of control, far above what was available to the entrepreneur who rode on horseback every few weeks to the cottage of a weaver or a spinner, can be provided by telephone or by on-line data monitoring. For people who get paid by commission or by the piece, for whom minute-by-minute supervision is unimportant, telecommunications can easily provide all the information on the results of their work that an organization wants. Brokers, salesmen, editors, interviewers, and typists, for example, can be supervised quite adequately from a distance.

To some degree, then, people will find it convenient to work from their homes, and entrepreneurs will find it advantageous to have them do so. It will permit part-time work by mothers, use of specialists in remote locations, and reduction of labor unions' ability to organize employees. People can be hired on an ad hoc basis for peak loads, and a business can get started faster without setting up a facility, saving the employer some capital costs of plant and saving the worker the costs of commuting and of moving when changing jobs.

Collection of all the people involved in a project under a single factory roof had some advantages besides supervision and communication. It allowed for better handling of materials and equipment. Tools could be shared by more people and stand idle less of the time. Raw materials would have to be moved less. Storage and inventory could be done more efficiently. All those considerations continue to apply in an age of telecommunications. Where material handling remains important, as in a supermarket or factory, people are still likely to come to the job, rather than the job to the person. Still, if a major reason for assembling people vanishes, the balance between centrally located and dispersed operations will shift.

Noncommercial institutions may see many of the same reasons for changing their way of doing things as may business. Today students' work in school is closely supervised by the teacher, but homework is totally unsupervised. Telecommunications permits the adoption of intermediate patterns in which the child can get assistance while doing homework, check answers, and/or be set back on the track from time to time.

Physicians have taken to the telephone avidly; perhaps data communications networks can be equally useful to them. It is clearly possible to monitor a patient's temperature, pulse, electrocardiogram, and various other measures by telecommunications. Perhaps person-to-person visits between doctor and patient will become less frequent, in which case the mutual accessibility of their location will become less important. If medical telemetry comes into common operation, perhaps there will be a need for groups of medical practitioners on 24-hour duty monitoring the remote sensor data. Perhaps patients will look on a wider geographic basis for the appropriate specialty, rather than considering location in the neighborhood.

In all these ways, in an era of good telecommunication, location becomes a less imperious constraint on human activities. To that extent, people can, if they wish, choose to live in remote locations at a tolerable price. Few occupations or other activities permit total disregard of location. Even most writers and artists occasionally have to be where the publishers and collectors are; only a few can assume that customers will beat a path to their door. And most people need not only face-to-face business contact but also external stimulation. Hermits are rarely motivated and productive. Everything we know about morale and productivity stresses the importance of the small group. So constraints of location are only matters of degree. If a larger proportion of the population does some of their work at home and needs to come to a specific job location less often, that in and of itself will have an effect on the overall pattern of population settlement. More people will be willing, on the few days when necessary, to commute further; more people will spend longer periods in a rural second home; for more people their second home may be their city home; more people will invest in larger houses, because they will need work space in them.

Perhaps reduced constraint on location will operate even more on institutions and groups than on individuals. Consider computer programmers. To simply write code they can be located anywhere where there is a terminal, but to remain on top of the profession they must pick up ideas informally, such as over coffee and sandwiches with colleagues. They may be remote from where the physical computer is, but they will want to be in a community where work of their kind is going on. Or consider stock brokers. To simply fill a customer's order can be done anywhere over telephone and terminal. But to be engaged in serious financing, they must be where they can have lunch with bankers and businessmen. They must be where there is a community of fellow professionals in related activity.

These communities need not all be in downtown Paris, or London, or Manhattan. Increasingly, the trend in megalopolis seems to be toward a more complex division of labor. University towns are an example of dispersed communities. Each has to have a certain critical size and be part of an extended intellectual community by communication. But they are located often in the less densely settled parts of megalopolis. We have seen such communities emerge in other fields, too. Corporate headquarters congregated in New York's Westchester County. Electronic companies sprang up along Route 128 around Boston and on the peninsula south of San Francisco. They form specialized communities within a larger structure.

That is not an antipolitan trend; it is an extension of the more dispersed and more complex division of labor of megalopolis. There is no reason to believe that cities will disappear nor even their great downtowns. Telecommunications does not eliminate the human need to associate physically with other people; it complements such association.

Some writers have suggested that telecommunications will create communities without contiguity; that means that people will begin to form communities with others who share a common interest but whom they hardly ever see. Amateur radio hams and computer hackers are examples of such a community. No doubt that can and will sometimes happen. There is much evidence, however, that even in a society with a highly sophisticated communication system that will not be a dominant pattern. When Bertil Thorngren plotted the places to which people make phone calls, he found that most calls are within two miles, with the frequency falling off with distance. People who see each other develop relationships that lead them to call each other. Mothers phone to ask a neighbor to send the children home. People phone to make appointments or as a prelude to dropping in. Neighbors become friends and then phone to chat. Face-to-face contacts serve different psychological functions than phone calls. In a close relationship these two kinds of contact reinforce and support each other.

That is one reason why it is sheer fantasy to believe that telecommunications will cause people to choose lives of physical isolation. It is also unrealistic insofar as a large portion of human activity is not just exchange of information but involves joint action on physical objects. The proportion of persons who are processing information has risen enormously, and these people could (conveniently or inconveniently, cheaply or expensively) operate from a distance via telecommunications. But there is still the other half of the working population who are handling things. Assembly lines will not be abolished by telecommunications, though computers reduce the number of people working on them. Truck drivers will still have to move products, even if they talk to each other by CB along the way. Haircuts are not going to be given by remote controlled automata—at least not in any period worth thinking about.

One can anticipate very large changes in the conduct of education, entertainment, science, the arts, government, finance, and administration in general. In all of those, the crucial activities can be carried out by remote communication. The effects of abundant and flexible electronic communication in other fields are less obvious, but they may be considerable. In manufacturing, for example, one may anticipate substantial further steps in the direction of individualization, exemplified by the emerging computer control of machines. Consider the possibility of an assembly line that does not produce items on speculation but according to individual orders. Now a customer either buys a mass-produced ready-made product or has a product custom made to order. Suppose there were a system by which a customer went into a shop to specify product type and material, and the specifications all were transmitted from a terminal to the factory, where the assembly line produced the item the same day. Suppose Detroit assembled cars after order under computer control in the same way. If something like that comes about, it would constitute a considerable impact of electronics and communication on the process of manufacture, but the basic process of physical production would still have much continuity with the way it is today. In housing, too, there is obvious continuity in the need for shelter, privacy, food, and sociability. Yet the design of housing might be substantially changed if families come to congregate in front of a six-foot television screen, and if both adults and children need work space by a terminal for their professional and school activities, presumably without interrupting one another.

Progress in communication is particularly important in a world in which there is increasing concern about scarcity of resources. There are no simple answers about how to keep improving the quality of life as energy becomes more expensive, or as the population/resource ratio becomes less favorable. However, it is clear that better use of information can increase the efficiency with which scarce resources are used. It is not clear what the optimal pattern of urban dispersion or concentration would be at different levels of energy cost and communication technology. There is a complicated interaction among those variables. Big cities are clearly more expensive places in which to do things than are smaller ones, but it is also more expensive to provide heat, food, and other goods to very sparsely settled populations. Somewhere in between there are optima to be found. In any case, compared with the present, it is certainly the view of virtually all planners that many activities ought to be dispersed out of the few overlarge cities that dominate most countries of the world.

In Japan, in France, indeed in most countries, planners want people to move out of the center to a more human and dispersed pattern of settlement. And most people when polled agree—that they would rather live in a place where they would have more room and less crowding if only certain kinds of jobs or services or educational opportunities were there. One untried way to achieve that would be to subsidize remote transmission of telecommunication services. In most countries in fact just the reverse is done. Long-distance telephone service is expensive and revenues from it subsidize cheap local phone service. We can speculate about what would happen if some government, as a policy, set a single uniform rate for phone calls (both local and long-distance) anywhere in the country, and set it at not much more than the present local call rate, perhaps if necessary, subsidizing the difference? It might markedly affect the pattern of settlement. Something like that was tried once before, forty years before the telephone, when the English reformer Rowland Hill proposed nationally uniform postal rates to achieve national integration and spur commerce. He pointed out that mail delivery was less distance sensitive than believed, since much of the cost was accounted for by local delivery, and only 1/10 of a penny per letter was chargeable to intercity transportation. The national penny-post was tried and became a great success. Very few countries today would permit the kind of penalization of remote areas that would occur if postal rates varied with distance.

Some economists object that such a subsidy to decentralization distorts resource allocations; but this is true for every subsidy. The argument for flat rates, in addition to their administrative simplicity, is that there are social costs in urban concentration. What or how extensive they are is an urban planning, not a communications, question. If in time such costs are found to be undesirable, an obvious way to offset them in part is by telecommunications pricing policy.

Communications, an Abundant Resource

There are no limits to the growth of ideas. If worry is justified about the scarcity of physical resources, all the more reason to improve the quality of life by use of resources that are not running out. Communications is one example.

If there is a reasonably well-working market, then those things that use scarce resources become more and more expensive; those that use less of them fall in price. Demand will respond to that financial incentive. The sacrifices forced by the rising prices of that which is scarce is offset by the ability to use more of that which is less so.

To have more energy, food, and transportation, one must use up some of the earth's scarce mantel and its elements. Communication, on the other hand, is a genuine addition to human welfare, but it does not necessarily use up physical resources. Granted, newspapers consume vast numbers of trees, but the information in the newspaper does not. It can be conveyed in resource-saving ways.

There is no obvious one-to-one relationship between the value of an idea and the cost of the medium that conveys it. A 50-million-dollar movie shown in large movie houses uses resources lavishly. Yet the same pleasure may arise from a school play. A thick Sunday paper with a magazine section is a large destroyer of trees and requires energy to dispose of it. A library book may produce as much enjoyment. In general the communicative activities in which people engage use much less by way of scarce resources than other major activities in their lives. That economy is measured by the small prices paid for using communications resources, be they broadcasts, periodicals, or conversation. Increasing the part communications plays in people's lives is a way of improving the quality of life with little drain on scarce resources.

Yet there are a few resources used by communication that one might worry about. Spectrum at once comes to mind, but we have already shown how that is an illusion due to the absence of prices. Four materials that might run short are copper, precious metals, plastics, and paper. But none of these, according to Michael Tyler and associates who studied material scarcities affecting British communications, poses a major threat to the expansion of telecommunications.[22]

The old AT&T used 15 percent of the copper consumed in the United States; in Britain, the corresponding number was between 5 and 10 percent of national copper consumption.[23] Copper prices, historically, have been notoriously volatile. The switch from iron to copper wires in the 1880s to accommodate the long-distance telephone system was itself made possible by a sharp decline in copper prices that had recently occurred with the opening up of new mines. The sharp rise in copper prices in the 1970s overlapped (when prices doubled) with the oil crises of 1974. In the long run copper is probably one of those scarce resources which must be partly replaced if its price is not to rise a great deal. It can be replaced. Aluminum is almost as good a conductor and is often used when copper prices rise. It is considerably more abundant than copper. And then there are glass fibers, made from silicon, one of the earth's most abundant materials.

Paper is a more critical material. It is pervasive in our society and a potential bottleneck for the phone system—not just the press. Tyler's conclusion is that with proper forest management, paper supplies can keep up with demand in the coming decades. There have, however, been short-term paper shortages. Paper prices have risen greatly at times, becoming, along with labor resistance to automation, the central financial problem of the press. As computer-controlled editing, setting, and printing has been introduced, production costs have dropped, while paper costs have been rising, making paper an ever more important item of cost in publishing.

In the 1830s, when, under the impact of faster presses, the penny paper went on the newsstands in the United States, the paper used was made from rags. In 1843 the United States imported two million pounds of ragstuffs, and by 1850 twenty-one million; that figure doubled by 1857. Rag and paper prices soared; European newspapers went into the red; export restrictions were put on rags. In the crisis that ensued experiments were made in producing paper from all sorts of alternative organic materials. By 1870, newspapers all across the country were using wood pulp paper instead of rags.[24] The price of paper dropped from 30 cents a pound to 31 cents a pound at the end of the century. From 1870 to 1880, 5,429 newspaper titles started up in the United States. The penny newspaper spread to England, and then in the 1890s came the halfpenny commercial paper—enlarged, incorporating many features from magazines, and dependent on advertising, which followed from the marketing of paper-wrapped branded consumer goods. Cheap paper is essential to the press as we know it, and it seems likely that improved technology of production and forest agriculture will continue to provide it.

However, if contrary to expectations there should be a massive run-up in either copper or paper prices, it would have an effect on communications. There would be a painful period of adjustment; but it would not stop the development of communications. It would increase the incentive to distribute information electronically (to CRT displays) using optical fibers and microwave circuits in place of copper cables and wires or paper. Substitutions for a scarce resource are almost always possible. Early in the century some observers worried about the exhaustion of the supply of trees for telephone poles.[25] In time, underground ducts, microwave circuits, cement poles, and scientific forestry took care of that. Thus, it seems unlikely that electronic communication will run into serious problems of scarcity—outside of spectrum issues—that will force up its real cost to society and therefore reduce its use. As far as one can see, a long era of boom lies ahead.

If so, one moral is clear. As some resources get scarcer and more expensive it becomes increasingly important to improve life by means of those services such as knowledge, culture, and communication that are less wasteful of scarce resources. Communications will become an increasingly significant element not only in total GNP but even more importantly in those parts of it that are not under pressure of scarcity of natural resources and which can, therefore, help to quench the specter of the limits of growth.[26]

Notes

1. "Telephone and the Doctor," *Literary Digest,* 44 (May 18, 1912): 1037. Before the telephone, it was practical for doctors to be in a single block; if the physician one sought was away, another was nearby. After the introduction of telephones one could locate the physician one wanted.
2. J. Alan Moyer, "Urban Growth and the Development of the Telephone: Some Relationships at the Turn of the Century," *The Social Impact of the Telephone,* Ithiel de Sola Pool, ed. (Cambridge: MIT Press, 1978). For London see P. M. Townroe, *Social and Political Consequences of the Motor Car* (Newton Abbot: David and Charles, 1974), pp. 52f: "One can illustrate this by looking at the development of the built-up area of London from the seventeenth century. One must realize that at that time there was no cheap public transport . . . In the eighteenth century a horse cost about three times as much as a motor car costs today in terms of working-class wages. A stage coach from Paddington into the city cost 2s 0d—single, which was 10 percent of a typical working-class wage. A taxi to do the same trip today would cost 2 percent of a typical working-class income, and a bus would be much cheaper. The only mass transport was on foot, and the cities were therefore very dense because a person had to live within a mile or so of his place of employment. In London, because there was some transport along the river, there was a linear development about a mile deep along the banks, particularly on the Northern side.

   As we go on 130 years to 1800, we find that the built-up area has not expanded much . . .

   By 1950, only 150 years later, the built-up area had grown by a linear factor of about 10. This . . . was the result of a whole series of transport improvements during the nineteenth century and the first half of the twentieth century. These started in 1829 with the horse buses . . .

These were followed by the urban railways from 1836 onwards. By 1880 there existed an area of central development, which was being served by horse buses and horse trams, plus development along the railway lines."

3. Jean Gottmann, "Megalopolis and Antipolis: The Telephone and the Structure of the City," in Pool, *Social Impact of the Telephone.*
4. H. G. Wells, *Anticipations* (New York: Harper Bros., 1902), p. 51.
5. Ibid., pp. 52f.
6. Ibid., p. 58.
7. Ibid., p. 66.
8. "Action at a Distance," *Scientific American,* 77 (1914): 39.
9. While in general the phone company supported the zoning movement, on one point it was at odds with the urban planners. Zoning was often used to set height limits, restricting the construction of tall buildings. Those, however, were heavy telephone users. Hence, telephone companies opposed height limitations.
10. See President's Research Committee, *Recent Social Trends* (New York: McGraw-Hill, 1933), pp. 197ff.
11. Arthur Pound, *The Telephone Idea: Fifty Years After* (New York: Greenberg, 1926).
12. Theodore K. Noss, *Resistance to Social Innovations* (Chicago: University of Chicago, Dept. of Sociology, 1944).
13. Vol. 52, Sept. 20, 1879, pp. 1187–1188.
14. "The Telephone in Modern Business," *Telephony,* 2, no. 5, pp. 65f.
15. Herbert N. Casson, "The Telephone as It Is Today," *World's Work,* 19 (Apri1 1910): 12777.
16. See John Short, Ederyn Williams, and Bruce Christie, *The Social Psychology of Telecommunications* (London: John Wiley, 1976), pp. 10–11, 166–167. They report an interesting example of synergistic effect in the reverse direction. "The opening of the Severn Bridge, a road bridge linking Southwest England with South Wales, was soon followed by the jamming of the telephone trunk routes between those areas" (p. 11).
17. See "Sociological Effects of the Telephone," *Scientific American,* 94 (1906): 500.
18. Bertil Thorngren, "Silent Actors: Communication Networks for Development" in Pool, *Social Impact of the Telephone.*
19. Alphonse Chapinis et al., "Studies in Interactive Communication: The Effects of Four Communication Modes on the Behavior of Teams during Cooperative Problem Solving," *Human Factors,* 14 (1972): 487–509.
20. Alex Reid, "Comparing Telephone with Face-to-Face Contact," in Pool, *Social Impact of the Telephone.* See also Short, *Social Psychology of Telecommunications.*
21. Thorngren, in Pool, *Social Impact of the Telephone.*
22. *Long Range Intelligence Bulletin 7,* Post Office Telecommunications, January 1976.
23. Ibid., p. 35.
24. Pauline Wingate, "Newsprint: From Rags to Riches—And Back Again," in Anthony Smith, ed., *Newspapers and Democracy* (Cambridge: MIT Press, 1980), p. 67; David C. Smith, "Wood Pulp & Newspapers, 1867–1900," *Business History Review,* 3 (Autumn 1964): 328–345.
25. "To Prolong Life of Telephone Poles," *Telephony,* 10, no. 1 (July 1905).
26. GNP is based upon market price. As a resource becomes more scarce, its contribution to GNP goes up; as it becomes plentiful, its price and contribution to GNP goes down. The information sector has been growing so rapidly that even despite falling prices its contribution to GNP has gone up. In the future, real information and communication may grow; whether its contribution to GNP will grow is another question.

The Blurring of Public and Private Behaviors

Joshua Meyrowitz

The impact of electronic media on social situations is not limited to a simple reshuffling of already "public" information. Electronic media do much more than take the information that was once available only in print and deliver it to new audiences. Print media and electronic media also differ in the *type* of information they convey.

Electronic media further integrate information-systems by merging formerly private situations into formerly public ones. In terms of the dramaturgical concepts discussed in earlier chapters, the shift from print media to electronic media is a shift from formal onstage, or front region, information to informal backstage, or back region, information, a shift from abstract impersonal messages to concrete personal ones. And this shift is not related directly to what "subjects" are discussed or to which particular people are placed in front of microphones or cameras, but to basic differences in the *form* of the information presented.

Information Forms

The differences between the types of information conveyed by electronic media and print media can be explained in terms of three dichotomies: communication vs. expression, discursive vs. presentational, and digital vs. analogic.

In one of his analyses of social interaction, Erving Goffman makes a distinction between "expression" and "communication." Although Goffman uses this distinction to suggest what is special about face-to-face encounters, this dichotomy is helpful here because it points to the differences in the type of information conveyed by different media.

"Expression," as defined by Goffman, refers to gestures, signs, vocalizations, marks, and movements produced by the *mere presence* of a person in an environment. The expressions exuded by the individual do not present any explicit "statement" or "argument" about things in general, but they do provide information about the expressing organism. The tone of voice and facial expressions of a person being interviewed, for example, give us some sense of the way the person is feeling and is responding to the interviewer's questions. The information received from expressions is always tied to the expressor and to the specific ongoing situation. Individuals' expressions were once accessible primarily to those in their immediate physical presence.[1]

In contrast, Goffman narrowly defines "communication" to mean the use of language, or language-like symbols, for the intentional transmission of a "message." In this special sense, "communication" involves purposeful symbolic utterances or written statements about things, events, and ideas. Unlike expression, communication can be abstract. A person writing a book, for example, can communicate about events in other countries, activities that took place in the past or are planned for the future, or unobservable abstractions such as "freedom" and "happiness."

In terms of this distinction, one can start and stop communicating at will, but one cannot stop expressing. Communications are consciously given, but expressions are unconsciously "given off."

Expressions are constant and much less controllable than communications. Further, while communications can be about anything, expressions, in one very important sense, are always "about" the individual giving them off. Expressions are personal and idiosyncratic; we look to them to discover how a person feels and what he or she is "like." For these reasons, most important decisions concerning relationships rely much more heavily on expressions than on communications. Few companies will hire new executives without meeting them first, and it is the rare person who chooses to marry someone merely on the basis of a résumé or a writing sample. Communications may provide important "facts," but expressive information is used to form basic "impressions."[2]

Although Goffman makes no explicit connection between the communication vs. expression dichotomy and the distinction between back and front regions, they are clearly related. Communications, because they are more controllable and manipulable, may be said to have a "front region bias," while expressions, because they reveal personal and private feelings, may be said to have a "back region bias."

One of the major differences between print media and electronic media is that print media contain only communications, while most electronic media also convey personal expressions. Electronic media make public a whole spectrum of information once confined to private interactions. Electronic media reveal information once exchanged only among people under each other's direct and close observation. In this sense, print media have a "front region bias," while electronic media have a "back region bias." A number of "dating services" have capitalized on the personal nature of electronic media by videotaping their clients for each other's consideration.

For the purpose of comparing print and electronic media, there is a strong parallel between Goffman's distinction between expression and communication and another dichotomy described by philosopher Susanne Langer: discursive vs. presentational symbols.

Langer distinguishes between discursive symbols, such as language, and presentational symbols, such as pictures. The relationship between a discursive symbol and what it stands for is always different from the relationship between a presentational symbol and its referent. Discursive symbols are abstract and arbitrary; they bear no physical resemblance to the objects and events they describe. The word "dog," for example, does not look like a dog. In addition, discursive symbols are composed of discrete units, such as words, each of which has a meaning independent of the particular arrangement of the elements. The same words can be arranged in different sequences to mean different things ("The dog bit the cat" or "the cat bit the dog"). Making sense of any arrangement of words also involves an agreed upon grammar that has little to do with the things described. If, for example, I try to describe in words two things that have happened simultaneously, I must still mention one of them after the other ("The cat bit the dog just as the man walked out of the house"). There is a structure to discursive symbols that has nothing to do with the structure of reality.

Langer contrasts the form of discursive symbols with the form of presentational (or nondiscursive) symbols, such as pictures or photographs. The presentational symbol has a more direct link with the thing it is about. A picture of something looks like the object; there is a direct physical resemblance. The picture imitates aspects of the real world: shapes, proportions, shadings, colors, and textures. Even stick figures of people look more like people than the word "people" does. Pictures of people and the word "people" are not equivalent symbols.

The elements that make up a picture are much more numerous than the elements of language, but the elements of a picture have no independent meanings. In isolation, for example, dark and light specks of photographic grain are only "blotches." The light and dark specks gain "meaning" only when they are arranged in a pattern that reflects the shape of *real* objects or people. Sentences, in contrast, have meaning only when they are arranged according to the *arbitrary* rules of a given grammatical code.

Because presentational symbols must take a form that resembles reality, they are much less "editable" than discursive symbols. One can describe a landscape in many different words and sentences, but one would not normally cut up a picture of a landscape and rearrange it in different patterns in order to describe it in different ways. Because a photograph is not composed of discrete units strung out in a linear row of meaningful pieces, we do not understand it by looking at one element after another in a set sequence. The photograph is understood in one act of seeing; it is perceived in a gestalt.

The meanings of words and sentences can be explained in other words and sentences, but the "meaning" of one picture cannot be conveyed by viewing another picture. Words can describe other things (including pictures and other words); pictures merely present themselves. The Declaration of Independence can be paraphrased and even translated into other languages, but the essence of pictures such as the Mona Lisa, or of the American Marines placing the flag at Iwo Jima, or of little John Kennedy, Jr. saluting the passing casket at the funeral of his father, or, for that matter, any family snapshot, can be only weakly conveyed in words. "Paraphrasing" is a peculiarity of language. One can express the same thoughts in different words. But it is difficult to express the same picture in different ways. Each picture conveys a unique message. Although I can use my own words to explain Langer's words, it would be impossible to present my own picture to convey an image of Langer.

A presentational symbol cannot be used to "discuss" a specific abstract idea. "Respect," "patriotism," or the theory of gravity cannot be pictured easily. In viewing a picture, we may be reminded of such ideas (i.e., the pictures may bring certain *words* to mind), but language is needed to conceive of them in the first place.

Langer contends that discursive language is not our only articulate means of communicating. Nondiscursive symbols are another equally important form of mental activity and another "intellectual" means of describing reality. Yet, notes Langer, presentational symbols are often devalued because they cannot be translated directly into discursive language. Symbols such as pictures, sounds, and music simply "present" feelings and emotions without the benefit of words.

The discursive/presentational dichotomy offers another means of distinguishing print and electronic media. Print strips messages of most of their presentational forms; it conveys only discursive information. But most electronic media convey a rich range of presentational information along with discursive symbols. Radio conveys sounds and vocalizations. Television adds visual forms. Like expressions, presentational information is directly tied to specific objects or people in specific contexts. More than print, electronic media tend to unite sender and receiver in an intimate web of personal experience and feeling.

The presentational/discursive distinction is similar to another dichotomy—digital vs. analogic—discussed by Paul Watzlawick, Janet Beavin, and Don Jackson in *Pragmatics of Human Communication*. Digital symbols are discrete units such as numbers or words. Digital systems are based on discontinuities—either/or. A number is either a 3 or a 4, but not both. A word is either one word or another. Analogic messages, in contrast, are continuous; they are based on the principle of more or less.

The difference between digital and analogic forms is exemplified in the two different types of watches and clocks now available. A digital watch presents the time in discrete units. It is either 12:57 or 12:58 (or some discrete unit in between, such as 12:57 and 10 seconds as opposed to 12:57 and 11 seconds). With an analogic watch, however, time is continuous. A person who sees the time as "12:57 and 10 seconds" on a digital watch may see the time as "about one o'clock" on an analogic watch. A similar difference exists between a digital calculator and the analogic slide rule. The slide rule is excellent for making "approximations," but it cannot yield precise digital data.

Human communication can also be broken down into digital and analogic systems. The words we speak or write in a letter are digital messages. Words and sentences are discrete units with relatively specific meanings. The blank in the sentence, "I ____________ eating fish" might be filled with the word "like" or "hate," one word *or* another word. But a hug or a handshake or a smile involve analogic expressions. There are no isolated units with independent meaning within a smile; there are no discrete degrees of sincerity in a handshake.

Watzlawick and his colleagues suggest that digital communications convey "content" messages, while analogic expressions convey "relationship" messages. That is, digital communications can be about things in general, while analogic messages tend to reveal how the persons emitting them feel about people and things around them or about the digital messages they are speaking or hearing.

This dichotomy also helps to distinguish between the form of information conveyed in print and electronic media. Print conveys digital information; most electronic media convey both digital and analogic information.

Personal vs. Impersonal Response

The three dichotomies summarized above offer a means of explaining why our responses to print media and to electronic media are so different. Print's communicative-discursive-digital messages and electronic media's expressive-presentational-analogic messages have different possibilities and limitations.

The messages in print and electronic media are about different things. Television, for example, can "present" an object or person in a manner in which verbal descriptions cannot. Thus, if the police hope to capture a fugitive, a verbal description would be much less helpful in identifying the fugitive than a videotape of the fugitive. Similarly, few people would trade an only photograph of a lost loved one for a poem about the loved one. But presentational messages are severely limited in their own way because they cannot convey abstract ideas or concepts. Thus the *idea* of "fugitive" or "justice" cannot be expressed in a videotaped image. It is possible to photograph two people eating together, but it is impossible to picture the concepts of "kindness," "friendship," or "love." It is possible to present a picture of *a* specific woman, but it is impossible to picture "femininity." One can present a picture of a ball or of a wedding ring, but the ideas of "toy," "roundness," or "continuity" must be discussed in words. The abstract idea of "great leader" can be spoken about in language, but a picture of a President is only a picture of a specific person.

Further, unless words are used to explain expressions and presentational symbols, their meanings are often ambiguous. Expressive messages have no qualifiers to specify which of several possible interpretations is the correct one. "There are tears of sorrow and tears of joy, the clenched fist may signal aggression or constraint, a smile may convey sympathy or contempt, reticence can be interpreted as tactfulness or indifference." Similarly, people often rely on a picture's verbal caption to determine the picture's specific meaning.

Ironically, electronic media's analogic messages are both more direct and more ambiguous, more "natural," yet less precise. The analogic message comes directly from the sender/subject; it suggests how the sender/subject "really feels." Yet one cannot base a contract on a smile and a wink. Words must be exchanged if the exact meaning of the message is to be understood.

Electronic media's presentational messages have no equivalents to linguistic grammar or syllogistic logic. A person's gestures or tone of voice cannot be rearranged and edited the way words can be manipulated because gestures and tone of voice are not divided into discrete units. Expressive messages have no logical links, such as "if-thens" or "either-ors." Expressive messages cannot be used to distinguish between the past, the present, and the future, and they cannot be used to make propositional statements that are capable of being proven true or false ("It will rain tomorrow" or "I have three hands"). The thesis of this book could never be developed in a series of television images. Ideas live through language.

Yet the discursive means through which authors convey their messages are highly "impersonal." The words on this page, for example, tell you very little about my personal existence. I may speak with a lisp or a heavy accent. I may have shifty eyes and I may constantly mispronounce words. I may be tall or short, young or old, mischievous or sober. But you cannot tell any of this from the words I write. Even an awkward phrase or a misspelled word cannot be attributed directly to me. Perhaps an editor or a typesetter made the mistake. Print offers shared responsibility, and a printed page is a shield behind which human idiosyncrasy and frailty may hide. After thirty seconds of face-to-face interaction between us (or after seeing me on television), however, you would know things about my *personal* being that you could not discover from reading everything I have ever written.

The impossibility of translating many of the presentational messages on television into discursive language may explain why television is often looked down upon by many "intellectuals." Yet while most of the messages in television entertainment are hardly "realistic" or "educational" in conventional (discursive) terms, they do present viewers with realistic images and sounds of people and places. These presentational symbols feed our store of "experience" rather than our store of discursive knowledge, arguments, and ideas.

Discursive and presentational forms are so distinct that they are apparently produced and perceived primarily by different hemispheres of our brains. In most right-handed people, the left hemisphere is the language center and the right hemisphere deals with tasks related to visual and spatial patterns. The left hemisphere is dominant in the handling of linear, sequential, and discrete symbols, but the right hemisphere is the primary realm of images, shapes, feelings, insights, and intuitions.

As a result of all these differences, the messages in television tend to be "about" something different from those of print. Our response to a verbal message (such as a speech printed in the newspaper) is often to ask whether it is true or false, whether the arguments are logical, and whether it is well-written. Print messages can be "issue"- and concept-related. But expressive messages on television often lead to a different response. They are more personal, more tied to the speaker. Expressive messages constantly tell us something about the expressors' orientations to the situation, how they are feeling, and what their emotional response to their own behavior is. A common response to a speech on television, therefore, is to think about the speaker rather than the speech. While watching a politician on television, it is not unusual for people to think or say: "He seems nervous"; "She looks like she's going to cry"; "I don't trust him"; or "She's tough!"

One can choose, of course, to write about very personal matters—as in Augustine's *Confessions* or in letters to *Penthouse* magazine—but unlike presentational media, one's physical self is completely absent. Even when printed words are about intimate matters, the form in which the message is conveyed is impersonal and abstract. Relatively little about the flesh and bones *person* is ever revealed in print. On television, in contrast, one can speak about very impersonal matters, and yet be unable to escape sending, along with the abstract "topic," a broad range of personal cues.

Much more than print media, television thrusts the personal, private realm into the public arena. Television provides the type of information we are accustomed to responding to when we are with family members and friends. Compared to print, television provides a rich personal profile of the communicator. The separation between private emotion and public communication is blurred. The sudden loss of breath, the welling of tears in the eyes, the voice that cracks with emotion or moves steadily through a difficult passage—all these convey information that is not discursive, digital, or communicative, but presentational, analogic, and expressive. When presidential candidate Edmund Muskie stood on the steps of the Manchester *Union Leader* in 1972 to denounce a newspaper story attacking his wife, the response to the televised message was based primarily on his expressive behavior rather than on his communicated verbal message. Most people remember that Muskie cried; few can remember what he said.

A good example of the differing importance of verbal and nonverbal messages in different media comes from George E. Reedy, who served as Press Secretary for President Johnson. In a 1977 article on Jimmy Carter's use of television, Reedy suggests that, "at the very least, [Carter] has demonstrated that one need not resort to words to put across rather sophisticated points." Reedy suggests that the best example of this comes from Carter's "fireside chat" on the energy crisis.

> *I have discussed this with individual Americans in many parts of the country and, except for news reporters who had to write stories about it or public officials who had to react to it, I have found no one who can remember anything he said. The fact that he wore a sweater while delivering the statement, however, has become legend and, significantly, everyone to whom I have talked ascribed a complex meaning to the act.*

The presentational nature of television affects our response to presidential debates and elections. The campaigns on television are rarely about issues; they are about personality and style. Media advisors often play more important roles in campaigns than political advisors and speech writers. In general, it is important for politicians to avoid getting upset or looking angry during television appearances. A politician's demeanor on television is often more important than his political programs. On the night of the Carter-Reagan debate, Reagan aide James Baker passed Reagan an index card with one word of advice: "Chuckle!" When most people think back on such televised debates they tend to remember the general style and emotional state of the participants rather than their specific verbal arguments.

There is no great mystery to the focus on the expressive aspects of television's messages. It is not as if television deadens people to verbal messages to an "abnormal" degree. The same focus on expressive information exists in many interpersonal encounters. Anthropologist Albert Mehrabian's studies of nonverbal behavior suggest that the relative weight people give to messages in face-to-face encounters is 7% to the verbal, 38% to vocal inflection, and 55% to facial expression. If these figures are accurate, more than 90% of the meaning of a message is derived from expressions rather than "communications." This explains why, when two people go out on a first date, they carefully monitor each other's expressions to get a sense of how good a time the other person is having. The formulaic verbal messages ("That was a very good movie" or "I had a very nice time") provide relatively little pertinent information.

It is primarily in written language that the words themselves become all-important. Without nonverbal cues, careful construction and rewriting are needed to capture the correct "feeling." Sarcasm, teasing, and other nuances of meaning are easily lost in letters and books. This is why even intimate love letters tend to have a "formal" quality, while many business meetings or telephone conversations seem quite intimate and personal.

In face-to-face encounters, the significance of the verbal message is inversely related to the amount of available vocal and gestural information. When nonverbal information is weak or absent, the verbal information becomes more dominant. The attention to a speaker's verbal message, therefore, varies with interpersonal distance. The closer the distance between people, the less attention paid to the verbal message. A group of friends tightly clustered around a restaurant table rarely produces a conversation that would be fully understandable from a transcript of the words they have spoken. In intimate encounters, sentences are not completed, there are changes in tense, and numerous grammatical errors are made. Less care is taken to construct and to listen to words when other sources of information are available.

There are several reasons why the focus on expressive messages may be even greater on television than in face-to-face interaction. Although very close physical proximity is the exception rather than the rule in live encounters, the opposite is true on television. There are few, if any, people to whom a viewer is supposed to respond who are not seen extensively in close-up. Further, the one-way direction of information flow allows the viewer to do something that is generally considered impolite in face-to-face interaction—*stare*. Finally, in live interactions, a person cannot devote all his or her energy to monitoring the behavior of others because he or she is also deeply involved both in producing expressions and communications and in thinking about how other people are monitoring those behaviors. But with television, the viewer can slump in a chair and focus exclusively on the expressive behavior of those televised. The viewer remains unobserved and relatively inexpressive. (Indeed, for all these reasons, television may have been a stimulus to the recent growth of the field of nonverbal communication.)

Because of the rich expressive dimensions of television, interviews on television are completely different from print interviews. On television, even a "no comment" or a long pause is a meaningful and significant "answer." It is rich in expressive information. It shows the speaker's response to the situation and to the content of the question, even though the response is essentially nonverbal. One of the most popular programs on television, "60 Minutes," often uses footage of interviewees pausing between words or stammering. Very tight close-ups are often used, so tight, in fact, that any move of the head makes it appear that the subject is trying to escape scrutiny. Such shots seem to suggest that the interviewee is hiding something or is dishonest. (In a sense, the use of such shots may itself be a kind of deception, because sudden head movements or the expressive signs of discomfort do not necessarily mean deception; they can simply signify excitement, nervousness in front of a camera, tiredness, a lapse in memory, or an upset stomach.)

Although some analysts attribute the new "personal" coverage and exposure of political leaders to a conscious decision on the part of journalists and editors, it is more likely that the new bias is supported by the ability of electronic media to "capture" personal attributes. It now seems fair (and necessary) literally to "expose" the personalities of leaders because the revealing behavior emanates from the leaders themselves, and unlike quotes in a newspaper, expressive behavior cannot be easily rearranged or edited.[3] If a politician cries while speaking a certain sentence, it is not necessary to mention it in print, but you cannot play a tape of the words without also conveying the crying. It is fitting that in the jargon of broadcast journalism, taped reports and interviews are called "actualities"; the expressions are allowed to "speak" for themselves.

For a number of reasons, the decisions surrounding the broadcast of a videotape of a person are different from the decisions to quote someone in a newspaper. If I overhear a politician swearing, and I print the words, *I* seem to take some responsibility for them as well. If, however, I record the words and expressions on audio or videotape and then broadcast them, the words and expressions seem to come more directly from the speaker, and my responsibility (though arguably the same) seems greatly diminished.

In this sense, when magazine editors printed a picture of Vice President Nelson Rockefeller making an obscene gesture in response to hecklers in 1976, the editors deferred responsibility for the gesture's appearance in their magazine to Mr. Rockefeller. Yet had no photographer caught the event on film, it is unlikely that the same editors would have printed an artist's rendering of the gesture or a detailed verbal description. The ability to capture personal expressions through cameras and microphones increases the extent of their incorporation into the public record.

The expressive nature of electronic messages offers one explanation of why the public is no longer offended by the idea of political ghost writers. Presidents and politicians once wrote (or claimed that they wrote) all their own speeches. After all, if most people in the country can only *read* what you have to say, they will want to believe that you really wrote the words yourself. If someone else wrote the words, there would be no message directly from or about you. The expressions accompanying a speech over radio or television, however, are so personal and revealing in themselves, that the authorship of the words is now less significant.[4] Indeed, a politician speaking on television could probably please an audience even by reciting a collection of quotes from others' essays and poems that reflected his or her philosophy. Conversely, a politician might alienate an audience by having someone else *deliver* a media speech—even if the words were written by the politician. On television, expressions usually dominate words.

The personal and emotional response to expressive cues on television explains some otherwise baffling findings in political polls. In recent years, pollsters have found a split between the voters' political views and their personal feelings toward politicians. In February 1982, for example, only 47% of the national population approved of "the way Ronald Reagan is handling his job as President," yet 70% approved of him "as a person." Another poll found that nearly a third of those interviewed disapproved of Ronald Reagan's policies, yet personally "liked him." In an electronic age, therefore, it is quite possible that many people will vote for candidates with whom they largely disagree, or vote against candidates who share their political philosophy. Such voting behavior would be unthinkable if the voters only had access to transcripts of the candidates' speeches and policy statements; it only makes sense when the voters feel they "know" the candidates personally.

The programs that thrive on television depend heavily on expressive rather than communicative content. The transcripts of talk shows such as the "Tonight" show and of game shows such as "Family Feud" would make boring and unenlightening reading. Talk shows do not succeed because of talk. And in spite of the common belief that quiz shows are "educational" in a traditional sense, little or nothing is "communicated" in these shows. Game shows, however, are rich in expressions. People are placed in situations of stress and they react and respond. (Indeed, potential contestants are screened for their expressiveness.) The "learning" in these programs comes from the "truth" and "reality" inherent in human behavior and experience.

The most expensive content on television—advertising—also relies much more heavily on expression than on communication. Few campaigns for nationally advertised products present a formal, logical, verbal "argument" for buying their product. Instead, they present people in highly expressive situations: smiling, laughing, hugging, kissing, singing, running, playing, or diving into water (often in slow motion). The implicit nonverbal arguments are that buying the advertised product will make you feel good, make you more huggable and lovable. Even when verbal messages are used, they often push discursive words as far as possible toward nonlinear presentational messages ("Have a Coke and a smile"; "Me and my RC"; "Coke is it!"). The expressive messages of advertising are almost impossible to regulate. As nondiscursive presentations, they cannot be proven true or false. If a diet soft drink commercial *said,* "Drink this and you'll be beautiful and have beautiful friends to play volleyball with on the beach," they could be prosecuted for fraud. But commercials rarely make such verbal claims. They do not appeal to rationality, but to the emotions and the senses.

It is not surprising that advertisers would want to circumvent rationality, but it is upsetting to realize that television news and documentaries often do the same. Transcripts of television newscasts often contain surprisingly little "information," and they seem strangely informal compared to newspaper reporting. The style of television news is often more suited to the neighborhood coffee shop than to the newspaper column.[5] Videotaped segments on the news frequently contain much more expression than communication. A microphone and camera thrust in the face of an airplane crash survivor or a mother who has just heard that her son has been killed yield very little communicated information, but offer viewers a rich source of expression.

The informality of television news content is often supported by a chatty style and informal interaction among newscasters. (This is especially true for local news programs.) Indeed, the perceived quality of the relationship among television news "teams" is often the key to the success or failure of a news program. The quips and interchanges are generally low-level communications, but they are meaningful expressions. They are intended to make the viewer feel that the newscasters are "nice people" and that they are friends off the set.

One of the oddities of television news is the fact that the newscasters become part of the news in a way that newspaper reporters rarely do. Broadcast reporters and interviewers are also personally exposed on television. Because we experience their personal expressions and responses to events, they are now as prominent as the public figures and stories they report on. After revealing the personalities and homes of many world leaders and celebrities, veteran interviewer Barbara Walters felt it only fair and sensible that she do a television program on her own home and personal life. Similarly, Walter Cronkite's retirement from the "CBS Evening News" was itself a major news story.

If, as discussed above, expressions are ambiguous and imprecise, why do so many television programs, including news shows, rely heavily on them, and why are viewers apparently so accepting of them? Expressions are powerful and effective because they seem more natural and real than words. Expressions come directly from a human being, and the sender has significantly less control over them than over communications. A person can lie verbally much more easily than he or she can "lie" nonverbally. An unethical person could say or write the false statement: "I love you." But it is much more difficult to *express* the same feeling nonverbally—especially over an extended period of time—unless it is true. Even beyond the issue of truth and falsity, however, is the basic sense that one has "met" or "knows" the persons one sees on television because one experiences their personal expressions.

It is also very difficult to have people change or manipulate their expressive behavior on demand. A threat or a twisted arm might force someone to write something he or she does not believe (such as a confession to a crime), but it is much more difficult to coerce someone into "sounding sincere." Further, expressive messages are so "personal" that it is very difficult for most people to "forge," or accurately mimic, another person's expressions. As an April Fool's prank, you might type a letter to a friend and sign it with the name of the friend's employer. If you do a good job, your friend may take the letter at face value. Yet if you call your friend on the telephone and pretend to be the employer, your task is much more difficult because you must capture the vocal style of the employer. And if you come to the door of your friend's house (or send a videotape) and pretend to be the employer, your chances of fooling your friend are very slim.

While words carry more precise meanings than expressions, they are often untrustworthy or insufficient when they are isolated on a page. When words are seen on a printed page, it is impossible to tell if they are composed out of fear or confidence, spoken with humility or conceit. It may be difficult to ascertain the true authorship of words, and it is impossible to check the sincerity of a statement by looking to expressions for evidence of deception. Even accurate quotes in a newspaper may be denied later by their speakers.

For all these reasons, expressive "evidence" on videotape is very powerful. Many police departments have found that videotaped confessions yield higher rates of conviction than written confessions because it is more difficult for defendants to deny such a confession or claim that it was forced out of them. Similarly, the FBI's Abscam videotapes of U.S. Congressmen accepting bribes presented much stronger evidence than would have been provided by verbal accounts of witnesses. For the same reasons, the "living will"—a videotape of a person speaking his or her own last will and testament—is growing more popular. It is difficult to contest a will in which your wealthy Uncle Harold leaves you only a spare tire, when you see and hear Harold himself bestowing that gift on you.

The relative difficulty of controlling or forging expressive messages explains why electronic media are more intimate, personal, and revealing than print media. The rich range of expressive information in television suggests a reason for the claim by many Americans that they find television the most "trustworthy" of all sources of news.[6] For even though the news on television is highly packaged, with much information coming from the same sources used by newspapers and radio, at least there is someone you "know personally" speaking it to you. The power of expressive messages is also probably the source of the credibility of "hidden camera" interviews, person-on-the-street interviews, and such shows as "Real People."

Different electronic media vary in terms of the range of expressive cues they transmit. Radio, for example, conveys only vocal expressions. Those who are capable of managing their vocalizations well may be able to use the medium very effectively—even if they cannot control other expressions simultaneously. A nervous speaker may be able to control his or her voice, but not the nervous shaking of a leg or swaying of the body. Radio is more revealing than print, but radio remains an active filter of any expressive cues that are visual.

The trend from print media to electronic media (and from radio to television), therefore, is a trend from formal and social front region messages to informal and personal back region messages. Print media present crafted, stylized messages, while electronic media present spontaneous and "natural" messages. Even when the "content" of news and entertainment programs is tightly scripted and planned, the expressions of the individuals in them present a "slice of life" view that is not comparable to printed messages. All other things being equal, electronic media tend to merge personal and public spheres.

The merger of personal and public has a dual effect on the information-systems associated with people of different ages, professions, and educational backgrounds. The expressive nature of electronic media of communication both homogenizes *access* to information and merges the general types of images *projected* by different people into the public realm.

Expressive messages are more equally accessible than verbal and printed messages. As Watzlawick and his colleagues note, young children, mental patients, and animals can accurately assess the general meaning of expressions, even when the linguistic cues are not meaningful to them. Similarly, two researchers of nonverbal behavior, Paul Ekman and Wallace Friesen, note that a lay person's interpretation of expressive messages often coincides with those of experts.

Although people could never learn a strange foreign language from reading a book written in that language, they can generally make some sense of the information presented by foreigners interacting on television. The "language" of action, vocalization, and interaction is rich in meaningful expressions. Lech Walesa, leader of the Polish workers' union, Solidarity, became an international hero through television even though many of his admirers could neither understand the words he spoke nor comprehend the complexity of the issues his union dealt with. Similarly, the average American may not understand all the implications of a politician's policy statements, but anyone can grasp the "meaning" of a campaign advertisement that shows a politician eating breakfast with his or her family, and anyone can sense the implications of an FBI sting film that shows a politician stuffing money in his or her pockets.

The expression/communication distinction sheds more light on the differences between reading habits and television viewing habits. A book on a mental hospital may contain specialized discursive data which can be understood only by those with training in the subject. But while a television show may *discuss* similar specialized topics, the program usually *presents* very nonspecialized information as well. In one sense, all television programs that present the behavior of people are about the same thing: human gesture, feeling, and emotion. The expressive quality of television makes almost any television program accessible to the average viewer.

The heavy reliance on expressions in electronic media also tends to equalize the types of messages sent by different people. While some people have more pleasant and persuasive expressive styles than other people, there are no clear graded levels of expressive competence. One does not learn "to express" by going to Harvard. Even animals and children can express "effectively"—though not necessarily to the effect desired. There may be ways to teach people to express more pleasingly, but our educational system takes that realm of behavior for granted.[7]

Every organism exudes expressions, and children and illiterates may be, at times, more pleasing expressors than renowned scholars and poets. The writing styles of people of different ages and educational backgrounds vary much more predictably than their presentational styles. While essays, articles, and books must be composed by trained "authors," the content of a video "essay" can be provided by almost any passing man, woman, child, or animal. In these terms, there is an underlying similarity among the messages presented by different people on television—a similarity that does not exist in any other form of communication except face-to-face interaction. Groups of people who were excluded from the public forum created by print because they lacked the requisite entry skills are now able to participate in the public arena created by electronic media.

There are, of course, expressive "superstars," people whose personal styles are pleasing to large sections of the population, people who are not only *able* to express (as we all are), but able to express in a way that many people find exciting. Talk show hosts such as Phil Donahue obviously have exceptional expressive abilities that can make them public "personalities," but, ironically, part of the appeal of Donahue's programs is that almost any member of his show's studio audience seems capable of providing some comedy or drama.

Further, those people who are excellent expressors are not necessarily excellent "communicators." The charismatic Lech Walesa, for example, is a powerful expressor in person and on television; but he would not command much attention in printed speeches or books. As *Time* magazine reports:

> *His working-class Polish is rough and often ungrammatical. . . . His speeches frequently are riddled with mixed metaphors and skewed analogies; Solidarity's leaders admit that Walesa . . . is more intuitive than intellectual. He rather defiantly claims that he has never read a serious book in his life.*
>
> *Yet Walesa got through his message of hope to his countrymen. Said a Warsaw journalist: "Sometimes he doesn't even make any sense, but he is always reassuring."*

The accessibility of expressions compared to the elitism of discursive communications may explain why most people feel that they do not need political experts to help them interpret and understand televised political "speeches." And the "truth" and "reality" of all people's expressive messages explains, in turn, why the televised reactions of the person-on-the-street are so acceptable in news programs and documentaries.

The information form of electronic media is changing our perceptions of the world and its people. And through electronic "documents," future generations will experience a new sort of past. "History" was once a discursive script written and acted by the rich, the powerful, and the educated. Because of their discursiveness, historical sources were also filled with arguments, propositions, ideas, and ideals. The growing archives of audio and videotape, however, thrust the common person into history; simultaneously, they reveal what is common even about our leaders. And the "language" of these records is the expressive-presentational-analogic form of gesture, feeling, and experience. It is not surprising that in the last three decades many historians have discovered a new vacuum in our understanding of the past—the everyday experience of the common people who did not write down their thoughts and who did not make history, but who only occupied it. A whole new historical industry has developed that involves analyzing parish documents, gravestones, and other traces of "the world we have lost."

"Imprint of" vs. "Report on"

All media act as filters that exclude aspects of "reality." Each medium, however, is a different type of filter. The more abstract, obtrusive, and slow the encoding process, the less messages resemble the things they are about. And the less messages resemble reality, the more the medium can be used to present stylized and idealized images of events. Conversely, the more the medium simply "captures" what is happening in a situation, without any particular effort on the part of those present, the more messages will closely resemble the behavior they are about. And the more messages are tied to real behavior and events, the less they can be manipulated and controlled. In this sense, slow, obtrusive media have a "front region bias," while fast, unobtrusive media have a "back region bias."

Written and printed messages tend to be more formal than electronically conveyed messages. Sitting down to write a letter involves an abstract, obtrusive code. It is difficult to forget about the intervening medium because you must "translate" information into its form. But sitting down in front of a microphone or television camera does not involve a complex recording activity. You simply "behave," and a "message" is automatically "captured."

The informality of electronic messages is heightened when the medium is bidirectional and interactive. Speaking to someone on the telephone, for example, is so natural that we almost forget about the intervening medium. (We often say "I spoke to ________________ last night," not even thinking to add that it was a telephone conversation. We rarely say "I spoke" when we mean that we wrote someone a letter.) Because of the unobtrusiveness of the medium and its code, the telephone conversation is much more spontaneous and much less contrived than most letters. Letters to a close friend are often more stylized than telephone conversations with a stranger. People often rewrite letters several times, but rarely does one say on the telephone: "Let me start this conversation again."

This analysis also explains why one post-print medium—the telegraph—tends to be even more formal and less expressive than print. The telegraph's Morse Code is a code of a code. A student's letter asking for money from his or her parents would contain less expressive information than a telephone call involving the same request. But a telegram would be the least expressive of all three. It, like a hieroglyph, deals with stock communicative phrases ("IN TROUBLE STOP SEND FIVE HUNDRED DOLLARS STOP LOVE BOB"). For the same reason, computer messages can be rather "stuffy" when the computer is used for discursive communication rather than presentational information.

The striking difference in formality of the telegraph compared to the telephone is suggested by the first messages that have commonly been attributed to these media. Morse's first telegraph message is said to have been either "What Hath God Wrought?" or "Attention Universe!" Both messages are highly formal and abstract; they are clearly front region messages. They convey little personal expression or affect. The first telephone message, in sharp contrast, is reported to have been "Watson, come here, I need you"—a personal, back region message.

If the content of a medium requires complex encoding, then the encoder needs time and effort to formulate messages. The encoding energy and time pull the sender out of live, ongoing situations. In live interactions, one can compliment a cook while eating a meal, but to write about a delicious meal, one must usually wait until the meal is over.

The burden of writing, however, can also be a blessing. The time a person *must* take to write something can be used to facilitate the more careful crafting and molding of the final message. The time can be used to eliminate mistakes, resolve ambiguities, and hide the fact that one needs time to construct and edit the message. Thus, we may feel that a postcard written to hosts a week after a visit may better convey our thoughts about the visit than the spontaneous thank you's mumbled as farewells were said.

The great traumas of writing are hidden by the delays inherent in the medium. Just transcribing and proofreading an essay take a great amount of time. Anxieties, revisions, doubts, sweating, and swearing can all be concealed within this buffer time zone. Writing not only excludes all expressive behavior, it also conceals the physical process of producing the message. A final essay, therefore, can have a tone of confidence and formality foreign to the first fumbling drafts.

The more slowly messages are encoded, sent, and received, the more formal their typical content. Slow messages must "summarize" events; instantaneous messages can present a "transcript" or an "imprint" of events. A moment-by-moment account is less organized, involves many insignificant details, and contains many aspects of behavior that role performers would often prefer to keep secret. An instantaneous message is "messier" and more back region in flavor.

In writing a letter to a close friend, for example, you might summarize events of the last few weeks or months. You might tell your friend how the weather has been, that your child has begun to talk, or that you have been growing your own vegetables. Yet if in the middle of your letter writing, the same friend calls you on the telephone, the type of information you convey often changes. While the letter provides a general summary of the past, the telephone conversation often deals with the specifics of the present. You may look out the window and comment on the wind blowing your lawn chairs over, or you may have to get off the phone for a moment to tell your child to stop screaming in the next room. On the phone, your tone of voice might reveal that you are tired today, and you may feel obligated to explain that your stomach is a bit upset from eating a bad tomato. Even when the subject matter is the same—you and your life—the letter tends to present a more formal front region view, while the telephone conversation includes informal, moment-to-moment, back region experiences.

In the same manner, television's live coverage of events is more back region in style than newspaper accounts. Television reports of hostage situations, assassination attempts, union negotiations, and airport accidents include the stream of moment-to-moment rumors, suggestions, indecisions, anxieties, and incorrect assumptions. The choppy, disorganized, and mostly uninformative live coverage of a man's threat to blow up the Washington Monument in 1982 demonstrated how little police and reporters actually knew about what was going on. The newspaper stories the next day presented a much cleaner image.

Similarly, when John W. Hinckley, Jr., attempted to kill President Reagan, the videotape recording showed the disorganized scuffle around the assassin and his victims. Television news then reported that Press Secretary James Brady was dead and carried an announcement by Secretary of State Alexander Haig saying that, according to the Constitution, he was in charge at the White House, pending the return of the Vice President. Then, television news reported that James Brady was alive and that Haig was mistaken in his claim of authority. Then, the misinformation and mistakes on the part of reporters and officials became news stories themselves.

Most television news is not "live," but unlike other media, television has the constant potential for instantaneous visual and aural coverage. In times of crisis, therefore, we expect television to use this potential. The videotape of the Hinckley assassination attempt went on the air four minutes after the first shot was fired. The speed of the spread of news forced other actions to occur quickly. According to later reports, Alexander Haig rushed to speak to the press because he feared that silence from the government would be interpreted at home and abroad as a sign of confusion and leaderlessness. The result was that he literally ran to the press room, and came before the cameras out of breath and sweating, thereby contributing to the impression he hoped to forestall. Similarly, television reporters felt the pressure to confirm or deny the rumor of Brady's death quickly. In one instance, CBS correspondent Lesley Stahl saw White House aides "dissolved in tears," which she suggested to viewers probably meant that Brady *had* died. Unknown to her at the time, the White House aides had just heard the false report of Brady's death on another television network!

In a sense, the instantaneous coverage of television created some of these "events." In an earlier time, the slower reporting would have filtered out many of these stories. The rumor that James Brady was dead, for example, might never have been reported, and then, certainly, the fact that there were mistaken reports concerning Brady's death would not have become a significant news event. The small details, mistakes, and confusions would have been hidden and forgotten in the invisible backstage fostered by the slower print media (just as the letter to your friend would probably not mention overturned lawn chairs, a screaming child, a bad tomato, and tiredness on a particular afternoon). Further, because government officials would have been unable to give an instantaneous response to the nation, a few hours delay could not have been interpreted as a sign of confusion.

Slowness, of course, is not a guarantee of accuracy (witness the headline "Dewey Defeats Truman"), but the speed of encoding in television, combined with its wide spectrum of nonverbal information, leads to a new degree of exposure of the many details, fluctuations, and uncertainties that were traditionally filtered out in newspaper reports. What was once part of the backstage area of life is now presented as "news."[8]

Thus, while much of the *content* of news is shaped by journalistic decisions and by political and economic factors, the *form* of the news is often greatly affected by the nature of the medium. For whether one believes that television news in the United States is biased or unbiased, controlled or free, censored or uncensored, its style remains very distinct from print news.

The speed with which messages can now travel from senders to receivers limits the rehearsal and preparation time for many social performers (and makes whatever backstage time *is* taken more apparent). Even when one sits down to write a letter "immediately," several hours may be spent preparing the message. But if one is interviewed on the telephone, radio, or television, the message may have to be developed on the spot. The beginning of a sentence must be spoken before the end of the sentence is fully formed. The apparent inarticulateness of many of our leaders today may be related more to this change in media than to a decline in our leaders' basic intelligence.

Regardless of the specific symbol system used, all written messages separate behavior from reports of behavior. Electronic media, however, present behavior and experience much more directly. Many of us are willing to share with our friends detailed verbal or written descriptions of our experiences. But we would not necessarily want our friends to be present during all the situations we describe. Similarly, few of us would want even our closest friends to monitor constantly our behaviors and interactions through a video and/or sound system.

A written or verbal account of an event can be "packaged" and "cleaned up." An executive may tell his or her spouse that a training session for employees went "very well." The executive may forget to mention that at one point he or she tripped over a chair. And the executive may be largely unaware of the degree to which he or she made speaking errors, paced back and forth, or left out steps in a sequence of arguments. Clearly, whatever the executive later *says* about the event is not equivalent to the spouse having experienced the event directly. Thus, even in a technically "public" performance, such as a talk before one's employees, there may be many "back region" idiosyncrasies and mistakes that performers do not care to have publicized beyond the situation of their occurrence. A videotape of the executive's performance, however, would tend to carry such situation-specific behaviors beyond the live setting.

Electronic media, of course, do not simply "present" an event. Whenever you aim a camera or microphone in one direction, for example, you inevitably ignore things in every other direction. But electronic imprints of events do not allow for a clear separation between the "stream of behavior" in a situation and the information about the situation made available—later or simultaneously—to those not physically present. (Indeed, a videotape of an event may even shock and upset participants in the event who are not accustomed to seeing themselves "as others see them.") Cameras and microphones tend to force people to have memories and accounts of events that resemble the electronically captured images and sounds. As many politicians have discovered recently, it is impossible to deny quotes once they have been preserved on audiotape.

For these reasons, no doubt, many judges and legislators are hesitant to permit the televising of trials and legislative sessions. A written statement or a brief public appearance allows for a highly controlled performance. But it is much more difficult for someone to sustain a good "onstage" performance during many grueling hours of a trial or legislative debate. A camera at such events may capture public officials looking tired, getting angry, yawning, scratching an itch, picking their noses, or not paying attention. It is interesting to imagine how our image of great leaders of the past would change if we had thousands of hours of videotape of them instead of having only transcripts of their speeches and glowing (and sanitized) descriptions of their actions.

The live, ongoing nature of most electronic communications makes it much more difficult (and at times, impossible) to separate the public thread of experience from the private one. A wide spectrum of behavior is conveyed through electronic media. Emotions and feelings "leak out" through television. A depressed person may be able to write a cheerful note, but unable to make a cheerful telephone call. Not only does the nature of electronic media make it simple to forget that one's behaviors are mediated, but it also leaves one few options even if one remembers. For when the camera captures how you behave, then how you choose to act *for the camera* is also how you *really* behave in the situation, and how you behave in the situation is how the camera captures you. When a wedding is videotaped, or when the reunion of hostages and their loved ones is televised, the record of the event is an imprint of the behaviors themselves. Clearly, there is a relationship between private behavior and public message in electronic media that does not exist in print media.

This discussion offers another reason why the widespread use of television and other electronic media may be having a profound effect on social behavior. To the extent that traditional distinctions among various group identities, different stages of socialization, and different ranks of authority were dependent on pre-electronic distinctions between public messages and personal, expressive behaviors, we would expect significant changes in a wide spectrum of social roles.

This analysis has suggested that electronic media lead to a form of interaction that was once found exclusively in immediate physical interaction. . . . For when electronic media begin to provide the information once available only when participants met in the same place at the same time, then electronic media begin to reshape the meaning of "place" itself.

Notes

1. A few exceptions to the need for direct physical presence are the scent left by an animal or person, a pattern of footprints in the sand, or the style of handwriting in a letter—all of which may provide information about an organism's emotional state.
2. Most communication theorists use the term "communication" to include both of these types of information. Thus, although I adhere to Goffman's special, limited notion of communication in this discussion, I use the term in the more typical, generic sense elsewhere in the book.
3. Films and videotapes can, of course, be "edited" in the sense that strips of action can be arranged in any sequence or cut to any length. But one cannot manipulate a person's actions and speech on videotape or film quite as freely as one can edit words on a page. In print, an editor may add words, delete words, change punctuation, and perhaps alter the spellings from American to British (in a sense, changing the author's "accent"). With film and tape, however, the deletion of a word is often obvious, the addition of a word almost impossible, and a sentence cut off in the middle sounds odd. To edit out a whole phrase or sentence, videotape editors must often cut to another shot (such as an interviewer nodding) so as to avoid a "jump cut" in which a person's head, body, and hands suddenly jump to a new position. Even the smoothest editing job, however, cannot change the cadence of a person's speaking or edit out a tone that smacks of conceit or uncertainty.
4. Although I have focused on media such as radio and television, the loudspeaker also has had a tremendous impact on the style of public address. With a microphone and amplifier, for example, a speaker could speak more softly and intimately, and more nuances of expression could emerge. It is probably no coincidence that the first use of amplifiers for presidential addresses,

the birth of broadcasting, and the formal addition of a "speechwriter" to the President's staff all took place in the same year.

5. I still remember my surprise the first time I read a transcript of the "CBS Evening News with Walter Cronkite." The news program was aired on October 25, 1972, and began as follows: "Good evening. North Vietnam today stepped up its propaganda war over the subject of peace. Hanoi said President Nixon is tricky, and crafty, and the real stumbling block to peace in Indo-China, not President Thieu of South Vietnam. In Paris meanwhile, the Viet Cong said a cease-fire could come within hours, but not until Mr. Nixon either dumps Thieu, or gets him to accept a coalition government." These words *sound* fine when read out loud, but they are very different from the reporting style of respected newspapers. Note the informal constructions and the use of words such as "tricky," "crafty," "dumps," and "gets." Later in the same broadcast, Cronkite opened another serious story with the line, "More of those deadly little letter bombs turned up today." The next day's *New York Times* reported the same events quite differently. The mention of the Communists' accusations was buried deep within the story on the cease-fire and began as follows: "The Communists' statements from various quarters concentrated . . . on warning that they would fight on to victory if a settlement was not reached. There were renewed accusations that peace reports were a 'political trick' by the United States designed to 'dupe public opinion.'" One of the two newspaper stories on the letter bombs began, "Explosive devices sent by mail critically injured an Arab postman and a secretary here today, and other letter-bombs exploded or were intercepted in Algeria, Libya and Egypt."
6. According to the Roper surveys, television has been viewed as the most believable source of news since 1961. Since 1968, it has held better than a two-to-one margin over the next most believable source, newspapers.
7. Actors are among the few people who consciously study expressive skills. In contrast to the stage, however, television "acting" often means learning how to behave "naturally" and pretending that no camera is present. Many business leaders now take training in speaking into television cameras so that they can address their employees through videotapes (or star in their companies' TV commercials). But again, a major part of their training is learning how to be comfortable about the presence of a camera. We know from the television series "Candid Camera" how good many unselfconscious "performers" can be.
8. Motion pictures stand between print and electronic media in terms of this variable. Film must be developed before it is projected. Filmed newsreels and interviews can be screened and edited in the backstage created by such necessary delays. But the telephone, radio, and television permit (and therefore, in a sense, demand) instant "live" coverage and transmission. If John Hinckley's attack on Ronald Reagan had been filmed instead of videotaped, for example, there would have been time for the Secret Service and the FBI to request a screening of the film before it played on television. There would have been time for authorities to ask for a delay in its showing or to decide that parts were too graphic to display on television. And the public need not have been informed of such control and censorship since films can *never* be shown immediately. Because electronic images can be transmitted instantly, any attempts at control or censorship become difficult and apparent. Such "biases" of electronics sometimes operate even in countries where the press is tightly controlled by the government. When the Pope visited Poland in 1983, for example—in part to criticize the government—the *possibility* of covering his trip "live" seemed to demand at least some live coverage. As a result, the Polish people heard serious criticism of their government on the government's own radio and television.

Remembrance of Films Past: Cable Television and Classic Hollywood Cinema

Barbara Klinger

The procedure which today relegates every work of art to the museum . . . is irreversible. It is not solely reprehensible, however, for it presages a situation in which art, having completed its estrangement from human ends, returns . . . to life. . . . [Museums] have actually transformed works of art into the hieroglyphics of history and brought them a new content while the old one shrivelled up. No conception of pure art, borrowed from the past and yet inadequate to it, can be offered to offset this fact.

Theodor W. Adorno, "Valery Proust Museum," 1967

As we enter the second century of great American filmmaking, AMC will, through its commitment to preserving this unique portion of our cultural heritage, continue to be the Museum of Classic Hollywood.

American Movie Classics promotion, 1994

Deanna Durbin smiles an impossibly sweet smile as she convinces the famous Leopold Stokowski to conduct an orchestra of down-on-their-luck musicians in *One Hundred Men and a Girl* (1937). Lupe Velez prevails, through sheer charisma, over a society matron trying to break up Velez's marriage to the matron's nephew in *Mexican Spitfire* (1939). The Ritz Brothers get hopelessly drunk toasting all of the French kings named Louis in their version of *The Three Musketeers* (1939). Cornell Wilde, square jaw firmly set in place, flees through the African bush from a tribe bent on killing him in *The Naked Prey* (1966). These scenes may not have the mythic resonance of the burning of Atlanta from *Gone with the Wind* (1939) or Kane's utterance of the word *Rosebud* at the beginning of *Citizen Kane,* but they are, nonetheless, part of the legion of classic-era Hollywood films shown every day in the United States.

While vintage films appear periodically in various high-profile venues, especially when they are rereleased on the big screen or materialize on network television during the holidays, they have become a staple of cable television. Venues that once acted as primary recycling centers for Hollywood features in the post–World War II era, such as network television and retrospective movie houses, either no longer regularly program this fare or, in the case of retrospective houses, have largely gone out of business. Beginning in the 1980s, VHS and cable TV displaced the networks as providers of all manner of Hollywood films, making old films, including the most obscure and forgotten titles, an even more intimate part of everyday life. With its round-the-clock programming, cable has become a repository for what is popularly referred to as "classic Hollywood cinema," that is, studio films made between the 1920s and 1960s. From A&E (Arts and Entertainment) to USA, cable stations have made classic films and television series a significant part of their offerings. In the process, they have become major purveyors of the media past. As they persistently re-present classic films to contemporary home viewers, cable stations help to create a home film culture devoted to the commemoration of vintage American cinema.

The resurrection of artifacts is, of course, an integral activity of all institutions associated with the arts. Indeed, it would be hard to imagine the U.S. cultural landscape without museum exhibitions of the Old Masters, performances of classical music, bookstores featuring canonical novels, DVD reissues, TV reruns, and "golden oldie" radio stations. We expect and even rely on the ready availability of forms from the past long after their original appearance. Since both media industries and audiences are substantially invested in reestablishing contact with past forms, it is worth examining the cultural forces that influence how we remember and value these artifacts.

Memory studies devoted to film typically analyze either the emotionally potent connection between visual media and individual histories or how certain texts (e.g., *Schindler's List* [1993]) attempt to construct the past and its public memory.[1] Each of these approaches provides valuable information about film's relationship to memory; however, far less attention has been paid to the practices of media companies committed to showcasing the media past. Yet, the visibility and centrality of Hollywood studios and media conglomerates, coupled with their immense capacity to revive bygone texts, suggest that they should be considered, along with museums and government memorials, as institutions of memory par excellence. Like museums and memorials, Hollywood acts as a custodian of the past, orchestrating meaningful and influential confrontations with its archive for viewers. And like these other sites, Hollywood has its share of aesthetic treasures and historic monuments; in a sense, *Gone with the Wind, Casablanca* (1942), and *Citizen Kane* have become as much markers of the nation's cultural heritage as Plymouth Rock is. How, then, do media companies repackage reissues for public consumption? In the process of reinventing "old" products for contemporary audiences, how do they reconstruct the past? What implications do their strategies of display have for understanding the connections between media industries and the social production of memory?

To pursue these questions, I examine the home presentation of classic films by one of the chief self-defined custodians of U.S. film culture today: cable movie channels devoted to the exhibition of vintage cinema. In this chapter, I concentrate on American Movie Classics (AMC), the oldest and most successful of these dedicated channels.

In a 1997 *New York Times* article, "Teaching New Generations the Joys of Old Movies," Stephen Henderson commends AMC, Turner Classic Movies (TCM), and Movies from Fox (FXM) for playing "a huge role in keeping the nation's film history alive." He writes that because they are "just about the only places movie fans can find their old favorites . . . the movies these networks choose to show will become the ones known by generations to come. . . . Without channels showing 30's hits and John Wayne westerns, millions might grow up film-illiterate." Henderson suggests that, while profit-motivated, AMC, TCM, and FXM have achieved the status of central archives of the past, promoting the viewer's awareness of film history, selecting films that will be remembered by future generations, and helping to prevent film illiteracy. The executives of these channels have taken it upon themselves to "simultaneously entertain and educate . . . citing a need for raising awareness of America's film heritage."[2] AMC's role as a showcase for the Hollywood past has been especially widely recognized. Cable executives and audiences alike have referred to AMC as the "the Metropolitan Museum" and "the Masterpiece Theater" of classic movies.[3]

As a part of mass culture that is zealously commemorative, AMC is an instructive example of how America's film past is preserved in the realm of domestic leisure. As we shall see, the channel is very much a part of what Michael Bommes and Patrick Wright define as "national heritage culture." National heritage culture involves a "public articulation or staging of the past . . . of immense extent, variety, and ubiquity."[4] In its public presentation of the past, national heritage culture selects aspects of the past for commemoration, supplying historical context to explain their significance to audiences. Heritage culture often also involves the care, preservation, and management of the selected past, ensuring that it will continue to represent the nation's achievements across generations. The acts of selection, contextualization, and preservation are necessarily ideological. Whether these acts seek to promote and maintain one vision of the past over possible others or to present diverse viewpoints, they enter into the cultural fray about the meaning of the nation's history.[5]

Typically, we tend to regard civic, state, government, and other official organizations as being responsible for preserving the physical aspects of the nation's past (e.g., its buildings and landscapes) as well as the memories of the persons, events, and rituals that compose its history (e.g., through monuments and public pageantry such as reenactments and parades). Bommes and Wright provide an important revision to such assumptions. They argue that reminders of a nation's heritage occur through a "whole battery of discourses and images" that echo throughout the culture's representations, including such things as advertisements and other mass cultural ephemera (253). National heritage culture, then, is not only the preserve of the museum or government memorial; it can also be vividly expressed and maintained in even the most quotidian forms of representation. Aspects of nationalism become part of the language of mass culture through strategies designed to heighten a product's visibility, legibility to its audiences, and resonance as a specifically American good.

With their elaborate exhibition strategies and consciousness of preserving and promoting the nation's film heritage, cable movie channel "cine-museums" provide an opportunity to examine the specifics of the traffic in classic film titles as well as the larger cultural and historical implications of resurrecting vintage texts. To begin, these cine-museums allow us to examine the classic label itself. Classic films are not born; they are made by various media, educational, and other agencies interested in revitalizing old properties within contemporary taste markets. Further, as a means of locating Hollywood relics within these markets, exhibition venues forcefully rehistoricize them, giving them new sellable, historical identities. Not surprisingly, the classic film is surrounded by discourses that emphasize a nostalgic view of the past, particularly through a lavish attention to "old" stars. Not only do stars act as agents of nostalgia; classic movie channels also define them as potent representatives of gender and race capable of supplying verities about America, past and present. Thus, the recycling of movies on cable involves a juggling of aesthetic and historical values that provide insight into the place of the classic within mass cultural hierarchies and its function as a platform for the popular writing and rewriting of U.S. history.

Certainly, the classic movie channel's historical revisionism is part of a public articulation or staging of the past, as Bommes and Wright say. But its participation in a national heritage culture devoted to film is also more explicitly realized. The rise of classic movie channels took place within a highly charged cultural moment in which saving the film past was freighted with concerns about the health of the nation. AMC and other such channels that have featured "preservation festivals" of various kinds in their programming are part of a larger film preservation movement, fueled in the late 1980s by government legislation and the support of certain social agencies and organizations. In addition to having significant ecological value and virtue, historical preservation is also a political mission, subject to perceived national concerns and crises. As I will argue, the contemporary film preservation movement was articulated in relation to multiple challenges to the sovereignty and identity of America at the end of the twentieth century. Within the politics and ideologies of this time, the classic cable movie channel participated in a discursive reconstruction of the nation spurred by a variety of perceived internal and external economic threats.

By studying this home film culture's aesthetic and historical "makeover" of the classic film and its prizing of film heritage, we can consider which memories of the cinematic and American past enter the highly valued market of the home, literally reaching viewers where they live. Given this film culture's preoccupation with history, we can also grasp far-reaching connections between the province of the home and institutions involved in the management and circulation of the past, while meditating on the nature of mass-produced memory itself. Before we engage with these issues, however, it is necessary to provide some background on AMC's history as the first cable station to program exclusively and without commercial interruption feature films from Hollywood's classic era.

About AMC

Launched in 1984 as a pay service to 250,000 subscribers, AMC gained a more solid footing in the industry three years later when it became part of basic cable, an arrangement that allowed subscribers to have the channel as part of a flat monthly fee. In 1991, Rainbow Media Holdings, Inc. (a subsidiary of Cablevision Systems Corporation and NBC) and TCI (Tele-Communications, Inc.) each owned half of AMC.[6] TCI and Cablevision were ranked, respectively, first and sixth among U.S. multisystem operators (or MSOs, companies that own more than one cable system), associating AMC with two top cable providers early in its history. At the time, TCI was also one of the largest media companies in the United States, just behind such other giants as Time Warner Inc., the Disney Company, and the Sony Corporation.[7] Presently, AMC is fully owned and operated by Rainbow Media Holdings.

Given its auspicious beginnings, by the early 1990s AMC was well situated in terms of two integral components of cable success: number of subscribers and programming rights acquisition. In the years following the channel's introduction, subscribers rose from 3.2 million in 1985 to 24 million by decade's end to 39 million in 1991.[8] By the mid-1990s, 61 million subscribed to the channel, making AMC sixteenth among the top twenty cable networks, just below CNN Headline News (at 62,619,000) and MTV (at 65,900,000). In 2004, AMC reached 84 million homes. As Douglas Gomery remarks, for both cable operators and viewers alike, AMC represents "one of the great success stories in the development of cable TV."[9]

Certainly, AMC's early part ownership by TCI was influential in its growth; this connection alone substantially enhanced AMC's audience, because, by 1991, TCI possessed nearly one-quarter of all cable subscribers. More subtly, Ted Turner's purchase of MGM/UA in the mid-1980s had considerable impact on AMC's success. When Turner bought MGM/UA Studios and its library of films, he found to his dismay that many of the films were under contract with AMC and other organizations. Taking legal action, Turner challenged AMC's cable rights to the films, but he eventually paid the channel fifty million dollars in order to reclaim older Warner Bros. and MGM titles. AMC used some of this money to secure new movie rights to enhance its own library of films, helping to build a solid and diverse foundation for programming.[10] By 2000, AMC had more than five thousand films in its library as well as licensing agreements with, among others, Samuel Goldwyn, RKO, 20th Century Fox, Paramount, Warner Bros., Universal, Columbia/Tristar Television Distribution, MGM/UA Domestic Television Distribution, and several independent distributors.[11] It should be noted that, despite a reputation for showing uncut versions of films, AMC often shows "TV-ready" prints edited for content such as nudity and offensive language.

At its inception, AMC's programming was aimed at a category of viewers long considered to be "cable resisters" by the industry. By showing older films—uninterrupted and uncolorized—AMC attracted the "gray demographic," the middle-aged and elderly audience.[12] AMC's early choice of hosts reflects this priority. The main host of the show, Bob Dorian, was part of the older demographic himself, as were many of the celebrity guest hosts who appeared on the channel in the 1980s and 1990s, including such stars of Hollywood's golden years as Debbie Reynolds, Shirley Jones, and Douglas Fairbanks Jr. The end of the 1990s found AMC trying to broaden its appeal to include younger, "hipper" viewers. At this point, AMC began to hire slightly more youthful hosts (such as John Burke and Neal Gabler) and to showcase guest celebrity hosts from contemporary films (such as Dennis Quaid, Darryl Hannah, and Samuel L. Jackson). AMC also featured original programming designed to put "a new spin on classic movies," to render them "more palatable to a wider audience."[13] These original programs, such as *Knockout: Hollywood's Love Affair with Boxing* and *Hollywood Goes to Court,* attempted to revitalize the classics by showing clips that trace the relevance of old films to fundamental aspects of American culture—in these cases, sports and the judicial system, respectively. In 2002, AMC modernized its offerings yet again in a stronger bid for the youth market. Along with some older fare, the channel regularly screens more contemporary films and has launched special programming in tune with the times, including reality-based shows and DVD-TV, a series that exhibits films with accompanying supplemental features, directors' commentaries, and other extras associated with DVD. To generate more revenue, AMC's film showings are now interrupted by commercials.

As part of these most recent changes, AMC's Web site (www.amctv.com) features images of twenty-somethings, ads for upcoming movie releases, and information about the AMC Movie Academy, a filmmaking program aimed at youth.

Through most of its history, AMC has organized its programming through "branded blocks," groupings of films according to certain themes, a concept that has long been a popular way for television networks and other programmers to repackage and promote classic Hollywood fare. These blocks have been thematized according to stars, directors, genres, historical periods, and special topics (such as the femme fatale and movies showcasing the Big Band sound). The channel's original programming has often taken the form of documentaries tied to the thematic festivals of the month (e.g., "Stars and Stripes: Hollywood and World War II" and "Between Heaven and Hell: Hollywood Looks at the Bible"). Further, since 1993 AMC has run a series of film preservation festivals in association with major nationwide organizations devoted to film preservation and restoration. In addition to its on-air strategies of presenting the past, beginning in 1988 AMC published a monthly magazine, *American Movie Classics,* which contained articles about Hollywood along with the channel's monthly screening schedule. AMC no longer publishes the magazine, but the magazine's content is available free of charge on the channel's Web site and in its newsletter.

AMC's 2002 transformations are interesting for what they indicate about changing strategies of classic film exhibition in a shifting and increasingly competitive multichannel cable universe. AMC appears to be leaving most of the market for pre-1950s Hollywood films to TCM, a cable channel that continues to court older viewers and die-hard classic movie fans. However, this chapter focuses on AMC in the 1980s and 1990s, when the channel's commitment to older Hollywood classics was key to its phenomenal success and to the role it played in American culture. During this era's political ferment, especially with respect to anxieties about globalization, establishing the concept of film heritage became an urgent and widespread national preoccupation. Cinema's past, perhaps, had never mattered more to so many.

Among the different facets of the classic cable movie channel, one of its most important operations is to assert the value of the aged artifact. An obvious goal perhaps, but the heritage industry's way of achieving it involves a vigorous reassessment of the relationship between past and present, forging an intimate link between the attribution of aesthetic worth and historical revisionism. With this in mind, we can examine three interrelated strategies AMC has used to affect the value of classic Hollywood films: the expansion of the classic film canon to make it more inclusive; the definition of film classics as vital historical documents; and the establishment of a crisis in our cinematic heritage accompanied by the need to rescue it from the ravages of time. While AMC's practices do not apply to all institutions that resurrect Hollywood's "Golden Era," its case presents an instructive view of the internal dynamics involved in contemporary film heritage discourse and the economic and political forces that interact with this discourse as it enters the viewer's domain.

What's in a Name?

We might expect that only those films that have been ranked in "best of cinema" lists, such as *Citizen Kane* and *Casablanca,* or, at the very least, those that were made during Hollywood's classic era (roughly from the 1920s through the 1950s) would be labeled as classics. However, AMC's use of this term is not so selective. Like many other media companies in the business of selling Hollywood films, AMC refers to nearly all of its properties as classics, no matter when they were made or how humble their reputation. Revealing the broad usage of the term, the channel has shown films as part of categories such as "Early Morning Classics," "Matinee Classics," "Saturday Classics," and "Classics Collection." In this context, it is entirely possible to see *The Ghost and Mr. Chicken* (1966), starring Don Knotts, hailed as a "comedy classic" and Bo Derek touted as a "legendary screen star."[14] Thus, AMC's use of the classic designation is inflated, a rhetorical gambit designed to canonize all Hollywood products, even those that have long flown under the aesthetic radar, in the hope that their instant value will translate into instant profit.

Although it may be driven by the most basic commercial motives, the attribution of classic status has substantial cultural implications. It represents a bid by what Pierre Bourdieu calls "agents of consecration" to create or enhance the legitimacy of works—an effort that circulates in culture as part of the identity of these works. Such agents are interested in conserving "the capital of symbolic goods," that is, in producing or maintaining the prestige value of artifacts within a system of cultural relations.[15] This act of conservation serves commercial interests but also attempts to secure status for the agents and their goods in the social hierarchy. Certain social institutions, such as museums and the educational system, as well as more informal groups, such as reading circles and fan clubs, routinely strive to establish or confirm the value of certain artists and their works. No matter whether their acts of consecration take place in what are traditionally considered high or low cultural arenas, these entities become tastemakers. They acquire the sheen of discerning readers or viewers within their particular territories. Such activities are omnipresent in the media, with the industry, critics, talk show hosts, fan clubs, and other groups generating lists of top books, CDs, films, television shows, and so on.

In Bourdieu's schema, additional agents—"agents of reproduction"—educate consumers to attain competency, to become "good" readers or viewers, thus enlisting them to support and to continue to reproduce the status of these works (121). Like a museum, AMC engages in both of these activities. The channel asserts the value of its artifacts as timeless classics and, through various means such as the commentary of hosts and the monthly magazine (which often featured articles by film academics), guides viewers in the art of classic film appreciation. Educators who appear on the channel give its films the respectability associated with literate, academic tastes.

In the process of manufacturing an aesthetic preserve for old Hollywood movies, media companies demonstrate the permeability of boundaries between high and low culture. Bourdieu writes that certain institutions operate in relation to a "field of restricted production," an aesthetic sphere primarily identified with high art. This sphere is tightly knit in terms of producers and audiences—for example, classical musicians and avant-garde artists performing for an edified cognoscenti. It also does not overtly identify itself as concerned with economics; rather, its concern appears to be with notions of art. Such fields are, at first glance, utterly opposed to what Bourdieu refers to as "large-scale production," or what is commonly called mass culture. Commercial productions in TV, film, radio, and literature are driven by the bottom line and aimed at the "average" consumer. Because of its obsession with profit and with reaching general audiences, large-scale production is, by definition, middle-brow. It is this middle-brow audience that "producers of this type of art and culture *explicitly* assign themselves, and which determines their technical and aesthetic choices" (125–26). Even when large-scale producers seek a targeted audience (e.g., women, men, teenagers), they still participate in an overall industry strategy of reaching increasing numbers of consumers.

However, because both the restricted and the large-scale spheres of production strive for crossover appeal, they are not as diametrically opposed as they seem. For example, art museums demonstrate their desire for the financial success associated with mass commercial enterprises by screening Hollywood films, embracing blockbuster exhibits, and devoting ever-increasing floor space to museum shops, where art-related paraphernalia is sold. Similarly, the producers of mass culture often aspire to escape their perceived debased status by defining themselves in relation to so-called legitimate culture. Media industries have long struggled to achieve the kind of symbolic capital or accumulated prestige that characterizes high art. Hollywood's history is marked by repeated efforts to align itself with respectability, whether through the European-inspired architecture of the silent era's motion picture palace, the adaptation of "serious" literary works, or the ceremonial bestowing of Oscar nominations and awards.

The classic designation becomes one means among many, then, of associating film producers and consumers with the symbolic capital of refined culture. Further, as such a marker of value, the classic label serves strategically to defend the status of the old product in the contemporary world of media consumption. When media companies define vintage films as classics, they help to create a cultural orthodoxy around Hollywood cinema of yore, protecting it from forces that could detract from or obliterate its significance. In a time that rabidly privileges new media products and technologies, the classic label wards off the appearance of obsolescence for the reissued film, allowing it to find a place in the market. To this end, AMC frequently characterizes classic-era movie stars as surpassing any subsequent achievements by contemporary celebrities who might appear to be more accomplished. Thus, Bing Crosby's casual, crooning singing style—the rage during the 1940s and 1950s—"changed the sound of popular music forever," while his song "White Christmas," which has sold thirty million to forty million copies, is "by far the best-selling single of all time. . . . Nobody has sold more records. Not the Beatles. Not Michael Jackson. Nobody." Similarly, an AMC magazine writer reports that it is "all but impossible to find an actor today who can match Cary Grant." In an article entitled "The Complete Cary: It Takes Five Modern Leading Men to Make One Classic Grant," the writer finds that Grant's versatility as an actor could be matched only by building him, à la the Frankenstein monster, out of five contemporary actors, each of whom represents one part of Grant's multifaceted persona. Thus, we would need Hugh Grant, Pierce Brosnan, George Clooney, Harrison Ford, and Sean Connery to be able to re-create Cary Grant's talents.[16]

Insisting on the superiority of Hollywood's Golden Era to the present, this kind of discourse shows the continuing relevance of old films to today's mass culture. As we shall later see, the classic label's defensive function also includes a cultural dimension: it helps to deflect criticisms that the vintage film could be subject to in the present, especially regarding dated, stereotypical portraits of gender and race.

In sum, although the classic label may not be a reliable indicator of aesthetic value (indeed, it demonstrates the term's slipperiness), it nonetheless strives to give an important status to aged forms as well as to their exhibitors and viewers, to create responsive audiences in a highly competitive media world, and to avoid the threat of devaluation. Situated squarely within large-scale production, AMC is an enterprise that nonetheless uses the language of so-called legitimate culture to bless the resurrection of old Hollywood, transforming the profane into the sacred. Once considered by some critics to be the home of mammon (just as contemporary Hollywood often is now), the Golden Era, with its crusty studio moguls and corporate backing, appears as more artistic, more innocent of the bottom line, than today's multinational media conglomerates. Removed from the taint of late capitalism, all kinds of old films—the noted, the forgotten, and the critically decried—can achieve aesthetic status by virtue of their historical vintage.

As AMC portrays the value of old Hollywood, it seeks to affect the home viewer's film literacy in order to sustain and reproduce the aesthetic and entertainment worth of the classic label. Often involving a comparison between past and present, the channel's exhibition strategies and appeals to its audiences are underwritten by a strong component of nostalgia. AMC presents Hollywood's stars, directors, and films as either evidence of a better past or a vital reminder of the best tendencies of the present. Because the baby boomers and their parents, who once constituted AMC's sole demographic, grew up watching many of these films in theaters or on network television, the audience is primed for reminiscence. As the classic is extolled as an emblem of the good old days, the viewer is treated not only to fond recollections of Tinseltown but also to pointed reassessments of the American past and present.

Seeing Stars

Some of AMC's invitations to experience classic cinema nostalgically are overt. For example, advertisements for its Big Band–era festival, which included *The Glenn Miller Story* (1954) and *The Benny Goodman Story* (1955), asked viewers to take "a sentimental journey down musical memory lane . . . put on your dancing shoes, roll up the rugs, turn up the volume and enjoy the sights and sounds of the Big Band Era." Similarly, in a series entitled "America's Movie Palace Memories," AMC let its audience participate in "the magic of a Saturday matinee." This series re-created the matinee experience by showing a block of cartoons, shorts, serials, newsreels, coming attractions, and feature films to take viewers "back to the glory days of Hollywood." The cablecasts were also set in "landmark movie theaters across the nation. . . . 'Great American Cinemas' [that] . . . serve as opulent reminders of the part neighborhood theaters played in our film heritage."[17]

Invariably historical in nature, these kinds of celebrations of the past are filled with a wistful yearning for yesteryear that is tinged with a view of bygone eras as superior to the present day. Nostalgia tends to arise at times of social change as a means of lamenting progress; thus, nostalgic presentations often show the past as a simpler, more harmonious time that can serve as an antidote to what is perceived as the messy complexity of the modern. Anxieties about the direction in which progress will take the country result, then, in indulgences in reveries meant to deny or offset the impact of change.[18] Certainly, through its "America's Movie Palace Memories" and other series, AMC defines itself as a place where such reveries can occur. However, the brand of Hollywood nostalgia on display is not simply concerned with the depiction of an earlier, rosier time as a means of reacting against change. Nostalgic enterprises also construct the relationship between the past and the present in a way that stress continuities rather than ruptures in a nation's history—a different means of dealing with perceptions of historical discontinuity. By showing a linear, rational progression in the nation's narrative, such enterprises reassure citizens that certain traditions and values that have always held the nation together will continue to do so. In the process, the backward gaze produces historical accounts that enter the field of public memory.

The problem with nostalgia in either its reactive or its inclusive variations is that, once empowered to tell the nation's story, it operates through what historiographers call "selective memory"—a pointedly partial retelling that often results, as Fredric Jameson argues, in a "chronological laundering" and neutralization of the past.[19] Nostalgic discourses tend to whitewash the past, repressing or minimizing conflicts that marked the nation so as to provide an affirmative picture of its political order and way of life. This is especially true of organizations that, like AMC, are commercially invested in resurrecting the past. As Michael Kammen argues, because it seeks to sell history to tourists and consumers, to make history into a "feel good" entertainment experience, this type of "entrepreneurial mode of selective memory" is as interested in forgetting unseemly aspects of the past as it is in commemorating its high points (535). While we can agree that nostalgia often produces unreliable narratives, nostalgic accounts are nonetheless significant because they represent popularizations of the past that vie for dominance in mass culture, circulating specific conceptions of history to citizens in public and private space.

AMC's particular mode of popularizing the past relies on establishing parallels between Hollywood and the nation. For Hollywood films to be able to signify the national story, they must first be intimately and transparently identified with U.S. history. AMC's strategies of display strive to deepen the value of old Hollywood products by repeatedly emphasizing their significance as unmediated documents of American life. As an AMC staff writer matter-of-factly comments, "AMC's commitment to classic movies runs deep, because classic movies are deeply rooted in American culture." Similarly, on the occasion of one of the channel's film preservation festivals, Martin Scorsese states, "Film is history. With every foot of film that is lost, we lose a link to our culture, to the world around us, to each other, and to ourselves."[20]

This sense of Hollywood's historical status informs the presentation of various film series on the channel. For example, in 1993 AMC programmed "Shots Seen 'round the World," a series designed to "celebrate the heights of motion picture achievement." The title of the series plays off one of the founding moments of the American Revolution, when, in 1775, farmers in Concord, Massachusetts, fired upon British soldiers who were going to destroy their guns and other means of waging resistance, an event commemorated as the "shot heard 'round the world." In its version of shots *seen* 'round the world, AMC presents film stills that have become a familiar part of the public's recognition of Hollywood. These shots are "moments audiences will cherish always. They've had an effect on our shared culture." The moments included in the first installment of the series are King Kong's battle with navy biplanes from the top of the Empire State Building; the scene from *Citizen Kane* in which Kane (Orson Welles) expires, uttering the word *Rosebud*; the foggy airport runway in *Casablanca* with Humphrey Bogart and Ingrid Bergman; Will Kane's (Gary Cooper) climactic walk alone down the deserted street in *High Noon* (1952); the beach scene with Deborah Kerr and Burt Lancaster's embrace in the surf in *From Here to Eternity* (1953); and Roger Thornhill's (Cary Grant) attempt to escape the crop-dusting assassins in *North by Northwest* (1959).[21] Through its presentation of these iconic images, this promotion accomplishes several rhetorical feats: it equates cinema and U.S. history, distinguishes the importance of cinematic imagery to the nation, and uses the language of heritage ("they've had an effect on our shared culture") to further stress cinema's historical significance. While few would deny film's close relationship to society or the need to preserve it as part of a culture's heritage, we should not overlook the significance of the way in which these issues are explained. The explanation allows the historical enterprise not only to stake certain claims about the importance of classic cinema but also to create a specific narrative of the nation's history.

In another kind of parallelism that relates cinema seamlessly to American life, AMC's film festivals are routinely aligned with holidays and special occasions, including Thanksgiving, Christmas, Black History Month, Memorial Day, and the Fourth of July. In addition, these festivals often coordinate a particular star with a holiday or event. Classic-era stars, the most visible, recognized, and promoted component of old Hollywood, become a central means of commemorating both the cinematic and the national past. Stars are everywhere in AMC's representations; besides appearing in films, they adorn advertisements for memorabilia in AMC's various venues, grace every cover of *American Movie Classics Magazine,* and are prominently featured on the Web site. They promote AMC's programming and act as co-hosts as well as interviewees in on-air programming. Even more than film directors or the films themselves, stars are the major subject of the channel's discourse. Their omnipresence allows the channel to draw from the cultural authority that celebrities have as representatives of the media industry while banking on them as a currency that has long driven the market in Hollywood memorabilia. Without stars, the world of film nostalgia would be a wasteland. Their familiar presence alone enables them to act as gateways to the American past for home viewers who grew up with them. Their gateway function, in turn, gives them a measure of historical authority that can be deployed within histories focused both on celebrating the grandeur of the past and on elaborating continuities between past and present that affirm contemporary life as well.

John Wayne, for example, who as late as 2001 ranked in the Gallup Poll as America's all-time favorite movie star (ahead of Julia Roberts, Tom Hanks, and Al Pacino), has also been a tremendously popular actor with AMC's viewers. On Memorial Day in 1992, one of AMC's many film festivals dedicated to his work featured some of his war films and Westerns. On this occasion, AMC magazine writer Nat Segaloff observes, "If there is one individual who symbolizes America to the rest of the world, it is John Wayne. 'The Duke' . . . portrayed such a range of heroes throughout his 50-year career that his face, voice, gait, and ethics may say more about this country than any scholar, industrialist or elected official." Segaloff acknowledges that in the 1960s Wayne became a controversial figure because of his right-wing politics, specifically his support of the Vietnam War. Any controversy was put to rest, however, during Wayne's illness and eventual death from cancer in 1979. This led "his fans, young and old, liberal and conservative" to send him "an outpouring of love and support. Few of them knew John Wayne the man; they were responding, instead, to Nathan Brittles, Ethan Edwards, Rooster Cogburn, the Ringo Kid, and an unequaled screen legacy that not only enriched our lives but helped to define them as well."[22] Wayne's symbolic power and screen legacy are so influential, Segaloff suggests, that his passing caused generational and political differences to collapse. In this case, cinema appears as more real than reality, capable of displacing and perhaps even resolving historical conflicts because of the audience's emotional attachment to characters and actors who represent American identity. Further evidence that Wayne, once referred to as "an Extra Star on the American Flag," is still considered a "true American hero" is amply provided by fan Web sites dedicated to the actor's accomplishments and memory.[23]

AMC's programming strategies underscore this identification of actor, role, and history. By staging a festival of an actor's films in association with a holiday, a double action occurs: the star comes to represent a ceremonial moment in American history, and that moment is observed in relation to the actor and the films in which he or she appears. In this case, Memorial Day is an occasion when war dead are remembered by loved ones and honored by the nation for sacrifices made to maintain the American way of life. Wayne films shown during the AMC commemoration include *Tall in the Saddle* (1944), *Flying Leathernecks* (1951), and *Fort Apache* (1948), in which he plays, respectively, a cowboy, a Marine, and a cavalry officer. These roles not only represent the classic U.S. masculine types of Westerner and soldier; they also portray episodes of national and global conflict that defined the nation's development and status in the world—the settling of the West and World War II. The synergy between film and occasion casts the actor as a player in American martial history and uses the classic film as the site of commemoration, reinforcing the sense that Hollywood is, as Scorsese insisted, history. Hollywood thus appears as a purveyor of the national narrative, telling stories that remember both the gains and losses of war, with Wayne personifying the struggles involved in building and maintaining the nation's security.

Henry Fonda is another AMC mainstay likewise associated with America, although to slightly different ends. To celebrate the Fourth of July in 1995, the channel programmed a festival of his films, including *Young Mr. Lincoln* (1939) and *The Grapes of Wrath* (1940). Writing for the channel's magazine, Gerald Peary remarks that Fonda is "an actor whose roles are so closely identified with the nation's history, he has become an American icon. . . . [His] movie legacy trailblazed an unforgettable path through our nation's history." Peary argues that Fonda, particularly in his roles as the title character in *Young Mr. Lincoln* and as Tom Joad in the adaptation of John Steinbeck's *The Grapes of Wrath,* "provided hallmarks of our national identity . . . [representing] the social conscience's unbending need that justice be served."[24]

This characterization of Fonda resonates suitably with Independence Day. The Fourth of July recognizes the success of America's victory over British forces in the 1770s—that is, the ability of patriots to pursue the aims of democracy in the face of oppressive powers. Fonda's portrayals of Lincoln and Joad likewise represent moments in U.S. history when individuals had to fight against overwhelming odds to further the cause of democracy. In these cases, the inequities exist within the democratic system itself in the forms of slavery and poverty. Although critical of the system, Fonda's characterizations suggest that democracy can rectify these wrongs through the actions of heroic individuals who oppose injustice. Unlike Wayne, who in his AMC biographies represents a relatively untroubled nationalism, Fonda embodies a kind of patriotism that exhibits faith in America's ability to resolve crises through the spirit of social conscience—democracy's mechanism for righting social wrongs. Thus, each actor's persona depicts a different way to accomplish the same thing: confirming the essential rightness of the American way. The strong association of each with director John Ford, a filmmaker often regarded as the representative recorder of the American experience, only adds to the nationalistic meaning attached to their star personas.

AMC, then, makes little or no distinction between the actor's real and reel personas. Making actors and their roles seem indistinguishable suggests that actors are playing on the stage of American history rather than on a sound stage in Hollywood. To blur this distinction further, AMC's discussions of stars often emphasize their offstage contributions to the nation: for example, Jimmy Stewart's stint as a bomber pilot and Clark Gable's enlistment as an army private in World War II. Rarely in the discourse on classic stars is there any sense that these actors were involved in a fictional, fanciful, or misleading rendition of American history. Yet films, like all texts, maintain complex, highly mediated relationships to historical developments that must be decoded.

Moreover, because stars' historical representativeness requires their idealization, this type of recollection generally avoids acknowledging aspects of a star's persona that might threaten his or her ability to radiate the resilient American character. Problems that characterize the industry or the star's life may be mentioned but only briefly. In this sense, the classic movie channel is the opposite of the "kiss-and-tell" celebrity biography or autobiography that revels in unsavory details (as does Kenneth Anger's *Hollywood Babylon,* for example). The scandalous exposé remains a bête noire among these more upbeat visions of the Hollywood past. The "clean" reromanticizing of stars far exceeds the world of AMC, defining other contemporary manifestations of celebrity worship, from A&E biographies to Web sites for fans of classic film, even sites devoted to such legendarily troubled stars as Judy Garland.[25]

This is not to say that the process of commemorating classic cinema is entirely unequipped to handle issues more difficult to square with a celebratory narrative. In fact, part of nostalgic revisionism involves the selective confrontation and handling of just those issues. In this vein, while male celebrities such as Wayne and Fonda may serve as relatively easy props for nationalistic rhetoric, women and people of color present greater challenges to representations of the past.

On AMC, old Hollywood appears to be stocked with sturdy stars, both male and female, whose performances testify to the system's ingenuity. Like their male counterparts, some female stars are recognized as vital national symbols. Katharine Hepburn, for example, is subject to a particularly rhapsodic account of her national representativeness. She is "the Statue of Liberty, Eleanor Roosevelt, and Lady Luck. The American ideal personified: the image of discipline, drive, the Puritan ethic fulfilled. She has been liberated since birth. There is nobody like her. She's *Kate*." Hepburn's Americanness is tied to her individualism, which in turn is composed of strange bedfellows: Puritanism (linking her to the nation's origins but also to the austerity associated with the Puritans) and personal liberation with a tinge of feminism, as evoked through the analogy to Eleanor Roosevelt. Her strength and rebelliousness left "an indelible mark on the film world with a body of work that serves as a testament to her indomitability. How lucky for us that the strengths, weaknesses, spirit, humor, and conviction of this American original were captured on film."[26]

Rosalind Russell, too, appears as an "indomitable" Hollywood figure. AMC held a festival of the actress's films on American Business Women's Day (September 22), including *Flight for Freedom* (1943) and *She Wouldn't Say Yes* (1945), in which Russell plays, respectively, an aviatrix and a psychiatrist. The actress is credited with giving "working women a tremendous boost." The festival's commentator remarks, "Long before *Working Women* magazine, sneakers with suits and flex time, there was Rosalind Russell, one of the career-minded leading ladies." Russell's "mastery of the female go-getter paved the way for strong professional women roles in the decades to follow, as working women gained equal footing at the office." Russell, it appears, was tough offstage as well, "a match for Tinseltown's most hard-boiled movie moguls."[27]

However, Russell's depictions were often compromised. The actress complained that the formula of her films began with business but ended with a "negligee and a desire for marriage." The AMC commentator acknowledges the limitations imposed on actresses by the convention of compulsory marriage but insists that the films still presented women's success in professions usually closed to them at the time, showing that they could earn a living on their own. With less qualification, Hepburn, too, was tamed. Her strong individualism was "tempered by love in the nine films she and [Spencer] Tracy made together. Her acting style became more naturalistic and her aquiline features seemed to soften . . . the taming of the shrew was about to begin. [She] was tough, but conquerable."[28] Whether in the form of Hollywood conventions or real-life relationships, love manages to domesticate the independent actress and the characters she plays.

Although love stories typically tame female characters, commemorative discussions of "strong" female stars must negotiate the consequences of this taming for the stars' personas as well as for its impact on Hollywood's legacy. In Russell's case, AMC depicts her characters as transcending the romantic conventions of the industry. Therefore, Russell is still able to represent the resourceful American working woman and, by extension, Hollywood's ability to foster and feature such images. In Hepburn's case, romance saves the star from extremism. It mitigates her independent streak, resulting in an improvement in her acting style and a greater sense of approachability in her characters. Thus, the felicitous influence of the relationship with Spencer Tracy on- and offscreen preserves the already legendary dimensions of this pairing and saves Hepburn from a too-shrill independence.

By proposing that compromise via the love story either has a salutary effect or is superseded by the strength of a female star's persona, AMC updates its images of women to conform better to contemporary gender ideals in a postfeminist era. That is, classic Hollywood actresses appear as forerunners of feminism, feminists before the second-wave women's liberation movement took place. Through actresses such as Russell, who represented strong independent professionals, Hollywood appears as prescient, anticipating the later women's movement. The industry may have even helped to inspire the movement, as women could see their own lives and ambitions reflected on the screen. Certainly, this recollection of classic Hollywood acknowledges difficulties with stereotypes that militated against fully liberated representations. Ultimately, however, these accounts portray old Hollywood as not only relevant to understanding the course of events in American history but also, through its embodiment of progressive elements, able to anticipate the liberation movements of the 1960s.

Just as it has attended to postfeminist attitudes, the channel has also responded to more contemporary thinking about race. AMC runs stories on African American actors and actresses, often resurrecting little-seen examples of their work in Hollywood. A story on Bill "Mr. Bojangles" Robinson in a 1991 issue of the AMC magazine relates how he was not only the "first black solo dancer to star on the white vaudeville circuits" but also the "world's most famous tap-dancer." In fact, he was so popular by 1930 that he "could sell out the largest and most prestigious houses in the land." Similar praise for another artist accompanies the April 1999 centennial celebration of Duke Ellington's birth. On this occasion, AMC host Nick Clooney interviewed John Hasse, Ellington's biographer, who contended that Ellington was "the greatest all-around musician this country has produced . . . more brilliant and original than any other—including Gershwin, Copland or Bernstein." Hence, Ellington has been "a towering influence on our culture, transcending his time, the jazz form and, eventually, even race."[29]

Beyond extolling the talents and centrality of African Americans to show business, AMC exhibits their film work to situate it within U.S. race histories. Thus, although movies represented opportunities for "Mr. Bojangles," his first film for RKO, *Dixiana* (1930), ran into problems with southern censors who "cut his numbers . . . a common fate then for any scenes in which blacks appeared as anything but servants or African 'savages.'"[30] In 1935, however, Robinson appeared with Shirley Temple in *The Little Colonel,* assuring his place in movie legend. Race also made "Ellington's connection with Hollywood superficial for most of his life," but "that doesn't mean his film appearances don't please."[31] In April 1999, AMC presented two of his rarely seen filmed shorts, *Black and Tan* (1929) and *Symphony in Black: A Rhapsody of Negro Life* (1935), as well as a number of feature films for which he had composed the score, including *Cabin in the Sky* (1943), Vincente Minnelli's all African American musical.

Similarly, for a celebration of Black History Month in February 1994, AMC unveiled such relatively unknown films as *Anna Lucasta* (1958), a race melodrama; *St. Louis Blues* (1958), a biography about father of the blues W. C. Handy (played by Nat King Cole); and *The Slender Thread* (1965), starring Sidney Poitier as a crisis line volunteer. As in the discussions of Robinson and Ellington, the festival's commentary acknowledges that "through Hollywood's lens, the image of African-Americans is distorted and disturbing." Writing for AMC, Thomas Doherty chronicles the stereotypes and limitations imposed on African American actors and actresses by racial prejudice, including southern censorship. At the same time, he points out that the past was not completely bankrupt in its depictions: "Even when the movies, like the nation, operated under a strict Jim Crow regime, an extraordinary assemblage of musicians, comedians, and actors surmounted the subservient status assigned by the scripts and lent dignity to roles designed to demean." With contemporary directors such as Robert Townsend and Spike Lee and actors such as Denzel Washington and Danny Glover, "successful filmmakers and stars have taken the reins of motion-picture production," allowing "the stories and talents of black artists [to enrich] the American cinema. . . . Today, the motion picture industry is experiencing a renaissance of African-American talent. No longer Hollywood's servant class, African-Americans now project their own images."[32]

In its celebration of African American contributions to entertainment and its presentations of rare or underplayed examples of race films, the channel depicts the integrity of African American stars as able to transcend social prejudice. Acknowledging that Hollywood itself was fraught with racial difficulties, the channel also insists that the industry nonetheless registered and preserved the performances of a Robinson or an Ellington. In this way, AMC joins other organizations in numbering these celebrities among America's national treasures and in regarding their work as part of the nation's cultural heritage.[33] At the same time, the cable channel's admission of prejudice in its chronicle of American history allows it to maintain credibility with its audiences, many of whom experienced the liberation movements of the 1960s firsthand.

The passing acknowledgment of prejudice can serve, however, to inoculate against a more penetrating assessment of race relations in the industry's past. Heritage narratives that feature racial inclusion often substantially deemphasize the depth of racial prejudice and the extent of the challenges it presented to actors and actresses of color.[34] These narratives also tend to offer a vision of history in which the problems of the past are well on their way to being resolved, ignoring problems that continue to exist in the industry and in society by presenting an untroubled vision of present-day gender and race relations. The absence of reflection on contemporary debates about the relation of Hollywood, women, and people of color (including their lack of representation in executive industry professions and the continuing use of stereotypes in film) and the accompanying depiction of contemporary life as free of gender and racial tensions (excluding developments such as the backlash against affirmative action) enable such simple assertions of progress, even as they render these narratives seriously suspect. Certainly, such assertions should be tempered by the public record;[35] but to treat American history more fully would necessitate a substantive critique of the gender and racial politics of Hollywood and of the nation—then and now—thus disturbing the ultimately affirmative goals of the heritage narrative.

It is important to point out that the commemorative activities of AMC are not unique or extraordinary in this regard. In fact, just the opposite is true. AMC provides one example of the rampant utopian historicism characteristic of institutions involved in the commercial management of memory. In such histories, painful aspects of a nation's past must be recognized to a certain extent, so that audiences will not be alienated by outright sanitization. Nonetheless, the stories finally emphasize the triumph of progress, the righting of past wrongs. As Michael Wallace argues in relation to another corporately sponsored revisitation of the past (Epcot Center's "American Adventure" in Disney World), through a selective reconstruction of the past, history gets redeployed "within a vision of an imperfect but still inevitable progress."[36] In the process of recording such a history, accounts emphasize consensus and unity over radical dissent and difference. While acknowledging inequities, these accounts highlight significant moments of the past that demonstrate the presence of enlightened social attitudes. The inexorable movement of history toward justice resolves cultural problems, essentially affirming the political and social arrangements of the present. The inevitability of progress secures a sense of the self-correcting rightness of the democratic system and the humane clairvoyance of the national character. The nostalgic and utopian commemorations offered by AMC and other organizations provide a vision of the past that is, in Kammen's words, "essentially history without guilt" (688).

Within these remembrances, films and stars appear as national treasures, as vital parts of the past that allow immediate access to the nation's history. This memorializing of Hollywood cinema not only emphasizes the value of its films and personnel but also symbolizes the nation's greatness. As Dipesh Chakrabarty argues, history writing is best envisioned as a practice of monumentalizing objects, in which making a "'heritage' out of assorted objects is essential to the politics of both nationalism and the nation-state." Through this process of monumentalizing, history becomes, "the business of the citizen, the subject of the grand narratives of Freedom and Progress that, ultimately, legitimize both the nation-state and the modern market."[37] Although only one voice in an immense sea of discourse devoted to defining the nation's history, the classic cable movie channel demonstrates how the promotion of recycled films and stars as indelible classics participates in the everyday confirmation of ideal notions of U.S. democracy. AMC's contribution to understanding the circulation of popular history, then, lies only partly in the traditional brand of nostalgia it offers. It also vividly illustrates the ability of neotraditional narratives of nation to continue to thrive, not only in the public space of theme parks, monuments, and museums, but on television and in the home, where such narratives are presented to millions of viewers.

Shouldering the weight of U.S. history, such entrepreneurial modes of selective memory are of course vulnerable to unintended revelations.[38] By raising past representations of women and African Americans to the surface and by resolving the issue of prejudice so expediently, AMC's stories risk revealing as much as they try to conceal about minority experience. We must grant the potential presence of such ideological instabilities in any memory project. At the same time, it is important to acknowledge the overt aspirations of heritage businesses and their place within larger cultural narratives devoted to memory. Built on an "artificially constructed past," tradition-driven nostalgia strives toward an "illusion of social consensus" that affirms the present order (Kammen, 4–5) and, consequently, attempts to quiet and displace views disruptive to the more agreeable history on offer.

In the case of agencies involved in fostering film memories, these attempts are bolstered by affiliations between a single voice such as the classic cable movie channel and larger social forces that support and amplify its particular take on classic Hollywood.

Certainly, the channel's conflation of Hollywood and American history, in which history appears to be directly expressed in film narratives, presents the past in a way that resonates with other popular manifestations of history in late twentieth-century U.S. culture. As historians have chronicled, during his presidency in the 1980s, Ronald Reagan improvised revisions of history that rearranged the past for conservatives, who were bent on restoring a national identity impervious to radical self-doubt (hence, Reagan's transformation of the Vietnam War from stigma to a source of national pride). In his "reign of error" he uttered factually incorrect statements, misremembered historical events, and confused movie events (including those from his classic Hollywood days) with actual historical occurrences. Hollywood provided the ex-movie star president with a "major source of the mythic iconography" to be deployed in his presentations and policies.[39] Whether he used movie references to bolster his campaigns or programs, from "Win One for the Gipper" to "Make My Day," or actually substantiated a historical event through its cinematic representation, Reagan's presidency was suffused with Hollywoodiana. In the process, Hollywood became one instrument in a revisionist toolbox, helping to renew traditional patriotic feeling in the nation.

Granting the affinities between AMC's presentation of the Hollywood past and Reagan's brand of historicism, the national narrative available on the cable movie channel also has a significant relation with a less obvious historical coordinate: the film preservation movement of the late 1980s and 1990s. During this time, the view that classic films were an endangered species that the U.S. government, national media organizations, and the public should rally to preserve gained particular momentum. Conservation, not just commemoration, was necessary to prevent the loss of precious objects from the past and the subsequent impact their loss would have on the nation's identity. With this identity at stake, discussions of film preservation on AMC and in other venues continued to magnify the nation's progressive accomplishments. However, preservation discourse concentrated additionally on technology's impact on the relationship between past and present. This emphasis on technology ultimately reveals an association between commemorative efforts directed at preservation and government laws, policies, and debates in the 1980s and 1990s that centered around a perceived crisis in U.S. sovereignty and hegemony due to the growing multinational global economy.

Saving the Past

AMC is not alone in its interest in film preservation—a salvage operation aimed at locating films in peril, physically repairing them, and seeing to their proper storage. Other cable movie channels, notably TCM, share AMC's activism in relation to preservation, showing documentaries on the subject *(The Race to Save 100 Years)* and attempting to raise public awareness and money for the cause in other ways. Film preservation is also part of larger industry trends as well as efforts by U.S. film archives to lobby for national recognition of the importance of rescuing the film heritage. If we understand the relatively recent rise of concern about preservation to include a broad range of efforts to save or renovate old films, from explicit preservation activities to restorations and digital reconstructions, then maintenance of the cinematic past emerges as a central facet of contemporary film culture.[40]

The industry has not always had an archival consciousness. In the past, media industries often discarded or misplaced films and television shows or left them to disintegrate in poor storage conditions. Because these industries felt that many of their products were valuable only in their initial runs, an enormous number of films (particularly from the silent era) and TV programs have been lost forever.[41] Over the last two decades, however, the economic rationale of repurposing has helped change the attitude of media companies toward their warehoused products. As we have seen, as ancillary exhibition sites and new media technologies have developed, old films and TV series (many rejuvenated with digitally remastered sound and image) have provided valuable content for home venues from cable TV to DVD.

Consequently, studios have come to see the importance of having preservation facilities on site. For example, for most of its history Columbia Pictures had no preservation laboratory. This situation changed after the Sony Corporation purchased the studio in 1989. In the process of developing high-definition television, Sony saw the possibility of rereleasing films and television programs in Columbia's library in ancillary formats. They spent millions of dollars constructing a preservation laboratory. For these products to be sales worthy in a world of new technologies, especially digital technologies, they had to be preserved, repaired, and updated so that they could appear in good form to new audiences accustomed to quality images and sound. The cost of making preservation masters, creating computer inventories, and using cool and dry storage vaults thus came to be seen as a necessary expense for protecting valuable assets.

As this example suggests, while the cause of film preservation has an important ecological dimension, it is also deeply influenced by economic imperatives. Like many preservation movements, it is motivated by the double concerns of conservation and commerce—that is, by both the commitment to safeguard a resource and the desire to find a profitable use for it.[42] These concerns are often opposed to one another, but at times preservation efforts succeed only when commercial benefits can be foreseen. In the case of cinema, it is arguable whether ecological considerations alone could have prevailed if new technologies and the accompanying need for programming content had not persuaded media conglomerates that there was literally gold to be found in the old Hollywood hills.

To return to cable television, the dynamics of preservation and profit in AMC's coverage of the cinematic past become particularly clear through three examples: its coverage of motion picture palaces, documentaries on the subject of preservation, and preservation festivals, conducted in cooperation with major film archives across the country.

At one time, *American Movie Classics Magazine* ran a monthly feature story on motion picture palaces. These circa-1920s theaters represent an important moment in film exhibition history, when exhibitors built extravagant theaters to attract audiences to films by offering them luxurious surroundings. AMC provided information about the current state of many of these theaters, including the Wang Center in Boston, Massachusetts (formerly the Metropolitan Theater); the Stanley Performing Arts Center in Utica, New York (formerly known as the Stanley); and the Paramount Theater in Oakland, California. The stories of all these theaters are remarkably similar: they were saved from the wrecking ball (a fate for most motion picture palaces) by private interests, civic-minded citizens, and/or local governments; they were then renovated for contemporary use, usually as performing arts centers featuring ballet companies, symphony orchestras, classic film series, and/or Broadway theater.

As AMC tells us, each motion picture palace represents the glories of the past and the threat posed by urban modernization. But the resolution of this conflict is a happy one: once properly renovated for contemporary use, these theaters can serve as cultural centers for cities while still exuding the exceptional character of the past. Boston's Metropolitan Theater, for example, was a "sumptuous entertainment palace that extended for a full city block and five stories into the sky . . . adorned with gold and marble fit for royalty." When it was being refurbished as the Wang Center, the process revealed "a vision of grandeur and extravagance hardly imaginable in a day when movies mean television or tiny black-box theaters."[43]

The description of the renovation process constructs a view of the past as well as a relationship between past and present. As we saw in the case of classic movies, old Hollywood is depicted as glorious—so glorious, in fact, that the experiences of cinema it offered remain unparalleled in today's viewing venues. The sumptuous art and architecture of the motion picture palace, coupled with its frequent contemporary function as a city center, help such buildings to gain admittance to the National Register of Historic Places. Nostalgia, combined with a sense of civic purpose, underwrites this particular heritage narrative. At the same time, these monuments of the past require extensive overhaul to operate in the present.

Preservationists refer to this kind of overhaul as adaptive reuse. The term *adaptive reuse* refers to a recycling of the past that integrates it with the present, preferably for profit. Historically, this strategy has provided a way to satisfy both preservationists and developers. Neither left in "sterile isolation" as a historical artifact nor destroyed in the name of progress, the past is saved and renewed to serve a contemporary function (Wallace 177).[44] Adaptive reuse provides the means for the past, languishing in profit limbo, to be revived by various social agencies willing to gamble on its born-again marketability. In the case of motion picture palaces, renovation allows the theater to be transformed into an integral part of today's civic society. Contemporary resources are necessary to save the splendid past from the more savage impulses of capitalism (which would see it in ruins) and to realize its potential for continued contributions to commercial as well as aesthetic and civic arenas.

The principles and rhetoric of adaptive reuse are also visible in the restoration of specific classic films. Alfred Hitchcock's *Vertigo,* for example, was restored and rereleased on the big screen as a special event in 1996. In 1995, Robert Harris and James Katz, film preservation experts who had previously restored *Spartacus* (1960) and *My Fair Lady* (1964), went to work on *Vertigo* with more than one million dollars of support from Universal Pictures. While *Vertigo* had been a critical disappointment when it first came out, its theatrical reappearance in 1984 (along with other Hitchcock titles kept out of distribution for many years) changed its reputation; critics began to regard it as one of the director's masterpieces. Thus, Universal's financing of its restoration was a fairly safe investment. As it turned out, audiences lined up in urban areas to see the restored *Vertigo* on the big screen, and its widescreen VHS counterpart sold out its first 250,000 units in a few weeks.[45]

When Harris and Katz initially examined the negative and color separations of the film, they found them faded, scratched, and shrunken. How they proceeded after this discovery is the subject of an AMC original documentary, *Obsessed with "Vertigo": New Life for Hitchcock's Masterpiece.* This documentary aired on the channel in the mid-1990s and is also available on the special collector's editions featuring the restored film on VHS, laser disc, and DVD. In *Obsessed with "Vertigo,"* the restorers agree that "time does things to movies . . . along the way *Vertigo* was almost lost to us forever . . . a film that looked nothing like what Hitchcock had created." But Universal, "at the forefront today of film preservation," stepped in to rescue the film "from the ravages of time."

Through researching such elements as the original colors of 1950s automobiles and the wardrobe for the film, designed by Edith Head, Harris and Katz sought to restore the film's original color schemes. They also worked on sound, digitizing the original orchestral recording sessions so that the film music would play well over today's theatrical and home sound systems and reinventing the foley and effects tracks, which Paramount Pictures had long ago discarded after it lost the rights to the film. In addition, the original version of *Vertigo,* shot in a widescreen process called VistaVision, was reduction-printed at 35mm in the 1950s because many theaters were not equipped to project this particular widescreen format. The film's restoration produced a true version of VistaVision in 70mm, regaining the quality that had been lost previously to reduction printing.[46]

Thus, the restored *Vertigo* reappeared to the public in theaters in 70mm VistaVision and DTS stereo sound, followed by widescreen special editions in THX sound for the home market. As Dave Kehr of the *New York Daily News* wrote, "For those who have never seen *Vertigo,* here is evidence that movies can occupy the highest plane of artistic expression. If you have seen it, you owe it to yourself to see it like this."[47] "To see it like this" intimates that there is something about the restored *Vertigo* that not only brings the film back from the dead but does so in style. Although discussions of the film's renovation emphasize how assiduously Harris and Katz aimed at reproducing Hitchcock's original, equal attention is paid to the role that technology and other contemporary variables played in delivering the "new, improved" version of the film. For example, *Vertigo*'s 1996 trailer tells us that it is "presented for the first time in 70mm and DTS Digital Stereo . . . fully restored and remastered," in a "stunning . . . version featuring enhanced picture and sound." These enhancements are depicted, however, as maintaining the spirit of the original. The AMC documentary and other forums define technical progress as being able both to re-create "the precise visual texture Hitchcock intended for every shot" and to improve on his vision in ways he would endorse. *Vertigo* was "seen as Hitchcock could only have dreamed it would look and sound . . . now for the first time people can see *Vertigo* the way that it was intended to be seen."[48]

The discourse of restoration includes numerous appeals to originality and authenticity, while leaving no doubt that the present improves on the past.[49] Along with other innovative elements involved in the process, restored and digitally enhanced visuals and sound signify the superiority of contemporary resources. The commentary surrounding *Vertigo*'s reissue reconciles whatever problems might arise concerning authenticity under these circumstances by assuring us that the new film can be re-viewed in a way that ultimately *realizes* the director's original intent. Because technology was inferior decades ago, Hitchcock could not achieve his full artistic vision. Now *Vertigo* can be seen in real VistaVision, remastered with improved image and sound. In this way, the aesthetic demand for the authenticity of the original is balanced with the marketability of technologically advanced effects.

Here, the principle of adaptive reuse meshes harmoniously with repurposing: the past is indeed saved, while being regenerated specifically in relation to the profit-conscious standards and demands of the present. Universal and the restorers rescue the past from the ravages of time as well as from any imperfections in its original design owing to dated technologies. Maintaining the cachet of the past as the repository for film classics, contemporary Hollywood provides a better past, inevitably underscoring the achievements of the present and the promise of the future. Again, as in AMC's star-driven historical accounts, these discussions of technology provide a progressive narrative, complete with the depiction of a slightly problematic past followed by a better today. Even a trace of heroic discourse is present in this commentary, particularly in relation to the ability that capital and technology have to triumph over the various limitations and liabilities of time. From this perspective, Hollywood and its personnel can be seen as protagonists in an epic historical drama, in which skill and powerhouse technological capabilities enable the preservation of the past in spite of what seem like insurmountable odds. In the visions of progress that accompany such preservation commentaries, the success of capital and technology in this mission ultimately testifies to the power of both the media industries and the nation.

Certainly part of the audience's excitement at re-viewing reissues is the anticipation that they have been "spruced up," restored, and/or digitally remastered to achieve a better-than-original status. In fact, a large part of the reissue market for "oldies," from CDs to DVDs, promotes these products under the banner of the "remaster." Nostalgic accounts may insist that the "good old days" are superior to the present, but the artifacts of those days must nevertheless be modernized. At the same time that audiences are dazzled by enhanced visual and audio delights, the promotion of the remastered version encourages them to be impressed by the technical prowess of the film industry, by its virtuoso performance as an entity that has the ability to ward off the effects of time while providing audiences with the latest in digital wizardry. Pleasures in re-viewing vintage cinema, then, are linked to an awe of the institution itself. The remembrance of films past is an experience permeated by the spectacle of Hollywood as the embodiment of advanced technoculture and therefore as exemplary of the capital enterprise that helps to define the exceptional power of the nation. This spectacle is present, then, not only in Hollywood's special effects extravaganzas such as *Titanic* (1997), which was often seen worldwide as representing the extraordinary capabilities of U.S. capitalism and technological advancement; it is amply portrayed in the promotions for recycled Hollywood products as well.

In this sense, media industries attempt to channel our memories of old films so that we regard them as monuments to the accomplishments of Hollywood, much like the Parthenon testifies to the glories of ancient Greece. Such self-referential activities help to define Hollywood as an important cultural institution devoted to preserving the past. Further, the industry characterizes its vintage titles as landmarks of cinematic and cultural achievement not simply by declaring their classic status but by carefully framing them within a technocentric narrative that updates and elaborates their appeal as examples of the industry's technical ingenuity and, ultimately, the nation's grandness.

AMC's film preservation festivals help to clarify more explicitly what is at stake in these testimonials about the nation's achievements. Efforts by high-profile activists such as film director Martin Scorsese and film archives across the country, as well as economic and political factors associated with globalization, have helped to bring about congressional legislation designed to salvage remaining American films from neglect and to bring the cause of film preservation into greater public awareness.

Conducted as fund-raisers for the cause, AMC's annual film preservation festivals have occurred in partnership with the Film Foundation (which includes Martin Scorsese, Francis Ford Coppola, George Lucas, Sydney Pollack, Robert Redford, and Steven Spielberg) and the foundation's Archivists Council. The council's members include the International Museum of Photography at George Eastman House, the Library of Congress, the Museum of Modern Art, the American Film Institute's National Center for Film and Video Preservation, and the UCLA Film and Television Archive.[50]

In 1993, AMC's first preservation festival focused on *Citizen Kane* and other well-known classics, with subsequent festivals organized according to themes. For example, the second annual festival was devoted to "keeping alive America's Western movie heritage"; the third, to Buster Keaton and film comedy; and the fourth, to the musical.[51] AMC made arrangements with several of the above archives to show their preservation prints of such films as "The Great Train Robbery" (1903), *The General* (1927), *The Iron Horse* (1924), *His Girl Friday* (1940), and *My Darling Clementine* (1946). This is only a small sample of the more famous classics screened by AMC. The channel has also shown more obscure films and listed many lesser-known titles that are in the process of being preserved by various archives.

Commentary accompanying these festivals cites statistics that demonstrate the need for intervention to save the film heritage. These statistics are most likely drawn from a widely cited 1990s congressionally mandated report on the state of film preservation in the United States. Researchers found that less than half of U.S. films made between 1895 and 1950 survive. Specifically, fewer than 20 percent of feature films from the 1920s and only 10 percent of features from the 1910s exist in complete form. Films made after the 1950s on acetate stock are also endangered; they face deterioration from color fading, "vinegar syndrome" (an irreversible film base decay), and other factors of age. In addition, many American films survive precariously in foreign archives and need to be reclaimed. Meanwhile, funding for film preservation has fallen to less than half of its 1980 levels (when adjusted for inflation).[52]

Writing about the preservation festivals for AMC magazine, Robert Moses emphasizes the urgency of saving Hollywood's past: "Fred Astaire's flying feet, the frosty allure of Greta Garbo, Groucho's manic rambles . . . these are among the dreams in the dark that American moviemakers shared with the world. As the 1900s became the American century, our country gave form to the first medium that could capture the dynamics of a rapidly changing world, to an art in motion and light that became the literature of the common man." Because of the loss, disintegration, and fading quality of the old films, "our entire motion picture heritage . . . may not be available to enrich the lives of future generations." Web sites devoted to film preservation concur: "Over the last hundred years, America has been documented by a nation of filmmakers—professionals and amateurs alike—working in every corner of the country. These filmmakers have recorded our traditions, captured the events of the day, and expressed our ambitions. Their work is the collective memory of the twentieth century." With the film heritage at risk, the alliance between film industry leaders and cinema archives and museums will "turn back time's threat to movie history." Indeed, AMC's preservation festivals help to celebrate this "triumph of preservation technology over time."[53]

The threat of loss clearly elevates cinema's importance as an exceptional record of history: its unique modernity allowed it to tune in to the rhythms of the twentieth century. Further, preservation commentary claims cinema as a uniquely American art form connected to a uniquely "American century." Without cinema, it is argued, present and future generations would lack the keys to the twentieth century and hence lose a sense of historical continuity that helps to maintain the nation's identity. While there is no doubt that films should be preserved, the promotion of film preservation continues to show the substantial investment of some social and media institutions in defining classic films as unmediated signifiers of American history. Moreover, cinema comes to embody American exceptionalism—the sense of the nation's sovereignty and superiority, particularly as it rose to economic and political prominence during the 1900s.

As I mentioned, the growth of discourse about film preservation is linked to a flurry of activity that has taken place on this subject in Congress, especially since the 1980s. A series of legislative efforts, including the passage of the National Film Preservation Acts of 1988, 1992, and 1996, has formally recognized film as part of the nation's heritage. The language arguing for the protection and preservation of film varies from bill to bill, but generally Congress has found that "(1) motion pictures are an indigenous American art form that has been emulated throughout the world; (2) certain motion pictures represent an enduring part of our Nation's historical and cultural heritage; and (3) it is appropriate and necessary for the Federal Government to recognize motion pictures as a significant American art form deserving of protection."[54]

While it is beyond the bounds of this chapter to detail the complex nature of this legislative history, several major outcomes are relevant here. In 1988, along with other heritage bills concerning funding for such things as the American Revolution Bicentennial Administration, the National Film Preservation Act became law. The act established the National Film Registry and the National Film Preservation Board. Administered and overseen by the Library of Congress, the National Film Registry selects films "that are culturally, historically, or aesthetically significant." The librarian provides "a seal to indicate that the film has been included in the National Film Registry as an enduring part of our national cultural heritage." This seal may be used "in the promotion of any version of such film that has not been materially altered."[55] The National Film Preservation Board, consisting of members from many units, including the Academy of Motion Picture Arts and Sciences, the Director's Guild of America, the Writer's Guild of America, the National Society of Film Critics, the Society for Cinema and Media Studies, and the American Film Institute, selects films for the National Film Registry.[56] Each year the board chooses twenty-five films to add to the registry.[57]

The National Film Preservation Act of 1992 reauthorized the earlier law.[58] In addition, this act asked the Library of Congress to study the current state of film preservation in the United States. It also authorized the library to establish procedures for including films in the National Film Registry, to help to create public consciousness of film heritage and the importance of preservation, and to provide reasonable access to prints for scholarly and research purposes. The study of the current state of film preservation and restoration activities produced under the directives of this act was entitled "Film Preservation 1993." Cited previously in relation to AMC's preservation efforts, this study found that America's film heritage was at serious risk. The National Film Preservation Board went to work to develop a plan that addressed archivists' concerns about these issues. The resulting plan, published in 1994 under the title "Redefining Film Preservation," represents the consensus that emerged from this process.[59] The board's response was to propose the formation of the National Film Preservation Foundation to raise grant money for preservation projects and programs.

The National Film Preservation Act of 1996 authorized the National Film Preservation Board for another seven years and formally created the National Film Preservation Foundation.[60] Starting operations in November 1997 and awarding its first grants in 1998, the foundation is a federally chartered, independent private-sector, nonprofit organization "designed to save America's film heritage." The foundation regards its mission as part of "our shared national responsibility to guarantee that the broadest range of America's film heritage can be seen by future generations." It raises private monies, matched with federal funds, to provide grants to "nonprofit film archives, historical societies, and other nonprofit institutions with film collections throughout the nation." Specifically, grants are to focus on so-called orphan films—films that would not survive without public support because they lack "commercial protectors" or "preservation benefactors" such as Hollywood. Orphans include "public domain films, silent films, documentaries, independent films, films of historical and regional importance, and films by or documenting minorities."[61]

Contemporary preservation causes run the gamut, then, from orphan films to more commercial Hollywood ventures. The motives and issues involved in the politics of the film preservation movement are as extensive and varied as this spectrum implies. Certainly, the imprimatur of President Clinton and Congress, as it supported the creation of new organizations devoted to the study and funding of film preservation, has provided official public recognition of silent and classic cinema's importance to the national heritage. This recognition has given commercial venues such as AMC a dignified rationale for pursuing high-profile conservation activities.

However, Congress's actions also reflect a different set of concerns. As the history of preservation in the United States shows, preservation tends to become an important issue during times of crisis. Between 1850 and 1920, conservationists helped to develop the national park system (with the support of federal legislation) in response to a multitude of factors. Among them was a perceived crisis in American national identity and purpose, fostered in part by the popularity of the "Turner Thesis." Advanced by historian Frederick Turner, the thesis defined the pioneer encounter with the Western frontier as the defining characteristic of American identity. Turner concluded that, because of successful settlement and the encroachments of industrialization on nature, the frontier was "closed." Thus, preserving the wilderness became an important means of maintaining American identity at a time when radical shifts in American life helped to transform a pastoral, agrarian ideal into the reality of a "tamed" industrialized space.[62]

A similar reaction occurred in the post–World War II era. During this time, suburbanization, the dissolution of inner-city historic neighborhoods through acts of urban renewal, and the construction of vast interstate highway systems dramatically changed the nation's landscape. To conservationists, the geographical conformity promoted by these developments and the forces of unrestrained growth precipitated something of a national emergency. As these changes severed the country's connections to the past, they threatened the sense of historical continuity that underpinned national identity. Without history—without a sense of the rootedness of national traditions—conservationists feared that the citizenry would lose its social cohesiveness, its sense of what it means to be American. In documents such as *With Heritage So Rich* (1966), conservationists argued that the nation was in a precarious situation that it could not afford to tolerate during the Cold War—an era when democracy was being challenged worldwide by the Soviet Union and other communist governments.[63] At the same time, the 1950s and 1960s saw the tremendous growth of mass tourism to historic sites (due in part to the new highways), showing entrepreneurs that history could be a "cash crop" (Wallace 173–76). The combination of politics and profit resulted in the creation of the modern preservation movement.

During the last two decades of the twentieth century, national identity was again called into question by an array of powerful forces. The emergence of media conglomerates, global capitalism, the explosion of new technologies in the information age, and the piracy of American media products confused and challenged traditional notions about the status of U.S. commerce and subsequently the nation itself.

The 1980s and 1990s saw a succession of takeovers and mergers in the communications and media industries that blurred the identities of companies that have long been major players in American business. For example, Viacom's purchase of CBS in 1999 consolidated a host of television, cable, movie, home video, publishing, and other businesses to make Viacom the world's second largest media company after Time Warner, Inc., combining the identity of these businesses under the ever-expanding umbrellas of corporate ownership. For some, such fusions of business interests in the world of communications called into question the survival of a democratic system. When AOL purchased Time Warner in January 2000, marrying the interests of the world's largest Internet service provider with one of the titans of the media business, news commentators raised concerns about one conglomerate's control of information in so many forums (the Internet, magazines, newspapers, movies, etc.).[64] A huge corporation's potential to exercise so much power over public access to information suggests the possibility of a totalitarian-like monopoly. Since freedom of information is one of the foundations of American democracy, this kind of venture represents a possible threat to the nation's principles, self-image, and future.

That the activities of media companies figured strongly in debates about the state of the nation becomes clearer in the case of the foreign purchase of U.S. businesses and other concerns. When the Japanese-owned Sony Corporation bought Columbia Pictures, the Gubar-Peters Entertainment Company, and CBS Records (the largest record company in the world) in the 1980s and Matsushita bought MCA/Universal in 1990, these particularly visible deals caused substantial controversy and emotional debate about the loss of major U.S. media to foreign interests. The purchases occurred at a time when politicians and citizens were already alarmed about the inroads that international companies were making onto American soil. The 1980s saw a significant growth in the sales of American farmland, real estate, factories, and banks to foreign companies. Adding fuel to the fire was the Japanese purchase of U.S. landmarks, particularly Rockefeller Center (including its skating rink and Christmas tree) and Radio City Music Hall in New York City. As one commentator put it, "What's next? . . . Mickey Mouse, Mickey Mantle, and the N.Y. Yankees?"[65] Thus began what David Morley and Kevin Robins have called a "Japan Panic," hysteria rooted in long-established ideologies about the invasion of the West by forces of the Orient, bolstered further by Japan's technological proficiency and growing superiority to the United States in a prized area of "superpower" strength.[66]

The acquisition of well-known American businesses, particularly media concerns, and major American landmarks resulted in legislative activity that sought to limit foreign ownership of entertainment companies and landmarks. The fear, shared by citizens as well, was both that foreign interests would dominate the nation's culture industries—industries with tremendous influence over citizens' everyday lives—and that foreign ownership of landmarks would put the custodianship of the nation's past into alien hands.[67] In addition, economic success in the twenty-first century was seen as strongly linked to dominance in telecommunications and electronics.

Thus, in 1989 Congress held a hearing with politicians and media industry representatives to discuss the impact of the global economy—the massive flows of money and capital across political boundaries—on American media. In these proceedings, politicians marshaled evidence and rhetoric to demonstrate the threat of foreign influence, while media representatives tried to assure them that, despite foreign ownership, business operations would be conducted in autonomous and traditionally American ways. Nonetheless, these proceedings tended to prompt a panicked response to globalization that envisioned the foreign takeover of American hearts, minds, and pocketbooks.

A number of politicians at the hearing offered evidence of the decreasing American presence in the media and communications industries. Some mentioned that only two decades before, U.S. companies had held an influential position in consumer electronics and in the record industry, making electronics companies such as RCA, Magnavox, General Electric, and Motorola into household words and more than a dozen U.S. record companies into worldwide competitors. By 1989, only Zenith remained a name in consumer electronics, and only two U.S.–owned companies, Warner and MCA, were among the top record companies. In these and other industries (such as the auto industry), Japanese and other foreign investors had made huge gains in U.S. markets; conversely, U.S. companies had not been able to purchase controlling shares of foreign companies because of indigenous laws against such arrangements.

These kinds of circumstances helped to create a general ideological climate of protectionism about American business and underscored the sense that the media embody the spirit of America. Because the media industry generated a trade surplus of three billion dollars in 1988, legislators concerned about the U.S. trade deficit and deficit spending in general took special notice. In the course of the hearing, Majority Leader Richard A. Gephardt argued, "The current international media industry is very much a child of America. Film, for example, is a uniquely American art form: we brought it to life, we made it talk, we used it to address our deepest social concerns. Now, we see our media industry on the global auction block" (Congress, House, 1989, 3). This nationalistic take on the history of cinema is echoed in other political statements and, as we have seen, in preservation discourses that identify cinema as essentially American. At the same time, the communication industry in general is viewed as "most vital to our national well-being. The lifeblood of democracy is communications, because it gives our citizens access to the marketplace of ideas . . . in the age of 'media without frontiers,' it's even more important that Americans maintain their First Amendment rights in the electronic media free from outside influence" (Congress, House, 1989, 5).

This nationalizing of media goods is spurred by at least one other factor: the rise in piracy of American films. Technologies from video to the Internet have made the piracy of U.S. media at home and abroad an extremely lucrative enterprise. Annually, the major movie studios lose approximately $250 million to domestic piracy and more than $3 billion overseas.[68] The use of the Internet as a venue for pirated films has only exacerbated the studios' immense concern about the illegal traffic in Hollywood films. In particular, the loss of revenue to overseas pirates enforces the sense that the media are an important economic and symbolic battleground in an era when the global flow of goods and information defines world economies.

Besides obvious economic ramifications, the question of how globalization affects national sovereignty lies at the heart of congressional discussions—how, that is, the purchase or theft of U.S. goods by foreign interests impinges on the definition of America as an autonomous entity that can exercise supreme authority. This is, of course, a key question in larger debates on this topic. Certainly, many scholars view globalization as meaning the "Coca-colonization" or McDonaldization of the world—the rise, rather than the fall, of Western hegemony. Concerns about this hegemony have been additionally fueled by what the *New York Times* refers to as "Net Americana," the global reach of U.S. companies involved in the Internet business.[69] For other scholars, however, globalization promises the potential demise of "*Pax Americana,* of American hegemony" by dispersing and dismantling American business concerns via multinational corporate ownership, the economic rise of other international powers, and other forces.[70] Yet, as Robert Holton argues, the massive changes associated with globalization "dispose neither of the idea of sovereignty as state autonomy from external coercion nor that of sovereignty as a bargaining resource that political elites may use in negotiation with external interests."[71] One of the challenges globalization presents to a nation is exactly how to maintain "the ideology of sovereignty in the face of conditions that could seriously undermine its existence. At the very least, this ideology serves to perpetuate the idea of the nation in order to project state power and economic viability in a rapidly changing world environment.

The views promulgated during the congressional hearing demonstrate the aggravated relationship that globalization has to concepts of sovereignty: in short, globalization makes defining America a problem. Part of this problem is addressed through the promotion of nationalism. The more American businesses become dispersed, dissolved, or unrecognizable through economic ventures, the more discourses turn urgently to the tasks of emphasizing self-determination, of clarifying principles that have traditionally defined the nation, of preserving aspects of the heritage that are threatened. Further, as the cases of films endangered by time and piracy show, maintaining the nation's hold on its own products is increasingly threatened by both lack of attention to film as heritage and illicit capitalists. Thus, as Congress put it, the nation must try to "repatriate 'lost' American films from international archives"; meanwhile, the Motion Picture Association of America's Internet piracy division works with the FBI to shut down piracy Web sites.[72]

Popular cultural artifacts have often served as insignias of the American character. The late twentieth century saw this association greatly amplified by concerns about the relationship between globalization and the self-determination of American businesses, between globalization and national identity. In the 1980s and 1990s, the impulses of nationalism materialized in claims made about objects from the past, whether they were the movies or the Rockettes. These objects were especially important insofar as they represented a time in the nation's history when American businesses appeared to be free from outside economic influences or threats. They thus could more easily embody the concept of America as a unified, hegemonic power.

With this in mind, one role that the film preservation movement played during this time was to protect the nation's heritage in the face of proliferating corporate mergers and other developments that increasingly complicated a product's identification as specifically American. Part of the rescue mission was to save a past that seemed to be clearly American—a past that appeared never to have experienced the complexities of the contemporary economy or to have been subject to compromise in the arena of national identity. In preservation discourse, Hollywood maintained its cachet as the site of distinctive American products, just as it helped to recall an era when the industry reigned supreme at both home and abroad. Retaining such an image of the past responded to threats against the sovereignty of American identity that some felt characterized the developing global economy. This is, then, part of the stake in saving classic Hollywood as part of the national heritage: holding on to artifacts that proclaim their American origins without apparent complication. In this way film preservation, driven by a spectrum of ecological and economic motives, gains an ideological dimension.

Postmodernist and other studies have suggested that global extension, international flows of capital and information, and the disintegrative forces of separatism, among other things, have deeply complicated any monolithic notions of nation. I have argued, conversely, that these very forces have helped to make nostalgia and other evidence of tradition more compelling. In times of crisis, social institutions work overtime to produce a historical "glue," often built on evocations of the past, with which to bind citizens to a reassuring sense of individual and collective identity.[73] In this vein, the recycling of classic Hollywood helps to demonstrate that master narratives of nation are alive and well in today's polymorphous cultural mix. They continue to be repackaged in highly visible ways by media conglomerates and dispersed to millions of viewers not only in public but also in the everyday environ of the home. The employment of such historical visions by key cultural institutions suggests that the confrontation with the cinematic past is often at the same time a confrontation with what Homi Bhabha has called that "prose of power that each nation can wield within its own sphere of influence."[74] Memories of old Hollywood appear to have far less stake in public memory debates than, say, memories of the Vietnam War or the U.S. bombing of Hiroshima and Nagasaki; in fact, the old film's quaintness and lack of apparent connection to contemporary politics make it seem quite distant from the agonistics of such debates. All the same, strolling down the cinematic memory lane involves encounters with highly charged images of America that are just as invested in placing indelible truths about the nation's history on the cultural stage.

Film and Memory: A Coda

Vintage artifacts may often be forgotten and doomed to decompose in a lonely corner of someone's basement, but when they are resurrected as events for public consumption, they can become, as Adorno suggests in the epigraph to this chapter, hieroglyphics of history. They appear as ciphers, characterized by an outdated system of writing set in a faraway past. To achieve sense in the present, they must be both decoded to make their past legible to contemporary audiences and recoded, that is, otherwise prepared for life in new circumstances of exhibition. Cine-museums and companies devoted to the commercial management of memory step in to perform this work of coding by recirculating meanings these artifacts once had for audiences and by inventing fresh meanings. This is part of the standard repackaging of "ancient relics" as they reappear to various audiences over time.

However, it is important to reiterate that the work of the cine-museum does not monopolize the social production of memory; memory is a contested terrain. Since old Hollywood films are bountifully recycled in culture, we can expect that their meanings are mobilized to support diverse, conflicting accounts of the past, depending on the contexts that reintroduce them to audiences. In addition, viewers will have differing relations to the historical visions produced by classic movie channels. Part of this different orientation arises from the special place films from the past may have in the recollections of individual viewers, bringing the personal strongly into the mix. As Annette Kuhn and Jackie Stacey have shown in their respective studies of individual memories of films, old features can have deep autobiographical resonances for audience members, shaping the manner in which these films signify history.[75] The potential for films to be repeated endlessly in the private space of the home makes them especially liable to the process of personalization. Further, as Lipsitz argues, viewers' responses, as they are informed by personal experience, may be at such strong variance with "hierarchically prepared and distributed mass culture" that oppositional interpretations of texts result. If these interpretations are sustained and integrated into a community—especially, for example, minority communities often left out of or misrepresented by grand historical narratives—they generate "counter memories," memories rooted in the local and immediate that force revision of totalizing dominant accounts.[76]

At the same time, a memory experienced locally or individually is not necessarily less problematic ideologically than one offered by a master narrative. Moreover, it is unlikely that private and public memories can be neatly separated. As Marita Sturken points out in her study of the impact of films and docudramas on Vietnam vets' memories of their own encounters with war, media representations of past moments can become extremely powerful, "weaving themselves into experiences and memories . . . [becoming] part of cultural memory." Her research finds that those who had participated in the war have difficulty at times in recalling their experiences *without* the mediations of film or television. Hence, the industries' historical revisions are not necessarily at odds with individual recollections; they can, in fact, gain a substantial foothold in the personal, the private, and ultimately within the collective memory of certain social groups. In other words, as the Popular Memory Group writes, "Private memories cannot . . . be readily unscrambled from the effects of dominant historical discourses. It is often these that supply the very terms by which a private history is thought through."[77] The home exhibition of media exemplifies this complex relationship between public and private memories. The images of classic Hollywood and television that enter the home on cable TV as part of each and every day's programming magnify the possibilities that individuals will construct personal relationships with the classics, just as these images enhance the presence of media companies in negotiating how that relationship will be formed or re-formed.

The classic movie channel cine-museum cannot unearth all of the elements that make up the "strangely composite constructions" of memory in such circumstances. But it does provide a concrete demonstration of how a particular vision of the past is produced and privileged in an attempt to construct an official account of the past within certain contemporary social contexts. While the complexities involved in creating public memories must be acknowledged, we should not minimize the salience and persistence of "processes of domination in the historical field" or these processes' ability to become a seamless part of private reveries.[78] Given the diffuse presence of Hollywood's commemorative activities, we should consider film memory as partly negotiated by agencies involved in the business of media recycling. Any Proustian model of memory, then, must be refracted through the lens of mass culture to capture how the remembrance of films past is shaped by the industries that revive the past as part of their capital enterprise.

Notes

1. For example, Lynn Spigel and Henry Jenkins have written on personal media recollections in "Same Bat Channel, Different Bat Times: Mass Culture and Popular Memory," in *The Many Lives of Batman: Critical Approaches to a Superhero and His Media,* ed. Roberta E. Pearson and William Uricchio (New York: Routledge, 1991), 117–48. For work on film texts and memory, see, for example, Miriam Bratu Hansen, "*Schindler's List* Is Not *Shoah:* The Second Commandment, Popular Modernism, and Public Memory," *Critical Inquiry 22,* no. 2 (1996): 292–312.
2. Stephen Henderson, "Teaching New Generations the Joys of Old Movies," *New York Times,* June 8, 1997, 29–30.
3. Les Luchter, "AMC, Bravo Uphold Cable's Original Promise, Sapan Says," *Multichannel News,* November 16, 1987, 11; Ron Alexander, "AMC, Where the Movie Never Ends," *New York Times,* November 12, 1991, 52.
4. Michael Bommes and Patrick Wright, "'Charms of Residence': The Public and the Past," in *Making Histories: Studies in History Writing and Politics,* ed. Richard Johnson et al. (Minneapolis: University of Minnesota Press, 1982), 264. Hereafter, this work is cited parenthetically in the text.
5. Thomas F. Gieryn has written about the contested nature of commemorations in "Balancing Acts: Science, *Enola Gay,* and History Wars at the Smithsonian," in *The Politics of Display: Museums, Science, and Culture,* ed. Sharon MacDonald (London: Routledge, 1998), 197–228.
6. Janet Wasko, *Hollywood in the Information Age: Beyond the Silver Screen* (Austin: University of Texas Press, 1994), 87. For more on AMC's origins and history, see Douglas Gomery, "American Movie Classics," in *The Cable Networks Handbook,* ed. Robert G. Picard (Riverside, CA: Carpelan Publishing, 1993), 8–14.
7. Patrick R. Parsons and Robert M. Frieden, *The Cable and Satellite Television Industries* (Needham Heights, MA: Allyn and Bacon, 1998), 123.
8. *Channels Field Guide '85* (December 1984): 62; and Richard Katz, "Reaching Classic Heights," *Channels,* December 17, 1990, 16–18, 22.

9. *Broadcasting Yearbook 1991* (Washington, DC: Broadcasting Publications, 1991), D-3; National Cable Television Association, *Cable Television Developments* (Fall 1996): 16–17; Richard Campbell, *Media and Culture: An Introduction to Mass Communication* (New York: St. Martin's Press, 1998), 164; "About AMC," www.amctv.com (accessed on October 10, 2004); and Gomery, "American Movie Classics," 8.
10. "AMC Reaches Agreement with 5 MSOs," *Cablevision*, December 9, 1985, 28; and Gomery, "American Movie Classics," 9–10.
11. "About AMC," www.amctv.com (accessed on January 6, 2003).
12. As one executive remarks, "TV viewers over the age of 45 are especially underserved because of cable programmers' preoccupation with pleasing advertisers. This also damages cable systems . . . because this age group dominates the cable-resister category that systems should be trying to reach." Les Luchter, "AMC, Bravo Uphold Cable's Original Promise, Sapan Says," *Multichannel News*, November 16, 1987, 11. See also Katz, "Reaching Classic Heights," 16–17; and Gomery, "American Movie Classics," 11. For academic studies on elderly media audiences, see John Tulloch, "Approaching the Audience: The Elderly," in *Remote Control: Television, Audiences, and Cultural Power*, ed. Ellen Seiter et al. (New York: Routledge, 1989), 180–203; and Nancy Wood Bliese, "Media in the Rocking Chair: Media Uses and Functions among the Elderly," in *Inter/Media: Interpersonal Communication in a Media World*, ed. Gary Gumpert and Robert Cathcart (New York: Oxford University Press, 1986), 573–82.
13. Katz, "Reaching Classic Heights," 22.
14. Advertisement, *American Movie Classics Magazine* (October 1996): 23.
15. Pierre Bourdieu, *The Field of Cultural Production: Essays on Art and Literature* (New York: Columbia University Press, 1993), 120–21. Hereafter, this work is cited parenthetically in the text.
16. Gerald Peary, "A Singular Voice: Bing Crosby's Style Changed the Sound of Popular Music Forever," *American Movie Classics Magazine* (December 1994): 4; and Eve Golden, "The Complete Cary: It Takes Five Modern Leading Men to Make One Classic Grant," *American Movie Classics Magazine* (August 1996): 7–8.
17. Joseph Varsalona, "Battle of the Big Bands: A Salute to the Finest Films of the Swing Era," *American Movie Classics Magazine* (February 1992): 4–6; and Ann McGuire, "America's Movie Palace Memories," *American Movie Classics Magazine* (April 1993): 3, 6–7.
18. On nostalgia's reactionary nature, see Michael Kammen, *Mystic Chords of Memory: The Transformation of Tradition in American Culture* (New York: Vintage Books, 1993). Hereafter, this work is cited parenthetically in the text.
19. Fredric Jameson, *The Geopolitical Aesthetic: Cinema and Space in the World System* (Bloomington: Indiana University Press, 1992), 120–21.
20. AMC brochure, *Second Annual Film Preservation Festival.*
21. "Shots Seen 'round the World," *American Movie Classics Magazine* (November 1993): 8.
22. Nat Segaloff, "An American Legend: A Memorial Day Tribute to the Man Who Came to Define the American Persona," *American Movie Classics Magazine* (May 1992): 8–9.
23. www.members.aol.com/fortscott/ and www.jwplace.com/patriot.html (accessed on February 20, 2003). A Google search on John Wayne turns up approximately five million results. Even with the repetition built into these searches, along with incidental inclusions (such as John Wayne Gacy and John Wayne Airport), this is still an impressive return, indicating Wayne's continuing place in the celebrity firmament.
24. Gerard Peary, "American Hero," *American Movie Classics Magazine* (July 1995): 4–7. The Fonda festival is called "Fireworks and Fonda: An All American Tribute to a National Icon."
25. Steven Cohan, "Judy on the Net: Judy Garland Fandom and 'the Gay Thing' Revisited," in *Keyframes: Popular Cinema and Cultural Studies*, ed. Matthew Tinkcom and Amy Villarejo (London: Routledge, 2001), 119–36.

26. Ann McGuire, "The Great Kate: She Has a Style All Her Own," *American Movie Classics Magazine* (July 1992): 8–9.
27. Alicia Potter, "Rosalind Russell: The Original Working Girl," *American Movie Classics Magazine* (September 1996): 11.
28. McGuire, "The Great Kate," 9.
29. Ed Hazell, "Classic Debut: Mr. Bojangles," *American Movie Classics Magazine* (December 1991): 7; Nick Clooney, "Ellington Onscreen: A Word from Duke's Biographer," *American Movie Classics Magazine* (April 1999): 7.
30. Hazell, "Classic Debut," 7.
31. Clooney, "Ellington," 7.
32. Thomas Doherty, "Black and White Hollywood," *American Movie Classics Magazine* (February 1994): 6–7.
33. Although popularizing the Hollywood past involves social inclusiveness to respond to contemporary developments, we should ask whether this inclusiveness disturbs existing hierarchies or ultimately functions to reassert white dominance over the now included social "others." For example, in the 1980s and 1990s, African American stars (such as Lena Horne and Sidney Poitier) appear on the cover of *American Movie Classics Magazine* primarily in relation to Black History Month. However, these February covers tend to feature more white than black stars. This suggests that African American stars are compartmentalized within a designated area of commemoration and that, even within this area, the memory of white stars can appear as a more commercially viable gateway to the past.
34. Thomas Cripps has chronicled Hollywood's compromised attempts to produce more liberal racial fare in *Making Movies Black: The Hollywood Message Movie from World War II to the Civil Rights Movement* (New York: Oxford University Press, 1993).
35. Numerous histories gauge the impact of the Reagan and Bush (George H. W.) years on minorities. See, for example, Robert R. Detlefsen, *Civil Rights under Reagan* (San Francisco: Institute for Contemporary Studies, 1991); and Steven A. Shull, *A Kindler, Gentler Racism? The Reagan-Bush Civil Rights Legacy* (Armonk, NY: M. E. Sharpe, 1993). Shull argues that, although this period saw some improvements, women and African Americans continued to be overrepresented in lower-level jobs. Further, the latter group experienced increased unemployment, a rise in segregated schooling and housing, and, generally, a poorer standard of living.
36. Michael Wallace, "Mickey Mouse History: Portraying the Past at Disney World," *Radical History Review* 32 (1985): 52.
37. Dipesh Chakrabarty, "The Death of History? Historical Consciousness and the Culture of Late Capitalism," *Public Culture* 4, no. 2 (1992): 63.
38. For example, in *Time Passages: Collective Memory and American Popular Culture* (Minneapolis: University of Minnesota Press, 1990), 39–75, George Lipsitz focuses on the inability of early network television to shoulder completely the ideological burden resulting from the clashing juxtaposition of ethnic, working-class communities rooted in pre–World War II traditions and the new suburban consumer order in the post–World War II era. Tensions and contradictions arose, questioning the implicit rightness of capitalist progress and threatening the smooth operation of the hegemonic enterprise.
39. Michael Wallace, "Ronald Reagan and the Politics of History," *Tikkun 2,* no. 1 (1987): 16–18.
40. The meanings of the terms *preservation, restoration,* and *digital reconstruction* can vary depending on context; further, they are not necessarily mutually exclusive. That said, preservation generally refers to the repair and/or safe storage of original celluloid; restoration returns a film negative to a form congruent with its original release condition and/or artistic intentions; and, without physically altering the celluloid version, digital reconstruction repairs defects of a degenerated negative to simulate original quality. For more on these distinctions, see Mark-Paul Meyer, "Work in Progress: Ethics of Film Restoration and New Technologies," http://evora.omega.it/~demos/faol/digital/digital.htm#1 (accessed on February 17, 2004).

41. Steve Bryant discusses the extensive neglect of old TV shows as well as events that led to a greater archival sensibility in Britain, in *The Television Heritage: Television Archiving Now and in an Uncertain Future* (London: British Film Institute Publishing, 1989). On the recycling of TV shows in Britain, see Tim O'Sullivan, "Nostalgia, Revelation, and Intimacy: Tendencies in the Flow of Modern Popular Television," in *The Television Studies Book,* ed. Christine Geraghty and David Lusted (London: Arnold, 1998), especially 201–4. For more on the subject of television heritage in a U.S. context, see Derek Kompare, *Rerun Nation: How Repeats Invented American Television* (New York: Routledge, 2005), especially 101–68.
42. Michael Wallace has written on the subject of the relationship of capitalism and conservation, in "Reflections on the History of Historic Preservation," in *Presenting the Past: Essays on History and the Public,* ed. Susan Porter Benson et al. (Philadelphia: Temple University Press, 1986), 165–99. Hereafter, this work is cited parenthetically in the text. The *Stanford Humanities Review*'s special issue on film preservation (Winter 2000) details the relationship of studios and archivists.
43. In *American Movie Classics Magazine:* "The Wang Center" (December 1991): 2; see also "The Stanley Performing Arts Center" (February 1992): 2; "The Kentucky Theater" (May 1992): 2; "The Circle Theater" (September 1992): 2; "The Orpheum Theater" (June 1992): 2; and "The Paramount Theater" (August 1992): 2. All of these articles were written by Ann McGuire.
44. Philip Rosen has written on the tension among preservation, restoration, and nationalism in architecture, in *Change Mummified: Cinema, Historicity, Theory* (Minneapolis: University of Minnesota Press, 2001), especially 44–78.
45. Bill Desowitz, "Only the Good Die and Return," *Los Angeles Times,* May 18, 1997, 21.
46. On *Vertigo*'s restoration, see also Royal S. Brown, "Back from among the Dead: The Restoration of Alfred Hitchcock's *Vertigo,*" *Cineaste* 23, no. 1 (1997): 4–9.
47. Liner notes for the 1996 widescreen video edition of *Vertigo.*
48. James Katz and Robert Harris, quoted in *Obsessed with "Vertigo"* and in the liner notes for the 1996 wide screen video edition of *Vertigo.*
49. Paul Grainge discusses politicized attempts to reconcile the issue of authenticity with the impact of new technologies in "Reclaiming Heritage: Colourization, Culture Wars, and the Politics of Nostalgia," *Cultural Studies* 13, no. 4 (1999): 621–38.
50. For more on organizations devoted to film preservation, see online sites such as the National Film Preservation Board of the Library of Congress (www.loc.gov/film/); the National Film Registry (www.loc.gov/film/titles.html); the National Film Preservation Foundation (www.filmpreservation.org/); the Association of Moving Image Archivists (www.amianet.org/); and the American Film Institute National Center For Film and Video Preservation (www.afi.com/about/preservation/preservation.aspx).
51. See *American Movie Classics Magazine,* issues for March 1993, October 1994, October 1995, and July 1996, for coverage of the first four preservation festivals.
52. For more on these findings, see www.filmpreservation.org or www.loc.gov/film/.
53. www.filmpreservation.org; Robert Moses, "Film Preservation: Making History, by Saving It," *American Movie Classics Magazine* (March 1993): 4–6; and *American Movie Classics Magazine* (October 1994): 8.
54. *National Film Preservation Act of 1988,* Public Law 100-446, 100th Cong., 2nd sess. (September 27, 1988). The complete text for each preservation act is available in LexisNexis Congressional (www.lexisnexis.com/academic/universe/congress/).
55. Ibid.
56. Although the board's membership changes, some additional members include the Department of Theatre, Film and Television, UCLA; the Department of Cinema Studies, NYU; the University Film and Video Association; the Motion Picture Association of America; The Association of Motion Picture and Television Producers; the Screen Actors Guild of America; the National

Society of Film Critics; the National Association of Theater Owners; the American Society of Cinematographers; and the International Photographers Guild.

57. The films honored by the registry since 1989 are listed on www.loc.gov/film/titles.html. In an effort to promote a grass-roots awareness of film preservation, a selection of the registry films periodically tours the nation.
58. *National Film Preservation Act of 1992,* Public Law 102-307, 102nd Cong., 2nd sess. (June 26, 1992).
59. For information on these findings, see the National Film Preservation Foundation's Web site, www.filmpreservation.org, or that of the National Film Preservation Board, www.loc.gov/film/.
60. *National Film Preservation Act of 1996,* Public Law 104-285, 104th Cong., 2nd sess. (October 11, 1996), HR 1734.
61. The site www.filmpreservation.org offers a complete description of the National Film Preservation Foundation; it also lists orphan films already preserved through the foundation's auspices. Further, in cooperation with the National Film Preservation Foundation, collections of rare films have been released on video and DVD.
62. For factors influencing the growth of the U.S. conservation movement, see the Library of Congress's conservation site, http://memory.loc.gov/ammem/amrvhtml/conspref.html.
63. Albert Rains et al., *With Heritage So Rich* (New York: Random House, 1966).
64. See, for example, Laurence Zuckerman, "As Media Influence Grows for a Handful, Can That Be a Good Thing?" *New York Times,* January 13, 2000, sec. C, 6.
65. U.S. Congress, House, Committee on Energy and Commerce, *Globalization of the Media: Hearing before the Subcommittee on Telecommunications and Finance,* 101st Cong., 1st sess., November 15, 1989, 8. Hereafter, this work is cited parenthetically in the text.
66. David Morley and Kevin Robins, *Spaces of Identity: Global Media, Electronic Landscapes, and Cultural Boundaries* (London: Routledge, 1995), 147–73.
67. Wasko, *Hollywood in the Information Age,* 67–68.
68. Steve Wilson, "On-Line Piracy Turns from Music to Movies," *New York Times,* July 29, 1999, sec. D, 1, 6.
69. Steve Lohr, "Welcome to the Internet, the First Global Colony," *New York Times,* January 9, 2000, sec. 4, 1, 4.
70. For a discussion of both positions on globalization, see Robert J. Holton, *Globalization and the Nation State* (London: Macmillan Press, 1998); and Ian Clark, *Globalization and Fragmentation: International Relations in the Twentieth Century* (New York: Oxford University Press, 1997).
71. Holton, *Globalization and the Nation State,* 106.
72. Wilson, "On-Line Piracy Turns from Music to Movies," sec. D, 6.
73. On this point, see Robert Hewison, quoted in David Harvey, *The Condition of Postmodernity: An Enquiry into the Origins of Cultural Change* (Oxford: Basil Blackwell, 1989), 86.
74. Homi K. Bhabha, ed., *Nation and Narration* (London: Routledge, 1990), 1.
75. Annette Kuhn, *Family Secrets: Acts of Memory and Imagination* (London: Verso, 1995); Kuhn, *Dreaming of Fred and Ginger: Cinema and Cultural Memory* (Washington Square, NY: New York University Press, 2002); and Jackie Stacey, *Star Gazing: Hollywood Cinema and Female Spectatorship* (London: Routledge, 1994).
76. Lipsitz, *Time Passages,* 67, 213.
77. Marita Sturken, *Tangled Memories: The Vietnam War, the AIDS Epidemic, and the Politics of Remembering* (Berkeley: University of California Press, 1997), 94–97; and the Popular Memory Group, "Popular Memory: Theory, Politics, Method," in *Making Histories,* ed. Johnson et al., 211.
78. The Popular Memory Group, "Popular Memory," 211, 207.

MODULE THREE

Understanding Media Convergence and Transmediation

Personal Networks and the Personal Communication System: Using Multiple Media to Connect

Jeffrey Boase

In contrast to technologically deterministic approaches that focus on how communication technology affects social relationships, this paper examines how individuals draw on a variety of commonly used communication media in conjunction with in-person contact to stay connected to their personal networks. I term this use of multiple communication media the 'personal communication system'. Findings are based on a random sample telephone survey of 2200 adults living throughout the continental USA. Descriptive statistics show that despite the popularity of email and mobile phones, in-person and landline phone contact are still the most common ways of connecting with personal networks. Multivariate analysis reveals a more complex picture of media use, showing that the extent to which each medium is used varies to differing degrees with the size and diversity of personal networks. Hierarchical cluster analysis is used to explore the possibility that individuals may have different types of personal communication systems. Results show only two distinct clusters: those who draw heavily on all types of media to connect with their personal networks and those who draw less heavily on all types of media. Heavy communicators typically have larger and more diverse personal networks than light communicators. When taken together, the results presented in this paper suggest that rather than radically altering relationships, communication technology is embedded in social networks as part of a larger communication system that individuals use to stay socially connected.

Keywords: community, social networks, personal networks, computer mediated communication, Internet, telephone, mobile phone, email, communication networks, social capital, personal relationships, informatics

Introduction

The development of communication technology during the past century has altered how individuals connect with their personal networks. With the rapid diffusion of the telephone, mobile phone, and email, people now have the ability to contact friends, family and workmates outside of their immediate vicinity, both synchronously and asynchronously. But however profound communication technology has been for *how* social interaction occurs, the empirical work reviewed below suggests something even more profound: personal networks remain robust in spite of their newly mediated existence.

How then to understand the social significance of communication technology? Rather than viewing relationships as directly impacted by communication technology, I propose a new framework that emphasizes how purposeful individuals draw on multiple types of communication media—including in-person contact—to stay connected to their personal networks. I call this purposeful use of multiple communication media the 'personal communication system'. Using this framework I analyze data collected from a random telephone survey of 2200 Americans to examine the extent to which individuals draw on communication technology and in-person contact, how the use of each of these media vary with personal network composition, and how these media are combined to maintain contact with personal networks.

Communication Technology and the Personal Communication System

By focusing on whether or not using communication technology weakens relationships by replacing in-person contact with less rich contact, researchers have often assumed that technology is an external force which affects social relationships. This assumption is typically guided by concern that using communication technology may weaken social ties, increasing the prevalence of social isolation throughout society (e.g. McPherson *et al.* 2006). However, with the exception of a few studies conducted using unrepresentative samples in contrived settings (Kiesler & Sproull 1991; Nie & Hillygus 2002), a great deal of empirical research has shown that telephones and the internet have not weakened relationships by acting as insufficient substitutes for in-person contact (Fischer 1992; Dennis & Kinney 1998; Anderson & Tracy 2001; Robinson *et al.* 2002; Walther 2002; Gershuny 2003; Hampton & Wellman 2003; Cummings *et al.* 2006; Shklovski *et al.* 2006; Hampton 2007; di Gennaro & Dutton 2007). To the contrary, these studies suggest that individuals draw on various media to fit their social needs and lifestyles.

In this section I will propose a new approach that builds on existing literature by viewing the use of various communication technologies as a set of tools used in conjunction with in-person contact to maintain contact with personal networks. This approach has four important elements that call attention to the significance of: (a) multiple communication media, (b) social affordances of communication technology, (c) personal networks, and, (d) the cognition and behavioural elements of social ties.

Multiple Communication Media—The Personal Communication System

I use the term 'personal communication system' to convey the idea that individuals draw on multiple kinds of communication technologies together with in-person contact for two reasons.[1]

First, the word 'personal' connotes that the use of communication media is centred on the individual. The variety of communication media currently available—as evident in the wide adoption of landline phones, mobile phones and email—means that individuals can combine different media differently, to suit their needs and personal preferences. Paying attention to how individuals use a variety of communication technology stems from Fischer's (1992) 'hermeneutic approach' that empathizes the purposeful user as being the main force that accounts for the use of communication technology. According to Fischer, this approach helps avoid billiard ball metaphors that overemphasize the causal influence of technology and ignore how individuals can intentionally use communication technology to suit their own personal needs.

Second, the word 'system' implies that individuals combine various communication media together to connect with their own personal networks, rather than using them separately to connect with separate social worlds. Although it is true that individuals may sometimes form new social ties through one medium such as the internet, as these ties become close they are often contacted through other means, either in person or by phone (Parks & Floyd 1996; Haythornthwaite & Wellman 1998; Haythornthwaite 2001; Hampton & Wellman 2003; di Gennaro & Dutton 2007; Hampton 2007). Although existing research clearly shows that individuals use multiple media together to maintain their personal networks, how and the extent to which they are used together requires further research. The term 'personal communication system' should serve as a useful conceptual starting point from which to begin studying how this system operates.

The Social Affordances of Communication Technology

The term 'affordance' was first used by psychologists to refer to the human ability of understanding how the intrinsic properties of objects allow them to be used for a variety of purposes (Gibson 1977, 1986). When applied to communication technology, the social affordance approach posits that individuals will choose to use particular kinds of communication media when there is a congruency between opportunities that they provide and the characteristics of the ties with whom they are used to communicate (Wellman *et al.* 2003; Boase & Wellman 2006). For example, the telephone affords individuals the ability to communicate with distant ties that would be difficult to contact otherwise. When individuals have numerous distant ties that they would like to quickly get hold of, they use the space-transcending affordance of the telephone to meet their communication demands. The social affordance approach allows researchers to take into account how the intrinsic properties of communication technologies may factor into their adoption and use, without making the technological deterministic assumption that these properties directly impact on users.

Personal Networks

A personal network is the set of social ties that an individual knows and communicates with. In contrast to social network researchers who define the boundaries of a particular group and then focus on its interconnected relationships, those who study personal networks focus on ties that are defined as significant by a particular individual (see Wellman 2007). The personal network approach is ideally suited for understanding the personal communication system because it takes the individual's relationships as its primary focus of analysis.

Lack of personal network measures in communication technology studies has made it difficult to evaluate the social significance of these technologies for at least two reasons. First, existing research tends to examine only frequency or duration of contact using communication technology rather than examining the number of personal ties actually contacted using technology. This makes it difficult to know if additional communication made using telephones, mobile phones and email is used to connect more intensely with just a few core ties or more broadly with a large number of ties throughout an individual's personal network. Second, personal network measures have been typically limited to contact with unspecified numbers of friends and family. This has made unclear whether people using these technologies typically have large and diverse or small and dense networks.

The Cognitive and Behavioural Sides of Personal Networks

Social ties have two important dimensions: cognition and behaviour. The cognitive dimension includes the belief that a social tie exists, as well as feelings of closeness, memory of past interaction and knowledge about that tie. The behavioural dimension is interaction that typically occurs by way of mediated and unmediated communication. When thinking of ties in this way it is clear that the social significance of communication technology lies not in how it alters ties as an external force, but rather in the fact that it is embedded in behaviour that is essential to the existence of ties. Much as Licoppe & Smoreda (2005) described it, it is not necessarily the content of the communication but rather the fact that it is being made which is often the most significant part of technologically mediated communication. The more that communication technology is used to maintain the active connection that is vital to the existence of ties, the greater its social significance.

Thinking of social ties in this way does not imply that there is no interaction between cognition and behaviour. Indeed, it is likely that there is a reciprocal process through which communication is necessary to form and maintain feelings of connection, and feelings of connection motivate continued communication. However, the point being made here is that feelings of connection alone are just one side of social ties, and that understanding the significance of communication technology means understanding how—and to what extent—it is embedded in personal networks as the means by which they are made real.

Research Questions

As argued, the social significance of communication media is evident in the extent to which they are used to facilitate communication. Thus, my first research question is as follows:

1. *To what extent do individuals use landline phones, mobile phones, email and in-person communication to maintain contact with their personal networks?*

Although understanding the extent to which communication media are typically embedded into personal networks is an important start, it only gives a basic sense of their social significance. The extent to which individuals draw on communication technology may depend heavily on the extent to which the affordances that it offers fit the social demands of their personal networks. Individuals with large and diverse personal networks may be more prone to use mobile phones, landline phones and email because each of these technologies affords the ability to switch quickly between ties (Wellman 1999, 2001). Moreover, some of these technologies may be more useful than others, depending on the particular network property being considered. The asynchronous nature of email may make it particularly well suited to help avoid scheduling conflicts that naturally arise as networks become large and diverse (Gibson 2005; Boase 2006). By contrast, landline and mobile phone contact may be suited for distant network ties and for those ties that are strong enough that individuals either know what time is best to call or are not concerned that the possible interruption of their phone call will be considered offensive (Boase & Wellman 2006).

Given the potentially strong relationship between the personal network properties and the use of these various communication media, the second research question is:

2. *To what extent does the use of each communication medium vary with personal network composition?*

Finally, although it is clear that individuals draw on multiple communication media in conjunction with in-person contact, it is unknown if there are common types of communication systems that are prevalent throughout the general population. It is possible, for example, that some individuals draw heavily on email in conjunction with in-person contact, while others mostly use telephone and in-person contact to maintain contact with their personal networks. Further, if different communication systems exist, they may reflect the prevalence of different kinds of personal networks. Accordingly, my third research question is:

3. *Are there different types of personal communication systems? If so, does personal network composition vary between each type of system?*

Data and Methodology

The findings presented here are based on data collected from the Pew Internet & American Life Project's Social Ties Survey, a random digit dial telephone survey of 2200 adults living in the continental USA. Interviews were conducted from 17 February to 17 March 2004, and lasted an average of 19 minutes per individual. All adults sampled had landline telephones (telephones that use physical outlets connected in one place) in their households, and all interviews were conducted in English. The response rate was 35 per cent, and approximately 96 per cent of those individuals who began the survey completed it in full. A comparison of sex, age, race, employment and education variables from the Social Ties Survey with the same variables in the US Census Bureau's 2003 American Communities Survey indicates that the Social Ties sample is similar to the general American population in its demographic composition (Boase 2006).

Measuring Personal Networks

Personal Network Size

The number of ties in an individual's personal network may affect the extent to which they use communication technology. Large-scale ego-centred surveys have typically used name generators to collect personal network data—for example, asking respondents to give the names of ties with whom they discuss important matters (Burt 1984; Marsden 1987; McPherson *et al.* 2006). Although these methods have the advantage of collecting detailed information about a few select ties, they are generally not good indicators of network size because they focus only on relatively small numbers of alters that are not representative of the large number of ties that make up an individual's personal network (Marin 2004; Marin & Hampton 2007). Rather than using the name generator method, the Social Ties Survey uses the summation method, which was specifically designed to measure personal network size. This method calculates network size by having respondents report the number of ties that they know in different relational categories—such as friends, workmates, kin, etc.—and then adding those together to give a single estimate of size (McCarty *et al.* 2001). Breaking down network size into a number of relational roles eases the cognitive burden of recall. McCarty *et al.* use comparative data and repeated measures to argue that the summation method is a valid and reliable proxy for network size.

To further reduce cognitive burden and give more detailed information about tie strength, a variation of the summation method used in the Social Ties Survey asked respondents to report the number of people with whom they 'feel very close' in a variety of relational roles, and then to report the number of people with whom they 'feel somewhat close' in the same relational roles. Closeness was chosen because Marsden and Campbell (1984) found it to be a fundamental property of tie strength, and pre-testing showed that thinking of ties that were very close and somewhat close was a relatively easy task for respondents.

Personal Network Diversity

The space transcending and asynchronous social affordances of communication technologies may also be useful when connecting with diverse personal networks, because tie diversity increases scheduling conflict and travel demands. Connecting with geographically diverse personal networks often requires substantial travel time, connecting with sparsely knit ties increases communication demand because ties must be contacted serially rather than all at once in group settings, and connecting with ties that work in diverse occupations increases scheduling conflict because people who work in different occupations often have different schedules. The measures used in the Social Ties Survey to operationalize these three types of diversity will be discussed in turn.

Geographical diversity is measured by asking respondents to report on the number of core and significant ties who live more than one hour's travel from their home (see Wellman & Tindall 1993; Fischer 1992; Mok & Wellman 2007).

Having sparsely knit personal networks is measured by asking respondents to report approximately how many of their core and significant ties know each other (0 = 'none', 1 = 'some', 2 = 'about half', 3 = 'many', and 4 = 'all'). Although this measure may be less accurate than the name generator method that asks respondents if specific ties know each other, time constraints and the potentially large numbers of core and significant ties collected using the summation method make using the name generator method unfeasible.

Another indicator of social diversity used in the Social Ties Survey is a variation of Lin's (2001) position generator method. Although Lin's original method asks respondents if they know *anyone* in occupations of varying prestige, this variation asks respondents if they know *active ties* in occupations of varying prestige. The measure is modified in this way because the Social Ties survey focuses only on active ties, rather than all possible ties in an individual's personal network.

This paper also analyses the data collected using this method somewhat differently than Lin. While Lin adds the results of these questions together to give a single scale indicating network diversity, this analysis adds together the results to give two scales: one indicating the diversity of ties in high prestige occupations, and the other indicating the diversity of ties in low prestige occupations. This scale is broken into two scales because I have theorized that diversity is associated with the use of email, and people in high prestige occupations are more likely to have email access than those in low prestige occupations. Social diversity can only matter for the use of email if the diverse ties that an individual would like to communicate with have email access. The first scale includes occupations that were selected at roughly equal intervals from a list of occupations with prestige scores higher than 50 points, while the second scale includes occupations selected at equal intervals from a list of occupations with prestige scores lower than 50 points (prestige scores came from Ganzeboom and Treiman 1996).

Measuring Personal Communication Systems

Understanding the extent to which these media are used to connect with personal networks is not captured well through measures that focus on frequency or duration of contact through various media. The Social Ties data include a measure of how many core and significant ties are typically contacted at least once a week by way of landline phone, mobile phone, email and in-person media. These core and significant tie measures are then summed together to give the number of active ties that are contacted at least weekly through each of these communication media.

Preliminary analysis shows only a few minor discrepancies between core and significant ties in terms of the associations between personal network composition and the use of communication media. Hence, the remainder of this paper reports only on active ties—the sum of core and significant ties.[2]

Control Variables

Answering research questions 2 and 3 involves the use of regression models that take into account a number of factors that may affect the association between personal network composition and the use of different communication media. Gender, education, type of occupation, age and organizational involvement have all been implicated as factors in the use of various communication media, and they have also been associated with differing types of personal network composition. For a more detailed discussion of how each of these factors may influence the use of different communication media, see Boase (2006) and Boase and Wellman (2006).

Contact through one communication medium may sometimes lead to contact through other communication media—such as when a person emails a friend about getting together for coffee. To control for the potentially spurious association between network properties and the particular communication medium under consideration, the other communication media used are treated as control variables.

Findings

Personal Networks in America

I will begin by describing the basic characteristics of personal networks in America using the Social Ties data. Overall, the results show that Americans maintain substantial numbers of active ties: a mean of 51 active ties, 24 core ties and 27 significant ties. The median measures show somewhat different results, with medians of 35 active ties, 15 core ties and 16 significant ties. The discrepancy between the mean and median numbers of ties is due to a positive skewing of active, core and significant tie distributions (skewness = 6.6 for active ties, 6.0 for core ties, and 7.0 for significant ties).

The relational distribution of personal networks shows that active personal networks are composed of a variety of ties. The mean percentage of ties in each relational category is as follows: 30 per cent are friends, 22 per cent are immediate kin, 20 per cent are extended kin, 18 per cent are known from work, and 10 per cent are neighbours.

Research Question 1: Connecting to Personal Networks By All Means

Question 1 asks:

1. *To what extent do individuals use landline phones, mobile phones, email and in-person communication to maintain active contact with their personal networks?*

Overall results indicate that Americans draw on a variety of media to stay connected to their personal networks. However, the role of communication technologies should not be overstated, as in-person contact still plays a leading role in personal communication systems. While cell phone and email technology are widespread, they have yet to be adopted by some Americans. At the time of the survey approximately 63 per cent of the respondents reported using email in the past month and 74 per cent reported using a mobile phone in the past month. Moreover, when examining the number of ties contacted using all of these media, these results indicate that despite the recent popularity of email and mobile phones, in-person and landline phone media are still the main ways in which individuals maintain contact with their active networks. The approximate mean number of active ties contacted at least weekly by each medium is as follows: 17 in-person, 12 by landline phone, 10 by mobile phone, and 10 by email. As with the distribution of network size, the number of ties contacted using each medium is positively skewed. The median number of active ties contacted at least weekly by each medium follow a similar order, but show substantially less use of mobile phones and email: 10 in-person, 7 by landline phone, 5 by mobile phone, and 4 by email.

Research Question 2: Communication Media and Personal Network Composition

Examining the typical use of these media in the general population only gives a general sense of their social significance. However, as discussed above, their significance may depend heavily on the size and diversity of personal networks. For example, the asynchronous nature of communication afforded by email implies that email may by more socially significant for people coping with the communication demands of large and diverse networks than those maintaining contact with small and densely knit personal networks. Question 2 asks:

2. *To what extent does the use of each communication medium vary with personal network composition?*

Zero inflated negative binomial count regression is used to understand the extent to which personal network composition and other potentially intervening factors account for in-person, landline phone, mobile phone and email contact.[3]

Overall, the extent to which each medium is used varies significantly with personal network composition (Table 1, 'in-person' 'landline' 'mobile' and 'email' models).

Personal Network Size

Generally these results suggest that people draw more heavily on in-person and email contact as the numbers of friends, work ties and kin in their active personal networks become large. By contrast, landline and mobile phones are not typically important ways of maintaining contact with large numbers of friend and work ties. Considering only those associations where $p < 0.05$, the results suggest that:

- In-person contact is made when active personal networks include large numbers of friend, work, neighbour and kin ties;
- Landline phones are used when active personal networks include numbers of neighbour and kin ties;
- Mobile phones are used when active personal networks include large numbers of kin ties;
- Email is used when active personal networks include large numbers of friend, work and kin ties.

TABLE 1 Event count and binary logistic regression of media use.

	In-person [1]	Landline [1]	Mobile [1]	E-mail [1]	Heavy communicators [2]
Control variables					
female	0.00	0.17**	0.02	0.33**	0.44**
college degree	–0.05	–0.01	–0.07	0.12	0.07
professional	0.09*	0.07	0.05	0.06	0.10
occupation age	0.03**	–0.01	0.03*	–0.01	0.01
age squared	–0.46**	0.05	–0.54**	0.05	–0.58
organizational involvement	0.03*	–0.02	0.04*	0.08**	0.16**
in-person	NA	0.03**	0.01**	0.00	NA
landline phone	0.02**	NA	0.02**	0.01**	NA
mobile phone	0.01**	0.01**	NA	0.01**	NA
e-mail	0.01*	0.01*	0.02**	NA	NA
Network size					
friends	0.02**	0.00	0.00	0.02**	0.06**
work ties	0.03**	0.00	0.01	0.02**	0.08**
neighbors	0.02**	0.02**	0.00	0.00	0.02
kin	0.01**	0.01**	0.01**	0.01**	0.03**
Network diversity					
high prestige diversity	–0.01	0.04**	0.04*	0.08**	0.15**
low prestige diversity	0.03**	0.04**	0.03*	0.02	0.08*
1+ hours travel	–0.01	0.00	0.00	0.01**	0.00
core tie density	0.06**	0.08**	0.09**	–0.01	0.16*
significant	0.11**	0.03	0.03	0.03	0.20**
tie density constant	2.62**	1.07*	3.06**	0.52	–1.91
Inflate					
constant	–4.33**	–3.32**	–2.29**	–1.82**	NA
N	2153	2153	1595	1359	2153
Number of zero observations	112	171	254	338	NA

*p ≤ 0.05
**p ≤ 0.01
[1] zero inflated negative binomial event count regression
[2] binary logistic regression
0 = light communicators
1 = heavy communicators

Personal Network Diversity

These results show that the relationship between active personal network diversity and the use of various communication media is complex. Considering only those associations where $p < 0.05$, the results suggest that:

- In-person contact is made when active personal networks include diverse ties in low prestige occupations, ties that are physically close, and dense groups of core and significant active ties;
- Landline phones are used when active personal networks include diverse ties in both low and high prestige occupations, and dense groups of core ties;
- Mobile phones are used when active personal networks include diverse ties in low and high prestige occupations, and dense groups of core ties;
- Email is used when active personal networks include diverse ties in high prestige occupations, and ties that are geographically distant.

Research Question 3: The Personal Communication System

The results presented so far only show how the use of each communication medium varies with personal network composition. To explore the ways in which these media may be combined together and how those different combinations vary with personal network composition, research question 3 asks:

> 3. *Are there different types of personal communication systems? If so, does personal network composition vary between each type of system?*

Because this question is exploratory in nature and emphasizes the importance of understanding how communication media are combined to form a complex whole, I use hierarchical cluster analysis to look for different patterns in the ways that in-person, landline phone, mobile phone and email are combined.

Although hierarchical cluster analysis has the advantage of avoiding assumptions about the number and kinds of different personal communication systems that may exist in the general population, it has been criticized for not providing any standard way of deciding if clusters differ from each other significantly. In this analysis I first use Ward's Method to cluster in-person, landline phone, mobile phone and email contact.[4] A Calinski and Harabasz index is then used to determine an optimum number of clusters.[5]

The Calinski and Harabasz index shows that two clusters are more distinctive than three or more clusters. Moreover, breaking communication media use into more than two clusters creates at least one cluster that is too small to be included in the analysis that follows ($N < 10$ per cent). For these reasons I use two clusters in the analysis that follows.

The two kinds of personal communication systems showing in these two clusters differ mostly by the number of active ties contacted at least weekly by each medium rather than by the way that media are combined. Because of the similar ways communication media are combined in both clusters and because they are only distinguishable by the extent to which these media are used, I will refer to these two clusters simply as the 'heavy communicators' and the 'light communicators'. For light communicators, the median number of ties contacted at least weekly is as follows: 9 in-person, 7 by landline phone, 4 by mobile phone, and 2 by email. For heavy communicators, the median number of ties contacted weekly is as follows: 31 in-person, 21 by landline phone, 13 by mobile phone, and 12 by email.

To answer the second part of Research Question 3, binary logistic regression analysis is used to examine how personal network composition varies between light and heavy communicators (Table 1, model 'heavy communicators'). Considering only those associations where $p < 0.05$, the results suggest that:

- Heavy communicators have greater numbers of friend, work, and kin ties in their active personal networks than light communicators;
- Heavy communicators know a greater diversity of ties working in both high and low prestige occupations than light communicators;
- Heavy communicators have more dense core and significant ties than light communicators.

Discussion and Conclusions

This paper uses a new approach to understanding the social significance of communication technology that focuses on how individuals use multiple communication technologies in conjunction with in-person communication to maintain contact with their personal networks.

The Social Ties data show that in-person and landline phones are the most common ways that Americans maintain contact with their active personal networks, but that mobile phone and email contact still plays a significant role in this regard. By helping people stay connected to their personal networks, communication technology plays a critical role in acting as the means by which the behavioural dimension of social ties is fulfilled.

The extent to which in-person, landline phone, mobile phone and email communication occurs varies in complex ways with personal network size and diversity. Although each of these media may be used to a greater extent as personal networks become large, each medium has a slightly different usage dependent on the kind of tie being examined. In-person communication and email are used to contact more ties as the numbers of friend, work and kin ties in active personal networks become large. By contrast, landline phones are used to contact more ties as the numbers of neighbours and kin in active personal networks become large, and mobile phones are used to contact more ties as the number of kin ties becomes large. These results also show that the relationship between active tie diversity and the number of ties contacted using various communication media is complex. More ties are contacted in-person when the diversity of low prestige ties, number of physically close ties, and density of core and significant ties increase in active personal networks. Somewhat differently, more ties are contacted by way of landline phone and mobile phone as the diversity of low and high prestige ties, number of distant ties, and density of core ties increase in active personal networks. Moreover, more ties are contacted by email when the diversity of high prestige ties and the number of geographically distant ties increase in active personal networks.

The social affordance approach can be used to make sense of these different patterns by suggesting that they are due to a congruency between personal network composition and the different opportunities that each medium affords. Although having large numbers of friend, work, neighbour and kin ties in one's active personal network may require substantial time and effort to communicate with in-person, the rich amount of information and emotional exchange that in-person communication affords may make it a necessary way of staying connected to these ties. Despite the importance of in-person communication, these results also show that email is still useful for connecting with large numbers of friends, kin and work ties, even though it lacks verbal and visual feedback. This is probably because the asynchronous nature of email makes it well suited to the scheduling conflicts that arise when maintaining contact with large numbers of these ties. Moreover, these results show that landline phones can also be useful for maintaining contact with large numbers of kin and neighbouring ties. Because landline phones afford the ability to connect from one household to another, they are well suited for maintaining contact with large families of kin living together and with large number of neighbourhood based ties (Wellman 2001). These results also show that mobile phones can be useful for maintaining contact with large numbers of kin ties. Because mobile phones afford person-to-person contact they may be especially useful for those who need to connect with many kin ties organizing everyday activities. Unlike friend, work and neighbouring ties, individuals may be more aware of what their kin ties are doing at any given time, making it easier for them to know when it is possible to interrupt their activity with a call to their mobile phone.

Social affordance can also be used to explain the complex patterns that emerge between personal network diversity and the media use. That ties in low prestige occupations may not have email access helps to explain why having these kinds of ties in active tie networks is positively associated with in-person, landline and mobile phone, but not email contact. By contrast, ties working in high prestige occupations are more likely to use a variety of technologies, which helps to explain why having a diversity of ties working in high prestige occupations is positively associated with email, mobile phone and landline phone contact but not in-person contact. The findings showing that tie distance is negatively associated with in-person contact and positively associated with email contact can also be well explained using the social affordance approach. In-person contact does not easily afford distant contact, while email does. The results regarding tie density also can be explained by way of social affordance. Dense groups of core ties often involve high expectations of commitment, making intense synchronous contact that occurs in-person, by landline and mobile phone both acceptable and perhaps expected. By contrast, the opportunities for asynchronous text based communication email may be less useful for connecting with dense groups.

Only two kinds of personal communication systems are prominent among the respondents: those that use all media to maintain contact with many ties—'heavy communicators'—and those that do not—'light communicators'. Heavy communicators have significantly more friend, work, kin, high and low prestige diversity, and greater amounts of density in their active personal networks than light communicators.

When taken together, these results indicate that Americans actively draw on their personal communication systems to stay connected to complex personal networks. The more complex the personal network, the more people draw on all of these media. These findings contradict the idea that communication technology detracts from personal relationships and leads to social isolation, as suggested by McPherson *et al.* (2006). Instead, these results show that directing attention to the ways in which communication technology is used as part of a personal communication system that is necessary for the continued existence of personal networks can make the social significance of communication technology more apparent.

Although this paper reveals a great deal about the role of communication media in personal networks, more work needs to be done in order to better understand the processes that underlie these general patterns. For example, closely examining the process that occurs when individuals choose to use one communication over another would reveal the extent to which media selection is based on active awareness of how using the affordances they offer fits with the kinds of personal networks being contacted. By better understanding the micro processes that underlie these macro patterns, we can better understand how and why communication media are used together to form personal communication systems.

About the Author

Jeffrey Boase is currently a postdoctoral researcher in the Department of Social Psychology at the University of Tokyo. His research focuses on how individuals use the internet and mobile phones to maintain and build their personal networks, and how they integrate these media into their daily lives. To investigate these issues Dr Boase has co-designed several large-scale surveys in America, Canada and Japan. His most recent work examines the social utility of web-enabled mobile phones in Japan, with a focus on how personal network dynamics shape the extent to which this technology is used to bridge and bond with social ties. Dr Boase has published over a dozen scholarly articles, and he has been awarded fellowships from the Social Sciences and Humanities Research Council of Canada and the National Center for Digital Government at Harvard University. *Address:* Department of Social Psychology, The University of Tokyo, 7-3-1 Hongo, Bunkyo-ku, Tokyo 113-0033, Japan. [email: jeff.boase@gmail.com]

Notes

1. Although 'personal communication system' is sometimes used in the technology industry to refer to cellular technology in America, it is not widely used in academic writing about the social significance of communication technology.
2. Core and significant tie results are kept separate for density variables in this analysis because the nature of these measures does not allow them to be summed together.
3. Zero inflated count regression is more suitable than regular OLS regression because the dependent variables are positive count numbers with a strong positive skew and a substantial number of zero values. Negative binominal count regression was used instead of Poisson count regression because the standard variation of each dependent variable was greater than its respective mean, indicating an over-dispersion that is better handled by negative binomial count regression than Poisson count regression. STATA's likelihood-ratio test confirmed the suitability of this choice.
4. Ward's Method is an efficient way of sorting data that uses an analysis of variance approach to minimize the Sum of Squares between separate clusters.
5. A Calinski and Harabasz index gives a value reflecting the overall distinctiveness of clusters at each stage of a cluster analysis. Larger values reflect more distinct clustering.

References

Anderson, B. & Tracey, K. (2001) 'Digital living: the impact (or otherwise) of the internet on everyday life', *American Behavioral Scientist,* vol. 45, no. 3, pp. 456–475.

Boase, J. (2006) *America Online and Offline: The Relationship of Personal Communication Media to E-mail and Other Communication Media,* Dissertation, Graduate Department of Sociology, The University of Toronto.

Boase, J. & Wellman, B. (2006) 'Personal relationships: on and off the internet', in *The Cambridge Handbook of Personal Relationships,* eds D. Perlman & A. L. Vangelisti, Blackwell, Oxford, pp. 709–723.

Burt, R. S. (1984) 'Network items and the General Social Survey', *Social Networks,* vol. 6, no. 1, pp. 293–339.

Cummings, J. N., Lee, J. B. & Kraut, R. (2006) 'Communication technology and friendships during the transition from high school to college', in *Computers, Phones, and the Internet: Domesticating Information Technology,* eds R. Kraut, M. Brynin & S. Kiesler, Oxford University Press, Oxford, pp. 265–278.

Dennis, A. R. & Kinney, S. T. (1998) 'Testing media richness theory in new media: the effects of cues, feedback, and task equivocally', *Information Systems Research,* vol. 9, no. 3, pp. 256–274.

Di Gennaro, C. and Dutton, W. H. (2007) 'Reconfiguring friendships: social relationships and the interne', *Information, Communication & Society,* vol. 10, no. 5, pp. 591–561.

Fischer, C. (1992) *America Calling: A Social History of the Telephone to 1940,* University of California Press, Berkeley.

Ganzeboom, H. B. & Treiman, D. J. (1996) 'Internationally comparable measures of occupational status for the 1988 International Standard Classification of Occupations', *Social Science Research,* vol. 25, no. 1, 201–239.

Gershuny, J. (2003) 'Web use and net nerds: a neofunctionalist analysis of the impact of information technology in the home', *Social Forces,* vol. 82, no. 1, pp. 141–168.

Gibson, J. (1977) 'The theory of affordances', in *Perceiving, Acting, and Knowing: Toward an Ecological Psychology,* eds R. E. Shaw & J. Bransford, Lawrence Erlbaum Associates, Hillsdale, NJ, pp. 67–82.

Gibson, J. (1986) *The Ecological Approach to Visual Perception,* Lawrence Erlbaum Associates, Hillsdale, NJ.

Gibson, D. (2005) 'Concurrency and commitment: network scheduling and its consequences for diffusion', *Journal of Mathematical Sociology,* vol. 29, no. 4, pp. 295–325.

Hampton, K. & Wellman, B. (2003) 'Neighboring in netville: how the internet supports community and social capital in a wired suburb', *City & Community,* vol. 2, no. 4, pp. 277–311.

Hampton, K. (2007) 'Neighborhoods in the network society: The e-Neighbors Study', *Information, Communication & Society,* vol. 10, no. 5, pp. 714–748.

Haythornthwaite, C. (2001) 'Exploring multiplexity: social network structures in a computer supported distance learning class', *The Information Society,* vol. 17, no. 3, pp. 211–226.

Haythornthwaite, C. & Wellman, B. (1998) 'Work, friendship and media use for information exchange in a networked organization', *Journal of the American Society for Information Science,* vol. 49, no. 12, pp. 1101–1114.

Kiesler, S. & Sproull, L. (1991) *Connections,* MIT Press, Cambridge, MA.

Licoppe, C. & Smoreda, Z. (2005) 'Are social networks technologically embedded? How networks are changing today with changes in communication technology', *Social Networks,* vol. 27, no. 4, pp. 317–335.

Lin, N. (2001) *Social Capital: A Theory of Social Structure and Action,* Cambridge University Press, Cambridge.

Marin, A. (2004) 'Are respondents more likely to list alters with certain characteristics? Implications for name generator data', *Social Networks,* vol. 26, no. 4, pp. 289–307.

Marin, A. & Hampton, K. (2007) 'Simplifying the personal network name generator: alternatives to traditional multiple-and single-name generators', *Field Methods,* vol. 19, no. 2, pp. 163–193.

Marsden, P. (1987) 'Core discussion networks of Americans', *American Sociological Review,* vol. 52, no. 1, pp. 112–131.

Marsden, P. & Campbell, K. (1984) 'Measuring tie strength', *Social Forces,* vol. 62, no. 2, pp. 482–501.

McCarty, C., Killworth, P. D., Bernard, H. R., Johnsen, E. C. & Shelley, G. A. (2001) 'Comparing two methods for estimating network size', *Human Organization,* vol. 60, no. 1, pp. 28–39.

McPherson, M., Smith-Lovin, L. & Brashears, M. (2006) 'Social isolation in America: changes in core discussion networks over two decades', *American Sociological Review,* vol. 71, no. 3, pp. 353–375.

Mok, D. & Wellman, B. with Basu, R. (2007) 'Did distance matter before the internet?', *Social Networks,* vol. 29, no. 3, pp. 430–461.

Nie, N. H. & Hillygus, D. S. (2002) 'Where does internet time come from? A reconnaissance', *IT & Society,* vol. 1, no. 2, pp. 1–20.

Parks, M. R. & Floyd, K. (1996) 'Making friends in cyberspace', *Journal of Communication,* vol. 46, no. 1, pp. 80–97.

Robinson, J. P., Kestnbaum, M., Neustadtl, A. & Alvarez, A. (2002) 'Information technology and functional time displacement', *IT & Society,* vol. 1, no. 2, pp. 21–36.

Shklovski, I., Kiesler, S. & Kraut, R. (2006) 'The internet and social interaction: a meta-analysis and critique of studies, 1995–2003', in *Computers, Phones, and the Internet: Domesticating Information Technology,* eds R. Kraut, M. Brynin & S. Kiesler, Oxford University Press, Oxford, pp. 251–264.

Walther, J. B. (2002) 'Time effects in computer-mediated groups: past, present, and future', in *Distributed Work,* eds P. Hinds & S. Kiesler, MIT Press, Cambridge, MA, pp. 235–257.

Wellman, B. (1999) 'From little boxes to loosely-bounded networks: the privatization and domestication of community', in *Sociology for the Twenty-first Century: Continuities and Cutting Edges,* ed. J. Abu-Lughod, University of Chicago Press, Chicago, pp. 94–114.

Wellman, B. (2001) 'Physical place and cyber-place: changing portals and the rise of networked individualism', *International Journal for Urban and Regional Research,* vol. 25, no. 2, pp. 227–252.

Wellman, B. (2007) 'The network is personal: introduction to a special issue of Social Networks', *Social Networks,* vol. 29, no. 3, pp. 349–356.

Wellman, B., Quan-Haase, A., Boase, J., Chen, W., Hampton, K., Isla de Diaz, I. & Miyata, K. (2003) 'The social affordances of the internet for networked individualism', *Journal of Computer-Mediated Communication,* vol. 8, no. 3.

Wellman, B. & Tindall, D. (1993) 'Reach out and touch some bodies: how telephone networks connect social networks', *Progress in Communication Science,* vol. 12, no. pp. 63–94.

Localizing the Internet Beyond Communities and Networks

John Postill

As the numbers of internet users worldwide continue to grow, the internet is becoming 'more local'. This article addresses the epistemological challenge posed by this global process of internet localization by examining some of the conceptual tools at the disposal of internet researchers. It argues that progress has been hampered by an overdependence on the problematic notions of community and network whose paradigmatic status has yet to be questioned by internet scholars. The article seeks to broaden the conceptual space of internet localization studies through a ground-up conceptualization exercise that draws inspiration from the field theories of both Pierre Bourdieu and the Manchester School of Anthropology, and is based on recent fieldwork in suburban Malaysia. This exploration demonstrates that a more nuanced understanding of the plural forms that residential sociality can take is needed in order to move beyond existing binaries such as 'network sociality' versus 'community sociality'.

Keywords: banal activism, community, field theory, internet localization, Malaysia, network, residential sociality, suburbia

Introduction

Until the mid-1990s the number of internet users worldwide was small and most users could not help but communicate with others at great distances. However, as numbers continue to grow, the internet is gradually becoming 'more local' (Davies and Crabtree, 2004). This global process of internet localization poses a set of logistic, methodological and conceptual challenges to researchers. Logistically, it demands that they spend sufficient time in a local setting in order to get to know, both online and offline, those who live, work and/or play there. Methodologically, it renders the very distinction of online versus offline social domains even more problematic than it has been hitherto (Hine, 2000), more so as internet and mobile technologies continue to converge. Conceptually, the challenge is how to keep track of the fast pace of technological change while avoiding the default position whereby a seemingly stationary 'local community' (or local college, local council, etc.) is assumed to be impacted upon by 'global' technologies.

This article concentrates on the third challenge. How can we conceptualize the relationship between technological and social change at the local level? More specifically, what conceptual tools do we have at our disposal to study the emergence of new internet-related forms of local sociality? To address these questions, first this article reviews the existing literature on internet localization, suggesting that the progress of this research area has been hampered by an overdependence on two key notions: community and network. Both notions have had uneven careers as social scientific terms, careers that are yet to be critically debated in the internet literature, and which are outlined below. However, more important than their strengths and limitations is the unrivalled paradigmatic status that these paired notions currently enjoy among scholars of internet localization. This article suggests the need to think beyond the community/network paradigm by broadening our analytical lexicon to do some justice to the plethora of forms of sociality that anthropologists and sociologists have identified down the decades.

This article carries out this conceptual broadening through ethnographic examples from recent fieldwork in suburban Kuala Lumpur, Malaysia. It explores the potential uses of concepts that lie outside the community/ network paradigm (e.g. field, arena, forum) by means of a fine-grained analysis of emergent forms of residential sociality, arguing that this kind of ground-up conceptualization reveals the inadequacies of overly general notions such as 'community sociality' or 'network sociality' (Wittel, 2001). It closes with a discussion of the wider theoretical implications of this investigation for the study of internet localization and with suggestions for further research.

The Community/Network Paradigm

Two approaches to the study of internet localization stand out in the existing literature, both hinging on the notions of community and network. First, there is the 'community informatics' approach, in which researchers study a 'local community' and assess its specific technological needs.[1] Researchers of this persuasion regard local communities as 'the bedrock of human development' (Gurstein, 2004). To withstand the onslaught of capitalism and become empowered, they suggest, local communities must take control of the very networked technologies that threaten their survival (Gurstein et al., 2003).

Second, there is the 'networked individualism' approach led by the Toronto network analyst Barry Wellman. Wellman has denounced the smuggling of obsolete notions of community from an earlier period of North-American community studies into internet localization studies. Where the old communities had 'streets and alleys', internet researchers are now imagining communities bound 'by bits and bytes' (Hampton and Wellman, 2003). For Wellman, this is an analytical cul-de-sac, for the crucible of North-American sociality has long ceased to be the local neighbourhood (Wellman and Leighton, 1979). However, as Wellman insists, this does not mean that communities have disappeared; rather, they have survived in the form of geographically dispersed personal communities, i.e. personal networks (Pahl, 2005). The internet merely reinforces a global trend towards networked individualism that was already well underway (Castells, 2001; Wellman et al., 2003).

Despite their differences, both perspectives on internet localization share a heavy reliance on the inseparable notions of community and network. Although ever popular among internet scholars, policymakers and activists, both terms have had troubled careers as theoretical concepts. These conceptual difficulties, it is suggested, demand closer attention if we intend to further our understanding of how the internet is becoming 'more local'. Here these will be merely sketched out, beginning with the notion of community.

The anthropologist Vered Amit (2002) has reviewed 'the trouble with community' as a theoretical concept. Amit argues that the term's strong emotional resonance makes it an ideal choice in public rhetoric, even though its empirical referent is seldom specified or, indeed, specifiable. Amit cautions that expressions of community always 'require skeptical investigation rather than providing a ready-made social unit upon which to hang analysis' (2002: 14). Relying on emotionally-charged, bounded notions such as community (or diaspora, nation, ethnic group, etc.) is unwise, she adds, for there are numerous sets of social relations that cannot be brought under these banners. Such sets include neighbours, co-workers and leisure partners—people who nevertheless may share 'a sense of contextual fellowship' that can be 'partial, ephemeral, specific to and dependent on particular contexts and activities' (Rapport and Amit, 2002: 5). Countering the often heard idea that community remains a valid term because it is a notion dear to millions of people around the world, Amit urges us not to conflate cultural categories with actual social groups.

This point has strong implications for the study of internet activism and other forms of social mobilization, since 'the assignation of membership in a particular cultural category does not tell us, in itself, which categories will actually be drawn on for the mobilization of social relations' (Amit, 2002: 18). For example, members of a local organizing committee may assure a researcher that all revellers at a street party are one 'community'. It does not follow that the same set of people will mobilize against the building of an airport in their vicinity. In sum, community merits attention as a polymorphous folk notion widely used both online and offline, but as an analytical concept with an identifiable empirical referent it is of little use. As one internet researcher aptly put it during a recent discussion on the ontological status of community: 'Fears of goblins do not goblins make' (Godard, 2006).

Turning now to 'network', in the 1950s and 1960s this notion appeared to offer anthropologists a way out of the entrapments of structural functionalism (Sanjek, 1996). By following individuals across social fields they hoped to be able to capture the open-ended nature of social life, particularly in the urban settings where numerous anthropologists were now finding themselves (Amit, 2007; Mitchell, 1969). However, they came to an impasse as they pursued ever more systematic 'morphological calculations' within increasingly small units of analysis (Amit, 2007). Consequently, social network analysis was all but abandoned by social anthropologists by the 1970s (Sanjek, 1996).

Interestingly it was precisely in the 1970s, as computers became more widely available, that social network analysis became popular with other social scientists (Freeman, 2007). One milestone was Granovetter's (1973) 'strength of weak ties' thesis, in which he showed that jobseekers in Boston found their 'weak' connections (e.g. with friends of friends) to be more useful in the job market than the 'strong' bonds of close friendship and kinship. This work helped to popularize social network analysis among North-American sociologists and economists (Knox et al., 2006). Today, social network analysis is used in a vast range of research areas, including mental illness, the spread of diseases and information, the sociology of organizations and internet studies (Freeman, 2007).

In recent years we have seen renewed anthropological interest in social networks, but anthropologists are still wary of this concept. Thus, in a recent review article, Knox et al. (2006) warn about the lack of critical reflection on key social network analysis notions such as 'whole network'. While whole populations are extremely difficult to research, drawing an arbitrary boundary around the network to be investigated in order to overcome this problem (e.g. by limiting the study to children in a few school classes) contradicts the fundamental idea of networks being unbounded and cutting across enduring groupings and organizations. In a similar vein, Amit (2007) urges us to reclaim the original promise of network as a notion that offers researchers the freedom to explore interpersonal links without any prior assumptions about what kinds of links or collectivities are more worthy of study (Hannerz, 1980).

However, this article is concerned here not with the limitations or potentialities of community and network as theoretical concepts, rather with their unchallenged dominance in the local internet studies literature. This paradigmatic dominance blinkers our view of the ongoing adoption of internet technologies by local authorities, companies and residents around the globe. It argues that one effective way to loosen the hold of community and network is to broaden our sociality lexicon. In the sections that follow, such broadening is carried out in two directions. First, by introducing a set of field theoretical notions such as 'social field', 'social drama', 'field law' and 'arena'; and second, by conceiving of sociality as being inherently plural and context-dependent (Amit and Rapport, 2002; Jean-Klein, 2003), rather than in the overly general terms that we find in the existing new media literature, e.g. 'community sociality' versus 'network sociality' (Wittel, 2001, see below).

Field, Network and Interaction

One available exit from the community/network bottleneck is the Habermasian concept of 'public sphere'. Holub defines the public sphere as:

> *An arena, independent of government [and market] . . . which is dedicated to rational debate and which is both accessible to entry and open to inspection by the citizenry. It is here . . . that public opinion is formed. (Holub, in Webster, 1995: 101–2)*

Despite Habermas' insistence that the public sphere (*Öffentlichkeit*) was a phase in European history, not a universal phenomenon, most new media scholars have used it as a normative, democratic ideal that all modern societies should aspire towards (Benson, 2007; Chadwick, 2006). Thus, Dahlberg (2001) has evaluated the citizen-led initiative Minnesota e-Democracy, built around an email list forum, against five predefined public sphere criteria: autonomy from state and market, reciprocal critique, reflexivity, sincerity and discursive inclusion. The problem with Dahlberg's strategy is that it prescribes what counts as a domain worthy of investigation. As with community, public sphere is used both as a 'rhetorical token' (Benson, 2007: 3) and a normative notion that guides research away from what is and towards what ought to be. A further difficulty with this concept is that its advocates, starting with Habermas (1989), have failed to explore how public spheres are internally differentiated (Peters, 2002, following Benson, 2007).

> *[W]hatever its qualities, any public sphere is necessarily a socially organised field, with characteristic lines of division, relationships of force, and other constitutive features. (Calhoun, 1992, quoted in Benson, 2007: 4)*

In view of these difficulties with public sphere, this article wishes to propose instead the concept of 'social field' as one possible way of overcoming the community/network impasse.[2] Put simply, a social field is a domain of practice in which social agents compete and cooperate over the same public rewards and prizes (Martin, 2003). One advantage of 'field' is that it is a neutral, technical term lacking the normative idealism of both public sphere and community. Field theorists have developed a sophisticated vocabulary that is being recruited increasingly to the study of media (Benson, 2007; Benson and Neveu, 2005; Couldry, 2007; Hesmondhalgh, 2006; Peterson, 2003). More pertinent to the case at hand, field theory offers us a framework with which to analyse the internet-mediated relations between local authorities and residents by treating these two parties not as discrete entities, but relationally, as two sectors of a porous, conflict-prone 'field of residential affairs' (see Epstein, 1958;Venkatesh, 2003).

Today we associate field theory with Pierre Bourdieu (1993, 1996), yet this theory has a far longer history originating in physics and Gestalt psychology (Martin, 2003). Bourdieu was critical of social network analysis for what he regarded as its naive commitment to interaction as the basis of human life, developing his field theory in opposition to social network analysis. He argued that by concentrating on people's visible interactions and ties, social network analysis practitioners fail to grasp the invisible network of objective relations binding human agents within a common cultural space (e.g. France) and its fields of practice (art, sociology, photography, etc.). For Bourdieu, social network analysis conflates structure with interaction, exaggerating the importance of 'social capital', i.e. the capital that accrues from social connections, while neglecting other species of capital such as cultural and symbolic capital (Knox et al., 2006). For example, two Parisian artists who have never met may possess similar amounts of symbolic capital (prestige, renown, etc.) and occupy neighbouring positions within the field of art. In Bourdieu's field theory, it is the agents' relative positions and amounts of field-specific capital that matter, not with whom they interact.

Although this article draws partly on Bourdieu's field lexicon, it finds his dismissal of interaction unhelpful on two accounts. First, it is hard to envisage how one could study the internet without considering its interactivity, e.g. the ease with which mailing list users can reply to posts (on media and interaction, see Thompson, 1995). Second, Bourdieu's opposition to the interactionism of social network analysis conceals the fact that within the social network analysis tradition there has always existed a tension between its interactionist (or connectionist) and its structural (or field) strands. While network analysts who adopt a contact approach indeed do map interactions and ties onto 'sociograms' (e.g. de Nooy, 2003), field-oriented social network analysis practitioners are more interested in 'structural relations usually opaque to actors' (Knox et al., 2006: 117).

In fact, a number of scholars have managed to graft successfully the notion of interaction onto their field analyses. Wouter de Nooy (2003: 323) has shown how the interactions of literary critics and female authors in the 1970s helped to establish and naturalize the category of 'feminist literature' within the Dutch literary field. For this theorist, a field of practice is shaped by objective power relations 'insofar as they influence the interaction within the field' (de Nooy, 2003: 323). Similarly, Victor Turner's (1974) reconstruction of a failed uprising in colonial Mexico, the Hidalgo Insurrection, tracks the interactions that took place in a series of 'arenas' over a period of six months. Turner understands the Insurrection to have been a 'social drama' unfolding across a rapidly shifting political field made up of the people, institutions and other resources mobilized to assist or thwart the rebellion (cf. Zald and McCarthy, 1988).

Suburban Frontiers

Victor Turner was a leading exponent of the Manchester School of Anthropology whose members were keenly interested in social change, particularly in the urbanizing regions of Central and Southern Africa during decolonization (Evens and Handelman, 2006). The situation there was curiously analogous to that faced today by suburbanites in many parts of the world. As with rural migrants in the booming urban areas of post-war Africa (Epstein, 1958), many present-day suburbanites find themselves in densely populated settlements with inadequate social and public facilities. The result is the mushrooming of ad hoc initiatives seeking to resolve the more pressing problems.

Newly-built suburbs are ideal settings in which to rethink our current dependency on community and network as the paradigmatic sociality notions in the study of internet localization. These are frontiers where newly-arrived people, technologies and ideas shape one another in unforeseeable ways. Over time new forms of residential sociality arise out of this flux as residents, private companies, local authorities and other human agents strive to 'produce locality' (Appadurai, 1996). In such unsettled conditions, any attempt at positing an existing 'local community' as impacted upon by a globalizing 'network logic' is doomed. New suburbs are particularly well suited to the study of emerging forms of residential sociality linked to 'banal activism'—the activism of seemingly mundane issues such as traffic congestion, waste disposal and petty crime.[3] Banal activism has been neglected by internet scholars, particularly in East and South-East Asia, where attention has centred on the 'serious' cyberactivism of the intelligentsia and the actual and potential uses of new digital technologies for political reform at the national level (e.g. Abbott, 2001, 2004; Gan et al., 2004; George, 2003; Hachigian, 2002; Hill and Sen, 2000; Holmes and Grieco, 2001; Loo, 2003; Nain, 2004; Rafael, 2003; Uimonen, 2003).

To date, the two main suburban internet studies to discuss residential sociality and banal activism provide useful entry points but are marred by their adherence to the community/network paradigm. The better-known study was conducted by Keith Hampton in the Toronto suburb of 'Netville' (a pseudonym) in 1997–9. Hampton combined survey research with participant observation in this new 'wired-up' locality to study the impact of the internet on 'local community' (Hampton and Wellman, 2003). He found that the internet helped Netville's settlers to make new friends and acquaintances both in their own immediate neighbourhoods and across the suburb, as well as being able to maintain older ties with geographically-dispersed friends and relatives. Residents with the most online contacts also tended to have the most offline contacts in the suburb. In accordance with Granovetter's 'strength of weak ties' dictum, local residents drew on their new contacts to make further contacts for information, socializing, mutual aid, etc., in the process increasing their local social capital. The web of social ties thus created also had important political implications, for it allowed residents to mobilize effectively when the developers attempted to withdraw the very technologies that had facilitated the collective production of sociality (Hampton, 2003; Hampton and Wellman, 2003).

A more recent study was conducted by Yael Levanon in the Tel Aviv suburbs of Ramat Beit-Shemesh and Modiin: the former settled by orthodox Jews, the latter by both religious and secular families. Levanon's starting point was, like Hampton's, the North-American literature on the reported decline in community social capital (Putnam, 1995; see also Putnam, 2000). Her aim was to study 'community networking' and its effect on local ties. On the basis of a questionnaire delivered to users of two local mailing lists, Mesch and Levanon (2003) argue that the internet allowed residents to find like-minded others across their suburb with whom to exchange information, socialize and cooperate—a finding echoing the Netville study. Another similarity was the use of the internet for banal activism: in the Israeli case, to oppose the building of a new shopping mall that would open on Saturdays and offer non-kosher food. Yet in contrast to their North-American counterparts, the Tel Aviv settlers had little need for the internet at the immediate neighbourhood level, for in Israel the neighbourhood remains a fulcrum of sociality.

These two studies further our understanding of internet localization in suburban settings in a number of ways. First, they point at cross-cultural similarities as well as contrasts in the internet-shaped making of suburban socialities. In both countries, suburban families with young children and dual-career parents are driven by the imperative to find and maintain a social environment conducive to family-building and class reproduction (see Miller, 1995); an imperative that shapes their use of internet technologies. However, the specific 'banal' issues that matter to residents can vary greatly from one locale to another, even within the same country. For example, plans to build a non-kosher restaurant were resisted by orthodox, not secular, Jews in suburban Tel Aviv. Second, the two studies demonstrate the continued usefulness of Granovetter's theory of weak ties in contexts other than Boston's 1970s job market (see also Amit, 2007; Haythornthwaite, 1998), enabling their authors to correct the overemphasis on strong, affective ties found in the community informatics literature (Hampton, 2003). Third, these studies shed light on the critical importance of two specific 'internet affordances' (Wellman et al., 2003), namely its interactivity and asynchronicity, to suburban residents who are able to engage with local issues despite their work and childcare commitments.

However, these studies are not without their shortcomings. First, they are both examples of the connectionist strand of social network analysis discussed earlier. This weakens their explanatory power when it comes to structural or 'field' questions. Murali Venkatesh (2003: 344) has broached such field-related questions with reference to Hampton's Netville research and suggested, following Melucci (1996), that collective action is always tethered to relational structures (or fields) that constrain action, although 'breakthrough social agency is always possible'. However, this relational line of inquiry is not pursued in Hampton's own work. Furthermore, both studies are caught up in the community/network semantic tangle, for example, by making contradictory use of the term 'community'. Thus, in Mesch and Levanon's (2003) analysis, community is used in places to refer to a pre-existing, unspecified collectivity ('the local community'), and in others to the future outcome of an ongoing effort ('community-building'), yet still in others to the suburb in its entirety ('the extended community') as opposed to the neighbourhood. As was argued earlier, community is a vague notion favoured in public rhetoric, not a sharp analytical tool with an identifiable empirical object. Amit puts it well:

> *Invocations of community . . . do not present analysts with clear-cut groupings so much as signal* fields *of complex processes through which sociality is sought, rejected, argued over, realized, interpreted, exploited or enforced. (2002: 14; emphasis added)*

This article turns now to an examination of such a 'field of complex processes' and its internet-related socialities by drawing on ethnographic research in suburban Malaysia.

A Field of Residential Affairs

Subang Jaya and its twin township, USJ, form a largely middle-income suburb of Kuala Lumpur, Malaysia. In 1998 the Subang Jaya municipality as a whole had an official population of 480,000, consisting of some 60 percent Chinese, 25 percent Malays and 15 percent Indians and 'Others'—predominantly immigrant workers from poorer Asian countries.[4] The local Creole is Malaysian English (see Nair-Venugopal, 2001). Anthropological fieldwork in Subang Jaya (mostly in USJ) was conducted for 12 months in 2003 to 2004, followed by intermittent online research since the present author's return to Britain. The aim was to study whether the internet was making any significant difference to the governance of this multiethnic locality. Subang Jaya-USJ is renowned in Malaysian information and communication technology (ICT) policy circles for its rich diversity of 'e-community' initiatives, ranging from a federally-funded 'smart township' project to a municipal 'cybermosque' and multimedia libraries to a self-funded residents' web forum, among numerous other projects. It was this vibrant internet scene that attracted this author to the locality.

Following the same suburban imperative that drove the Tel Aviv and Toronto suburban settlers described in the previous section, most Subang Jaya residents arrived in the 1990s in search of a green, safe environment to raise their young families while pursuing careers in the private sector. The omens were auspicious for the award-winning township dotted with small neighbourhood parks (*padang*). Yet as the decade drew to a close, familiar urban problems that incomers thought they had left behind began to surface, including traffic congestion, petty crime, degraded civic amenities and a shortage of Chinese schools. As a result of growing discontent, a manner of banal activism emerged around these issues in the late 1990s, one in which the internet played an important part.

It would be entirely possible to discuss Subang Jaya's residential activism in its own right, as its own 'field of striving' (Martin, 2003), but it is perhaps more productive to cast the net wider so as to encompass the local authorities and their grass roots initiatives (see Epstein, 1958). This is because the activities of local residents, politicians and municipal staff are inextricably entwined. To this end, this article will examine Subang Jaya's 'field of residential affairs': that is, the field of organized striving in which residents, politicians, municipal staff, journalists and other social agents compete and cooperate over matters of concern to local residents, often by means of the internet. In addition to bringing under one analytical umbrella both residents and local authorities, the coined phrase 'residential affairs' has the virtue of avoiding the sort of tacit normative commitment to a cohesive 'community' found in formulations such as 'community building'.

An invisible line divides Subang Jaya's field of residential affairs into two main sectors: a nongovernmental sector led by internet activists, and a governmental sector led by the local authorities. The 'fundamental laws' (Bourdieu, 1991) of the field differ from one sector to the other. Thus, residents wishing to amass symbolic capital are expected to volunteer their valuable time freely for the greater good of 'the community'. This 'interest in disinterestedness' (Bourdieu, 1993, 1998) can be described as the unwritten law of selfless volunteerism. In contrast, agents from the governmental field sector must abide by a different fundamental law, *turun padang*—Malay for 'to go down to the ground'. To solve local issues, however banal they may seem to an outsider, local politicians and civil servants must first go down to the ground. It is only by doing this, the belief goes, that powerful outsiders will be able to resolve concrete issues, thereby gaining residents' trust and support.

Field Socialities

Fields of striving do not necessarily exhibit a homogenous 'field sociality'. This is certainly the case with Subang Jaya's field of residential affairs, where a plurality of socialities has arisen over time. The perils of reducing sociality to totalizing categories such as 'field sociality' or 'community sociality' are apparent in Andreas Wittel's (2001) discussion of the new media industry in London. Wittel distinguishes two broad kinds of sociality: community versus network sociality. Community sociality is the pre-modern, sluggish sociality of physically-localized collectivities. By contrast, network sociality is fast-paced and based on fleeting, instrumental encounters (e.g. speed dating) with a large set of 'contacts'. Young urban professionals working in new media industries epitomize this latter form of sociality, which Wittel (following Castells, 1996) regards as the defining sociality of our era.

Wittel's notion of network sociality glosses over notable differences in how people interact with one another within the same field of practice, e.g. the new media industry in London. Surely the quality of a social interaction within a speed-dating session differs markedly from that in the office canteen or in a board meeting? How do media industry workers in London navigate these different micro-social settings while pursuing their goals (advancing their careers, socializing, mating, etc.)? These questions cannot be answered unless we develop a more nuanced understanding of sociality.

This section unpacks the notion of sociality ethnographically, suggesting that it cannot be reduced to a community/network binary. Instead, it suggests that researchers need to approach this question with an open mind, with the expectation that sociality may take on plural forms even within a single universe of practice. The three distinct forms of field sociality to be discussed are committee sociality, patrol sociality and web forum sociality (other emergent forms that cannot be discussed here would include ritual sociality, street party sociality and sporting sociality). It will outline their specificities in terms of the type of interaction, mode of discourse and field articulations. It is suggested that it is only through these fine-grained distinctions derived from empirical research on the ground that we can begin to theorize the elusive relationship between internet usage and the emergence of new forms of residential sociality.

Committee Sociality

Despite the prominence of committees within social and political activism worldwide, 'the sociality of committees themselves has remained a largely unseen ethnographic object' (Jean-Klein, 2003: 557). The account of committee sociality presented here is based on participant observation at a number of meetings in Subang Jaya and USJ. The Jawatankuasa Penduduk (JKP; residents' committees system) was launched in February 2001 by the municipal council. The then council president, Ahmad Fuad, linked the pioneering scheme to Local Agenda 21, a United Nations programme aimed at improving local governance around the globe. Fuad used the public rhetoric of community to call on 'the involvement of the community as a whole' in the new initiative. In practice, however, most committee members are political appointees drawn from the subfield of residential activism. Wishing to 'delegate some authority to the people', Fuad launched a plan to allocate RM100,000 (US$29,000) annually to each of the 24 committees 'for small and immediate projects' such as drain repairs or tree-cutting (Yeoh, 2005).

The residents' committee sociality is the co-present, synchronous sociality of monthly meetings devoted to discussing local issues (cf. Jean-Klein, 2003). Meetings are held at night in air-conditioned rooms and attendance is restricted to committee members and their guests. The bodily orientation is primarily face-to-face, although this depends, of course, on interlocutors' seating positions relative to one another (Pink, nd). Food and drinks are usually provided and consumed in the room during the break. These 'ephemeral items . . . live on in the form of the social relations that they produce, and which are in turn responsible for reproducing the comestibles' (Gell, 1986: 112). The discourse is oral, polylogical and gesturally rich but also mediated by texts (e.g. agenda, emails, letters, websites, etc.), that is, meetings are 'literacy events' (Street, 2000). Albeit a largely sedentary activity, attendants undertake a metaphorical journey whose itinerary is the agenda. It is the task of the chair to 'bring to order' participants who stray too far from this discursive itinerary (Pink, nd). These sessions are deliberative in that participants aim to reach consensual decisions (see http://www.thefreedictionary.com/deliberate). Even though meetings are held behind closed doors, minutes are in principle open to all local residents for inspection.

As regards the committees' articulation with the rest of the field of residential affairs, these hybrid organizations are located on the border between the field's governmental and non-governmental sectors. This ambiguous location is a perennial source of friction. The predominant mood at these sessions is neither one of open conflict nor of fellowship. Instead, there prevails an atmosphere of frustration with the council's apparent inability to solve but a fraction of the myriad issues brought to the table. As one chair put it once: 'But what is it we're doing? We cannot just come to meetings!'

Patrol Sociality

This form of USJ sociality springs from a deep-seated fear of crime in the severely underpoliced suburb. In 1999, a neighbourhood watch committee was formed in the precinct of USJ18. Each member was entrusted with organizing night patrols for a single street. In its heyday, the scheme boasted 330 volunteer patrollers—virtually all of them middle-aged men—guarding the precinct's 536 houses. Night patrol sociality is the side-by-side (not face-to-face), outdoors sociality of the night beat, when volunteer patrollers take time out to walk the streets in pairs. Patrollers carry torches, whistles, batons or long sticks, mobile phones, pen and paper (Pengawasan Seijiran Neighbourhood Watch, nd). The discourse is oral, informal and agenda-free as well as gesturally poor on account of patrollers' collateral bodily orientation and the nocturnal conditions. Because of the physical and cognitive constraints of verbal discourse (Hutchins, 1995), the two partners take it in turns to carry forth the single conversational thread. In keeping with the dyadic nature of patrols, the discourse is dialogical rather than polylogical (see Mulkay, 1985). Unlike the discourse of committees, patrol discourse is non-deliberative, unrecorded and private.

Regarding its field articulations, patrolling is a result of the collaboration between local residents, the police and the National IT Council, which seed-funded a neighbourhood watch web portal under an e-community scheme. Despite this governmental connection, patrolling itself is largely free from the political tensions and frustrations that beset committee meetings. The mood is one of camaraderie and fellowship—the shoulder-to-shoulder fellowship of the beat. While committee members live off meagre extrinsic rewards (i.e. the occasional resolution of a pending issue), patrolling is both a rewarding activity in its own right as well as epitomizing the fundamental field law of selfless volunteerism (on the rewards of practice, see Warde, 2005).

Web Forum Sociality

A unique quality of social intercourse has evolved on Subang Jaya's thriving web portal, USJ.com.my. Founded in 1999, this portal is proudly independent from the government. At the time of writing, the USJ.com.my forums had clocked close to 15,000 threads (topics of discussion), 232,000 posts and had some 13,000 registered members. The main forum is devoted ostensibly to 'issues related to USJ Subang Jaya Community', although in practice it is open to any topic so long as participants tread carefully on 'sensitive' subjects such as Islam. Thread starters compete to attract posters to their own threads by choosing topical issues. In contrast to the gesturally-rich committee meetings, web forum sociality relies on emoticons and avatars to compensate for the reduced bodily cues of online communication. As noted in connection to the Israeli mailing lists, the asynchronicity of most internet communication allows busy suburbanites to participate at their own leisure. The discourse is polylogical, non-deliberative, recorded and public.

How does web forum sociality articulate with Subang Jaya's field of residential affairs? First, its independence from the state places the forum firmly in the non-governmental sector of the field. However, it is important, once again, not to conflate our folk and technical terms. If during calm periods the forum lives up to its name as a cordial informational market, the collective mood can change very rapidly from one of conviviality to another of confrontation and inflammatory language ('flaming'), sometimes leading to collective action. At such times the forum morphs into an arena. In Victor Turner's field theory, an arena is a 'bounded spatial unit in which precise, visible antagonists, individual or corporate, contend with one another for prizes and/or honour' (1974: 132–3). Arenas are 'explicit frames' in which 'nothing is left merely implied' and major decisions are taken in public view (1974: 134). With his notion of arena, Turner was distancing himself from game theory and other rational actor models which had dominated political anthropology since the 1960s. Turner emphasized that an arena is neither a marketplace nor a forum, although both can become an arena 'under appropriate field conditions' (1974: 134). These conditions have arisen a number of times throughout USJ.com.my's short history, for example, residents have been pitted against the local authorities over an issue of general concern.

Conclusion

The field of local internet studies appears to suffer from semantic agoraphobia—a fear of open semantic spaces. Yet researching local settings should not necessarily limit one's conceptual space to one or two familiar notions, especially if these are of questionable sociological value, as is the case with community. This is particularly noticeable in suburban studies, where a reliance on community and network is strangely at odds with a frontier-like scenario in which people, technologies and other cultural artefacts are co-producing new forms of residential sociality in unpredictable ways.

By drawing on the field theory lexicon of both Bourdieu and the Manchester School, this study was able to bring a set of concepts that lie partly outside the community/network paradigm (field, interaction, sociality, arena, etc.) to bear on the ethnographic analysis. Like Eric Klinenberg (2005) in his study of US youth media, it has found field theory to be 'a useful way of seeing patterns within a messy domain of social action' (Couldry, 2007: 211)—in the case presented here, the domain of local governance in a Malaysian suburb. This ground-up theoretical exploration (Hesmondhalgh and Toynbee, 2006) revealed not a homogenous 'field sociality', rather an internally differentiated field of striving with various forms of sociality distinguished by the nature of their interactions, discursive practices and field articulations. Subang Jaya residents and local authorities are not appropriating the internet wholesale. In common with countless people around the globe, they are appropriating internet technologies selectively for specific purposes (Miller and Slater, 2000) and within fields of self-organized striving. The present analysis suggests that the internet technologies which are adopted will depend not only on their cost and technical affordances (Wellman et al., 2003) but also on the adopting field's inner differentiation. Thus, all three initiatives analysed above have sought to foster online discourse and conviviality through web forums, but only USJ.com.my has succeeded. Such success can be attributed in part to this portal's symbolic location at the heart of the non-governmental field sector (a location that attracts civic-minded residents), and in part to the virtual lack of thematic restrictions, which gives it an edge over the neighbourhood watch forum, with its narrow crime remit. The analysis also showed that certain field socialities are more prone than others to fluctuations in their 'moods'. Careful diachronic attention to these mood swings can teach us about a field's uneven patterns of sociotechnological change and continuity, with some field niches enjoying better insulation from external pressures than others (Bourdieu, 1996; Epstein, 1958).

There was much ground that could not be covered in this brief discussion, including questions about internet localization and social identity, other internet technologies besides web forums, or the ongoing convergence of internet and mobile technologies. Further empirical research and ground-up theorization is needed on these matters, particularly in new suburban settlements, where internet-related socialities can be studied as they emerge and compared with analogous processes unfolding in other suburbs. Other approaches besides field theory may prove fruitful as well (e.g. actor–network theory, structuration theory and material culture studies), especially if they are undertaken in the spirit of openness towards new forms of sociality that is the hallmark of all frontiers.

About the Author

John Postill is a senior lecturer in media at Sheffield Hallam University. He is the author of Media and Nation Building (Berghahn, 2006) and Grounding the Internet (Berghahn, forthcoming). He has published widely on the anthropology of media and is the founder of the EASA Media Anthropology Network.

Address: Sheffield Hallam University, Unit 9, Science Park, Howard Street, Sheffield S1 1WB, UK. [email: j.postill@shu.ac.uk]

Acknowledgment

This research was undertaken while a research fellow at the University of Bremen in 2002–2005. The project was part of Netcultures, an international comparative study funded by the Volkswagen Foundation.

Notes

1. For reviews of this literature see Loader and Keeble (2004), Pigg and Crank (2004) and Taylor (2004).
2. A variant of field theory is being used as the theoretical thrust by a media research group at Goldsmiths College, University of London, see http://www.goldsmiths.ac.uk/media-research-programme/

3. Personal communication from Alexander T. Smith, 22 May 2006, who independently coined the term 'banal activism' after anthropological fieldwork among Conservative Party activists in Scotland.
4. Personal communication from Arpah Bt. Abdul Razak, Subang Jaya Municipal Council, May 2003.

References

Abbott, J.P. (2001) 'Democracy@internet.asia? The Challenges to the Emancipatory Potential of the Net: Lessons from China and Malaysia', *Third World Quarterly* 22(1): 99–114.

Abbott, J.P. (2004) 'The Internet, *Reformasi* and Democratisation in Malaysia', in E.T. Gomez (ed.) *The State of Malaysia: Ethnicity, Equity and Reform*, pp. 79–104. London: Routledge.

Amit, V. (2002) 'Anthropology and Community: Some Opening Notes', in V. Amit and N. Rapport, *The Trouble with Community*, pp. 13–25. London: Pluto Press.

Amit, V. (2007) 'Globalization through "Weak Ties": A Study of Transnational Networks Among Mobile Professionals', in V. Amit (ed.) *Going First Class? New Approaches to Privileged Travel and Movement*, pp. 53–71. Oxford: Berghahn.

Amit, V. and N. Rapport (2002) *The Trouble with Community*. London: Pluto Press.

Appadurai, A. (1996) 'The Production of Locality', in *Modernity at Large*, pp. 178–99. Minneapolis, MN: University of Minnesota Press.

Benson, R. (2007) 'After Habermas: The Revival of a Macro-sociology of Media', paper presented at the American Sociological Association Annual Conference, NewYork, 11 August, URL (consulted December 2007):
http://steinhardt.nyu.edu/scmsAdmin/uploads/000/671/Benson_ASA.pdf

Benson, R. and E. Neveu (eds) (2005) *Bourdieu and the Journalistic Field*. Cambridge: Polity Press.

Bourdieu, P. (1991) 'The Peculiar History of Scientific Reason', *Sociological Forum* 6(1): 3–26.

Bourdieu, P. (1993) *The Field of Cultural Production*. Cambridge: Polity Press.

Bourdieu, P. (1996) *The Rules of Art: Genesis and Structure of the Literary Field*. Cambridge: Polity Press.

Bourdieu, P. (1998) *Practica Reason*. Oxford: Polity Press.

Calhoun, C. (1992) 'Introduction: Habermas and the Public Sphere', in C. Calhoun (ed.) *Habermas and the Public Sphere*, pp. 1–48. Cambridge, MA: MIT Press.

Castells, M. (1996) *The Rise of the Network Society*. London: Blackwell.

Castells, M. (2001) *The Internet Galaxy: Reflections on the Internet, Business, and Society*. New York: Oxford University Press.

Chadwick, A. (2006) *Internet Politics: States, Citizens, and New Communication Technologies*. Oxford: Oxford University Press.

Couldry, N. (2007) 'Bourdieu and the Media: The Promise and Limits of Field Theory (Review of Benson and Neveu, 2005)', *Theory and Society* 36(2): 209–213.

Dahlberg, L. (2001) 'Extending the Public Sphere through Cyberspace: The Case of Minnesota e-Democracy', *First Monday* 6(3), URL (consulted December 2007): http://www.firstmonday.org/issues/issue6_3/dahlberg/

Davies, W. and J. Crabtree (2004) 'Invisible Villages: Technolocalism and Community Renewal', *Renewal* 12(1), URL (consulted January 2008): http://www.renewal.org.uk/issues/2004%20 Voulme%2012/Invisible%20Villages.htm

de Nooy, W. (2003) 'Fields and Networks: Correspondence Analysis and Social Network Analysis in the Framework of Field Theory', *Poetics* 31(5): 305–27.

Epstein, A.L. (1958) *Politics in an Urban African Community*. Manchester: Manchester University Press.

Evens, T.M.S. and D. Handelman (eds) (2006) *The Manchester School: Practice and Ethnographic Praxis in Anthropology*. Oxford: Berghahn.

Freeman, L. (2007) 'The Study of Social Networks', URL (consulted December 2007): http://www.insna.org/INSNA/na_inf.html

Gan, S., J. Gomez and J. Uwe (eds) (2004) *Asian Cyberactivism: Freedom of Expression and Media Censorship*. Bangkok: Friedrich Naumann Foundation.

Gell, A. (1986) 'Newcomers to the Worlds of Goods: Consumption among the Muria Gonds', in A. Appadurai (ed.) *The Social Life of Things: Commodities in Cultural Perspective*, pp. 110–38. Cambridge: Cambridge University Press.

George, C. (2003) 'The Internet and the Narrow Tailoring Dilemma for "Asian" Democracies', *Communication Review* 6(3): 247–68.

Godard, E. (2006) 'Community and Our Conceptual Lexicon', URL (consulted December 2007): http://listserv.aoir.org/pipermail/air-l-aoir.org/2006-August/010366.html

Granovetter, M. (1973) 'The Strength of Weak Ties', *American Journal of Sociology* 78(6): 1360–80.

Gurstein, M. (2004) 'Editorial: Welcome to the Journal of Community Informatics', *Journal of Community Informatics* 1(1), URL (consulted December 2007): http://ci-journal.net/index.php/ciej/article/view/193/148

Gurstein, M., M.J. Menou and S. Stafeev (eds) (2003) *Community Networking and Community Informatics: Prospects, Approaches, Instruments*. St Petersburg: Centre of Community Networking and Information Policy Studies.

Habermas, J. (1989) 'The Structural Transformation of the Public Sphere. An Inquiry into a Category of Bourgeois Society'. Cambridge, MA: The MIT Press.

Hachigian, N. (2002) 'The Internet and Power in One-party East Asian States', *Washington Quarterly* 25(3): 41–58.

Hampton, K.N. (2003) 'Grieving for a Lost Network: Collective Action in a Wired Suburb', *The Information Society* 19(5): 417–28.

Hampton, K.N. and B. Wellman (2003) 'Neighboring in Netville: How the Internet Supports Community and Social Capital in a Wired Suburb', *City and Community* 2(3): 277–311.

Hannerz, U. (1980) *Exploring the City: Inquiries Toward an Urban Anthropology*. New York: Columbia University.

Haythornthwaite, C. (1998) 'A Social Network Study of the Growth of Community Among Distance Learners', *Information Research* 4(1), URL (consulted December 2007): http://informationr.net/ir/4-1/paper49.html

Hesmondhalgh, D. (2006) 'Bourdieu, the Media and Cultural Production', *Media, Culture and Society* 28(2): 211–31.

Hesmondhalgh, D. and J. Toynbee (2006) 'A Successful Conference on Media Change and Social Theory', *CRESC News* 4, December, URL (consulted December 2007): http://www.cresc.ac.uk/publications/documents/CRESCNewsIV.pdf

Hill, D. and K. Sen (2000) 'The Internet in Indonesia's New Democracy', in P. Ferdinand (ed.) *The Internet, Democracy and Democratization*, pp. 119–36. London: Frank Cass.

Hine, C. (2000) *Virtual Ethnography*. London: Sage.

Holmes, L. and M. Grieco (2001) 'The Internet, Email, and the Malaysian Political Crisis: The Power of Transparency', *Asia Pacific Business Review* 8(2): 59–72.

Hutchins, E. (1995) *Cognition in the Wild*. Cambridge, MA: MIT Press.

Jean-Klein, I. (2003) 'Into Committees, Out of the House? Familiar Forms in the Organization of Palestinian Committee Activism during the First Intifada', *American Ethnologist* 30(4): 556–77.

Klinenberg, E. (2005) 'Channeling into the Journalistic Field: Youth Activism and the Media Justice Movement', in R. Benson and E. Neveu (eds) *Pierre Bourdieu and the Journalistic Field*, pp. 174–92. Cambridge: Polity Press.

Knox, H., M. Savage and P. Harvey (2006), 'Social Networks and Spatial Relations: Networks as Method, Metaphor and Form', *Economy and Society* 35(1): 113–40.

Loader, B.D. and L. Keeble (2004) *Challenging the Digital Divide? A Literature Review of Community Informatics Initiatives*. York: Joseph Rowntree Foundation/YPS.

Loo, E. (2003) 'Opening Windows to 'e-Democracy' in Malaysia', paper presented at Hawaii International Conference on Arts and Humanities, Honolulu, 12–15 January, URL (consulted December 2007): http://www.hichumanities.org/AHproceedings/Eric%20Loo. pdf

Martin, J.L. (2003) 'What Is Field Theory?', *American Journal of Sociology* 109(1): 1–49.

Melucci, A. (1996) *Challenging Codes: Collective Action in the Information Age*. Cambridge: Cambridge University Press.

Mesch, G.S. and Y. Levanon (2003) 'Community Networking and Locally based Social Ties in Two Suburban Locations', *City and Community* 2(5): 335–52.

Miller, L. (1995) 'Family Togetherness and the Suburban Ideal', *Sociological Forum* 10(3): 393–418.

Miller, D. and D. Slater (2000) *The Internet: An Ethnographic Approach*. Oxford: Berg.

Mitchell, J.C. (ed.) (1969) *Social Networks in Urban Situations*. Manchester: Manchester University Press.

Mulkay, M. (1985) *The Word and the World*. London: Allen and Unwin.

Nain, Z. (2004) 'New Technologies and the Future of the Media in Malaysia', in *Communicating the Future: Proceedings from the National Conference on the Future of Media in a Knowledge Society. Rights, Responsibilities and Risks*, pp. 101–16. Kuala Lumpur: Kinibooks.

Nair-Venugopal, S. (2001) *Language Choice and Communication in Malaysian Business*. Bangi: Penerbit UKM.

Pahl, R. (2005) 'Are all Communities Communities in the Mind?', *Sociological Review* 53(4): 621–40.

Pengawasan Seijiran Neighbourhood Watch (nd) 'About Night Patrol', URL (consulted December 2007): http://usj18.nwatch.net.my/article.cfm?id=68

Peters, B. (2002) 'On Public Deliberation and Public Culture: Reflections on the Public Sphere', Working Paper 7/97, Institut für Interkulturelle und Internationale Studien (InIIS), Universität Bremen, Germany.

Peterson, M.A. (2003) *Anthropology and Mass Communication: Myth and Media in the New Millennium*. Oxford: Berghahn.

Pigg, K.E. and L.D. Crank (2004) 'Building Community Social Capital: The Potential and Promise of Information and Communications Technologies', *Journal of Community Informatics* 1(1), URL (consulted December 2007): http://ci-journal.net/misc/JoCI%20 Vol1%20Issue1.pdf

Pink, S. (nd) '"It Brought us Together": from Communities to Activist Socialities', unpublished paper. Loughborough: Loughborough University, Loughborough.

Putnam, D.R. (1995) 'Tuning In, Tuning Out: The Strange Disappearance of Social Capital in America', *Political Science and Politics* 28(4): 664–71.

Putnam, D.R. (2000) *Bowling Alone: The Collapse and Revival of American Community*. New York: Simon and Schuster.

Rafael, V. (2003) 'The Cell Phone and the Crowd: Messianic Politics in the Contemporary Philippines', *Public Culture* 15(3): 399–425.

Rapport, N. and V. Amit (2002) 'Prologue: The Book's Questions', in *The Trouble with Community*, pp. 1–12. London: Pluto Press.

Sanjek, R. (1996) 'Network Analysis', in A. Barnard and J. Spencer (eds) *Encyclopedia of Social and Cultural Anthropology*, pp. 396–7. London: Routledge.

Street, B. (2000) 'Literacy Events and Literacy Practices', in M. Martin-Jones and K. Jones (eds) *Multilingual Literacies: Comparative Perspectives on Research and Practice*, pp. 17–29. Amsterdam: John Benjamin.

Taylor, W. (2004) *Community Informatics in Perspective*. Hershey, PA: Idea Group.

Thompson, J.B. (1995) *The Media and Modernity: A Social Theory of the Media*. Cambridge: Polity Press.

Turner, V.W. (1974) *Dramas, Fields and Metaphors: Symbolic Action in Human Society*. Ithaca, NY: Cornell University Press.

Uimonen, P. (2003) 'Mediated Management of Meaning: On-line Nation Building in Malaysia', *Global Networks* 3(3): 299–314.

Venkatesh, M. (2003) 'The Community Network Lifecycle: A Framework for Research and Action', *The Information Society* 19: 339–47.

Warde, A. (2005) 'Consumption and Theories of Practice', *Journal of Consumer Practice* 5(2): 131–53.

Webster, F. (1995) *Theories of the Information Society*. London: Routledge.

Wellman, B. and B. Leighton (1979) 'Networks, Neighborhoods and Communities', *Urban Affairs Quarterly* 14(3): 363–390.

Wellman, B., A. Quan-Hasse, J. Boase, W. Chen, K. Hampton, I.I. de Diaz and K. Miyata (2003) 'The Social Affordances of the Internet for Networked Individualism', *Journal of Computer-Mediated Communication* 8(3), URL (consulted December 2007): http://jcmc.indiana.edu/vol8/issue3/wellman.html

Wittel, A. (2001) 'Toward a Network Sociality', *Theory, Culture & Society* 18(6): 51–76.

Yeoh, P.C. (2005) 'The Jawatankuasa Penduduk (JKP) Covering Subang Jaya/USJ and Sunway', URL (consulted December 2007): http://nwatch.net.my/topicOpen.cfm?start=1&count=10&id=25960A05-1584-4BE5-875EF359762CD9D3&

Zald, M.N. and J.D. McCarthy (1988) *The Dynamics of Social Movements: Resource Mobilization, Social Control and Tactics*. Lanham, MD: University Press of America.

Multimodal Computer-Mediated Communication and Social Support among Older Chinese Internet Users

Bo Xie

Using the grounded theory approach, this ethnographic study examines the relationship between the use of three different modes of computer-mediated communication (CMC) and the types of social support exchanged within an online community of older Chinese (age 50 or older). The findings indicate that the voice chat room is best suited for companionship while the online forum is used primarily for informational support. The exchange of emotional and instrumental support is most likely to occur via instant messaging (IM). These findings suggest that each of these particular modes of CMC, due to its inherent features, may be specialized to sustain certain interactions and relationships. Yet, because older Chinese in this study employ multiple modes of CMC to interact and develop relationships, the overall relationships developed through CMC are multidimensional.

Introduction

Computer-mediated social support (CMSS) or social support exchanged through computer-mediated communication (CMC) has been a major topic of recent research on the social implications of information and communication technologies (ICTs). Earlier research on CMSS focused on online support groups that were formed via "older" modes of CMC such as Usenet newsgroups, electronic bulletin boards, and listserv/mailing lists that are characterized by text-based, asynchronous, and anonymous interactions (Herring, 2002; White & Dorman, 2001). Newer modes of CMC or "multimodal CMC" (Herring, 2002; Soukup, 2000) that combine text-, voice- and/or video-based interactions (e.g., instant messaging or IM) are beginning to receive attention in the literature (e.g., Bryant, Sanders-Jackson, & Smallwood, 2006; Hu, Wood, Smith, & Westbrook, 2004; Nastri, Peña, & Hancock, 2006).

Compared with the vast amount of CMC and CMSS research on many social groups, to date there is relatively little research on CMC and CMSS among the older population (with a few exceptions that focused on text-based CMC, which will be discussed below). Further, at least in the English literature, even less, if anything, is known about CMC and CMSS among the older population in developing countries. Yet, as many scholars have pointed out, the impact of ICTs greatly depends on individual motives, goals, lifestyles, and life stages (Bargh & McKenna, 2004; Haythornthwaite, 2001) and existing social and power structures and relations (DiMaggio, Hargittai, Neuman, & Robinson, 2001; Holloway & Valentine, 2003; S. Jones, 1999; Wilson & Peterson, 2002). Without empirical evidence, one cannot know for sure if the findings generated from one particular age group (e.g., college students or teenagers) in a particular setting (e.g., developed countries) can be generalized to other age groups (e.g., older adults) in another setting (e.g., developing countries).

As one of the first attempts to address these limitations of the literature, this study examines CMSS in an Internet community of older Chinese and explores the types of social support exchanged in three different modes of CMC: voice chat, forums, and IM.

The Aging and Internet Trends in China

Similar to many other countries around the world, the Chinese population is aging. The 2000 Chinese census data showed that in November 2000, 6.96 percent of the Chinese population (88 million) were age 65 or older (National Bureau of Statistics of the People's Republic of China, 2001). It is estimated that in 2007, approximately 7.9 percent of the Chinese population (104 million) were age 65 or older (Central Intelligence Agency, 2007). Another important recent development in China has been the wide spread adoption of the Internet. Between October 1997 and December 2006, the number of Chinese Internet users increased from 620,000 to 137 million (China Internet Network Information Center, 2007).

Although the percentage of Chinese Internet users above age 50 has been consistently lower than 5% of the total Chinese Internet population, because the size of the total Internet population has increased so much, the total number of older Chinese Internet users has also increased significantly. In June 1998, there were only 14,400 Chinese Internet users aged 50 or above. Yet by the end of 2006, more than 4.25 million older Chinese were using the Internet (China Internet Network Information Center, 2007).

Despite the rapid development of ICTs in China, studies of Internet use in China have primarily focused on the political and economic aspects of ICT adoption (Kluver & Chen, 2005; McMillan & Hwant, 2002). Many important areas (e.g., how ICTs can be used by older Chinese to form and sustain supportive social relationships) are significantly under studied. Yet studying the impact of ICTs on the social relationships and social support of older Chinese has unique importance in the context of contemporary China. This is because during the past several decades, the forces of industrialization, capitalization, and globalization have resulted in dramatic social changes in Chinese society, including the everyday lives of older Chinese (Bian, 2002; Ikels, 1996; Mjelde-Mossey, Chi, & Lou, 2006; Mjelde-Mossey & Walz, 2006; Perry & Selden, 2000; Sun, 2004; Tang & Parish, 2000; Walder, 1989; Warner, 2001; Yoon & Hendricks, 2006; Zhou & Hou, 1999). Research shows that recent social changes in China have had negative effects on the social support networks and relationships of older Chinese (Cheng & Chan, 2006; Chong, 2007; Joseph & Phillips, 1999; Lam, 2006; Lee & Hong-kin, 2005; Ng, Phillips, & Lee, 2002; Price & Fang, 2002). Within this context, an important question is: Can ICTs help older Chinese overcome these social changes to maintain supportive social relationships in later life?

Modes of Computer-Mediated Communication (CMC)

Earlier CMC theories argue that, compared with face-to-face communication, CMC has a number of limitations resulting from its inherent characteristics (for a recent review and critiques, see Walther, Gay, & Hancock, 2005). The information richness/media richness model (Daft & Lengel, 1984, 1986; Daft, Lengel, & Trevino, 1987) holds that text-based CMC is "lean" media that is more suitable for simple, straightforward tasks such as scheduling, while "rich" media such as face-to-face communication is better suited for complicated tasks that require intensive interactions and negotiations. Similarly, the social presence theory (Short, Williams, & Christie, 1976; Spears & Lea, 1992) and the social context cues theory (Kiesler, Siegel, & McGuire, 1984; Kiesler & Sproull, 1992; Sproull & Kiesler, 1991) suggest that the text-based CMC environment reduces or filters out physical and contextual social cues and, as such, makes it difficult, if not impossible, to develop close, intimate relationships. Other researchers, however, argue that the relative lack of visual, aural, and contextual cues in CMC (as compared with face-to-face interactions) could only slow down, rather than completely prevent, the development of intimate online relationships; over time, users would be able to overcome the relative lack of social contextual cues and develop supportive online social networks (Walther, 1992, 1995; Walther, Anderson, & Parks, 1994).

These earlier CMC theories targeted predominantly text-based modes of CMC (e.g., Usenet newsgroups, electronic bulletin boards, or listserv/mailing lists). With the development of ICTs, newer modes of CMC (e.g., voice chat, IM, 3D online games, and, more recently, social networking sites such as Facebook, YouTube, and Blogs) that allow the exchange of not only textual but also audio, visual, and/or graphical information are increasingly challenging earlier theories (boyd & Ellison, 2007; Ellison, Steinfield, & Lampe, 2007; Hu et al., 2004; Kim, Kim, Park, & Rice, 2007; Lange, 2007; Stefanone & Jang, 2007; Valkenburg & Peter, 2007). Compared with older modes of CMC that are generally anonymous, asynchronous, publicly accessible, group-based, and persistent (interactions are automatically documented unless users choose to delete), newer modes of CMC are more likely to be identifiable, synchronous, private, dyadic, and ephemeral (evidence of interaction vanishes by default unless users choose to save) (Herring, 2002). These different features of different modes of CMC might have different effects on online interactions and relationships and thus should be examined thoroughly (Herring, 2002; Soukup, 2000). Based on their empirical findings, Swickert and colleagues (2002) caution that the findings based on one particular mode of CMC may not be generalizable to interactions that occur using other modes of CMC.

Furthermore, CMC is by no means limited to one particular mode. In fact, several earlier studies reported that users employed additional modes (e.g., private email, postal mail, telephone, and face-to-face meeting) to communicate and develop relationships with those whom they first met via one particular mode of CMC (Kollock & Smith, 1996; Ogan, 1993; Parks & Floyd, 1996; Rheingold, 1993; Salem, Bogat, & Reid, 1997; Turner, Grube, & Meyers, 2001). More recent research on newer modes of CMC such as IM also suggests that frequent CMC promotes the desire to interact face-to-face (e.g., Hu et al., 2004). Therefore, it is important to pay attention to not only the individual but also the collective effects of different modes of CMC.

Computer-Mediated Social Support (CMSS)

Although the precise definition of social support varies from study to study, it is widely acknowledged that social support has multiple dimensions: emotional, instrumental, appraisal, and informational (Barrera, 1986; House, 1981; Vaux, 1988). Some researchers have cautioned that Internet use could decrease social ties and social support and increase social isolation (Kiesler & Kraut, 1999; Kraut et al., 1998; Nie, 2001; Sanders, Field, Diego, & Kalplan, 2000). Yet, a growing number of studies provide concrete evidence that CMC can indeed help form and maintain online relationships that can facilitate the exchange of social support (Constant, Sproull, & Kiesler, 1996; Eastin & LaRose, 2005; Ellison et al., 2007; Hu et al., 2004; Kim et al., 2007; Lange, 2007; Parks & Floyd, 1996; Ridings & Gefen, 2004; Salem et al., 1997; Shaw & Gant, 2002; Stefanone & Jang, 2007; Tichon & Shapiro, 2003; Turner et al., 2001; Valkenburg & Peter, 2007; Walther & Boyd, 2002; Wellman et al., 1996).

There is preliminary evidence that different modes of CMC may have different effects on perceived social support. For instance, a survey study by Swickert and colleagues (2002) found a "marginally significant" ($p = 0.08$) relationship between social support and interaction via IM and online game-playing. Yet the same study found no significant correlations between social support and interaction via other modes of CMC including electronic bulletin boards, chat rooms, MUDs, and email. Morgan and Cotton (2003) found that increased use of email, chat rooms, and IM—all of which serve communicative functions—was associated with decreased depressive symptoms, while increased use of online shopping, game-playing, and doing research—all of which serve non-communicative functions—was associated with increased depressive symptoms. In a study that explored CMSS in support-oriented Usenet newsgroups, researchers failed to find any significant association between online support seeking activity and perceived social support (Eastin & LaRose, 2005). However, the "online support seeking activity" in this study was measured in a way that lumped together various types of Internet activities (e.g., email, group listings, chat rooms, and discussion boards) and thus could not distinguish whether or how users exchange different types of social support through different types of Internet activities. In a study that compared the impact of IM and that of chat in public chat rooms on adolescent well-being, researchers found that IM was primarily used to interact with existing friends and facilitated the development of existing friendships. In comparison, chat in public chat rooms was more often used to form new relationships, and had no effect on the quality of existing friendships (Valkenburg & Peter, 2007). The findings of these studies suggest that, in studying CMSS, it is necessary to differentiate different modes of CMC because the latter may have different effects on CMSS.

CMSS in the Older Population

Currently, the majority of social research on the intersection of older adults and ICTs is limited to the individual level, and many important issues, including how CMC affects older adults' interpersonal relationships and social support networks, are significantly understudied (Xie, 2006a). The few available studies on this subject, however, have shown promising results. In an early study that was originally designed to explore older adults' interactions with computer games, researchers unexpectedly found that participants favored interacting with other people via computer networks, an unplanned activity, more than playing single-user computer games (Danowski & Sacks, 1980). A survey conducted among new members of the SeniorNet online community—hosted by American Online at the time—revealed that, although the most frequently reported reason for becoming involved with the online community was to gain "information access," the most popular activity in this online community was actually communicating with others via email, forums, and conferencing (Furlong, 1989). Another SeniorNet online community study reported that participants engaged in more online companionship, or interaction that is sought for purely social, enjoyable purposes, than supportive relationships that are often oriented toward solving a particular problem or achieving a specific goal (Wright, 2000). An ethnographic study of older Japanese's online interactions in a senior-oriented electronic mailing list community found that members of the online community were willing, and indeed encouraged by other members, to disclose personal information about their past experiences and present situations via CMC. Such self-disclosure greatly facilitated their development of both companionship and supportive relationships in the online community (Kanayama, 2003).

The Present Study

The handful of previous studies on—text-based—CMSS among older adults in developed countries provides preliminary knowledge about CMC among these older adults. Yet little is known about how older adults in developing countries may employ CMC for CMSS, which presents a pressing need for research. Due to the scarcity of available data on this subject, the researcher chose to use ethnographic methods—including semi-structured, open-ended interviewing and participant observation—to collect data. These methods can help to generate data that would be difficult to obtain through quantitative methods such as survey (Bernard, 1998). Grounded theory (Glaser & Strauss, 1967; Strauss & Corbin, 1998), a powerful tool widely used to analyze qualitative data and build theory that is grounded in data, was used to guide data analysis and theory building of this study.

Method

Research Site: The OldKids (*lao xiaohai*, a Chinese phrase that refers to active seniors and can be literally translated as "Old Kids") online community is maintained by a Shanghai-based organization that provides face-to-face computer training to older Chinese. First launched in early 2000, this online community now has over 10,000 registered members. It provides free CMC services including voice chat rooms, online forums, and IM (for a detailed description of the history and development of the OldKids organization and community, see Xie, 2005).

Interviewees: Interviewees were recruited from members of the OldKids community. A total of thirty-three older Chinese—all were retirees—were interviewed. Interviewee ages ranged from 50 to 79. Nineteen of them were female, and fourteen were male.

Interview Topics: Semi-structured open-ended interviews were conducted in May and October of 2004 (for a methodological discussion on the interview techniques used in this study, see Kazmer & Xie, in press). Based on the concepts and findings present in the existing literature, the initial interview guide, which was used to guide the first several interviews, covered a number of key topics; among them, the ones most relevant to the focus of this article included: participation and activities in the OldKids online community; experience of age-related stigma in the online community; subjects that are typically addressed—and *not* addressed or deliberately avoided—when interacting in the OldKids online community; and feelings about online relationships. Later interviews incorporated new topics emerged from the first several interviews. See the Data Analysis section below.

Participant Observation: Participant observation was conducted in the OldKids voice chat room and online forum during the period from May of 2004 to December of 2004. A non-active participant approach (i.e., "lurking") was taken to ensure that the normal patterns of online interactions would not be interrupted or changed by the presence of the researcher. This approach overcomes the major limitation of active participation: i.e., observer bias that may lead to a loss of validity (Fox & Roberts, 1999). Following Jones' (1994) guidelines, all information that may disclose participants' (online and offline) identities and locations has been removed to protect participants' privacy.

Data Analysis: Data analysis for this research project was guided by grounded theory (Glaser & Strauss, 1967; Strauss & Corbin, 1998), such that data collection and analysis occurred simultaneously to ensure the co-evolution of data and theory. Following the constant comparative method provided by Glaser and Strauss (1967), immediately after each interview, the first step was to write a short descriptive and analytic summary to record general impressions of the interview process and the interviewee. After each interview, audio data was transcribed into text and then translated into English as soon as possible. The text interview data were then coded and memoed on a computer using the qualitative data analysis software Atlas/Ti.

Microanalysis or "detailed line-by-line analysis" (Strauss & Corbin, 1998, p. 57) was conducted on the first several interviews to generate initial concepts, categories (higher level, more abstract concepts), and possible relationships among categories. Specifically, open coding was conducted to identify salient concepts and their properties (i.e., characteristics) and dimensions (i.e., "the range along which general properties of a category vary, giving specification to a category and variation to the theory"; Strauss & Corbin, 1998, p. 101). Once the identified concepts were beginning to merge into categories, axial coding was conducted to systematically explore the properties and dimensions of those categories, which led to the formation and further development of dense, related categories. Results of the microanalysis of the initial interviews guided the selection of remaining interviewees, a process called *theoretical sampling* or "sampling on the basis of emerging concepts, with the aim being to explore the dimensional range or varied conditions along which the properties of concepts vary" (Strauss & Corbin, 1998, p. 73).

In keeping with the grounded theory approach, results of this initial stage microanalysis were also used to improve the interview guide for the remaining interviews. Some topics that did not appear to be salient were dropped (e.g., stigma) and new, emerging salient themes added (e.g., seeking, providing, and receiving emotional and instrumental support via the voice chat room, and switching from one mode of CMC to another to continue the interaction). In this way, the researcher was able to detect, examine, and focus on emerging and important concepts and categories, which were further examined and tested during participant observation of the chat room and online forum. New concepts identified through participant observation were added and integrated into the interview guide for the remaining interviews (e.g., sharing personal feelings in the online forum).

While the microanalysis of the first several interviews focused on detecting new concepts and thus relied more on open coding than axial coding, the latter was used more heavily during the middle stage of this project. Open coding, however, was still conducted during the middle stage of this study to detect new concepts and categories. Thus during this stage the researcher was able to focus on developing and enriching existing categories and their properties, dimensions, and relationships while at the same time continuing to be sensitive to emerging concepts and categories.

The remaining stage of data analysis relied primarily on selective coding to integrate and refine categories to form theory. At this stage, commonalities and differences in categories were further sought, core or central categories were identified, and then explanations were formulated. This procedure allowed the researcher to cut down the original list of categories and to be more focused on categories that could better explain the underlying phenomena. Finally, after developing these explanations and refining the categories, the theory that could best interpret the data was written.

Key Concepts and Categories

Open coding of the interview data identified two sets of concepts: a) voice chat room use, IM use, and online forum use; and b) companionship, emotional support, and informational support. These concepts quickly merged into two categories: while the first set of concepts clearly fit into the category of *modes of CMC,* the second set of concepts was categorized as *types of social support.* Further analysis revealed interesting relationships among the concepts of these categories: while the voice chat room was primarily associated with the exchange of companionship, IM and the online forum featured the exchange of emotional and informational support, respectively. Thus three new categories were generated: voice chat room-companionship, IM-emotional support, and online forum-informational support.

Through axial coding, the properties, dimensions, and relationships of these three new categories were systematically explored. The voice chat room-companionship category was shown to characterize singing/talking, "nothing personal," and "having fun"/enjoyment, while the IM-emotional support category featured communication of intimate subjects, exchange of deep, personal feelings, and emotional support. Characteristics or properties of the online forum-informational support category included communication of computer-related subjects, "very general level" interactions, and learning/education.

Variations of these categories were also discovered: personal, sensitive subjects were occasionally discussed in both the voice chat room and online forum, in which cases emotional support was sought and provided. The exceptional cases thus suggest that these categories vary along the dimensions of general-sensitive talks and no-emotional-emotional support. Further, the overall relationship between the categories of modes of CMC and types of social support was by no means restricted by the boundaries (along the properties and dimensions) of the three new categories. Rather, because older Chinese in this study employed multiple modes of CMC to interact and develop relationships, the overall relationships developed and social support exchanged through CMC were multidimensional.

The following three sub-sections provide more details about these new categories.

Voice Chat-Companionship

Voice chat is especially popular among OldKids members because it does not require typing (the OldKids chat room also provides a text-based chat function; so members can type if they would like to, but they don't have to) and thus helps older Chinese bypass one of the main barriers of older Chinese's use of computers: inputting Chinese characters (for a detailed discussion about the major factors that inhibit older Chinese's computer typing/inputting, see Xie, 2005, p. 192). Most interestingly, a number of OldKids members have been creatively using the voice chat room to display their musical talent by singing in the voice chat room. By using special software, registered users can even play Karaoke and sing along with music. In fact, singing instead of talking has become the most common activity in the OldKids voice chat room. Although not everyone enjoys singing, for those who do the voice chat room has become the most frequently visited part of the OldKids website.

The data indicate that the primary reason for users to visit the OldKids chat room is to spend time together and have fun. Topics not directly related to singing or music are rarely addressed in the chat room. For instance, an active user of the voice chat room states:

> *In general, we go to the [voice] chat room to sing, to hear others to sing, or just to greet each other. Nothing personal. We don't normally talk about personal issues. We like to talk about happy things*—having fun is the primary purpose for us to visit this chat room.

Although OldKids members in general perceive and use the voice chat room primarily as a place for singing and having fun, there are exceptional cases where the voice chat room can be used to express deeper feelings or concerns for others. For instance, a member recalls how, one time, when a net-friend (*wang you*) whom he met online got sick, he used the online voice chat room in a unique way to show his caring and support:

> *One time, one net-friend had to go to a hospital to have a surgery, because the doctor suspected that she had cancer. Before she went to the hospital, she came to our voice chat room to say goodbye to us. It sounded like she was saying farewell to us. So I told her that I truly believed that she would be back [to the OldKids chat room] very soon. I also promised her that I would sing a song—"missing you so much" in the chat room for her everyday. I believe that if you are truly sincere, it will happen as you wish. So I had been singing that song for her in the chat room everyday, as I promised . . . As it turned out, she came back within a month! I was very happy.*

This story suggests that, when necessary, OldKids users are not hesitant or shy to use the voice chat room to provide emotional support to one another. Even if that support was not immediately received by the targeted receiver, it could still have positive effects on the support provider and other members of the chat room and therefore help create a friendly and supportive atmosphere in the chat room.

This member also tells a story about how the voice chat room has even facilitated the exchange of tangible aid between two OldKids members:

> *One net-friend, who lives in Wuhan, has a son working in Beijing. One time the son of this net-friend got really sick and had to be hospitalized. Because he was there alone, his mother was really worried that nobody would take care of him. So she asked a net-friend who lived in Beijing to take care of her son for her. The Beijing net-friend was very nice; she took care of the son just like taking care of her own child. When the son recovered, the Beijing net-friend took him to her house and logged into our OldKids voice chat room so that the son could talk to his mother and tell her that he recovered. It was so touching . . . That was really something. Even now when I think about it I can still feel it . . .*

In this case, although the exchange of instrumental support did not actually take place in the online world, the online chat room nevertheless clearly contributed to the exchange of instrumental support in the physical world, which generated emotional support in both the online and offline settings. However, it is important to point out that the cases where online voice chat room facilitates the exchange of emotional and instrumental support are likely to be the exception rather than the rule for the following reasons. First, during the 8-month period of participant observation in the voice chat room, no similar cases were observed. Rather, what was observed in the chat room was simply the seeking, providing, and receiving of companionship and enjoyment. Second, the exchange of emotional and especially instrumental support in the chat room was only detected in a few interviews. Although the stories reported by those interviewees evidently had involved a larger number of participants, without further evidence, it is difficult to draw conclusions on how many or who had actually been involved in the exchange of emotional or instrumental support and why. To answer these questions, it is necessary to conduct further research, which will be addressed in the Conclusion.

Online Forum-Informational Support

When interacting in the online forum, the most common subjects that members talk about are computer-related ones. In particular, members have found that the online forum is a great place to learn about computers and the Internet from peers—in other words, the online forum is used primarily to seek, provide, and receive informational support. As a member states:

> *There are many people online who are willing and eager to help you. If you don't know how to do something [on the computer], you can simply post your question in the online forum. Almost immediately, someone will come and teach you how to do it.*

Although it is not unusual for some people to discuss other subjects (e.g., family) in the online forum, those subjects are usually limited to "a very general level." As one OldKids member states:

> *Learning about computers is the primary subject we talk about in the online forums. We talk about computer knowledge, skills, and tips; we ask questions about computers and the Internet; and we share news about recent advances in Internet technology. We also greet each other during the holidays, including posting greeting cards in the forums. Sometimes we talk about our families too, but only at a very general level.*

Further exploration suggests that there are two main reasons for not talking about personal or sensitive subjects in the online forum: first, members realize that the discussions posted in the online forum are available to the public. Therefore, they generally avoid discussing personal or sensitive issues in the public forum. Second, most members of the OldKids online community are students or former students of one or several OldKids computer classes. Not surprisingly, they perceive the online forum as an extension of the (offline) OldKids computer classes where they can continue to learn about new ICTs (Xie, 2006a, 2006b, 2007a). The OldKids online forum, therefore, is used primarily for educational purposes.

Interestingly, although in general OldKids members discuss only computer-related subjects in the online forum, the observation in the OldKids online forum did identify one case where a very personal and sensitive subject was brought up and many older Chinese actively participated in the discussion: In August 9, 2004, a message was posted in an OldKids forum telling the story of a 75-year old man who has been the primary caregiver for his wife—who is now in a vegetative state—for 13 years. This man said in the post that he had been very lonely and depressed, and had been thinking about divorcing his wife and finding another woman to remarry. In the post, he asked: would this be legal? Would it be ethical? What should I do? What if my children have objections?

The man asked other older people for advice on an extremely complicated issue. And others responded actively: in the following four days, sixteen deliberately written messages were posted to the forum addressing various aspects of the situation. The discussion came to a halt when the discussion leader announced that, at the end of the month, OldKids and eastday.com (a Shanghai-based Internet service company) were going to co-organize an online discussion in an eastday.com maintained online chat room with the old man himself. On August 30, the online (text-based) discussion was held in an eastday.com online chat room as scheduled, and the old man—who was then introduced as "Old Q"—discussed his deep, personal feelings and concerns with other older adults over the Internet. After the two-hour online discussion with Old Q in this eastday.com chat room, OldKids members continued the discussion on the OldKids online forum until the end of 2004. Various aspects—legal, ethical, cross-generational conflicts, altruism versus selfishness, etc.—were brought up, discussed, and debated.

This "Old Q" case suggests that, although the OldKids online forum is primarily used to exchange computer-related informational support, it can also help participants seek, provide, and receive advice and support on extremely personal, sensitive, and complicated issues.

Instant Messaging (IM)-Emotional Support

Interviews with OldKids members revealed that older Chinese Internet users were much more likely to share deep, personal, and sensitive feelings and thoughts via IM than via voice chat or the online forum. An important reason is that, unlike the public voice chat room and online forum, IM allows private interaction, thus ensuring privacy. One member of the OldKids online community, who insists that net-friends are "just like brothers and sisters," gives this example:

> *We talk about personal and sensitive topics [online]. For example, one net-friend told me that her mother was very ill and that she was really depressed, because she's concerned about her mother's health, and because she had to spend most of her time taking care of her mother. She told me all about her situation and feelings.*

When further asked how he and his net-friends talked about those issues online, and if it was in the online forum or something else, he responded immediately:

> *No, no, no, no. Not posting messages in the forum. How could we do that in the forum? Everybody could see it. It was via IM, where we could type, voice chat, send pictures, and see each other through our digital video cameras.*

Another member agrees that the online forum is used for public discussions while IM is used for private conversations. Further, she points out that IM can be an effective tool to *continue* a conversation that was initiated in the online forum:

> *Sometimes we do talk about personal or sensitive subjects, but not in the online forums—we usually talk about those subjects via IM where we can have one-to-one private conversations . . . If, when we interact in an online forum, it looks like the conversation might go into a more personal direction, we would switch to IM to continue.*

This member gives an example of how, after a signal to have a private conversation was given and received in a public online forum, she and her net-friend immediately switched to IM to have a very personal conversation:

> *One year, on Mother's Day, a good net-friend of mine posted a message in an OldKids online forum to wish me happy Mother's Day. I posted a message to reply, in which I said, thanks for your greetings; but my own daughter didn't send me anything on this day. I'm unhappy. That net-friend then posted another message and suggested that, if I would like, we could have a conversation via IM, which we did . . . We had a really deep and personal conversation via IM. I would never discuss things like this in the online forums.*

These OldKids members' stories convincingly suggest that older Chinese Internet users not only have good understandings of the strengths and limitations of different modes of CMC but also are quite capable of taking advantage of the opportunities the technology provides while avoiding the weaknesses of the technology.

Discussion

This study reveals that companionship, information, and emotional support are the most common supportive activities among the older Chinese Internet users who participated in this study. The findings of this study seem to provide preliminary evidence that the positive association between social support and CMC as reported in many previous studies (that were conducted mostly among younger age groups in developed countries) could be generalized to the older population in developing countries such as China. Different from previous research that focused on text-based CMC, however, this study shows that different modes of CMC (voice chat room, online forum, and IM) are used by these older Chinese Internet users for different purposes and activities. As a result, the main types of social support exchanged within those online settings are also different.

Specifically, the voice chat room is primarily a place where participants spend time together and have fun chatting and especially singing. Older Chinese usually participate in the voice chat room to obtain a sense of belonging, social connectedness, and enjoyment. The voice chat room, therefore, facilitates the development of companionship relationships. Occasionally, however, the voice chat room is also used to provide emotional support for those experiencing life stresses. In other words, similar to findings reported in previous studies on older adults' relationship development via text-based modes of CMC (e.g., Kanayama, 2003; Wright, 2000), the voice chat room also facilitates the development of online companionship as well as supportive relationships among older adults. These findings suggest that CMC modes that are characterized by synchronous, group interaction, and public access, whether they are text- or voice-based and whether or not they allow anonymity, are well suited for social interaction and relational communication. However, because voice-based CMC helps overcome the typing/inputting barrier to older adults' but especially older Chinese's use of new ICTs (Xie, 2005), it may be a more efficient way for older adults to communicate and construct online relationships than text-based CMC.

While the voice chat room serves the purpose of constructing online companionship as well as supportive relationships, the online forum primarily provides informational support to aid participants to solve computer problems. This is due in large part to the inherent features of text-based CMC, especially asynchronous interaction, which allows deliberate thinking and writing that are required for solving complex problems. This feature is especially advantageous for older adults in that it allows older adults to respond to others' posts at their own pace (Kanayama, 2003). Furthermore, text-based CMC ensures that problem-solving-related information is generated by a group of individuals with complimentary expertise and experience, and is persistent and publicly accessible. Although previous research has reported that the relatively low level of social accountability associated with anonymity—which is possible via text-based CMC—might result in untrustworthy or junk information (Herring, 2002), this is not the case in the OldKids online forums, where members generally enjoy and value providing useful information for others.

Problem-solving-oriented support facilitated by text-based CMC is important in that it helps older adults overcome major barriers to their adoption of new ICTs. Research shows that a lack of knowledge about the technology and inadequate training prevent older adults from using computer technology (Xie, 2003). The online forum provides rich opportunities for older adults to get necessary information and knowledge from peers, and therefore greatly facilitates their adoption of new ICTs.

However, this does not mean that the education-oriented online forum cannot be used to address non-computer-related subjects or exchange other types of social support. As demonstrated in the "Old Q" case, the OldKids online forum was also used to seek, provide, and receive advice on personal, sensitive, and potentially embarrassing issues. Here *anonymity* resulting from text-based modes of CMC is a crucial factor that explains why "Old Q" was willing to disclose such information in the public online forum. This finding echoes earlier research which suggested that a purported disadvantage of text-based CMC—i.e., the relative anonymity associated with the limited social presence or cues—could actually facilitate the sharing of personal, sensitive, embarrassing, or potentially risky information, feelings, and thoughts and, consequently, increase self-disclosure and intimacy of interpersonal relationships (Ferguson, 1997; Galinsky, Schopler, & Abell, 1997; Klemm & Nolan, 1998; Madara, 1997).

Although both the voice chat room and the online forum are only occasionally used to share deep and personal feelings and thoughts and exchange emotional support, this is not the case for the IM mode. In fact, it is common for OldKids members to disclose private and sensitive information and exchange emotional support via IM. This use of IM is consistent with the findings reported in other IM studies which also showed that IM use facilitated positive relationships and fostered intimacy (Gross, Juvonen, & Gable, 2002; Hu et al., 2004; Valkenburg & Peter, 2007). An important reason for these similar findings is that IM ensures private, dyadic, and ephemeral interactions and thus protects privacy. Interestingly, this study also finds that OldKids members consciously use the IM mode to continue conversations that they have started (but do not wish to continue) in the online forum. This finding supports other research that also shows that online relationships are not confined to one particular mode of CMC; rather, individuals often use multiple channels/modes to communicate, interact, and develop relationships (Hu et al., 2004; Kim et al., 2007; Kollock & Smith, 1996; Ogan, 1993; Parks & Floyd, 1996; Rheingold, 1993; Salem et al., 1997; Turner et al., 2001; Valkenburg & Peter, 2007).

Parks and Floyd (1996) suggested that, with advances in ICTs that would enable the exchange of not only text-based but also audio- and video-based information, the reduced-cues belief that was widely circulated in the CMC literature might "simply become a theoretic antique" (p. 93). This study provides empirical evidence contributing to the finding that CMC is indeed no longer limited to the text-based mode that features asynchrony and anonymity. Rather, voice- and video-based modes have enabled more synchronous and less anonymous communication and interaction. In this sense, the argument that CMC is purely based on the limited social presence or contextual cues is not valid any more. However, it is important to keep in mind that advances in technology do not eliminate the advantages of conventional text-based CMC. In situations where anonymous and asynchronous interactions are more desirable (e.g., in the "Old Q" case), individuals can still choose to use conventional mode of CMC. In other words, technological advances have only provided more options from which users can choose.

Wellman and Gulia (1999) speculated that "while people can find almost any kind of support on the Net, most of the support available through one relationship is rather specialized" (p. 171). The findings of this study support the first half of Wellman and Gulia's speculation: as reported above, older Chinese Internet users exchange companionship and various types of social support through CMC. However, the findings of this study do not seem to agree with the second half of Wellman and Gulia's speculation because online relationships as reported by older Chinese participants are generally not specialized to the exchange of one or another particular type of social support. Rather, the relationship between two individuals of the OldKids online community usually provides multiple types of social support—except that some types of support are exchanged in one specialized mode of CMC, while other types of support are exchanged in another specialized mode of CMC. Because individuals use multiple modes of CMC to communicate and interact, they can develop relationships that are capable of providing multiple types of social support. Thus, *although the major type of support sustained in one mode of CMC may be rather specialized, the overall relationships can be multidimensional because relationship development can cross multiple modes of CMC.*

These findings suggest an important factor that had been previously ignored: i.e., the particular mode of CMC through which an online relationship is formed and sustained. Each CMC mode has its unique, inherent functions that may be used to facilitate one instead of another particular type of online interaction and social support. More importantly, it is common that individuals use multiple modes of CMC to interact and develop relationships. Yet research on CMSS so far has largely lagged behind advances in Internet technology. Future research needs to pay more attention to different modes of CMC and explore thoroughly the implications of modes of CMC on interpersonal relationships and social support.

Studying the impact of ICTs on older Chinese's social relationships and social support has unique importance in the context of contemporary China. Due to recent social changes, in contemporary Chinese society filial piety may be changing, perhaps eroding; although the family has traditionally been and is still an important source of support for senior parents and grandparents, perhaps it is no longer the primary source of support for elderly Chinese (Joseph & Phillips, 1999; Ng et al., 2002). In the meantime, the social networks and relationships that older Chinese have developed during their work years are increasingly threatened by recent social changes: in recent years many older Chinese are forced to take early retirement because their work units have not been doing well since recent economic reforms (Price & Fang, 2002). As a result, more and more older Chinese retire—and cut off from their work-related social networks and relationships—in their early 50s or even late 40s. In the context of these social changes, the findings of this study present promising signs for using ICTs to help overcome negative impact of social changes on the lives of older Chinese. Facilitating older adults' relationship development through new ICTs is important to all nations experiencing a graying of the population but especially important to nations like China, where recent social changes, in addition to age-related ones, have increasingly weakened the social networks and relationships of older Chinese, including those who participated in this study (Xie, 2005, 2006a, 2006b, 2007a, 2007b, in press-a, in press-b; Xie & Jaeger, 2008). Within this context, it is necessary to try to make use of the full potential of new ICTs—although technology cannot be the sole solution, it certainly can be a solution to the problems and issues faced by aging societies and individuals.

Implications

The findings of this study have important theoretical implications for online social support. First, with the development of computer technology, CMC is no longer limited to text-based approaches that—arguably—reduce or filter out physical and contextual social cues and thus impede the development of close, intimate relationships. CMC nowadays takes place more and more often in voice- and video-based modes that enable richer social presence and contextual cues and thus facilitates the formation and development of close, intimate relationships. Earlier CMC theories that argue CMC is not as "rich" as face-to-face communication are no longer valid. Second, each mode of CMC, due to its inherent technological features, may be specialized to sustain one particular type of social interaction and relationship. As such, it will be important to find the right match between each mode of CMC (and its technological features) and each type of social relationship and support. Finally, although the type of social relationship and support sustained by one particular mode of CMC is quite specialized, because individuals use multiple modes of CMC and can freely—and often deliberately—switch from one to another when interaction online, the overall relationships developed through CMC can be multidimensional.

The findings of this study also have implications for policy-making agencies and non-government organizations that serve older adults. In a time when the population is aging rapidly (not only in developing countries like China but also in developed countries like the U.S.), it is imperative that government and non-government agencies develop effective and efficient interventions that may help both society and individuals better cope with aging-related changes (e.g., retirement, relocation, deteriorating health condition, loss of loved ones, and the resulting diminishing social relationships and interactions). Because the aging of the population is coincident with the dramatic development of new ICTs worldwide, it is only logical to envision that, in an increasingly technological world, at least some of those interventions would involve new ICTs.

The findings of this study expand the literature on older adults' use of ICTs by detecting and reporting the specific ways that older adults use newer and more complicated technical applications to construct and develop supportive relationships. As such, policy-makers and administrators of senior-oriented organizations should be further convinced of older adults' potential and be more compelled to design interventions that may facilitate older adults' relationship development through new ICTs. This is important to all nations experiencing a graying of the population but especially important to nations like China where recent social changes, in addition to age-related ones, have increasingly weakened the social networks and relationships of older Chinese (Xie, 2006a, 2006b, 2007a, 2007b, in press-a; Xie & Jaeger, 2008). Within this context, it is necessary to make use of the potential of new ICTs—although technology cannot be the sole solution, it certainly can be a solution to the problems and issues faced by aging societies and individuals.

Limitations and Future Directions

This study detected interesting relationships among the three new categories, including the phenomena that both public modes of CMC—the chat room and online forum—were occasionally used to talk about personal, sensitive subjects and exchange emotional support, and that older Chinese participants switched from one mode of CMC to another in order to exchange a different type of subjects and social support. Yet, due to the qualitative nature of this study, it was not possible to determine how often those phenomena occurred in the chat room and online forum, and how often the switch between different modes of CMC took place. Future research will benefit from examining, comparing, and contrasting the frequencies of both exchanging emotional support in each of these modes of CMC and switching from one mode to another. It will also be necessary to examine how different frequencies of occurrences and switches affect the development of overall social relationships and in what ways.

This study was only able to investigate three modes of CMC because those were the only ones available in the selected research site, the OldKids online community. In future research it will be interesting to examine how other modes of CMC (e.g., multi-user online games, email) sustain relationship development and social support in ways that may be different from (or similar to) those examined in this study. Perhaps more importantly, it will be necessary to examine how different *combinations* of particular modes of CMC might have different effects on the construction and maintenance of social relationships and the types of social support that arise from those relationships.

Another limitation of this study was that it did not specifically look at the relationship between individual characteristics (e.g., computer skills, gender) and choices to use which modes of CMC. For instance, did older Chinese who used the voice-based mode of CMC differ significantly in terms of typing skills from their age peers who used text-based modes of CMC such as the online forum or IM? Were older Chinese who used the public modes of CMC for the exchange of personal, sensitive subjects and emotional support more likely to be older men or women? These questions deserve systematic examination in future research.

About the Author

Bo Xie is an assistant professor in the College of Information Studies at the University of Maryland, College Park. Her research focuses on the intersection of older age, information and communication technologies, and health.

Address: 2117G Hornbake Building, South Wing, University of Maryland, College Park, MD 20742, USA

Acknowledgments

The findings reported in this article were part of a large research project that examined the civic engagement, social relationships, and psychological well-being of older Chinese and older American Internet users. The larger research project was supported by the National Science Foundation under Grant No. 0431373. The author would like to thank Ken Fleischmann for his insightful editorial assistance.

References

Bargh, J. A., & McKenna, K. Y. A. (2004). The Internet and social life. *Annual Review of Psychology*, 55, 573–590.

Barrera, M. (1986). Distinctions between social support concepts, measures, and models. *American Journal of Community Psychology*, 14(4), 413–445.

Bernard, H. R. (Ed.). (1998). *Handbook of Methods in Cultural Anthropology.* Walnut Creek, California: AltaMira/Sage.

Bian, Y. (2002). Chinese social stratification and social mobility. *Annual Review of Sociology*, 28, 91–116.

boyd, d. m., & Ellison, N. B. (2007). Social network sites: Definition, history, and scholarship. *Journal of Computer-Mediated Communication*, 13(1), article 11. http://jcmc.indiana.edu/vol13/issue1/boyd.ellison.html

Bryant, J. A., Sanders-Jackson, A., & Smallwood, A. M. K. (2006). IMing, text messaging, and adolescent social networks. *Journal of Computer-Mediated Communication*, 11(2), article 10. http://jcmc.indiana.edu/vol11/issue2/bryant.html

Central Intelligence Agency. (2007). *The world factbook*. Washington DC: Central Intelligence Agency. Retrieved November 16, 2007, from www.cia.gov/library/publications/the-world-factbook/index.html

Cheng, S.-T., & Chan, A. C. M. (2006). Filial piety and psychological well-being in well older Chinese. *Journals of Gerontology: Series B: Psychological Sciences and Social Sciences*, 61B(5), 262–269.

China Internet Network Information Center. (2007). *The 19th Survey Report on Internet Development in China (January 2007)*. Beijing, China: China Internet Network Information Center. Retrieved June 18, 2007 from http://www.cnnic.com.cn/uploadfiles/pdf/2007/1/23/113114.pdf

Chong, A. M.-l. (2007). Promoting the psychosocial health of the elderly—the role of social workers. *Social Work in Health Care*, 44(1–2), 91–109.

Constant, D., Sproull, L., & Kiesler, S. (1996). The kindness of strangers: The usefulness of electronic weak ties for technical advice. *Organization Science*, 7(2), 119–135.

Daft, R. L., & Lengel, R. H. (1984). Information Richness: A New Approach to Managerial Behavior and Organization Design. *Research in Organizational Behavior*, 6, 191–233.

Daft, R. L., & Lengel, R. H. (1986). Organizational Informational Requirements, Media Richness and Structural Design. *Management Science*, 32, 554–571.

Daft, R. L., Lengel, R. H., & Trevino, L. K. (1987). Message Equivocality, Media Selection, and Manager Performance: Implications for Information Systems. *MIS Quarterly*, 11, 355–368.

Danowski, J. A., & Sacks, W. (1980). Computer communication and the elderly. *Experimental Aging Research*, 6(2), 125–135.

DiMaggio, P., Hargittai, E., Neuman, W. R., & Robinson, J. P. (2001). Social implications of the Internet. *Annual Review of Sociology*, 27, 307–336.

Eastin, M. S., & LaRose, R. (2005). Alt.support: Modeling social support online. *Computers in Human Behavior*, 21, 977–992.

Ellison, N. B., Steinfield, C., & Lampe, C. (2007). The benefits of Facebook "friends:" Social capital and college students' use of online social network sites. *Journal of Computer-Mediated Communication*, 12(4), article 1. http://jcmc.indiana.edu/vol12/issue4/ellison.html

Ferguson, T. (1997). Health care in cyberspace: Patients lead a revolution. *Futurist*, 31(6), 29–34.

Fox, N., & Roberts, C. (1999). GPs in cyberspace: The sociology of a "virtual community". *Sociological Review*, 47(4), 643–671.

Furlong, M. S. (1989). An electronic community for older adults: The SeniorNet network. *Journal of Communication*, 39(3), 145–153.

Galinsky, M. J., Schopler, J. H., & Abell, M. D. (1997). Connecting group members through telephone and computer groups. *Health and Social Work*, 22, 181–188.

Glaser, B. G., & Strauss, A. L. (1967). *The Discovery of Grounded Theory: Strategies for Qualitative Research*. Chicago: Aldine.

Gross, E. F., Juvonen, J., & Gable, S. L. (2002). Internet use and well-being in adolescence. *Journal of Social Issues*, 58(1), 75–90.

Haythornthwaite, C. (2001). Introduction: The Internet in everyday life. *The American Behavioral Scientist*, 45(3), 363–382.

Herring, S. C. (2002). Computer-mediated communication on the Internet. In B. Cornin (Ed.), *Annual Review of Information Science and Technology* (Vol. 36, pp. 109–168). Medford, NJ: Information Today.

Holloway, S. L., & Valentine, G. (2003). *Cyberkids: Children in the Information Age.* London & New York: RoutledgeFalmer.

House, J. S. (1981). *Work Stress and Social Support.* Reading, MA: Addison-Wesley.

Hu, Y., Wood, J. F., Smith, V., & Westbrook, N. (2004). Friendships through IM: Examining the Relationship between Instant Messaging and Intimacy. *Journal of Computer-Mediated Communication,* **10**(1), article 6. http://jcmc.indiana.edu/vol10/issue1/hu.html

Ikels, C. (1996). *The Return of the God of Wealth: The Transition to a Market Economy in Urban China.* Stanford, California: Stanford University Press.

Jones, R. A. (1994). The ethics of research in cyberspace. *Internet Research,* 4(3), 30–35.

Jones, S. (Ed.). (1999). *Doing Internet Research: Critical Issues and Methods for Examining the Net.* Thousand Oaks, London, & New Delhi: Sage.

Joseph, A. E., & Phillips, D. R. (1999). Ageing in rural China: Impacts of increasing diversity in family and community resources. *Journal of Cross-Cultural Gerontology,* **14**, 153–168.

Kanayama, T. (2003). Ethnographic research on the experience of Japanese elderly people online. *New Media & Society,* 5(2), 267–288.

Kazmer, M. M., & Xie, B. (in press). Qualitative interviewing in Internet studies: Playing with the media, playing with the method. *Information, Communication and Society.*

Kiesler, S., & Kraut, R. (1999). Internet use and ties that bind. *American Psychologist,* **54**, 783–784.

Kiesler, S., Siegel, J., & McGuire, T. (1984). Social Psychological Aspects of Computer-Mediated Communication. *American Psychologist,* **39**(10), 1123–1134.

Kiesler, S., & Sproull, L. (1992). Group Decision Making and Communication Technology. *Organizational Behavior and Human Decision Processes,* **52**(1), 96–123.

Kim, H., Kim, G. J., Park, H. W., & Rice, R. E. (2007). Configurations of relationships in different media: FtF, email, instant messenger, mobile phone, and SMS. *Journal of Computer-Mediated Communication,* **12**(4), article 3. http://jcmc.indiana.edu/vol12/issue4/kim.html

Klemm, P., & Nolan, M. T. (1998). Internet cancer support groups: Legal and ethical issues for nurse researchers. *Oncology Nursing Forum,* **25**, 673–676.

Kluver, R., & Chen, Y. (2005). The Internet in China: A meta-review of research. *The Information Society,* **21**, 301–308.

Kollock, P., & Smith, M. (1996). Managing the virtual commons: Cooperation and conflict in computer communities. In S. C. Herring (Ed.), *Computer-mediated Communication: Linguistic, Social, and Cross-cultural Perspectives* (pp. 109–128). Philadelphia, PA: John Benjamins.

Kraut, R., Patterson, M., Lundmark, V., Kiesler, S., Mukopadhyan, T., & Scherlis, W. (1998). Internet paradox: A social technology that reduces social involvement and psychological well-being? *American Psychologist,* 53(9), 1017–1031.

Lam, R. C. (2006). Contradictions between traditional Chinese values and the actual performance: a study of the caregiving roles of the modern sandwich generation in Hong Kong. *Journal of Comparative Family Studies,* **37**(2), 299–313.

Lange, P. G. (2007). Publicly private and privately public: Social networking on YouTube. *Journal of Computer-Mediated Communication,* **13**(1), article 18. http://jcmc.indiana.edu/vol13/issue1/lange.html

Lee, W. K. M., & Hong-kin, K. (2005). Older women and family care in Hong Kong: differences in filial expectation and practices. *Journal of Women and Aging,* **17**(1–2), 129–150.

Madara, E. J. (1997). The mutual-aid self-help online revolution. *Social Policy,* **27**, 20–26.

McMillan, S. J., & Hwant, J. S. (2002). Nailing jell-o to the wall and herding cats: A content analysis of Chinese and U.S. newspaper coverage of the Internet in China. *Journal of Intercultural Communication Research*, **31**(2), 107–125.

Mjelde-Mossey, L. A., Chi, I., & Lou, V. W. Q. (2006). Relationship between adherence to tradition and depression in Chinese elders in China. *Aging and Mental Health*, **10**(1), 19–26.

Mjelde-Mossey, L. A., & Walz, E. (2006). Changing cultural and social environments: implications for older East Asian women. *Journal of Women and Aging*, **18**(1), 5–20.

Morgan, C., & Cotten, S. (2003). The relationship between Internet activities and depressive symptoms in a sample of college freshmen. *CyberPsychology & Behavior*, **6**(2), 133–142.

Nastri, J., Peña, J., & Hancock, J. T. (2006). The construction of away messages: A speech act analysis. *Journal of Computer-Mediated Communication*, **11**(4), article 7. http://jcmc.indiana.edu/vol11/issue4/nastri.html

National Bureau of Statistics of the People's Republic of China. (2001). *Report on the 5th Census.* Beijing, China.

Ng, A. C. Y., Phillips, D. R., & Lee, W. K.-M. (2002). Persistence and challenges to filial piety and informal support of older persons in a modern Chinese society: A case study in Tuen Mun, Hong Kong. *Journal of Aging Studies*, **16**, 135–153.

Nie, N. H. (2001). Stability, interpersonal relationships and the Internet: Reconciling conflicting findings. *American Behavioral Scientist*, **45**, 420–435.

Ogan, C. (1993). Listserver communication during the Gulf War: What kind of medium is the electronic bulletin board? *Journal of Broadcasting and Electronic Media*, **37**, 177–196.

Parks, M. R., & Floyd, K. (1996). Making friends in cyberspace. *Journal of Communication*, **46**, 80–97.

Perry, E. J., & Selden, M. (Eds.). (2000). *Chinese Society, Change, Conflict and Resistance.* London: Routledge.

Price, R. H., & Fang, L. (2002). Unemployed Chinese workers: the survivors, the worried young and the discouraged old. *International Journal of Human Resource Management*, **13**(3), 416–430.

Rheingold, H. (1993). *The Virtual Community: Homesteading on the electronic frontier.* Reading, MA: Addison-Wesley.

Ridings, C. M., & Gefen, D. (2004). Virtual community attraction: Why people hang out online. *Journal of Computer-Mediated Communication*, **10**(1), available: http://jcmc.indiana.edu/vol10/issue1/ridings_gefen.html

Salem, D. A., Bogat, G. A., & Reid, C. (1997). Mutual help goes on-line. *Journal of Community Psychology*, **25**, 189–207.

Sanders, C. E., Field, T. M., Diego, M., & Kalplan, M. (2000). The relationship of Internet use to depression and social isolation among adolescents. *Adolescence*, **35**, 237–242.

Shaw, L. H., & Gant, C. S. W. (2002). In defense of the Internet: The relationship between Internet communication and depression, loneliness, self-esteem, and perceived social support. *CyberPsychology & Behavior*, **5**, 157–171.

Short, J., Williams, E., & Christie, B. (1976). *The Social Psychology of Telecommunication.* London: John Wiley.

Soukup, C. (2000). Building a Theory of Multimedia CMC: An Analysis, Critique and Integration of Computer-mediated Communication Theory and Research. *New Media & Society*, **2**(4), 407–425.

Spears, R., & Lea, M. (1992). Social Influence and the Influence of the "Social" in Computer-mediated Communication. In M. Lea (Ed.), *Contexts of Computer-mediated Communication* (pp. 30–65). London: Harvester Wheatsheaf.

Sproull, L., & Kiesler, S. (1991). *Connections: New Ways of Working in the Networked Organization.* Cambridge, MA: MIT Press.

Stefanone, M. A., & Jang, C.-Y. (2007). Writing for friends and family: The interpersonal nature of blogs. *Journal of Computer-Mediated Communication*, 13(1), article 7. http://jcmc.indiana.edu/vol13/issue1/stefanone.html

Strauss, A. L., & Corbin, J. (1998). *Basics of Qualitative Research: Techniques and Procedures for Developing Grounded Theory* (2nd ed.). Thousand Oaks, CA: Sage.

Sun, R. (2004). Worry about medical care, family support, and depression of the elders in urban China. *Research on Aging*, **26**(5), 559–585

Swickert, R. J., Hittner, J. B., Harris, J. L., & Herring, J. A. (2002). Relationships among Internet use, personality, and social support. *Computers in Human Behavior*, **18**, 437–451.

Tang, W., & Parish, W. L. (2000). *Chinese Urban Life under Reform: The Changing Social Contract.* New York: Cambridge University Press.

Tichon, J. G., & Shapiro, M. (2003). The process of sharing social support in cyberspace. *CyberPsychology & Behavior*, **6**, 161–170.

Turner, J. W., Grube, J. A., & Meyers, J. (2001). Developing an optimal match within online communities: An exploration of CMC support communities and traditional support. *Journal of Communication*, **51**(2), 231–251.

Valkenburg, P. M., & Peter, J. (2007). Online communication and adolescent well-being: Testing the stimulation versus the displacement hypothesis. *Journal of Computer-Mediated Communication*, **12**(4), article 2. http://jcmc.indiana.edu/vol12/ issue4/valkenburg.html

Vaux, A. (1988). *Social Support: Theory, Research, and Intervention.* New York: Praeger.

Walder, A. G. (1989). Social change in post-revolution China. *Annual Review of Sociology*, **15**, 405–424.

Walther, J. B. (1992). Interpersonal effects in computer-mediated interaction: A relational perspective. *Communication Research*, **19**(1), 52–90.

Walther, J. B. (1995). Relational aspects of computer-mediated communication: Experimental observations over time. *Organization Science*, **6**(2), 186–203.

Walther, J. B., Anderson, J. F., & Parks, D. W. (1994). Interpersonal effects in computer-mediated interaction: A meta-analysis of social and antisocial communication. *Communication Research*, **21**, 460–487.

Walther, J. B., & Boyd, S. (2002). Attraction to computer-mediated social support. In C. A. Lin & D. Atkin (Eds.), *Communication Technology and Society: Audience Adoption and Uses* (pp. 153–188). Cresskill, NJ: Hampton Press.

Walther, J. B., Gay, G., & Hancock, J. T. (2005). How Do Communication and Technology Researchers Study the Internet? *Journal of Communication*, **55**(3), 632–657.

Warner, M. (2001). The new Chinese worker and the Challenge of Globalization: An overview. *International Journal of Human Resource Management*, **12**(1), 134–141.

Wellman, B., & Gulia, M. (1999). Virtual communities as communities: Net surfers don't ride alone. In M. A. Smith & P. Kollock (Eds.), *Communities in Cyberspace* (pp. 167–195). London & New York: Routledge.

Wellman, B., Salaff, J., Dimitrova, D., Garton, L., Gulia, M., & Haythornthwaite, C. (1996). Computer networks as social networks: Collaborative work, telework, and virtual community. *Annual Review of Sociology*, **22**, 213–238.

White, M., & Dorman, S. M. (2001). Receiving social support online: Implications for health education. *Health Education Research*, **16**(6), 693–707.

Wilson, S. M., & Peterson, L. C. (2002). The anthropology of online communities. *Annual Review of Anthropology*, **31**, 449–467.

Wright, K. B. (2000). Computer-mediated social support, older adults, and coping. *Journal of Communication,* **50**, 100–118.

Xie, B. (2003). Older adults, computers, and the Internet: Future directions. *Gerontechnology,* 2(4), 289–305.

Xie, B. (2005). Getting older adults online: The experiences of SeniorNet (USA) and OldKids (China). In B. Jaeger (Ed.), *Young Technologies in Old Hands—An International View on Senior Citizens' Utilization of ICT* (pp. 175–204). Copenhagen, Denmark: DJOF Publishing.

Xie, B. (2006a). *Growing Older in the Information Age: Civic Engagement, Social Relationships, and Well-being of Older Internet Users in China and the United States.* Unpublished Dissertation, Rensselaer Polytechnic Institute, Troy, New York.

Xie, B. (2006b). Perceptions of computer learning among older Americans and older Chinese. *First Monday,* **11**(10), URL: http://www.firstmonday.org/issues/issue11_10/xie/index.html

Xie, B. (2007a). Information technology education for older adults as a continuing peer-learning process: A Chinese case study. *Educational Gerontology,* 33(5), 429–450.

Xie, B. (2007b). Older Chinese, the Internet, and well-being. *Care Management Journals: Journal of Long Term Home Health Care,* 8(1), 33–38.

Xie, B. (in press-a). Civic engagement among older Chinese Internet users. *Journal of Applied Gerontology.*

Xie, B. (in press-b). The mutual shaping of online and offline social relationships. *Information Research.*

Xie, B., & Jaeger, P. T. (2008). Older adults and political participation on the Internet: A cross-cultural comparison of the United States and China. *Journal of Cross-Cultural Gerontology,* 23, 1–15.

Yoon, H., & Hendricks, J. (Eds.). (2006). *Handbook of Asian Aging.* Amityville, NY: Baywood Pub. Co.

Zhou, X., & Hou, L. (1999). Children of the Cultural Revolution: The state and the life course in the People's Republic of China. *American Sociological Review,* **64**(1), 12–36.

Buying into American Idol: How We Are Being Sold on Reality Television

Henry Jenkins

Who would have predicted that reality television series, such as *Survivor* (2000) and *American Idol* (2002), would turn out to be the first killer application of media convergence—the big new thing that demonstrated the power that lies at the intersection between old and new media? Initial experiments with interactive television in the mid-1990s were largely written off as failures. Most people didn't want to stop watching television just to buy the clothes one of the *Friends* (1994) was wearing. Few were interested in trivia quizzes flashing up at the bottom of the screen during sportscasts or James Bond movies. Critics argued that most of us simply wanted to sit back and watch television rather than interact with it. The current success of reality television is forcing the media industry to rethink some of those assumptions. The shift is one from real-time interaction toward asynchronous participation.

Few can argue with *American Idol*'s success. By the final weeks of its second 2003 season, FOX Broadcasting Company was receiving more than 20 million telephone calls or text messages per episode casting verdicts on the *American Idol* contestants.[1] This made the phone companies happy because they have been trying to find a way to get Americans more excited about text messaging, which hasn't taken off in the United States the way it has in Asia and northern Europe. Of the 140 million mobile phones in the United States today, only 27 million are being used for text messaging.[2] AT&T Wireless reported that roughly a third of those who participated in *American Idol* through text messaging had never sent a text message before.[3] As an AT&T spokesman explained, "Our venture with FOX has done more to educate the public and get people texting than any marketing activity in this country to date."[4]

American Idol commanded two of the top five time slots throughout the important May 2003 sweeps period. More than 40 million people watched the final segment of the final episode of *American Idol*'s second season. By the third season, FOX devoted 13.5 hours to *American Idol* during the crucial May sweeps period, representing nearly one quarter of their total prime-time schedule for the month.[5]

This made advertisers happy. As MediaCom chief executive Jon Mandel explains, "We know when people are watching a show they care about, they tend to watch commercials more. Unfortunately, there aren't that many shows people care about."[6] *American Idol,* based on the successful British series *Pop Idol,* was sold to FOX through an aggressive campaign by the Creative Artists Agency, which saw the series as an ideal match for their client, Coca-Cola, and its 12–24-year-old target audience.[7] And what a match it has been. For those of you without a television or a teenage offspring, *American Idol* is a showcase of unknown singers—some good, some very bad—from around the country. Each week, the finalists perform and the audience votes out one contestant. In the end, the surviving performer gets a record contract and a promotion deal. *Forbes* ranked *American Idol* as the most profitable of all reality series, estimating that it had netted the network more than $260 million in profits by the end of its third season.[8]

All of this really made the networks happy. Reality television programs hold their own during the summer months, when network viewership is traditionally at its lowest ebb. And, as importantly, reality television has been a savior as broadcast networks try to resist cable television's attempts to siphon away their core audience. In 2002, for the first time, the cable networks' combined share outstripped that of the broadcast networks. No given cable channel has had the power and reach of a CBS, NBC, or ABC, but year by year broadcast networks become less central to their viewership. Overall, television audiences in summer decline 8–10 percent, but the major networks lost 30 percent of their market in the summer of 2002.[9] Cable networks like Showtime or HBO use the summer months to launch new episodes of their hot sitcoms (such as *Sex and the City,* 1998) and dramas (such as *Six Feet Under,* 2001), pitting them against broadcast network reruns. Viewers tend to remain with cable once the fall season starts. So, the broadcast networks are countering by offering more original programming in the summer, with the less-expensive reality television programs becoming their best weapon. When they are successful, reality series generate as much or more buzz than the cable shows they are competing against and thus slow viewership erosion. Even if a reality series doesn't make ratings history, as the first seasons of *Survivor* and *American Idol* did, its lower returns are almost always better than the network would get on a rerun. The trade-off is that reality programs have a shorter shelf life and limited life-after-syndication, though they can represent significant sales when sold directly to consumers on DVD.

And this makes the media conglomerates happier still, since *American Idol* was from the start not simply a television program but a transmedia franchise. The show's first season winner, Kelly Clarkson, signed to RCA Records and had an immediate number 1 hit single on the Billboard Hot 100, "A Moment Like This." The song went on to become the top-selling U.S. single for 2002. Kelly's initial singles got played more than 80,000 times on radio stations in 2002. An *American Idol* book made the bestseller list,[10] and the *American Idol* contestants played to sold-out houses on their nationwide concert tour. Production began immediately on a feature-length movie, *From Justin to Kelly* (2003), though the film ultimately generated low box-office returns.

Not everyone, however, was enchanted with *American Idol*'s success. Speaking for many critics of reality television, Karla Peterson ranted in the *San Diego Union-Tribune:*

> American Idol *was not a dumb summer fling, but a conniving multimedia monster. Shameless product placement. Bloodless nostalgia. Incestuous corporate hype. Like the show's Stepford divas—who dutifully parroted every shriek, quiver and growl from the Mariah Carey catalog—*American Idol *has absorbed the sins of our debauched culture and spit them out in a lump of reconstituted evil. And because we were so dazzled by its brazen lack of redeeming qualities, we stepped over the mess and happily followed it over the abyss.*[11]

Peterson is correct that *American Idol* was shaped at every level by blatant commercial calculations. Yet, her moral outrage doesn't take us very far toward understanding its appeal to the network, advertisers, or consumers.

To understand *American Idol*'s success, we need to better understand the changed context within which American broadcasting is operating and the changed model of consumer behavior shaping programming and marketing strategies. We need to know more about what I am calling "affective economics." By affective economics, I mean a new configuration of marketing theory, still somewhat on the fringes but gaining ground within the media industry, which seeks to understand the emotional underpinnings of consumer decision-making as a driving force behind viewing and purchasing decisions. In many ways, affective economics represents an attempt to catch up with work in cultural studies over the last several decades on fan communities and viewer commitments. There is a crucial difference, however: the cultural studies work sought to understand media consumption from the fan's point of view, articulating desires and fantasies that were ill-served by the current media system; the new marketing discourse seeks to mold those consumer desires to shape purchasing decisions. While they are increasingly interested in the qualities of audience experience, the media and brand companies still struggle with the economic side of affective economics—the need to quantify desire, to measure connections, and to commodify commitments—and perhaps most importantly of all, the need to transform all of the above into return on investment (ROI). These bottom-line pressures often deflect attempts to understand the complexity of audience behavior even when such knowledge is desperately needed by companies that want to survive in the coming decades. Rather than rethinking the terms of their analysis, they are struggling to fit these new insights into familiar economic categories. It is still a world where what can be counted is what counts most.

Arguably, fans of certain cult television shows may gain greater influence over programming decisions in an age of affective economics. From time to time, networks reprioritize certain segments of their audience and the result is a shift in program strategies to more fully reflect those tastes—a shift from rural to urban viewers changed television content in the 1960s, a renewed interest in minority viewers led to more Afrocentric sitcoms throughout the 1990s, and a shift toward an emphasis on loyal viewers has been changing what reaches the air in the early twenty-first century. Fans are seeing more shows reflecting their tastes and interests reaching the air; those shows are being designed to maximize elements that appeal to fans; and those shows that fans like are apt to remain on the air longer because they are more likely to get renewed in borderline cases. Here's the paradox: to be desired by the networks is to have your tastes commodified. On the one hand, to be commodified expands a group's cultural visibility. Those groups that have no recognized economic value get ignored. That said, commodification is also a form of exploitation. Those groups that are commodified find themselves targeted more aggressively by marketers and often feel they have lost control over their own culture, since it is mass produced and mass marketed. One cannot help but have conflicted feelings because one doesn't want to go unrepresented—but one doesn't want to be exploited, either.

For years, fan groups, seeking to rally support for endangered series, have argued that networks should be focused more on the quality of audience engagement with the series and less on the quantity of viewers. Increasingly, advertisers and networks are coming to more or less the same conclusion. Marketers seek to shape brand reputations, not through an individual transaction but through the sum total of interactions with the customer—an ongoing process that increasingly occurs across a range of different media "touch points." They don't simply want to get a consumer to make a single purchase, but rather to build a long-term relationship with a brand. New models of marketing seek to expand consumers' emotional, social, and intellectual investments with the goal of shaping consumption patterns. In the past, media producers spoke of "impressions." Now, they are exploring the concept of audience "expressions," trying to understand how and why audiences react to the content. Marketing gurus argue that building a committed "brand community" may be the surest means of expanding consumer loyalty and that product placements will allow brands to tap some of the affective force of the affiliated entertainment properties. For this reason, shows such as *American Idol* are being watched closely by advertisers, marketing companies, television networks, and trade press reporters, all eager to understand how corporate convergence strategies may be reshaping the branding process. Early evidence suggests that the most valuable consumers are what the industry calls "loyals," or what we call fans. Loyals are more apt to watch series faithfully, more apt to pay attention to advertising, and more apt to buy products.

For the moment, I want readers to bracket their anxieties about consumerism and their fear of Madison Avenue. I do not intend this chapter to be in any simple sense an endorsement of or apology for the changes that are taking place. My own view is that this emerging discourse of affective economics has both positive and negative implications: allowing advertisers to tap the power of collective intelligence and direct it toward their own ends, but at the same time allowing consumers to form their own kind of collective bargaining structure that they can use to challenge corporate decisions. I will return to this issue of consumer power in the concluding chapter of this book. Even if you want to criticize the way American capitalism works, you need to recognize that the models of marketing depicted in classic accounts, such as Vance Packard's *Hidden Persuaders* (1957), no longer adequately describe the way the media industries are operating.[12] Even if you believe that fan and brand communities lack the clout to significantly alter corporate behavior, you still need to understand the way participation works within this new affective economy so that you can direct criticisms at the actual mechanisms by which Madison Avenue seeks to reshape our hearts and minds.

At industry gatherings around the country, corporate visionaries and brand gurus are promoting what I am calling affective economics as the solution to a perceived crisis in American broadcasting—a crisis brought about by shifts in media technology that are granting viewers much greater control over the flow of media into their homes. Affective economics sees active audiences as potentially valuable if they can be courted and won over by advertisers. In this chapter, we will be looking more closely at the ways that advertisers and networks think about their audiences in the age of media convergence and the ways those assumptions about branding, audience commitment, and social viewing are shaping series such as *American Idol*. *American Idol* offers up a fantasy of empowerment—"America" gets to "decide" upon the next Idol. This promise of participation helps build fan investments, but it may also lead to misunderstandings and disappointments as viewers feel that their votes have not been counted.

"Impress Me"

An advertisement created several years ago for Apple Box Productions, Inc., depicts the new youth consumer: his straggling dishwater blonde hair hangs down into his glaring eyes, his mouth is turned down into a challenging sneer and his finger posed over the remote. "You've got three seconds. Impress me," he says.[13] One false move and he will zap us. No longer a couch potato (if he ever was), he determines what, when, and how he watches media. He is itinerant—free of commitments to particular series, going where his fancy takes him.

The word "impress" serves double duty here, depending if it is read from the consumer's point of view or the marketer's. It refers to the consumer's search for something so "impressive" that he pauses his relentless search for novelty. It also refers to the "impression" that is the unit of measurement historically deployed by the networks in their conversations with potential sponsors—the raw number of "eyeballs" watching a television program at a specific moment in time. What interests me here is the way the cultural and economic, and the consumer and corporate, meanings intersect. How does the viewer's search for compelling content translate into exposure to sponsored messages?

Much fuss was made a few years back about the ineffectiveness of banner advertising on the Web because the "click through" rate was so low. Relatively few people who saw the banner were following its link and purchasing the product. If television advertising had been judged by that same standard, it would have been found to be equally ineffective. The impression is not a measurement of how many people buy the product or even comprehend the message; it is purely a measurement of how many people have the set tuned to a particular channel. The impression is an even looser measurement when applied to other media. For example, the impressions created by a billboard are measured in terms of the sheer number of cars that drive by a particular intersection. According to marketing researcher Robert Kozinets, "It is not only that the impression is a clumsy way to track media insights. . . . The impression is a symptom of the larger business misunderstanding about what can be tracked, understood, and related back to particular investments."[14] Advertisers, however, are increasingly demanding accountability from media outlets for the degree of actual exposure they receive and for the quality of relationship this creates with their consumers. They want to understand how effective different media are about getting their messages before their potential buyers.

Just as the clumsiness of audience measurements have been exposed, the networks have also witnessed a breakdown in viewer loyalty—the problem posed by our shaggy-haired young friend. First, there has been a proliferation of media options—a move from three major networks to a cable environment with hundreds of more specialized channels and the introduction of alternative forms of home entertainment, including the Internet, video, DVD, and computer and video games. Initially, the amount of time people spent consuming media each day expanded as the range of media options grew, but this expansion could only go so far given what a large portion of time the average consumer spent engaged with entertainment content outside of work, school, or sleep. Confronted with seemingly infinite variety, the average consumer settled into a pattern of watching ten to fifteen different media outlets. Broadcast programming still commands a higher degree of loyalty than cable programming, but the major broadcast networks are attracting a smaller slice of the pie as audience fragmentation continues. In the 1960s, an advertiser could reach 80 percent of U.S. women with a prime-time spot on the three networks. Today, it has been estimated that the same spot would have to run on one hundred TV channels to reach the same number of viewers.[15]

As advertisers grow anxious about whether network programming can reach audiences, they are diversifying their advertising budgets and looking to extend their brands across multiple distribution outlets that they hope will allow them to target a diverse selection of smaller niche markets. As Sumner Redstone, chairman of Viacom, told *Businessweek,* "What advertisers buy is platforms to get their brand promoted, and we've got four platforms for them. We're everywhere, because in this day and age you have to be where advertisers need to be."[16] A researcher for Forrester Research summarized the trends: "Monolithic blocks of eyeballs are gone. In their place is a perpetually shifting mosaic of audience microsegments that forces marketers to play an endless game of audience hide-and-seek."[17]

Next-generation technologies—especially the digital video-recorder (DVR)—are enabling more and more consumers to skip commercials. Right now, 43 percent of VCR-using households are skipping adverts and many in the media industry are terrified of what's going to happen when technologies such as TiVo, which Nielsen Media Research president Susan Whiting calls "the VCR on steroids," become more widespread.[18] Current users of digital video recorders scan through commercials about 59 percent of the time.[19] This doesn't mean that 59 percent of users skip commercials; it means that the average consumer watches about 41 percent of the aired advertisements. *Advertising Age* reporter Scott Donaton explains: "As advertisers lose the ability to invade the home, and consumer's minds, they will be forced to wait for an invitation. This means that they have to learn what kinds of advertising content customers will actually be willing to seek out and receive."[20]

Rishad Tobaccowala, president of the media-buying group Starcom MediaVest, sparked panic at a gathering of television executives in 2002 when he made what turned out to be the premature prediction that the thirty-second commercial would be dead by 2005. FOX Television chairman Sandy Grushow argued that the networks are nowhere near prepared for such a development: "Not only will everyone have to get drenched, but struck by lightning before significant progress is made."[21] As network executives search for their umbrellas, product placements are the most oft-discussed alternative, though no one really believes they can replace the $8 billion spent each year on commercials. For this transformation to occur, Lee Gabler, co-chairman and partner of Creative Artists Agency, argued, "The biggest hurdle we have to go over . . . is the integration of the networks, the studios, the ad agencies, the advertisers, the talent agencies, and anybody else that's involved in this space. We must be able to sit down collectively and cooperatively to come up with a solution, Right now, the ad agencies are frightened about anybody getting in their space, the networks are in denial, and the advertisers don't have a solution."[22]

In this context, the American viewing public is becoming harder and harder to impress. The television industry is increasingly focusing on understanding consumers who have a prolonged relationship and active engagement with media content, who show a willingness to track down that content across the cable spectrum and across a range of other media platforms. Such consumers, they believe, represent their best hope for the future. This next-generation audience research focuses attention on what consumers do with media content once it has passed across their eyeballs, seeing each subsequent interaction as valuable because it reinforces their relationship to the series and, potentially, its sponsors. Responding to that demand, Initiative Media, a company that advises many of the Fortune 500 companies about their advertising placement, advocates an alternative approach to audience measurement they call "expression."[23] Expression charts attentiveness to programming and advertising, time spent with the program, and the degree of viewer loyalty and affinity to the program and its sponsors. Their concept of expression emerged through collaboration with the MIT Comparative Media Studies Program. Expression may start at the level of the individual consumer, but by definition it situates consumption within a larger social and cultural context. Consumers not only watch media; they also share media with one another, whether this consists of wearing a T-shirt proclaiming their passion for a particular product, posting a message on a discussion list recommending a product to a friend, or creating a parody of a commercial that circulates on the Internet. Expression can be seen as an investment in the brand and not simply an exposure to it.

Lovemarks and Emotional Capital

Delivering the keynote address at *Advertising Age*'s Madison + Vine conference on February 5, 2003, Coca-Cola president Steven J. Heyer outlined his vision for the future relations between the advertising ("Madison") and the entertainment industries ("Vine"). His speech offers a glimpse into the thinking of one of *American Idol*'s primary sponsors.[24] Heyer opened by identifying a range of problems that "demand a new approach to connecting with audiences" and force a rethinking of the old mass media paradigm: "The fragmentation and proliferation of media, and the consolidation in media ownership—soon to be followed by a wholesale unbundling. The erosion of mass markets. The empowerment of consumers who now have an unrivaled ability to edit and avoid advertising and to shift day parts. A consumer trend toward mass customization and personalization." Confronting profound shifts in consumer behavior, Heyer then outlined what he saw as his "convergence" strategy—the greater collaboration between content providers and sponsors to shape the total entertainment package. The focus, he argued, should be less on the content per se than on the "why, where, and how" the various entertainment media are brought together and the relationship that gets brokered with the consumer. As he explained, "Imagine if we used our collective tool kit to create an ever-expanding variety of interactions for people that—over time—built a relationship, an ongoing series of transactions, that is unique, differentiated and deeper" than any the entertainment industry has offered before.

Heyer's speech evokes the logic of brand extension, the idea that successful brands are built by exploiting multiple contacts between the brand and consumer. The strength of a connection is measured in terms of its emotional impact. The experience should not be contained within a single media platform, but should extend across as many media as possible. Brand extension builds on audience interest in particular content to bring them into contact again and again with an associated brand. Following this logic, Coca-Cola sees itself less as a soft drink bottler and more as an entertainment company that actively shapes as well as sponsors sporting events, concerts, movies, and television series. This intensification of feelings enables entertainment content—and brand messages—to break through the "clutter" and become memorable for consumers: "We will use a diverse array of entertainment assets to break into people's hearts and minds. In that order. . . . We're moving to ideas that elicit emotion and create connections. And this speeds the convergence of Madison + Vine. Because the ideas which have always sat at the heart of the stories you've told and the content you've sold . . . whether movies or music or television . . . are no longer just intellectual property, they're emotional capital."

Kevin Roberts, the CEO Worldwide of Saatchi & Saatchi, argues that the future of consumer relations lies with "lovemarks" that are more powerful than traditional "brands" because they command the "love" as well as the "respect" of consumers: "The emotions are a serious opportunity to get in touch with consumers. And best of all, emotion is an unlimited resource. It's always there—waiting to be tapped with new ideas, new inspirations, and new experiences."[25] Arguing that only a small number of customers make purchase decisions based purely on rational criteria, Roberts urges marketers to develop multisensory (and multimedia) experiences that create more vivid impressions and to tap the power of stories to shape consumer identifications. For example, Coca-Cola's corporate Web site (http://www2.coca-cola.com/heritage/stories/index.html) includes a section where consumers can share their own personal stories about their relationship with the product, stories that get organized around themes such as "romance," "reminders of family," "childhood memories," "an affordable luxury," "times with friends," and a "memory of home." These themes merge core emotional relationships with core promotional themes, helping people not simply to integrate Coca-Cola into their memories of their lives, but also helping to frame those memories in terms of the marketing pitch.

American Idol wants its fans to feel the love or, more specifically, the "love marks." Audience participation is a way of getting *American Idol* viewers more deeply invested, shoring up their loyalty to the franchise and its sponsors. This investment begins with the turnout of millions of would-be contestants at auditions held in stadiums and convention hotels around the country. Many more people watch the series than try out; many more try out than make the air; many more make the air than become finalists. But, at every step along the way, the viewers are invited to imagine that "it could be me or someone I know." From there, the weekly votes increase the viewer's engagement, building a strong allegiance to the individual performers. By the time the records are released, many of the core consumers have already endorsed the performers, and fan clubs are already involved with grassroots marketing. For example, fans of Clay Aiken, the runnerup on Season Two, turned their disappointment into a campaign to ensure that his album, *Measure of a Man* (2003), outsold first-place finisher Ruben Studdard's *Soulful* (2003). Clay's album sold more than 200,000 more copies than Studdard's in its opening week on the charts—though one suspects that the record executives would have been happy whichever way the sales contest went.[26] Coca-Cola, in turn, brands key series elements: contestants wait in the "red room" before going on stage; judges sip from Coca-Cola cups; highlights get featured on the official program Web site surrounded by a Coca-Cola logo; soft drink promotions reward tickets to the finales; Coca-Cola sends *Idol* performers to NASCAR races and other sporting events that it sponsors; and Coca-Cola's sponsorship figures prominently at the *American Idol* finalist's national concert tour.[27]

Heyer spoke of a shift "away from broadcast TV as the anchor medium" and toward "experience-based, access-driven marketing" as the ideal means of reaching the emerging generation of consumers. Cokemusic.com further aligns the soft drink company with people's enjoyment of popular music, allowing for a range of different participatory and interactive options. Members can pay for downloads of popular songs or redeem coupons that allow them to download songs for free. Members can create their own music mixes, share them with one another, and receive ratings from other site visitors. Ratings points reward "decibels" that can be redeemed to purchase virtual furnishing for their "pads," allowing further customization and a deeper sense of belonging in the world of Coca-Cola. "Performers" develop reputations and followings, which provide emotional incentives for them to spend even more time working on their "mixes." More casual site visitors can participate in a range of quizzes, games, and contests. Cokemusic.com has become the third most popular Web site among teens, registering more than 6 million users who spend an average of forty minutes per visit. As Carol Kruse, the director of interactive marketing for the company, explains, "They're having fun, they're learning about music, they're building a sense of community . . . and it's all in a very safe and friendly Coke environment."[28]

Brand loyalty is the holy grail of affective economics because of what economists call the 80/20 rule: for most consumer products, 80 percent of purchases are made by 20 percent of their consumer base. Maintaining the allegiance of that 20 percent stabilizes the market and allows them to adopt an array of other approaches to court those who would make the other 20 percent of purchases.[29] Corporations are turning toward active consumers because they must do so if they are going to survive; some have learned that such consumers can be allies, but many still fear and distrust them, seeking ways to harness this emerging power toward their own ends.

Something of this ambivalence can be seen in Roberts's description of what he calls "inspirational consumers" or others call "brand advocates": "They are the ones who promote and advocate for the brand. The ones . . . who suggest improvements and refinements, who create websites and spread the word. They are also the people who act as moral guardians for the brands they love. They make sure the wrongs are righted and hold the brand fast to its stated principles."[30] Roberts acknowledges that these "inspirational consumers," individually and collectively, place demands on corporations, citing the example of the outcry when Coca-Cola sought to replace its classic formula with "New Coke" and was forced within two months to back off from that decision. Roberts argues that companies need to listen closely when these inspirational consumers speak—especially when they

Product Placement and *The Apprentice*

Mark Burnett, the executive producer of *Survivor* and *The Apprentice* (2004), has been on the cutting-edge of experiments in brand integration. After finding networks highly resistant to his initial *Survivor* proposal, the producer agreed to help offset the anticipated costs of the production by preselling sponsorship, convincing companies such as Reebok to pay $4 million apiece for product placements during the series.* His second series, *The Restaurant* (2003), was fully funded by product placements from Mitsubishi, American Express, and Coors Brewing Company.† With *The Apprentice,* Burnett charged up to $25 million per company for a significant product placement, and in the process the series became a test site for a range of different approaches of linking brands and series content.‡

How many different ways is *The Apprentice* involved in branding?

1. *The Brand as Protagonist:* Program host Donald Trump casts himself and his corporate empire as the series protagonists. In the course of the series, we visit his many different companies, meet his staff (as well as his fiancée), visit his apartment, and learn about his business philosophy. The contestants are vying for a chance to help run one of his projects, which is presented as if it were the greatest opportunity any young business person might aspire toward.
2. *The Brand as Taskmaster:* In the second season (Fall 2004), contestants were asked to design and market test toys for Toys 'R' Us and Mattel, to develop new ice cream flavors for Ciao Bella, to redesign the bottle for a new Pepsi product, to sell a new M&M candy bar on the streets, and to market a new Vanilla Mint toothpaste for Procter & Gamble. Procter & Gamble spokesman Bryan McCleary commented, "Having an entire episode dedicated to selling the benefits of this new product was a very appealing story line—the viewer actually ends up rooting for the brand to succeed."§
3. *The Branding Process as Entertainment:* On the September 23, 2004, episode, contestants demonstrated ways of linking brands and entertainment (circus acrobats and clowns, the New York Mets) to create buzz for the new Crest product. Other story lines centered on their efforts to produce spots that would promote recruitment by the New York Police Department or to peddle household goods on the Home Shopping Network.
4. *The Brand as Helper:* Frequently, the contestants consult with a range of smaller companies (such as the Alliance Talent Agency) who aid them in their tasks in return for exposure.
5. *The Brand as Prize:* In many cases, Trump rewards contestants with access to himself and his "things" or to luxury meals and services (such as a caviar feast at Petrossian's or jewelry from Graff's).
6. *The Brand as Tie-In:* Following an episode in which the contestants designed ice cream, viewers at home were able to order samples of the flavors online to serve at their next *Apprentice* theme party. Similarly, although they had not planned on producing a tie-in with the series, Mattel was so excited with the results of their episode that they eventually marketed

* Ted Nadger, "The End of TV 101: Reality Programs, Formats, and the New Business of Television," in Susan Murray and Laurie Ouellette (eds.), *Reality Television: Remaking Television Culture* (New York: New York University Press, 2004).

† Wade Paulsen, "NBC's *The Restaurant* Funded Solely by Product Placement," *Reality TV World,* July 18, 2003, http://www.realitytvworld.com/index/articles/stor.php?s=1429.

‡ Michael McCarthy, "Also Starring (Your Product Name Here)," *U.S.A. Today*, August 12, 2004.

§ "Sponsors Buy into Reality TV," Product Placement News, *ITVX*, December 6, 2004, http://www.itvx.com/iPageCount,2,ppnews.asp.

the Mighty Morpher toy car that was designed by the contestants. J. C. Penney distributed a catalog of Levi jeans that was designed by one of the teams during another challenge.

7. *The Brand as Community:* Through a tie-in between *The Apprentice* and Friendster, fans asserted their affiliation with specific contestants and the producers collected real-time data about audience response.
8. *The Brand as Event:* Trump launched a sweepstakes competition with Yahoo! Hot Jobs, whose 25k award is designed to encourage new initiatives. A sign for the service was mounted on top of the cab that took the dejected contestants away, and in one comic interstitial, the colorful Raj appeared as a taxi driver.
9. *The Contestants as Brand:* The female contestants were showcased modeling lingerie as "The Women of *The Apprentice*" in *Maxim* magazine.
10. *The Brand as Judges:* As the second season neared its finale, Trump allowed a range of executives from other companies, including Unilever HPC, PepsiCo, Bear Stearns, and the New England Patriots, to help him to winnow down the finalists.

These examples scarcely exhaust the roles brands play in the series (and doesn't include the various ways NBC is using the series to revise its own brand identity).The temptation among media-savvy people is to dismiss *The Apprentice* as nothing but one big product placement, but this would not adequately explain its popularity. *The Apprentice* is popular because it's a well-made show, and the brand tie-ins work because they are linked to its core emotional mechanics. We care about the brands because they become the focus of contests or because they shape our identifications with the characters. But, as a general rule, the reality shows that have gotten the highest ratings have been those that have had the most original and compelling formats.

criticize a company decision. A company that loses faith with its "inspirational consumers," he argues, will soon lose its core market: "When a consumer loves you enough to take action, any action, it is time to take notice. Immediately."[31] Roberts praises companies that actively court such fans through, to continue the Coca-Cola example, hosting events and conventions where their collectibles are appraised and showcased. The first fan club for Coca-Cola formed in 1974, a grassroots effort by a small group of enthusiasts. Today, fan clubs operate in twenty-eight different countries around the world and host a global network of local and national conventions that the company uses to bring together and address its most dedicated consumers.

Roberts's advice about courting "inspirational consumers" is echoed across a range of other business best-sellers, such as Marc Gobé's *Emotional Branding: The New Paradigm for Connecting Brands to People* (2001), Matthew W. Ragas's *The Power of Cult Branding: How 9 Magnetic Brands Turned Customers into Loyal Followers (and Yours Can, Too)* (2002), and John Hagel III and Arthur G. Armstrong's *Net.Gain: Expanding Markets through Virtual Communities* (1997).[32] They point toward a world where the most valued consumer may be the one who is most passionate, dedicated, and actively engaged. Far from marginal, fans are the central players in a courtship dance between consumers and marketers. As one noted industry guide explains, "Marketing in an interactive world is a collaborative process with the marketer helping the consumer to buy and the consumer helping the marketer to sell."[33] This search for "inspirational consumers" is starting to impact the way television audiences are appraised and the ways advertisers think about selling products.

Zappers, Casuals, and Loyals

Industry insiders often deploy the distinction among zappers, casuals, and loyals: this distinction manages to blur together how, why, and what consumers watch. Zappers are people who constantly flit across the dial—watching snippets of shows rather than sitting down for a prolonged engagement. Loyals actually watch fewer hours of television each week than the general population: they cherry pick those shows that best satisfy their interests; they give themselves over fully to them; they tape them and may watch them more than one time; they spend more of their social time talking about them; and they are more likely to pursue content across media channels. Loyals watch series; zappers watch television. Loyals form long-term commitments; zappers are like the folks at cocktail parties who are always looking over their shoulders to see if someone more interesting has just entered the room. Casuals fall somewhere in between; they watch a particular series when they think of it or have nothing better to do. They generally watch it from start to finish but are more apt to wander away if it starts to bore them. They may be more likely to conduct conversations or do other household activities over the show rather than give it their full attention.

No given viewer is exclusively a loyal, a causal or a zapper; most watch television in different ways on different occasions. The most discriminating viewer will zap around the dial in a hotel room or at the end of a hard day. And sometimes zappers get hooked into a series and watch it every week. Nobody knows for sure yet whether the new media environment has produced more zappers, casuals, or loyals. For one thing, A. C. Nielsen's continued focus on entire program blocks rather than more microscopic units of time means that they have no real way of measuring zapping or, indeed, the fluctuating loyalties of more casual viewers.

Throughout much of the 1990s, industry analysts overstressed the significance of the zappers. For example, Phillip Swann asserts in his book, *TV.Com: How Television Is Shaping Our Future:* "Few viewers today can sit through an entire program without picking up the remote and checking out another channel. . . . Today's viewer needs constant gratification: if she's not entertained or intrigued for any stretch of time, she will flip the dial."[34] Swann thinks interactive television should and will be designed for zappers. In Swann's future, variety and magazine shows will almost entirely displace dramas, and the few remaining series will be shrunk to thirty minutes or less. According to Swann, "[There will be] fewer occasions where people sit down and watch a show from beginning to end without interruptions. People will start watching TV shows the way they read books: a little at a time. . . . The concept of 'appointment television'—arranging to be home at a precise time to watch a particular program—will soon be a thing of the past."[35] Refusing to bow out just yet, the networks want to hold on to appointment viewing by constructing new forms of programming that demand and reward immediate attention, and they want to build up viewer loyalty by intensifying the affective appeal of their programs.

Industry research now suggests that loyals are much more valuable than zappers. According to a study done by Initiative Media, the average network program was identified as a "favorite series" by only 6 percent of its viewers. But, in some cases, as many as 50 or 60 percent of viewers may rank a program as their favorites. Early evidence suggests that these loyals have a higher rate of brand recall (a key concern of advertisers) and are much less likely to be lured away from the networks toward competing cable content (a key concern of programmers). Loyals are twice as likely to pay attention to advertisements and two to three times more likely to remember product categories than more casual viewers. And they are between 5 to 20 percent more likely to recall specific sponsors—not huge numbers, perhaps, but big enough that they can give a competitive edge to advertisers who consistently target shows with a high degree of viewer loyalty. Historically, networks ignored those fan bases in making decisions about renewing series, seeing fans as unrepresentative of the general public; but advertisers are increasingly realizing that they may be better advised investing their dollars behind shows that have a high favorability than shows that have high ratings. As this research impacts programming decisions, the media industry is trying to generate content that will attract loyals, slow down zappers, and turn casuals into fans.

At first glance, *American Idol* looks like it was designed for zappers, Each episode breaks down into bite-size units of only a few minutes' duration as each of the competing performers sings and is judged. To some degree, reality series are built up of "attractions," short, highly emotionally charged units that can be watched in or out of sequence. But the series is designed to support and sustain multiple levels of engagement.

America's Army

In 1997, the National Research Council, acting as an adviser to the U.S. Defense Department, issued its own vision for convergence culture, which they called "Modeling and Simulation: Linking Entertainment and Defense." Recognizing that the consumer electronic entertainment sector was outpacing defense research in developing simulation and artificial intelligence techniques, the defense department sought ways to collaborate with industry to develop games that could help them to recruit and train a next generation fighting force: "The DOD is interested in this capacity for large scale training exercises; the games industry is interested in networked games that would allow hundreds or thousands of players to participate."* Some have seen this report as representing a major first step toward the establishment of what is being called the military/entertainment complex. Yet, the report acknowledged many of the same challenges to collaboration we have identified elsewhere in convergence culture: "The entertainment industry and DOD are two different cultures, with different languages, different business models, and separate communities of constituents. . . . Success will rely on sustained commitment from both sides—and from a shared belief that the benefits of collaboration are worth the costs."†

Responding to this report, people in the U.S. military began to explore how games could be used to speak to younger Americans who were alienated or bored by traditional approaches to recruitment. The military also wanted to tap the communities that emerged around games as a means of rebuilding the social connections between military and civilians at a time when most military volunteers came from a relatively narrow sector of the population. Colonel E. Casey Wardynski, the man who originated the *America's Army* project, explains:

> *Whereas in the past a young American could gain insights into military service by listening to the recollections or the advice of an older brother, an uncle, a father, or perhaps a neighbor, today opportunities for such insights are relatively scarce. To the extent that information about military service shapes the career plans of young Americans today, these decisions are influenced by movies, magazines, books, and advertising. . . . Consequently, it is not surprising that young Americans with little or no contact with Soldiers are less likely to include Soldiering as a potential career. To counter this situation, the game's originator reasoned that the Army would reduce search costs by framing information about Soldiering within the entertaining and immersive context of a game. . . . A game would provide virtual experiences and insights into the development, organization, and employment of Soldiers in* America's Army.‡

The *America's Army* project has the ambitious vision of developing itself as a general popular culture brand for all kinds of media, hoping to extend outward to include comic books, television series, youth organizations, perhaps even feature films, though the game will continue to be the hub for this brand identity. When they launched the project, they made a decision

---

* Unless otherwise noted, my discussion of America's Army draws upon Zhan Li, "The Potential of America's Army: The Video Game as Civilian-Military Public Sphere," Master's thesis, Comparative Media Studies, MIT, Summer 2003.

† All quotes in this paragraph are taken from National Research Council, Committee on Modeling and Simulation, "Modeling and Simulation: Linking Entertainment and Defense," Washington, D.C., 1997, http://www.nap.edu/html/modeling/.

‡ E. Casey Wardynski, "Informing Popular Culture: The America's Army Game Concept," in Margaret Davis (ed.), *America's Army PC Game: Vision and Realization* (San Francisco: Yerba Buena Art Center, 2004), pp. 6–8.

to brand *America's Army* separately from the U.S. Army brand, allowing it to develop independently as an entertainment property. What they have produced are hard-core fans of *America's Army* who may or may not be supporters of the U.S. military. *America's Army* rapidly became one of the most popular games on the market. By August 2003, almost 2 million registered users had played over 185 million ten-minute missions. In 2004, the marketing firm i to i Research surveyed high school and college students and found them overwhelmingly pro-military. Asked the source of their favorable impressions, 40 percent cited recent combat operations in Afghanistan and Iraq. Nearly as many—30 percent—cited their experiences playing *America's Army.*[§]

The U.S. Department of Defense wanted to use the game not simply to simulate military processes but also to inculcate values. Players are never to be rewarded for killing virtual American soldiers. Each player sees the members of his or her own team as American and the opposing team members as enemies. Fragging teammates would be grounds for immediate expulsion from the game. On the other hand, the game was designed to reward players with increases in rank and access to more advanced missions when they respect the military codes of conduct.

The game has attracted international interest—42 percent of visitors to the official *America's Army* Web site log in from outside the United States (though some of these are probably service personnel and their families stationed overseas). There are organized groups of players and brigades, representing a range of different nationalities, including some from parts of the world that have traditionally been regarded as enemies. The game's designers advocated successfully for the suspension of many military regulations restricting the expression of opposing ideas to create a robust forum—which they call a "Virtual Community of Interest in Soldiering." There, civilians and service men and women could talk openly about the values of military service and even debate the merits of current military conflicts. The forum site initially offered links to a diverse array of alternative sources for information about the war, including the controversial Arabic news station, Al Jazeera. A site design in 2004 seemed calculated, at least in part, to reduce the overtly political content, cutting back on news links and discussion forums that had become centers for debates about America's continued military presence in Iraq.

Seeking to deploy them as ambassadors, the military issues gold stars to any veteran or active soldier who joined the game, and these players enjoy such a high level of prestige in the community that some players try to pass themselves off as veterans just to grab a little of that respect. The veterans take great pleasure in exposing such hoaxes, drilling them with questions that only someone who had served would know, and through this process, they assert the importance of real world experience over game-play fantasies.

As the Iraqi war began, some players said they were playing the game and watching the news at the same time, trying to achieve through fantasy what they hoped would happen in reality. When several Americans were taken prisoners by the Iraqi army, many organized efforts staged hostage rescue scenarios that engaged in Hollywood-style fantasies of how one might get these men and women back safely. Some members of the veterans' clans shipped out to serve in the Middle East but remained in contact with the other players in their community, sending back home front-line perspectives on what was unfolding. As the death tolls mounted, some of the veterans and service-personnel groups gathered together in the game world not to fight, but to talk through their feelings of anxiety and loss. *America's Army,* thus, may be more effective at providing a space for civilians and service folk to discuss the serious experience of real-life war than as a vehicle for propaganda.

[§] Wagner James Au, "John Kerry: The Video Game," *Salon,* April 13, 2004, http://www.salon.com/tech/feature/2004/04/13/battlefield_vietnam/.

American Idol is designed to pull in every possible viewer and to give each a reason not to change the channel. Many elements loyals find repetitive ensure the program's continued accessibility to casuals—things like the recaps of the previous episodes, the recurring profiles of the contestants, the rereading of key quotes from the judge's assessments. Each of these segments reorients casuals to the contest's basic mechanics or provides the background that's needed to appreciate the dramatic conflicts in that night's episode. As they move into their final weeks and more casuals are drawn into the snowballing phenomenon, *American Idol* and many other reality shows may devote an entire episode to the season's highlights, designed to provide an easy entry point. Beyond this, each episode is constructed to allow a satisfactory entertainment experience. In *American Idol,* each Tuesday night episode includes performances by all of those contestants still in the competition. Each episode also includes a cliffhanger, so *American Idol* viewers are encouraged to tune in the following night to see how the voting went. These unresolved elements are intended to pull casuals toward a more committed relationship.

As for loyals, perhaps the single most important factor separating reality from other kinds of nonfiction programming is serialization. Talent contests are a well-established genre in American broadcasting, going back at least as far as *Major Bowles' Original Amateur Hour* on radio in the 1930s. What *American Idol* added to the mix, however, was the unfolding of the competition across a season, rather than in the course of a single broadcast. Or to be more accurate, serialized talent competitions had already sprung up on cable networks, such as MTV and VH1, but FOX brought them over to the major networks and made them prime-time entertainment. In serializing the talent competition, *American Idol* is simply following a trend that runs across all contemporary television—a movement away from the self-contained episodes that dominated broadcasting for its first several decades in favor of longer and more complicated program arcs and more elaborate appeals to series history. Serialization rewards the competency and mastery of loyals. The reason loyals watch every episode isn't simply that they enjoy them; they need to have seen every episode to make sense of long-term developments.

Every reality series starts out with a cast larger than most audience members can grasp and most of those characters will receive relatively limited airtime. As the winnowing process occurs, however, certain characters will emerge as audience favorites, and a good producer anticipates those interests and rewards them by providing those characters with more airtime. Viewers move from thinking of the characters as generic types toward thinking of the characters as particular individuals. Viewers get to know the contestants, learn their personality, their motives for competing, their backgrounds, and, in some cases, other members of their families. In *American Idol,* viewers watch them improve or crash and burn. This may be why *American Idol* has become such a powerful marketing tool for launching the careers of young performers compared to earlier televised talent competition.

Talk among Yourselves!

There has historically been a tendency within industry discourse to focus either on mass, undifferentiated audiences (of the kinds that get measured by the ratings system) or individual consumers. Marketing researchers now speak about "brand communities," trying to better understand why some groups of consumers form intense bonds with the product and, through the product, with fellow consumers. In one study that helped to define the concept of "brand community," marketing professors Albert M. Muniz Jr. and Thomas C. O'Guinn concluded: "Brand communities carry out important functions on behalf of the brand, such as sharing information, perpetuating the history and culture of the brand, and providing assistance [to other users]. They provide social structure to the relationship between marketers and consumer. Communities exert pressure on members to remain loyal to the collective and to the brand."[36] These brand ethnographers research specific groups of highly committed consumers (such as Harley-Davidson riders, Apple Computer users, or Saturn drivers) or what they call "brandfests," social events (either commercially sponsored or grassroots) that pull together large numbers of consumers.

As these brand communities move online, they are able to sustain these social connections over long periods and thus to intensify the role the community plays in their purchasing decisions; they expand the number of potential consumers who interact with the community and help to move casual consumers into a more intense engagement with the product. Marketing professor Robert Kozinets sees these online consumption communities, whether focused on a single product or a cluster of related products (coffee, wine, cigars), as places "where groups of consumers with similar interests actively seek and exchange information about prices, quality, manufacturers, retailers, company ethics, company history, product history, and other consumption related characteristics."[37] In short, they are something like Pierre Lévy's knowledge communities applied to consumer decision making. Participation within such communities does not simply reaffirm their brand affiliation but also empowers these groups to assert their own demands on the company. As Kozinets explains, "Loyal consumers are creating their tastes together as a community. This is a revolutionary change. Online, consumers evaluate quality together. They negotiate consumption standards. Moderating product meanings, they brand and rebrand together. Individuals place great weight on the judgment of their fellow community of consumption members. . . . Collective responses temper individual reception of marketing communications. . . . Organizations of consumers can make successful demands on marketers that individual consumers can not."[38]

Just as the social dynamic of these online communities reaffirm and/or redefine their individual members' brand loyalties, a similar social dynamic shapes the ways people consume media and products within their families or with friends. A team of researchers from MIT's Comparative Media Studies Program and Initiative Media joined forces to document audience response to the second season of *American Idol*.[39] The MIT team sent researchers into homes and dorm rooms to observe people watching television; we did one-on-one interviews with a range of different consumers; we did surveys through the official *American Idol* Web sites; and we monitored discussions within the fan community. The Initiative team ran large-scale surveys and focus groups and collected aggregate data from the official *Idol*-on-FOX home page. We wanted to better understand how people integrated the experience of watching *American Idol* into the rest of their social interactions.

The Initiative Media/MIT research team found that in almost every social space where *American Idol* was watched, viewers of different degrees of commitment were present.

In one family setting, for example, the two youngest children (girl, 9; boy, 7) were among the first into the family room for the start of *American Idol*; they generally watched the last few minutes of *Lizzie McGuire* (Disney, 2001) before turning the channel to FOX. As the program started, both routinely raised their voices to announce to those throughout the house that "it's starting!" As the opening segments began, they recalled last week's performances and made comments about what the judges are wearing or how their hair looks. By the end of the first segment, the mother had usually arrived and stood at the door. Typically, she would come and go for the first thirty minutes of the program, working in the kitchen or running up and down the stairs. Such distracted viewing is fairly common for women; even those fairly committed to the program must respond to competing demands on their attention in the early evening hours.[40] The father would typically enter the broadcast in progress and the older daughter would only occasionally view the program, allowing other family members to bring her up to speed as needed. The son's attention would wander during the commercial break and he would begin to zap around the dial and they might miss the start of the next program segment unless the mother and daughter called him back to FOX on time. Finally, as the entire family was sitting in front of the television during *American Idol*'s recap in the last five minutes of the program, they would debate the contestant before the mother would call in the vote. The conversation never ended in disagreement, although in later weeks the young girl became more vocal about whom she expected to win, waffling between Ruben and Clay. The father took it all in and endorsed the family's choice based on the snippets of performance he saw in the recap.

Through family conversation, the loyals pulled the casuals into the fold and held on to the attention of the would-be zapper; they announced when the program was on and updated those who have missed segments. In the absence of such reinforcement, some of the family members probably would not return each week, yet even the most casual participants see watching the series as a family ritual. One of the effects of expanding audience participation is to give such families more different ways to engage with the content; discussing who to vote for becomes part of the viewing experience and provides an incentive for everyone to watch the recap if not the individual performances. Researchers have found that such shared rituals or mutual evaluations are central to the sense of affiliation members feel to the group, and it makes sense that similar rituals would be played out in individual households.[41] *American Idol* can become family entertainment because it lies at the intersection between youth and adult tastes, allowing everyone to show some expertise. Most of the contestants are in their teens or twenties. To broaden its focus, the show brings in aging pop stars as guest judges and coaches: Burt Bacharach, Billy Joel, and Olivia Newton-John appeal not to the kids who compete but to their parents'—or even their grandparents'—generation.

The researchers who observed college students watching *American Idol* in a dorm common area found similar patterns: different students had different investments in the various contestants and carried on debates from week to week about their relative merits. Catch phrases might be dropped ironically into the conversation. People who had missed some episodes could reenter the series with the help of their friends because they knew the rules of the competition and had some passing familiarity with the contestants. Some people wandered into the commons area with no prior plans of watching the series and got sucked in. The number of committed viewers grew week by week as the competition mounted and as watching the series became more central to the social life of the dorm community. Interestingly, the final episodes conflicted with final exams, so the group taped them, made a pact not to look at the results, and scheduled time to watch them together.

Across a range of studies, Initiative Media has found that different genres of entertainment provoke different degrees or kinds of social interactions. Drama viewers are the most likely to view alone, comedy to view with family members, and reality viewers to view with friends. Demographically, 18–34-year-old viewers have the most varied habits, depending on program genre. Adults over 50 either view alone or with family, but rarely with friends, while 35–49-year-olds are most consistent, viewing all genres primarily with family members. People watching in groups pay more attention to the program content, are less likely to shift channels mid-broadcast, and are more likely to access program-related Web sites. Of course, as those viewers move onto the Web, some are choosing to discuss their interpretations and assessments of the programs via online fan communities. Social viewing, then, would appear to be an important driver behind brand and content extension.

A survey of 13,280 *American Idol* viewers, conducted through the official FOX Web site, found that the majority of fans discovered the series on the basis of word-of-mouth and watched it regularly because other people they knew were also watching it. (On the zapper side of the equation, the same study found that significantly more viewers stumbled onto the series while channel surfing than tuned in consciously on the basis of prior awareness.) While historically men make the viewing decisions during the primetime hours, only slightly more women (32 percent) than men (31 percent) said they started watching the show because other family members were watching it. Altogether, 78 percent of *American Idol* viewers surveyed said that they watched the show with family or friends, and 74 percent reported that they talked about the show with friends during the week between episodes.

Such conversations extended beyond the initial viewing group to friends, workmates, or more distanced relatives. As one respondent told us, "My mother lives in Africa, my aunt lives in Russia, but they are able to watch the show on the Armed Forces Network over the weekend. My other aunts, scattered about the country, will create tests and drop stupid hints that all come clear when they finally get to watch the show. It's a family viewing occurrence, which I don't usually have the opportunity to do." Even if they missed individual episodes, study participants made a conscious effort to keep up with developments to participate in casual conversations with peers and coworkers. Consequently, many more people know about *American Idol,* follow its development, and even get exposed to its marketing message than actually sit down and watch.

Phone companies, across the board, observed a marked increase in traffic on Tuesday nights following the broadcasts. In the third season's final week, Verizon reported an increase of 116 million calls, a 7 percent increase over a typical Tuesday, and SBC saw 100 million more calls, an 80 percent increase over a normal weeknight.[42] In all likelihood, this increase was not simply stimulated by the large number of people who were voting, but also indicates the number of people chatting about the program content.

How Gossip Fuels Convergence

One of the survey respondents captures the spirit of these conversations: "[Watching *American Idol*] helps me to relax because it gives me something to talk about with friends that doesn't effect our lives in any big way; therefore, it is an easy thing to discuss." Historically, gossip has been dismissed as "worthless and idle chatter," but over the past several decades, feminist scholars have begun to reappraise the place of gossip in women's community, and subsequent writers have extended it to talk about interactions within a much broader range of communities. Writing in 1980, Deborah Jones described gossip as "a way of talking between women in their roles as women, intimate in style, personal and domestic in topic and setting."[43] Gossip, she argued, allowed women to talk about their common experiences, share expertise, and reinforce social norms. While the fluidity of gossip makes it difficult to study or document, Jones suggests that gossip is an important resource that women historically have used to connect their personal experiences within a larger sphere beyond their immediate domestic environment. The specific content of gossip is often less important than the social ties created through the exchange of secrets between participants—and for that reason, the social functions of gossip hold when dealing with television content. It isn't who you are talking *about* but who you are talking *with* that matters. Gossip builds common ground between participants, as those who exchange information assure one another of what they share. Gossip is finally a way of talking about yourself through critiquing the actions and values of others.

As cyberspace broadens the sphere of our social interactions, it becomes even more important to be able to talk about people we share in common via the media than people from our local community who will not be known by all of the participants in an online conversation. Into that space step the complex, often contradictory figures who appear on reality television. Reality television provides consumers with a steady stream of ethical dramas, as contestants are forced to make choices about whom to trust and what limits to set on their own behavior. Viewers can argue about whether Joe Millionaire picked the right woman or The Donald fired the right apprentice, whether it's OK to lie your way to success on *Survivor,* and whether Clay, Ruben, or Kimberley sang best on *American Idol.* In a focus group study of reality television viewers conducted by Initiative Media, 60.9 percent of the respondents said that the ethical conduct of the contestants was a central topic of their discussions around such series. By way of contrast, 67 percent discussed the outcomes, 35 percent discussed strategies, and 64 percent discussed personalities. Rather than being morally debasing, ethically dubious on-air conduct frequently encourages a public discussion of ethics and morality that reaffirms much more conservative values and assumptions. In a multicultural society, talking through differences in values becomes a mechanism by which different social groups can learn more about how they each see the world, so there is a real value in gossip that extends into virtual rather than face-to-face communities. The reality contestants put themselves forward to be judged by their audience; and through their judgments, the audience reaffirms their own shared values by expressing outrage over their social transgressions and they learn about their differences by sharing notes about how they respond to shared ethical dramas.

American Idol viewers debate whether the contest should be decided on the basis of "pure talent," or whether it is legitimate to draw on other factors, such as personality or appearance, which are often key to defining commercial success. Consider, for example, the self-righteous indignation expressed by one *American Idol* viewer who wrote to the CMS/Initiative Media research team convinced that we had a direct pipeline back to the producers. This message followed an episode where fan favorite and ultimate winner Ruben Studdard unexpectedly ranked near the bottom:

> *Do you really think the American public believes for one second that Ruben could possibly be voted into the bottom? Ruben has never had one negative comment said about him nor has he ever not been excellent. He has never missed a tune. . . . It is very cruel to mislead these young people into trusting you are going to be honest and fair. This is a talent show isn't it? Hence the word TALENT SHOW. . . . So do the right thing and seriously look at who is counting those votes and maybe check to see if they actually know how to count. If you are rigging the show you will probably burn in hell for being so stupid. (A true and honest American)*

What was striking to us about this comment, apart from its cynicism about the voting process, was its moral intensity; its firm belief that the outcome of a talent competition should be read in terms of questions of justice, honesty, and equity. Another respondent referenced her "responsibility" to monitor the results to see "whether America has chosen and voted fairly . . . whether or not America really votes based on real talents or just tits and asses."

Assessment is a two-part process: first, discussions about the performances, and then discussions about the outcomes. Among the most committed and socially linked members of a consumption community, these evaluative standards are arrived at collectively, just as the members of the family we described pooled their individual tastes to make a collective decision about who should win. Such a process tends to pull toward a consensus over time, and then, over a longer time, the consensus no longer seems to be something that was disputed or haggled over; it is the commonsense outcome. We can see this as part of the process through which collective intelligence generates shared knowledge. Some critics, such as Cass Sunstein, argue that this process of consensus formation tends to decrease the diversity of perspectives that any community member encounters; people tend to flock toward groups that share their existing biases, and over time they hear less and less disagreement about those core assumptions.[44] At the same time, this consensus-forming process increases the likelihood that these brand and fan communities will speak up when corporate interests cross the group's consensus. In the course of a season of *American Idol,* total consensus may not have been achieved, but most members of the online community saw season 2 as a contest between Clay and Ruben, making the buildup toward the season finale that much more powerful. The community expressed outrage, however, when voting went against the perceived consensus, as occurred one week when Ruben was almost bumped (the incident provoking the above response) or as occurred frequently throughout the third season.

Because the characters are real people whose lives continue beyond the series borders, viewers are left feeling like there is more and more they might know about them, which provides an incentive to track down additional information through multiple media channels. This search for the hidden "truth" of reality television is what motivated the spoiling described in chapter 2. The Initiative Media survey found that 45 percent of *American Idol*'s loyals went onto the Web in search of more information about the program, and it is generally agreed that reality television is one of the primary drivers of traffic to network Web sites.

These networks build upon synergies within the entertainment corporations to ensure that talk about their hit reality series continues throughout the week. The contestants are featured prominently on morning and late-night talk shows and on network affiliated chatrooms. The results of at least the top-rated reality series are seen as news events that will be covered even by rival networks. In the case of *American Idol,* for example, *USA Today, Entertainment Weekly,* and AOL each conducted its own independent audience surveys designed to second-guess the likely results before they are aired in the network. Online gossip magazines, such as *The Smoking Gun,* have sought to tap public interest of the series, digging out old criminal records or divorce proceedings involving contestants. In some cases, reality series market access to exclusive content, which further expands the viewing experience. Given the pervasiveness and diversity of such publicity, any given fan is apt to know something his friends do not, thus creating an incentive to share knowledge. This publicity also has the effect of making some viewers more apt to want to watch the episodes as they are aired so they can avoid finding out the outcomes in a less dramatically compelling fashion. For other consumers, such coverage keeps their interest alive even if they are unable to watch some installments and makes it more likely they will tune in for the final episodes of a particular run.

Contesting the Vote

So far, we have focused our discussion on those factors that ensure viewer loyalty to *American Idol,* but as Heyer's speech suggests, sponsors are seeking to transfer viewer loyalty from entertainment properties onto their brands. The majority of the people our research team interviewed were acutely aware that *American Idol* was serving as a testing ground for branding strategies and were eager to offer their opinions about the experiments as they unfolded. Product placements and program-themed commercials became an acknowledged part of the *American Idol* phenomenon, something people, in some cases, tuned in to see—much as the Super Bowl has become as much a showcase for advertising as a sporting event. Coca-Cola spoofed the uncompromising honesty of judge Simon Cowell, depicting him as forced by a mobster to read an endorsement for Vanilla Coke; Ford created new musical segments each week featuring the program contestants; AT&T created a campaign that mimicked *Legally Blonde* (2001) and showed an airheaded teenager going around the country encouraging people to participate in the call-in voting process. Sponsors are not simply seeking the chance to advertise their products; they are seeking to brand the content so that the red of the *American Idol* set becomes inseparable from Coca-Cola's sponsorship of the series, so that the Ford spots featuring the contestants becomes part of the evidence fans mobilize in support of their favorite performers, and so that AT&T's text-messaging system becomes the preferred vehicle for voting.

Viewers are more accepting of product placements in reality programming than in any other genre (they are least comfortable with product placements in drama, news, and children's programming). Some are turned off by this hypercommercialism, but for others, recognizing marketplace interventions has become part of the "game": "I find myself trying to pick out products placed in shows and get an a-ha moment when I find one." Even those who claim not to watch commercials are drawn toward series-targeted spots: "You know what I do in the commercial breaks? Refill my popcorn bowl. Go to the bathroom. Bake a cake. Sing a song. Dance a dance. I refuse to be made to sit through that crap!!! However, I really like the Simon/Vanilla Coke commercial." Even many of those who refused to watch the show because it was overcommercialized still accurately named its sponsors. In some cases sponsors improved public perception of their brand, whereas others potentially damaged their standing. As one regular viewer told our researchers, "Now I know for sure that AT&T Wireless and Ford and Coca-Cola advertise with them, but it's to the point of being annoying and I want nothing to do with those particular brands now." Others couldn't separate consumerism from their participation in the series: "*Sigh* Yes, I purchased a sweatshirt from Old Navy because Aiken wore it in the studio recording of 'God Bless the USA' and I loathe Old Navy. Normally, I hate that kind of stuff." The early evidence, however, suggests that as a general rule, the more invested viewers became in *American Idol,* the more committed they became to its sponsors.

Such a tight integration of advertising and content is not without its risks, since the credibility of the sponsors became closely linked with the credibility of the competition itself. Kozinets warns that participating in a consumption community heightens one's awareness of the consumption and marketing process and reaffirms feelings of resentment if a company exploits that relationship. The collective voice speaks louder and often more decisively than its individual members. Such expressions reach the ear not only of the companies being challenged but also the mainstream media; consumer backlashes are increasingly being covered as "scandals," which puts further pressure on the companies to respond. In some cases, Kozinets notes, the corporations, angry over their loss of control, threaten or punish their most loyal consumers, undermining valuable relationships. Kevin Roberts argues that companies need to see such scandals as opportunities to listen to and learn from their most hard-core consumers, building up greater loyalty through their responsiveness rather than tearing it down through indifference or overreaction.

The down-to-the-wire contest between Clay Aiken and Ruben Studdard turned out to be almost as close as the 2000 presidential election, with the two finalists separated by a little more than a hundred thousand votes out of 24 million votes cast. The text message votes all got through and were counted—several million worth—whereas millions of telephone callers faced endless busy signals. As one fan explained in our survey, "Hanging chads in Florida is nothing compared to this stupid voting procedure." Clay supporters were particularly vocal about the degree to which clogged phone lines made it impossible to get an accurate count, and some argued that the lines may have been arbitrarily restricted to ensure a close race. The *American Idol* producers had raised expectations about responsiveness to audience feedback and thus faced a backlash when they failed to meet those expectations. By the third season, inconsistencies in the voting made headlines in national newspapers, with the network forced to acknowledge that significant numbers of callers were not able to register their votes because local phone lines were being flooded. The result was an uneven counting of votes from one region to another. For example, viewers in Hawaii, an area where there were relatively few people competing for access to the local connection, could have cast as much as a third of the total votes third season, an imbalance which some have argued accounted for the prolonged run of a Hawaiian contestants.[45] As the controversy intensified, they expanded voting hours and added more phone lines to try to lower public disappointment. An editorial in *Broadcasting & Cable* warned: "Viewer loyalty is hard to build and tough to maintain. . . . With AT&T one of the show's sponsors, FOX needs to go out of its way to avoid the appearance that it could be in cahoots with the phone company to drive as many calls as possible, whether or not they get through."[46] Despite such scrutiny, FOX has refused to release the actual vote counts, offering only partial information on a selective basis during the broadcasts. Many fans argue that such selective reporting makes it hard for them to put much faith in the reliability and impartiality of the process.

Complaints went beyond the voting mechanism to include concerns about how particular contestants were "pushed" by the judges and the producers, given higher prominence on the show and more supportive comments, or, in some cases, intentionally attacked to inspire audience backlash against the judging. Cynics saw the producers as more interested in generating controversial and compelling programming than in recognizing talent. Much as the spoilers sought to thwart Mark Burnett's efforts to keep the *Survivor* outcome a secret, the *American Idol* online community took pleasure in trying to read through the "mechanisms" by which the producers "engineer" the results. As one fan explained, "I like seeing Simon trying to figure out the power of saying evil things to create a backlash, and saying to a so-so performer that they are amazing."

For many, such efforts to shape public response were seen as an extension of the sponsor interference into program content. The performers, they argued, were becoming so "packaged" that they were no different from the other products being advertised. In some cases, the "Idols" became models who displayed new fashions, new makeup, and hair-styling products. Fans suggested this refashioning of their images was simply the first step in what would result in overprocessed versions of their performances when their albums were released.

This degree of anger suggests that product placements might be a double-edged sword—on the one hand, higher consumer awareness and, on the other, higher consumer scrutiny. Virtually every research participant had some criticism of the ways that commercialism tainted the series, complicating arguments which might see media-savvy marketers manipulating naïve and gullible consumers. Even loyals complained that the series was sometimes nothing more than a "merchandise machine."

These online communities gave "inspirational consumers" a place to talk about their resistance to these new forms of commercialism. In critiquing the results, fans often focused on the corporate interests they saw as shaping the outcome. Through this process, more economically conscious participants could educate others about the commercial structures shaping American broadcasting. In some cases, *American Idol* fans used the resources of these online communities to identify flaws in the voting system. This summary from one fan site suggests the sophistication with which they were collecting information:

> *Most text messages go through—according to message board posts, websites maintained by the text messaging company, and through news articles. But viewers pay a small fee to send a vote in text—so paying for the vote does give you leverage. Yet last year when Ruben Studdard won, texters on an* American Idol *message board reported some of their messages didn't go through. Hours after the calls, people said their phone companies send back text error messages from their phone carriers saying some messages failed. Up till that point, texters were reporting 100 percent completion.*[47]

American Idol fans discussed voting strategies they felt would counter such distortions in the competition. Their efforts might be aimed at supporting the best singer, balancing out negative comments, or undermining "heavily marketed" contestants. The producers had, from day 1, sought to position the third season as a battle between three black "divas," and the judges had all but proclaimed Fantasia Barrino the likely winner. As the other two black "divas" went down and as Fantasia ranked near the bottom of the vote counts across several weeks, guest judge Elton John denounced America's voting patterns as "incredibly racist."[48] Such seemingly erratic voting patterns, however, make more sense if we see them in the context of a growing backlash among the most hard-core viewers to what they saw as open attempts to take away their right to choose the Idol.

Sponsoring such a show ensures that companies will get talked about, but it doesn't guarantee what the audience is going to say about them. Much of this chapter has looked at *American Idol* in terms of the behind-the-scenes calculations of media companies such as FOX, consumer brands such as Coca-Cola, and marketing researchers such as Initiative Media. Yet, we must also take seriously the backlash of Roberts's "inspirational consumers." Who wins *American Idol,* in the end, doesn't matter that much in the great scheme of things. But, the debates about *Idol* voting are debates about the terms of audience participation in American media. At a time when networks and sponsors are joining forces to shape the emotional context through which we watch their shows, then consumers are also scrutinizing the mechanisms of participation they are being offered. If the rhetoric of Lovemarks emphasizes the audience's activities and investments as a central source of value in brands, then the consumption community may well hold the corporations accountable for what they do in the name of those brands and for their responsiveness (or lack thereof) to consumer demands. Such disputes generated considerable "heat" around the series, drawing in many new viewers, but they also alienated and disenfranchised many of the most dedicated ones.

Too much backlash can damage ratings or hurt sales of the products. The Initiative Media study found that AT&T, the company that had branded the voting mechanism, was damaged by the public backlash and that the other key sponsors—Coca-Cola and Ford—may have been hurt as well. No one would imagine that viewers might translate bad will to one advertiser in a traditional segment of commercials toward another advertiser. Yet in a world where sponsors are more closely associated with the content, all of the hosting companies may be negatively affected by any negative perceptions that emerge around the series. It is through struggles that the relationship between media producers and consumers will get redefined in the coming decades.

Understanding when audience backlash hurts companies—or, for that matter, how far companies can go in shaping the nature of audience participation—is central to what I have been calling affective economics. If a program is going to become, in Heyer's terms, the "emotional capital" of its consumers, then we can expect consumers to make different investments in the program than the producers do, and for the love behind the Lovemarks to turn into hate when producers alter something that the brand community sees as fundamental to their experience. For the moment, the marketing industry still has a long way to go if it wants to understand the complexity of audiences' emotional investments in entertainment properties and brands. And audiences have a long way to go if they are going to exploit the points of entry that affective economics offers them for collective action and grassroots criticism of corporate conduct.

Notes

1. Jefferson Graham, "Idol Voting Strained Nerves, Nation's Telephone Systems," *USA Today,* May 27, 2003, http:www.usatoday.com/life/television/news/2003-05-26-idol_x.htm.
2. Jeff Smith, "Getting the Mssg: U.S. Wireless Carriers Mining the Airwaves for Ways to Profit from Text Messaging," *Rocky Mountain News,* May 19, 2003.
3. Ibid.
4. "AT&T Wireless Text Messaging Takes Center Stage with Unprecedented Performance on Fox's American Idol," *PR Newswire,* April 16, 2003.
5. Scott Collins and Maria Elena Fernandez, "Unwanted Wrinkles for Idol," *Los Angeles Times,* May 25, 2004, p. 1.
6. Stuart Elliott, "The Media Business: Some Sponsors Are Backing Off to Fine-Tune the Art of Blending Their Products into Television Shows," *New York Times,* January 22, 2003.
7. Jennifer Pendleton, "Idol a Standard for Integration," *Advertising Age,* March 24, 2003.
8. Penelope Patsuris, "The Most Profitable Reality Series," *Forbes,* September 7, 2004, http://www.forbes.com/home_europe/business/2004/09/07/cx_pp_0907realitytv.html.
9. Gary Levin, "No Summer Vacation on TV: Networks Aggressively Chase Audiences with Reality, Original Series," *USA Today,* June 3, 2004, p. 1D.
10. Carla Hay, "Idol Ups Stakes for TV Talent," *Billboard,* April 26, 2003.
11. Karla Peterson, "False Idols: How to Face Down a Media Monster So We No Longer Worship Moments Like This," *San Diego Union-Tribune,* December 16, 2002.
12. Vance Packard, *The Hidden Persuaders* (New York: Bantam, 1957).
13. Applebox Productions, Inc., marketing postcard, circa 2000.
14. Personal correspondence with the author, December 31, 2004.
15. Anthony Bianco, "The Vanishing Mass Market," *Businessweek,* July 12, 2004, p. 62.
16. Ibid., p. 64.
17. Ibid., p. 62.
18. Susan Whiting, remarks at the MIT Communications Forum, April 17, 2003. Streaming audio of the session can be found at http://web.mit.edu/comm-forum/forums/nielsen.html#audiocast.
19. Stacey Koerner, remarks at Media in Transition 3 Conference: Television, MIT, Cambridge, Mass., on May 3, 2003. Streaming audio of this session can be found at http://cms.mit.edu/mit3/.
20. Scott Donaton, *Madison and Vine: Why the Entertainment and Advertising Industries Must Converge to Survive* (New York: McGraw-Hill, 2004), pp. 10–11.
21. Michael Schneider, "Fox Revs Ford for Blurb Free 24," *Variety,* July 21, 2002.
22. Donaton, *Madison and Vine,* p. 18.
23. Stacey Lynn Koerner, David Ernst, Henry Jenkins, and Alex Chisholm, "Pathways to Measuring Consumer Behavior in an Age of Media Convergence," presented at the Advertising Research Foundation/ESOMAR Conference, Cannes, France, June 2002.
24. Steven J. Heyer, keynote remarks delivered before Advertising Age's Hollywood + Vine Conference, Beverly Hills Hotel, Beverly Hills, California, February 5, 2003. For a transcript of the remarks, see http://www.egta.com/pages/Newsletter%20-%20Heyer.pdf. All subsequent references to Heyer refer to these remarks.
25. Kevin Roberts, *Lovemarks: The Future Beyond Brands* (New York: Power House Books, 2004), p. 43.
26. Joe D'Angelo, "Ruben Debuts at #1 but Can't Match Clay's First-Week Sales," VH1, December 17, 2003, http://www.vh1.com/artists/news/1482928/12172003/aiken_clay.jhtml.

27. Theresa Howard, "Real Winner of 'American Idol': Coke," *USA Today,* September 8, 2002; Wayne Friedman, "Negotiating the American Idol Product Placement Deal," *Advertising Age,* September 29, 2003, accessed at http://www.adage.com/news.cms?newsId=38800.
28. Sara Wilson, interview with Carol Kruse, IMedia Connection, October 2, 2003, http://www.imediaconnection.com/content/1309.asp.
29. Robert V. Kozinets, "E-Tribalized Marketing? The Strategic Implications of Virtual Communities of Consumption," *European Management Journal,* 17 (3) (1999): 252–264.
30. Roberts, *Lovemarks,* p. 170.
31. Ibid., p. 172.
32. Marc Gobe, *Emotional Branding: The New Paradigm for Connecting Brands to People* (New York: Allworth Press, 2001); John Hagel III and Arthur G. Armstrong, *Net.Gain: Expanding Markets through Virtual Communities* (Cambridge, Mass.: Harvard University Press, 1997).
33. Don Peppers, "Introduction," in Seth Gordon, *Permission Marketing: Turning Strangers into Friends and Friends into Customers* (New York: Simon and Schuster, 1999), p. 12.
34. Phillip Swann, *TV.Com: How Television Is Shaping Our Future* (New York: TV Books, 2000), pp. 9–10.
35. Ibid., p. 31.
36. Albert M. Muniz Jr. and Thomas C. O'Guinn, "Brand Community," *Journal of Consumer Research,* March 2001, p. 427.
37. Kozinets, "E-Tribalized Marketing?" p. 10.
38. Ibid., p. 12.
39. Preliminary findings were reported in David Ernst, Stacey Lynn Koerner, Henry Jenkins, Sangita Shresthova, Brian Thiesen, and Alex Chisholm, "Walking the Path: Exploring the Drivers of Expression," presented at the Advertising Research Foundation/ESOMAR Conference, June 2003.
40. David Morley, *Family Television: Cultural Power and Domestic Leisure* (London: Routledge, 1996).
41. James H. McAlexander, John W. Schouten, and Harold F. Koenig, "Building Brand Community," *Journal of Marketing,* January 2002, pp. 38–54.
42. Deborah Starr Seibel, "*American Idol* Outrage: Your Vote Doesn't Count," *Broadcasting & Cable,* May 17, 2004, p. 1.
43. Deborah Jones, "Gossip: Note on Women's Oral Culture," *Women's Studies International Quarterly* 3 (1980): 194–195.
44. Cass Sunstein, *Republic.Com* (Princeton, N.J.: Princeton University Press, 2002).
45. Wade Paulsen, "Distorted *American Idol* Voting Due to an Overtaxed American Power Grid?" *Reality TV World,* http://www.realitytvworld.com/index/articles/story.php?s=2570.
46. Staff, "The Right Fix for Fox," *Broadcasting & Cable,* May 24, 2004, p. 36.
47. Joan Giglione, "What's Wrong with the *American Idol* Voting System," May 24, 2004, no longer on the Web.
48. Wade Paulson, "Elton John Calls *American Idol* Voting 'Incredibly Racist,'" *Reality TV World,* April 28, 2004, http://www.realitytvworld.com/index/articles/story.php?s=2526.

MODULE FOUR

Digital and Network Aesthetics

Digital Film

Stephen Keane

The digital has been theorized in immensely fluid terms. This is based on the practical understanding that digital processes work through intricate sequences and seeming endless combinations of binary code. Or, founded upon what Nicholas Negroponte has termed the 'DNA of Information' (Negroponte 1995: 11–20), digital processes have entered into not only the body but also the genetic makeup of existing media forms. The terms most commonly applied from out of this effervescent reformation include 'manipulability', 'conversion', 'versatility' and 'interactivity'. But one of the main effects of the digital is that it has also given retrospective strength to the familiar properties of analogue in terms of 'permanence', 'authenticity', 'depth' and 'truth' (see Usai 2001; Dixon 2003).

In this chapter, I will begin by looking at the basic advantages and disadvantages of the digital image as applied to photography and film. This will be followed by consideration of digital practices in key areas of film production such as sound and editing. In the second section, I will go on to look at recent arguments surrounding 'independent' versus 'mainstream' production, beginning with the ways in which digital video has primarily been said to have benefited experimental and independent filmmaking. The most recent development, however, has been that of high-definition (HD) cameras, and filmmakers now have a variety of choices regarding which format they choose—whether film, video, digital video or high-definition. In my case studies, I will be looking at Mike Figgis's experimental *Timecode* and the DV horror films, *My Little Eye* and *28 Days Later.*

Going Digital

As Timothy Binkley argues, three of the main advantages of digital media are: 'conversion', 'resilience' and 'interactivity'. First of all, analogue media work through 'transcription'. Whether ink on paper, sound on vinyl or light on film, once transcribed, analogue signals remain permanently imprinted on the receiving medium. Data from a digital source, however, is effectively stored and available for an array of uses within the receiving medium or, indeed, to be further converted to other media (Binkley 1993: 94–101). Second, one chip in the wrong place on a sculpture or too many scratches on a vinyl record, and analogue media can suffer permanent physical damage. Even where the same could happen with digital media—the computer going wrong or a scratch on a DVD—the data doesn't simply disappear, the essential source unrecoverable. Similarly, while analogue media often work through the ability to tape music and films, the more the copies, the more the reduction in quality. Data is there to be endlessly manipulated and converted in digital media, and doesn't suffer the same 'generation loss' (Binkley 1993: 101–3). And finally, Binkley sees in the spatial properties of digital media new forms of representation and involvement. It is in this sense that Binkley regards digital media in terms of 'dialogue' rather than 'monologue', the play-and-receive properties of analogue media replaced by a greater interplay between the user-creator and digital forms (Binkley 1993: 116–19). This can be applied to a number of activities from graphic design to playing videogames and downloading music compilations.

The Digital Image

The relative pros and cons of the digital have found particular focus as applied to the digital image. These will be explored in relation to still images and then the moving image. As Janet Harbord argues, the shift from analogue to digital can be theorized via further distinctions between 'modernism and postmodernism; the bounded and the fluid, the fixed and the ephemeral' (Harbord 2002: 139). Digital images provide us with practical and relatively 'settled' examples of digital processes in that they are the impermanent results of those processes. To begin with digital photography: as well as the usual analogue viewfinder, digital cameras contain an LCD screen. Allowing the user to hold the camera at a distance, the digital screen also allows instant viewing of the photos taken. This leads to relatively little wastage as pictures can be viewed and chosen on-screen. Stored on the camera's memory card, digital photos can then be converted onto the PC. This allows the user to edit and manipulate the images on screen: alter their size, clean them up in terms of brightness and contrast, or add effects and patterns. Once stored on the PC, digital photos can be set as screensavers, sent by email or posted on webpages. And, ultimately, the user might also want to print the pictures, whether through a photo-quality printer, emailed to their service provider's picture service or through a traditional photo-processing outlet.

The possible disadvantages of the digital might already be apparent from the above sketch of what many regard as being the hyper-convenient nature of digital cameras (and, even more so, mobile phone cameras). Digital photographs can be flexible bordering on disposable, the amount of pictures taken providing for a quantity rather than quality of images. This is further compounded by the virtual nature of so many digital photos that are not made into actual prints and what many regard to be the artificial nature of digital images. Lacking material depth and infinitely manipulable, the 'art' and 'aura' of the analogue photograph are lost along with longstanding notions of photographic 'authenticity' and 'truth'. Memories are now more like snatches of activity and can, in any case, be cleaned up after the event. And where photography has traditionally been regarded as amateur and democratic, it is now restricted to 'prosumers' (see Mitchell 1992; Lister 1995; Stallabrass, 1996: 13–39).

When applied to film, the binary of analogue and digital follows in similar terms of fixity and flux but as applied to the ongoing history of the moving image. Having established some of the main advantages and disadvantages of the digital image, the continuity approach to both photography and film looks at the digital image more in terms of historical evolution rather than the shock of the new. This is particularly manifest in Philip Rosen's account of the history and historicity of cinema; that is, the ways in which cinema has developed and the ways in which those developments have been theorized. In Rosen's account, claims for the radical novelty of the digital image can be put into perspective by pulling back on ecstatic accounts of the free-floating image favoured by advocates of the 'digital utopia'. Where theorists of the analogue—or 'indexical'—image such as Andre Bazin and Roland Barthes fixated on light-sensitive chemicals, proponents of the digital image fetishize maths (Rosen 2001: 305). If the numerical bases of the digital represented a complete rescuing of the image from physical referents, however, the digital image would remain purely abstract. Hence, Rosen argues that the digital, in effect, *simulates* the analogue and *mimics* photography, film and television. Or, to put it another way, the digital image still has to be an image of something. Rosen goes on to look at overenthusiastic discourses surrounding the digital at all levels of film production, distribution and exhibition. With regard to the supposed 'infinite malleability' (Rosen 2001: 340) of the digital image, for example, Rosen quite rightly points out that, once released, digital films become just as permanent as analogue films.

Digital Production

In its purest form, the digital image translates to what we actually see in print and on screen. Before we look at the ways in which this applies to the digital video image and then spectacular special-effects images, there are also the ways in which digital practices work within film production.

Two of the foremost digital film practices have been in the areas of sound and editing. Sound designers such as Walter Murch and Ben Burtt have been working on 'electronic' recording and mixing since the 1970s. Developing an extensive library of sound effects, the storage capacities of the digital mean that all of these sounds are available in the one place at the same time. The sound effects can then be manipulated and combined, and filtered and channeled to give them depth and direction. Principally, digital sound practices have been said to enhance the 'worldizing' properties of contemporary film, creating more complete, three-dimensional environments (see Ondaatje 2002). The *spatial* advantages of digital processes have also come to transform the temporal art of editing. As sound designer and film editor Walter Murch states, 1995 was a watershed year in editing, the last time the number of mechanically edited films equalled the number of digital (Murch 2001: xi). The three main digital editing systems are Avid, Apple's Final Cut Pro and Microsoft's Lightworks. Murch highlights two of the main advantages of digital editing systems as being 'virtual assembly' and 'instant random access'. Unlike the 'destructive' analogue, the virtual assembly allows editors to edit sequences continuously because it is a process that separates images and information (Murch 2001: 81–2). Instant random access means that the editor can work on any part of a film at any time. Combining both of these elements, digital editing brings production and postproduction closer together in that the editing can take place as soon as the day's rushes are available, the process remaining open, as always, until the final print (Murch 2001: 107–9).

Digital editing is also referred to as non-linear editing (NLE). Digital editing's ability to work out of sequence has been used to further arguments surrounding the loss of cohesion of contemporary film narratives. As Monica Mak argues, however, digital editing still ends up 'continuity editing' when it comes to the final film text and even more so in the case of blockbusters that make use of separate production units (Mak 2003: 43–5). The most recent development in digital editing has been that of '3D editing', the ability to edit within rather than solely across frames. This has resulted in an 'Avid aesthetic' (Griffiths 2003: 19) evidenced in films such as *Requiem for a Dream* (2000), *The Rules of Attraction* (2002) and even *Hulk* (2003), each of which makes notable use of split-screen techniques. The overall 'layering' effect of 3D editing also means that actors can be moved to another part of the frame rather than entire scenes having to be re-shot. And it is also used in special-effects sequences where further perspective is required between virtual backdrops and upfront spectacle. Blurring the distinction between production and postproduction, director and editor, in total, digital editing enhances the claim that films are either made or undone in the editing suite.

Regarded in such practical terms, Janet Harbord distances herself from overarching 'historical' and 'theoretical' accounts of digital processes by looking at 'cultural' and 'industrial' applications; that is, at particular aspects of worldwide film production. She does so by focusing on two films released in 1999, the assumed blockbuster *The Matrix* and the ostensible independent film, *Run Lola Run*. While both have been read extensively in terms of digital culture—in particular their use of videogame-style techniques—up close and beyond superficial consideration of genre, they are of course very different films. As Harbord argues:

> *[T]here are significant differences between the texts that locate them within different film cultures, an argument that problematizes the concept of digital culture as transforming film production in general terms. Technological transformations may be used across different contexts of production, but each context brings to bear the history of that paradigm. (Harbord 2002: 143)*

Harbord regards *The Matrix* as 'hypertextual' and *Run Lola Run* as 'intertextual'. Providing for polished special effects, continuity editing and a linear narrative, *The Matrix* is wholly designed to take viewers into its world. Playing with different film and video formats *Run Lola Run* has sharp visible cuts between live action and animation; and continually stopping and restarting, we never settle into one consistent narrative. The film is intertextual in that it is well aware of its philosophical content and its own textual status in this respect. *The Matrix* may well comment on notions of reality but also presents viewers with a much more attractive hyperreality (Harbord 2002: 143–6). Harbord's choice of films is useful in that they also lead to complications in the ostensible distinction between 'independent' and 'mainstream' films. An assumed Hollywood product, *The Matrix* was also an Australian co-production and more of an unheralded rather than calculated blockbuster. And while *Run Lola Run* is an ostensibly independent German film, it was also an international—or 'crossover'—success, a love story and crime thriller that combines elements of animation, music and videogames.

From DV to HD

'Independent' and 'mainstream' are, of course, loaded terms, the former privileged in terms of talented filmmakers working on low budgets to establish a distinct artistic vision, in opposition to commercial studio products that work through predictable narratives driven by stars and spectacle designed to appeal to large audiences. In the US context, independence refers to those films that are produced without financial backing from the major studios (see Lewis 1998; King 2005). This is complicated somewhat by independent films that go on to be distributed by major Hollywood studios and the assumption that all non-US and English-speaking films automatically qualify as independent or 'arthouse' films. More consistent, however, are the ways in which independent films have been characterized in terms of their use of formats other than 35mm film: namely video, 16mm and digital video. Of these, 35mm became the accepted standard for major feature film production after the 'lower quality' 16mm had been sidelined to documentaries and newsreels in the 1930s and 1940s (see Winston 1996; Enticknap 2005). With its 'truthful' speed of 24 frames per second and 'correct' aspect ratio of 1.37, 35mm is the accepted Academy format for feature film production. With its slower frame rates and television-style aspect ratio of 1.33, 16mm film became the favoured alternative format of independent directors in the 1950s and 1960s. This was particularly evident in the 'nouvelle vague' of the period. Supremely characterized by Jean-Luc Godard's *A Bout de Souffle (Breathless) (1959)* and François Truffaut's *Jules et Jim* (1962), the French New Wave opposed mainstream Hollywood production as particularly manifest in big-budget widescreen epics. Not adverse to borrowing from the pulp verve of American film noir, the New Wave represented a back-to-basics filmmaking which eschewed the 35mm Technicolor standard in favour of 16mm black-and-white. Rejecting expensive sets and industrial-strength hardware, this was filmmaking on a more immediate and intimate scale, often taking place on the streets and favouring naturalistic performances shot from numerous unspecified perspectives. Challenging the tidy narratives and composed frames of classical Hollywood cinema, the New Wave also made use of semi-improvised techniques such as irregular angles and jump-cut editing (see Bordwell 1985; Marie 2002; Neupert 2003).

Since the late 1960s, independent filmmakers have also had access to the further reality format of video. With speeds of anywhere between 60 and 25 frames per second, since the 1990s digital video has worked in simultaneously strengthening and refining the independent-versus-mainstream argument. Which is to say that, regarded in opposition to the increasing use of digital processes in contemporary Hollywood cinema, DV opened up the potential for filmmaking of a 'comparative' yet noticeably lowbudget nature. As Holly Willis suggests, from digital video onwards, the distinction between film and video has become hard to maintain (Willis 2005: 4). This refers not only to Hollywood's appropriation of DV but also to our own ability to make 'desktop' films.

Dogme

Accounts of DV tend to begin with the Dogme 95 movement. In what has been loosely termed the New New Wave, Dogme's famous 'Vow of Chastity' makes clear the similarity in intent but differences in approach to the original nouvelle vague: 'In 1960 enough was enough! The movie had been cosmeticised to death, they said; yet since then the use of cosmetics has exploded' (von Trier and Vinterberg 1995a). Where the French New Wave is said to have failed because of its 'bourgeois romanticism', the Denmark-based Dogme 95 saw itself as strictly avant-garde, redirecting the technological democratization of cinema through the strength of the group. The Vow of Chastity has ten commandments:

1. Filming must be done on location.
2. Sound and image must be produced together rather than through recourse to post-production.
3. The camera must be handheld.
4. The film must be in colour and special lighting is not acceptable.
5. Optical work and filters are forbidden.
6. The film must not contain 'superficial action'.
7. Temporal and geographical alienation are forbidden, the films taking place in the 'here and now'.
8. Genre movies are not acceptable.
9. The film format must be 35mm.
10. The director must not be credited. (von Trier and Vinterberg 1995b)

The movement was unleashed upon the world through the success of the first two Dogme films, Thomas Vinterberg's *Festen (The Celebration)* and Lars von Trier's *Idioterne (The Idiots),* at the Cannes film festival in 1998. Followed by contributions from the two other founding members, Soren Kragh-Jacobsen's *Mifune's Sidste Sang (Mifune's Last Stand)* (1999) and Kristian Levring's *The King is Alive* (1999), further films have come from Argentina, Korea and all over Europe. The first American entry was Harmony Korine's *Julien Donkey-Boy* (1999) and to date there have been thirty-seven Dogme films, the most recent being the British entry, Jan Dunn's *Gypo* (2005).

The advantages of DV, the format of choice for the Dogme filmmakers who quickly ignored their own 35mm rule, relate to production and effect. Where mainstream cinema is said to take full advantage of digital's ability to fake and manipulate the perceived reality of the photographic image, DV brings immediacy and authenticity back into the process. Location shooting releases Dogme films from the restrictions of the custom-built studio and the handheld nature of the cameras means that they can be anywhere, at any distance and often in multiple locations. This also impacts upon performance in the sense that the actors are no longing playing to camera. Two particular elements of DV filmmaking are worth looking at in more detail: film form and aesthetics. Where the digital image is often said to provide for clarity, it is also extremely versatile in being able to pick up on natural light in an almost impressionistic manner. The digital video image has been characterized in some of the same ways as the video image, in terms of 'blur' and 'grain' but also in terms of this extra-sensitivity, the camera either pre-programmed to catch the light in certain ways or enhanced after the fact in order to convey a particular aesthetic effect. This can be seen in the opening to *Festen.* The film's central character walks down a sunny country road. The effect is one of a yellow, bleached heat. This is subsequently countered by the clinical light and tomb-like darkness of the family gathering in the mansion, a celebration that becomes more like a wake. The darkness later takes an impressionistic turn as a haunted atmosphere is evoked through candlelight and mirrors. With particular regard to formal composition, DV filmmakers are prone to avoid traditional framing and take advantage of the mobile properties of digital video cameras. This is more like catching the action on the run. Although relatively contained in the mansion scenes within *Festen,* the camera combines form and content by having the camera 'sneak' around; roaming rooms and peeking round doors as the viewer assumes the position of an invisible invitee, as curious as the other guests to unearth the family's hidden secrets. This takes much more of an anarchic turn in *Idioterne* as the viewer becomes more of a participant in the group's activities. In effect, the camera also 'spazzes out' and on occasions spirals out of control the more the 'normals' in the film forget who they are (see also Schepelern 2000; Chaudhuri 2005).

Variety

Like 16mm and analogue video before it, a number of alternative orthodoxies have come to surround the use of DV. Where Hollywood's reality has traditionally resulted from colour, clarity and careful composition, the reality of independent film is that of the shaky camera and the black-and-white or blurred image. In short, where Hollywood ultimately provides for a sort of glamorous hyper-reality, the 'true' reality favoured by independent filmmakers is messy and off-colour. Such is the dominance of mainstream film, however, that the reality fostered by independent film is regarded as alternative and artistic. But are we truly at the stage where DV can be said to have replaced 16mm and analogue video as the low-budget alternative to 35mm film? This is where we start to qualify the situation. The Internet Movie Database (imdb.com), for example, currently lists over 800 DV entries (January 2005). Notable DV films outside the Dogme movement include *Chuck and Buck* (2000), *Open Water* (2003) and *9 Songs* (2004). Characterized in terms of immediacy and authenticity, DV has also become the main format for recent documentary films such as *Touching the Void* (2003) and *Super Size Me* (2004). The number of IMDb entries relating to TV and short films, however, makes it difficult to ascertain an official number of feature films. DV has proven versatile rather than dominant. From a strictly amateur perspective, the availability of DV, mini DV and software packages such as Avid and Windows Movie Maker have provided for more complete production systems than Super 8 in the 1960s, Hi 8 in the 1980s and camcorders in the 1990s. And, from this amateur base, DV has come to be used in numerous other forms and outlets, whether short films, music videos, internet films or the 'new avant-garde' of multimedia installation artists and video DJs. There is a definite DV culture at work here, fostered by initiatives such as Next Wave Films and festivals such as ResFest (see Griffiths 2003; Hanson 2003; Willis 2005).

With particular regard to feature films, therefore, DV has more correctly added the *choice* of another film format. That is, we are currently at an exciting transitional phase where filmmakers are making use of video, film or digital video, and often in combination with each other. Although this might appear to pull back on the distinctive qualities of DV, it also locates digital video within the current context of film itself becoming increasingly digital. We are at the stage right now, for example, where analogue film can be committed to a digital intermediate which allows it to be run through digital editing and colour grading before the transfer back to 35mm. It is in this sense that while certain films such as *The Last Broadcast* (1998), *Pi* (1998) and *Run Lola Run* are erroneously used in accounts of DV, it's understandable why (see, for example, Griffiths 2003). *The Last Broadcast* and its more successful cousin, *The Blair Witch Project* (1999), are generally caught up in the reality effect attributed to DV, an assumption that arises from the fact that DV is also, essentially, video (both of these films made use of a combination of Hi-8 video and 16mm film). Similarly, the 16mm *Pi* and the combined 16mm and 35mm *Run Lola Run* are films that made extensive use of digital postproduction.

High-Definition

The most recent development in digital cinematography is high-definition. Although the two terms are often confused and simply conflated, high-definition would not be possible without digital *compression*. This applies to television as much as film in that compression refers to the quantity, quality and speed of data—leading to clarity of image in terms of film, for example, and speed and clarity in terms of television transmission. Read mainly in terms of Hollywood fighting back against the use of digital video by independent filmmakers, high-definition is the generic name for professional-standard cameras that take advantage of the versatility of digital film but enable a resolution and depth of field similar to 35mm. The progressive systems have been: HDDV, DVCAM, DVC Pro, HDCAM and, most recently, HDTV. The development from HDDV to HDTV has been that from specific cameras to the overall cinematographic process. But to find filmmakers as commercially and ideologically opposed as Lars von Trier and George Lucas using the same systems also goes some way to tracing the *convergent* developments at work here. Post-Dogme, von Trier has worked with

35mm film in *Breaking the Waves* (1996), DVCAM on *Dancer in the Dark* (2000) and HDTV on *Dogville* (2003). Justified in terms of technical innovation, Lucas made use of HDCAM equipment for *Star Wars* Episode II and the HDTV process for Episode III. The shift from CAMs to TV also demonstrates the further synergy between film and television. DV is now used regularly in television production. And HDTV, first used in the Spielberg-produced miniseries *Band of Brothers* (2001), has come to be utilized in recent feature films such as Robert Rodriguez's *Once Upon a Time in Mexico* (2003) and *Sin City,* and Michael Mann's *Collateral* (2004). This allows for the speed of television production but also the versatility and clarity of the digital image.

As British director, Mike Figgis, has assessed the situation from an independent perspective:

> *Hollywood are terrified. They are afraid, they're absolutely terrified, because if I can prove that I can make a film for, say, £100,000—a feature film with stars—and the equipment that I will use will go in the boot of my car; that is a terrifying prospect for an industry that's over-invested in hardware . . . And so what Hollywood has now done, which I think is really dumb, is they're pretending that the digital revolution is this thing called High-Definition and Lucassound and Lucas this and Lucas that. What they're really trying to do is divert the onslaught by saying we're really on the same side as you. However, what they're saying is please don't take our big things away from us.*[1]

Leading to a new dominance and uniformity to a certain extent, as things currently stand, however, along with analogue video and film, DV and HD more correctly provide for a *range* of options and aesthetics. Consider the use of thirty DV cameras in *Julien Donkey-Boy,* for example, and the clarity of HD in the *Star Wars* prequels. And then consider the use of DV and HD in the moody, ambient night-time aesthetics of *Collateral.*

In the following case studies, I will be focusing on the use of digital video in terms of film form, aesthetics and genre. Mike Figgis's *Timecode* represents a particular experiment in conventional film form, progressing as it does entirely through split-screens. Independent filmmaking criteria remain in terms of low budgets leading to the necessity of experimentation but how do the alternative modes of digital video translate into a genre such as horror? In the second case study, I will explore this question in relation to Marc Evans's *My Little Eye* and Danny Boyle's *28 Days Later.* In the first example, I will be looking primarily at film form and, in the second, DV aesthetics.

Case Study 1

Timecode *(2000)*

Mike Figgis's *Timecode* presents us with a linear, split-screen narrative, the principal distinctions being that the story progresses in real time over four screens, our attention drawn to the main characters who inhabit each and ultimately every quarter of the screen through variances in sound. Lev Manovich and Keith Griffiths take leave to question exactly how experimental we should regard the film. 'Although it adopts some of the visual conventions of computer culture,' states Manovich, 'it does not yet deal with the underlying logic of a computer code' (Manovich 2002: 217). And, while valuing its 'playful storytelling and narrative order', Griffiths argues that 'in terms of altering the actual surface or the palette of the film, *Timecode* remains strictly conventional' (Griffiths 2003: 20). As a digitally experimental feature film, however, *Timecode* remains an excellent introduction to the ways in which digital video can be used to subvert conventional aspects of film form and narrative.

First of all, could *Timecode* have been produced as an analogue, 35mm film? As Figgis states, he was particularly inspired by the multi-character, multi-narrative films of Robert Altman. Similar to Altman's *The Player* (1992) and *Short Cuts* (1993) in terms of setting and situations, the principal way in which *Timecode* distinguishes itself from Altman's films is that we can see all of the characters at once through use of the split-screen technique in what Figgis terms a 'four dimensional' multi-narrative film.[2] The first familiarity of the film is that it is set in the film world of Los Angeles.

The second familiarity is how the film plays out with regard to genre. As Lev Manovich argues, the problem with the 'real time' technique of DV films such as *Festen* and *Timecode* is that we are often presented with *more* incidents than 'the artificially compressed time of traditional film narrative' (Manovich 2002: 214–15). Multi-generic in terms of mixing comedy, drama, sex and murder, the compression resulting out of both temporal flow and spatial organization within *Timecode* presents us with clichés that are also deliberate. As the banal sex drama of the casting sessions and the pretentious boardroom presentation suggest, both mainstream and arthouse filmmaking are variously taken to task in the film.

Based around a behind-the-scenes story that can also be regarded as metafictional, the ways in which that story is presented to us is where the real experimentation lies. As we have already discussed, one of the main advantages of DV filmmaking is that production and postproduction are brought closer together. In order to think in four dimensions, the musically trained Figgis wrote the narrative of the film on music manuscript, allowing the essential construction and development to be written on four lines of parallel action over a six-page script or 'score'. The actors were then involved in one week of improvisation to flesh out the characters and dialogue. Tightly synchronized in terms of the dance of cameras, fifteen continuous 93-minute takes were performed over the next three weeks, the cast and crew immediately able to watch each version of the film and develop accordingly throughout the overall process. Absent of editing and special effects, postproduction was accomplished almost entirely through sound mixing.

These production elements are only useful insofar as they go on to inform the actual film text. Film producer Alex (Stellan Skarsgard) is having a relationship with actresses Emma (Saffron Burrows) and Rose (Salma Hayek). This would be a conventional love triangle were it not for the fact that Rose is also involved with her agent, Lauren (Jeanne Tripplehorn). Shaped by the form of the film, this love quadrangle plays out across four screens. The opening fifteen minutes of *Timecode* work in terms of a conventional introduction but also allow for formal orientation as each of the main characters is established in a separate screen. Prefaced by the earthquake that affects all of the characters, however, the rest of the film will see the main players intersecting as the initially very formal split-screen structure is given fluidity. The congress of characters starts to play through screen breaks: Emma from the top-right Screen 2 and then Rose from top-left Screen 1 individually meeting up with Alex in bottom-right Screen 4 and bottom-left Screen 3. In accordance with a developing distinction, the top half of the screen shows characters in transit feeding into and out of the main events in the film production world at the bottom half of the screen. The first screen break occurs when Emma arrives at the Red Mullet offices. We see her entering from her own screen, Screen 2, and she simultaneously appears in Screens 2 and 4 as she kisses Alex, which also involves a symbolic split after the join as a reverse shot shows Alex in Screen 2 and Emma in Screen 4. Each has entered into the other's screen worlds and, significantly, Alex is the only character to appear in all four screens. Further, promiscuous, screen breaks occur during the sex scene between Rose and Alex, which takes place in and across Screens 3 and 4. A singular focus in mainstream films, while our attention is drawn to the two lovers, it is also illicit as they make love behind the screen test of a love scene being projected on Screens 1 and 2. In tight frames whose only light and sound are provided by the screen test, such as Lauren listening in on the act, we too are left trying to distinguish the real sex from the film-within-a-film sex. This segment of *Timecode* ends with all four characters alone with their thoughts, Emma and Rose looking in the mirror in Screens 2 and 4, and Lauren and Alex unaware of each other having a cigarette outside the offices in Screens 1 and 3.

As well as subverting traditional film conventions, *Timecode* also makes use of aspects of new media. It does so through both form and content; that is, what we see on screen and what the characters do on screen. In his analysis of what he calls 'DV realism', Lev Manovich regards the immediacy of spatially aware real time DV films as being the result of tapping into today's 'reality media', principally reality television and webcams. Manovich regards *Timecode* as utilizing formal aspects of modern telecommunications but, again, built upon relatively traditional narrative bases (Manovich 2002: 215–16). Using new technologies of seeing and surveillance is not the sole preserve of DV realism, having been used in films and television series like *The Matrix* and *24*, and in non-narrative video and performance art. Essentially, however, *Timecode* can be located outside such high-tech plotting and the denial of narrative in what is, after all, a character-driven film. If this is part of its so-called conventional narrative, the point is that much more reliance on new technologies would marginalize the film's characters, who remain our principal focus throughout.

Beginning with the split-screen technique, *Timecode* echoes the windowed properties of contemporary media, whether the internet, DVDs or even multiplayer videogames (see also Hanson 2003: 70–1; Verevis 2005). The four quarters also have the effect of chopping this piece of cinema down to television or computer monitor size, a reduction of potential scale in favour of intimacy and a surveillance-style invasion of privacy. Surveillance and voyeurism are principally conveyed through the roving cameras. Where one camera is usually enough for the innate voyeurism of film viewers, here the four cameras cover events from all distances and angles, often when hiding behind objects (the literal result of each cameraman not wanting to be exposed by the other cameras). This verité style, which also involves out-of-focus shots and imperfect framing, is followed through to the sound mixing in terms of tuning in and out of snatches of conversation.

If most of this relates to what the film viewer sees, the characters within *Timecode* are also connected by various technologies. The security guard would be the closest to providing an omniscient perspective if he weren't so distracted from his CCTV monitors when peddling drugs. All of the main characters are connected by the film-production world they inhabit and it's significant that technology is only explicitly foregrounded here when Anna (Mia Maestro) presents Red Mullet with a discourse on digital film complete with DJ and a live soundtrack demonstration. Sound technologies play a part in the developing narrative of the film in terms of surveillance, connection and disconnection, principally Lauren's bugging of Rose and assorted arrangements made on mobile phones. *Timecode* ends with a film camera and the mobile phone. Avant-garde director Anna films the death of mainstream producer Alex, her ambition for pure verité given its ultimate demonstration. And the final scene of the film involves mobile phone calls across all four screens: Rose phoning Lauren, who doesn't answer the call; and the dying Alex phoning Emma, his final moments caught on film and his last words transmitted over a phone line inaudible to the film viewer. Overall, then, *Timecode* focuses on two developments: the recent potential for digital filmmaking and the ways in which film will have to adapt to new media technologies. It concludes with ambivalence, the cold substantial art of film and the final connection made on something as seemingly insignificant as a mobile phone.

Case Study 2

My Little Eye *(2002) and* 28 Days Later *(2002)*

From *The Night of the Living Dead* (1968) and *The Texas Chainsaw Massacre* (1974) to *The Blair Witch Project* and current retrograde slasher films, the horror genre progresses by continually going back to basics. As Matt Hanson states: 'Cinematic horror has always been at its eeriest and its most convincing when stripped down. The genre is at its best flayed of all pretensions or extraneous details—any hint, in fact, that what is happening has any root in artifice' (Hanson 2003: 13). Although it is too soon—or possibly too late—to talk about a definitive wave of DV horror films, a new post-*Blair Witch* aesthetic has developed through various uses of film and video. Hence, for example, *Wendigo* (2001), which makes use of 16mm, and *Session 9* (2001), which was one of the first films to use HDDV (see Aloi 2005). The same 'authenticity' can be found in *The Descent* (2005), which makes use of film, video and camcorder formats, and *Wolf Creek* (2005), which uses HDTV. The purest examples of DV horror are Marc Evans's *My Little Eye* and Danny Boyle's *28 Days Later.* Enabling a back-to-basics approach to some of the most effective horror conventions, digital video provided these films with a much more versatile aesthetic and also allowed for a formal rearrangement of the more expected material of the genre.

My Little Eye combines the two main outlets of today's 'reality' media: reality television and the internet. In the first instance, the film is grounded in many of the recent conventions of *Big Brother*-style television shows: audition tapes; a group of participants isolated in a house for a determinate period and ultimate prize; all-seeing cameras; and the occasional delivery or surprise designed to breathe life or breed more contempt into contestants so that viewers will keep watching. Beyond this, however, the live internet format chosen for the film makes it much more speculative in that current webcasts are not as advanced or organized. In a sense, therefore, the film takes us into what could have been had reality television not proven a more acceptable version of the sort of voyeuristic material privately available on the internet. But this, of course, is exaggerated as appropriate to make it into a horror film.

My Little Eye creates much of its atmosphere through the combination of sound and images. Opening with a pixellated glaze and the sound of a modem, the camera pulls back to reveal a computer screen. The first of many associations between passive film voyeur and interactive computer user, the anonymous user clicks on a 'Wanted!' pop-up which results in the following text:

> *5 contestants for REALITY WEBCAST*
> *Spend 6 months in a house for $1 MILLION*
> *If anyone leaves, everyone loses!*
> *R U the perfect housemate?*
> *CLICK TO ENTER*

Following on from the actual film title, we are then introduced to the main characters via their audition tapes. Having themselves clicked to enter, these participants soon find themselves unwitting victims of a 'snuff' webcast through which subscribers bet on their fluctuating odds of survival.

Rarely breaking the point-of-view perspective of the adopted webcast format, *My Little Eye* works in simultaneously 'framing' the characters and directing what we see on screen. Horror cinema has always been about positioning, the viewer made to identify with the victim or even the killer. Here, the webcast format allows for a further level, the director Evans in complete control but on a parallel with 'The Company' in the film. In the opening sequence, we are provided with a four-quartered split-screen montage of the participants' arrival and first months in the house. Resembling *Timecode* in terms of spatial organization, this initial technique allows for the compression of time similar to the ways in which reality television gives us edited highlights or the internet subscriber might look through archive material. The first uneventful weeks in the house bypassed and, moving on to full screen, what would otherwise be presented as five stock teenagers chatting their way through exposition becomes a way of familiarizing the viewer to the film's defamiliarizing form. Shot mostly from the top corners of the room and progressing through quick snatches and occasional zooms, we are effectively seeing the action of the film through the myriad surveillance cameras that pervade the house. All movement of the cameras is accompanied by electronic clicks and whirrs, and the film takes on further distance and involvement when the participants are shot via two-way mirrors, for example, or from outside-in. Privacy is obviously not an option in the house and this is made clear when the participants go to bed in the subsequent scene. Feeling safer in the darkness, the participants are nevertheless illuminated by the green haze of infra-red cameras and monitored by further sound devices. Talking about the males in the house, for example, the female characters whisper. But similar to the way in which the cameras focus in to avoid blurred images, the sound devices work in clarifying muffled sounds. Throughout the film, the more the assumed privacy, the more the probing intensity of techniques.

As producer Jon Finn states, *My Little Eye* is a 'low-tech' film that uses camera equipment bought from high street retailers.[3] Compare this with two other internet-themed horror films from 2002—*FearDotCom* and *Halloween: Resurrection*—and the overall digital effects of the film become much more pronounced. One of the subsequent values of using consumer-grade digital cameras was exactly the ability to explore surveillance, which becomes most illicit and paradoxically evident when small cameras are attached to pens, torches, a joypad and a shower nozzle. Talking further about the challenges and advantages of shooting *My Little Eye* in this way, Finn claims it a piece of 'anti-filmmaking'. Which is to say that the film has to follow the webcast format throughout and this also determines the ways in which it works as a horror film. Viewing the majority of the action through locked-on cameras means that using familiar horror film conventions such as point-of-view shots or overtly skewed angles is not a viable option. Working within this small screen format, however, has its advantages with regard to horror. Exaggerating the characteristic complicity of horror film viewers and further restricting the participants' movement and actions, the format shapes both the participants' unease and ours. It is in this respect that we can see the variety of camera shots as *extending* horror film conventions. As Marc Evans states, the main difference is that instead of 'anticipating' the action, the cameras are 'behind' the action. Referring to specific moments where the cameras take advantage of how exposed the participants are by locking on to the backs of their heads and necks, this also refers to the way in which the film viewer follows the action. In effect we find ourselves positioned *between* The Company and the participants, essentially powerless but with a wider perspective than the participants. The final revelation might still come as something of a surprise but at this point we're fully complicit in the game.

Freed from the formal requirements of *My Little Eye,* Danny Boyle's *28 Days Later* allows us to look much more at DV aesthetics. A postapocalyptic zombie film that works through urban desolation, it makes stark use of contrasts, principally day and night on a visual level, and peace and rage on an allied aural level. Following the necessary horror mechanics to a certain extent (the Infected only come out at night), these contrasts are first accentuated and then blurred by the use of digital video. Following on from the dark, claustrophobic laboratory and the hit-and-run pace of the opening sequence, the film immediately progresses to a barren London hospital where Jim (Cillian Murphy) wakes up bathed in glaring light. As Jim becomes accustomed to the light, it takes on more of a dreamlike hue which is continued as he leaves the hospital and walks out into a deserted London. Throughout the London sequence as a whole, the DV cameras work through emphasis and de-emphasis. The camera angles both frame and cut across the straight lines of the buildings, advertising hoardings and familiar landmarks. This is at once a city laid bare—an architectural grid without the usual organic mass of people—and a city in an unfamiliar situation where the lack of people is exactly the thing that allows our attention to be drawn to the skewed geometry of the buildings. Long shots and close-ups follow Jim's isolation through what becomes a montage sequence and the dreamlike hue takes on more of an environmental blur. As the sun beats down on silver structures through the rising haze, it's a hot day but it also looks like post-nuclear.

For Danny Boyle and Anthony Dod Mantle, cinematographer of *Festen, Julien Donkey-Boy* and *Dogville,* this London sequence proved to be the biggest advantage with regard to using digital video. Prepared beforehand with digital photographs of the city and given one morning to shoot the sequence, the use of several cameras allowed for unlimited footage from numerous angles to be edited together in postproduction. Speaking further of the use of DV in *28 Days Later,* Boyle highlights two other benefits:

> *We wanted it to feel different in texture from normal film. Because it's an apocalypse, you can use a different hue, because nobody knows what things will look like if everybody's killed or there are no cars. So . . . we would tickle the color of the film occasionally to create a slightly strange universe. Against that, I wanted this enormous energy from those who are infected, which I was going to get through this particular menu on the camera, which allows you to alter the frame rate; things appear to be speeded up but actually it's real time. So you kind of snatch at fast images, like falling rain or a man running, snatching at them in a slightly unreliable way. The idea is that you can't quite trust your usual sense of judgment about perception, depth and distance when dealing with the infected. (Quoted in Hunter 2003: 2)*

Boyle has also justified his choice of digital video in terms of the predominantly urban nature of the film. Speaking in particular about the British context, we are indeed very used to seeing the pervasive coverage and grainy aesthetic of CCTV cameras. When applied to the film this can be taken very literally in that, if there were an apocalypse of some sort, then there wouldn't be a film crew to record the after-events.[4]

Far from presenting us with omniscient camera shots surveying empty streets, the action scenes in *28 Days Later* are conveyed through abrupt editing and often presented in terms of sudden environmental effects. Hence, when Selena (Naomie Harris) and Mark (Noah Huntley) rescue Jim by petrol-bombing a service station, the impact is also that of light exploding into dark. Conversely, smashing through windows or scrambling over barriers, when the Infected attack it is presented in terms of the dark trying to break through receding sources of light. From the sunny South to the rainy North, the climax to the film takes place in a baroque country house. And, moving on from the previous urban effects, the digital aesthetics now work in conveying a gothic atmosphere. If we were to isolate the action elements of the film from the atmospheric, it would principally be a matter of looking at the moments of physical violence. From individual attacks to the final onslaught, the staccato movement of the Infected is the most noticeable action effect within the film. Filmed with the camera shooting at variances between 24 and 16,000 frames per second, the Infected run and attack through blinks of the eye. It is only when the same technique is applied to environmental effects, however, that the action sequences represent a full onslaught as it is illuminated by speckled flashes of gunfire, lightning and searchlights. Rain takes on the appearance of liquid metal and, more so than ever, the cameras can barely isolate the Infected as they smash through doors and windows. And Jim is himself filmed in the same manner as he moves between the soldiers and zombies, ultimately leading us to question whose side he is on and whether he too has become Infected.

The digital composition and aesthetics of *28 Days Later* take us into a world divorced from our own, and to the extent that any flashes of the familiar appear in stark relief. There is something too perfect about the impressionistic and almost hyperreal blue skies and green fields on the road from London to Manchester, for example. But it is entirely fitting that moments of comfort in the film use different, familiar, film stocks. When Jim visits his family home in the suburbs, for example, his thoughts become a Super 8 flashback. And the fact that the final scene of the film is shot using 35mm film is also significant in that we, the audience, are very familiar with that look. Framed by a jet scouring the countryside, the three survivors are rescued and we too are released, re-conditioned, back into widescreen clarity and out of the film.

Notes

1. Quoted from the television documentary, *Power to the Pixel* (UK, Channel 4, 9/10/01).
2. See Figgis's Video Diary and Audio Commentary on the *Timecode* DVD. The DVD also includes the first test version of the film and an Interactive Audio Mix that allows viewers to isolate the sound on each of the four screens.
3. Taken from the DVD Commentary by Evans and Finn. The DVD also offers an Interactive Mode that presents the film as if displayed on a computer screen and is most effective when played on a laptop. Adopting a webpage format, the main window follows the film and the five 'player' icons are presented down the right-hand side of the screen, their odds of survival continually fluctuating. The viewer-user can also access Archive Footage, the full Audition Tapes and two further special features are the Multi Camera option, which enables the choice of camera angles on four selected scenes, and a chilling Audio Commentary by 'The Company'.
4. See the DVD documentary, *Pure Rage: The Making of 28 Days Later.* See also the Audio Commentary by Boyle and writer Alex Garland.

Call and Response: Storytelling in the Neverwinter Vault

Robin Woods

In her essay "Towards a Game Theory of Game," Celia Pearce suggests that the expansion of academic disciplines, especially narrative, into the territory of game theory is a kind of "theoretical imperialism," and notes that "If you have a hammer, everything looks like a nail"—that is, if you're a narrative theorist, everything looks like a story (Pearce, *First Person* 144). Pearce here is taking part in a familiar debate: nearly every discussion of the narrative aspects of computer gaming begins with a description of the hotly contested ground between narrative theorists interested in gaming and the self-named ludologists, who focus on the mechanical—that is, physical—aspects of gaming as a constitutive part of the computer gaming experience. The ludologists—most particularly Espen Aarseth, in his important work *Cybertext*—call for a game theory that focuses on and arises from the gaming experience, rather than importing models from other disciplines and superimposing them on games and game theory. For such critics, the spread of literary theory into the gaming world might indeed look like an attempt to annex new ground, and to grind its resources in the same old academic mills that have operated for decades and longer.

Certainly academe has a tendency to expand its territory in sometimes disquieting ways, and tempers flare in critical debates as much as anywhere else. But even revolutionary forms do not appear in a vacuum, and anything that sparks widespread change throughout a culture will necessarily bump up against a number of different theoretical structures. New forms reconfigure the existing forms around them, and new theoretical possibilities allow us to look at older forms and theories in new ways. Though clearly not all games are narratives, the study of narrative has much to contribute to game theory, and the development of the computer game can teach us a good deal about the way narrative works.

There remains a significant challenge, however, in finding a model that addresses both the elements that are individual (one might even say unique) to gaming as well as what are clearly narrative structures and techniques. In order to create such a model, I would propose looking to performance theory, specifically to the oral performance theories of Albert Bates Lord, Richard Bauman, and John Miles Foley. Though the study of oral transmission and oral performance seems far removed from the world of computer games and hypertext, it provides possibilities for discussing both the physical aspects of gaming as well as its storytelling aspects. Such studies are useful because they focus on what Foley calls the "performance arena" (Foley 8), the entire storytelling scenario including teller, tale, and listener. They take particular interest in the interaction among these participants in the storytelling event, and in story itself as something that develops out of that interaction, to form what Bauman calls an "emergent" text (Bauman ff). That is, interactivity takes place on a different level of discourse: not between game and gamer, but between gamer and designer, and the "emergent text" Bauman refers to is the narrative that drives the game.

The term "emergent" is well known to students of computer game theory, and of other sciences as well: emergent behavior is behavior that evolves unpredictably from a program—for example, Janet Murray, in *Hamlet on the Holodeck,* describes artificial agents that "are able to act in ways that go beyond what they have been explicitly programmed to do" (Murray 239–40). What is at stake here—and what is at stake for Bauman as well—is the much-vexed question of interactivity. Interactivity is the desideratum of computer gaming, or at least of computer role-playing games. Ideally, the player in an interactive game collaborates with the designer in producing a tale that bears the player's stamp, that is actualized and to some degree co-authored by her. Murray identifies three categories that comprise an interactive gaming experience: immersion, in which the player feels herself to be transported to and functioning in an alternative environment not corresponding to her present reality; agency, or the ability to take meaningful action that leads to perceivable results; and transformation, in which the player either adopts an identity not her own for the duration of the game, experiences variations of the gaming experience, or is personally changed by the experience. As Michael Mateas notes, these are all—agency included—phenomenological categories, relating to the gamer's experience of the game (Mateas, *First Person* 22). In other words, the gaming experience may well change the player, but it does not change the game, and thus cannot be described as truly interactive. Critics note that the "free" choices a player seems to make in any game are circumscribed by choices the designer has already made, and even games that provide alternate endings according to the player's decisions can only offer something that has already been written. Aarseth, in fact, goes so far as to say that "interactive fiction is best understood as a fiction: the fiction of interactivity" (51).

But oral performance theory gives us a way to account for what is interactive in gaming, and for ways in which the emergent nature of an orally performed text coincides with the emergent behavior of a computer system that evolves unpredictably. Oral performance theory first arose from the research of Milman Parry and his student, Albert Bates Lord. Parry and Lord did fieldwork in the 1930s in what was then Yugoslavia, where traditional singers, called guslars, still performed centuries-old Slavic epics. From observing these performances Parry developed his seminal theory of oral composition, positing that oral performers compose their texts using an array of traditional phrases, from which they select according to the metrical requirements of the song, the content of the song, and the audience's response to the song. These performers learn their songs from other performers, but each performance is not a verbatim redaction of the text that has been learned. Instead, the singer puts together a text in the very act of recounting it, leading Parry to his famous observation that "the moment of composition is the performance" (Lord 1960:13). That is, the text is created anew with each performance, and each performance is an original version of the text—not a debased copy descending from a mythopoetic golden age, but a new composition in itself.

What gives oral performance both its traditional resonance and its creative power are fluidity and immediacy: the exact nature of the text is unfixed, undetermined, until the action of performance and reaction of audience response call it into specific form. Such an arena is very different from that in which designer and gamer interact. A designer will not (ordinarily) meet with the gamer while he or she creates, and gamers who play in multiplayer may never meet each other face to face—in fact, *will not* see each other face to face even if they are playing in the same room, for each will communicate only through the medium of the computer, and then only through the further mediation of an avatar. It is hard to imagine a more mediate, less immediate interaction than that which takes place between any person and any other person engaged in computer gaming.

But while the computer game is not itself an immediate arena, it is part of a larger arena in which designer and gamer converse, discuss game play, and modify aspects of the text. For although the interaction is not immediate, the computer itself provides the means for direct and public discussion. Gamers communicate with designers and with one another to give feedback and consider what makes a game most effective. Not every game designer responds positively—or at all—to commentary from players, and certainly not all fan feedback is positive or even civil. Gamesite forums frequently reflect frustrations from both players and designers who find each other unresponsive. But such forums can provide swift and responsive communication between player and designer, and in so doing they comprise a part of the storytelling arena that Foley describes. And unlike a printed text, but like an oral text, a computer game can be endlessly altered to suit the expectations of players, with version 3.5b of any given game representing a new text in itself, one that has been altered according to the wishes of its audience. Such a text is "emergent" in Bauman's sense of the word: it is a story that develops out of the interaction between storyteller and audience. And it is emergent in Murray's terms as well, as a system that grows unpredictably. It is, in fact, an arena in which the game is changed as well as the player.

Seen in this way, it is clear that both game and oral text fit Espen Aarseth's definition of a cybertext, one which "focuses on the mechanical organization of the text, by positing the intricacies of the medium as the integral part of literary exchange. However, it also centers attention on the consumer, or user, of the text, as a more integrated figure than even reader-response theorists would claim" (Aarseth 1). As we have seen, it is the physical and mechanical interaction of the user with the text, and the user's potential for making changes to that text, that determines a cybertext, or what Aarseth elsewhere calls the ergodic text. Looking at both codex and computer text, Aarseth, in *Cybertext,* maps out selected works—including such examples as Michael Joyce's *Afternoon,* Herman Melville's *Moby-Dick,* and the *I Ching*—according to such markers as time functions, user's point of view, and types of engagement opportunities, in order to determine ways in which different texts are manipulable by players, and susceptible to changes arising from player behavior—or, once again, emergent. Though Aarseth, as we can see, examines a wide range of texts, his dismissal of the "proponents of secondary orality" (108) leads him to ignore the performed text, which in fact conforms in all ways to his criteria for the configurative text. Though one might shudder to call an orally performed traditional tale a cybertext, it is a text which requires a physical presence for both speaker and listener (or text and user), involves a physical response from the listener, and offers the listener an opportunity to modify the text. Like the ergodic texts that Aarseth considers, the orally performed text must have an audience even to exist: if there is no audience, there is no performance; if there is no player to perform—the gaming action—the ergodic action—no game is being played. While no one would claim that an oral tale is the same thing as a computer game, the crucial role played by the audience is, we might argue, constitutive of both genres, and provides us with new opportunities to see where and how interactivity truly takes place.

A particularly salient instance is provided by the game *Neverwinter Nights,* released by Bioware in 2002. *Neverwinter Nights* is a CD-ROM roleplaying game set in the Dungeons and Dragons world of Forgotten Realms, and has its roots in the pen and paper gaming experience. Included in the *Neverwinter Nights* package is the Bioware Aurora Toolset which enables amateur game designers—even rank amateurs—to build modifications of the original campaign. Additionally, Bioware provides server space on which these amateur builders can post their modules, and make them accessible to other players. It also provides discussion line space for each module for player feedback and comments, in addition to general forum space. *Neverwinter Nights* is a hugely successful game by all standards: it has earned 86 awards (including several Game of the Year and Roleplaying Game of the Year awards); it has sold, along with its two expansion packs, more than two million copies in the US and abroad; and it has an enormous fan base of gamers who not only buy and play the games, but who are actively engaged on the Neverwinter Vault forums and discussion lines (currently more than 650,000 postings), in guilds, multiplayer games and persistent worlds.

Most significant in all of this is the fan content and the interest it generates. According to the *Neverwinter Nights* website, there are nearly 4000 amateur modules posted on the site; the top ten modules alone have been downloaded more than a million times. Bioware provides tutorials for builders learning to use the toolset, and it also provides forum space in which builders can talk to each other and share ideas. In addition, it supports guides and tutorials which the players write and post, thus giving its fans a large presence in the feedback loop. The fan-based tutorials in particular blur the distinction between company and community, as gamers take roles as designers and teachers.

And among the fans themselves, players and amateur builders are widely overlapping categories; builders show up on one another's gamesite discussions lines just as they appear in the forums. And builders rely on all players, builders and non-builders alike, to provide feedback on their modules. Much of this feedback concerns bugs the players have found, but players also provide criticism of game content, and discuss elements that make the module they're playing, as well as computer role-playing games in general, effective or ineffective.

Not all players, of course, participate in such discussions or provide feedback; the first chapter of builder Stefan Gagne's highly regarded "Penultuma" series, for example, registers 104,704 downloads, but only 105 postings. But all players engage with a game that has been modified by player feedback, and which reflects, to some extent, community standards as to what makes a game good or bad. The importance of such feedback is reflected in Aulslime's "Builder's Manifesto," posted as an editorial. Aulslime advises builders how to recruit playtesters from the Neverwinter community, and adds a warning: "It is INCREDIBLY difficult not to reply to or explain to playtesters how they missed the point of the module or how they didn't get the idea of the plot. Don't do that. The reality is that if a player didn't 'get it,' it is always your fault. Your module doesn't say what YOU think it says, it says what a guy who has never seen it before thinks it says. Your adventure has to tell the story without any help or guidance from you or else it isn't finished." Here Aulslime is saying what we all know, that a fresh pair of eyes can make us see our work in different ways. But he is also voicing the need for games to be player-centered, and giving the gamer final authority for what the module is and does.

This view of game creation permeates the discussion lines for *Neverwinter*'s amateur game modules, and the resulting interactions suggest that an engagement is taking place between gamer and designer that is similar to that between oral storyteller and audience. Richard Bauman, in *Verbal Art as Performance,* notes that "performance as a mode of spoken verbal communication consists in the assumption of responsibility to an audience for a display of communicative competence. This competence rests on the knowledge and ability to speak in socially appropriate ways. Performance involves on the part of the performer an assumption of accountability to the audience for the way in which communication is carried out, above and beyond its referential content" (Bauman 11). Although Bauman is speaking only of oral performance, his point regarding a performer's accountability to her audience mirrors precisely the gaming arena in which players discuss different aspects of the games they play, debate the criteria for what makes a game good or bad, and consider how well or how badly the designer has risen to such standards. Bauman further says, "From the point of view of the audience, the act of expression on the part of the performer is thus marked as a subject for evaluation for the way it is done, for the relative skill and effectiveness of the performer's display of competence." In other words, by the very act of performance, the oral storyteller enters into a kind of social contract with her audience by which they are licensed to criticize the performance and give voice to normative standards in response. Similarly, in posting a module on the Bioware site, the amateur game builder is automatically situated in an evaluative arena in which the very fact of her module's public exposure invites public critique.

And this critique is crucial to the designers. One builder identified as KnightMayor, writing on one of the Vault's forums, notes that "While I can't speak for all mod builders, I find the feedback and support from players is sometimes the only thing that can keep me going. With long mods there are large gaps of time between releases. When mods are first released, I get a lot of feedback and this gets me really motivated for the next one. Over time the comments become less and less frequent. With this decreased feedback/support accompanied by the exhaustion, repetition and frustration of not having finished yet, it can make working on a mod feel more like a chore than a hobby." KnightMayor is right not to speak for other builders, since many designers have been discouraged by negative feedback from fans. But his comments point out the crucial dynamic between players and builders. As we have seen, without the gamer there is no reason for the game, and indeed, in ergodic terms, no game at all. Gamers rely heavily on audience response, both in the gametesting phase and after a module's release, just as the oral storyteller relies on her audience to tell her where her tale succeeds and where it fails.

We find an abundance of such responses for most of the modules that generate any interest at all. A nice example can be found in the works of Dave Mason. A member of the *Neverwinter Nights* Hall of Fame (the Hall of Fame also being provided by Bioware, based once again on player feedback), Dave Mason has at present produced five modules in two separate series. Mason is among the more loquacious of the *Neverwinter* amateur designers, a frequent participant both on the forums and on his own games sites, and like other builders, his conversation reflects both a concern with the playability of his own modules, and an interest in what makes for a good gaming experience in general.

Mason's first module, "Where Am I?," is a case in point. Mason himself begins the discussion thread for this game, writing on 8/26/02, "Just a couple of comments I wanted to add here. I marked this as the final version, but that is in no way indicative that I intend never to work on it again. I openly welcome all comments and criticisms. I will try to make it better if the interest is there for me to do so." Mason here makes clear what is true of the oral texts Lord and Foley describe: there is no fixed text, only instances of the text. For Mason, there is no such thing as a final version at all, only the most recent version. We might also say that Mason's comment suggests that the word "final" affixed to his module means that he has finished building the mod in isolation. The "final" product is all his own, but the further, future instances of his game must be driven by the interest of his audience. If version 1.0 of "Where Am I?" is Mason's, then version 1.5 belongs to Mason and his gaming audience.

As "Where Am I?" begins, the player character awakes to find she has been kidnapped by a gang of gnome bandits during a drinking binge, and brought to an enchanted town which no one is able to leave. (The logical flaw in this scenario, that the gnomes must have left the town in order to kidnap the hero, has been pointed out by a number of the game's players.) Despite a crippling hangover, the hero discovers the reason for her kidnapping (pecuniary) as well as the source of the town's magical imprisonment: a dragon who is keeping the townspeople trapped and then tormenting them with summoned monsters for his own amusement. The hero kills the dragon, frees the town, and goes on her way—ideally, to another Dave Mason adventure just up the Sword Coast.

"Where Am I?" is a short module by Neverwinter Vault standards. It is aimed at high-level player characters, is rich in treasure, and features several particularly difficult encounters with high-level monsters. All these aspects are hallmarks of Mason's modules, and all garner comment from players. The high-level pitch of Mason's games is especially important, since it means that players are more likely to import pre-existing characters into the game than to generate new ones. High-level modules are by no means rare, but many designers prefer to pitch their games at first level characters, so that the newly-generated character will fit into a module's storyline better. A previously generated character brings with her a kind of personal history—and often a kind of relationship with the player—that can clash with the role the games asks her to assume. And these competing elements do clash, as we see from the site postings. One player named Paladin writes, "An excellent modul[sic]! I enjoyed it very much, even though the story might seem a little bit . . . strange . . . who has ever heard of a drunken paladin of Torm??" (qtd. in Mason). In fact, the player character's hangover, along with a general recognition of her past acts of heroism (a given for any high-level character), are the only things Mason imposes on the game's protagonist. But a hangover is far from an empty index of character, and the player behind a saintly paladin character may well wonder why her character has been suddenly living life in the adventurers' fast lane.

Another poster named Calabar makes a similar complaint a few days later: "The only negative experience I had was with the dialog. I'm not sure if I can really explain it, but it just didn't feel right. I often had to choose dialog I felt my character would never dream of saying. I think the problem was that the dialog was too modern and didn't feel right in that setting. And I think I often came off as a bit of a jerk. *shrugs* Perhaps its [sic] just me, but the dialog detracted a great deal from my enjoyment of the game" (qtd. in Mason). Dialogue options are, of course, the central way in which personality and personal history are imposed on a player character, and they determine the voice with which the character—and to some degree the player behind the character—speaks: note that Calabar begins by referring to what his character says, and ends by referring to himself.

Mason's response is a typical one, at least for him: "Thanks very much for both your kind words and criticisms. I must say, you've inspired me and pointed out something about my mod that I really hadn't thought about. . . . I believe for my next update, I will be providing more 'good' character dialog paths, or at least, comments that don't seem so 'jerky.'" Though the cranky, hungover condition of the player character is important to the story of "Where Am I?," Mason here decides to alter the character's in-game identity in response to the prompts of his audience. Once again we see that Mason regards his own text as perfectly fluid, perfectly alterable, rather than as a fixed object, and in fact he expects to alter it as a matter of course: in his "next update."

We also see that Mason believes his players have something to tell him about his own production. Like Aulslime, who maintains that "Your module doesn't mean what YOU think it means," Mason acknowledges that Calabar has something to teach him about his game, and cedes some part of his authority—of his authorship—to the player.

But before Mason can alter the game, another poster named Endgamer writes in with a different opinion: "I don't think you should change the dialog at all. It was one of the strengths of your mod in my opinion. I totally thought the things you had the PC say were consistent with a grumpy character with a hangover" (qtd. in Mason). Mason's response is again typical: "Thanks for the compliment. Now I don't know what to do! I still appreciate Calabar's comments as valuable feedback. I guess I'll leave it as is, and live with the fact that you just can't please everyone."

The standoff over the hero's dialogue in "Where Am I?" reminds us that opinions on gameplay are likely to vary, and that tastes among gamers—even gamers of a certain subset, as with the *Neverwinter Nights* following—must necessarily differ widely. And yet, in practical terms it appears that there are certain normative standards for games and gameplay, and that those standards are voiced in and to some degree enacted on the game sites and forums, as we can see from Mason's "A Harper's Tale" series. As we've noted, among the hot-button (so to speak) issues that elicit comments from Mason's players are the frequency and difficulty of combat encounters, and the amount of treasure available within the game. Such complaints arise in particular in the discussion lines for "A Harper's Tale 1," Mason's third module. "A Harper's Tale" is a highly rated series, and the first chapter has a rating of 9.28 out of ten, ranking 63 out of 551 ranked modules in the Vault standings. Player comments on the discussion line run heavily in the game's favor, but some who like the mod, and all of those who do not, complain about the balance of encounters and treasure. One poster, Seryn, notes, "Far too much fighting! Maps are overcrowded with monsters, and they even respawn" (that is, come back to life once the player character leaves and then re-enters the scene. One player refers to the experience as "jailbreak at the zoo" (qtd. in Mason), and another calls the encounters "pointless and repetitive" (qtd. in Mason).

In his earlier responses, Mason defends the respawning practice, as in the following posting: "I just do the best I can to make something I'll find fun to play, and feel grateful that there are a few others that feel the same way I do. I understand not everyone's going to like it." But later in the postings for "A Harper's Tale 1," Mason suggests that he may change the game's respawn rate: "Hi bradkay, yeah, I'm thinking of an update to completely remove all the respawning encounters (make them single shot only) BTW, some encounters are infinite, and some are fixed. What can I say Sometimes, I like to go hunting! :)." Mason is saying here what he frequently says elsewhere: that he builds the modules that he himself would like to play. But he is also showing his willingness to respond to the players' suggestions. And in fact, Mason's most recent module, "A Harper's Tale 3," features no respawned monsters. Though there are still plenty of opportunities to "go hunting," combat encounters are sharply reconfigured, and feature fewer battles against much more difficult foes. By accommodating his players in this (and other) matters, Mason is tempering an idiosyncratic view of gaming (I build what I want to play) with a community-oriented view that leads him to change his game according to the response of his players.

The engagement of these players, and the overlap of players and builders which marks *Neverwinter Nights* in particular, remind us of Brenda Laurel's oft-quoted statement that "people who are participating in the representation aren't audience members anymore. It's not that the audience joins the actors on the stage; it's that they *become* actors—and the notion of 'passive' observers disappears" (Laurel 17). We might take Laurel's formulation one step farther and say that the players become not actors but playwrights—or rather, both actors *and* playwrights. Unlike Howard the Duck, the 1970s comic book character who is "trapped in a world he never made," the *Neverwinter Nights* player travels through worlds she has helped create, through gamescapes that have formed partly under her hand. The games she plays are ergodic, because they require her physical presence and mechanical action to move forward; they are engaged, because the choices she makes affects her immediate experience of gameplay; and they are interactive, because she participates in a community-wide discourse that effects changes on the games she plays. As with traditional culture, or any culture for that matter, the product resulting from such a dynamic reflects an agreement between designer and gamer, speaker and listener, as to what makes a story worth telling or a game worth playing. In a coffeehouse in Serbia, or on a computer screen anywhere in the world, the speaker calls to us, and our response determines the nature of the culture we have agreed to share.

References

Aarseth, Espen. *Cybertext.* Baltimore and London: Johns Hopkins University Press, 1997.

Aulslime. "Builder's Manifesto." Online posting. 1 Jan 2004. Neverwinter Vault: Features: Editorials. 14 October 2004 <http://nwvault.ign.com/features/editorials/012504.shtml>.

Bauman, Richard. *Verbal Art as Performance.* Prospect Heights, Illinois: Waveland, 1977.

Bioware. *Neverwinter Nights.* 2002.

Foley, John Miles. *The Singer of Tales in Performance.* Bloomington: Indiana University Press, 1995.

KnightMayor. "Making Modules Bigger. Longer, and Uncut." Online posting. 21 Jan. 2004. Neverwinter Vault: Players: NWN Modules. 14 October 2004 <http://nwn.bioware.com/forums/viewtopic.html?topic=313370&forum=66&highlight=feedback>.

Laurel, Brenda. *Computers as Theatre.* Menlo Park, California: Addison, 1991.

Lord, Albert Bates. *The Singer of Tales.* Cambridge and London: Harvard University Press, 1960.

Mason, Dave. "Where Am I?" Online posting. 26 Aug. 2002. Neverwinter Vault: Modules. 14 Oct. 2004 <http://nwvault.ign.com/Files/modules/data/1030339290713.shtml>.

———. "A Harper's Tale 1." Online posting. 30 Sept. 2003. Neverwinter Vault: Modules. 14 Oct. 2004 <http://nwvault.ign.com/Files/modules/data/1054436694285.shtml>.

Murray, Janet. *Hamlet on the Holodeck: The Future of Narrative in* Cyberspace. 1997; rpt 1998, Cambridge: MIT Press.

Wardrip-Fruin, Noah and Harrigan, Pat. *First-Person: New Media as Story, Performance, and Game.* Cambridge: MIT Press, 2004.

From Barbie to Mortal Kombat: Further Reflections

Henry Jenkins

It is early summer, 2001. I am walking through the great exhibition hall at the Electronic Entertainment Expo (E3), the most important trade show in the games industry. It is hard to imagine many activities that would pack in so many men per square foot of exhibition space and so few women. Many of the women we see here are dressed in scant clothing—leather-bound Valkyries, bikini-clad space bunnies, or women in tight jumpsuits prepared for motion-capture demonstrations. In what is the loudest, most visually oversaturated place I have ever been—with the possible exception of the streets of Hong Kong—women are used as bait to lure dazed buyers into the booths. One understands instantly why most of the games on the market look and sound the way they do. It has little to do with consumer tastes and everything to do with the competition for attention at the floor show. In this environment, subtle games—nuanced games—disappear without a trace. The Sims or Black & White, for example, commanded far less space at the Electronic Arts booth than the more action-packed games, and Majestic debuted off the main floor altogether in a corporate suite.

One booth captures my attention. A female gamer is taking on all challengers. There is a long line of male gamers, game designers, and game executives, each pumping each other up, in the expectation of "whipping her sweet little butt," as I overhear one man to say. And one by one they return from the encounter with their tails between their legs, their masculinity in tatters. She is *that good!* She is one with the machine—the closest thing to a cyborg I've ever encountered—with absolute economy and precision of movement. She knows the game so well that she anticipates every move and compensates for it. And she slaughters all of the best gamers.

Later that same summer, I am walking down the streets of downtown Melbourne and stumble past the window to a game arcade. In the window of the arcade, there are two teenage Moslem women—dressed in flowing robes, headdresses, and veils—and they are dancing barefoot in front of Dance Dance Revolution. They seem absolutely free of all inhibitions, totally in touch with the rhythms of the machine. And people along the streets, like myself, are stopping and gaping at the spectacular dance performance unfolding before us.

I could be optimistic and suggest that these two chance encounters suggest the increased visibility of female gamers. After all, the Interactive Digital Software Association is reporting dramatic increases in the number of women playing games in the United States (43 percent overall, with women constituting the majority of online gamers). The same research shows that, while the female Quake clans get lots of publicity for breaking across the gender barriers in gaming, more women play traditional card games such as bridge, board games such as chess, or trivia games such as Jeopardy, and only a small number play the kinds of games most often produced by the IDSA's member companies.

I fear that what these stories really suggest is that the sight of a female gamer remains a remarkable spectacle within a commercial and cultural space still dominated by male designers and male consumers.

From Barbie to Mortal Kombat: Gender and Computer Games was conceived and coedited with Justine Cassell in the midst of an era of limited but very real optimism about the ability of female-run startup companies to transform the game market, creating new kinds of games that might appeal to a broader range of female consumers.[1] Barbie Fashion Designer had been a top seller for the Christmas season, 1996, and had continued throughout the year to outsell industry standards, such as Quake or Myst, establishing that there was potentially a rather large market for female-centered software titles. Brenda Laurel, who was one of the most respected women in the computer industry, had established Purple Moon games with the explicit goal of designing products that reflected her sociological and ethnographic research into young girls' play patterns. And a number of other companies—Girl Games, Her Interactive—were producing girl-targeted titles; independent artists, such as Theresa Duncan (Chop Suey), were designing playful interactive works with a distinctly feminine sensibility; the major games companies were being forced to reconsider their marketing and design decisions to factor in female consumers more fully; and the introduction of Tomb Raider's Lara Croft character represented a significant new era for the female action hero in games.

The girl's game movement took shape around a series of competing goals and expectations:

1. *Economic.* The platform game market had entered into an age of heightened competition at a time when, in fact, 90 percent of American boys were already playing computer games. To survive, these game companies understood that they would need to expand their market and thus, then as now, there were three major targets—casual gamers, older gamers, and women. Any product that could succeed in attracting one or more of those prized demographics might hold the key for the company's long-term viability.
2. *Political.* The plight of preteen girls had become a central focus of feminist concern, following the publication of a series of books that suggested serious self-esteem issues that were impacting their ability to learn and grow within the classroom. Further research suggested that the gender gap in technological fields was growing rather than shrinking despite decades of feminist intervention, and the computer was more and more being coded as a "masculine" technology within the culture. Some felt that computer games might hold a key to getting girls engaged with computers at an earlier age, a head-start program for technological literacy. By the time they encountered computers in the classroom, boys had spent many more hours playing games than girls and often shoved them off the hardware to show them how it should be used.
3. *Technological.* The introduction of the CD-ROM as a staple of the home computer opened up a new opportunity for female-centered games to find their market. As long as platform games ruled the roost, there was little chance of building up a girl's market because consumers would have to buy the hardware before they could play games at a time when there was not going to be a critical mass of relevant product available. Once the home computer became the locus of game play—either through CD-ROM games or web-based games—then people who had bought the computer for other purposes could take a chance on buying software for girls or playing an online game. Moreover, while three major companies determined what products would be available for their platforms, the CD-ROM and web game markets were open to competition. Technological changes had lowered the barriers of entry into the market place.
4. *Entrepreneurial.* A growing number of women had tried working within the mainstream industry, enjoyed some degree of success, but had wanted to develop independence so they could create products that more fully reflected their perspectives and experiences. The girl's game movement caught the rising tide of female entrepreneurship in American culture at a time when women were starting new businesses at a rate significantly higher than men and in doing so were introducing new kinds of products, opening new kinds of markets, and developing new forms of business management and new kinds of customer relationships.

5. *Aesthetic.* The girl's game movement promised new kinds of content, new models of play and interactivity, new visual aesthetics, and new approaches to the soundtrack. The movement set a goal of making games radically different from those on the market, so that they could attract new kinds of consumers. This made the movement a hotbed for innovation and experimentation, with a strong push toward more psychologically nuanced characters, softer color palettes, more richly-layered soundtracks, new interface designs (including those, such as Barbie Fashion Designer, that helped bridge between the computer and real-world play), and more complex stories.

Our book tried to document the complex interweaving of ideological idealism and market calculations that shaped the development of the girl's game movement. We were interested in the ways that the girl's game movement brought together feminist academics and female CEOs in a collective effort to transform the current state and the future direction of a sector of the entertainment industry which is drawing in an annual income roughly equivalent to the gross domestic box office returns for Hollywood movies.

In that context, the book's focus shifted from whether there was going to be software for girls (which seemed almost a given) to *what kinds* of software for *what kinds* of girls. We asked, for example, whether games should be designed to reflect girl's existing tastes and interests or to transform them, and whether this question represented a contradiction within feminist entrepreneurship. Feminism has always sought to critique and reinvent gender roles, whereas entrepreneurship has had to start where the market was. We pointed, for example, to a controversial statement that Brenda Laurel had made, "I agreed that whatever solution the research suggested, I'd go along with. Even if it meant shipping products in pink boxes."[2] In many ways, this was a challenge that feminist critics had never had to confront before—how to write about cultural production in a context where women now exercised a considerable degree of power. If women were heading up the games company, their efforts could no longer be called marginal, but at the same time they were now forced to respond to the same economic realities which shaped corporate decision making in what remained a largely patriarchal society. Earlier generations of feminists had celebrated filmmakers like Dorothy Arzner who produced films from the margins of the studio system as "engaged in a kind of "countercinema." Yet, a game company executive like Laurel had control over (and thus was forced to take responsibility for) the products she shipped under the Purple Moon logo.

In hindsight, it is troubling how quickly Laurel found herself caught in a feminist backlash against the reproduction of feminine stereotypes in her work when so little attention was being given to her explicit goal of using popular culture to empower young girls. We might compare the feminist backlash against Purple Moon to, say, the much more friendly reception given to *Buffy the Vampire Slayer, Xena: Warrior Princess,* or *The Power Puff Girls.* As Laurel writes in her new book, *Utopian Entrepreneur,* "By trying to do anything socially positive at all, the utopian entrepreneur opens herself up to the endless critique that she is in fact not doing enough. . . . I am reminded of the old saw: the one who attacks you is likely to be the one closest to you on the road. . . . I wondered, did anyone notice that this wasn't Barbie—that Rocket struggled mightily to be ethical and self-defined? . . . Our characters exhibited loyalty, honor, love, and courage. They also struggled with gossip, jealousy, cheating, lipstick, smoking, exclusion, racism, poverty, materialism, and broken homes. When we had to choose, we sacrificed political correctness in order to meet the girls where they were, in the realities of their own lives."[3]

What bothers me even more, however, is the degree to which Purple Moon has been treated not only within the industry but also within academia as a "failed experiment," as if the importance of a political and cultural intervention could be or should be measured only by market standards. As *Utopian Entrepreneur* makes clear, however, even by market standards, the company could scarcely be viewed as a failure, Simply because it couldn't compete effectively for the girl's market against Barbie, the most successful girls franchise of all time, or American Girl, another successful franchise with brand loyalties arranged well before it sought to introduce its software line. Purple Moon sought to use a new set of characters and a new set of stories in order to introduce a new kind of product to a new kind of market. In doing so, it had notable success on the web, where it became the third most popular site for children, though its success came at a time when the business models for turning a revenue stream from the web still had not jelled (and—guess what, folks—they still haven't, several years later). As Laurel explains, "Here's one of the perversities of dot-capitalism: if Purple Moon had not actually produced any real products, I'd probably be 'post-economic' today. Just as the dot-economy started spinning straw into gold, Purple Moon was spending real money to make real products to go into real shelves in real stores. In investment terms, this was a big mistake. Even though we had an extremely popular Website, the embarrassing detail of real goods prevented us from passing for a dot-company in the venture community. And so, instead of the wild valuations that made some of our younger friends multimillionaires, the valuation of Purple Moon could never exceed some small multiple of our revenues—because we actually had revenues."[4]

What killed Purple Moon wasn't the lack of a girl's market for games or the invalidity of her core model about what a girl's game might look like, but rather the impatience of the company's core investors to make the kinds of turnaround that they were seeing elsewhere in the digital economy. In the end, Purple Moon suffered the fate that befalls most startups—it was acquired by a larger and more firmly established company, an old-economy company—in this case, Mattel. In trying to break into a market, startups are often forced to try riskier new methods or to test alternative approaches or to innovate new kinds of products, whereas established companies tend to be more conservative, taking a wait-and-see attitude toward innovation since they can afford to make adjustments down the line. Larger companies absorb startups for two reasons—to kill competition, and to absorb the innovation back into the mainstream of the industry. Mattel apparently absorbed Purple Moon with an eye toward killing off competition to Barbie with the result that it made only a halfhearted attempt to extend the Rocket franchise beyond those products that Laurel's company had produced. Mattel never understood what it had acquired. Rocket's ultimate disappearance from the marketplace, then, may have less to do with the viability of its model and more to do with the serious corporate forces that Purple Moon was challenging.

One of the most frequently asked questions when our book first appeared was whether it made sense to gender segregate—that is, to create a girls' market rather than expanding the existing boys' market to include more gender-neutral material. We were told, for example, that no one designed games specifically for boys. I would suggest that the release of a major piece of hardware known as the Gameboy suggests that the industry *did* identify its products along gender lines. Many felt that it was time to consciously signal to girls that there might be some forms of digital entertainment that reflected their tastes and interests. Without such a gesture, a gender-neutral play space would remain simply a de facto masculine space. The argument for more "feminine" content lay in the recognition that a significant number of girls were playing games—those who were already comfortable searching for and finding common ground with boys—whereas those girls most apt to be left behind were those whose tastes were the most traditionally feminine.

Suppose we take at face value the claim that game designers aren't designing for boys, that they are simply designing games they would like to play. The existing employee pool for the games industry is overwhelmingly male, so the games designed appeal overwhelmingly to men. One important step would be to significantly expand the number of women working in the industry and give them enough room to develop products that would reflect their tastes and sensibilities. Another would be to make a conscious effort to broaden the range of game content—based on research, if not firsthand experience—in order to expand the number of women who get excited about the medium and thus help to develop the next generation of game designers. What this situation doesn't justify is complacency on the part of the games industry.

The decline of the girl's game companies has taken some of the pressure off the major games companies to respond to this gender gap, buying them time to take a more conservative approach. We are not seeing what we might have anticipated five years ago—a growing "pink" aisle at the software stores. Instead, we are seeing the major companies absorb those lessons that they can from girl games companies' experimentation and apply them to existing franchises that have already proven successful with their predominantly male consumers. So, for example, Barbie Fashion

Designer enables girls to design Barbie's clothes, style her hair, select her theme music, choreograph her movements, and have her walk down the fashion runway. The new World Wrestling Federation games would then enable boys to design the wrestlers' clothes, style their hair, select their theme music, choreograph their movements, and then have them walk down the aisle into the ring, where they beat the crap out of each other. Is this new focus on designing your own characters at least partially a product of the success of the Barbie franchise? Hard to tell, given the fact that hardcore gamers have long traded in "skins," but the parallels between the two interface designs are striking.

Ironically, this more conservative strategy of integrating selected traits from girl's game models into existing boy's game genres and franchises may be leading us toward something approaching the "gender-neutral" play spaces some of us were advocating. (Let's bracket for the moment the nagging question of whether any of us really knows what it would look like to live in a *gender-neutral* society and thus whether gender-neutral doesn't still mean second-class citizenship for girls.)

It is interesting to look at The Sims, one of the most successful games in recent years, in light of the distinctions that Brenda Laurel mapped between the classic boy game and her vision for the ideal girl game:

> GIRLS—Leading characters are everyday people that girls can easily relate to, and are as real to girls as their best friends.
> BOYS—Leading characters are fantasy-based action heroes with "super-power" abilities.[5]

In the case of The Sims, the default set of characters are, indeed, "everyday people," the setting is domestic, and the challenges confronted are familiar ones. Because The Sims supports the development and sharing of "skins," artifacts, and architecture, some players have developed larger-than-life protagonists and spaces that can be retrofit into the game universe—including the kinds of superhero or wrestler characters found in more traditional games. What is striking, however, is the degree to which boys are forced to adjust the game to accommodate their interests, whereas so often in our culture, women are expected to read their interests from the margins of popular culture texts that center on male interests.

> GIRLS—Goal is to explore and have new experiences, with degrees of success and varying outcomes.
> BOYS—Goal is to win, and the play is linear. Outcome is black and white; die and start over; one "right" solution.

The Sims lends itself to a broad array of different goals and outcomes. Often, players use the game to do psychological experiments. Friends have told me that they have used the game to see if they are compatible with a proposed set of roommates or learn what adjustments they would need to make in order to move in with their girlfriends. A single mother described the different ways that she and her son used the game to work through the issues surrounding the recent divorce—the son to create a perfect nuclear family, the divorced mother to imagine what it would be like to reenter the dating world. The features of the game are familiar and evocative, encouraging us to use them to simulate our own experiences and thus test different social strategies, much as the design of the Purple Moon games was designed to allow pre-adolescent girls to rehearse social strategies for responding to the challenges of high school life.

> GIRLS—Play focuses on multi-sensory immersion, discovery, and strong story lines.
> BOYS—Speed and action are key.

The exploratory nature of The Sims differs dramatically from the fixed goals and rapid-fire action associated with traditional boy's games. The pace of this game is slower, dealing more with gradual processes and repeated routines, rather than rapid-fire challenges and conflicts. The Sims almost immediately developed a fan culture surrounding the production of scrapbooks documenting the experiences of the Sims characters and over time, these scrapbooks became more and more narrativized. To read those scrapbooks is to learn more and more about the emotional lives and motives of these characters. Indeed, while the scrapbooks were intended to record game actions, many players report that they play the game to get the images they need to complete narratives they want to tell. The Sims has thus found a way to encourage players to create more narratively compelling experiences. While Purple Moon authored stories within its games (while providing a web space in which girls could expand upon their understanding of those characters), The Sims was designed as a "sandbox" or "dollhouse" where players could play out their own narratives.

> GIRLS—Feature everyday "real life" settings as well as new places to explore.
> BOYS—Features non-realistic, larger-than-life settings.

In *From Barbie to Mortal Kombat,* I contrasted Sim City with the "play town" that Harriet constructed for her imaginary life in the classic children's book, *Harriet the Spy*: "Sim City embraces stock themes from boys' play, such as building forts, shaping earth with toy trucks, or damming creeks, playing them out on a much larger scale. For Harriet, the mapping of the space was only the first step in preparing the ground for a rich saga of life and death, joy and sorrow, and those are the elements that are totally lacking in most simulation games." The Sims, however, is Harriet's "play town." The shift in the scale of the game from Sim City's objectifying top-down view of urban renewal to The Sims' subjective, eye-level perspective on individual characters, reflects this openness to new kinds of narratives, which center around characters and their emotional lives. The interface is designed to foreground emotional responses—rendered all the more powerfully when abstracted from the specific verbal content of their speech. The characters weep when other characters die; they make passionate love; they flirt; they get depressed and mope; they get tired and cranky. As The Sims moves from a stand-alone game to a multiplayer online game, the universe of spaces and characters expands enormously, creating many more opportunities for both spatial and social exploration.

> GIRLS—Success comes through development of friendships.
> BOYS—Success comes through the elimination of competitors.

Consider Laurel's descriptions of the different ways that girls and boys compete: "Girls and boys are equally competitive—they are just different kinds of competition. Girls assert social influence and structure relationships while boys seek to dominate and defeat." Girl's competition, she asserts, reflects a "desire to establish relationships/friendships," to extend their "social status" through processes of affiliation and exclusion, and to "figure out" opponents. The Sims can be seen as a game that rewards precisely these kinds of social and cultural competition, including the development of a rewards system based on the player's ability to establish strong social bonds with a broad array of other players and characters. Much of the game play consists of social interactions within the family or with friends, and the expansion packs offer more options for hosting parties and thus expanding the circle of one's social contacts. The game excludes almost altogether those forms of competition that are traditionally associated with boy's games, where "social status tends to be by achievement and physical domination" and where the goal is to "outwit" opponents. Indeed, in the absence of pre-structured goals or pre-determined competition, the element of direct physical combat is excluded from The Sims almost altogether.

I have spoken to Will Wright and others at Maxis and I am reasonably convinced that they were not directly modeling The Sims on girl's game–movement products. Rather, the decisions they made came out of a context where there were more female designers and more highly ranked female designers than I have seen at any other mainstream game studio. In such a context, even if there is no conscious goal of expanding the female market, the unconscious decisions made by men and women working together is likely to produce a product that is very different from one where the intuitive decisions were made by an all or predominantly male team of designers. Not surprisingly, then, The Sims has proven to be highly successful in attracting female players while at the same time expanding the range of play experiences available to boys.

I concluded my essay in *From Barbie to Mortal Kombat* with these words of caution:

> *[Sociologist Barrie] Thorne finds that aggressive 'border-work' is more likely to occur when children are forced together by adults than when they find themselves interacting more spontaneously, more likely to occur in prestructured institutional settings like the school yard than in the informal settings of the subdivisions and apartment complexes. All of this suggests that our fantasy of designing games that will provide common play "spaces for girls and boys may be illusive and as full of its own complications and challenges as creating a 'girls only' space or encouraging girls to venture into traditional male turf. We are not yet sure what a gender-neutral space will look like. Creating such a space would mean redesigning not only the nature of computer games but also the nature of society. The danger may be that in such a space, gender differences are going to be more acutely felt, as boys and girls will be repelled from each other rather than drawn together.*
>
> *There are reasons why this is a place where neither the feminist entrepreneurs nor the makers of boys' games are ready to go, yet as the girls' market is secured, the challenge must be to find a way to move beyond our existing categories and to once again invent new kinds of virtual play spaces.*

The Sims is an important step in that process—one that shares many of the traits associated with the girl's game movement without calling attention to them as such, and embeds them within a successful preexisting franchise with a known appeal to male consumers: The Sims, of course, in reproducing domestic space, reproduces many of the ideological assumptions that currently shape the contemporary family, yet it also gives us the tools to rescript and restage those relationships in new terms. As such, it would seem to me to represent the logical culmination of the girl's game movement and the beginning of the process of developing a more "gender-neutral" play space. As we move in that direction, however, we need to be aware of the kind of "border-work" that demarks places where boys and girls can and cannot play together and try to understand what it really means to provide a common ground for both genders.

Notes

1. Justine Cassell and Henry Jenkins, eds., *From Barbie to Mortal Kombat: Gender and Computer Games* (Cambridge: MIT Press, 1998). This essay was originally presented at Playing by the Rules: The Cultural Policy Challenges of Video Games Conference, hosted by the Cultural Policy Center, University of Chicago, October 26–27, 2001.
2. Brenda Laurel, as quoted in G. Beato, "Girls Games: Computer Games for Girls Is No Longer an Oxymoron," *Wired,* April 1997.
3. Brenda Laurel, *Utopian Entrepreneur* (Cambridge: MIT Press, 2001).
4. Ibid., 3.
5. This formulation of the differences between girls and games, which run throughout the rest of this essay, was formulated by Brenda Laurel. It was widely reproduced in internal documents and press releases from Purple Moon in the late 1990s. She discusses these ideas in detail with Jennifer Glos and Shari Goldin, "An Interview with Brenda Laurel (Purple Moon)," in Cassell and Jenkins, eds., *From Barbie to Mortal Kombat,* 118–135.

Image Future

Lev Manovich

Today the techniques of traditional animation, cinematography, and computer graphics are often used in combination to create new hybrid moving image forms. I discuss this process using the example of a particularly intricate hybrid Universal Capture method used in the second and third films of The Matrix *trilogy. Rather than expecting that any of the present "pure" forms will dominate the future of visual and moving image cultures, I suggest that the future belongs to such hybrids.*

For the larger part of the twentieth century, different areas of commercial moving image culture maintained their distinct production methods and distinct aesthetics. Films and cartoons were produced completely differently and it was easy to tell their visual languages apart. Today the situation is different. Computerization of all areas of moving image production created a common pool of techniques, which can be used regardless of whether one is creating motion graphics for television, a narrative feature, an animated feature, or a music video. The ability to composite many layers of imagery with varied transparency, to place still and moving elements within a shared 3D virtual space and then move a virtual camera through this space, to apply simulated motion blur and depth of field effect, to change over time any visual parameter of a frame—all these can now be equally applied to any images regardless of whether they were captured via a lens-based recording, drawn by hand, created with 3D software, etc.

The existence of this common vocabulary of computer-based techniques does not mean that all films now look the same. What it means, however, is that while most live action films and animated features do look quite distinct today, this is the result of a deliberate choices rather than the inevitable consequence of differences in production methods and technology. At the same time, outside of the realm of live action films and animation features, the aesthetics of moving image culture has dramatically changed during the 1990s.

What happened can be summarized in the following way. Around middle of the 1990s, the simulated physical media for moving and still image production (cinematography, animation, graphic design, typography), new computer media (3D animation), and new computer techniques (compositing, multiple levels of transparency) started to interact within a single computing environment—either a personal computer or a relatively inexpensive graphics workstation affordable for small companies and even individuals. The result was the emergence of a new hybrid aesthetics that quickly became the norm. Today this aesthetics is at work in practically all short moving image forms: TV advertising and TV graphics, music videos, short animations, broadcast graphics, film titles, music videos, Web splash pages. It also defines a new field of media production—motion graphics—but it is important to note that the hybrid aesthetics is not confined to this field but can be found at work everywhere else.

This aesthetics exists in endless variations but its logic is the same: juxtaposition of previously distinct visual languages of different media within the same sequence and, quite often, within same frame. Hand-drawn elements, photographic cutouts, video, type, 3D elements are not simply placed next to each but interweaved together. The resulting visual language is a hybrid. It can be also called a metalanguage as it combines the languages of design, typography, cell animation, 3D computer animation, painting, and cinematography.

Along with special effects features, the hybrid (or meta) aesthetics of a great majority of short moving images sequences which surround us today is the most visible effect of computerization of moving image production. In this case animation frequently appears as one element of a sequence or even a single frame. But this is just one, more obvious, role of animation in contemporary post-digital visual landscape. In this article I will discuss its other role: as a generalized technique that can be applied to any images, including film and video. Here, animation functions not as a medium but as a set of general-purpose techniques—used together with other techniques in the common pool of options available to a filmmaker/designer.

I have chosen a particular example for my discussion that I think will illustrate well this new role of animation. It is a relatively new method of combining live action and CG. Called "Universal Capture" (U-cap) by their creators, it was first systematically used on a large scale by ESC Entertainment in *Matrix 2* and *Matrix 3* films from *The Matrix* trilogy. I will discuss how this method is different from the now standard and older techniques of integrating live action and computer graphics elements. Universal Capture also creates visual hybrids—but they are quite different from the hybrids found in motion graphics and other short moving image forms today. In the case of Universal Capture, different types of imagery are not mixed together but rather *fused* to create a new kind of image. This image combines "the best" qualities of two types of imagery that we normally understand as being ontologically the opposites: live action recording and 3D computer animation. I will suggest that such image hybrids are likely to play a large role in future visual culture while the place of "pure" images that are not fused or mixed with anything is likely to diminish.

Uneven Development

What kinds of images would dominate visual culture a number of decades from now? Would they still be similar to the typical image that surrounds us today—photographs that are digitally manipulated and often combined with various graphical elements and type? Or would future images be completely different? Would photographic code fade away in favor of something else?

There are good reasons to assume that the future images would be photograph-like. Like a virus, a photograph turned out to be an incredibly resilient representational code: it survived waves of technological change, including computerization of all stages of cultural production and distribution. The reason for this persistence of photographic code lies in its flexibility: photographs can be easily mixed with all other visual forms—drawings, 2D and 3D designs, line diagrams, and type. As a result, while photographs truly dominate contemporary visual culture, most of them are not pure photographs but various mutations and hybrids: photographs which went through various filters and manual adjustments to achieve a more stylized look, a more flat graphic look, more saturated color, etc.; photographs mixed with design and type elements; photographs which are not limited to the part of the spectrum visible to a human eye (nigh vision, x-ray); simulated photographs done with 3D computer graphics; and so on. Therefore, while we can say that today we live in a "photographic culture," we also need to start reading the word "photographic" in a new way. "Photographic" today is really photo-GRAPHIC, the photo providing only an initial layer for the overall graphical mix. (In the area of moving images, the term "motion graphics" captures perfectly the same development: the subordination of live action cinematography to the graphic code.)

One way is which change happens in nature, society, and culture is inside out. The internal structure changes first, and this change affects the visible skin only later. For instance, according to Marxist theory of historical development, infrastructure (i.e., mode of production in a given society—also called "base") changes well before superstructure (ideology and culture in this society). In a different example, think of technology design in the twentieth century: typically a new type of machine was at first fitted within old, familiar skin: for instance, early twentieth century cars emulated the form of horse carriage. The familiar McLuhan's idea that the new media first emulates old media is another example of this type of change. In this case, a new mode of media production, so to speak, is first used to support old structure of media organization, before the new structure emerges. For instance, first typesets book were designed to emulate hand-written books; cinema first emulated theatre; and so on.

This concept of uneven development can be useful in thinking about the changes in contemporary visual culture. Since this process started fifty years ago, computerization of photography (and cinematography) has by now completely changed the internal structure of a photographic image. Yet its "skin," i.e., the way a typical photograph looks, still largely remains the same. It is therefore possible that at some point in the future the "skin" of an image would also become completely different, but this did not happen yet. So we can say at present our visual culture is characterized by a new computer "base" and old photographic "superstructure."

The Matrix films provide us with a very rich set of examples perfect for thinking further about these issues. The trilogy is an allegory about how its visual universe is constructed. That is, the films tell us about *The Matrix,* the virtual universe which is maintained by computers—and of course, visually the images of *The Matrix* which we the viewers see in the films were all indeed assembled with the help of software (the animators sometimes used Maya but mostly relied on custom written programs). So there is a perfect symmetry between us, the viewers of a film, and the people who live inside *The Matrix*—except while the computers running *The Matrix* are capable of doing it in real time, most scenes in each of *The Matrix* films took months and even years to put together. (So *The Matrix* can be also interpreted as the futuristic vision of computer games at a point in a future when it would become possible to render *The Matrix*-style visual effects in real time.)

The key to the visual universe of *The Matrix* is the new set of computer graphic techniques that over the years were developed by a number of people both in academia and in the special effects industry including Georgi Borshukov and John Gaeta.[1] Their inventors coined a number of names for these techniques: "virtual cinema," "virtual human," "virtual cinematography," "universal capture." Together, these techniques represent a true milestone in the history of computer-driven special effects. They take to their logical conclusion the developments of the 1990s such as motion capture, and simultaneously open a new stage. We can say that with *The Matrix,* the old "base" of photography has finally been completely replaced by a new computer-driven one. What remains to be seen is how the "superstructure" of a photographic image—what it represents and how—will change to accommodate this "base."

Reality Simulation versus Reality Sampling

Before proceeding, I should note that not all of special effects in *The Matrix* rely on Universal Capture and, of course, other Hollywood films already use some of the same strategies. However, in this text I decided to focus on the use of this process in *the Matrix* because Universal Capture was actually developed for the second and third films of the trilogy. And while the complete credits for everybody involved in developing the process would run for a number of lines, in this text I will identify it with Gaeta. The reason is not because, as a senior special effects supervisor for *The Matrix Reloaded* and *The Matrix Revolutions* he got most publicity. More importantly, in contrast to many others in the special effects industry, Gaeta has extensively reflected on the techniques he and his colleagues have developed, presenting it as a new paradigm for cinema and entertainment and coining useful terms and concepts for understanding it.

In order to understand better the significance of Gaeta's method, lets briefly run through the history of 3D photo-realistic image synthesis and its use in the film industry. In 1963 Lawrence G. Roberts (who later in the 1960s became one of the key people behind the development of Arpanet but at that time was a graduate student at MIT) published a description of a computer algorithm to construct images in linear perspective. These images represented the objects' edges as lines; in contemporary language of computer graphics they can be called "wire frames." Approximately ten years later computer scientists designed algorithms that allowed for the creation of shaded images (so-called Gouraud shading and Phong shading, named after the computer scientists who create the corresponding algorithms). From the middle of the 1970s to the end of the 1980s the field of 3D computer graphics went through rapid development. Every year new fundamental techniques were created: transparency, shadows, image mapping, bump texturing, particle system, compositing, ray tracing, radiosity, and so on.[2] By the end of this creative and fruitful period in the history of the field, it was possible to use combination of these techniques to synthesize images of almost every subject that often were not easily distinguishable from traditional cinematography.

All this research was based on one fundamental assumption: in order to re-create an image of reality identical to the one captured by a film camera, we need to systematically simulate the actual physics involved in construction of this image. This means simulating the complex interactions between light sources, the properties of different materials (cloth, metal, glass, etc.), and the properties of physical film cameras, including all their limitations such as depth of field and motion blur. Since it was obvious to computer scientists that if they exactly simulate all this physics, a computer would take forever to calculate even a single image, they put their energy in inventing various short cuts which would create sufficiently realistic images while involving fewer calculation steps. So in fact each of the techniques for image synthesis I mentioned in the previous paragraph is one such "hack"—a particular approximation of a particular subset of all possible interactions between light sources, materials, and cameras.

This assumption also means that you are re-creating reality step-by-step, from scratch. Every time you want to make a still image or an animation of some object or a scene, the story of creation from The Bible is being replayed.

(I imagine God creating Universe by going through the numerous menus of a professional 3D modeling, animation, and rendering program such as Maya. First he has to make all the geometry: manipulating splines, extruding contours, adding bevels . . . Next for every object and creature he has to choose the material properties: specular color, transparency level, image, bump and reflexion maps, and so on. He finishes one set of parameters, wipes his forehead, and starts working on the next set. Now on defining the lights: again, dozens of menu options need to be selected. He renders the scene, looks at the result, and admires his creation. But he is far from being done: the universe he has in mind is not a still image but an animation, which means that the water has to flow, the grass and leaves have to move under the blow of the wind, and all the creatures also have to move. He sights and opens another set of menus where he has to define the parameters of algorithms that simulate the physics of motion. And on, and on, and on. Finally the world itself is finished and it looks good; but now God wants to create the Man so he can admire his creation. God sights again, and takes from the shelf a particular Maya manuals from the complete set which occupies the whole shelf . . .)

Of course we are in somewhat better position than God was. He was creating everything for the first time, so he could not borrow things from anywhere. Therefore everything had to be built and defined from scratch. But we are not creating a new universe but instead visually simulating universe that already exists, i.e., physical reality. Therefore computer scientists working on 3D computer graphics techniques have realized early on that in addition to approximating the physics involved they can also sometimes take another shortcut. Instead of defining something from scratch through the algorithms, they can simply *sample* it from existing reality and incorporate these samples in the construction process.

The examples of the application of this idea are the techniques of texture mapping and bump mapping which were introduced already in the second part of the 1970s. With texture mapping, any 2D digital image—which can be a close-up of some texture such as wood grain or bricks, but which can be also anything else, for instance a logo, a photograph of a face or of clouds—is mathematically wrapped around a 3D model. This is a very effective way to add visual richness of a real world to a virtual scene. Bump texturing works similarly, but in this case the 2D image is used as a way to quickly add complexity to the geometry itself. For instance, instead of having to manually model all the little cracks and indentations which make up the 3D texture of a concrete wall, an artist can simply take a photograph of an existing wall, convert into a grayscale image, and then feed this image to the rendering algorithm. The algorithm treats grayscale image as a depth map, i.e., the value of every pixel is being interpreted as relative height of the surface. So in this example, light pixels become points on the wall that are a little in front while dark pixels become points that are a little behind. The result is enormous saving in the amount of time necessary to recreate a particular but very important aspect of our physical reality: a slight and usually regular 3D texture found in most natural and many human-made surfaces, from the bark of a tree to a weaved cloth.

Other 3D computer graphics techniques based on the idea of sampling existing reality include reflection mapping and 3D digitizing. Despite the fact that all these techniques have been always widely used as soon as they were invented, many people in the computer graphics field (as far as I can see) always felt that they were cheating. Why? I think this feeling was there because the overall conceptual paradigm for creating photorealistic computer graphics was to simulate everything from scratch through algorithms. So if you had to use the techniques based on directly sampling reality, you somehow felt that this was just temporary—because the appropriate algorithms were not yet developed or because the machines were too slow. You also had this feeling because once you started to manually sample reality and then tried to include these samples in your perfect algorithmically defined image, things rarely would fit exactly right, and painstaking manual adjustments were required. For instance, texture mapping would work perfectly if applied to strait surface, but if the surface were curved, inevitable distortion would occur.

Throughout the 1970s and 1980s the "reality simulation" paradigm and "reality sampling" paradigms co-existed side-by-side. More precisely, as I suggested above, sampling paradigm was "imbedded" within reality simulation paradigm. It was common sense that the right way to create photorealistic images of reality is by simulating its physics as precisely as one could. Sampling existing reality now and then and then adding these samples to a virtual scene was a trick, a shortcut within overwise honest game of simulation.

Building *The Matrix*

So far we looked at the paradigms of 3D computer graphics field without considering the uses of the simulated images. So what happens if you want to incorporate photorealistic images into a film? This introduces a new constraint. Not only every simulated image has to be consistent internally, with the cast shadows corresponding to the light sources, and so on, but now it also has to be consistent with the cinematography of a film. The simulated universe and live action universe have to match perfectly (I am talking here about the "normal" use of computer graphics in narrative films and not the hybrid aesthetics of TV graphics, music videos, etc. which deliberately juxtaposes different visual codes). As can be seen in retrospect, this new constraint eventually changed the relationship between the two paradigms in favor of sampling paradigm. But this is only visible now, after *The Matrix* films made the sampling paradigm the basis of its visual universe.[3]

At first, when filmmakers started to incorporate synthetic 3D images in films, this did not have any effect on how computer scientists thought about computer graphics. 3D computer graphics for the first time briefly appeared in a feature film in 1980 in *Looker*. Throughout the 1980s, a number of films were made which used computer images but always only as a small element within the overall film narrative. (Released in 1982, *Tron* can be compared to *The Matrix* since its narrative universe is situated inside computer and created through computer graphics—but this was an exception.) For instance, one of *Star Trek* films contained a scene of a planet coming to life; it was created using the very first particle system. But this was a single scene, and it had no interaction with all other scenes in the film.

In the early 1990s the situation has started to change. With pioneering films such as *The Abyss* (James Cameron, 1989), *Terminator 2* (James Cameron, 1991), and *Jurassic Park* (Steven Spielberg, 1993) computer generated characters became the key protagonists of feature films. This meant that they would appear in dozens or even hundreds of shots throughout a film, and that in most of these shots computer characters would have to be integrated with real environments and human actors captured via live action photography (called in the business "live plate.") Examples are the T-100 cyborg character in *Terminator 2: Judgment Day,* or dinosaurs in *Jurassic Park.* These computer-generated characters are situated inside the live action universe that is the result of sampling physical reality via 35 mm film camera. The simulated world is located inside the captured world, and the two have to match perfectly.

As I pointed out in my *The Language of New Media* in the discussion of compositing, perfectly aligning elements that come from different sources is one of fundamental challenges of computer-based realism. Throughout the 1990s filmmakers and special effects artists have dealt with this challenge using a variety of techniques and methods. What Gaeta realised earlier than others is that the best way to align the two universes of live action and 3D computer graphics was to build *a single new universe*.[4]

Rather than treating sampling reality as just one technique to be used along with many other "proper" algorithmic techniques of image synthesis, Gaeta and his colleagues turned it into the key foundation of Universal Capture process. The process systematically takes physical reality apart and then systematically reassembles the elements into a virtual computer-based representation. The result is a new kind of image that has photographic/cinematographic appearance and level of detail yet internally is structured in a completely different way.

Universal Capture was developed and refined over a three-year period from 2000 to 2003.[5] How does the process work? There are actually more stages and details involved, but the basic procedure is the following.[6] An actor's performance in ambient lighting is recorded using five synchronized high-resolution video cameras. "Performance" in this case includes everything an actor will say in a film and all possible facial expressions.[7] (During the production the studio was capturing over 5 terabytes of data each day.) Next special algorithms are used to track each pixel's movement over time at every frame. This information is combined with a 3D model of a neutral expression of the actor created using cyberscan scanner. The result is an animated 3D shape that accurately represents the geometry of the actor's head as it changes during a particular performance. The shape is mapped with color information extracted from the captured video sequences. A separate very high resolution scan of the actor's face is used to create the map of small-scale surface details like pores and wrinkles, and this map is also added to the model.

After all the data has been extracted, aligned, and combine, the result is what Gaeta calls a "virtual human"—a highly accurate reconstruction of the captured performance, now available as a 3D computer graphics data—with all the advantages that come from having such representation. For instance, because actor's performance now exists as a 3D object in virtual space, the filmmaker can animate virtual camera and "play" the reconstructed performance from an arbitrary angle. Similarly, the virtual head can be also lighted in any way desirable. It can be also attached to a separately constructed CG body.[8] For example, all the characters which appeared the Burly Brawl scene in *The Matrix 2* were created by combining the heads constructed via Universal Capture done on the leading actors with CG bodies which used motion capture data from a different set of performers. Because all the characters along with the set were computer generated, this allowed the directors of the scene to choreograph the virtual camera, having it fly around the scene in a way not possible with real cameras on a real physical set.

The process was appropriately named Total Capture because it captures all the possible information from an object or a scene using a number of recording methods—or at least, whatever is possible to capture using current technologies. Different dimensions—color, 3D geometry, reflectivity and texture—are captured separately and then put back together to create a more detailed and realistic representation.

Total Capture is significantly different from the commonly accepted methods used to create computer-based special effects such as keyframe animation and physically based modeling. In the first method, an animator specifies the key positions of a 3D model, and the computer calculates in-between frames. With the second method, all the animation is automatically created by software that simulates the physics underlying the movement. (This method thus represents a particular instance of "reality simulation" paradigm I already discussed.) For instance, to create a realistic animation of moving creature, the programmers model its skeleton, muscles, and skin, and specify the algorithms that simulate the actual physics involved. Often the two methods are combined: for instance, physically based modeling can be used to animate a running dinosaur while manual animation can be used for shots where the dinosaur interacts with human characters.

In recent years, the most impressive achievement in physically based modeling was the battle in *The Lord of the Rings: Return of the King* (Peter Jackson, 2003) which involved tens of thousands of virtual soldiers all driven by *Massive* software.[9] Similar to the Non-human Players (or bots) in computer games, each virtual soldier was given the ability to "see" the terrain and other soldiers, a set of priorities and an independent "brain," i.e., a AI program which directs character's actions based on the perceptual inputs and priorities. But in contrast to games AI, Massive software does not have to run in real time. Therefore it can create the scenes with tens and even hundreds of thousands realistically behaving agents (one commercial created with the help of *Massive* software featured 146,000 virtual characters.)

Universal Capture method uses neither manual animation nor simulation of the underlying physics. Instead, it directly samples physical reality, including color, texture and the movement of the actors. Short sequences of the actor's performances are encoded as 3D computer animations; these animations form a library from which the filmmakers can then draw as they compose a scene. The analogy with musical sampling is obvious here. As Gaeta pointed out, his team never used manual animation to try to tweak the motion of character's face; however, just as a musician may do it, they would often "hold" particular expression before going to the next one.[10] This suggests another analogy—editing videotape. But this is a second-degree editing, so to speak: instead of simply capturing segments of reality on video and then joining them together, Gaeta's method produces complete virtual recreations of particular phenomena—self-contained micro-worlds—which can be then further edited and embedded within a larger 3D simulated space.

Animation as an Idea

The brief overview of the methods of computer graphics that I presented above in order to explain Universal Capture offers good examples of the multiplicity of ways in which animation is used in contemporary moving image culture. If we consider this multiplicity, it is possible to come to a conclusion that "animation" as a separate medium in fact hardly exists anymore. At the same time, the general principles and techniques of putting objects and images into motion developed in nineteenth and twentieth century animation are used much more frequently now than before computerization. But they are hardly ever used by themselves—usually they are combined with other techniques drawn from live action cinematography and computer graphics.

So where does animation start and end today? When you see a Disney animated feature or a motion graphics short it is obvious that you are seeing "animation." Regardless of whether the process involves drawing images by hand or using 3D software, the principle is the same: somebody created the drawings or 3D objects, set keyframes and then created in-between positions. (Of course in the course of commercial films, this is not one person but large teams.) The objects can be created in multiple ways and inbetweening can be done manually or automatically by the software, but this does not change the basic logic. The movement, or any other change over time, is defined manually—usually via keyframes (but not always). In retrospect, the definition of movement via keys probably was the essence of twentieth century animation. It was used in traditional cell animation by Disney and others, for stop motion animation by Starevich and Trnka, for the 3D animated shorts by Pixar, and it continues to be used today in animated features that combine traditional cell method and 3D computer animation. And while experimental animators such as Norman McLaren refused keys/inbetweens system in favor of drawing each frame on film by hand without explicitly defining the keys, this did not change the overall logic: the movement was created by hand. Not surprisingly, most animation artists exploited this key feature of animation in different ways, turning it into aesthetics: for instance, exaggerated squash and stretch in Disney, or the discontinuous jumps between frames in McLaren.

What about other ways in which images and objects can be set into motion? Consider for example the methods developed in computer graphics: physically based modeling, particle systems, formal grammars, artificial life, and behavioral animation. In all these methods, the animator does not directly create the movement. Instead it is created by the software that uses some kind of mathematical model. For instance, in the case of physically based modeling the animator may sets the parameters of a computer model which simulates a physical force such as wind which will deform a piece of cloth over a number of frames. Or, she may instruct the ball to drop on the floor, and let the physics model control how the ball will bounce after it hits the floor. In the case of particle systems used to model everything from fireworks, explosions, water and gas to animal flocks and swarms, the animator only has to define initial conditions: a number of particles, their speed, their lifespan, etc.

In contrast to live action cinema, these computer graphics methods do not capture real physical movement. Does it mean that they belong to animation? If we accept that the defining feature of traditional animation was manual creation of movement, the answer will be no. But things are not so simple. With all these methods, the animator sets the initial parameters, runs the model, adjusts the parameters, and repeats this production loop until she is satisfied with the result. So while the actual movement is produced not by hand by a mathematical model, the animator maintains significant control. In a way, the animator acts as a film director—only in this case she is directing not the actors but the computer model until it produces a satisfactory performance. Or we can also compare her to a film editor who is selecting among best performances of the computer model.

James Blinn, a computer scientist responsible for creating many fundamental techniques of computer graphics, once made an interesting analogy to explain the difference between manual keyframing method and physically based modeling.[11] He told the audience at a SIGGRAPH panel that the difference between the two methods is analogous to the difference between painting and photography. In Blinn's terms, an animator who creates movement by manually defining keyframes and drawing inbetween frames is like a painter who is observing the world and then making a painting of it. The resemblance between a painting and the world depends on painter's skills, imagination and intentions. Whereas an animator who uses physically based modeling is like a photographer who captures the world as it actually is. Blinn wanted to emphasize that mathematical techniques can create a realistic simulation of movement in the physical world and an animator only has to capture what is created by the simulation. Although this analogy is useful, I think it is not completely accurate. Obviously, the traditional photographer whom Blinn had in mind (i.e., before Photoshop) chooses composition, contrast, depth of field, and many other parameters. Similarly, an animator who is using physically based modeling also has control over a large number of parameters and it depends on her skills and perseverance to make the model produce a satisfying animation. Consider the following example from the related area of software art, which uses some of the same mathematical methods. Casey Reas, an artist who is well-known both for his Processing programming environment and his own still images and animations, told me recently that he may spend only a couple of hours writing a software program to create a new work—and then another two years working with the different parameters of the same program and producing endless test images until he is satisfied with the results.[12] So while at first physically based modeling appears to be opposite of traditional animation in that the movement is created by a computer, in fact it should be understood as a hybrid between animation and computer simulation. While the animator no longer directly draws each phase of movement, she is working with the parameters of the mathematical model that "draws" the actual movement.

And what about Universal Capture method as used in *The Matrix*? Gaeta and his colleagues also banished keyframing animation—but they did not used any mathematical modes to automatically generate motion either. As we saw, their solution was to capture the actual performances of an actor (i.e., movements of actor's face), and then reconstruct it as a 3D sequence. Together, these reconstructed sequences form a library of facial expressions. The filmmaker can then draw from this library editing together a sequence of expressions (but not interfering with any parameters of separate sequences). It is important to stress that a 3D model has no muscles, or other controls traditionally used in animating computer graphics faces—it is used "as is."

Just as it is the case when animator employs mathematical models, this method avoids drawing individual movements by hand. And yet, its logic is that of animation rather than of cinema. The filmmaker chooses individual sequences of actors' performances, edits them, blends them if necessary, and places them in a particular order to create a scene. In short, the scene is actually constructed by hand even though its components are not. So while in traditional animation the animator draws each frame to create a short sequence (for instance, a character turning his head), here the filmmaker "draws" on a higher level: manipulating whole sequences as opposed to their individual frames.

To create final movie scenes, Universal Capture is combined with Virtual Cinematography: staging the lighting, the positions and movement of a virtual camera that is "filming" the virtual performances. What makes this Virtual Cinematography as opposed to simply computer graphics? The reason is that the world as seen by a virtual camera is different from a normal world of computer graphics. It consists from reconstructions of the actual set and the actual performers created via Universal Capture. The aim is to avoid more manual process usually used to create 3D models and sets. Instead, the data about the physical world is captured and then used to create a precise virtual replica.

Ultimately, ESC's production method as used in *Matrix* is neither "pure" animation, nor cinematography, nor traditional special effects, nor traditional CG. And this is typical of moving image culture today. When the techniques drawn from these different traditions are fused together in a computer environment, the result is not a sum of components but a variety of hybrid methods such as Universal Capture. I believe that this how different moving image techniques function now in general. After computerization virtualized them—"extracting" them from their particular physical media to turn into algorithms—they start interacting and creating hybrids. Which means that in most cases, we will no longer find any of these techniques in their pure original state.

For instance, what does it mean when we see depth of field effect in motion graphics, films and television programs which use neither live action footage nor photorealistic 3D graphics but have a more stylized look? Originally an artifact of lens-based recording, depth of field was simulated in a computer when the main goal of 3D compute graphics field was to create maximum "photorealism," i.e., synthetic scenes not distinguishable from live action cinematography.[13] But once this technique became available, moving image artists gradually realized that it can be used regardless of how realistic or abstract the visual style is—as long as there is a suggestion of a 3D space. Typography moving in perspective through an empty space; drawn 2D characters positioned on different layers in a 3D space; a field of animated particles—any composition can be put through the simulated depth of field.

The fact that this effect is simulated and removed from its original physical media means that a designer can manipulate it a variety of ways. The parameters which define what part of the space is in focus can be independently animated, i.e., set to change over time—because they are simply the numbers controlling the algorithm and not something built into the optics of a physical lens. So while simulated depth of field can be said to maintain the memory of the particular physical media (lens-based photo and film recording) from which it came from, it became an essentially new technique which functions as a "character" in its own right. It has the fluidity and versatility not available previously. Its connection to the physical world is ambiguous at best. On the one hand, it only makes sense to use depth of field if you are constructing a 3D space even if it is defined in a minimal way by using only a few or even a single depth cue such as lines converging towards the vanishing point or foreshortening. On the other hand, the designer can be said to "draw" this effect in any way desirable. The axis controlling depth of field does not need to be perpendicular to the image plane, the area in focus can be anywhere in space, it can also quickly move around the space, etc.

Coming back to Universal Capture, it is worthwhile to quote Gaeta who himself is very clear that what he and his colleagues have created is a new hybrid. In 2004 interview, he says: "If I had to define virtual cinema, I would say it is somewhere between a live-action film and a computer-generated animated film. It is computer generated, but it is derived from real world people, places and things."[14] Although Universal Capture offers a particularly striking example of such "somewhere between," most forms of moving image created today are similarly "somewhere between," with animation being one of the coordinate axes of this new space of hybridity.

"Universal Capture": Reality Re-assembled

The method which came to be called "Universal Capture" combines the best of two worlds: physical reality as captured by lens-based cameras, and synthetic 3D computer graphics. While it is possible to recreate the richness of the visible world through manual painting and animation, as well as through various computer graphics techniques (texture mapping, bump mapping, physical modeling, etc.), it is expensive in terms of labor involved. Even with physically based modeling techniques endless parameters have to be tweaked before the animation looks right. In contrast, capturing visible reality through lens on film, tape, DVD-R, computer hard drive, or other media is cheap: just point the camera and press "record" button.

The disadvantage of such recordings is that they lack flexibility demanded by contemporary remix culture. *This culture demands not self-contained aesthetic objects or self-contained records of reality but smaller units—parts that can be easily changed and combined with other parts in endless combinations.* However, lens-based recording process flattens the semantic structure of reality—i.e., the different objects which occupy distinct areas of a 3D physical space. It converts a space filled with discrete objects into a flat field of image grains or pixels that do not carry any information of where they came from (i.e., which objects they correspond to). Therefore, any kind of editing operation—deleting objects, adding new ones, compositing, etc.—becomes quite difficult. Before anything can be done with an object in the image, it has to be manually separated by creating a mask. And unless an image shows an object that is properly lighted and shot against a special blue or green background, it is impossible to mask the object precisely.

In contrast, 3D computer generated worlds have the exact flexibility one would expect from media in information age. (It is not therefore accidental that 3D computer graphics representation—along with hypertext and other new computer-based data representation methods—was conceptualized in the same decade when the transformation of advanced industrialized societies into information societies became visible.) In a 3D computer generated worlds everything is discrete. The world consists from a number of separate objects. Objects are defined by points described as XYZ coordinates; other properties of objects such as color, transparency and reflectivity are similarly described in terms of discrete numbers. This means that the semantics structure of a scene is completely preserved and is easily accessible at any time. To duplicate an object hundred times requires only a few mouse clicks or typing a short command; similarly, all other properties of a world can be always easily changed. And since each object itself consists from discrete components (flat polygons or surface patches defined by splines), it is equally easy to change its 3D form by selecting and manipulating its components. In addition, just as a sequence of genes contains the code that is expanded into a complex organism, a compact description of a 3D world that contains only the coordinates of the objects can be quickly transmitted through the network, with the client computer reconstructing the full world (this is how online multi-player computer games and simulators work).

Beginning in the late 1970s when James Blinn introduced texture mapping,[15] computer scientists, designers and animators were gradually expanding the range of information that can be recorded in the real world and then incorporated into a computer model. Until the early 1990s this information mostly involved the appearance of the objects: color, texture, light effects. The next significant step was the development of motion capture. During the first half of the 1990s it was quickly adopted in the movie and game industries. Now computer synthesized worlds relied not only on sampling the visual appearance of the real world but also on sampling of movements of animals and humans in this world. Building on all these techniques, Gaeta's method takes them to a new stage: capturing just about everything that at present can be captured and then reassembling the samples to create a digital (and thus completely malleable) recreation. Put in a larger context, the resulting 2D/3D hybrid representation perfectly fits with the most progressive trends in contemporary culture which are all based on the idea of a hybrid.

The New Hybrid

It is my strong feeling that the emerging "information aesthetics" (i.e., the new cultural features specific to information society) has or will have a very different logic from what modernism. The later was driven by a strong desire to erase the old—visible as much in the avant-garde artists' (particularly the futurists) statements that museums should be burned, as well as in the dramatic destruction of all social and spiritual realities of many people in Russia after the 1917 revolution, and in other countries after they became Soviet satellites after 1945. Culturally and ideologically, modernists wanted to start with "tabula rasa," radically distancing them from the past. It was only in the 1960s that this move started to feel inappropriate, as manifested both in loosening of ideology in communist countries and the beginnings of new post-modern sensibility in the West. To quote the title of a famous book by Robert Venturi and et al (published in 1972, it was the first systematic manifestation of new sensibility), *Learning from Las Vegas* meant admitting that organically developing vernacular cultures involves bricolage and hybridity, rather than purity seen for instance in "international style" which was still practiced by architects world-wide at that time. Driven less by the desire to imitate vernacular cultures and more by the new availability of previous cultural artifacts stored on magnetic and soon digital media, in the 1980s commercial culture in the West systematically replaced purity by stylistic heterogeneity. Finally, when Soviet Empire collapsed, post-modernism has won world over.

Today we have a very real danger of being imprisoned by new "international style"—something which we can "global international." The cultural globalization, of which cheap airline flights and Internet are two most visible carriers, erases certain cultural specificity with the energy and speed impossible for modernism. Yet we also witness today a different logic at work: the desire to creatively place together old and new—local and transnational—in various combinations. It is this logic, for instance, which made cities such as Barcelona (where I talked with John Gaeta in the context of Art Futura 2003 festival which led to this article), such a "hip" and "in" place today. All over Barcelona, architectural styles of many past centuries co-exist with new "cool" spaces of bars, hotels, museums, and so on. Medieval meets multi-national, Gaudy meets Dolce and Gabana, Mediterranean time meets Internet time. The result is the incredible sense of energy which one feels physically just walking along the street. It is this hybrid energy, which characterizes in my view the most interesting cultural phenomena today.[16] The hybrid 2D/3D image of *The Matrix* is one such hybrid.

The historians of cinema often draw a contrast between the Lumières and Marey. Along with a number of inventors in other countries all working independently from each other, the Lumières created what we now know as cinema with its visual effect of continuous motion based on the perceptual synthesis of discrete images. Earlier Maybridge already developed a way to take successive photographs of a moving object such as horse; eventually the Lumières and others figured out how to take enough samples so when projected they perceptually fuse into continuous motion. Being a scientist, Marey was driven by an opposite desire: not to create a seamless illusion of the visible world but rather to be able to understand its structure by keeping subsequent samples discrete. Since he wanted to be able to easily compare these samples, he perfected a method where the subsequent images of moving objects were superimposed within a single image, thus making the changes clearly visible.

The hybrid image of *The Matrix* in some ways can be understand as the synthesis of these two approaches which for a hundred years ago remained in opposition. Like the Lumières, Gaeta's goal is to create a seamless illusion of continuous motion. In the same time, like Marey, he also wants to be able to edit and sequence the individual recordings.

In the beginning of this article I evoked the notion of uneven development, pointing that often the inside structure ("infrastructure") completely changes before the surface ("superstructure") catches up. What does this idea imply for the future of images and in particular 2D/3D hybrids as developed by Gaeta and others? As Gaeta pointed out in 2003, while his method can be used to make all kinds of images, so far it was used in the service of realism as it is defined in cinema—i.e., anything the viewer will see has to obey the laws of physics.[17] So in the case of *The Matrix,* its images still have traditional "realistic" appearance while internally they are structured in a completely new way. In short, we see the old "superstructure" which stills sits on top of "old" infrastructure. What kinds of images would we see then the superstructure" would finally catch up with the infrastructure?

Of course, while the images of Hollywood special effects movies so far follow the constraint of realism, i.e., obeying the laws of physics, they are also not exactly the same as before. In order to sell movie tickets, DVDs, and all other merchandise, each new special effects film tries to top the previous one showing something that nobody has seen before. In *The Matrix* it was "bullet time"; in *The Matrix 2* it was the Burly Brawl scene where dozens of identical clones fight Neo; in *Matrix 3* it was the Superpunch.[18] The fact that the image is constructed differently internally does allow for all kinds of new effects; listening to Gaeta it is clear that for him the key advantage of such image is the possibilities it offers for virtual cinematography. That is, if before camera movement was limited to a small and well-defined set of moves—pan, dolly, roll—now it can move in any trajectory imaginable for as long as the director wants. Gaeta talks about the Burly Brawl scene in terms of *virtual choreography*: both choreographing the intricate and long camera moves impossible in the real word and also all the bodies participating in the flight (all of them are digital recreations assembled using Gaeta's method as described above).

According to Gaeta, creating this one scene took about three years. So while in principle Gaeta's method represents the most flexible way to recreate visible reality in a computer so far, it will be years before this method is streamlined and standardized enough for these advantages to become obvious. But when it happens, the artists will have an extremely flexible hybrid medium at their disposal: completely virtualized cinema. Rather than expecting that any of the present pure forms will dominate the future of visual culture, I think this future belongs to such hybrids. In other words, the future images would probably be still photographic—although only on the surface.

And what about animation? What will be its future? As I have tried to explain, besides animated films proper and animated sequences used as a part of other moving image projects, animation has become a set of principles and techniques which animators, filmmakers and employ today to create new methods and new visual styles. Therefore, I think that it is not worthwhile to ask if this or that visual style or method for creating moving images which emerged after computerization is "animation" or not. It is more productive to say that most of these methods were born from animation and have animation DNA—mixed with DNA from other media. I think that such a perspective which considers "animation in an extended field" is a more productive way to think about animation today, especially if we want our reflections to be relevant for everybody concerned with contemporary visual and media cultures.

Notes

1. For technical details of the method, see the publications of Georgi Borshukov: www.virtualcinematography.org/publications.html.
2. Although not everybody would agree with this analysis, I feel that after the end of 1980s the field has significantly slowed down: on the other hand, all key techniques which can be used to create photorealistic 3D images have been already discovered; on the other hand, rapid development of computer hardware in the 1990s meant that computer scientists no longer had to develop new techniques to make the rendering faster, since the already developed algorithms would now run fast enough.
3. The terms "reality simulation" and "reality sampling" are made up by me for this text; the terms "virtual cinema," "virtual human," "universal capture" and "virtual cinematography" come from John Gaeta.
4. Therefore, while the article in *Wired* which positioned Gaeta as a groundbreaking pioneer and as a rebel working outside of Hollywood contained the typical journalistic exaggeration, it was not that far from the truth. Steve Silberman, "*Matrix 2,*" *Wired* 11.05 (May 2003) <http://www.wired.com/wired/archive/11.05/matrix2.html>.

5. Georgi Borshukov, "Making of *The Superpunch,*" presentation at Imagina 2004, available at www.virtualcinematography.org/publications/acrobat/Superpunch.pdf.
6. The details can be found in George Borshukov, Dan Piponi, Oystein Larsen, J. P. Lewis, Christina Tempelaar-Lietz, "Universal Capture—Image-based Facial Animation for 'The Matrix Reloaded,' SIGGRAPH 2003 Sketches and Applications Program, available at http://www.virtualcinematography.org/publications/acrobat/UCap-s2003.pdf.
7. The method captures only the geometry and images of actor's head; body movements are recorded separately using motion capture.
8. Borshukov et al, "Universal Capture."
9. See www.massivesoftware.com.
10. John Gaeta, presentation during a workshop on the making of *The Matrix,* Art Futura 2003 festival, Barcelona, October 12, 2003.
11. I don't remember the exact year of SIGGRAPH conference where Blinn has spoke but I think it was end of the 1980s when physically based modeling was still a new concept.
12. Casey Reas, private communication, April 2005.
13. For more on this process, see the chapter "Synthetic Realism and its Discontents" in *The Langauge of New Media*.
14. Catherine Feeny, "'The Matrix' Revealed: An Interview with John Gaeta," VFXPro, May 9, 2004 <www.uemedia.net/CPC/vfxpro/article_7062.shtml>.
15. J. F. Blinn, "Simulation of Wrinkled Surfaces," *Computer Graphics* (August 1978): 286–92.
16. Seen in this perspective, my earlier book *The Language of New Media* can be seen as a systematic investigation of a particular slice of contemporary culture driven by this hybrid aesthetics: the slice where the logic of digital networked computer intersects the numerous logics of already established cultural forms. Lev Manovich, *The Language of New Media* (The MIT Press, 2001).
17. John Gaeta, making of *Matrix* workshop.
18. Borshukov, "Making of *The Superpunch*."

PART II

Examining Mediation Themes and Theaters of Action

MODULE FIVE

Identity Online

Revealing the 'Real' Me, Searching for the 'Actual' You: Presentations of Self on an Internet Dating Site

Monica T. Whitty[1]

This paper considers the presentation of self on an internet dating site. Thirty men and 30 women were interviewed about their online dating experiences. They were asked about how they constructed their profiles and how they viewed other individuals' profiles. Which types of presentations of self led to more successful offline romantic relationships were also investigated. Additionally, gender differences were examined. In line with previous research on presentation of self online, individuals were quite strategic in their online presentations. However, important differences between initiating a relationship on an internet dating site and other spaces (online and offline) included the type of self disclosed as well as the depth of breadth of information individuals self-disclosed about themselves before any one-on-one conversations took place.

Keywords: online dating, Internet dating, Internet relationships, possible selves, self-presentation, identity

1. Introduction

Previous research has demonstrated that individuals can be quite strategic in their presentations of self in cyberspace (e.g., Walther, Slovacek, & Tidwell, 2001). Others have contended that the success of moving an online relationship offline may be dependent on the type of self that is presented in the cyber-world (e.g., Bargh, McKenna, & Fitzsimons, 2002; McKenna, Green, & Gleason, 2002). This study was interested in how men and women presented themselves on an online dating site. It does so by considering theories on 'possible selves' and Goffman's 'performed self' theory.

1.1. Presentation of Self-Online: Possible Selves

It has for some time been recognized that online relationships do initiate online and can move successfully offline (Whitty, 2007; Whitty & Carr, 2006). In more recent times researchers have been interested in how these relationships progress. Some, for example, have been interested in how individuals go about presenting themselves in cyberspace and which presentations of self lead to more successful relationship development.

Theorists have argued that individuals can be quite selective in their self-presentations online (e.g., Bargh et al., 2002; McKenna et al., 2002; Walther et al., 2001). For example, Walther et al. (2001) believe that while of course individuals do tend to be strategic in their presentation of self-offline, in CMC impression management is more controllable and fluid. They claim that "online communicators may exploit the capabilities of text-based, nonvisual interaction to form levels of affinity that would be unexpected in parallel offline interactions" (Walther et al., 2001, p. 110). As a consequence in some situations CMC users idealize their virtual partners. Bargh and colleagues (Bargh et al., 2002; McKenna et al., 2002) have also focused on presentation of self-online. In particular, they have focused on which presentation of self on the internet is more likely to lead to closer relationships. These researchers have drawn from Rogers' and Higgins' work on personality to come up with two aspects of self that they believe are important to consider when focusing on the development of relationships online—'true' selves and 'actual' selves.

As stated above, Bargh et al. (2002) and McKenna et al. (2002) have drawn from Rogers' and Higgins' work on personality to arrive at two aspects of the self that they believe are important to consider when focusing on the development of relationships online, these being the *'true self'* and *'actual self'*. These theorists drew from Rogers' (1951) work to define the true self (or what they also refer to as the *'Real Me'*) as traits or characteristics that individuals possess and would like to but are not usually able to express. Rogers (1951) developed a humanistic personality theory where the 'self' is a central construct. Rogers believed that the self developed through interactions with others, and that the point of therapy was to help people to discover their true selves. He understood the 'true self' to represent one's inner core—who they really are. Theoretically, an individual can do this if they experience 'unconditional positive regard'. In contrast, drawing from Higgins' (1987) research, these theorists defined the *actual* self as traits or characteristics that individuals possess and express to others in social settings. Higgins (1987) made a clear distinction between three aspects of the self: the *'actual self'*, *'ideal self'*, and *'ought to self'*. The 'actual self' is the representation of how you or another actually believes you are; the 'ideal self' is the representation of how you or another would like to see yourself, including hopes and wishes for you; and the 'ought to self' represents the attributes that you believe you should possess. In line with Higgins, these researchers claim that as one develops trust and intimacy with one's partner they are more likely to disclose aspects of themselves that are not widely known to others. They have argued that individuals who are more likely to express their true self online will consider the relationships they form in this space to be more identity-important compared to those individuals who are more likely to express their true selves in non-internet relationships.

To test the above ideas, Bargh, McKenna and their colleagues conducted a number of experiments. Bargh et al. (2002) measured true and actual selves by asking participants to list a maximum of 10 traits or characteristics that participants believed they actually possessed and expressed to others in social settings, as well as what characteristics individuals possessed and would like to but are typically unable to express to others. Their series of experiments revealed that the individuals' true selves were more accessible in memory after interacting with a stranger online compared to face-to-face. Moreover, they found that participants tended to like each other more when they meet first online compared to face-to-face.

Taking this work a step further, McKenna et al. (2002) were interested in whether individuals who are better able to disclose their 'true' selves online than offline were more equipped to form close relationships online and then take these relationships offline successfully. They randomly selected 20 Usenet newsgroups to include in their study. Over a 3-week period, questionnaires were emailed to every fifth poster in each of the newsgroups (excluding spam). Their first study found that when people convey their 'true' self online they develop strong internet relationships and bring these relationships into their 'real' lives. Two years after this initial study 354 of the 568 participants were emailed a follow-up survey (the remainder of the sample had email addresses that were no longer valid). In line with these researchers' prediction, these relationships remained relatively stable and durable over the 2-year period; however, one has to wonder how the 38% of the sample that were not followed up faired. In this same research, McKenna et al. (2002) found that participants who were more socially anxious and lonely were somewhat more likely to believe they could express their true selves with others online than they could with people they knew offline. McKenna et al. (2002) conclude from this research that:

> *rather than turning to the Internet as a way of hiding from real life and from forming real relationships, individuals use it as a means not only of maintaining ties with existing family and friends but also of forming close and meaningful new relationships in a relatively nonthreatening environment. The Internet may also be helpful for those who have difficulty forging relationships in face-to-face situations because of shyness, social anxiety, or a lack of social skills. (p. 30)*

1.2. Presentation of Self-Online: Goffman

Goffman was very interested in the ways people present themselves in their everyday face-to-face encounters. In his book *The Presentation of Self in Everyday Life*, Goffman (1959) argued for a dualistic image of the self. He described the self as both a performer and a character. According to Goffman (1959) the 'self-as-performer' is not merely a social product, but also has a basic motivational core. In contrast, the 'self-as-character' represents an individuals' unique humanity. It is this part of the self which is a social product; that is, performed outwardly in social life. The 'self-as-character' is one's inner self.

Goffman believed that individuals need to present themselves as an acceptable person to others. He stated that "the impressions that the others give tend to be treated as claims and promises they have implicitly made, and claims and promises tend to have a moral character" (Goffman, 1959/1997, p. 21). He argued that individuals can be strategic in their impression formation. In particular Goffman was interested in distinguishing between expressions 'given' (e.g., spoken communication) and expressions 'given off' (e.g., nonverbal cues) in a face-to-face interaction.

Researchers have applied Goffman's theory to online presentations of self. Miller (1995) claims that although depth and richness of self-presentation might not seem immediately apparent online that nonetheless "the problem of establishing and maintaining an acceptable self remains, and there is a range of expressive resources available for this end". Miller and Arnold (2001) applied Goffman's theory to explain how woman academics construct their own webpages. Drawing from his theory they suggest that women academics struggled with establishing a credible presence on the websites.

1.3. Offline Presentations and Relationship Development

Of course long before the internet ever existed psychologists were interested in how romantic relationships initiated offline and what types of presentation of self typically leads to budding relationships. One of the most popular theories to explain relationship development is '*Social Penetration Theory*'(Altman & Taylor, 1973). Social Penetration Theory is an incremental theory which argues that relationships move to greater levels of intimacy over time. According to this theory, how greater intimacy is achieved is typically through depth and breadth of self-disclosure. Breadth of self-disclosure refers to discussing a range of topics, such as information about one's family, career, and so forth. Depth refers to the more central core of one's personality; that is, the more unique aspects of one's self. The timing of how much one self-discloses is crucial to determining whether a relationship will continue to proceed. Rushing self-disclosure in the early stages of a relationship can seem unnatural and desperate and can lead to an abrupt end.

1.4. Study Purpose

This present study was interested in how individuals present themselves in cyberspace. Given that cyberspace is not one generic space (Whitty & Carr, 2006) it is important to investigate how individuals present themselves in different spaces online. In particular, this study was interested in how individuals present themselves on an online dating site, as well as their judgments on how others present themselves on this site. Online dating sites have increased in popularity and will no doubt continue to do so (Brym & Lenton, 2003; Whitty, 2007; Whitty & Carr, 2006). However, there is a dearth of research available on how individuals use this space to initiate and develop relationships.

Online dating sites are set-up very differently to other places online, such as newsgroups, chat rooms, and the like. On online dating sites individuals are required to construct a profile. On this profile they can upload photographs and videos of themselves and are given the opportunity to write a description of who they are. The way individuals contact each other on the site varies depending on the way the site is set-up. The online dating site which this study focused on is one of the largest Australian online dating sites. Individuals contact each other on this dating site by firstly sending someone a 'kiss'. The 'kiss' is a sentence sent through the site to an undisclosed email indicating that the individual is interested in the person they contacted. The contacted person can then respond to three options, including (a) they are interested and are requesting that the other person spend money on sending an email through the site, or (b) they are interested and they themselves will spend money on emailing the person through the site, or (c) they were flattered but uninterested. Next, individuals pay for '*stamps*' which would enable them to send emails through the site. This is how the company makes their money. Individuals can write anything in these emails. It is within this email that individuals can disclose their personal email so that contact can be made off the dating site.

This present study examined the types of selves individuals are more likely to present on an online dating site, and how they view other people's presentation of self. Additionally, it considered what types of presentations of self will lead to the development of a successful romantic relationships offline. Moreover, because previous research on offline attraction has found that men and women are attracted to different qualities in the opposite sex, gender differences were also taken into account.

2. Method

2.1. Participants

Sixty online daters were interviewed for this study (30 men and 30 women). The ages ranged from 23 to 60 years, with an overall mean age of 43.40 years (SD = 8.70). This is very close to the average mean age for the total number of individuals who use the site (M = 45 years). The mean age for men was 42.63 years (SD = 10.57) and for women was 44.97 (SD = 6.27) years. All participants resided in Australia. Forty-three percent of the sample stated that their relationship status was single, 46% were divorced or separated, 8% had a girlfriend or a boyfriend, and 3% were married or in a cohabiting relationship. All of the participants self-reported as being heterosexual. Individuals reported that they had been using the online dating site for an average of 27.42 months (SD = 16.36). Interestingly, about a third (27%) of individuals stated that they were using online dating as the only method of finding a potential date.

In considering the highest level of education achieved by this sample, 34% had completed high school, 28% had a diploma, 28% had a degree and 10% had achieved postgraduate qualifications. The Australian 2001 census data revealed that 3% of Australians held postgraduate qualifications, 8% had a degree, and 22% had received a diploma (Australian Bureau of Statistics, 2001). Hence, this sample was slightly skewed towards a better educated group of individuals.

2.2. Procedure

Participants were recruited from the largest Australian online dating site. For ethical reasons (to keep the online daters' identity confidential) the manager of the online dating company carried out the recruitment of participants. She randomly selected 300 participants from the database inviting them to participate in a telephone interview about their online dating experiences. There was a response rate of 20%, which is reasonable given the personal nature of the study; the possibility that many of the participants might have discontinued using the site and hence had no interest in participating in the study; and some of the emails might have no longer been operable. The individuals were emailed from the manager the details of the study, including the aims of the study, what was involved, and contact details of the principal investigator and the ethics board at the University of Western Sydney (where the research was conducted). They were told that if they were interested in being interviewed that they were to either phone or email the principal investigator to organize a time for a telephone interview.

A structured interview schedule, which mostly consisted of open-ended questions, was designed for this study. Participants were initially asked to report basic demographic details about themselves (e.g., age, relationships status, and socio-economic status). Next they were asked to explain their motivations for using the site. Participants were then asked to explain in detail how they went about using the site, including questions on how they constructed their own profiles, what sort of profiles they were attracted to, and how they went about contacting other individuals on the site. They were then asked to explain how the relationship progressed from online to offline, including questions on the pace of the relationship, how well others matched up to their profiles, as well as what kind of personal information they disclosed to their online potential date. They were asked to describe how their face-to-face dates were typically set-up and why they were set-up in this way. They were finally asked to compare online dating with other forms of dating.

The interviews were all conducted on the telephone. Male participants were interviewed by a male research assistant and female participants were interviewed by the author of this paper. They typically went for 40–50 min. All participants granted their permission to have the conversations audio taped. Participants were assured anonymity and were told that pseudonyms would be used to replace their real names. Participants were reminded that they could withdraw from the study at any time up until the completion of the study and were given a number for a free telephone counseling service in case the study raised any issues for them that they wanted to speak to a counselor about.

2.3. Analysis

Grounded theory methodology (Glaser & Strauss, 1967) guided the procedures for this research. As Willig (2001) points out "grounded theory involves the progressive identification and integration of *categories of meaning* from data. Grounded theory is both the process of category identification and integration (as *method*) and its product (as *theory*)" (p. 33). In this study themes emerged from the participants' responses rather than a priori categories. Although strictly speaking grounded theory puts aside any previous theories in order to develop new theories (Glaser & Strauss, 1967), this analysis did take into account previous work on presentation of self online and offline when considering the data. Previous researchers have also conducted qualitative research in this way (Willig, 2001). Once themes were decided upon each transcript was re-examined and coded for whether that theme was absent or present. To ensure inter-coder reliability two scorers separately coded the entire data set. In the very few cases where there was disagreement the researchers discussed their disagreements until consensus was reached. Moreover, given that this study was also interested in gender differences, logistic regressions were carried out using the Backward Wald procedure, which produces a Chi-square statistic. Because logistic regression cannot always obtain accurate calculations with zero cells, Fisher's exact test was carried out when there was a zero cell.

3. Results and Discussion

The following analysis details how the participants presented themselves on this internet dating site, what aspects they misrepresented, as well as what features they found attractive in others. Next, participants' views on how others presented themselves on this dating site is discussed as well as how well these profiles matched with the person when they met them face-to-face. Extracts from the interviews are presented here as illustration of the themes. Pseudonyms are used in place of the participants' real names.

3.1. Constructing Their Own Profile

The online daters interviewed for this study stressed the importance of crafting an attractive profile. Table 1 presents a list of aspects that individuals considered in constructing their own profiles. Participants typically elected to have a photo (many selecting the most flattering photo they could find), some even going as far as having a glamour shot. Descriptions of favorite interests and activities, one's personality, occupation, and hopes and dreams were also deemed important by some of the participants. Others tried to make their profile stand out by adding humor or finding a way to make themselves appear unique. More women than men stated that they included a photo (and more women than men said that this was a glamour photo). Moreover, more women than men said that they wrote about their interests.

It is noteworthy that participants did discuss in detail the importance of having a photograph on their profile. This is nicely illustrated by Joan:

I Did you put a photograph up of yourself?

J Yes.

I And how did you decide what type of photograph to put up of yourself?

J Well obviously you want something decent and I didn't have a lot of things around. So the photograph that I have got was a studio glamour photograph where they did your makeup and your hair and made you look gorgeous, but that is the only decent photograph I had. (Joan)

A number of previous theorists have downplayed the importance of the body online and the need to appear physically attractive. For example, Levine (2000) wrote that "the beauty of the virtual medium is that flirting is based on words, charm, and seduction, not physical attraction and cues" (p. 565). Similarly, Rollman, Krug, and Parente (2000) comment that "by eliminating time, distance, and body, the architects of the Internet have created an unhindered medium that connects the mind and spirit" (p. 161). However, in stark contrast, this study revealed that participants believed that the need to present a good physical image of themselves was more important than any other characteristic.

The results from this study highlight some of the similarities between online and offline attraction. Offline, physical characteristics play a critical role in attractions for both men and women.

TABLE 1 Aspects individuals included in their own profiles.

Constructing own profile	Men *f* (%)	Women *f* (%)	Total *f* (%)	χ^2	β	Exp β
Photo	18 (60.0%)	26 (86.7%)	44 (73.3%)	5.03* Women > men	1.47	4.33
Interests/activities they enjoy doing	11 (36.7%)	21 (70.0%)	32 (53.3%)	6.43* Women > men	1.39	4.03
Description of personality	10 (33.3%)	11 (36.7%)	21 (35.0%)	ns		
Made it humorous	7 (23.3%)	3 (10.0%)	10 (16.7%)	ns		
Occupation	4 (13.3%)	4 (13.3%)	8 (13.3%)	ns		
Demonstrate their intelligence	4 (13.3%)	3 (10.0%)	7 (11.7%)	ns		
Wrote it to sound different/unique	4 (13.3%)	3 (10.0%)	7 (11.7%)	ns		
Played down own beauty	1 (3.3%)	5 (16.7%)	6 (10.0%)	ns		
Glamour shot	0 (0.0%)	5 (16.7%)	5 (8.3%)	5.46* Women > men		
Hopes and dreams	1 (3.3%)	3 (10.0%)	4 (6.7%)	ns		

*$p < 0.05$, **$p < 0.01$, ***$p < 0.001$.

Note: Fisher's exact test was carried out for glamour shot.

Moreover, research on face-to-face attraction has found that men are typically more readily aroused sexually than women by visual stimuli (e.g., Ellis & Symons, 1990; Townsend, 1993). Hence, it is no surprise, that more women than men included photos in their profiles, and more women than men choose to have 'glamour' photos of themselves. Townsend and Wasserman (1997) also contend that women are more interested in men that demonstrate a willingness to invest in them. This perhaps explains why the women put more effort into detailing their interests—to be certain that they are attracting someone truly interested in them.

Participants talked about the importance of writing a profile that successfully attracted others. Some described this as a process of 'selling themselves'. This is explained by Wayne below:

> W The other thing for me personally is I'm great at writing trade manuals for someone, but when it comes to writing about yourself and trying to sell yourself it's a very different story. I don't know whether that's more of a male trait than a female trait. It depends how good you want to try selling yourself too isn't it? (Wayne)

Individuals did admit to misrepresenting themselves on their profiles. This they stated was not for malicious reasons, but rather as a way to attract others. Mostly they saw their misrepresentations as exaggerations rather than blatant lies. About half the sample (51%) admitted to misrepresenting themselves on the site. As demonstrated in Table 2, individuals admitted to lying about their looks, their current relationships, age, weight, socio-economic status, and interests. Interestingly, the women lied about looks or used outdated photos more than men did. Again this can be explained by men placing more emphasis on being attracted to a partner who is physical attractive. With this being the case women have more reason to misrepresent their looks.

As stated above, most participants rationalized that their misrepresentations were not out-and-out lies, but rather mere exaggerations, and often exaggerations that they thought others were probably also doing in their profiles. To give some examples:

> S Actually, it's quite funny that there is, you can pick, there is a thing for body type and you can pick 'slim, average, athletic, a bit overweight'. Do you know any chick that is going to tell you that they are a bit overweight?
>
> I I wouldn't know, I guess not?
>
> S That is right, so I just say average.
>
> I It's like the idea of perception of what is 'average', and average is you know probably a bit overweight anyway. (Suszi)

TABLE 2 Participants misrepresentations on their own profiles.

Misrepresentations on own profiles	Men *f* (%)	Women *f* (%)	Total *f* (%)	χ^2	β	Exp β
Looks: photo over a year out of date	2 (6.7%)	14 (46.7%)	16 (26.7%)	9.38** Women > men	2.51	12.25
Details about their own relationship/children (e.g., having them, who they live with, etc.)	4 (13.3%)	2 (6.7%)	6 (10.0%)	ns		
Age	3 (10.0%)	1 (3.3%)	4 (6.7%)	ns		
Weight (e.g., said average when a bit overweight)	2 (6.7%)	2 (6.7%)	4 (6.7%)	ns		
SES (occupation, being employed, professional, income)	2 (6.7%)	1 (3.3%)	3 (5.0%)	ns		
Interests	3 (10.0%)	0 (0.0%)	3 (5.0%)	ns		

$^*p < 0.05$, $^{**}p < 0.01$, $^{***}p < 0.001$.

M Well it's all perception of reality isn't it. Everyone's view of the world is different, my view of myself may be very different to what someone else thinks it is. But you know, my view is that you had to sell yourself without being. I mean the last thing you want is reality shock, when people get there and go 'that is not what he said he was about', so you have got to make the most of what you have got without exaggerating it so much that they never see you again. So, I would say it was probably 90% accurate with a few little embellishments you know.

I Tell us about the embellishments.

M Oh gosh . . . you know things like portraying things in the most positive light. You know things like interests. I don't think I really have definite music interests or anything, but I just said, 'I am on a first name basis with people at HMV' but I am not really but it doesn't matter. You know that sort of thing. (Matthew)

The interviewees explained that constructing a profile was a dynamic process. As demonstrated in some of the quotations above, people experimented with what photos and descriptions of themselves would be more successful at attracting others to their profile. They also re-wrote profiles to attract the partner they were searching for. However, there was a third concern that participants considered in constructing their profiles. Participants were also aware that relationships were not developed online but rather offline. Over half of the participants (57.4%) stated that they met their date within a week or two after initial contact on the site. Another 10.3% stated that they met their date within a month. Hence, it is fair to say that individuals use online dating sites as a means to identify a potential date and that cyberspace is not, in the main, utilized as a medium to get to know the person. In fact, participants were adamant about the importance of the first meeting, with 67.6% of the participants stating that the first meeting determined if the relationship would progress. They also stated that any emailing or telephone calls that were made prior to the first date were more to organize the date and verify information about a person, rather than disclosing any further information about themselves. As expressed by Andrew:

A We exchanged at least two emails each after the first contact email I guess and then probably phoned up to organize a time to meet and get together. Not so much as getting on the phone and speaking for 3h or anything like that. . . . (Andrew)

Given this, individuals were concerned that their profiles reflect how they 'actually' are in everyday situations. In order to ensure this, individuals asked their friends and family to check if their profiles were a true reflection of themselves, as Crystal explains:

I What about your own profile?

C I tried to vary it I suppose. . . . As far as myself goes, I tried to just put who I am but I must admit it is really hard because other people may see you in a different way, so I actually think it's a good idea to have somebody else look at your profile, like a friend because as I said you may see yourself differently than everyone else does. (Crystal)

The findings reported here differ somewhat to the results yielded in the work carried out by Bargh, McKenna, and colleagues (e.g., Bargh et al., 2002; McKenna et al., 2002). As summarized earlier in this paper, these researchers believe that individuals who are able to express their 'true' self-online (traits or characteristics that individuals possess and would like to but are not usually able to express in their everyday lives) are more likely to develop strong internet relationships and bring these relationships into their 'real' lives. However, the participants in this current study instead appeared to be trying to create a balance between keeping their profiles real (actual self) as well as selling themselves (or describing how they would like to be). The motivation for this was based on not wanting to disappoint the date once they met them face-to-face, but at the same time trying to attract a decent number of individuals to choose from. Therefore, cyberspace is partly perceived by these participants as a safe space to identify their 'true' self or play around with presentations of themselves (Whitty, 2003) but at the same time how much people do this is restricted if they want to ensure their date is not disappointed when they meet face-to-face.

3.2. Considering Other Online Daters' Profiles

Individuals might have felt the need to present a profile that presented a mix of their actual and true self; however, is that what they, in turn, expected of their potential dates? The interviewees were very clear about what they hoped for in a partner. In fact, when asked they presented a *'shopping list'* of characteristics they were seeking out. Table 3 presents a list of the attractive qualities people were looking for in a partner.

Contrary to some theorists' views (e.g., Levine, 2000; Rollman et al., 2000) these online daters did not de-emphasize physical attractiveness as an important quality. Noticeably, looks actually topped the list. The importance placed on physical characteristics may be greater for online daters than for individuals developing relationships in other places online. This is for two reasons: (a) when individuals first 'meet' the person online, they are presented with a photo and not simply text, (b) there are a plethora of choices (and unlike places where 100s of people are interacting online, such as MOOs, online daters 'know' that everyone on the site is seeking romance) and so they can bypass the less attractive profiles and make a play for the more attractive ones.

Interestingly, the traditional gender differences for what men and women are attracted to did not emerge in this study. Unlike previous studies on offline attraction (see for example, Ellis & Symons, 1990; Townsend, 1993), men did not state significantly more than women that they were looking for an attractive looking person. Moreover, women did not state more than men that they were looking for someone with high socio-economic status. This is contrary to what researchers have found in regards to offline relationships (see for example, Buss & Barnes, 1986; Kenrick, Sadalla, Groth, & Trost, 1990; Townsend & Wasserman, 1997) where women are typically more interested than men in seeking out a partner with high SES. Physical attractiveness (mentioned by 90%) and socio-economic status (mentioned by 73.3%) were highly sought out qualities by both men and women in this study. Therefore, what might be happening on these dating sites is that given the numerous perceived choices available to individuals (with 1000s of profiles to sought through), these online

TABLE 3 Aspects participants were looking for in a partner.

Attractive qualities	**Men *f* (%)**	**Women *f* (%)**	**Total *f* (%)**	χ^2
Looks	29 (96.7%)	25 (83.3%)	54 (90.0%)	ns
Similar interests/values	25 (83.3%)	26 (86.7%)	51 (85.0%)	ns
SES	22 (73.3%)	22 (73.3%)	44 (73.3%)	ns
Personality	19 (63.3%)	24 (80.0%)	43 (71.1%)	ns
Honest/genuine/real people	17 (56.7%)	15 (50.0%)	32 (53.3%)	ns
Attracted to a certain age group	11 (36.7%)	17 (56.7%)	28 (46.7%)	ns
Height	9 (30.0%)	16 (53.3%)	25 (41.7%)	ns
Proximity	11 (36.7%)	13 (43.3%)	24 (40.0%)	ns
Size/weight	14 (46.7%)	8 (26.7%)	22 (36.7%)	ns
Non-smokers	11 (36.7%)	6 (20.0%)	17 (28.3%)	ns
Different/unique people/people with different interests to their own	7 (23.3%)	10 (33.3%)	17 (28.3%)	ns
Humor	6 (20.0%)	8 (26.7%)	14 (23.3%)	ns
Someone with no children/off their hands	5 (16.7%)	7 (23.3%)	12 (20.0%)	ns
Someone who has children	5 (16.7%)	5 (16.7%)	10 (16.7%)	ns
Opening line	5 (16.7%)	2 (6.7%)	7 (11.7%)	ns
Star sign	1 (3.3%)	5 (16.7%)	6 (10.0%)	ns
Someone who wants children	1 (3.3%)	3 (10.0%)	4 (6.7%)	ns

*$p < 0.05$, **$p < 0.01$, ***$p < 0.001$.

daters are consequently, adding more to their 'wish list' for a potential partner. This is a very different playing field when compared to traditional methods of meeting people in pubs and clubs where one perceives far fewer potential dates.

Despite admissions of their own exaggerations in their profiles, the participants in this study were often outraged to find when they meet face-to-face that their date had misrepresented themselves in their profiles. Table 4 displays some of the characteristics that the participants' dates misrepresented about themselves. Interestingly, male online daters were said to misrepresent their height and their relationships status more than women.

The most common way that individuals misrepresented themselves was in regards to their physical appearance. Again, this is perhaps not surprising given this was the most highly valued characteristic. Some exemplars of the ways the participants' online dates misrepresented themselves are presented below:

I In what way do they lie about their profession?

L Most guys make you understand on their thing [profile] that they are in management, but when you talk to them they are not really in management, they are not even middle management some of them.

I So they kind of exaggerate their role considerably?

L Yes definitely. (Lisa)

C A lot of people say they are looking for a long-term relationship or a friendship and what I think they are after is a one night stand. (Christine)

As demonstrated in the extracts above, some of the misrepresentations were blatant lies; however, it could also be argued that some of the misrepresentations were individuals' representations of their 'true' selves. As illustrated by Kim:

K I think also some people have got a very different grasp of who they are compared to what the rest of the world sees, so I don't think it is so much that people blatantly lie, I haven't experienced that. One guy who I found, he was relying on how he has got a very funny personality, or a good disposition sort of thing. He turned out to be one of the most depressing people I have ever met in my life. (Kim)

TABLE 4 How other online daters misrepresented themselves on the online dating site.

Other online dater's misrepresentations	**Men *f* (%)**	**Women *f* (%)**	**Total *f* (%)**	χ^2
Looks (including describing themselves as better looking than what they really are; out-dated photos; a photo of a different person)	24 (80.0%)	17 (56.7%)	41 (68.3%)	ns
Weight/size	9 (30.0%)	16 (53.3%)	25 (41.7%)	ns
Personality	10 (33.3%)	9 (30.0%)	19 (31.7%)	ns
Age	9 (30.0%)	6 (20.0%)	15 (25.0%)	ns
Height Men > women	10 (33.3%)	0 (0.0%)	10 (16.7%)	12.00**
Their intentions (e.g., just for sex not relationship)	5 (16.7%)	5 (16.7%)	10 (16.7%)	ns
SES (education, occupation, income, being professional)	6 (20.0%)	3 (10.0%)	9 (15.0%)	ns
Relationship status (did not admit on profile that they are married) Men > women	8 (26.7%)	0 (0.0%)	8 (13.3%)	9.23**
Having children	2 (6.7%)	3 (10.0%)	5 (8.3%)	ns
Being a smoker	1 (3.3%)	2 (6.7%)	3 (5.0%)	ns

$^*p < 0.05$, $^{**}p < 0.01$, $^{***}p < 0.001$.

Note: Fisher's exact test was carried out for height and relationship status.

In addition to sorting through the lies presented on profiles, interviewees were often weary of profiles that contained 'cheesy' clichés. They mostly avoided these types of individuals, as they believed these people to be far less 'real'. To some, clichés became a 'turn off', rather than the 'turn on' that the person writing the profile had hoped for. To give an example of a clichéd profile:

G ... and everyone says that they like walking on the beach, and anyone will tell you that it is so goddamn boring (laugh). And when they say that I think 'oh my God, not another one, it is almost a turn off'.

I They are all on that beach walking (laughing).

G Oh my God, if there are so many on that beach walking, why don't they run into each other.

I So a bit cliché that one?

G Terrible, it is almost a turn off now. (Grace)

When we examine which characteristics individuals are attracted to in this particular online space, again Bargh, McKenna, and colleagues' (e.g., Bargh et al., 2002; McKenna et al., 2002) theory does not hold true. Rather interviewees stated that they were attracted to genuine and honest people and they hoped that an individual's profile presents something about who the individual 'really' is—rather than a stereotypical, clichéd self or a self that the individual would like to possess but typically do not express in their day-to-day settings. In fact, over half of the participants said that an attractive person was one who was genuine and real (see Table 3). It appears that these individuals were more attracted to individuals who expressed their 'actual' self. That is, online daters perceived honest and genuine people to be those who included in their profiles the traits or characteristics that they typically express in everyday offline social settings.

These online daters were most concerned with what their date looked like and how they behaved in their first face-to-face meeting. In fact, 67.6% claimed that the first face-to-face meeting was a screening out process—one that determined if there was a possibility for a relationship to develop. This is nicely expressed by Phil:

P Well I got sick of meeting girls and everything's based on this, even though it's not supposed to be looks, everything's based on the first meeting, that's the big shock compared to meeting someone in a pub or someone introducing you. (Phil)

Given the importance placed on that first meeting and making the best first face-to-face impression, if the individual did not match up to the profile, as described earlier, the online dater was highly disappointed with their date—often judging them as dishonest people. These online daters learnt fairly quickly that people might present a different self-online than they do offline. Therefore, unlike McKenna et al.'s (2002) sample, individuals who used the online dating site did not want to spend time getting to know one another in cyberspace. In fact, they were quite the opposite, expressing a desire to meet the person face-to-face as soon as possible. Hence, there was little opportunity or desire for people to get to know individuals 'true' selves or for individuals to gradually express their true selves.

3.3. Developing a Theory for the Presentation of Self on Internet Dating Sites

This study revealed both similarities and differences between dating initiated on an internet dating site when compared to other forms of dating both online and offline. Similar to dating face-to-face, individuals stressed the importance of physical looks as an attractive quality they were looking for in a partner. Also, as other previous researchers have contended (e.g., Turkle, 1995; Whitty, 2003), these participants found that cyberspace was a place where they could be somewhat creative with their identity and presentation of self. However, there were limits to how imaginative individuals could be in their presentations of self.

In line with Walther's theories (e.g., Walther, 1995), these online daters were very strategic in their presentation of self. However, the ways these individuals presented themselves in cyberspace differs to what previous researchers have found. Although online daters were tempted to present a 'true' self or a more attractive self than what they typically are in face-to-face encounters, they were nonetheless mindful of the importance of presenting an 'actual' self. Although further research is required to test this claim, it seemed apparent that, contrary to McKenna et al.'s (2002) study, those who presented their 'actual' selves in this study were more likely to successfully progress a relationship offline.

As mentioned in Section 1 of this paper, some theorists have applied Goffman's theory to explain presentation of self in cyberspace. His theory might also be of use to explain individuals' expectations of how others ought to present themselves on an online dating site. On an online dating site there is very little opportunity to witness expressions 'given off'. Possible ways to 'read between the lines' or look for the expressions 'given off' is evident when participants skeptically viewed clichéd profiles. Moreover, online daters checked to see if their date matched up to their profile when they meet face-to-face. In Goffman's terms they are able to see how well the 'real character' matches up to the performance. When there is a discrepancy, as Goffman would predict, the online daters in this study judge their dates are immoral, believing they had an obligation to match the impressions created in their profile.

Gender differences that are evident in offline attraction were only partly evident in this study. Women were more likely than men to present a photo of themselves and made greater attempts at ensuring an attractive image of themselves was presented. However, when it came to what attractive qualities men and women preferred there were no significant differences. However, rather than deemphasizing qualities, online daters appeared to hope for more from their potential date. Arguably, this is because online daters have a greater number of potential mates to select from than in typical face-to-face situations. Given this perception of choice, online daters might be raising their hopes and expectations as to what potential mate they might attract. However, while on the surface this might seem a strength of online dating, future research needs to investigate if one does indeed attract a better quality mate, or if this raise in expectations leads to greater disappointment.

What also emerged from this study is the amount of information individuals initially self-disclose to their potential mates is very different to other forms of dating both online and offline. As highlighted in the introduction, when it comes to offline relationships the social penetration theory initially proposed by Altman and Taylor (1973) and modified by others (e.g., Morton, Alexander, & Altman, 1976) essentially argues that relationships move from less intimate to more intimate involvement over time. The process has been described using an onion analogy, arguing that people self-disclose deeper and deeper aspects about themselves as the relationship progresses. This theory discusses depth and breadth. According to social penetration theory, in the early phases of relationship development one moves with caution, discussing less intimate topics and checking in the conversations for signs of reciprocity. Gradually one feels safer to admit to other aspects of themselves. This process arguably happens in chat rooms, discussion boards, IM and the like. Moreover, as mentioned earlier, researchers have found that this process often takes place quicker, since people are potentially in a safer environment to reveal core aspects of themselves (e.g., Cooper & Sportolari, 1997; Walther, 1995; Whitty, 2003, 2004).

There is far less opportunity for relationships to develop on an internet dating site in the way proposed by the social penetration theory. On an internet dating site the profiles are set-up in such as way to reveal both depth and breadth. For instance, within the profiles, individuals typically have to provide information about surface levels aspects of themselves, such as, eye color, drinking and smoking habits, relationship status, number and types of pets and occupation. In addition, they are given space to write more in-depth about themselves, where they are asked to describe their personality, interests (what they read, music they listen to and so forth), their ideal date, their political persuasion. They are encouraged on these sites to open up about all aspects of themselves online—so that they will attract the most appropriate person. Given the amount of information they are presented with it is no surprise the conversations that take place via email, telephone and so forth prior to the first meeting are more to clarify information about the person as well as to arrange the meeting.

Therefore, online dating is arguably even more removed from what people are used to when it comes to developing a relationship. There is not any real opportunity to test the waters gradually and check for reciprocity, instead, reciprocity is determined prior to communication with the individual. Therefore, as others have found when it comes to presentation of self online—this can be quite strategic. In a way though, individuals have more control over their presentation on an internet dating site than any other space online. However, in turn, because the profile compiles all the information about the person in one chunk, it is easier to check back to. Hence, others are less forgiving when there is a mismatch between the person they meet face-to-face and the person they were presented with in the profile.

3.4. Limitations

There are some limitations to this study. Given that this was a small sample size the results are difficult to generalize. Moreover, different online sites are structured in slightly different ways and some are geared towards specific interests (e.g., sites set-up for certain religious orientations). Again, this makes it difficult to generalize these results to all online daters. Nonetheless, the rich data presented here does present some important insights as to how online daters present themselves and their expectations of how others ought to present themselves. Future research might test out the claims made here by developing a survey which can be distributed to a variety of types of online dating sites as well as a larger sample of individuals using them.

3.5. Conclusions

This research highlights that it is critical that social scientists should not develop one grand theory to explain how everyone presents themselves within cyberspace. Cyberspace is not one generic space. Rather it is important to consider how different spaces online are constructed. This research points out that individuals are strategic in how they present themselves on an internet dating site; however, the strategies employed here are different to other spaces like newsgroups. Although online daters found this to be a space where they can experiment with how they present their identity, in order for the relationship to continue offline successfully it would seem that presentations of how they appear in everyday face-to-face situations are crucial.

Acknowledgments

The author thank Lisa Lunnon for her assistance with coding, and The University of Western Sydney and the Australian online dating company who funded the project.

Note

1. Tel.: +44 28 90975654. E-mail address: monica.whitty@ntu.ac.uk. Division of Psychology, Nottingham Trent University, York House, Nottingham NG1 4BU, UK

References

Altman, I., & Taylor, D. A. (1973). *Social penetration: The development of interpersonal relationships.* New York: Holt, Rinehart & Winston.

Australian Bureau of Statistics. (November 2001). 2001 Census basic community profile and snapshot. Canberra, Australian Capital Territory: Australian Bureau of Statistics. http://www.abs.gov.au/Ausstats/abs@census.nsf/, Retrieved 05.06.05.

Bargh, J. A., McKenna, K. Y. A., & Fitzsimons, G. M. (2002). Can you see the real me? Activation and expression of the "true self" on the internet. *Journal of Social Issues, 58,* 33–48.

Brym, R. J., Lenton, R. L. (2003). Love at first byte: Internet dating in Canada. http://www.societyinquestion4e.nelson.com/Chapter33Online.pdf, Retrieved 25.03.05.

Buss, D. M., & Barnes, M. (1986). Preferences in human mate selection. *Journal of Personality and Social Psychology, 50,* 559–570.

Cooper, A., & Sportolari, L. (1997). Romance in cyberspace: Understanding online attraction. *Journal of Sex Education and Therapy, 22*(1), 7–14.

Ellis, B., & Symons, D. (1990). Sex differences in sexual fantasy. *Journal of Sex Research, 27*(4), 527–555.

Glaser, B. G., & Strauss, A. (1967). *The discovery of grounded theory: Strategies for qualitative research.* Chicago: Aldine Publishing Co.

Goffman, E. (1959/1997). In the presentation of self in everyday life. In C. Lemert & A. Branaman (Eds.), *The Goffman reader*. Massachusetts, USA, New York: Blackwell Publishers, Doubleday Anchor (original work published in 1959).

Higgins, E. T. (1987). Self-discrepancy theory. *Psychological Review, 94,* 1120–1134.

Kenrick, D. T., Sadalla, E. K., Groth, G., & Trost, M. R. (1990). Evolution, traits, and the stages of human courtship: Qualifying the parental investment model. *Journal of Personality, 58,* 97–116.

Levine, D. (2000). Virtual attraction: What rocks your boat. *CyberPsychology and Behavior, 3,* 565–573.

McKenna, K. Y. A., Green, A. S., & Gleason, M. E. J. (2002). Relationship formation on the Internet: What's the big attraction? *Journal of Social Issues, 58,* 9–31.

Miller, H. (1995). The presentation of self in electronic life: Goffman on the internet. Embodied Knowledge and Virtual Space, June 1995. http://ess.ntu.ac.uk/miller/cyberpsych/goffman.htm, Retrieved 26.07.06.

Miller, H., & Arnold, J. (2001). Breaking away from grounded identity? Women academics on the web. *CyberPsychology and Behavior, 4*(1), 95–108.

Morton, L., Alexander, J., & Altman, I. (1976). Communication and relationship definition. In G. R. Miller (Ed.), *Explorations in interpersonal communication* (pp. 105–125). Beverly Hills, CA: Sage.

Rogers, C. (1951). *Client-centered therapy.* Boston: Houghton-Mifflin.

Rollman, J., Krug, K., & Parente, F. (2000). The chat room phenomenon: Reciprocal communication in cyberspace. *CyberPsychology and Behavior, 3*(2), 161–166.

Townsend, J. (1993). Sexuality and partner selection: Sex differences among college students. *Ethology and Sociobiology, 14,* 305–330.

Townsend, J., & Wasserman, T. (1997). The perception of sexual attractiveness: Sex differences in variability. *Achieves of Sexual Behavior, 26,* 243–268.

Turkle, S. (1995). *Life on the screen: Identity in the age of the Internet.* London: Weidenfeld & Nicolson.

Walther, J. B. (1995). Relational aspects of computer-mediated communication: Experimental observations over time. *Organizational Science, 6,* 186–203.

Walther, J. B., Slovacek, C., & Tidwell, L. (2001). Is a picture worth a thousand words? Photographic images in long-term and short-term computer-mediated communication. *Communication Research, 28,* 105–134.

Whitty, M. T. (2003). Cyber-flirting: Playing at love on the internet. *Theory and Psychology, 13,* 339–357.

Whitty, M. T. (2004). Cyber-flirting: An examination of men's and women's flirting behaviour both offline and on the Internet. *Behaviour Change, 21,* 115–126.

Whitty, M. T. (2007). The art of selling one's self on an online dating site: The BAR approach. In M. T. Whitty, A. J. Baker, & J. A. Inman (Eds.), *Online matchmaking* (pp. 57–69). Houndsmills: Palgrave Macmillan.

Whitty, M. T., & Carr, A. N. (2006). *Cyberspace romance: The psychology of online relationships.* Basingstoke: Palgrave Macmillan.

Willig, C. (2001). *Introducing qualitative research in psychology: Adventures in theory and method.* Maidenhead: Open University Press.

The Chronicles of Me: Understanding Blogging as a Religious Practice

Pauline Hope Cheong, Alexander Halavais, and Kyounghee Kwon

Blogs represent an especially interesting site of online religious communication. Analysis of the content of 200 blogs with mentions of topics related to Christianity, as well as interviews of a subset of these bloggers, suggests that blogs provide an integrative experience for the faithful, not a "third place," but a melding of the personal and the communal, the sacred and the profane. Religious bloggers operate outside the realm of the conventional nuclear church as they connect and link to mainstream news sites, other nonreligious blogs, and online collaborative knowledge networks such as Wikipedia. By chronicling how they experience faith in their everyday lives, these bloggers aim to communicate not only to their communities and to a wider public but also to themselves. This view of blogging as a contemplative religious experience differs from the popular characterization of blogging as a trivial activity.

With the growth of the Internet, many religious believers and institutions have adopted technology for expressing their faith, reaching out to nonbelievers, and building stronger ties among believers (Dawson & Cowan, 2004). Blogs have emerged as a popular genre of online communication and have been utilized for religious expression, but scarce attention has been paid to understanding emergent blogging practices, particularly blogging as a religious practice. According to Buddenbaum's review of social science and the study of media and religion (2002), the tendency among researchers has been to "study the media and religion in isolation from each other and both at least somewhat separate from other institutions and from the surrounding culture" (p. 14). Subsequently, research in newer media needs to incorporate religion in the contemporary information society, including the study of religious blogging.

The emergence of blogs also raises important questions about storytelling and self-representation, motivations for media use, and the building of community online. Accordingly, the presence of religious discourse in blogs calls for fresh thinking about the representation of believers and their faiths, motivations for religious storytelling, the constitution of religious practices, together with the interplay of online and offline connections among religious believers, nonbelievers, and seekers. The rise of online scribes or religiously inspired blog authors poses interesting questions about the changing nature of personal media uses and gratifications and the role of new media in peoples' everyday lives, including their spiritual verve.

This study explores blogging as a religious practice in order to understand the nature of blogs, the motivations behind blog use for religious expression, and the implications for religious community building. We first discuss the debate concerning the rise of online religions, two views on religiosity and Internet use, the emergence of blogs as an alternative form of communication and with the potential to be shaped as "sacred spaces" for religious community. Following that, we deploy several levels of data to capture and analyze the ways and extent to which the blogosphere functions as a space for various religious expressions and practices associated with Christianity.

Online Religion: God in the Blogosphere

In recent years, online religious activity has been identified as popular and important. For example, according to the latest results of the Pew Internet and American Life Project study on "cyber-faith,"64% of "wired Americans" have used the Internet for various religious purposes, including activities such as seeking and giving religious information/advice online, downloading religious music, and purchasing religious items (Hoover, Clarke, & Rainie, 2004). However, there are concerns about the role of the Internet in fraying religious ties given how some commentators have noted that the momentum of religion is increasingly in the practices of "autonomous selves" (Giddens, 1991), as individuals choose to consume online religious experiences and withdraw from houses of worship (Lyon, 2000). Alternatively, the Internet is seen as a place of revival given the waning influence of religious institutions (Cobb, 1998).

Yet, given the nature of the Internet as a meta-medium (Agre, 1998), debate on the influence of the Internet on religious practice is largely dependent on the particular technology in use. Blogs represent one of the more recent incarnations of social software supported by networked computing (Tepper, 2003). They represent the culmination of two decades of software systems that support threaded, persistent conversations, including Usenet, Bulletin Board Systems, listservs, and Internet forums. The most recent uses of the Web—sometimes called Web 2.0—are marked not only by their openness to contributions by the user but also by their tendency to foment community. Definitions of blogs generally note that they are frequently updated sites, with dated postings, often linking to, and linked from, other blogs (Walker, 2005; Winer, 2003).

Recent data document the swelling growth of weblogs in the last three years. According to the Pew Internet and American Life Project's study on "content creation online," more than 53 million or about 44% of adult American Internet users have contributed material online. Eleven percent of Internet users have read or visited the blogs of other Internet users, and about one-third of these visitors have posted material to the blog (Lenhart, Horrigan & Fallows, 2004). Technorati, a blog search engine, reports it is currently tracking nearly 70 million blogs, with more than 1.6 million new posts daily (technorati.com, March 1, 2007), and that number continues to grow exponentially. These blogs support a range of issues, including political commentary (Matheson, 2003), scholarly research (Halavais, 2006), blawgs that cover legal matters (Oi, 2006), economic blogs (Quiggin, 2006), and kittyblogs of oversharing triviality (McFedries, 2003). Although recent books on blogging have documented blogs in different industrial and social contexts (e.g., Bruns & Jacobs, 2006; Tremayne, 2006), no one has published on religious-related blogs. Given the rise of online religious activity, bloggers may also act as online scribes and contribute to the rise of online religion, a form of many-to-many religious communication developed aside from the influence of religious institutions (Helland, 2002).

Religious Blogging: Different Paradigms on Religiosity and Internet Use

On the micro-level of individual Internet use, various commentators have expressed contrasting assessments for religiosity and Internet use that have implications for religious blogging. One view argues that the Internet presents dramatic challenges to religious adherents, as secularization theory posits an inverse relationship between the processes of modernization (technological, social, and economic) and the importance of religion and religious beliefs in society (McGrath, 2004). As traditional peoples began to understand science and technology, increasing educational levels lead to a more skeptical attitude toward religious belief, undermining religiously-oriented belief systems. Religious leaders lose their efficacy as other forms of authority (such as the state, educational leaders, scientists, etc.) teach and legislate in life domains that had previously been dominated by religious authorities (Norris & Inglehart, 2004).

Various communication scholars have applied the tenets of secularization theory to posit a significant negative relationship between religion and mass media use, as the mass media is perceived by some to be aligned with the values of modern, secular society (Buddenbaum & Stout, 1996). For instance, secularization theory has been used to explain how religious adherence may negatively predict older forms of media such as newspaper reading and television viewing since some people allegedly shun media that are perceived to be incompatible with their beliefs (Buddenbaum, 1986; Hamilton & Rubin, 1992). Recently, Armfield and Holbert (2003) examined the statistical relationship between religiosity and Internet use and concluded that results from their survey data analysis showed that "the more religious an individual is, the less likely he or she will use the Internet" (p. 139). Drawing upon a model of secularism as well as uses and gratifications, they argue that their results may be explained by the perceived ethos of the Internet since "Internet largely embodies a more secular worldview," religious persons are less likely to use it (p. 136). They also note that their application of secularization theory "mirrors the personal identity function of media outlined in uses and gratifications scholarship which states that individuals often turn to the media to reinforce some preexisting norm or value structure and shy away from those media outlets that do not reflect their values" (p. 130). From this viewpoint, believers who construe the Internet to be antithetical to their faith beliefs and practices may be less enthusiastic about the Internet and utilizing the blogosphere.

Another more recently articulated view emphasizes a positively synergistic relationship between religiosity and Internet use. Recent studies argue that Internet use by individual believers may be attributed to the extent to which religious communities spiritualize and socially shape the Internet in line with their discourse, worldviews, and practices (Barzilai-Nahon & Barzilai, 2005; Cheong & Poon, 2008). Kluver & Cheong (2007) argue against the secularization thesis in light of technological modernization by discussing how religious leaders of five world religions perceive the Internet as compatible in a highly wired context and mandate its use for institutional growth and community building. A socio-historical perspective on the religious use of mass media also highlights how the religious have strategically employed the newest technology throughout its history (Soukup, 2002). Starting with the acts of writing on paper to print images, radio, and television broadcasts, the Internet is the latest medium to be adopted in ways similar to how older forms of mass media have been utilized to advance religious missions. It is particularly interesting that Serfaty (2004) notes that the tradition of blogging finds some of its roots in the religious diaries of previous centuries. In contemporary society, according to Ammerman (1997), a voluntaristic mode of spiritual involvement is emerging where fluid and plural types of commitment, diverse worship approaches, and a bricolage of convenient, portable religious practices exist. She recommends future research to examine "how religious rhetorics and practices are enacted" (p. 205), given how traditional dichotomies such as sacred-secular and member-nonmember are increasingly irrelevant in a world where "actors are constantly choosing their ways of being religious" (p. 204).

The blogosphere serves as an interesting social laboratory to examine the emerging mode of religious connections, especially in light of recent evidence suggesting the Internet is the "new normal" increasingly woven into people's daily rhythms of work, play, and critical life decisions (Rainie & Horrigan, 2005). For instance, ethnographic research conducted on Christian email communities shows how Christians have perceived the Internet as a "sacramental space" and socially shape it for their information seeking, worship, and interaction with other believers (Campbell, 2005). More specifically, Murley (2005) points out how blogs may serve as an "extension of the Reformation concept of the priesthood of all believers" providing individuals the freedom to express their views on faith outside the confines of organized religion and mediation of religious leadership. Likewise, according to Murley (2005), blogs may provide ministers an alternative space to broadcast their beliefs on contemporary spirituality, including critiques of existing religious practices; an act of resistance akin to Martin Luther's nailing of the 95 theses to the door of the Wittenberg Church that sparked off the Protestant Reformation. Forbes & Mahan (2005) note that blogging "fully represents the paradox of the internet, creating greater individualism in religion at the same time that it forms new communities" (p. 131). Perhaps most striking in the material examined below is the degree to which conservative Catholics in the United States have embraced this role of direct interpretation. Several commentators have also argued that the Internet is the ideal space for religious imaginings as certain inherent characteristics of the Internet are conducive for cyber spirituality. For example, Henderson (2000) writes that the Internet is a fitting religious symbol as it has corresponding features to the image of God as being accessible at any place, at any time, and for anyone.

The motivations for religious blogging may also encompass less extreme needs for confrontation and opposition, and may include other personal needs. According to prior research on the uses and gratifications theory, people selectively use mass media to fulfill one or more of the following needs for information, personal identity, integration and social interaction, and entertainment (Blumer & Klatz, 1974; Palmgreen, 1984). Given how audiences can now actively interact with the Internet, the popularization of the Internet has revitalized interest in the uses and gratifications of newer media (Ruggiero, 2004). Blogs with their chronological entry and hyperlink connections may redefine the processes of need gratifications and media use, blurring interpersonal communication and mass media use as gratifications may come from the medium's content, from general exposure to blogs, from familiarity with the genre of religious blogs, and from the context in which it is used. Based on an analysis of interviews with 23 popular bloggers in the United States, Nardi, Schiano, Gumbrecht, and Swartz (2004) identified five motivations for blogging; documenting one's life, providing commentary and opinions, expressing deeply felt emotions, articulating ideas through writing, and forming and maintaining community forums. With regard to religion, blogs may also represent a reintegration of religious life into the practices of the everyday as even among blogs that are explicitly dedicated to religious topics, the breadth of entries spans a wide range of public and personal concerns. Likewise, on blogs that are not directly religious in nature, some entries are clearly marked by religious practice and influenced by spiritual beliefs.

In sum, given the dearth of research specific to religious blogging, we were interested to find out how individuals are utilizing the Internet for religiously related discourse, in particular to examine a) the content of religion-related blogs, b) the degree to which religious blogging is integrated into other spaces on the Web, and c) the motivations of bloggers who blog as a form of religious practice. As Herring et al. (2004) note, weblogs are "the latest genre of Internet communication to attain widespread popularity, yet their characteristics have not been systematically described." They report that results from their content analysis of 203 randomly selected weblogs show that media coverage, scholars and bloggers exaggerate the extent to which blogs are linked, interactive, and oriented toward external events, and underestimate the importance of blogs as individualistic, intimate forms of self-expression. Therefore, although a paucity of research exists on religiously oriented blogs, we expect that the blogosphere may serve as a growing locus for religious expression and an alternative arena to established religious organizations and practices. This study takes an "ecumenical" approach to the exploration of blogging as religious practice; an interdisciplinary and multi-method data collection and analysis approach recommended for the study of new fields of media and religion (Buddenbaum, 2002). In the next section, we describe our data collection steps and methods of analysis.

Examining Religion-Related Bogs

While the totality of blogs is often referred to as a seemingly unitary "blogosphere," that label covers a variety of sites created by different individuals and groups for different audiences, with varying intentions. Many blogs are started and then quickly abandoned, never having reached a continuing readership, or are rarely or never linked to by other bloggers (Henning, 2004; Herring, 2005; Li & Walejko, 2007; Trammell, Tarkowski, Hofmokl, & Sapp, 2006). Because we were interested in "public" blogs, those that were read by some form of audience and had a community following, however small, we collected a purposive sample of blogs from Technorati (http://www.technorati.com). Technorati provides a way of ranking blogs by citations and was used because blogs tracked by Technorati are integrated into the larger community of the blogosphere to some degree. Also, Technorati provides the ability to search for the most recently posted items that include a given keyword.

During the four days ending in Easter 2006, URLs were identified for 200 blogs that had recently posted something containing a keyword related to Christianity. First, a dictionary of terms was constructed, drawing on a previous content analysis dictionary (Stone, 1966) and a printed dictionary of religious terms (Kauffman, 1969). From this larger dictionary, 10 keywords were chosen that were most likely to yield Christianity-related posts: Bible, Christ, church, congregation, lord, pastor, pray, priest, salvation, and faith. Using these search terms several times each day, blogs were drawn as long as they were publicly accessible; blogs that required passwords or were otherwise not public were excluded. We chose only blogs that were predominantly written in English. Blogs that used the terms in nonreligious contexts (for example, "priest" used as surname or as a character-class in some games) were excluded.

Front pages of these blogs were archived and content analyzed by human coders, seeking information on items that indicate general topical coverage of the blogs, linkage patterns, and indications of the uses and gratifications afforded by the blog to the blogger. The coding frame was composed of 8 blog-level variables and 99 entry-level variables. Eight blog-level variables included basic information for self-introduction, the first five links on the blogroll (list of other blogs the author likes or reads), when it was present, and indications of the number of authors and whether they identified their blog or themselves with particular religions. An additional 99 variables were coded for the first four blog entries listed on the index page that were over 50 words in length. These included content-related indicators, a coding of the first three hyperlinks present in the entry, perceived intentions, objectives, or gratifications of the entry, and coding of the comments for the entry. Content-related themes coded on the entry-level were summed up later and re-analyzed for the blog-level exploration. For example, if content of religious news was coded as "1 (existing)" at least in one of four postings, it was re-coded as "1" on blog-level. It should be noted that as in other exploratory coding of blog content and intent of users (e.g., Lenhart & Fox, 2006; Papacharissi, 2006), we did not analyze content as a single mutually exclusive unit, as doing so potentially reduced the degree to which we could measure the co-occurrence of particular themes. As a result, we did not use factor analysis or other forms of statistical inference that depend on such mutual exclusivity. Ten percent of each of these frames was coded twice to test for inter-rater reliability. Cohen's kappa was more than 0.8 for all variables. Given the exploratory nature of this research, we triangulated our quantitative analysis with qualitative interviews with direct questions to the authors of a subset of the blogs in the sample regarding their blogging practices, intentions, and motivations. Of 200 bloggers, 49 responded to the e-mail interview conducted in October 2006.

Characteristics of the Sample

An archive of the front page of each of these blogs was collected, along with the contemporaneous authority rank on Technorati, when available. As noted, this rank indicates how many hyperlinks from other blogs link to a particular blog. The top-ranked blogs in Technorati routinely receive more than100,000 in-links from other blogs. The highest ranked blog in our sample was at number 995. A substantial number (64) were not ranked because they lacked a profile on Technorati, though their postings were available through a search. The median ranking of blogs with ranks on Technorati was 463,342 and is distributed heavily toward the tail of the rankings. While there is not necessarily a correlation between the Technorati rank and blog traffic, the Alexa service (www.alexa.com) estimates the reach of the top-ranked blog in the sample at 8.75 per million internet users, while a site that is ranked 9,811 on Technorati, Alexa estimates a reach of 0.1 per million. The majority of the blogs tracked in this sample receive too little traffic to be tracked by Alexa. In other words, the blogs that make up this sample are part of what has been called the "long tail" of blogs, representing the less visible, everyday practice of blogging.

The sample was collected based on the appearance of terms related to Christianity. This intentionally resulted in a sample that included both blogs devoted to religious topics and those that mentioned religious topics only in passing. Of the 200 blogs, 80 provided a clear indication that religion was an important part of the blog. Forty-six (23%) included religious terms in their title or the "motto" of the site (e.g., "A Catholic Guy"). Of the 72 blogs that included a statement about the purpose of the blog, 18 (25%) indicated that the blog had a religious focus, and of the 138 with some form of a biographical note, 56 (41%) provided a religious affiliation. Fifteen of the blogs were in some way affiliated with a local church or congregation. Most (173) did not clearly identify a denomination, but Catholics, Episcopalians, and Baptists made up half of those who did indicate a particular denomination.

A proportion of blogs in the sample (64%) was not "religious blogs" and made use of religious terms sporadically. A number of these posted complaints or criticisms of Christianity, its religious practices and complaints against clergy (e.g., "the pastor is a [expletive] and can't so much as return a phone call . . . "), or a criticism of the faithful (e.g., "I am far, far more frightened by the idea of a Christian running at me with a Bible than I am by an Arab approaching me with a bomb strapped to his chest"). Many, however, simply used religious terms only in passing (e.g., "it is just that when I see a chicken, I don't think 'oh, poor little helpless creature of God.'")

Results

Topics of Blog Posts

Given the dearth of research specific to religious blogging, we were interested to find what sorts of utterances were produced by bloggers who touched on religious topics. We sought out postings in the five broad themes listed in Table 1. Of the 200 blogs, the most frequent appearance of religious motifs was found in descriptions of personal experiences, emotional reactions, and thoughts about God (89 of the 200 blogs). This was followed by didactic content (78), religious and social criticism (45), news and information (31), and efforts to coordinate events, gatherings, and other "real life" social activities (14). As noted, these themes are nonexclusive so that a posting, for example, that celebrates a personal experience (rejoicing in the author's 26th year as a Catholic), might also connect with other church members, or provide a Christian perspective on a current event. A recent survey of bloggers by the Pew Internet and American Life project presents a similar set of motivations expressed by bloggers more generally (Lenhart & Fox, 2006). It finds that among the major reasons for blogging, authors list "documenting personal experiences" (50%) and "sharing practical knowledge or skills" (34%), which appear to be aligned with our categories for "personal religiosity" and "didactic content."

More detail is available by concentrating on the posts individually. A total of 789 individual posts were coded. Many of the postings lacked any content related to religion, as we will explore further below. Of those that did contain references to religion, many (89) were made only in passing. Some of these—requesting prayers or invoking the name of God—clearly identify the author as religious, but provide little more information than this. Others simply note that the author went to church or reference the upcoming holiday.

TABLE 1 Types of religious content appearing in blogs.

Type of content	Number of blogs
Personal religiosity	89 (44.5%)
Didactic content	78 (39.0%)
Criticism and social issues	45 (22.5%)
News and information	31 (15.5%)
Coordination of practices	14 (7%)

A number of different sorts of content make up the broader "personal religiosity" category, including observations of faith and descriptions of religious experiences. The majority of postings in this category were those in which bloggers made positive emotional statements of devotion, awe, appreciation, or belief (76 postings). In many cases, these were simple expressions ("I'm just so grateful for Christ's sacrifice"), but often these were tied to a description of everyday experiences of the blogger, which they indicated helped to cement their belief or demonstrate the existence of God in their own life ("I found myself so moved by the motions of the towel, wiping it dry."). Several posts (22) spoke against the church or the faithful. Some of these were also emotional ("What the hell is wrong with these people??????"), but many were merely skeptical or sarcastic ("And, for today only, he is dying for the consumption of chocolate eggs before noon.").

Many of the personal statements also represented more contemplative statements about the author's own faith. By its nature, much of this is intended to present a perspective on some particular element of faith, for example, the meaning of the crucifixion and resurrection (53 postings). These rarely were critical of a particular position but rather presented some form of a justification for the individual's personal faith. Personal religious experiences also included accounts of participation in rituals such as attending church (28 posts) or engaging in missionary or evangelistic work, pilgrimages, Bible study, meditation, and other religious activities (28 posts).

The second broad category was "didactic" or "educational" posts; those intended to inform the reader on the topic of faith from an authoritative source. While some of the critical posts might also be educational, the posts in the "didactic" category generally drew on existing authoritative claims. Some (21 postings) made reference to the history of Christianity and other religions, or important figures (e.g., hagiography). A larger number presented references from the Bible (59 postings) or by other authoritative religious or scholarly texts (72 postings), including sermons and catechism reference works. Indeed, while reference to web-accessible news is common among bloggers, the religious bloggers in our sample made reference to published works at an extraordinarily high rate.

While strictly didactic content and theological criticism may seem easy to confuse, we took the former to be based in authoritative reference, while the latter clearly indicated an area of disagreement. Because the categorization was nonexclusive, some postings fell into both of these categories, providing an argument and citing authoritative works. We identified three forms of criticism employed by bloggers. The first and most prevalent was criticisms of the institutions of the church, or of particular dogma or orthodoxies (26 postings). Many of these included theological arguments to convince the reader of a particular position, or recapitulated internecine conflicts. A related group of posts presented or argued against critiques of Christianity (17 postings). Finally, 19 postings tookup an argument over social issues—abortion, foreign affairs, and gay and lesbian issues, among others—from a religious perspective.

A number of the postings seemed to be largely informational, providing some form of news or information related to religious practice. For example, if news sites included a story relevant to the bloggers' own church or denomination, they might summarize it and link to it (34 posts). News items that treated broader social issues of religious interest were also noted (11 postings) even if they did not include an extensive discussion. The final category consisted of postings related to a particular church or religious organization, seeking to announce, organize, or coordinate the actions of the group (19 postings). Occasionally, a posting attempted to bring together the virtual community for a "meet-up" to continue the discussions already started online.

By understanding the composition of the typical religious blog post, we begin to form a picture of religious blogging as a practice. In some ways, religious blogging reflects the standard interests of bloggers everywhere. There is a focus on documenting personal experience, followed by interest in teaching and learning, as well as monitoring and making sense of the mainstream news. Beyond blogging, these are the kinds of functions Harold Lasswell (1948) identified as being at the core of mass communication in the middle of the last century. While religious blogging echoes the practices of blogging generally, it also emphasizes a particularly integrative function. While blog punditry and political blogs may rely on the news media to feed their conversations, religious bloggers seem more likely to gather from a combination of sources: their own lived experience, and the texts and beliefs of their community.

Integrating Religious Discourse

One of our central questions was the degree to which religious blogging was integrated into other forms of blogging and other spaces on the Web. That is, while a greater division between religious and secular discourse in the West may be dictated in part by the place-based institutions of each, does the blogosphere allow for greater integration of religious thought with activities in other public realms? What is the context of religious blogging? To answer that question we examined the degree to which topics, hyperlinks, and comments tended to be more integrated between religious and non-religious subjects, and what factors influenced this.

As noted, each of the first four entries was coded for topic. While some detail was extracted from religious postings, the general category of nonreligious postings was also coded. While much of this content was social commentary, the most frequent form of blogging was the relation of everyday activities and thoughts, which made up 323 of the 789 posts. Of these postings about everyday personal events, 122 also included some form of religious commentary. Thus, within a large number of posts, we saw mixtures of religious and other contexts where bloggers located the sacred within the profane and vice versa.

Many of these posts that mixed the everyday with the religious concerned the daily minutia of participating in a religious group, or engaging in religious interests. For example, one blogger asks for advice on where to buy Mormon art in Salt Lake City, and a number of readers respond with suggestions. Another remarks, "I woke up at 9:30 to sing in the church choir, completely sleep-deprived because I was up late getting drunk at a drag show. I like to keep balance in my life." As in other forms of blogging, this provides an interesting window into the everyday practices of religion. For example, some popular blogs track a pastor's process of writing a sermon, and provide a kind of transparency that might otherwise be absent. Others post about the spirituality of everyday events. A blogger writes of the vocation of nurses who must counsel a daughter who is watching her father die slowly. The same blogger writes, "Ice cream is one of my proofs of God's existence." In some ways, this encapsulates many of the ways religious bloggers discover grace in the joys and challenges of their everyday lives, and share these lessons with readers.

Devout bloggers are certainly not the only ones who blog about religion in their daily lives. One blogger writes about his agnosticism, while another blogger indicates a former priest convinced the blogger never to pray again. A number of the entries complained about the Easter holiday and friends and relatives bringing the bloggers to church. One blogger pleaded for Christians to "stop ambushing us honest heathens at the shopping malls in some bizarre attempt to drag us to church." No doubt, such an integration of everyday experience with spiritual reflection appears in private conversations, and occasionally within sermons, but blogging provides an unusual platform for the mixture of the profane with the sacred, the everyday with the eternal.

The Blogosphere Context

Much has been written about the structure of hyperlinks within the blogosphere (Park & Thelwall, 2003). While the blogosphere is in some ways an interconnected whole, it is clearly not a uniform space. What is the context of religious blogging within the larger network of blogs? We extracted all of the external hyperlinks on the index pages of our sample, and examined the most popular links. We also took the first five hyperlinks of each of the posts we examined and coded them for type and apparent reason for use.

Table 2 lists the domains present in hyperlinks on the index page of four or more blogs in each of two groups: blogs that provided some form of indication of being "religious blogs" and those that did not. A total of 8,409 hyperlinks were extracted from the first group (80 blogs), and 9,855 were

TABLE 2 URLs of sites linked to by four or more of the blogs in the sample.

Religious domain	# Linking	Nonreligious domain	# Linking
http://www.amazon.com	21	http://news.yahoo.com	14
http://en.wikipedia.org	18	http://en.wikipedia.org	13
http://creativecommons.org	18	http://www.nytimes.com	13
http://technorati.com	18	http://www.washingtonpost.com	12
http://www.truthlaidbear.com	15	http://www.imdb.com	12
http://www.bloglines.com	14	http://technorati.com	12
http://www.google.com	12	http://rpc.blogrolling.com	11
http://rpc.blogrolling.com	10	http://michellemalkin.com	10
http://www.ringsurf.com	9	http://www.bloglines.com	10
http://www.blogwise.com	9	http://www.technorati.com	10
http://www.nytimes.com	8	http://www.cnn.com	9
http://www.technorati.com	8	http://www.nationalreview.com	9
http://www.feedburner.com	7	http://www.amazon.com	9
http://www.biblegateway.com	7	http://www.truthlaidbear.com	8
http://www.blogstreet.com	7	http://maps.google.com	8
http://www.newsgator.com	7	http://www.google.com	8
http://news.bbc.co.uk	6	http://www.townhall.com	7
http://www.washingtonpost.com	6	http://www.latimes.com	7
http://www.blogcatalog.com	5	http://www.drudgereport.com	6
http://blogs.salon.com	5	http://www.guardian.co.uk	6
http://www.christianitytoday.com	5	http://news.bbc.co.uk	5
http://homepage.mac.com	5	http://www.blogarama.com	5
http://www.getreligion.org	5	http://www.blogrolling.com	5
http://www.ewtn.com	4	http://abcnews.go.com	5
http://amywelborn.typepad.com	4	http://creativecommons.org	5
http://www.blogarama.com	4	http://www.telegraph.co.uk	5
http://www.cafepress.com	4	http://www.foxnews.com	5
http://www.challies.com	4	http://www.newsgator.com	5
http://www.vatican.va	4	http://www.boston.com	5
http://www.prolifeblogs.com	4	http://www.opinionjournal.com	4
http://www.jordoncooper.com	4	http://www.usatoday.com	4
http://www.geocities.com	4	http://mediamatters.org	4
http://www.one.org	4	http://kevan.org	4
http://www.nationalreview.com	4	http://weblog.herald.com	4
http://www.evangelicaloutpost.com	4	http://www.makepovertyhistory.org	4
http://www.feedblitz.com	4	http://www.worldnetdaily.com	4
http://www.blogs4god.com	4	http://www.whitehouse.gov	4
http://www.thestar.com	4	http://www.newyorker.com	4
http://www.sojo.net	4	http://www.buzzmachine.com	4
http://www.newadvent.org	4	http://instapundit.com	4
		http://www.ringsurf.com	4
		http://www.dailykos.com	4
		http://www.khaleejtimes.com	4
		http://www.leftyblogs.com	4

extracted from the second (120 blogs). Excluded from these lists are links to sites that may be considered in some way infrastructural, such as blogger.com and livejournal.com, which, while they may not have been internal to the blog itself, were integral to its operation. Links that were added as advertising to free blog hosts were excluded. Finally, also excluded were links to sites that provided technical support like feed services, image and file hosting, and statistics generators.

The sites that prove to be most interesting to members of the two groups can provide some indication of collective interests. Turning first to the sites most frequently linked to from blogs that are not self-identified as religious, we find a distribution of favorite sites that is not far from what we might expect from the larger universe of blogs. Prominent in this list is the number of mainstream news sites, which have consistently been among the most-linked sites among blogs (Halavais, 2002). News sites are present among the religious bloggers as well, though not in as great a number. The religious bloggers supplement these mainstream news outlets with information sources specific to religious concerns. Christianity Today, for example, is a Web site and online magazine that features resources and stories of interest to Christians, EWTN is the Web site for a Catholic organization that produces religious broadcasting and other activities, and Sojourners (sojo.net) is an online magazine covering "faith, politics, culture."

The nonreligious bloggers also prove to be a good source of links to what are sometimes called the "A-list" blogs that garner a particularly large readership. These include blogs such as Michelle Malkin, Buzzmachine, Instapundit, and the Daily Koz, each of which figure on the top hundred popular blogs listed on Technorati. These blogs are less linked to among the religious bloggers. Naturally, some of this has to do with the difference in numbers of blogs in the two groups. Nonetheless, the Daily Koz, a popular liberal political blogger, for example, is only linked to from one of the index pages among the self-identified religious blogs. The religious bloggers clearly have their own A-list. Included among these are Get Religion, Amy Welborn, Tim Challies, Jordon Cooper, and the Evangelical Outpost. Technorati suggests that each of these blogs receive thousands of links from the blogosphere at large, reinforcing the idea that they represent a community A-list, an alternative to the sites most frequently linked to from bloggers whose focus is not primarily religious. The idea of an alternative is reinforced by the masthead of Get Religion, which includes a quotation from William Schneider, a political analyst: "The press . . . just doesn't get religion." In addition to the major blogs, two aggregators that collect and display posts from a variety of religious bloggers (prolifeblogs and blogs4God) are often linked from the blogs that self-identify as religious. Just as very popular blogs have risen to the level of mass media for the blogosphere as a whole, and thereby gained recognition from the traditional mass media, certain Christian bloggers have attracted a similar large audience from their own community. It would be interesting to discover the degree to which traditional Christian media are attending to this "Christian A-List."

Among the blog services frequently referenced by all the blogs in the sample was Technorati, which in addition to tracking blogs more generally provides the ability to include topic tags in individual blogs. These provide an avenue through which bloggers can engage in a dialog of postings across blogs. In total there were 104 such tags among the religious blogs and 211 among the nonreligious blogs. Perhaps unsurprisingly, only 16 of the tags in the nonreligious blogs were in some way religiously oriented (Arminianism, Baptism, Calvinism, Catholic, Catholics, Christian Politics, Christians, Christianity, Good Friday, Gospel, Irresistible Grace, Jesus, Kingdom of God, Pope Benedict, Spirituality, Temple Emmanuel). In other words, authors generally did not think of their religious posts as something that they should classify as part of a larger discussion of religious issues. On the other hand, more than half, or 58, of the 104 tags among the blogs that self-identified as religious were also religious in nature.

Another common link was to Wikipedia, a collaboratively edited online encyclopedia. Of the 104 links from self-identified religious blogs, only 10 were religious in nature. Only 41 links were made from the nonreligious blogs directly to Wikipedia articles, and only 2 of these were to an article with a religious topic ("Catharism" and "Pope Benedict XVI").

Other sites emerge as authoritative in each of the two groups. For example, while the official White House site is frequently linked among the nonreligious bloggers, the official Vatican site is among the most linked for religious bloggers. Likewise, a Bible gateway site that provides biblical quotations ranks highly among religious bloggers. To borrow the title of a well-known paper on hyperlink analysis, "You are what you link," (Adamic & Adar, 2001), and the community of religious bloggers shows one facet of its character in the makeup of the links it shares.

Table 3 outlines the types of sites that the bloggers in the sample linked to in their "blogroll," the list of sites they believe their readers would be interested in. Those blogs identified by their religiosity unsurprisingly are more likely to link to other religious blogs, and most frequently blogs that represent their own denomination of Christianity. In both the case of links more broadly, and the

TABLE 3 Target of hyperlinks within sampled blog entries, according to self-identification of religious intent of blog.

Type of target	From religious blogs	From nonreligious blogs
Personal Web site of same religion/denomination	34	8
Institutional Web site of same religion/denomination	40	12
Personal Web site of different religion/denomination	2	1
Institutional Web site of different religion/denomination	1	0
Secular source of religious information	35	15
Not religion-related	38	103

specific case of links in the blogroll, religious bloggers express their community and solidarity with those they are drawn into conversation with.

From the Voices of Religious Bloggers

Past research in the uses and gratifications of mass media have examined the reasons reflected in media use in various forms, including religious television and radio use (e.g., Abelman, 1988; Hamilton & Rabin, 1992). A growing number of studies have examined the uses of the Internet (e.g., Armfield & Holbert, 2003; Flanagan & Metzger, 2001; Laney, 2005), and a small number of studies have started to address blogging from a uses and gratifications perspective (Johnson & Kaye, 2004; Kaye, 2006; Trammell et al., 2006). Responses from interviews with religious bloggers among the sample analyzed here suggest that blogs are typically used to fulfill the following key needs, including the a) transmission of values, b) entertainment and escapism, c) integration and interaction. In some instances, these needs have been interpreted and expressed in religious terms to refer specifically to the faith factor (Laney, 2005) or the faith-based aspects of these media needs, for example, the motivation to sustain "fellowship" or to interact with and integrate into online and offline religious communities. In addition, according to most blog respondents, their blogs serve multiple purposes, as the motivation for blogging is, in the words of one blogger, "archival tinged with cathartic and communicative purposes" where the "public goal is to offer and argue my views, privately, [. . .] to blow off steam."

First, the data suggest that one major use of religiously oriented blogs is to convey and perpetuate the values of religious bloggers. Several bloggers emphasized their role as "online scribes" who chronicle their "spiritual journey via blogging." A few bloggers said they were particularly motivated to write about religious experiences in their daily lives. They stressed that they provided unique viewpoints from the "intersection of religious faith and the culture at large" as they "write about whatever is going on" and everyday occurrences.

For some bloggers, religious writings online reflected their personal mission to provide a prophetic voice, to "give others inspiration in their lives," "introduce the relevance of faith in Christ to current issues," and "to engage people in thoughtful and meaningful dialogue regarding the paradoxes of our faith and practice and the future of the Church." One subject's blog provided a means of "articulating a new paradigm" that integrates spirituality with political and cultural concerns. He said,

> *I am frustrated by polarization in both political and spiritual life. I feel as if our culture has torn apart the truth, like the mother in the King Solomon story who wanted to cut the baby in half. The political and cultural left has part of it, the right has another part, and they're fighting with each other over which incomplete and bleeding half is sufficient.*

A few bloggers also specifically mentioned their blogs as "apologetics," giving a defense of their faith to "warn and edify the body of Christ." Two bloggers explicitly cited biblical scripture as a reason for their blog, to "expose the unfruitful works of darkness Ephesians 5:11" and "to introduce or clarify the saving grace of Christ Jesus, John 3:16." These bloggers highlighted the need for them to "defend religion against anti-religious bigotry,' or to talk about religious bigots" as they felt that their voices were marginal and sometimes "distorted and misrepresented" in the blogosphere and mainstream media. In the same vein, some bloggers said that the purpose of much of their writing online was to share their beliefs with people in hopes of providing a fair religious perspective, to help others better understand their faith:

> *I have discussed religion as a topic many times and will continue to do so. Religion is a major social and political phenomenon. Unfortunately the mainstream media is actively hostile towards all religion except liberalism. Consequently, the views of a majority are metaphorically seated at the back of the bus. I live in a state with a large dynamic religious community, and I find their views extremely interesting and relevant to the national discourse. It seems 'good business sense' to give voice to the silent majority.*

Similarly, another subject noted:

> *The majority of what goes on my blog is religious content. The reason I do this is because I believe that modern Christianity has perverted my faith and I want to speak out against it. I also like to write things down when I am thinking about them, it helps me figure out what I am thinking. I also want to share my ideas about faith and God with other people.*

In this way, their sentiments reflected the findings of earlier research on the use of religious television, where usage was "reactionary" to what is perceived to be "secular" media (Abelman, 1987, 1988). Here, blogs serve as an alternative online space to convey their dissatisfaction with the ways that their faith is presented.

However, not all bloggers expressed an overt evangelistic passion, some commented that their blogs were not aimed to convince their readers but "simply to reflect" their own personal experiences and thoughts. One blogger commented:

> *I view religion as a private matter. I have faith and it does inform my viewpoints, but the purpose of my writing is not to evangelize or convert.*

While another noted:

> *I have expressed some religious content, mostly how I was raised, and how my religious views are ingrained into my personality, thus, my life. My views are not intended to offend or sway others to see my point of view, rather, offer another perspective, mine.*

Thus, blogging provides a venue for the transmission of values and culture, ranging from an explicitly evangelical mission to defend the faith, to goad readers to become seekers, to the subtle expressions of religious views ingrained in personal reflections. In these ways, blogs may serve as "an extension of the priesthood of all believers," providing believers an avenue of expression past the precincts of institutionalized religion (Murley, 2005).

Another key use of religious blogs that emerged from the bloggers interviewed was the need for a writing space to diminish stress. According to the bloggers, blogs were "an outlet for writing," a "ventilation system," "random thought catcher," a "public diary of sorts," and a "catharsis that can be accomplished by releasing whatever it is from within yourself, it becomes something that you'd like to keep doing on a higher level."

Prior literature highlights how religious media use may be seen as a form of leisure providing entertainment and psychological release (Abelman,1988; Hamilton & Rubin, 1992). Here, according to some bloggers, blogging was professed to be a form of pleasurable escape. However, contrary to the illusory and sometimes deceptive role identity plays online, one blogger commented on the need to present his authentic self, as his blog was a way to share his beliefs. He wrote, "it is my portal directly to my life, in order for people to see the real me." In addition, this form of cathartic writing is not necessarily crude or loosely constructed as one blogger commented on the form of her personal reflexivity,

> *I frequently blog about my Christian faith. I do it because it is the highest expression of who I am. I am a reader and thinker and I teach Reformed Christian doctrine to women, so it not only permeates the things I think about but more importantly the way I think.*

Hence, the conventional use of media for entertainment is noted by some religious bloggers to be important. Bloggers write to "vent" but the use of blogs for cathartic release is also interpreted more cautiously by some to comprise of an accurate portrayal of self and well-crafted expressions of faith in action.

Finally, the third key expressed by religious bloggers related to the interaction with other believers. Given the interactive capacity of the Internet, several bloggers said their blogs were an avenue for connecting to others and for "God to work" as "the body of Christ is not limited to being inside a church." One respondent, who was a pastor, mentioned that the blog was "an extension" of his ministry, saying "blogs are a good way to create a place for others to know me better and connect with them more then 1–2 times a week face-to-face."

The majority of the respondents also mentioned the hyperlinks, hits, and comments from other bloggers, including comments from family, friends, and even strangers who have since become their friends online. In this way, the blog functions as a form of social support, to open up new connections to others, as one blogger said, his blog

> *has opened up a whole new world to me. I have made connections with so many people all across the country, who I've never met, yet I consider them my friends. I journal randomly, yet I am a frequent commenter on many different journals. I enjoy reading about other people and what is going on in their lives. There are several blogs I read that are like a "fix" . . . if I don't read them first thing in the morning, I feel like I'm missing something. That has surprised me, wholly . . . I didn't think I'd care about people I don't "know". I feel I've become accepted in a non-judgmental community . . .*

Given that blogs are interactive social networking platforms, bloggers may participate actively to comment on other blogs, thus constituting religious networks across the blogosphere as one blogger points out, her blog represents,

> *. . . more of a network, not really an audience. Many of my blog visitors are from blogs I also read. Other visitors just stumble across it while searching for terms that show up in my tags, or are referrals from other bloggers. Part of the reward of having a blog, other than improving your writing skills and learning more about yourself as you write, is meeting other bloggers and getting to know them. Comments from readers spur you on to more creativity!*

Similarly, another blogger likened blogging to a "team sport" where "you give links, you get links. You grow by helping others grow. It's very cooperative." Hence, the integrative use of the blogosphere was perceived to be mostly positive as bloggers said these online connections enabled them to broaden their reach and strengthen the extended "body of Christ." One blogger also noted how comments and the e-mails that he receives also "creates a sense of accountability that was not expected" such that blogging becomes a "discipline and commitment" to produce "writing that delivers epiphany."

Conclusion

There are two movements that are sometimes affiliated with Christian blogging. The first is as a space for the "emergent church." The emergent church is a slippery idea, but it is generally associated with a theology accommodative to postmodern culture that does not rely upon a singular, rationalist narrative, draws on a diversity of practices, and generally appeals to a networked generation (Barna, 2005). We find some evidence of discourse in this sample that would tend to indicate work in building alternative frameworks for interpretation. Results from our analyses suggest that some bloggers are articulating critical discourse of the institutionalized character of the church, and debate over traditional norms and practices. Religious bloggers are also operating outside the realm of the conventional nuclear church as they connect and link to the mainstream news sites, other nonreligious blogs, and online collaborative knowledge networks like Wikipedia. However, major parts of the religious blogosphere examined still reflect the influence of traditional religious experiences, including church attendance, Bible study, and missionary work as well as connections to prominent online Christian informational sources, and self-identified Christian authors. While it is often difficult to judge age and approach, many of the bloggers in the sample appear to be older and more inclined to engage in traditional forms of rational theology.

The other frequently referenced group of religious bloggers is the conservative Catholic movement in the United States, an outgrowth of the scandals surrounding the Catholic Church, and what some Catholics in the United States perceive as an approach more liberal than that elsewhere in the world. Without a clear voice in the American Church, many of those who are seeking changes have turned to the Web. A significant number of the blogs in the sample engaged this larger conservative Catholic movement, and these sites tended to be more interlinked and engage in a broader conversation than other blogs in the sample. Interview data from bloggers illustrate why bloggers write, revealing a gradient of needs related to evangelism and sharing of their faith, as they take on varied roles of the online scribe, prophet, and evangelist online. In this way, data here provide further support for the spiritual shaping of technology perspective (e.g., Campbell, 2005; Kluver & Cheong, 2007), as some bloggers shape the blogosphere in alignment with their spirituality and faith practices.

While these two groups often draw public attention, they are not fully representative of either religious bloggers or religious content in the blogosphere. Religious bloggers represent a multiplicity of intents and practices. For some, this consists of seekers who are documenting a personal spiritual journey. For others, blogging represents little spiritual content. It is an opportunity to vent and represents a document of how the empyrean intersects the everyday. In both cases, there are examples of substantial integration of the religious with the secular, illustrating the dynamics of the nexus between religion, culture, and communication.

In conclusion, given the recent popularity of weblogs and revitalized interest in religious expressions, this study explored both the structural and the descriptive concerns of religion online by examining blog entries and blog links in the blogosphere, and by interviewing bloggers who write about religion. Using a triangulation of qualitative and quantitative methods, this paper investigated the content of religious related blogs, the degree to which religious blogging is integrated into other spaces on the Web and the motivations of bloggers who blog as a form of religious practice.

Findings here present both promising avenues for further research and demonstrate the limitations of the work presented, particularly in terms of the sample. The use of keywords to collect the sample raises two issues. First, in order to limit the number of keywords, it was necessary to restrict our study to Christian-related blogs. The range of religious practices present among bloggers, naturally, extends beyond Christianity. Bunt (2004) has written on the vibrant community of Muslims blogging in Farsi and Arabic in recent years, and others have explored how Buddhists may adopt cyberspace for building community (Prebish, 2004; Kim, 2005), raising the possibility of Buddhist blogs serving as meditation grounds for the exchange of discourse and contemplative psychotherapy (www.sangha.org). While other religions have not received as much scholarly attention, there are many opportunities for studying how religions with different discourse traditions engage this new medium. Second, the limited size of the sampling dictionary, a necessity of our collection method, may have biased the sample toward particular religious practices. Furthermore, the English-language sample privileged bloggers in English-speaking countries, particularly in the more populated United States. Sampling blogs has always been a complicated affair, and other approaches (e.g., snowballing a sample through hyperlinks from a small set of exemplars) come with different, but equally challenging, biases. Finally, although the window of time just before Easter provided us with a more intensive time of contemplation among Christians and those who would otherwise be at the margins of faith, the degree to which blogging during this period might differ in substance from other periods is not clear.

Future research should expand the investigation by delving into more posts that constitute blog texts within a religious tradition and investigate the links of individual blog posts. As blogs offer an alternative in religious self-expression, future research could investigate the implications of critical communicative shifts on traditional religious norms, including governing sanctions and concepts of religious authority and leadership. A comparative analysis of blog practices by denominations and faith would further illuminate blogging as a religious practice and the (re)integration of religious into everyday life and faith communities.

Note

Correspondence should be addressed to Pauline Hope Cheong, Hugh Downs School of Human Communication, Arizona State University, Stauffer Hall, Room A462, Tempe, AZ 85287. E-mail: pauline.cheong@asu.edu

References

Abelman, R. (1987). Religious television uses and gratifications. *Journal of Broadcasting & Electronic Media,* 31, 293–307.

Abelman, R. (1988). Motivations for viewing the 700 Club. *Journalism Quarterly, 65,* 112–118.

Adamic, L. A., & Adar, E. (2001). You are what you link. *Tenth annual International World Wide Web Conference,* Hong Kong. Retrieved August 1, 2006, from http://www10.org/program/society/yawyl/YouAreWhatYouLink.htm

Agre, P. (1998). The Internet and public discourse, *First Monday, 3*(3). Retrieved June 1, 1998, from http://www.firstmonday.dk/issues/issue3_3/agre

Ammerman, N. T. (1997). Organized religion in a voluntaristic society. *Sociology of Religion, 58*(3), 203–215.

Armfield, G. G., & Holbert, R. L. (2003). The relationship between religiosity and Internet use. *Journal of Media and Religion, 2*(3), 129–144.

Barna, G. (2005). *Revolution*. Wheaton: Tyndale House Publishers.

Barzilai-Nahon, K., & Barzilai, G. (2005). Cultured technology: The Internet and religious fundamentalism. *The Information Society, 21,* 25–40.

Blumer, J. G., & Katz, E. (1974). *The uses of mass communication.* Beverly Hills, CA: Sage.

Bruns, A., & Jacobs, J. (Eds.). (2006). *Uses of blogs.* New York: Peter Lang Publishing.

Buddenbaum, J. M. (1981). Characteristics and media-related needs of the audience for religious TV. *Journalism Quarterly, 58,* 266–272.

Buddenbaum, J. M. (1986). An analysis of religion news coverage in three major newspapers. *Journalism Quarterly, 63,* 600–606.

Buddenbaum, J. M. (2002). Social science and the study of media and religion: going forward by looking back. *Journal of Media and Religion, 1*(1), 13–24.

Buddenbaum, J. M., & Stout, D. A. (1996). Religion and mass media use: a review of the mass communication and sociology literature. In D. A. Stout & J. M. Buddenbaum (Eds.), *Religion and mass media: Audiences and adaptations* (pp. 12–34). Thousand Oaks, CA: Sage.

Bunt, G. (2004). "Rip. Burn. Pray.": Islamic expression online. In L. Dawson& E. Cowan (Eds.), *Religion online: Finding faith on the Internet* (pp. 123–134). London: Routledge.

Campbell, H. (2005). *Exploring religious community online: We are one in the network.* New York: Peter Lang.

Cheong, P. H., & Poon, J. P. (2008). 'WWW.faith.org': (Re)structuring communication and social capital building among religious organizations. *Information, Communication and Society, 11*(1), 89–110.

Cobb, J. (1998). *Cybergrace. The search for God in the digital world.* New York: Crown Publishers.

Dawson, L. L., & Cowan. D. E. (2004). *Religion online: Finding faith on the Internet*. New York: Routledge.

Flanagan, A., & Metzger, M. (2001). Internet use in the contemporary media environment, *Human Computer Research, 27*(1), 153–181.

Forbes, B. D., & Mahan, J. H. (2005). *Religion and popular culture in America*. Berkeley, CA: University of California Press.

Giddens, A. (1991). *Modernity and self-identity. Self and society in the late modern age*. Cambridge: Polity Press.

Halavais, A. (2002). Blogs and the "social weather." Paper presented at the Association of Internet Researchers 3.0, October 13–16, Maastricht, Netherlands.

Halavais, A. (2006). Scholarly blogging: Moving toward the visible college. In A. Bruns & J. Jacobs (Eds.), *Uses of blogs* (pp. 117–126). New York: Peter Lang Publishing.

Hamilton, N. F., & Rubin, A. (1992). The influence of religiosity on television viewing. *Journalism Quarterly, 69,* 677–678.

Helland, C. (2002). Surfing for salvation, *Religion, 32,* 293–302.

Henderson, C. (2000). The Internet as a metaphor for God? *CrossCurrents, 50,* 1–2.

Henning, J. (2004). The blogging iceberg: Of 4.12 million hosted weblogs, most little seen quickly abandoned. *Perseus Survey Solutions.* Retrieved August 1, 2007, from http://perseus.com/survey/resources/perseus_blogging_iceberg.pdf

Herring, S. C., Scheidt, L. A., Bonus, S., & Wright, E. (2004, January). Bridging the gap: A genre analysis of weblogs. *Proceedings of the 37th Annual Hawaii International Conference on Systems Sciences.*

Herring, S. C., Kouper, I., Paolillo, J. C., Scheidt, L. A., Tyworth, M., Welsch, P., Wright, E., & Yu, N. (2005). Conversations in the blogosphere: An analysis "from the bottom up." *Proceedings of the 38th Hawaii International Conference on Systems Sciences*.

Hoover, S. M., Clark, L. S., & Rainie, L. (2004). Faith online. *Pew Internet and American Life Project.* Retrieved January 12, 2006, fromhttp://www.pewinternet.org/topics.asp?pageD1&cD6

Johnson, T. J., & Kaye, B. K. (2004). Wag the blog: How reliance on traditional media and the Internet influence credibility perceptions of weblogs among blog users. *Journalism & Mass Communications Quarterly, 81*(3), 622–642.

Kauffman, D. T. (1969). *The dictionary of religious terms*. Westwood, NJ: Revell.

Kaye, B. K. (2006). Blog use motivations: an exploratory study. In P. Tremayne (Ed.), *Blogging, citizenship, and the future of media* (pp. 127–148). New York: Routledge.

Kim, M-C (2005). Online Buddhist community: An alternative religious organization in the information age. In M. T. Hojsgaard & M. Warburg (Eds.), *Religion and cyberspace* (pp. 138–148). London: Routledge.

Kluver, R., & Cheong, P. H. (2007). Technological modernization, the Internet and religion in Singapore. *Journal of Computer-Mediated Communication, 12*(3), article 18. Retrieved May 5, 2007, from http://jcmc.indiana.edu/vol12/issue3/kluver.html

Laney, M. J. (2005). Christian web usage: Motives and desires. In M. T. Hojsgaard & M. Warburg (Eds.), *Religion and cyberspace* (pp. 166–179). London: Routledge.

Lasswell, H. (1948). The structure and function of communication in society. In L. Bryson (Ed.), *The communication of ideas: A series of addresses* (pp. 37–51). New York: Institute for Religious and Social Studies.

Lenhart, A., & Fox, S. (2006). Bloggers: A portrait of the Internet's new storytellers. *Pew Internet and American Life Project*. Retrieved December 21, 2006, from http://www.pewinternet.org/PPF/r/186/report_display.asp

Lenhart, A., Horrigan, J., & Fallows, D. (2004). Content creation online. *Pew Internet and American Life Project*. Retrieved February 29, 2004, from http://www.pewinternet.org/PPF/r/113/report_display.asp

Li, D. (2005). Why do you blog: A uses-and-gratifications inquiry into bloggers' motivations. Unpublished master's thesis. Marquette University.

Lyon, D. (2000). *Jesus in Disneyland: Religion in postmodern times*. Cambridge: Polity Press.

Matheson, D. (2003). Weblogs and the epistemology of the news: Some trends in online journalism. *New Media & Society, 6*(4), 443–468.

McGrath, A. (2004). *The twilight of atheism: The rise and fall of disbelief in the modern world*. London: Rider.

Mcfedries, P. (2003). Technically speaking: Blah, blah, blog. *IEEE Spectrum, 40*(12), 60.

Murley, B. (2005). *The mediahood of all receivers: Blogging as an extension of the Reformation concept of the priesthood of the believer*. Paper presented at The Second Annual Faith and Communication Conference. Campbell University, Buies Creek, N.C.

Nardi, B. A., Schiano, D. J., Gumbrecht, M., & Swartz, L. (2004). Why we blog. *Communications of the ACM, 47*(12), 41–46.

Norris, P., & Inglehart, R. (2004). *Sacred and secular: Religion and politics worldwide*. Cambridge: Cambridge University Press.

Oi, I. (2006). Blogging the legal commons. In A. Bruns & J. Jacobs (Eds.), *Uses of blogs* (pp. 81–90). New York: Peter Lang Publishing.

Palmgreen, P. (1984). Uses and gratifications: A theoretical perspective, *Communication Yearbook, 8,* 20–55.

Papacharissi, Z. (2006). Audiences as media producers: Content analysis of 260 blogs. In P. Tremayne (Ed.), *Blogging, citizenship, and the future of media* (pp. 21–38). New York: Routledge.

Park, H., & Thelwall, M. (2003). Hyperlink analyses of the World Wide Web: A review. *Journal of Computer-Mediated Communication, 8*(4). Retrieved May 5 2004, from http://jcmc.indiana.edu/vol8/issue4/park.html

Prebish, C. S. (2004). The cybersangha: Buddhism on the Internet. In L. L. Dawson & D. E. Cowan. (Eds.), *Religion online: Finding faith on the Internet* (pp. 135–147). New York: Routledge.

Quiggin, J. (2006). Economic blogs and blog economics. In A. Bruns & J. Jacobs (Eds.), *Uses of blogs* (pp. 69–80). New York: Peter Lang Publishing.

Rainie, L., & Horrigan, J. (2005). Internet evolution. *Pew Internet and American Life Project*. Retrieved October 5, 2005, from http://www.pewinternet.org/PPF/r/148/report_display.asp

Ruggiero, T. (2000). Uses and gratifications theory in the 21st century. *Mass Communication & Society, 3*(1), 3–37.

Serfaty, V. (2004). *The mirror and the veil: an overview of American online diaries and blogs*. Amsterdam: Rodopi.

Soukup, P. (2002). Media and religion. *Communication Research Trends, 21*(2), 1–44.

Stone, P. J. (1966). *The general inquirer: A computer approach to content analysis*. Cambridge, MA: MIT Press.

Tepper, M. (2003, Sept.). The rise of social software. *NetWorker Magazine,* pp. 19–23.

Trammell, K. D., Tarkowski, A., Hofmokl, J., & Sapp, A. M. (2006). Rzeczpospolita blogów [Republic of Blog]: Examining Polish bloggers through content analysis. *Journal of Computer-Mediated Communication, 11*(3), Retrieved November 2, 2006, from http://jcmc.indiana.edu/vol11/issue3/trammell.html

Tremayne, P. (2006). (Ed.). *Blogging, citizenship, and the future of media*. New York: Routledge.

Walker, J. (2005). Weblog. In D. Herman, M. Jahn, & M. Ryan (Eds.), *Routledge encyclopedia of narrative theory* (p. 45). London and New York: Routledge.

Winer, D. (2003). What makes a weblog a weblog? Retrieved September 6, 2004, from http://blogs.law.harvard.edu/whatMakesAWeblogAWeblog

Feminist Sexualities, Race and the Internet: An Investigation of suicidegirls.com

Shoshana Magnet

This article analyzes representations of feminism and sexuality on Suicide Girls (www.suicidegirls.com), a commercial site which features the online journals, profiles and nude photographs of young, heavily tattooed, punk women. It highlights the ways in which this site attempts to subvert the male gaze by changing contemporary photographic practices. It also interrogates the way in which the feminist potential of this site remains constrained by its inclusion of only a limited number of women of colour and only as a marketing 'strategy' of diversity. It argues that rather than a critical race feminist commitment to inclusivity and structural change, this strategy of 'diversity' is reflective of the internet tenet which holds that 'content diversity is good business'. Thus, it concludes that rather than a feminist site which operates in the hope of broadening understandings of female sexuality, this site prioritizes profit to the detriment of feminist content.

Keywords: commodification, cyberfeminism, cybersex, race

Introduction

I recently overheard a heterosexual couple discussing their membership of the website Suicide Girls (www.suicidegirls.com). Remarking that they loved the 'feminist community' on the site, they commented that the 'queered sexuality' represented on Suicide Girls differentiated it from other commercial erotica they had seen. 'It's not exploitative,' they stressed to a third person standing with them, 'the women take the photos themselves.' The third person in the group described herself as a 'pro-sex' feminist and part of a feminist erotica collective. She vociferously interrogated their membership in the site.

This conversation parallels the discussions of cybersex found in academic texts concerning the liberatory potential of cybersex in the face of the increasing commodification of the internet. A simplistic rendering of the debate posits two versions of cyberfeminism: cybertopian feminism which understands cybersex as liberating, and cyberdystopian feminism which sees cybersex as exploitative. Although recent treatises on cybersex have attempted to break down this false binary, these two opposed positions remain—sometimes mirroring Second Wave feminist debates which occurred with respect to pornography (Segal, 2004).

This article considers the impact of cybersex on sexual expression and interrogates the contradictory ways in which women's bodies are constituted online. The analysis of the website Suicide

Girls demonstrates that 'utopian' and 'dystopian' cyberfeminist perspectives both diverge and parallel each other more closely than they acknowledge. They share common oversights: both tend to fail to consider the impact of racism and commodification on the liberatory potential of cybersex.[1] The analysis interrogates the way in which both utopian and dystopian cyberfeminist understandings of sexuality assume whiteness as normative and fail to emphasize the importance of race to feminist understandings of online sexuality. To do so, it considers the competing and conflicting discourses of cyberfeminism and the enriching scholarship of critical race feminist internet theorists. It also takes into account the intersection of feminist and political economic theories in order to interrogate practices of commodification on Suicide Girls.

Suicide Girls is a commercial site which features the online journals, profiles and nude photographs of young, heavily tattooed, punk women. This study examines cybersex from a feminist perspective that is committed to combating intersecting systems of oppression, including those resulting from racism, classism, ageism, sexism, ableism and heterosexism or homophobia. In keeping with the history of feminism and the critiques of exclusionary practices of Second Wave feminism and its initial failure to rely on and to make central women of colour, disabled, poor and queer women,[2] it defines feminist representations of sexuality as those which include the depictions of women in sexually empowered positions that do not endorse sexist, ageist, racist, ableist or heterosexist stereotypes. It concludes that some of Suicide Girls' unique online practices do open up feminist possibilities. However, rather than a location of liberation or of feminist sexuality, cybersex on Suicide Girls significantly reproduces the commodification of women and reinforces racialized hierarchies of sexual subordination.

Method

To interrogate the possibilities for the feminist representations of female sexuality offered by Suicide Girls, the text and photos of the models and members are analyzed. All the documents cited are public documents and may be accessed by any user. The names cited are adopted by the models themselves and thus do not reveal the user's identity. With respect to text, the methodology consists of a discourse analysis of the profiles of the models, as well as of the threads attached to the 'feminist' and 'equal opportunity'[3] groups within the 'politics and activism' category of the Suicide Girls website. Combining Lindlof and Taylor's (2002) method of discourse analysis with the methodology of Glenn Stillar (1998), this analysis attempts to determine whether and how feminist representations of sexuality occur on the Suicide Girls website and the understanding of these representations by the members of this online community. The purpose of interaction (how it might encourage or discourage online participation), the temporal context of the study, the form of computer-mediated communication (CMC) being studied and how it might alternately facilitate or constrain interaction, the characteristics of the participants and the external research context, are considered to determine how feminist constructions of sexuality are produced and constrained (Lindlof and Taylor, 2002). With respect to visual texts such as photos, Sturken and Cartwright's (2001) methodology for studying visual culture is combined with that of Nicholas Mirzoeff (1999), each of whom emphasizes the constructed nature of both visual texts and 'practices of looking' (Sturken and Cartwright, 2001).

Suicide Girls: Its Feminist Leanings and Liberatory Intentions

Site Description

Established in 2001 by site owners 'Spooky' (male) and 'Missy' (female), Suicide Girls is a for-profit site allegedly developed as a response to the plethora of oppressive images of plastic, blonde, fake-breasted women which have dominated pornography on the internet (Tomlin, 2002). Described as a site featuring erotica which 'challeng[es] social mores of beauty' (Trumbull, 2004), Suicide Girls is regularly lauded for being a site which focuses on both beauty and brains. As one female member describes the site: 'These are girls that I can respect' (http://suicidegirls.com/tour/). Models must apply for acceptance. If accepted, their obligations include posting photos of themselves and participating in an online journal. They are paid for their participation at a rate of $100 to $200 per photo shoot.[4] Both members and models may participate on bulletin boards and online journals, but models must participate regularly. Each model is introduced by a descriptive paragraph written by site owner Missy. Although some members join only to look at the photographs of the models, other members indicate that their reasons for joining consist of the opportunity to be part of a sex-positive community which includes women who are 'beautifully imperfect' (*Nerve* magazine, http://suicidegirls.com/tour/) and who, far from being the usual silent plastic pin-ups, are feisty and intelligent, as witnessed from their written participation on the site.

Feminism and Suicide Girls

While the owners of this site do not use the word 'feminist' to express its ideological affiliation, Suicide Girls is frequently described as a feminist website by the media which have reviewed the site, including publications ranging from *Bitch Magazine* to *Wired*. However, it is not the reviews of this site, or the popularity that it has attained among Third Wave feminist communities as a result of its inclusion of feminist content, that is responsible for the site's feminist identification. Rather, it is that the site explicitly attempts to interrupt the male gaze through the subversion of the standard photographic practices utilized by more traditional forms of pornography. Film theorists such as Laura Mulvey (1975) and E. Ann Kaplan (1983) have theorized the ramifications of this 'penetrative' and 'structuring' gaze. In her landmark article 'Visual Pleasure and Narrative Cinema', Mulvey defined 'to-be-looked-at-ness' as the way in which women are 'simultaneously looked at and displayed' in film and then coded for 'visual and erotic impact' (1975: 7). Mulvey concluded that the male gaze produced by cinematic texts tied 'woman down to her place as bearer of meaning, not maker of meaning' (1975: 6). Kaplan furthers Mulvey's assertion in arguing that while the gaze might 'not necessarily be male (literally), to own and activate the gaze, given our language and the structure of the unconscious, is to be in the 'masculine' position' (Kaplan, 1983: 30). Clearly, pornography too has been subject to this structuring gaze and its ramifications.

However, on Suicide Girls, the production of this gaze is by the objects themselves—helping the site to meet Kaplan's call for women to 'own the gaze' (1983: 24). In a deliberate effort to interrupt the structuring male gaze, the women themselves decide how revealing their photos will be, how each shot will be staged and which photos will be made available online. One of the liberatory potentials of textual cybersex is that 'the penetrative gaze of pornography is absent' (Gillis, 2004: 96). In taking the cameras back into their own hands through staging their photo shoots, the models on Suicide Girls are redirecting the objectifying male gaze described by Mulvey—transforming 'to-be-looked-at-ness'. Women become the subjects and objects of their own desiring gaze rather than its passive recipients. Furthermore, instead of being alienated from the means of production, on this website the models are the cultural producers of their images. This is a very important difference with respect to the agency that they are able to access and is a change meant to facilitate their empowerment. These models are not '[a]ssigned the place of object (lack)' only 'passively appearing rather than acting' (Kaplan, 1983: 26).

Neither do the Suicide Girls models construct their sexual pleasure 'only around [their] own objectification' (Kaplan, 1983: 26). Many models assert that they enjoy participating on Suicide Girls because it gives them a chance to subvert these very stereotypes about women in the sex industry: 'posing nude is a way to try to change inhibitions and ideas about nudity and female sexuality' (http://suicidegirls.com/groups/Feminists/topics/128470/). The women describe themselves as the *réalisateurs* of their fantasies. The photographic methods are identified as a key reason that many women are keen to become paid members of the site (despite the limited monetary compensation). It is a fantasy that is difficult to realize: for every potential model who is chosen, 350 apply (Tomlin, 2002).

The models on Suicide Girls sport extensive piercing, tattoos and dyed hair—all traditional signifiers of the female grotesque (Braunberger, 2000). The eroticization of these markers of 'deviant' femininity is meant to disrupt the conventional norms of female beauty. An overwhelming number of the photographs are of women's faces. Focusing on the face rather than the body is a highly unusual practice in pornography, and may be used as a deliberate photographic technique which visually indicates an emphasis on the mind (Valdivia, forthcoming).

This photographic practice is in accordance with an emphasis on the minds of the women on Suicide Girls, who must answer a number of questions in order to be selected for the site. As a result of the models' required contributions to the Suicide Girls community through journaling and posting on the bulletin boards, the models are emotionally as well as physically fleshed out (Tomlin, 2002). The models' journals deal with life events, ranging from their issues with sexual harassment at work (http://suicidegirls.com/groups/Feminists/topics/131311/), to their worries about schoolwork (http://suicidegirls.com/girls/Aya/), to their views on women in the sex industry (http://suicidegirls.com/groups/Feminists/topics/128470/). Rarely is the content of their journals erotic. Suicide Girls has a group devoted to feminist discussion, with threads ranging from discussions of *Roe* v. *Wade* (410 U.S.113 (1973)) to sexual harassment. This is a highly unusual, perhaps unprecedented, practice for a site featuring nude photos of women. It is important to note that Suicide Girls itself claims a subscription rate that is approximately 50 percent women. Although it is impossible to verify this claim, a strong female presence is evident on the site.

Before proceeding with a further analysis of Suicide Girls, both cyberfeminist and critical race literature about the possibilities for women's sexual expression in new media help to shed further light on the ambivalent position that Suicide Girls occupies with respect to scopophilic pleasure and the history of pornography.

Cyberfeminist Approaches to Cybersex: Utopian and Dystopian Perspectives

Utopian Cyberfeminist Perspectives on Cybersex

Sadie Plant is representative of those cyberfeminists who claim the potential for a utopian view of cybersex and its possibilities for women. Plant argues that cybersex is the 'epitome of disembodied pleasure, contact free sex without secretions in a zone of total autonomy' (2000: 460). Plant grounds her claim in the assertion that cybersex is free from the unpleasant 'side effects' of sex which disproportionately affect women, including 'conception, diseases and the sad obligations of emotional need' (2000: 460).

It is partially the potential of cybersex to create a post-corporeal world through the use of virtual bodies which has led cybertopian feminists to be laudatory about its emancipatory potential for women. These cyberfeminists challenge feminism's 'pathological need to worship goddess figures at the altar of the natural body', which 'if it ever existed to begin with, was no more' (Senft, 2002: 539). Certainly, the models on Suicide Girls, who resemble cyborgs[5] more than goddesses, attempt to interrupt the beauty mythology surrounding the 'natural' body. As noted, in order to become a model, it is necessary to have a number of 'body modifications' such as large tattoos, multiple piercings or dyed hair. A combination of all three is preferred (as made explicit by the women who choose the models: http://www.wweek.com/story.php?story=3716&page=3).

Utopian cyberfeminists argue that online, women are freed from the rigid norms that traditionally restrict female sexuality. Rather than being reduced to passive female objects that must conform to male desires, cybertopian feminists argue that in 'feminist-inspired virtual worlds' a female body is 'staged as active, intelligent and polymorphously sexual' (Steffensen, 2002: 217). Many of these characteristics emerge from the profiles of the models on Suicide Girls. In the 'Favourite Books and Films' section, both models and members list a wide range of explicitly feminist literature. Another section asks the models what 'makes them happy', what 'makes them sad' and what 'gets them hot'. Suicide Girl Ada refuses to respond to the 'makes me sad' section and instead says: 'I'm going to pretend that this says "pisses me off ".' Other models have changed this section to read 'things

I hate'. In the 'makes me hot' section, Ada lists 'sex with the lights on' and 'girls who look like me'. These responses may be read as discursive attempts to subvert notions of passive female sexuality, as may the fact that vibrators and other sex toys are commonly listed in the '5 Items I Can't Live Without' section of both the female members' and models' profiles.[6] Thus these profiles represent a form of CMC that facilitates the interruption of traditional notions of female sexuality, in large part because they are used as spaces where the models can subvert these narratives discursively by explicitly giving voice to both their anger and their lust. In a sense, the purpose of this interaction may be understood as an attempt on the part of the models to tell their viewers how they wish to be read (as subjects, not objects) and to encourage a certain kind of (feminist) participation on the site.

Cybertopian Perspectives: The Fluidity of Gender and Sexual Orientation Online

The number of models who cite their attraction to other women confirms the assertion of theorist Nina Wakeford (2000) who argues that women are able to experiment with sexual orientation more readily in cyberspace. Asserting that many women who self-identify as 'straight' in real life may play with masculinity, transsexuality, bisexuality or lesbianism online, Wakeford argues that this may help to facilitate the 'coming out' process for queer women in a safe space where they are not subject to the same heteronormative pressures which dominate real-life experience. Certainly, there are a number of models who refer to relationships with their boyfriends while simultaneously making reference to their desire for women. For example, suicide girl Ada writes in her journal about her attraction to the other suicide girls, but the 'current crush' section in her profile lists 'my boyfriend'. After new pictures of a model are posted, women[7] (both models and members) usually cluster on the boards to discuss and to praise the model. It is clear from their comments that the female models and members find the photos of women erotic. For example, Bracket (a model) commented on suicide girl Amina's photoset with the following:

> *OMGOMGOMG!!!!!!!!!!!!!!!!! [Oh My God] *fallin off the chair* *runing [sic] to exchange panties* omg. ☺ wow. ☺☺☺☺☺)'*
> *(http://suicidegirls.com/girls/multiple/photos/Two+of+Hearts/)*

This frank expression of queer sexual desire from a model or a female member is a common response to a new photoset.

Cybertopian feminists claim that the internet's anonymity enhances female agency. They argue that women are able to 'transcend their sexual and gendered particularities' by being able to say yes or no to sexual propositions 'without fear of reprisal' (Blair, 1998: 208; Foster, 2000: 445). A recurring theme on Suicide Girls is models who note that they feel limited by their real-life personas, but freed by their ability to 'shed' them on Suicide Girls. For example, in reply to why she decided to do Suicide Girls, model Ada says: 'Because I'm a shy exhibitionist.'

Orgasm in Cyberspace

Asserting that the emphasis on the orgasm achieved through vaginal penetration has undergone revision in cyberspace, utopian cyberfeminists argue that women who participate in cybersex describe a new possibility for sexual fulfillment: the cyberorgasm. In sharp contrast to the heteronormative understanding of an orgasm as requiring the insertion of the penis, the cyberorgasm does not involve physical stimulation. Rather, it is an orgasm that occurs in the mind as one types up one's sexual fulfillment. Discussions of cybersex have helped to open up discussions around female masturbation. It is accepted that this is the usual way that women achieve pleasure through cybersex. Certainly, female masturbation and sexual pleasure for women are common topics on Suicide Girls. There are several groups devoted to women's sexuality—including a 'gay girls' and a 'girls only' group. The discussions on these groups—which range from femme–femme relationships to finding a good lubricant—helpfully debunk many of the myths that surround female sexuality, where women are frequently understood to experience neither sexual desire nor sexual fulfillment. Interestingly, cybersex is often touted by utopian cyberfeminists as a catalyst which will motivate women to get online (Cumberland, 2002). The hope of cybertopian feminists is that these women will become part of a 'force that is changing computer environments from a masculinist "virile reality"' (Cumberland, 2002: 180) to a space that enables women to express and participate freely in the making of their own sexualities. Suicide Girls represents one online space that is making attempts to meet the above requirements.

More recent scholarship on online sexuality attempts both to problematize and temper this utopian perspective while retaining optimism about the feminist potential of cybersex. Kath Albury (2003) is one of Australia's premier researchers of pornography on the internet. In her work, Albury attempts to clarify the distinction between morality and ethics with respect to online pornography. While Albury concedes that a particular text might be 'immoral' for a number of reasons,[8] she asserts that morality is not a helpful framework from which to examine online pornography. It imposes a simplistic binary of 'good' versus 'bad' representations of sexuality. Instead, Albury proposes using ethics as a framework from which to understand online porn, as it allows for the celebration of the broadening of sexual experiences and sexuality online while understanding that not all spaces allow for this diversification of possibilities. Thus Albury facilitates an understanding of the internet as a space which 'is not necessarily utopian' (2003: 208) if it meets basic ethical standards, paving the way for critiques of online pornography if they are contextualized within a community-based or personal process, but refusing an external 'all bets are off' relativism in which all cybersex is uniformly condemned. This theoretical approach best describes most members' understanding of Suicide Girls. Suicide girl Aya sums it up when she says:

> *SG [Suicide Girls] isn't perfect—some critics ALMOST get it right. Let's be realistic—if SG was a forum for women to empower themselves by getting naked, SG wouldn't charge nor would they pay us.*
> *(http://suicidegirls.com/groups/Feminists/topics/20807/)*

Yet she also asserts that this site does have feminist potential, arguing that the proof is in the fact that models are participating for sheer enjoyment, not only for the money:

> *You see it where Bettina says she'd take naked pictures of herself just because. Where girls are doing it to proclaim or reclaim their sexuality.*
> *(http://suicidegirls.com/groups/Feminists/topics/20807/)*

In noting the complexity of Suicide Girls as a feminist site, this posting helps to set the stage for a closer look at more critical cyberfeminist considerations of the possibilities of cybersex, to which this article turns below.

Dystopian Cyberfeminist Perspectives

Perhaps the greatest difference between utopian and dystopian cyberfeminist perspectives is their attitude towards the internet's potential for disembodiment and the implications of this possibility for women's sexual emancipation. Cyberfeminists who are more suspicious of the liberatory possibilities of women's sexuality online, from Melanie Stewart Millar (1998) to Susan Hawthorne and Renate Klein (1999) and Donna Hughes (1999), argue that it is important to remember the lived realities in which women's bodies are grounded. In suggesting that the internet is postcorporeal, cyberfeminists such as Sadie Plant are giving voice to the idea that cyborgs have no history: 'Whatever avatar you select for your scene, you cannot resist becoming cyborg as well . . . it will be post human whatever it is. Suddenly it always was, you always were' (Plant, 2000: 466). In making this suggestion, Plant is conveniently able to ignore the history of sexual inequality in which women's bodies have been mired offline, asserting that these inequalities will melt in the heat of sexy new technology.

Although the Suicide Girls website is full of self-praise for its own 'liberating' sexual practices and considerable time and energy is invested in distinguishing the site from traditional pornography and in arguing for its feminist potential,[9] the models themselves acknowledge the complex ways in which they are positioned by the discourse of the site. The following was posted by Aya:

> *I view pornographic modeling [as an] . . . active, discourse-challenging activity. There has always been a notion of female pornography models as a vile entity that threatens [sic] society because of the fact that these women are actively sexual (or at least perceived that way). The fact of the matter is, I'm doing this mostly because I enjoy it. It's pleasurable to ME and I'm DOING—whatever else is secondary. That's progress.*
>
> *I can't speak for all the of [sic] the girls on the site.*
>
> *I'm not going to say that this site promotes a 'realistic' potrayal [sic] of women all the time, nor would I say that this site alone promotes positive images of women—there are other places that do similar things.*
>
> *But what it FACILITATES is what's important. The girls are known as sexual BEINGS and not just sex objects. The notion of community and the fact that I can at least pretend I have half a brain by posting this adds a certain depth—people may jerk off to my pictures but they can also plainly see that I have opinions, thoughts and beliefs and that I'm not just a dumb girl.*
> *(http://suicidegirls. com/groups/Feminists/topics/32259/)*

This post demonstrates the overlap between both utopian and dystopian cyberfeminist perspectives in that it acknowledges both the feminist potential of Suicide Girls and the way in which it challenges the traditional, objectifying discourses around pornography and female models, while admitting that the site sometimes may endorse these oppressive narratives. However, what Aya fails to address is the limitations of an agency that comes from being a fleshed-out 'sexual being' as opposed to a 'dumb' (read: voiceless) model.[10] She also ignores the bodily risks to which the models may be subject. Although the models may exert their individual agency in participating on the site and many of the models derive both payment and enjoyment from participating, some are clearly only modelling out of necessity. One of the profile questions is: 'Why did you do SG?' Many models reply, 'For the money.' Furthermore, Aya ignores the fact that participating on Suicide Girls may place one at risk.

Critical cyberfeminist theorists have suggested that more utopian cyberfeminists have conveniently ignored the implications of a medium that has led to the growth of large-scale trafficking in women (Doring, 2000). They argue that rather than protecting women, the anonymity of the internet has made it an effective tool to engage in the wholesale buying and selling of women and children. Critics dubious of the liberating potential of the internet for women's sexuality argue that the 'standards and values of the net are being set by the sex industry and its supporters' rather than by the average female user (Hughes, 1999: 158). Certainly, the internet has allowed for the unequalled objectification of women.[11] Websites exist where men can rate sex trade workers on everything from the firmness of their flesh to the scars on their bodies (Hughes, 1999). Many sites on the internet demonstrate 'an awareness of racism, colonization, global economic inequalities and sexism' (Hughes, 1999: 159) that is restricted to understanding how to profit from these systemic inequalities. For example, the site World Sex Guide (www.worldsexguide.com) is organized by country and men participate from all over the world in ranking how 'docile' or 'authentic' the sex workers of a particular country are.

In addition, the freedom associated with the anonymity of the internet has contributed to a deluge of online sexual harassment (Branwyn, 2000). Neither does online harassment end with unwanted come-ons; virtual violence and virtual rape also occur (Dibbell, 1998; Kendall, 1996) and real-life stalking and violence may follow. The 'bodily dimension' of many of these sites 'opens up a field for a new kind of harassment based on actions and objects rather than just on spoken discourse' (Woodland, 2000: 426).[12] New users are particularly vulnerable to online violence. Lori Kendall describes the experience of a woman who was trying out a multi-user domain (MUD; online video gaming) for the first time and who had lagged out temporarily. Another user took advantage of her being temporarily unable to issue commands and pretended that her virtual body was committing sexual acts. By the time 'she unlagged, the person in question was gone and she was left with the equivalent of an obscene phone call on her screen' (Kendall, 1996: 212). Online harassment and violence may not only cause women to avoid certain chat areas or to cancel their memberships to particular lists, but may deter them from going online in the first place (Branwyn, 2000: 401). These experiences can be particularly harrowing for women who are victims of violence, bringing back the memory of real-life trauma (Doring, 2000).

Online violence may be more subtle than virtual rape. For example, cybersex may occur with only the 'partial' or 'half-hearted' consent of women who are 'unable to safeguard the equality of their interests in hierarchical relationships' (Doring, 2000: 870). This type of behaviour is encouraged by gendered suggestions around cybersex, such as the ones found in a internet sex guide written by and for women:

> *If you need some ideas about what men want women to say or do to them, read a few* Penthouse *letters, or watch a X-rated video. Most women (if they're honest about it) have faked an orgasm or two. Just fake it on the screen. (Skrilloff and Gould, cited in Doring, 2000: 870)*

This is a severely limited understanding of cybersex which exemplifies Albury's assertion that all online spaces are not utopian places engaged in the process of transforming traditional understandings of sexuality.[13]

Certainly, anonymity is insufficient protection on a site such as Suicide Girls where community formation and contact between the members and the models is encouraged by the site owners. Suicide girls are regularly recognized on the street by members of the site (http://www.wweek.com/story.php?story=3716&page=4). The risks to the models were highlighted in an article about the site:

> *Bailey Maxwell, a 22-year-old SG model, admits that the blurred lines between girls and members admit potential trouble. 'I worry and I know Missy and Spooky do too, that the girls are too trusting of the members because of the strong sense of community the site has', she writes by email. 'I try not to post too much personal information on my journal.' (Phillips, 2002)*

However, such self-protective measures are limited in their impact. The web makes it increasingly easy to track down a user's personal information such as home telephone number and address and the threatening nature of some posts makes the vulnerability of the models clear. In response to model Aya's first photoset, one member wrote:

> *I am an instant fan. I want more photos and if I have to fly to Canada, hunt you down and just capture some candid stakeout pictures of you on my Sony digital, then goddamn it I will. (http://suicidegirls.com/girls/Aya/photos/Cotton+Candy/page2/)*

Other members, who have been absent from their journals, tend to resurface with apologies and explanations that they had temporarily 'disappeared' because they were being harassed by someone on the site. A particularly harrowing story concerned one of the models who had been having difficulty getting paid by a photographer, only to realize that the same photographer had been accused of murdering an actress in the making of a snuff film (http://www.livejournal.com/community/sgirls/2004/03/24/; NBC, 2004). The dangers of online female sexual expression need to be acknowledged further by utopian cyberfeminism as well as addressed by Suicide Girls.

Racism in Cyberspace

Before conducting further analysis of Suicide Girls, it is necessary to investigate the way in which neither of these cyberfeminist perspectives fully consider the impact of ethnicity and racism on cybersexuality. Interrogating the intersectional oppression of women of colour is integral to an understanding of the impact of cybersex on women's sexual autonomy. Critical race feminist scholars from Ella Shohat (1998) and Sherene Razack (2002) to Patricia Hill Collins (2000) have noted the existence of a matrix of domination in which systems of oppression work mutually to reinforce one another. The impossibility of fighting only one location of oppression is described by Sherene Razack: 'We fail to realize that we cannot undo our own marginality without simultaneously undoing all systems of oppression' (2002: 5). Any analysis of cybersexuality must attend both to the impact of ethnicity and racism and must not essentialize women either as completely sexually empowered or victimized.

The utopian theorizing of cyberfeminists who suggest that cyberspace is postcorporeal has been criticized by critical race internet theorists such as Lisa Nakamura. Unconvinced that the internet remains free from the discrimination that contaminates 'real life', Nakamura suggests that gender, class and race hierarchies have been carried onto the internet. Using the metaphor of the after-image to describe 'a particular kind of historically and culturally grounded seeing or mis-seeing' of the postcorporeal body (2002: 322), Nakamura asserts that although the body may be left behind upon entering cyberspace, the after-image of the body's ethnicity, gender and class lingers. This implies that 'fluid selves are no less subject to cultural hegemonies, rules of conduct and regulating cultural norms than are solid ones' (2002: 325). Nakamura gives a number of examples of the way in which online personas continue to be mired in bodily vulnerabilities. Her example of Kozmo is particularly striking. Kozmo, an online service that delivers food and an array of convenience items, refused to deliver its products to certain zip codes. It did so, not because they were not affluent, but because these zip codes represented affluent African American addresses. Thus, Nakamura argues that

> *in order to think rigorously about posthuman and virtuality, one must ground one's critique in the lived realities of humans . . . online identities can never be truly fluid if you live in the wrong zip code. (2002: 328)*

Beth Kolko reinforces Nakamura's argument in her article 'Erasing @race'. Here she notes that the default ethnicity is set to white on most MUDs. Asking how 'the interface of virtual worlds came to efface @race and create a default whiteness for the worlds' (2000: 216), Kolko notes that the absence of ethnicity is particularly surprising in a medium that dramatizes other aspects of identity, including gender. Kolko concludes that the fact that MUDs are missing the '@race verb . . . speaks volumes about the assumptions technology designers carry with them' (2000: 225). Thus, Kolko argues that the internet is far from a postcorporeal utopia, but is instead a space which has retained a 'cultural map of assumed whiteness' (2000: 225).

Internet sex is described regularly as the last place to have 'safe sex', a space where one never has to interrupt the flow with awkward questions. However, if the default user is assumed to be white, women of colour are often 'passing' online unless they explicitly articulate their racial identity. Therefore, cybersex where one user writes 'I run my tongue along your creamy back' is not necessarily experienced as 'liberating' by an African Canadian woman who must choose either to disrupt the spontaneous flow of sexual discourse by 'coming out' as a woman of colour, or engage in cybersex as a white woman.[14] Moreover, the possibility that a user of colour will be subject to racism upon disclosure means that, rather than a postcorporeal sexual haven, the internet may become a postcorporeal hell, replete not only with all the usual racism but with additional discrimination particular to the medium. For the global consumer who is depicted as white, the internet serves as a glass window from which 'users can consume the sights of travel as if they were tourists' (Nakamura, 2000: 19). The queer user may enjoy the 'anonymity of the closet' while remaining 'connected to the sociality of public spaces' (Woodland, 2000: 417).[15] Yet, for users of colour, there are few such spaces. The window in their closet serves primarily as a one-way mirror: although persons of colour may be able to see out, they are stuffed relentlessly into a space that makes them invisible—that is, until 'commerce and discourse come into play', at which point the racialized body is mercilessly outed (Nakamura, 2002: 329).

The Suicide Girls website is a prime example of an online space in which whiteness is made normative. Racialized models are largely excluded. When a woman of colour is profiled, her body inevitably becomes a signifier of the exotic. It is this 'othered' position that is put forward as the signifier of her desirability. For example, Aya is one of the few Asian Canadian models on the site. In

Missy's introductory paragraph for Aya, she is described as a 'delicate beauty' and is constantly praised for being 'graceful' and 'like a pretty little doll'—despite the fact that this explicitly contradicts her online persona, which is of a vocal, highly-politicized feminist. Comments which encapsulate orientalism follow Aya's photosets:

> *oh wow. i have a thing for Asian ladies. you are one of the cutest ever. (http://suicidegirls.com/girls/Aya/photos/Cotton+Candy/page3/)*

Neko, a model of colour who appears African American, is also described using radicalized metaphors. In her introductory paragraph for Neko, Missy asserts that she has the 'fullest lips and most soulful eyes here' (http://suicidegirls.com/girls/Neko/photos/Bathroom+Time/). It is Neko's racialized features which signify her position as a sex object. None of the white models are introduced in this way. White suicide girls tend to be described as 'strong' (http://suicidegirls.com/girls/Absinthe/photos/Listen+Up/), 'feisty' or ready to 'kick you in the teeth' (http://suicidegirls.com/girls/Sphinx/photos/Beat/). Thus, on Suicide Girls, 'images of displayed ethnic and racial difference' are used 'in order to bracket them off as exotic and irremediably "other"' (Nakamura, 2002: 89), failing to meet Ella Shohat's challenge that feminism 'seek a non-essentialist approach to aesthetics' (1998: 29).

Racist comments made by members are largely ignored. For example, comments such as 'so exotic' or 'just like a china doll' go unremarked, despite the fact that bulletin boards are usually a form of CMC that facilitates interaction and discussion. A further example of uncommented-on racism occurred in a response to the posting of photos for a South Asian woman—'India'. India's photos are interesting in that they generated not only the usual orientalism ('I LOVE Indian girls!!!11 LOVE LOVE LOVE!!!!' http://suicidegirls.com/girls/India/photos/Smily+Face/page2/, and 'Very exotic looking' http://suicidegirls.com/girls/India/photos/Smily+Face/page3/), but also explicitly racist comments. Generally, the response to new photos posted on Suicide Girls follows a particular format. After a model posts photos, there are a number of flattering posts commenting on the model's beauty and/or sexiness. The only negative comments that were found in response to a photoset occurred after India had posted her first shoot. One member posted:

> *How did she become a SGirl anyway, doesnt [sic] make the cut in my opinion. ☺ (http://suicidegirls.com/girls/India/photos/Smily+Face/page13/)*

What serves to distinguish India from the vast majority of other models on the site is her ethnicity. In raising the issue of who 'belongs' on the site, this post may be understood to be referencing India's racialization—which is construed as 'inappropriate' for the site. Although this comment was highly out of the ordinary and unsuitable according to the site conventions, it went unaddressed.

This desire to ignore references to ethnicity and racism has been documented by Kolko et al. in *Race in Cyberspace* (2000). Here, they note that usually, ethnicity is not only erased online, but any attempt to bring ethnicity out of the virtual closet is met with resistance. As an example, they cite the response to one of their postings on a scholarly listserv that argued that ethnicity was not a biological reality but was instead a social construct. This posting began a flame war where the author of the post was denounced as a 'dangerous relativist who didn't believe in the real world', a person capable of thinking that a 'white couple could produce a baby with brown skin and epicanthic folds' (Kolko et al., 2000: 2). Highlighting ethnicity in this case was grounds for brutal *ad hominem* attack.

Although racism on Suicide Girls's bulletin boards goes uncommented, both the members and the site owners are aware of the lack of diversity on the site. Member response may be noted partially from the comments which follow the posting of photos of women of colour. These tend to include '[i]t's so nice to see some different women of color' (http://suicidegirls.com/girls/India/photos/Smily+Face/page2/) and 'it is nice to see a little more diversity on this site' (http://suicidegirls.com/girls/India/photos/Smily+Face/page13/). Aware of the criticism of the site for the paucity of models of colour, Suicide Girls has posted the following 'affirmative action' notice:

> *Suicide Girls encourages women of color to apply. We aim to be a more diverse site and we need your help. (http://suicidegirls.com/model/)*

In interviews with site owner Spooky, he agrees that where the site 'may be failing' is in the endorsement of white femininity or, as he terms it, 'the idea of beauty that is [in] Western media' which he admits 'permeates how we choose girls' (Tomlin, 2002).

To the extent that the site claims to be 'committed to diversity', this is arguably evidence of the 'commodification of difference' by a corporation in order to sell a product (Shohat, 1998: 3), namely exotic women of colour. Rather than coming from a critical race feminist commitment to inclusivity and structural change, this strategy of 'diversity' is reflective of the internet tenet which holds that 'content diversity is good business'. It is relatively cheap to include more information online and 'diverse' content widens the net cast for consumers (Duggan, 2003: 268). This is an example of what Appadurai terms 'the principle that more difference is better' as opposed to a commitment to 'creating a habitus where diversity is at the heart of the apparatus itself ' (Appadurai, 1996: 26). As Ella Shohat reminds us in her introduction to *Talking Visions,* a commitment to 'diversity' through the limited strategy of affirmative action will never 'offer a fully adequate response to the dumb structures of domination' (1998: 6). Limited diversity (read: racialization) broadens the appeal of the site to a white target market. To date, Suicide Girls seems to fall into this category of commitment to inclusivity. The way in which race and ethnicity are commodified on the site will be dealt with in more detail below.

Racializing the Female Grotesque

As previously noted, traditionally tattoos, piercing and dyed hair have been signifiers of 'bodies in revolt' or of the 'female grotesque' (Braunberger, 2000). Mary Russo argues that the 'grotesque, particularly as a bodily category, emerges as a deviation from the norm' (1995: 11–12). The bodily differences sported by the models mark their bodies as deviantly female:

> *Normalization as it is enforced in what Teresa de Lauretis has referred to as the 'technologies of gender' . . . [is] harsh and effective in its highly calibrated differentiation of female bodies in the service of a homogeneity called gender difference. (Russo, 1995: 11–12)*

Both large tattoos and unusual piercings (such as the nipples or septum) are symbols of hypermasculinity and serve as a reminder of the unstable and performative nature of gender (Butler, 1999; Halberstam, 1998). Additionally, both disrupt traditional notions of demure female beauty and sexuality, causing them to be read as 'unfeminine' (Halberstam, 1998; Sturken and Cartwright, 2001). They transform the female body into a display of 'bodily excess' and resist the ever-present normalizing discourses which attempt to govern femininity:

> *In a culture built on women's silence and bent on maintaining silence as a primary part of the relationship between women's bodies and cultural writing, the rules have been simple. The written body may only speak from a patriarchal script that tries to limit women's voices and bodies to supporting roles and scenery. (Russo, 1995:13)*

Tattoos, piercing and punk hair reframe male/female, center/periphery relations. They subvert the gaze so often trained on women's bodies by both recognizing and inviting viewers to look. Whereas a 'woman's body is a sex object, a tattooed woman's body is a lascivious sex object' (Braunberger, 2000).[16] Moreover, in signifying 'female masculinity' (Halberstam, 1998) tattoos may be coded as a form of butch lesbian identity—thus implying a subversive form of femininity or sexuality.

It is these depictions of 'monstrous beauty' or subversive forms of femininity that partially have given rise to the excitement about the feminist potential of Suicide Girls. What has been neglected is the way in which tattoos are coded differently for white women and for women of colour. In part, the models are intriguing because they are largely depictions of sweet, white femininity which has been desecrated. While a tattoo coupled with white femininity that be an erotic depiction of transgressive female sexuality, a tattoo or an unusual piercing on a racialized body is more likely to be read as 'primitive' or 'tribal' on an African American or African Canadian body and as exotic on an Asian American, Asian Canadian or Latina body. A website such as Suicide Girls is read as a progressive form of representation of female sexuality because of the way in which these body alterations suggest a deviant form of (white) femininity. However, if it is only the white models who are depicted as transgressing the traditional boundaries of femininity, whereas the same marks on women of colour are used to endorse racialized narratives around 'primitive' or 'exotic' sexuality, the site remains highly problematic. Furthermore, as bell hooks (1992) has argued, the commodification of transgression and its signifiers strips them of their radical potential.

Commodification, Ethnicity, Gender and Suicide Girls

The internet is a space ripe for the commodification and depoliticization associated with the consolidation of corporate capital. In *Digital Capitalism: Networking the Global Market System* (1999), Daniel Schiller asserts that the utopian promise of new information technologies tends to be derailed by the capitalistic framework within which they are located. Schiller argues that new information technologies such as the internet are more likely to facilitate commercialization rather than democratization and that this trend has meant that a wide range of online practices are undergoing commodification. Schiller suggests that this is in large part because the internet has facilitated both the spread and consolidation of corporate rule. Here, this article is particularly concerned with the way in which the internet facilitates the commodification of difference. Below, a brief review is conducted of the way in which the commodification of difference has occurred with respect to the gay community[17] and communities of colour, and the implications more generally of this corporate tactic for political disengagement. Finally, this analysis is extended to an interrogation of how capitalism and the commodification of feminism and ethnicity are manifested on Suicide Girls.

Commodifying Queerness

Alexandra Chasin's book *Selling Out: The Gay and Lesbian Movement Goes to Market* (2000) asserts the dangers of the commodification of queerness for corporate profit. She argues that the capitalist market serves to depoliticize queer activist movements and assimilate their members—drawing 'social movements focused on winning rights' into 'market-based tactics and objectives' (2000: xvii). In this way, the market promotes assimilation to a homogeneous national culture, encouraging identity difference only to the extent that it serves as a basis for niche marketing' (Chasin, 2000: xvii). In a related vein, Katherine Sender's book *Business, not Politics* (2004) on the making of the gay market, traces the development of this newly-visible entity and the intensification of corporate efforts to secure this niche. Sender demonstrates that corporations pitched the expansion as a matter of 'business, not politics', identifying politics as the realm of the 'irrational . . . biased towards the interest of one group and . . . utterly incompatible with—even damaging to—the needs of a healthily functioning economy' (2004: 3). Despite disavowals about the political nature of its 'business decisions', this incorporation of a politicized community has significant ramifications. Sender argues that marketing directed towards the gay community is a discourse through which sexuality is both constructed and produced, fundamentally altering the meaning of gay/lesbian/bisexual/transsexual (GLBT) identity. While Sender makes the case that this market discourse is neither 'exploitative nor liberatory' (2004: 9), she acknowledges its potential to reduce an inclusive queer community to a monolithic white, male, affluent gay consumer.

Commodifying Ethnicity

A second illustration of the commodification of difference is documented by Ann duCille's article on the marketing of 'ethnic' Barbies by Mattel. duCille attributes the Mattel 'ethnic innovations' of the 1970s to its increasing 'awareness of the growing black and Hispanic middle classes, who [had] more disposable income than ever before and hence more money to spend on Barbie dolls and their accessories' (1999: 109). duCille is less interested in what motives drove the Mattel corporate strategy than with the question of how it managed to 'produce multicultural meaning and market ethnic diversity' (1999: 111). She concluded that Mattel decided that the 'cost of mass-producing dolls to represent the heterogeneity of the world would be far greater than either corporation or consumer would be willing to pay' and thus produced difference by 'reproducing stereotyped forms and visible signs of racial and ethnic difference'—signifying otherness, not by replicating the individual differences of real bodies but by mass marketing the discursively familiar, by reproducing stereotyped forms and visible signs of racial and ethnic difference' (1999: 111). That is, Mattel signified ethnicity with tinted plastic and a change of costume—simply representing their signature white doll in a racialized form but never going so far as to 'break the mould' and attempt to produce a diversity of representation.[18] On the strength of this example, duCille argues that

> *the toy industry is only one of many venues where, multiculturalism posed as an answer to critical questions about inclusion, diversity and equality, has collapsed into an additive campaign that augments but does not necessarily alter the Eurocentric* status quo. *Barbie 'gone ethnic' by way of dye job and costume changes seems to me but a metaphor for the way multiculturalism has been used as kind of a quick fix by both liberal humanism and late capitalism. Made from essentially the same mould as what Mattel considers its signature doll—the traditional blonde, blue-eyed Barbie—tawny tinted ethnic reproductions are both signs and symptoms of an easy pluralism that simply melts down and adds on a reconstituted other without transforming the established social order, without changing the mould. (1999: 112)*

This passage is worth quoting at length to demonstrate the ease and the breadth of the corporate 'strategy of diversity', as well its ability to depoliticize movements for social change. When difference is manipulated into a palatable commodified form, it succeeds only in reinforcing the norm rather than effecting substantive change: 'things fall apart, but the centre holds remarkably firm' (1999: 127). The case of 'ethnic' Barbie reveals that the 'other' dolls served only to reinforce the signature white doll as the standard, while her racialized sisters remained the exception.

Implications of the Commodification of Difference for Suicide Girls

How do these two examples of the commodification of difference inform our understanding of the contested position of Suicide Girls? Unlike the commodification of queerness, in which corporations attempt to distance themselves from a 'political' agenda, Suicide Girls explicitly attempts to capitalize on its apparent feminism. In addition to the inclusion of feminist content through groups, bulletin boards and profiling of feminist members, the homepage features quotes designed to illustrate the site's 'feminist potential'. These include quotes from female members saying that the models on the site are 'women they can respect', as well as reviews from publications claiming that the site breaks down conventional 'social mores' around beauty. Suicide Girls goes still further, featuring quotes that are designed to distance it from a purely business model and to correlate the site to a pseudo-activist movement or a 'grassroots phenom' (*TimeOut New York*). Its models are referred to as 'post-modern pin-up girls for the alternative nation' (*Boston Phoenix*) and the site is described as 'not just monetary but a cultural icon' (Chuck Palahniuk, author of *Fight Club*, http://suicidegirls.com/tour/).

In contrast to the corporate market for the gay community, which makes explicit its desire to distance itself from social change and to emphasize consumption, Suicide Girls is engaged in the important political work of subverting the characteristic male gaze of pornography. Although the site increases marketability by including the additional category of 'feminist' into the mix, it also modifies fundamental practices of looking through transformation of photographic norms. In this way, it does achieve one of the liberatory promises of cybersex, the claim that the structure of sex would be altered.[19] The photographic practices of Suicide Girls allow for a change in the objectifying, exploitative male gaze characteristic of so many other visual texts.

However, although this gaze may no longer be specifically misogynist, it remains a racialized gaze which maintains whiteness as normative and women of colour as othered. Although the website claims to transcend capitalist practices—its motto is 'SG is not a lifestyle brand, it's a lifestyle' (http://suicidegirls.com/tour/)—the website draws on traditional marketing practices of the gendered and racialized body. This results in the commodification of women of colour in order to diversify 'product offerings' and improve profits. The site adheres to racist media ideals which idealize white femininity—the safest and most reliable product. Moreover, when racialized models are included on the site, they are used to demonstrate the way in which ethnic difference may be shored up in order to stabilize whiteness as normative. Much like 'ethnic' Barbie, the answer to duCille's question 'how does difference look?' (1999: 114), is 'like familiar racist stereotypes', including a 'delicate' Asian Canadian model and a 'soulful' African American model. Suicide Girls understands multiculturalism as a call to sell a more diverse range of products with no broader commitment to changing the social order. Like Barbie, in which all the dolls are made from the same mould as is its 'signature' white doll, the models on Suicide Girls perpetuate the easy additive which poses little threat to the established order or to racist understandings of beauty. Ethnicity on Suicide Girls has fallen into 'the category of precious ready-to-wear difference' in which 'to be profitable, racial and cultural diversity—global heterogeneity—must be reducible to such common, reproducible denominators as colour and costume' (duCille, 1999: 116) without producing a diversity of representations. The site maintains these very narrow representations even though no additional manufacturing costs would be incurred. It would be no more expensive for Suicide Girls to include a diverse range of women, women whose bodies might challenge the normative standards of beauty to which the site adheres.

A final example illustrates the way in which the site emphasizes profit to the exclusion of a political agenda. This phenomenon is highlighted by the treatment on the Suicide Girls site of one its models, 'Frenchie'. Frenchie continually uses racist and homophobic rhetoric, including vicious racial epithets (http://suicidegirls.com/girls/Frenchie/), despite a policy discouraging racist or homophobic language on the site. Frenchie is a very popular model. Many of Frenchie's supporters are members who share the racist views that Frenchie freely expresses and that they cannot express for fear of being 'toaded' (i.e. being expelled from a site by erasing a person's user information). To date, the site owners have failed to remove Frenchie to avoid angering some of their paying members.

Conclusion

The problematic position occupied by Suicide Girls is clarified by the insights of both utopian and dystopian theoretical frameworks of cyberfeminism and the ways in which they overlap. Utopian cyberfeminism would emphasize the photographic practices of Suicide Girls, which allow the models to attempt to subvert the scopophilic gaze by staging their own photo shoots. Dystopian cyberfeminism would highlight the relatively low pay offered by Suicide Girls and the vulnerability of these young women to both online and real-life harassment. Both frameworks are restricted by their failure to account for ethnicity. Complicating the attitudes of utopian cyberfeminists and dystopian cyberfeminists by taking ethnicity into account reveals the way in which women of colour are either essentialized or neglected. Suicide Girls claims to eroticize deviant forms of femininity such as cyborgian sexuality or the female grotesque. However, an examination of the characteristics of the female bodies displayed on the site demonstrates the way in which racism prevents an understanding of the bodies of women of colour as deviantly female. Instead, their bodies are framed by racist stereotypes which result in their objectification, or they are simply excluded altogether. As a result, the ways in which both whiteness and commodification operate on Suicide Girls limit the site's transgressive potential. This limitation occurs because the website draws on traditional marketing practices of the gendered and racialized body to produce a profitable website.

Although Suicide Girls does fundamentally alter oppressive photographic practices which rely on the objectification of women for the male gaze and provides a space that acknowledges female desire, ultimately it belies its feminist potential. The site's ability to broaden understandings of female sexuality remain limited by 'the intertwining of patriarchy and capitalism' (Meehan and Riordan, 2002: ix), the commodification of difference for profit and the way in which the site maintains whiteness as normative. Even though Suicide Girls diverges from the corporate approach to the gay market, ultimately these strategies converge as both are commodified and, in this process, their ability to effect change is constrained. A capitalist emphasis on profit as a primary motive rather than on a critical feminist commitment to inclusivity and structural change means that the site uses new technologies to represent old inequalities.

About the Author

Shoshana Magnet is a PhD candidate, SSHRC Doctoral Fellow and video artist at the Institute of Communication Research, University of Illinois, Urbana-Champaign, and a researcher for the 'On the Identity Trail: Anonymity and Authentification in a Networked Society' project, University of Ottawa. Her published work has appeared in *Qualitative Inquiry, The Journal of Communication Inquiry, Canadian Women's Studies/Les cahiers de la femme, The Journal of Men's Health and Gender* and *Atlantis*. Her videos have screened at festivals in Canada, Europe and the USA. *Address:* University of Illinois at Urbana-Champaign, 228 Gregory Hall, 810 South Wright Street, Urbana, IL 61801, USA. [email: magnet@uiuc.edu]

Acknowledgments

The author would like to thank the anonymous reviewers for their excellent suggestions. She would also like to thank Kent Ono, Paula Treichler, Angharad Valdivia, Celiany Rivera Velazquez, Jillian Baez, Joan Chan, Aisha Durham and Himika Bhattacharya for their terrific comments and suggestions.

Notes

1. I include both textual and graphical depictions of sexuality within the category of cybersex.
2. Of course, these categories do overlap. The Second Wave of feminism initially emphasized only the concerns of middle-class, white women and failed to consider the implication of 'race' and class (Bulkin et al. , 1984; hooks, 1989; Lorde, 1984; Monture, 1986), sexual orientation (Cain, 1993; Lorde, 1984; Majury and Young, 1995) and disability (d'Aoust, 1994; Masuda, 1992).
3. 'Equal opportunity' is the name given to the group that discusses issues of ethnicity and racism.
4. Given that the models are required to participate in the online community through writing in their online journals and posting on the site (a good deal of extra work) one would expect their pay to reflect this extra effort. Yet, these models are not paid any more than the average, and sometimes less. The average dollar amount for a photo shoot is $200; suicide girls make from $100–$200.
5. Cyborg here refers to Donna Haraway's article entitled a 'Manifesto for Cyborgs' (1991), in which she defines cyborgs as mixtures of animal and machines, constructs which blur the boundaries between the natural and the real. In redefining the boundaries which formerly separated machines from humans and making them more fluid, Haraway argues that women are no longer tied to the natural body and thus can be pigeonholed no longer according to essentialized stereotypes which restrict them to rigid gender roles.
6. It should be noted that these references to sexuality may also be read as an attempt by the models to code themselves as sexual beings. In keeping with the terminology used on the site, if I refer to a 'suicide girl', this means a woman who is a model for Suicide Girls. If I refer to a member, this means someone who has paid to access the site.

7. Of course, online it is not possible to be certain that a stated gender is biological. In this article, I assume that the stated gender is the actual gender. If a comment is made by a model, the gender of the model is 'known', as they must post their own pictures and journal on the site.
8. Albury asserts that a particular pornographic text might be immoral if it induces the women involved to become '"alienated" from their own sexuality, since pornographic sex is performed as labour, rather than simply for the purposes of recreation or procreation' (2003: 198).
9. For an extended discussion of whether or not Suicide Girls is exploitative pornography, see http://suicidegirls.com/groups/Feminists/topics/20807/
10. Equating agency with the ability to speak has long been problematized by feminist disability theorists such as Susan Wendell (1996) and Rosemary Garland Thomson (1997).
11. An article in the Canadian newspaper *The Globe and Mail* discussed the 'Camgirls' phenomenon, where young women (mostly teenage girls) have personal websites which boast webcams, online diaries and wishlists of Amazon.com items. Those who enjoy viewing these sites can pick an item to send to the young women. Although the 'Camgirls see themselves as holding the controls', security officials believe that 'it's only a matter of time before a stalker tags a victim who has volunteered too much information'. The Camgirls phenomenon is seen by many as the logical extension of consumer culture: 'They have taken commodified girl power to its logical extreme—they have objectified themselves' (Anderssen, 2002: A3).
12. A well-known case of virtual rape is the example of 'Mr. Bungle' in *Lambdamoo,* where a user manipulated a sub-program known as a voodoo doll in order to virtually rape a female player (Dibbell, 1998).
13. The utopian perspective may be summed up by the now infamous MCI commercial: 'There is no race. There is no gender. There is no age. There are no infirmities. There are only minds. Utopia? No, internet' (cited in Nakamura, 2000: 15).
14. It would be different if this woman had chosen to represent herself as white as a form of identity exploration. However, many women are forced into passing not because they have decided explicitly to alter their ethnicity, but because of the normative assumptions that surround it on the internet.
15. It should be noted that the internet has been touted often as a utopia for queer-identified individuals (Foster, 2000).
16. This is in keeping with the photographic practices adopted by the site—where the women are understood to be inviting the camera's stare through the staging of each shot.
17. Following from Sender (2004), I use 'gay' community to designate the queer community, as companies are primarily marketing only to affluent gay men, rather than lesbian, transsexual/transgendered or other queer individuals.
18. However, this is not to suggest that a doll which had stereotypical signifiers of blackness would be more 'authentic'. This would be a highly problematic proposition. It is simply to note that producing white dolls in darker plastic is not a way of significantly transforming the norm or, as duCille notes in the case of Barbie, the mould.
19. As noted above, one example given by cyberfeminists as proof of the potential for cybersex to transform traditional understandings of pleasure is the cyberorgasm.

References

Albury, K. (2003) 'The Ethics of Porn on the Net', in C. Lumby and E. Probyn (eds) *Remote Control: New Media, New Ethics,* pp. 196–211. Cambridge: Cambridge University Press.

Anderssen, E. (2002) 'Camgirls: Empowerment or Predator Enticement?', *The Globe and Mail,* 19 October, p. A3.

Appadurai, A. (1996) 'Diversity and Disciplinarity as Cultural Artifacts', in C. Nelson and D.P. Gaonkar (ed.) *Disciplinarity and Dissent in Cultural Studies,* pp. 1–23. New York: Routledge.

Blair, C. (1998) 'Netsex: Empowerment through Discourse', in B.L. Ebo (ed.) *Cyberghetto or Cybertopia? Race, Class and Gender on the Internet,* pp. 205–17. Westport, CT: Praeger.

Branwyn, G. (2000) 'Compu_Sex: Erotica for Cybernauts', in D. Bell and B.M. Kennedy (eds) *The Cybercultures Reader,* pp. 396–402. London: Routledge.

Braunberger, C. (2000) 'Revolting Bodies: The Monster Beauty of Tattooed Women', *National Women's Studies Association Journal* 12(2): 1–23.

Bryson, M. (2004) 'When Jill Jacks In: Queer Women and the Internet', *Feminist Media Studies* 4(3): 239–54.

Bulkin, E., M.B. Pratt and B. Smith (1984) *Yours in Struggle: Three Feminist Perspectives on Anti-Semitism and Racism*. New York: Long Haul Press.

Butler, J. (1999) *Gender Trouble: Feminism and the Subversion of Identity*. New York: Routledge.

Cain, P.A. (1993) 'Feminist Jurisprudence: Grounding the Theories', in D.K. Weisberg (ed.) *Feminist Legal Theory: Foundations,* pp. 237–47. Philadelphia, PA: Temple University Press.

Chasin, A. (2000) *Selling Out: The Gay and Lesbian Movement Goes to Market*. New York: St Martin's Press.

Cumberland, S. (2002) 'The Five Wives of Ibn Fadlan: Women's Collaborative Fiction on Antonio Banderas', in M. Flanagan and A. Booth (eds) *Reload: Rethinking Women and Cyberculture,* pp. 175–94. Cambridge, MA: MIT Press.

d'Aoust, V. (1994) 'Competency, Autonomy and Choice: On Being a Lesbian and Having Disabilities', *Canadian Journal of Women and the Law/Révue Femmes et Droit* 7: 564–75.

Dibbell, J. (1998) *My Tiny Life: Crime and Passion in a Virtual World*. New York: Holt.

Doring, N. (2000) 'Feminist Views of Cybersex: Victimization, Liberation and Empowerment', *CyberPsychology and Behavior* 3(5): 863–84.

duCille, A. (1999) 'Black Barbie and the Deep Play of Difference', in M. Shiach (ed.) *Feminism and Cultural Studies,* pp. 106–32. Oxford: Oxford University Press.

Duggan, L. (2003) *The Twilight of Equality? Neoliberalism, Cultural Politics and the Attack on Democracy*. Boston, MA: Beacon Press.

Foster, T. (2000) 'Trapped by the Body? Telepresence Technologies and Transgender Performance in Feminist and Lesbian Rewritings of Cyberpunk Fiction', in D. Bell and B.M. Kennedy (eds) *The Cybercultures Reader,* pp. 439–59. London: Routledge.

Garland Thomson, R. (1997) *Extraordinary Bodies: Figuring Physical Disability in American Culture and Literature*. New York: Columbia University Press.

Gillis, S. (2004) 'Cybersex', in P.C. Gibson (ed.) *More Dirty Looks: Gender, Pornography and Power,* pp. 92–101. London: British Film Institute.

Halberstam, J. (1998) *Female Masculinity*. Durham, NC: Duke University Press.

Haraway, D.J. (1991) 'A Cyborg Manifesto: Science, Technology and Socialist-feminism in the Late Twentieth Century', in *Simians, Cyborgs and Women: The Reinvention of Nature,* pp. 127–48. New York: Routledge.

Hawthorne, S. and R. Klein (1999) *Cyberfeminism: Connectivity, Critique and Creativity*. Melbourne: Spinifex Press.

Hill Collins, P. (2000) 'What's Going On? Black Feminist Thought and the Politics of Postmodernism', in E. St Pierre and W.S. Pillow (eds) *Working the Ruins: Feminist Poststructural Theory and Methods in Education,* pp. 41–73. New York: Routledge.

hooks, b. (1989) *Talking Back: Thinking Feminist, Thinking Black*. Boston, MA: South End Press.

hooks, b. (1992) *Black Looks: Race and Representation*. Boston, MA: South End Press.

Hughes, D. (1999) 'The Internet and the Global Prostitution Industry', in S. Hawthorne and R. Klein (ed.) *Cyberfeminism: Connectivity, Critique and Creativity,* pp. 157–84. Melbourne: Spinifex Press.

Kaplan, E.A. (1983) *Women and Film: Both Sides of the Camera*. New York: Methuen.

Kendall, L. (1996) 'MUDder? I Hardly Know' Er? Adventures of a Feminist MUDder', in L. Cherny and E.R. Weise (eds) *Wired_Women,* pp. 207–23. Seattle, WA: Seal Press.

Kolko, B.E. (2000) 'Erasing @race: Going White in the (Inter)Face', in B. Kolko, L. Nakamura and G.B. Rodman (eds) *Race in Cyberspace,* pp. 213–32. New York: Routledge.

Kolko, B.E., L. Nakamura and G.B. Rodman (2000) 'Introduction', in B.E. Kolko, L. Nakamura and G.B. Rodman (eds) *Race in Cyberspace,* pp. 1–14. New York: Routledge.

Lindlof, T.R. (1995) *Qualitative Communication Research Methods*. Thousand Oaks, CA: Sage.

Lindlof, T.R. and B.C. Taylor (2002) *Qualitative Communication Research Methods*. Thousand Oaks, CA: Sage.

Lorde, A. (1984) *Sister Outsider: Articles and Speeches*. Trumansburg, NY: Crossing Press.

Majury, D. and Y. Young (1995) 'Lesbian Perspectives', in T.C. Collective (ed.) *Breaking Anonymity: The Chilly Climate for Women Faculty,* pp. 276–86. Waterloo: Wilfrid Laurier Press.

Masuda, S. (1992) *Meeting Our Needs: Access Manual for Transition Houses*. Vancouver: Disabled Women's Network Canada.

Meehan, E.R. and E. Riordan (2002) *Sex and Money: Feminism and Political Economy in the Media*. Minneapolis, MN: University of Minnesota Press.

Millar, M.S. (1998) *Cracking the Gender Code: Who Rules the Wired World?* Toronto: Second Story Press.

Mirzoeff, N. (1999) *An Introduction to Visual Culture*. London: Routledge.

Monture, P.A. (1986) 'Ka-Nin-Geh-Heh-Gah-E-Sa-Nonh-Yah-Gah', *Canadian Journal of Women and the Law/Révue Femmes et Droit* 2: 159–70.

Mulvey, L. (1975) 'Visual Pleasure and Narrative Cinema', *Screen* 16(3): 6–18.

Nakamura, L. (2000) 'Race in/for Cyberspace: Identity Tourism and Racial Passing on the Internet', in D. Bell and B.M. Kennedy (eds) *The Cybercultures Reader,* pp. 712–20. London: Routledge.

Nakamura, L. (2002) 'After/Images of Identity: Gender, Technology and Identity Politics', in M. Flannagan and A. Booth (eds) *Reload: Rethinking Women and Cyberculture,* pp. 321–31. Cambridge, MA: MIT Press.

NBC (2004) 'Photographer Denies He Killed Adult-film Actress: Man Charged with Murder', URL (consulted August 2005): http://www.nbc10.com/news/2944340/detail.html

Phillips, N. (2002) 'Cynical, Bitter, Jaded as Hell. Also Naked', *Citypages.com: The Online News and Arts Weekly of the Twin Cities* 23(1147), URL (consulted 3 April 2007): http://citypages.com/databank/23/1147/article10895.asp

Plant (2000) 'Coming Across the Future', in D. Bell and B.M. Kennedy (eds) *The Cybercultures Reader,* pp. 460–70. London: Routledge.

Razack, S. (2002) *Race, Space and the Law: Unmapping a White Settler Society*. Toronto: Between the Lines.

Russo, M.J. (1995) *The Female Grotesque: Risk, Excess and Modernity*. New York: Routledge.

Schiller, D. (1999) *Digital Capitalism: Networking the Global Market System*. Cambridge, MA: MIT Press.

Segal, L. (2004) 'Only the Literal: The Contradictions of Anti-pornography Feminism', in P.C. Gibson (ed.) *More Dirty Looks: Gender, Pornography and Power,* pp. 43–62. London: British Film Institute.

Sender, K. (2004) *Business, Not Politics: The Making of the Gay Market*. New York: Columbia University Press.

Senft, T.M. (2002) 'Shockingly Tech_splicit: The Performance Politics of Orlan and Other Cyborgs', in M. Flanagan and A. Booth (eds) *Reload: Rethinking Women and Cyberculture*, pp. 539–45. Cambridge, MA: MIT Press.

Shohat, E. (1998) *Talking Visions: Multicultural Feminism in Transnational Age*. New York/ Cambridge, MA: New Museum of Contemporary Art/MIT Press.

Steffensen, J. (2002) 'Doing it Digitally: Rosalind Brodsky and the Art of Virtual Female Subjectivity', in M. Flanagan and A. Booth (eds) *Reload: Rethinking Women and Cyberculture*, pp. 209–33. Cambridge, MA: MIT Press.

Stillar, G.F. (1998) *Analyzing Everyday Texts: Discourse, Rhetoric and Social Perspectives*. Thousand Oaks, CA: Sage.

Sturken, M. and L. Cartwright (2001) *Practices of Looking: An Introduction to Visual Culture*. Oxford: Oxford University Press.

Tomlin, A. (2002) 'Sex, Dreads and Rock 'n' Roll', *Bitch Magazine*, URL (consulted 3 April 2007): http://www.bitchmagazine.com/archives/12_02sg/sg.shtml

Trumbull, J. (2004) 'Sexy Strutting, Lip Studding: Balls-out Burlesque', *Independent Florida Alligator Online*, URL (consulted 3 April 2007): http://www.alligator.org/edit/detours/issues/stories/040122suicide.html

Valdivia, A. (forthcoming) 'Is Penélope to J-Lo as Culture Is to Nature? Eurocentric approaches to "Latin" beauties', in M. Mendible (ed.) *Bananas to Buttocks: The Latina Body in Popular Culture*. Austin, TX: University of Texas Press.

Wakeford, N. (2000) 'Networking Women and Grrrls with Information/Communication Technology: Surfing Tales of the World Wide Web', in D. Bell and B.M. Kennedy (eds) *The Cybercultures Reader*, pp. 350–59. London: Routledge.

Wendell, S. (1996) *The Rejected Body: Feminist Philosophical Reflections on Disability*. New York: Routledge.

Woodland, R. (2000) 'Queer Spaces, Modem Boys and Pagan Statues: Gay/Lesbian Identity and the Construction of Cyberspace', in D. Bell and B.M. Kennedy (eds) *The Cybercultures Reader*, pp. 416–31. London, New York: Routledge.

I'll Take My Stand in Dixie-Net: White Guys, the South, and Cyberspace

Tara McPherson

After a certain point, the subject of the public sphere must become a matter of local investigations into particular collectivities and practices.

—*Bruce Robbins,* The Phantom Public Sphere

I. Of Neo-Confederates and Cybertheory

I do not recall *exactly* how I first stumbled upon one of the many outposts of Dixie in cyberspace. I do know that one summer I was working on a manuscript on race and Southern femininity in the twentieth century, and, in an obvious (and perhaps narcissistic) ruse to avoid writing one day, I typed "tara" into a Web search engine and started surfing sites. I visited the Tara museum in Atlanta and several other Atlanta pages (many tied to the Olympics) and then, with one stray click of the mouse, found myself on the doorstep of the Confederate Embassy in Washington, D.C. Having lived the previous summer in D.C. without ever encountering such an edifice, I was surprised to learn that it existed as a real building in physical space in the nation's capital. Later, I began to ponder what purpose a Confederate embassy might serve. Would I go there to get a visa to travel to Mississippi for a conference? Perhaps I could visit it to plea on behalf of loved ones left behind in Louisiana (most of my family is still there)? At any rate, I spent many hours during the next few days following the neo-Confederate trail through cyberspace and then put these sites out of my mind.

Or so I thought. Throughout the past two years, as I read, thought, and taught about new technologies, I found myself drifting back below the virtual Mason-Dixon line, wondering how these neo-rebel sites might be reconciled with much of the new media theory I was reading. Such theories often maintain, in the words of Allucquère Rosanne Stone, that cyberspace functions as a kind of public theater, "a base . . . for . . . [the] cyborg" (39), suggesting that, in their play, these cyborgs are rewriting the standard of the bounded, embodied individual (43). Here, prosthetic communication enables computer users to overcome the self/body binary. Sherry Turkle likewise foregrounds cyberspace's ability "to give people the chance to express multiple . . . aspects of the self" making "possible the creation of an identity so fluid and multiple that it strains the limits of the notion" of identity (12). While I focus here on Turkle and Stone (whose books, in many respects, I quite like), I see their work as indicative of a larger movement within technotheory to focus on the tendency of "prosthetic living" to foster "rapid alternations of identity" in which computer users inevitably come to "cycle through different characters and genders" (Turkle, 174). In this scenario, prolonged exposure to cyberspace irrevocably produces multiple selves (or, at least more selves than one entered the Net with), though the racial identity of these "many selves" has not received much attention.

And stable, embodied identity is not the only thing transformed by cyberspace. For Turkle, "Life on the screen is also without origins and foundation" (47). Indeed, "our very rootedness to place is attenuated" (178) via virtual living, suggesting that collective as well as individual identities are uprooted when confronted by the postmodern geographies of cyberspace (though Turkle is far more interested in selves than in communities). While Turkle does briefly consider the downside to cyber-transformations of place, she generally sees this loss of rootedness as, if not positive, somehow inevitable. More utopian cybernauts like Howard Rheingold have celebrated the Internet's ability to overcome geographical boundaries, envisioning it as a kind of yellow brick road leading to a harmonious global village. Even those wary of the charms of the virtual world (like Michael Heim or Stephen Doheny-Farina) still worry about its ability to destroy one's sense of place.[1]

But the neo-Confederates guarding the portals of the Confederate embassy in cyberspace seem relatively unconcerned with the prosthetic nature of cybercommunication. Likewise, they busily continue to envision and build a very specific, if virtual, South without paying much mind to the Internet's potential to enable, in the words of Stone, "many selves to inhabit one body." Still, some of their concerns do overlap with those of contemporary cybertheory. For instance, Stone notes that the stories we tell ourselves about technologies (old or new) reveal "our deep proclivity, our deep need, to make self in the world" (83) and to "stabilize a sense of presence" in uncertain times (88). I think the work of the neo-Confederates in cyberspace reveals a very sincere attempt to make "self" in the world and to articulate a very particular (and racially naturalized) presence. Still, their endeavors are notably less marked by a celebration of multiplicity and of play than much of today's technotheory. For these cyberrebels, reconstructing Dixie and its citizens is not about play at all; rather, it is a very serious battle over the demands of place, race, and identity.

Cybercommunities like those of the neo-Confederates invoke specific registers of place, yet these places, like the majority of writing about cyberspace, evade precise discussions about race or racism. Certainly race is one of the nodal points around which public discourse on the South (if not the nation) has turned throughout the twentieth century, but in the post–Civil Rights era representations of race and racism proceed via different logics than they did before the 1960s. In the first half of the century, Southern images of blackness and whiteness were often mutually interdependent, sketching the contours of whiteness (particularly white womanhood) in contrast to blackness, if only to insist upon racial difference. (For an example of this, one need simply recall the intertwined relation of Mammy and Scarlett in Margaret Mitchell's 1936 epic *Gone With the Wind.*) By century's end, Southern racial representations are much more likely to elide race entirely or to figure whiteness as independent, as separate from blackness, particularly via a dismissal of the region's complex racial history. (In Alexandra Ripley's 1991 *Gone With the Wind* sequel, *Scarlett,* the white heroine flees to Ireland, thus skirting the issue of blackness altogether.)[2]

These two logics can be labeled "overt" versus "covert" and differ in that the former *brings together* figurations of racial difference in order to fix the categories while the latter *enacts a separation* that nonetheless achieves a similar end. Upon close examination, this separation is very much in evidence in the neo-Dixie of cyberspace, a place which is nothing if not white. This chapter will highlight how this "cyberwhitening" takes place via the creation of new regional identities that refigure white Southern masculinity by borrowing from the language of the civil rights struggle. It is equally important to think through how such covert strategies might also inform our own theorizations of cyberspace.

What, for instance, does it mean to champion cyberspace as placeless or as a breeding ground for new identities when talk about race is conspicuously absent from these conversations? Rather than simply "forgetting" or "overlooking" race, these theories enable the evasion of the race question, underwriting the whiteness of cyberspace. Put differently, they revolve around and facilitate an inability to articulate the place of race and the race of place in visions of the future. Thus, when theories of the multiple-selved, unrooted inhabitants of cyberspace do not directly and explicitly engage with issues of race, racism, and racial representation, they help to construct a raceless fantasy space, free of the contradictions of life at the end of the millennium, that shares many traits with cyber-Dixie. Whether this space is filled with MUDers happily swapping gender, freed from the constraints of embodied identity and geographic origin, or neorebels dreaming of cybersecession and new visions of old places, the default setting is still all too white. Our theories of cyberspace must tackle race full on if we are to short-circuit the covert racial representations which structure life in the U.S. at the end of the millennium.

Throughout the last year, when I mentioned to friends that I was doing research on neo-Confederates in cyberspace, the general reaction consisted of grimaces and inquiries about how I could stand investigating such rednecks and racists. Such an assessment is one these neorebels are fully aware of, and, as we shall see, it is an evaluation to which they try to respond. For my part, I did have to wonder why I was studying this group (beyond the recent academic fascination with white trash), especially since I enjoyed sites like "The Lipstick Librarian" so much more.[3] In part, I suspect it is because their version of what is best about the South differs so radically from my own. On another level, my friends' questions highlighted for me certain issues of taste and judgement regarding academic inquiries. In what follows, I try not to reinscribe our own academic binaries and categories (progressive/reactionary, racist/nonracist) when describing these neorebels but instead attempt to understand what their virtual battles have to tell us about the self, the citizen and the public performance of white masculinity at the close of the mechanical age. In other words, what insights can an investigation into these local collectivities provide for our attempts at theorizing the possibilities of structuring new (and newly hued) public spheres online?

Exploring Virtual Dixie

Though I thus far have been referring to these neo-Confederates as if they were a monolithic and self-named entity, the terms "neo-Confederate" and "neorebel" are not always used on the webpages I examined, though I suspect they are titles many in the groups would embrace. I use these terms to refer to the creators and inhabitants of a fairly broad cluster of both individual- and group-authored websites, sites primarily concerned with "preserving Southern heritage." They often refer to themselves as "Southern nationalists" or as "Southrons." For the most part, these pages are designed and frequented by white men of a wide range of ages and classes, though the ages seem to cluster in the eighteen-to-thirty-year-old and the forty-five-to-sixty-year-old groups. Most are native-born Southerners as well, though a good number no longer reside within the physical borders of Dixie, and many also participate in Southern heritage groups offline. In fact, these groups have received a substantial amount of press coverage in the past few years, largely because two recent books, Peter Applebome's *Dixie Rising* and Tony Horwitz's *Confederates in the Attic,* have brought news of the neo-Confederate movement to a wide-ranging audience.

No single group or individual website embodies every aspect or member of what I call the cyber-Confederacy, but most of the pages reference a long list of links to other Confederate websites and several umbrella sites serve to introduce and catalog the variety and number of Confederate webpages that exist (I have visited well into the hundreds). Many of these are updated weekly (if not daily), and some also link to listservs and newsgroups (though not many; cyberchat does not seem to be the primary goal of many of these sites.) Three major sites, which together cover pretty much all of the cyber-Confederacy, are Dixie-net (http://www.dixienet.org), The Confederate Network (http://www.confederate.net), and The Heritage Preservation Association (http://www.hpa.org). Each of these sites contains information (or links to information) on Confederate history, reenactments, Southern merchants, and what many of the sites refer to as "heritage violations," a term that most often refers to attempts to ban or remove symbols of the Confederacy, particularly Confederate flags. Many of the sites advocate Southern separatism or nationalism, sometimes via secession.

Each of these warehouse sites also includes some type of mission statement, often in the form of a bulleted manifesto stating the group's aims. For instance, Dixie-net, the official website maintained by the Southern League (which is spearheaded by Alabama history professor Michael Hill and is in some ways comparable to the conservative National Association of Scholars) asserts that the League's purpose is to "advance the social, economic, cultural and political independence of the

Southern people by all honourable and peaceful means."[4] A fourth warehouse site, Dixieland Ring (http://www.geocities.com/BourbonStreet/2757/index.htw), addresses similar themes but stages a more inclusive appeal, defining Southernness a bit more broadly. This site does labor to reconfigure the much-heralded rhizomatic nature of cyberspace by structuring its links in a circle, creating a kind of closed mobility. The page itself proclaims that, ideally, "the [user] keeps moving forward, [and] eventually he will end up back where he started." Of course, individual visitors to the site are free to abandon the ring structure, but the physical construction of the pages encourages an orderly progression through a list of pro-Southern sites that works against the celebrated anarchy of the Internet.

During a tour of these sites, it quickly becomes obvious that they trade heavily in what might be deemed the signifiers of the Confederacy, from their frequently animated Southern flags to their structuring of a sonic Southernness via countless (and varied) renditions of "Dixie" and other Confederate battle songs. Battle itself is frequently figured as a primary frame of reference and as an originary moment, and the facts of the "war between the states" (never the Civil War; in fact, several sites ban the use of the term) constitute a large part of the historical information the sites detail. The war itself becomes the ground upon which claims to heritage are waged, though here heritage clearly functions as a universal and naturalized category which only some can lay claim to and which all "real" (read "white") Southerners would die to defend. The South's complex racial history and its relationship to the Civil War disappear as the war is rewritten in univocal terms.

Furthermore, these sites construct a phantasmagoric South via several varieties of mapping. For instance, many sites revision U.S. geography, presenting maps of the eleven states of the Confederacy floating detached and separate from the rest of the United States. The Dixienet site actually maps and literalizes the drama of secession via an animated image in which the Southern region breaks free from a map of the union and hovers inviolate (and in rebel gray) at the top of a page. This page also narrates the story of "The South as Its Own Nation," underscoring the vast demographic and economic resources of the region through a series of graphs and charts that detail the viability of the South as a separate nation. On the page, the visitor learns that "in economic power, a Southern nation composed of the eleven States would have the fourth largest gross domestic product . . . after the remainder of the United States, Japan, and Germany," a precise example of the narration of nation as imagined community.

Such narratives highlight that, much like webpages, the South is still undergoing "reconstruction." They also visualize a fantasy of a new Confederacy and a virtual secession at precisely the moment that black Americans are moving to the South in greater numbers than they are leaving it for the first time since the Civil War. The carefully constructed graphs and charts of this imagined Southern nation do not offer the visitor any demographic information on the racial makeup of the South. One can only imagine that the African-American families recently relocated from Detroit or Chicago would feel none too comfortable taking up citizenship in such a Southern nation. That the neo-Confederate imagined community glosses over the heterogeneity and diversity of the actual South almost goes without saying, but it is important to highlight the elisions of these covert strategies.

Other mappings are less literal, calling upon imagery of the old South to transform the "unrooted" realms of cyberspace into particular cyberplaces that correlate to real and imagined landscapes of gentility. These cognitive mappings structure an imagined place of history and heritage by referencing the ways of the Old South. One such site is the Confederate Embassy, which first initiated me on this journey. This site presents a virtual photo album of white Southern fantasy, replete with images of modern-day dancing belles, smiling rebel gentlemen, and a luxuriant, entirely white plantation home.

To anyone familiar with Southern history, these hoopskirted images of the past, in concert with old black and white images of the Southern soldier, smack of Lost Cause sentiments. "The Lost Cause" refers to a widely-prevalent movement across the South in the late nineteenth and early twentieth centuries to "enshrine the memory of the Civil War." This included the formation of Confederate memorial associations, the celebration of Confederate Memorial Day, and the beginnings of the still-active organizations, the Sons and Daughters of the Confederacy. A crucial trope upon which

Lost Cause sentiments depended was an idealized and nostalgic recasting of Southern gentility, hospitality, and manners. Today's neo-Confederates are clearly linked to this historic Lost Cause ideology, both in their defense of the memorials and statues of Confederate soldiers erected during that earlier period, and in their figurations of Southern grace and manners, suggesting that origins and foundations are not always lost in cyberspace.

I have written elsewhere about how this Southern reverence for manners functioned historically as a very particular technology of location, designed precisely to "help" one know one's place, but this rebirth of the Lost Cause over one hundred years after its origins interests me primarily for the *differences* it displays from its original.[5] Neo-Confederate revampings of Lost Cause imagery depend heavily on their reconfiguration of two important visual registers as they rework iconic signs of both femininity and race. The Lost Cause ideology constructed a very particular figure of white Southern femininity, that of the Southern lady, and quickly installed her upon a pedestal where she served both to represent the gentility of the South and, via imagined threats to her safety, to justify a violent and prolonged attack on black masculinity. The idealized Southern lady underwrote the myth of the black male rapist and the horrors of lynching.[6]

Though crucial to the figurations of the Lost Cause at the turn of the last century, these icons have largely disappeared in the phantasmatic South the neo-Confederates have built for the new millennium, replaced as they are with a near obsession with the contours of white Southern masculinity. Several personal webpages, often created by younger white men and linked to larger clearinghouse sites like DixieNet, have names like "The Virginian Gentleman" and commit a great deal of space to delineating the Southern gentleman's finer qualities: he strives to forgo the use of power and "feels humbled himself when he cannot help humbling others." He is the descendant of the crusader and the champion of justice. But in these reconstructions, the Southern gentleman, though modeled on Lost Cause portraits of Robert E. Lee and Jefferson Davis, stands largely alone, unencumbered by the lofty version of femininity that once accompanied him.

Gone too is any overt imaging of blackness or explicit expression of racism. A concern with white masculinity and its preservation replaces representations of blackness, and almost all of the sites decry any racism, hate-mongering, or Klan or neo-Nazi activity. (Many of the webpages feature an "anti-Klan" logo, a crossed-out image of a Klansman's hood.) Often, the pages express dismay (some seemingly genuine, others less so) over the continued perceptions that protecting Southern heritage means one must be racist. Hence, these sites abandon the overt racism of the Lost Cause era for the more palatable covert racism characteristic of the post–Civil Rights era. While this turn from explicit racism can simply be an act of bad faith, understanding how covert racism works in many of these sites entails recognizing that, often, practitioners of this brand of racism (quite fervently) do not believe themselves to be racist. Labeling them and their cyberspaces "racist" does little to help us understand how they understand either whiteness or blackness. I will return to these questions shortly, but first I want to examine how the sites deploy a doubled and particular mode of address, an address expressly structured toward a public.

At one level the sites hail fellow white rebels, enunciating a very clear address to a "you" who resembles the creators of the sites. For instance, one page opens with the promise that "preserving OUR heritage is preserving YOUR heritage." Membership pages delineate quite specifically what the "you" visiting the site must believe in (or at least agree to) in order to join the organizations. Yet, the sites are not just for consumption by like-minded individuals. Unlike, say, Confederate conventions or reenactments, these sites also stage a carefully mediated public address designed both to dissociate their actions from overtly racist causes and to educate the public on issues "of Southern civil rights."

This mode of address carefully and selectively appropriates the discursive practices of both civil rights groups and a wide-range of nationalist struggles. Thus, page after page insists on the need to reclaim history in order to end the oppression of the Southern people and offers practical tips on how to "buy Dixie" or to boycott corporations that violate Southern heritage. The Heritage Preservation Association home page labels the group "the most successful Civil Rights Organization for Southern Heritage," noting their efforts at "guarding our future by preserving our past" and inviting the viewer to "help us tip the scales of Justice for Dixie."

Sites also speak of "the cultural genocide of the Confederacy" and of campaigns of ethnic cleansing against Southerners. Several groups include links to the Homelands Page, a clearinghouse for nationalist movements worldwide. (If you visit this page from the Confederate States of America network, you are in for one of the Net's surreal juxtapositions: it is possible to view the Chiapas homepage from beneath a border extolling the benefits of Confederate secession.) Frequent comparisons are also made between the Southern independence movement and struggles in Scotland, Northern Italy, Croatia, and Quebec. While we may rightfully worry about comparisons between independence for white Southerners and the Chiapas revolutionaries, this call for a virtual Dixie does signal an attempt at reconstituting a public sphere where the finer points of citizenship and its requirements can be interrogated, debated, and lived. Though neo-Confederate organizations have existed throughout the twentieth century, these Internet communities signal a new level of awareness about public perception and battles over public spaces, and a heightened awareness of the functions of publicity. These sites understand that successful publicity now requires an evasion of questions of race and racial representation.

What I am suggesting here is that these men are fashioning a "technologically mediated publicness" that sustains a desire for origin and homeplace. In other words, they can be seen as actively constructing what Nancy Fraser has called a "subaltern counterpublic," designed to mediate and reconfigure the boundaries between the private and public. Like other counterpublics, these men coin new languages and invest old signs (like the flag) with new meanings as they construct new access routes to the public. Fraser, of course, primarily discusses the subaltern counterpublic in terms of groups that challenge dominant ideology (her example is feminism). What does it mean that white, mostly middle-class men—the group we usually see as important players in the public sphere—feel the need to battle for alternative publics? After all, both Democrat and Republican white Southerners—from Ted Turner to Newt Gingrich to Bill Clinton—have emerged as key players on the national scene, and the neo-Confederates' own statistics indicate the South's economic growth and stability during the past few decades. Yet despite their improved economic position and the prominence of many white men from their region, these men clearly view themselves as marginalized because of their Southernness, and they actively construct spaces in which this origin can be discussed, celebrated, and protected from attacks, real or imagined.[7]

This virtual battle is being fought to defend a very specific Southern heritage, a heritage that is undeniably white. While "whiteness" itself is rarely mentioned in these webpages, Celtic, Anglo, and European ancestry often is. Thus the Southern league proclaims it will "affirm the legacy of our precious Anglo-celtic heritage." In White, Richard Dyer notes that the turn to situating whiteness as ethnicity (Celtic, Polish, etc.) "tends to lead away from a consideration of whiteness itself" (5). Put differently, an exploration of white Southernness couched in the terms of ethnic identity is less likely to produce an understanding of the privileges whiteness confers and often functions as yet another form of covert racism. This is not to say that we shouldn't attempt to understand ethnicity's relation to whiteness but that fetishizing ethnicity is not enough. Contemporary theories of race often posit whiteness as devoid of content, as, in the words of Toni Morrison, "mute, meaningless, unfathomable" (59). It is perhaps useful to think of these neo-Confederate appropriations of nationalist or civil rights struggles as precisely an attempt to give whiteness both voice and content. Without a discourse or images of blackness to delineate the contours of whiteness, these men struggle to find new ways of securing the meaning of whiteness.

Emotional defenses of the Southern flag or other Confederate symbols are closely tied to this attempt to carve out an embodied meaning for whiteness. One site included a variety of letters protesting Ole Miss's decision to ban Confederate symbols from the school's publications and sports events. Rich in sarcasm, one of the writers suggests renaming the football team (formerly the Rebels) "The Guilt" and changing the school colors to "blush red." While I was at first tempted to dismiss these letters as racist refusals to see the pain caused to many by these symbols, I recognized in this evocation of guilt a response shared by many of my white students when we study theories of whiteness or of race more generally. Many claim anger at being made to feel guilty about what they describe as a past during which they were not even alive: a response provoked as much, I think, by our theories of whiteness as by their naïveté. One of the limits of these theories is their tendency to slip between readings of whiteness as a dominant (and dominating) ideology and the various ways white people are positioned in relation to this spectrum of privilege and power. If the left is not willing to offer up a possible reading of a progressive or oppositional whiteness, conservative groups like the neo-Confederates will be happy to assign whiteness another content.

The neo-Confederate webpages replay a logic of separatism that I have elsewhere categorized as the lenticular or covert logic of post–Civil Rights discourse in the United States. Such discourses stage a new visibility for whiteness as an injured, wronged, violated whiteness and also underscore the degree to which we lack compelling narratives or theorizations of successful union (between North and South and between races).[8] This inability to think beyond separatism also permeates more "liberal" accounts of today's South. The previously mentioned popular books *Dixie Rising* and *Confederates in the Attic,* both authored by prominent, liberal journalists, offer up portraits of a still racially separate South, but they frame their tales in such a way as to suggest an equivalence between white and black separatism, overlooking the historic ways in which white separatism has been supported by state institutions and given access to and power over black bodies. In many ways, Horwitz and Applebome reinscribe the separatist attitudes about which they seem so dismayed by presenting present-day segregation as "simply the way it is," something that *all* Southerners (black and white) really want. Though these two authors hope for a South different from the South envisioned by Georgia Senator Newt Gingrich or the Southern League, the stories they provide of the South do little to challenge the vision of the neo-Confederates.[9]

Though I am clearly critical of the neo-Confederates and their narrow definition of Southern heritage as conservative, white, and mostly male, I do think that their cyber-South expresses a real desire for some type of social or community life that is less bound by a distant and seemingly unresponsive government and more fully engaged in the local and the regional. Though their response may be to call for "states' rights" and a decrease in federal power, I wonder how the sense of dissatisfaction with public life that they express might be differently mobilized? How could we narrate other versions of Southern history and place that are not bleached to a blinding whiteness? Richard Dyer writes of the need to make whiteness strange in order to reveal its limits and its constructedness. One way to achieve this in relation to Southern culture would be to continue to tell the stories of those "eccentric" and brave Southerners, white and black, who have struggled to build a different South less seeped in the sentiments of the Lost Cause. Such stories could appeal to many Southerners who perceive their region to be marginalized without replaying the covert racial politics that neo-Confederate sites so often deploy.

I imagine websites that reclaim Southern heritage for progressive causes by reminding us of other Southern sons and daughters, including the likes of Katharine DuPre Lumpkin, Angela Davis, Martin Luther King Jr. and Morris Dees. Such sites would offer alternatives to the webpages of the neo-Confederates, reflecting a more diverse South and redefining Southern heritage. They could also point the way beyond the binary poles of covert and overt racial logics. If overt racial representation brings together black and white in order to privilege whiteness, and covert strategies repress difference to the same end, what seems necessary is an overt representation of racial difference and *its interconnections* that explores the interrelation of this difference without privileging either term. We cannot understand or learn from the South's racial history by repressing it any more than we can champion the identity-shifting potential of cyberspace without studying its racial contours.[10]

Some Final Thoughts

Beyond suggesting new ways for thinking through and negotiating the demands of white identity, I think virtual Dixie also suggests the limits of conceptualizing cyberspace as both freed from place and origin and as a "playground for nontraumatic multiplicity." Both proponents and critics of online community and culture have noted the Internet's ability to free us from physical spaces, propelling us into, alternately, freer virtual spaces or bottomless abysses of placelessness, but the constructions of the cyberrebels illustrate the ways in which virtual life serves instead to forge new (and reactionary) relations to precise geophysical spaces.

If these virtual rebels are not willing to abandon the physical South in pursuit of virtual glory, neither do they seem very interested in experimenting with the hybridity of the self. Rather, the neo-Confederates are reformulating identity politics for the cyberrealm, intent on the construction of a fairly unified cybersubject. And he's not playing. The theories of multiplicity we have used to describe identity formation in cyberspace often rely on MUDs and game playing in sketching their models of community and identity, but such models make it hard to envision how other subaltern publics might be formulating more fixed identities via the Web, reconstructing "the old South, the new South, and the future South" as a fairly traditional and static place. They also evade questions of racial representation in cyberspace, suggesting that in the identity-swapping realms of the cyber, one identity is the same as the next. While I am as happy to engage in prosthetic communication as the next girl, I wonder what we might gain from shifting our theories of cyberspace away from tropes of "play," "multiplicity," and "theater" toward explorations of "citizens," "politics," and "publics." We might just find our way toward a Southern cybercitizen less bound by the nostalgic racial politics of Dixie and toward new theories of cyberspace less trapped in the covert racial representations that seem all too familiar today.

Notes

1. Howard Rheingold has recently become a vocal critic of the commercialization of the Internet. a development that has certainly dampened his early utopian portraits of cybercommunity. Accounts such as Doheny-Farina's also tend to suggest that Internet users leave behind their sense of history when entering the Web. He writes that virtual communities "encourage us to ignore, forget, or become blind to our sense of . . . place and community" as we cast off the cumbersome weight of our past lives. My examination of the neo-Confederates illustrates that the past is very much with them as they traverse cyberspace.
2. The different strategies of racial representation utilized in *Gone With the Wind* and its sequel, *Scarlett,* are further explored in my essay. "Seeing in Black and White: Gender and Racial Visibility from *Gone With the Wind* to *Scarlett*."
3. See for example, the anthology *White Trash,* edited by Matt Wray and Annalee Newitz. This recent academic turn to studying whiteness has been met with an understandable skepticism by some within the academy. Richard Dyer notes that, while it is crucial that we "come to see whiteness" (8). it is also important that our investigations of whiteness not become yet another excuse to do what we white folks have done all along, that is, talk and think only about ourselves.
4. Michael Hill has been a vocal defender of Southern nationalism and has been profiled in academic publications like *The Chronicle of Higher Education*. He describes the politics of the group as "paleoconservative" and claims that the organization's membership is "well over four figures" (Chronicle A9).
5. I discuss the ways in which Southern etiquette most often worked in the service of power and also examine multiple ways in which the tropes of the old South are reworked for the present day in my *Reconstructing Dixie: Race, Place and Femininity in the Deep South*.
6. Excellent discussions of the race and gender dynamics of Southern lynching campaigns can be found in the work of Angela Davis and Jacquelyn Dowd Hall.
7. As noted earlier, many of my academic peers grimace when I describe this research project, noting their distaste for "rednecks." This reaction does suggest that the neo-Confederate perception of "anti-Southern" attitudes is not entirely imagined. To view white Southerners as inherently and irredeemably racist is about as insightful as arguing that the Civil War had nothing to do with race.

8. A lenticular image is the type of image found on what are often called "3-D" postcards, composed of two separate but interlaced images. The very structure of the card makes it almost impossible to view the two images together, enforcing a structural separatism. For more on this trope, see my *Reconstructing Dixie*.
9. It is only fair to note that, while I here draw an equivalence between the goals of the neo-Confederates and the writings of Horwitz and Applebome, neither the journalists nor the neo-rebels would be happy with that equation. In fact, both authors have been denounced as Yankee scalawags by the neo-Confederates, and, in an interesting interview, the Pulitzer-prize winning Horwitz comments that the neocon's "continuing sense of aggrievement and of being put down is at odds with the reality of Southern life" now that the South "is the most economically vibrant part of the country" (24). Still, neither journalist successfully imagines what a harmoniously multiracial South might look like, and neither pays much attention to historical or contemporary figures who are struggling toward such a vision.
10. As much as I think we have to learn from the neo-Confederate websites, it seems unlikely that many of these sites could accommodate this wider view of the South as racially diverse. Overt references to blackness would destabilize the carefully naturalized (i.e., white) universality of "Southern heritage" that the sites construct. What I am suggesting is that there are many ways to define Southern heritage and that we need sites that narrate these other origin stories. One might, for instance, begin with sites like those of the Southern Poverty Law Center or the Griffith IN CONTEXT project currently underway at Georgia Tech.

References

Applebome, Peter. *Dixie Rising: How the South is Shaping American Values, Politics, and Culture.* New York: Harcourt, Brace and Company, 1996.

Davis, Angela. *Women, Race and Class*. New York: Vintage Books, 1981.

Dyer, Richard. *White*. New York: Routledge, 1997.

Doheny-Farina, Stephen. *The Wired Neighborhood*. New Haven, CT: Yale University Press, 1996.

Fraser, Nancy. "Rethinking the Public Sphere: A Contribution to the Critique of Actually Existing Democracy," in *The Phantom Public Sphere,* Bruce Robbins, ed. Minneapolis: University of Minnesota Press, 1993.

Hall, Jacquelyn Dowd. *Revolt Against Chivalry: Jessie Daniel Ames and the Women's Campaign Against Lynching*. New York: Columbia University Press, 1974.

Heim, Michael. *The Metaphysics of Virtual Reality*. New York: Oxford University Press, 1993.

Homburg, Mark W. "Tony Horwitz's Civil Wargasm: An Interview." *The North Carolina Review of Books,* Summer 1998, 8.

Horwitz, Tony. *Confederates in the Attic: Dispatches from an Unfinished Civil War*. New York: Pantheon Books, 1998.

McPherson, Tara. *Reconstructing Dixie: Race, Place, and Femininity in the Deep South.* Durham, NC: Duke University Press, forthcoming.

———. "Seeing in Black and White: Gender and Racial Visibility from *Gone With the Wind* to *Scarlett,*" in *Hop on Pop: The Politics and Pleasures of Popular Culture,* Henry Jenkins, Tara McPherson, and Jane Shattuc, eds. Durham, NC: Duke University Press, forthcoming.

Mitchell, Margaret. *Gone With the Wind*. New York: Macmillan, 1936.

Newitz, Annalee and Matt Wray, ed. *White Trash: Race and Class in America*. New York: Routledge, 1996.

Rheingold, Howard. *The Virtual Community: Homesteading on the Electronic Frontier.* Reading, MA: Addison-Wesley, 1993.

Ripley, Alexandra. *Scarlett*. New York: Warner Books, 1991.

Robbins, Bruce, ed. *The Phantom Public Sphere*. Minneapolis: University of Minnesota Press, 1993.

Shea, Christopher. "Defending Dixie: Scholars in the Southern League Want To Resurrect the Ideals of the Old South." *The Chronicle of Higher Education,* 42.11 (1995): A9.

Stone, Roseanne Allucquere. *The War of Desire and Technology at the Close of the Mechanical Age*. Cambridge, MA: MIT Press, 1995.

Turkle, Sherry. *Life on the Screen: Identity in the Age of the Internet*. New York: Simon and Schuster, 1995.

MODULE SIX

Digital Inclusion and Inequalities

Digital Inequality: Differences in Young Adults' Use of the Internet

Eszter Hargittai[1] and Amanda Hinnant

This article expands understanding of the digital divide to more nuanced measures of use by examining differences in young adults' online activities. Young adults are the most highly connected age group, but that does not mean that their Internet uses are homogenous. Analyzing data about the Web uses of 270 adults from across the United States, the article explores the differences in 18- to 26-year-olds' online activities and what social factors explain the variation. Findings suggest that those with higher levels of education and of a more resource-rich background use the Web for more "capital-enhancing" activities. Detailed analyses of user attributes also reveal that online skill is an important mediating factor in the types of activities people pursue online. The authors discuss the implications of these findings for a "second-level digital divide," that is, differences among the population of young adult Internet users.

Keywords: skill; self-perceived knowledge; Internet; Web use; online behavior; young adults; digital divide

As Internet use has diffused to a growing portion of the American population (Fallows, 2004; National Telecommunications and Information Administration [NTIA], 2002), scholars have started calling for refined approaches to the study of the so-called digital divide (Barzilai-Nahon, 2006; Hargittai, 2002, 2003a; Mossberger, Tolbert, & Stansbury, 2003; van Dijk, 1999, 2005). These researchers, when looking at Internet use statistics, have argued that we need to move past the binary classification of users versus nonusers and explore details about people's Web uses in order to get a deeper understanding of where inequalities may lie regarding new information technologies. Such work has focused on the types of uses for which people turn to the Internet, and a smaller body of literature has explored the importance of people's abilities in using the medium. We focus on both of these refined measures of Internet use to explore where the so-called "second-level digital divide" (Hargittai, 2002) may lie across users.

Work that has looked at the relationship of various socioeconomic status (SES) variables to Internet use has found that some differences among population segments are fading, in some cases considerably. Data about access statistics suggest that the gender divide has disappeared in the United States (Ono & Zavodny, 2003; Wasserman & Richmond-Abbott, 2005), yet scholars looking at more detailed measures of Internet use have noted that men and women are not equally exposed to the medium (Cooper, 2006; Cooper & Weaver, 2003; Kennedy, Wellman, & Klement, 2003; Ono & Zavodny, 2003). Men use the Web more frequently than do women and have higher levels of self-perceived Web-use skills despite similarities in actual skills (Hargittai & Shafer, 2006). Research among young people has shown gender may also play a role in media multitasking, with girls multitasking more than boys (Foehr, 2006). Others have emphasized the importance of factors such as experience with the medium (Loges & Jung, 2001) and the importance of one's social surroundings (Hargittai, 2003b) in explaining how people have incorporated the Internet into their everyday lives.

We take these observations about differential online experiences seriously and empirically examine how Internet users differ in their online activities. We look at the implications different online activities may have for the medium's implications for people's life chances, that is, people's ability to improve their human, financial, political, social, and cultural capital. In particular, we examine the online behavior of the age group that is the most connected: young adults (Fox, 2004). Because this population segment has the largest level of diffusion, considering refined measures of use differences is of particular importance with this group of people given that simple measures of access may wrongly suggest that inequalities have all but disappeared.

To establish our premise that the study of young adults is particularly relevant when it comes to refined measures of Internet use, we start by reviewing the literature on the relationship of age to Internet uses. We then review the existing literature on studies of skill as we consider that to be an especially important area for refined measures of online behavior. Next, we briefly discuss the types of activities for which people go online and do so by paying attention to differences across varying segments of the population. We then describe our data and method of analysis, followed by a discussion of our findings. We conclude by exploring the implications of our findings for which segments of the young adult Internet population are best poised to benefit from their Web use.

Age and Internet Use

According to the most recent figures from the Current Population Survey's Computer and Internet Use Supplement (NTIA, 2004), the most connected age group among Internet users is the segment of those between 18 and 24 who are in school, with 86.7% online in 2003. Past reports of the Pew Internet and American Life Project, a premier authority about Americans' Internet uses based on their regular nationwide surveys, have shown growth in the percentage of young people between 18 and 29 who go online. In 2004, they reported that 77% of people in that age group went online; in 2006, they reported that 88% of that same group were going online, percentages higher than for any other age group (Fox, 2004; Madden, 2006). These figures suggest that predictions about generational differences are on target regarding the diffusion of basic Internet use. Young adults are much more likely than their older counterparts to be online. By limiting our focus to this segment of the population, we concentrate on the age group that is the most connected among all Americans.

Age is associated with variations in Web access and Web use. Young adults lead the age groups with their use of communication tools, such as instant messaging and chatting, and they are also more likely to pursue hobby or entertainment activities, such as downloading music or surfing for fun (Fox & Madden, 2005; Madden, 2003). They also use the medium for getting information about leisure time activities more than others (Howard, Rainie, & Jones, 2001). It is noteworthy, however, that those who are between 29 and 59 tend to use the Internet more than the younger people to perform job research and to use government sites (Fox & Madden, 2005). Research on health information seeking has shown that individuals who seek health materials online are likely to be younger than those who report seeking health information only offline (Cotten & Gupta, 2004). This body of literature suggests that there are differences among age groups in what they do online; however, we cannot conclude from these studies that there is homogeneity in Internet use within age groups.

Many have speculated about the implications of information technology for youth. Jenkins (2004) identified the two most prevalent conflicting myths as "The Myth of the Columbine Generation" and the "Myth of the Digital Generation." Although the former exaggerates the risks of going online, the latter exaggerates the benefits. More research seems to focus on the "Myth of the Digital Generation," envisioning children using the Internet for homework, educational information seeking, and connecting with friends (Livingstone & Bober, 2004). Of those 9- to 19-year-olds who use the Internet on a daily basis, 90% use it for school, 72% use it for e-mail, 70% use it for gaming, and 26% use it to read the news. These figures suggest that although some activities are widespread among all young users, others are restricted to just a portion of them. Overall, researchers in this area have advocated that studies go beyond a focus on access to exploring in detail the nature and quality of use as well as the relevant social conditions (Livingstone, 2003).

Importance of Skill

Some scholars have suggested that access and ability are two disparate issues within the digital divide (DiMaggio, Hargittai, Celeste, & Shafer, 2004; Hargittai, 2003a; Latimer, 2001; Mossberger et al., 2003; Stanley, 2003; van Dijk, 1999; Warschauer, 2003), a difference some have referred to as the "second-level digital divide" (Hargittai, 2002). Research has endeavored to create new ways to measure both computer literacy and online fluency together in order to more fully assess information seeking and disseminating skills (Bunz, 2004). A few studies have started to empirically examine the ways in which ability, or lack thereof, intensifies basic differences in the extent to which people across different segments of the population are able to benefit from use of the medium (Freese, Rivas, & Hargittai, 2006; Hargittai, 2003a; Mossberger et al., 2003; van Dijk, 2005). Some studies have also looked at how specific professions value certain technological skills (Lowrey & Becker, 2001; Russial & Wanta, 1998), but there has been little work on the online abilities of the average user. We focus on differences in general digital literacy, arguing that the way in which people utilize the Internet is at least in part driven by their online skills.

Differences in skill may also increase the "knowledge gap." Although the original knowledge gap hypothesis attributed variation in information acquisition to the SES of population segments (Tichenor, Donohue, & Olien, 1970), scholars have since offered other possible explanations for the knowledge gap. Many of the alternative explanations for the knowledge gap are related to or contingent on SES characteristics. Motivation (Ettema & Kline, 1977; Viswanath, Kahn, Finnegan, Hertog, & Potter, 1993), education level (Brantgarde, 1983; Wanta & Elliott, 1995) and media use (Gaziano, 1983; Griffin, 1990; Kleinnijenhuis, 1991) are three such causal factors that may influence the knowledge gap. A person's ability to search online could also influence the kind of material he or she finds on the Web and thus influence the knowledge gap (Bonfadelli, 2002). Van Dijk (1997, 1999, 2005) similarly predicted that a "usage gap" would materialize between those who use digital technology for work and education and those who use it largely for entertainment.

In the same vein as the knowledge gap hypothesis, research found the Sesame Street effect for television use (Cook et al., 1975). Children from more privileged backgrounds were shown to be more likely viewers of educational programming. In terms of Internet and computer use, this would suggest that young people who are already advantaged gain more from the new technology and thus increase the gap between the haves and have-nots. Thus, even when access to home computers is equal, children who come from a higher SES background experience greater educational gains than do children from lower SES backgrounds (Attewell & Battle, 1999; Livingstone & Helsper, 2007). Although some work has explored this relationship for children, little related research has focused on young adults.

Young people usually consider themselves more skilled at using the Internet than their parents (Livingstone, 2001). However, a study comparing the information-seeking abilities of teens and adults found that teens completed assigned navigating tasks 55% of the time, whereas adults completed the tasks 66% of the time (Nielsen, 2005). Teens are likely to have less patience and poor research skills. Moreover, children between the ages of 9 to 19 lack skills in evaluating material they find online (Livingstone, 2003). They also know little about searching or search engines. Although some teens may be able to serve as the "experts" for Internet use in their households or social circles (Kiesler, Zdaniuk, Lundmark, & Kraut, 2000), children are often inhibited by their lack of social support in Internet use.

Ultimately, there are several ways in which skills can affect Web use, which in turn can affect how the user may be able to benefit from the medium. This discrepancy in beneficial payoffs may exacerbate digital inequality (DiMaggio et al., 2004). This article examines how measures of online skill relate to the Web uses of young adults.

Refined Studies of Internet Use

Some researchers have moved the literature on Web use forward by considering details about people's Internet uses when charting differences among them. Some have suggested more in-depth ways of studying the digital divide to include differences in equipment, autonomy of use, social support, skill, and the purposes for which one is using the Internet (DiMaggio et al., 2004; Hassani, 2006). Another approach to refined studies of differences in online behavior led to the creation of the Internet Connectedness Index (Jung, Qiu, & Kim, 2001), which measures the long-term inequalities of Web users, largely based on whether their Web use will increase their chances for upward mobility or increase certain types of capital. For example, if the Web is used as a toy rather than as a tool, it may not enhance the user's career (Jung et al., 2001), although some have argued that recreational uses can yield beneficial outcomes as well (Sandvig, 2001).

One particular aspect of Internet use that has been explored by several scholars recently concerns what we may call "autonomy of use," that is, freedom to use the technology when and where one wants without constraint from others such as lines of library patrons or employer supervision. In many cases, home access can be considered the most autonomous, especially when coupled with high-speed Internet connection that allows quick access of Web sites, many of which increasingly rely on resource-intensive presentations such as animated graphics or video. Livingstone and Helsper (2007) found that children in the United Kingdom who have access to the Internet at home have been users for more years and tend to spend more time online on a weekly basis than do those who do not have home access. Looking at the access locations of Americans, Hassani (2006) found that people who have more locations at which they can access the Internet tend to engage in activities from which they may benefit more than those with fewer access points.

Like Hassani, following the lead of DiMaggio and Hargittai (2002), we focus on "capital-enhancing" uses of the Web to explore who is more likely to engage in uses that may enhance one's life chances. By such uses we refer to activities that may lead to more informed political participation (seeking political or government information online), help with one's career advancement (exploring career or job opportunities on the Web), or consulting information about financial and health services. We argue, similarly to DiMaggio and Hargittai (2002), that engaging in capital-enhancing activities is more likely to offer users opportunities for upward mobility than certain other types of online activities (e.g., checking sports scores, reading jokes) and thus is of particular concern to our explorations of digital inequality.

Previous research has found that Internet use is associated with heightened, rather than diminished, social capital for users (Best & Krueger, 2006; Katz & Rice, 2002), knowledge of current events (Tewksbury, Weaver, & Maddex, 2001), and civic engagement (Wellman, Haase, Witte, & Hampton, 2002). But these studies do not compare differences among users; rather, they consider the possible beneficial outcome of online activities between users and nonusers. We know less about the extent to which some Internet users may benefit from use of the medium more than others depending on the particular ways in which they embrace the medium.

Refined data about average users' online behavior show that although some activities are nearly universal (e.g., sending or receiving e-mail is nearly universal), most activities are less widespread across all users (Madden, 2003). For example, more people go online to get news than to look for health information (Madden, 2003). We seek to establish even more refined measures of how people use the Internet in order to understand the intricate ways in which inequality becomes manifest in Web use. More specifically, there are differences along demographic lines in how people use the Internet, and there are also disparities in use according to people's experiences with using the medium measured both by number of years online and amount of time spent using the Web on a daily basis.

Research has found that education is a strong predictor of what types of online activities a person will pursue. Sending e-mail, searching for financial, political, or government information, and banking online are all associated with higher education (Howard et al., 2001). Those with a higher education and a higher household income are less likely than those with less education and income to use instant messaging or download music, but they are more likely to seek news and product information or arrange for travel online and to use the Internet for work (Madden, 2003). Users who have more connection points through which to access the Internet—more autonomy of use—are more likely to use the Internet for beneficial purposes, including seeking health information, researching products, purchasing products, and banking online (Hassani, 2006). Users who have a lower income are more likely to use the Internet for gambling (Spooner & Rainie, 2000).

Growth in basic user statistics does not necessarily mean that everybody is taking advantage of the medium in similar ways. Here, we explore how young adults utilize the Internet, which may have an impact on their life chances and social participation. More specifically, we distinguish among the different types of activities people do online, arguing that recreational Internet usage may not have the same capital-enhancing effects as certain other types of use. A specific type of Web use that can potentially influence a user's upward mobility is online job hunting. People who look for jobs online tend to be young adults and more highly educated, to be male (Boyce & Rainie, 2002), to be unemployed (Anderson, 2004), to have a higher household income (Madden, 2003), and to have more computer skills (Anderson, 2004).

Another way to examine which people conduct which types of activities online is to explore user typologies. Scholars have discovered differences among veterans and newcomers online (Horrigan, 2002; Howard et al., 2001). Veterans are defined as people who have been online for at least 3 years, whereas newcomers started using the Internet in the past year. Veterans are more likely to have managed their finances online and to have used the Internet in relation to work (Howard et al., 2001). They are also more likely to be from higher SES groups. Those who have been online longer and who do more online are also more likely to seek health information online (Rice, 2006).

A more specific type of veteran is the "netizen," who, in addition to being online for 3 years, goes online from home every day (Howard et al., 2001). Of netizens, 33% are between 18 and 29, whereas only 5% are 65 or older. Netizens, in comparison to less avid users, engage in more capital-enhancing activities online than do "utilitarians," "experimenters," or "newcomers." They are more likely to trade stocks online, do banking, get news, or do research or training for jobs. Newcomers, who are drawn to the Internet by the allure of fun activities, are relatively unwilling to use the Internet for financial transactions. The activities of the various groups suggest that those with the longest and most frequent use of the Internet are most likely to engage in activities from which they may benefit. Although length of experience and frequency of at-home logging on are said to be the most useful predictors of which activities people do online (Howard et al., 2001), the level of skill may also prove to be a significant predictor. It is this particular relationship of online abilities and types of uses that we aim to test in this article.

Hypotheses

The preceding literature review leads us to consider the following hypotheses regarding young adult Internet users' online skill and types of Web uses. It is important to remember that because we focus on Internet users only, we have already restricted our sample to households with higher levels of SES, a traditional predictor of basic Internet use (NTIA, 2004).

Hypotheses Regarding Skill

Hypothesis 1a: People with higher levels of education will exhibit higher levels of digital literacy.

Hypothesis 1b: Those with more autonomy in using the medium will exhibit higher levels of self-reported skill.

Hypothesis 1c: People with more experience using the Internet will have higher self-reported online skills.

Hypothesis 1d: Users who have high-speed connections will have better self-reported skills.

Hypotheses Regarding Visits to Capital-Enhancing Web Sites

Hypothesis 2a: Education will exhibit a positive relationship with the number of capital-enhancing sites a user visits.

Hypothesis 2b: Autonomy of use will be positively associated with visits to capital-enhancing Web sites.

Hypothesis 2c: Experience will show a positive relationship with users' likelihood of viewing capital-enhancing sites.

Hypothesis 2d: The availability of a high-speed connection will be positively associated with the number of capital-enhancing sites visited by the user.

Hypothesis 2e: Those with higher levels of self-reported skills will have visited more capital-enhancing Web sites than those with lower levels of self-reported skills.

Method

Design and Procedure

A telephone survey was conducted by the national survey firm Schulman, Ronca and Bucuvalas, Inc. using a sample of listed households of 18- to 26-year-olds supplied by the company Experian compiled from more than 3,200 original public and private sources including white pages and census information. Each household was screened to determine the number of adult Internet users between the ages of 18 and 26, and the nearest birthday sampling method was used to randomly select among eligible household members. The survey was conducted using computer-assisted telephone interviewing to reduce interview length and minimize recording errors. The study was administered between August 24 and September 6, 2004, yielding 270 completed interviews. The response rate was 21% (based on American Association of Public Opinion Research [AAPOR] RR3 calculations; see AAPOR, 2008, p. 33). The relatively low response rate limits the extent to which we can draw generalizable conclusions from the sample; nonetheless, given how unique it is to administer such questions to a national sample, the study offers new insights heretofore unexplored on such a population.

Sample

Table 1 presents descriptive statistics about the sociodemographic variables in the sample. In the second set of columns in Table 1, we present comparison figures from the nationally representative sample of the Current Population Survey's Computer and Internet Use Supplement conducted in October 2003. These are the figures most comparable in timing available from the Bureau of Labor Statistics on Internet users. A comparison of the two samples shows that our sample is similar to the national sample of Internet users. Although Internet users in our sample are more highly educated, on average, than users from the large comparison data set, this may be because of the time difference between the two surveys. If anything, this characteristic of our data is likely going to make our findings conservative with respect to any possible concerns raised by having lower levels of education as an Internet user. Nonetheless, given that we also have considerably fewer women in our sample than is representative of national Internet usage statistics, we have to take care in the extent to which we make generalizations based on our findings.

Although it is a relatively small sample, this data set is especially well suited to testing the proposed hypotheses. Unlike many existing sources, these data include detailed information about digital literacy, a measure rarely considered in other studies (perhaps precisely because of a lack of data in this domain). Moreover, the fact that the survey was administered to a national sample of U.S. adults makes it distinctive compared to many studies that often rely on less generalizable samples of users or even smaller samples than our own (e.g., Jung et al., 2001; Kerawalla & Crook, 2002; Loges & Jung, 2001; Wellman et al., 2002). Although the sample cannot be said to be representative

TABLE 1 Background of study participants.

	Study Sample 2004			Current Population Survey 2003		
	M	*SD*	*n*	*M*	*SD*	*n*
Age (18 to 26)	21.80	2.56	270	21.97	2.60	15,400
Gender (1 = *female*, 0 = *male*)	0.36	0.48	270	0.51	0.50	15,400
African American	0.06	0.24	270	0.11	0.31	15,400
Hispanic	0.09	0.29	267	0.16	0.37	15,400
Did not graduate from high school	0.13	0.34	268	0.16	0.37	15,400
High school, no college	0.21	0.40	268	0.33	0.47	15,400
Some college, no degree	0.42	0.49	268	0.37	0.48	15,400
College degree	0.24	0.43	268	0.12	0.32	15,400
Household income (logged)	10.73	0.80	208	10.43	1.00	12,896

Note: Unless otherwise indicated in parentheses after the name of the variable, values are 0 to 1, where the variable is coded 1 if the variable's title is in the affirmative (e.g., "high school, no degree"; 1 = *yes*, 0 = *no*). Current population survey results include only the same age group as this study's sample: 18- to 26-year-old young adults.

of American adults generally speaking, and so we must be cautious about generalizing too widely from our findings, the unique availability of data about the digital literacy of young adults makes it a sufficiently distinctive resource that its use for research is warranted even though future studies will want to expand the special measures to more representative and thus more generalizable samples.

Measures: Independent Variables

As noted in the hypotheses above, we are interested in the predictive power of users' education, autonomy of use, online experiences, and quality of equipment on their online abilities. We control for some demographic characteristics in the models, namely, age and gender. We use degree completed for level of education, where having a college degree serves as the base in the models. We exclude information about income because of a large number of missing values on this variable (23% of respondents did not report this information). Also, we argue that, on the aggregate, income information for this group is likely less reliable than that for other populations. The majority (59.3%) of our respondents lived with their parents, whose income they may not have been able to accurately report. An additional 10% lived with roommates, also complicating the utility of this measure.

We examine the importance of whether people use the medium at home, arguably the location offering the most autonomy for use. We also look at whether respondents have freedom in using a computer for Internet access at work. This variable takes into account whether people have a job and have access to a network-connected computer there in the first place. Only people who are in this situation and claim to be "somewhat" or "very" free in using their work computer for online activities got a value of 1 on this binary variable (others were assigned a 0). We measure users' experiences with the medium in two ways: (a) veteran status (Has the respondent been an Internet user for over 3 years?) and (b) daily online access (as compared to less frequent uses), to account for the frequency with which the respondent uses the Internet. Table 2 presents descriptive statistics for these variables. We include a dummy variable to signal whether the user has access to a high-speed connection.

Measures: Dependent Variables

Of interest here are the online skills of Internet users and the types of Web sites they visit. We measure the former by aggregating responses to five questions about participants' knowledge of Internet-related terms. Previous research has shown that these measures are better proxies for users' actual skill than users' self-perceived abilities (Hargittai, 2005). The five self-rated knowledge items asked respondents to indicate on a 5-point Likert-type scale their understanding of the following terms:

TABLE 2 Descriptive statistics of Web use variables used in the analyses as independent variables.

	Project Participants		
	M	*SD*	*n*
Free to use computer at work for nonwork	0.40	0.49	270
Uses Internet at home	0.80	0.40	270
Web user for more than 3 years	0.86	0.34	270
Goes online less often than daily	0.25	0.43	270
Has high-speed access	0.80	0.40	261

Note: Answer options are coded as follows: 0 = *no*, 1 = *yes*.

jpg, frames, preference settings, newsgroups, pdf. These variables were derived from a study that compared measures of actual skill to survey measures of skill and found a high correlation with the above items (Hargittai, 2005). The resulting index variable in our study ranges from 5 to 25, and its Cronbach's alpha is .84. We refer to this construct as "self-reported skill" and "skill" in this article to be consistent with literature on the topic, while recognizing that it is a measure of self-reported knowledge.

For types of uses, we look at users' tendency to visit Web sites—measured as a visit in the past month—about national and international news, the presidential elections, health and financial information, government services, and product information. This results in an index variable that takes on values from 0 to 7. Table 3 presents the descriptive statistics for these items.

Method of Analysis

We use ordinary least squares (OLS) regression as the method for analyzing the data. Because both of our outcome variables are interval level, this method is most appropriate. We checked the correlations among the independent variables, and none of them were prohibitively high, so they may be included in the analyses simultaneously (see Tables 4 and 5). We also performed other diagnostics to make sure our data meet the requirements of OLS regression analysis.

TABLE 3 Descriptive statistics of Web use variables used in the analyses as dependent variables.

	M	*SD*	*n*
JPEG (1 to 5)	3.2	1.6	263
Frames (1 to 5)	2.6	1.6	264
Preference settings (1 to 5)	3.7	1.4	265
Newsgroups (1 to 5)	2.8	1.4	268
PDF (1 to 5)	3.1	1.6	267
Knowledge item score (5 to 25)	15.2	5.9	260
Sum of capital enhancing sites visited (0 to 7)	3.9	2.0	270

Note: Minimum and maximum values are in parentheses.

TABLE 4 Correlation coefficients for sociodemographic variables used in the analyses.

	1	2	3	4	5	6
1. Age						
2. Female	.0295					
3. Less than high school	–.2881****	.0024				
4. High school	–.2772****	–.0521	–.2002****			
5. Some college	.0133	–.0233	–.3363****	–.4339****		
6. College	.4776****	.0744	–.2206****	–.2846****	–.4782****	

*$p < .1$. **$p < .05$. ***$p < .01$. ****$p < .001$.

TABLE 5 Correlation coefficients for Web use variables used in the analyses.

	1	2	3	4	5	6	7	8	9	10	11	12
1. Web user for more than 3 years												
2. Web rare	–.1658**											
3. Autonomy at work	.0835	–.1254**										
4. Use at home	.1508**	–.3072****	–.0454									
5. High speed	.0676	–.1511**	.1445**	–.1627***								
6. Cap sites	.2050****	–.3110****	.1974***	.1572***	.1503**							
7. JPEG	.2079****	–.3829****	.2032****	.2020****	.0891	.3610****						
8. Frames	.0783	–.2730****	.0777	.0857	.0749	.2716****	.2927****					
9. Preference settings	.2148****	–.2977****	.0433	.2806****	.0424	.3828****	.5605****	.4632****				
10. Newsgroups	.0930	–.2455****	.1534**	.1254**	.1233**	.2906****	.3985****	.4316****	.4784****			
11. PDF	.1893***	–.4169****	.1933***	.1486**	.1073*	.3850****	.6010****	.5659****	.4962****	.4774****		
12. Sum skills	.2061****	–.4317****	.1710***	.2201****	.1221*	.4462****	.8169****	.7903****	.7679****	.6979****	.8173****	

*$p < .1$. **$p < .05$. ***$p < .01$. ****$p < .001$.

Results

Explaining Differences in Self-Perceived Knowledge

As noted earlier, there is little previous work exploring predictors of average users' online skills. The first analysis we explore is to see what explains differences in people's ability to understand Internet-related terms, our measure of their online abilities. We present the results of two models in Table 6. First, we look at the relationship of the demographic variables to self-reported skill without the inclusion of other variables in the model. The results suggest that women are considerably more likely to report lower levels of understanding about Internet-related terms. We also find that respondents with less than a high school education, or only a high school education, or some college education are statistically significantly less knowledgeable about the Internet than those with a college degree. This is evidence of Hypothesis 1a, according to which a higher level of education is associated with higher-level self-reported skills.

Next, we add information about the details of people's Web use and experience to the model. Gender continues to show a statistically significant relationship to understanding Internet-related terms, although the magnitude of the association has decreased. Testing Hypothesis 1b, about the importance of autonomy to people's familiarity with Web-use terms, we find that those with access to a network-connected computer at work who are free to use the machine for surfing the Web report higher levels of knowledge about Internet-related terms than do those who do not have this type of work access. Also, those who use the Internet at home are more likely to be knowledgeable about the medium than those who report no home usage. Regarding Hypothesis 1c, about the relationship of experience to skill, the results suggest that those who have been users for fewer years and those who spend little time online—people who go online less than once daily—are less knowledgeable about the Web than more frequent users. The implication of this finding is that having more time to spend online likely allows people to familiarize themselves with an increasing number of features of the medium. We have to reject Hypothesis 1d, as the results do not show a statistically significant relationship between the quality of Internet access and self-reported skill.

Once we introduce information about the details of people's Internet uses into the model, we no longer find a statistically significant relationship between a user's level of education and his or her self-reported online abilities. This suggests that examining the relationship between self-perceived Internet know-how and education is important to do in the context of additional variables concerning a user's Internet experience.

TABLE 6 Ordinary least squares regression predicting knowledge level of Internet-related terms or "self-reported skill."

Variable	*B*	*SE B*	*B*	*SE B*
Age	–0.149	0.167	–0.036	0.156
Female (= 1)	–2.637***	0.741	–2.151**	0.680
Less than high school	–5.238**	1.415	–1.803	1.351
High school	–2.936**	1.206	–0.894	1.120
Some college	–1.939*	0.963	–1.032	0.881
Autonomy with computer at work			1.504*	0.681
Use at home			1.895***	0.891
Veteran user (more than 3 years)			2.410	1.020
Rare user (less than once per day)			–4.282****	0.828
High-speed connection			1.106	0.848
Intercept	21.495	4.084	13.650****	4.012
n	258		249	
R^2	.10		.28	
Adjusted R^2	.08		.25	

*p < .1. **p < .05. ***p < .01. ****p < .001.

Explaining Different Levels of Capital-Enhancing Online Behavior

As we discussed above, there are numerous ways in which users may take advantage of their Internet access. We have argued, following the work of others (DiMaggio et al., 2004), that not all online activities are equally important to enhancing one's human, financial, and social capital. If the Internet is to serve as a life-enhancing medium, we must examine in detail the purposes to which users put it. Here, we analyze data about our sample to explore which factors explain different levels of capital-enhancing online activities among young adult Internet users across the United States.

Table 7 presents the findings from OLS regression analyses on the number of capital-enhancing activities in which users engage. Again, we run two models, first looking at the relationship of demographic variables to types of uses on their own. We find evidence for Hypothesis 2a, given that the data display a strong relationship between level of education and types of Web uses. The less education a user has, the less likely he or she is to look at capital-enhancing sites. This finding suggests that social inequalities may be perpetuated online, given that those who are already in more privileged positions are more likely to use the medium for activities from which they may benefit.

In the full model, the association of education and visits to capital-enhancing sites remains strong, although it is weakened somewhat, suggesting, again, mediating effects of the other variables. We find some support for Hypothesis 2b as autonomy of use (measured as freedom to use one's work computer for Internet surfing) displays a statistically significant relationship with number of capital-enhancing Web sites visited at the $p = .1$ level. However, Hypothesis 2c is not supported as being a veteran, or a rare user does not relate to such site visits. We also do not find any effect of a high-speed connection, rejecting Hypothesis 2d as well.

An innovative component of this study is to use self-reported skill as a predictor of types of uses to which people put their Internet access. We argue that it is incorrect to assume that the types of sites people visit are simply a reflection of the types of sites they *prefer* to visit. Rather, we contend that online behavior is at least in part a reflection of people's online abilities. To test this hypothesis, we included information about people's self-reported Web-use skills in the model predicting capital-enhancing site visits. This variable exhibits the strongest statistical significance with our outcome measure. Those who report higher levels of knowledge about Internet-related terms are considerably more likely to visit capital-enhancing Web sites than their less knowledgeable counterparts.

TABLE 7 Ordinary least squares regression predicting number of capital-enhancing online activities.

Variable	**B**	***SE B***	**B**	***SE B***
Age	–0.024	0.056	0.009	0.054
Female (= 1)	–0.275	0.248	0.003	0.240
Less than high school	–2.039****	0.463	–1.009**	0.468
High school	–1.328***	0.409	–0.856**	0.387
Some college	–0.820**	0.327	–0.448	0.305
Autonomy with computer at work			0.395*	0.238
Use at home			0.231	0.311
Veteran user (over 3 years)			0.069	0.356
Rare user (less than once/day)			–0.320	0.302
High-speed connection			0.356	0.294
Self-reported skill (knowledge-item score)			0.119****	0.022
Intercept	5.393	1.365	1.782	1.418
n	268		249	
R^2	.09		.27	
Adjusted R^2	.08		.24	

*$p < .1$. **$p < .05$. ***$p < .01$. ****$p < .001$.

Our innovation to gather data about self-reported skill and to include information about it in the model has proved significant.

Discussion

Initial studies of the digital divide have solely explored differences in access statistics. This was a legitimate focus of inquiry in the early stages of the Internet's diffusion given that differences were considerable across various segments of the population and the medium was not yet evolved enough to offer numerous types of uses. As more and more people have gone online and as the Internet has become a source of an increasing number of activities, it is important that we include more nuanced measures of use in our studies of online inequalities.

In this study, our aim has been to focus on two as-of-yet mostly neglected refined measures of use: skill and visits to capital-enhancing Web sites. Although other researchers have considered some of these variables, they have rarely done so using data from a national sample of users. Given the relatively low response rate and small sample, we have to be cautious about generalizing too far from our sample. Nonetheless, given the unique nature of the data set for a national sample, this study is an important first step in recognizing some relationships between Internet users' background characteristics and their self-reported Internet skills and usage. In particular, for this sample of 18- to 26-year-old American adults, we have shown that women are less likely to claim knowledge about online terminology and features, and those who use the Web infrequently also report lower levels of know-how about it. Interestingly, we also found that educational background does not display a statistically significant relationship with self-reported online skill once we control for details about respondents' uses, such as their autonomy and experience with the medium.

An especially novel contribution of our project is that we have been able to use data on users' self-reported Web-use skill as an independent variable to examine its relationship to people's online activities. We have found that educational background influences people's likelihood to visit capital-enhancing sites. We have also found that those with higher levels of self-reported skill are more likely to visit the types of Web sites that may contribute to improving their life chances and from which their human and financial capital may benefit. Future studies should extend data collection to larger, more representative samples to test whether our results hold for more generalizable populations.

Overall, our findings suggest that information about access and use statistics are not refined enough measures to explore all the ways in which differentiated Internet use may contribute to social inequality. Insofar as Internet use can enhance people's life chances, it is the types of activities for which people use the medium that will be most important in examining potential divides. By focusing on the amount of capital-enhancing online activities in which people engage and uncovering the background characteristics that predict these differences, we have contributed to a more in-depth understanding of the possible sources of digital inequality among populations whose members are, for the most part, already on the "right" side of the digital divide. Of course, lacking longitudinal data, we cannot make causal arguments. Nonetheless, our findings suggest that simply being connected will not necessarily solve all potential sources of inequality, and so studies of more nuanced uses of the Web are important as Internet use spreads to an increasing portion of the population. Moreover, because we find that self-reported skill is an important predictor of whether people use the medium for capital-enhancing activities, a focus on training that augments people's self-confidence and knowledge about the Internet must be a component of initiatives that strive to encourage the diffusion of the medium's use across different population segments.

About the Authors

Eszter Hargittai (PhD, Princeton University, 2003) is associate professor of communication studies and faculty associate of the Institute for Policy Research at Northwestern University, where she heads the Web Use Project. Her research looks at the social and policy implications of information technologies, with a particular interest in questions of social inequality.

Amanda Hinnant (PhD, Northwestern University, 2006) is assistant professor at the Missouri School of Journalism, University of Missouri. Her interests include health journalism, information seeking, sociology of news, and magazine trends.

Note

1. The authors thank the reviewers and editor for comments from which the article greatly benefited. The authors are also indebted to Paul DiMaggio, Edward Freeland, and Peter Miller for their helpful input. Support from the Markle Foundation is kindly acknowledged.

References

American Association of Public Opinion Research. (2008). *Standard definitions: Final dispositions of case codes and outcome rates for surveys.* Lenexa, KS: Author.

Anderson, B. (2004). *Everyday research in the knowledge society: Who uses ICTs to find job and health information.* Essex, UK: University of Essex.

Attewell, P., & Battle, J. (1999). Home computers and school performance. *Information Society, 15*(1), 1–10.

Barzilai-Nahon, K. (2006). Gaps and bits: Conceptualizing measurements for digital divides. *Information Society, 22*(5), 269–278.

Best, S. J., & Krueger, B. S. (2006). Online interactions and social capital: Distinguishing between new and existing ties. *Social Science Computer Review, 24*(4), 395–410.

Bonfadelli, H. (2002). The Internet and knowledge gaps: A theoretical and empirical investigation. *European Journal of Communication, 17*(1), 65–84.

Boyce, A., & Rainie, L. (2002). *Online job hunting.* Washington, DC: Pew Internet and American Life Project.

Brantgarde, L. (1983). The information gap and municipal politics in Sweden. *Communication Research, 10,* 357–373.

Bunz, U. (2004). The Computer-Email-Web (CEW) Fluency Scale—development and validation. *International Journal of Human-Computer Interaction, 17*(4), 479–506.

Cook, T. D., Appleton, H., Conner, R. F., Shaffer, A., Tamkin, G., & Weber, S. J. (1975). *"Sesame Street" revisited.* New York: Russell Sage.

Cooper, J. (2006). The digital divide: The special case of gender. *Journal of Computer Assisted Learning, 22*(5), 320–334.

Cooper, J., & Weaver, K. (2003). *Gender and computers.* Mahwah, NJ: Lawrence Erlbaum.

Cotten, S. R., & Gupta, S. S. (2004). Characteristics of online and offline health information seekers and factors that discriminate between them. *Social Science & Medicine, 59*(9), 1795–1806.

DiMaggio, P., & Hargittai, E. (2002, August). *The new digital inequality: Social stratification among Internet users.* Paper presented at the American Sociological Association annual meetings, Chicago.

DiMaggio, P., Hargittai, E., Celeste, C., & Shafer, S. (2004). Digital inequality: From unequal access to differentiated use. In K. Neckerman (Ed.), *Social inequality* (pp. 355–400). New York: Russell Sage.

Ettema, J. S., & Kline, F. G. (1977). Deficits, differences, and ceilings: Contingent conditions for understanding the knowledge gap. *Communication Research, 4,* 179–202.

Fallows, D. (2004). *The Internet and daily life* (No. 131). Washington, DC: Pew Internet and American Life Project.

Foehr, U. G. (2006). *Media multitasking among American youth: Prevalence, predictors and pairings.* Menlo Park, CA: Kaiser Family Foundation.

Fox, S. (2004). *Older Americans and the Internet.* Washington, DC: Pew Internet and American Life Project.

Fox, S., & Madden, M. (2005). *Generations online.* Washington, DC: Pew Internet and American Life Project.

Freese, J., Rivas, S., & Hargittai, E. (2006). Cognitive ability and Internet use among older adults. *Poetics, 34*(4–5), 236–249.

Gaziano, C. (1983). Neighborhood newspapers, citizen groups, and public affairs knowledge gaps. *Journalism Quarterly, 16,* 556–566, 599.

Griffin, R. J. (1990). Energy in the eighties: Education, communication and the knowledge gap. *Journalism Quarterly, 67,* 554–566.

Hargittai, E. (2002, April). Second-level digital divide: Differences in people's online skills. *First Monday, 7*(4). Retrieved April 17, 2008, from http://firstmonday.org/issues/issue7_4/hargittai/index.html

Hargittai, E. (2003a). *How wide a Web? Inequalities in accessing information online.* Unpublished doctoral dissertation, Princeton University, Princeton, NJ.

Hargittai, E. (2003b). Informed Web surfing: The social context of user sophistication. In P. E. N. Howard & S. Jones (Eds.), *Society online* (pp. 257–274). Thousand Oaks, CA: Sage.

Hargittai, E. (2005). Survey measures of Web-oriented digital literacy. *Social Science Computer Review, 23,* 371–379.

Hargittai, E., & Shafer, S. (2006). Differences in actual and perceived online skills: The role of gender. *Social Science Quarterly, 87*(2), 432–448.

Hassani, S. N. (2006). Locating digital divides at home, work, and everywhere else. *Poetics, 34*(4–5), 250–272.

Horrigan, J. B. (2002). *Getting serious online.* Washington, DC: Pew Internet and American Life Project.

Howard, P. N., Rainie, L., & Jones, S. (2001). Days and nights on the Internet: The impact of a diffusing technology. *American Behavioral Scientist, 45*(3), 383–404.

Jenkins, H. (2004, September 3). The myths of growing up online. *Technology Review.* Retrieved April 16, 2008, from http://www.technologyreview.com/Biotech/13773/

Jung, J.-Y., Qiu, J. L., & Kim, Y.-C. (2001). Internet connectedness and inequality: Beyond the "divide." *Communication Research, 28,* 507–535.

Katz, J. E., & Rice, R. E. (2002). *Social consequences of Internet use: Access, involvement and interaction.* Cambridge, MA: MIT Press.

Kennedy, T., Wellman, B., & Klement, D. (2003). Gendering the digital divide. *IT & Society, 1*(5), 72–96.

Kerawalla, L., & Crook, C. (2002). Children's computer use at home and at school: Context and continuity. *British Educational Research Journal, 28,* 751–771.

Kiesler, S., Zdaniuk, B., Lundmark, V., & Kraut, R. (2000). Troubles with the Internet: The dynamics of help at home. *Human-Computer Interaction, 15*(4), 323–351.

Kleinnijenhuis, J. (1991). Newspaper complexity and the knowledge gap. *European Journal of Communication, 6*(4), 499–522.

Latimer, C. P. (2001). *The digital divide: Understanding and addressing the challenge.* New York: New York State Forum for Information Resource Management.

Livingstone, S. (2001). Children on-line: Emerging uses of the Internet at home. *Journal of the IBTE, 2,* 1–7.

Livingstone, S. (2003). Children's use of the Internet: Reflections on the emerging research agenda. *New Media & Society, 5*(2), 147–167.

Livingstone, S., & Bober, M. (2004). *UK children go online: Surveying the experiences of young people and their parents.* London: London School of Economics and Political Science. Retrieved April 17, 2008, from http://www.children-go-online.net

Livingstone, S., & Helsper, E. (2007). Gradations in digital inclusion: Children, young people, and the digital divide. *New Media & Society, 9*(4), 671–696.

Loges, W. E., & Jung, J.-Y. (2001). Exploring the digital divide: Internet connectedness and age. *Communication Research, 28,* 536–562.

Lowrey, W., & Becker, L. B. (2001). The impact of technological skill on job-finding success in the mass communication labor market. *Journalism and Mass Communication Quarterly, 78*(4), 754–770.

Madden, M. (2003). *America's online pursuits.* Washington, DC: Pew Internet and American Life Project.

Madden, M. (2006). *Internet penetration and impact.* Washington, DC: Pew Internet and American Life Project.

Mossberger, K., Tolbert, C. J., & Stansbury, M. (2003). *Virtual inequality: Beyond the digital divide.* Washington, DC: Georgetown University Press.

National Telecommunications and Information Administration. (2002). *A nation online: How Americans are expanding their use of the Internet.* Washington, DC: Author.

National Telecommunications and Information Administration. (2004). *A nation online: Entering the broadband age.* Washington, DC: Author.

Nielsen, J. (2005). *Usability of Websites for teenagers.* Fremont, CA: Nielsen Norman Group.

Ono, H., & Zavodny, M. (2003). Gender and the Internet. *Social Science Quarterly, 84*(1), 111–121.

Rice, R. E. (2006). Influences, usage, and outcomes of Internet health information searching: Multivariate results from the Pew Surveys. *International Journal of Medical Informatics, 75*(1), 8–28.

Russial, J., & Wanta, W. (1998). Digital imagine skills and the hiring and training of photojournalists. *Journalism and Mass Communication Quarterly, 75*(3), 593–605.

Sandvig, C. (2001). Unexpected outcomes in digital divide policy: What children really do in the public library. In B. M. Compaine & S. Greenstein (Eds.), *Communications policy in transition: The Internet and beyond* (pp. 265–293). Cambridge, MA: MIT Press.

Spooner, T., & Rainie, L. (2000). *African-Americans and the Internet.* Washington, DC: Pew Internet and American Life Project.

Stanley, L. D. (2003). Beyond access: Psychosocial barriers to computer literacy. *Information Society, 19*(5), 407–416.

Tewksbury, D., Weaver, A. J., & Maddex, B. D. (2001). Accidentally informed: News exposure on the World Wide Web. *Journalism and Mass Communication Quarterly, 78*(3), 533–554.

Tichenor, P. J., Donohue, G. A., & Olien, C. N. (1970). Mass media flow and differential growth in knowledge. *Public Opinion Quarterly, 34*(2), 159–170.

van Dijk, J. A. G. M. (1997). *Universal service from the perspective of consumers and citizens.* Brussels, Belgium: Information Society Forum.

van Dijk, J. A. G. M. (1999). *The network society: Social aspects of new media.* London: Sage.

van Dijk, J. A. G. M. (2005). *The deepening divide.* London: Sage.

Viswanath, K., Kahn, E., Finnegan, J. J. R., Hertog, J., & Potter, J. D. (1993). Motivation and the "knowledge gap": Effects of a campaign to reduce diet-related cancer risk. *Communication Research, 20,* 546–563.

Wanta, W., & Elliott, W. R. (1995). Did the "magic" work? Knowledge of HIV/AIDS and the knowledge gap hypothesis. *Journalism and Mass Communication Quarterly, 72*(2), 312–321.

Warschauer, M. (2003). *Technology and social inclusion*. Cambridge, MA: MIT Press.

Wasserman, I. M., & Richmond-Abbott, M. (2005). Gender and the Internet: Causes of variation in access, level, and scope of use. *Social Science Quarterly, 86*(1), 252–270.

Wellman, B., Haase, A. Q.,Witte, J., & Hampton, K. (2002). Capitalizing on the Internet: Social contact, civic engagement, and sense of community. In B. Wellman & H. Haythornthwaite (Eds.), *The Internet in everyday life* (Vol. 45, pp. 436–455). Oxford, UK: Blackwell.

The Young and Techless? Investigating Internet Use and Problem-Solving Behaviors of Young Adults in Singapore

Pauline Hope Cheong and Jessie Poon

This article critically examines the technologically-savvy image of young adults by investigating the digital divide issues underlying youth internet use, including their daily computer and internet problem-solving behaviors. The study draws on data from a web-based questionnaire and face-to-face interviews with young adults in Singapore, a country where internet adoption is pervasive. Contrary to popular conceptualizations of youths as a cohort of technically-savvy experts, the findings showed considerable variance in their internet expertise and problem-solving behaviors, with some demonstrating limited knowledge of internet use and awareness of troubleshooting strategies. The analyses also showed that internet skills and self-efficacy in internet-related problem-solving behaviors were significantly related to the internet practices of young adults. The findings suggest that in wired contexts, variations in post-adoption patterns may reflect more accurately the extent and presence of social stratification, extending the meaning and scope of the digital divide.

Keywords: digital divide, internet use, problem-solving behaviors, Singapore, young adults

Images of Singapore's young wired generation abound in popular media. They have been called the 'Internet Generation', the 'tech-whizzes' and 'always connected'. Accompanying these portrayals are often celebratory rhetoric regarding the e-lifestyle of young people who have grown up with computers, the internet and cellphones and are known to be using these technologies to communicate anytime, anywhere (Sng, 2001). According to Don Tapscott in *Growing Up Digital,* 'The Net Generation has arrived! . . . For the first time in history, children are more comfortable, knowledgeable and literate than their parents about an innovation central to society' (1998: 1–2). But how are contemporary young adults embracing the so-called 'wired lifestyle' in their daily lives, and are some of them lagging behind others in terms of their internet use?

Recent studies in the USA and abroad indicate high internet adoption rates among the younger generation. For example, reports compiled by the Pew Internet & American Life Project found that nearly 90 percent of young people aged 12 to 17 in 2004 had online access, up from about three-quarters of youths in 2000. By comparison, about 66 percent of American adults in 2004 used the internet (Lenhart et al., 2005). A recent 2007 survey showed that 85 percent of teens between 12 and 17 years engage at least occasionally in electronic personal communication via email, text messaging and posting comments on social networking sites (Lenhart et al., 2008). Research by the UK Children Go Online project found that most of the youths aged 9 and 19 years have access to the internet at home, and almost all have access at school (Livingstone et al., 2004). In the past decade, internet adoption has burgeoned in the economically developed cities of Asia (Internet World Statistics, 2008). In Singapore, the majority of youths have access to computers and are using broadband internet connections both inside and outside of their homes (Jung et al., 2005; Kuo et al., 2002). These high penetration rates have led some commentators to dismiss the digital divide 'myth' in confidence that adoption and usage figures will progressively rise due to declining costs of computers and information technology (Compaine, 2001).

The digital divide, defined by conventional access and computer ownership terms, seems to be a transient, even irrelevant, concept in this information age. However, this article argues that high penetration rates do not equate automatically to the demise of the digital divide. On the contrary, other aspects of the divide may endure and be deepening due to widening inequalities in economic resources, social support and technical skills for internet use. The internet is becoming so ubiquitous, even banal, that successful bridging of the digital divide is taken for granted often in most industrialized societies. Yet, paradoxically, notions of internet 'access' and 'use' have received less critical attention. As Stephen Graham reminds us, the need to address the invisibility of sociotechnical power is a 'critical challenge because of the growing importance of truly opaque and invisible systems of social control that are based on the continuous use of automated software in order to mediate all sorts of social relations' (2004: 24). In particular, researchers of the UK Children Go Online project have urged for attention to the variation in the quality of internet use among young people, since 'the blanket label of 'online experts' gives children confidence but may deny them the attention and guidance that could benefit them' (Livingstone et al., 2005: 5).

The purpose of this article is to examine critically the technologically-savvy image of youths by investigating the digital divide issues underlying the internet use of young adults in Singapore, including their computer and internet problem-solving behaviors. More than two-thirds of Singaporeans have adopted the internet, a proportion comparable to the USA and higher than Australia, the UK and many countries in the European Union (Internet World Statistics, 2005). These high penetration rates make Singapore an interesting context in which to examine the social variations associated with post-adoption digital divides. Insights from the study extend to other wired societies where internet adoption is pervasive, especially among young persons. Given that the internet is utilized increasingly to create and disseminate information, the digital divide phenomenon raises important concerns about social stratification for young adults seeking economic and political participation in mediated contexts (Norris, 2001).

First, this article reviews some of the relevant literature concerning the digital divide. Second, it presents the responses from survey research and a complementary analysis of responses from interview research among youths. Finally, it concludes with a discussion of the results and policy implications.

Literature Review

Investigating the Digital Divide: Beyond Access to Internet Use and Problem-Solving Behaviors

Over the past decade, digital inequality has been associated most commonly with the digital divide. The earliest phases of the debate on the digital divide, as popularly discussed in America, focused on the 'information haves and information have-nots' (Lynch, 2001) and between heavy and light internet users based on time spent online. However, calibrating the digital divide using computer ownership and time online indicators may be misleading, as people of different skills and motivations use the internet in varied ways (Kling, 2000). As a corrective sequel, another area of scholarship has examined the multidimensional nature of the digital divide and highlighted the influence of social context and relationships on internet use (e.g. Cheong, 2007; Cheong and Wilkin, 2005; Jung et al., 2001; Van Dijk and Hacker, 2003). After a review of the literature on internet access and inequality, DiMaggio et al. (2004) stressed that future research should focus on inequalities in technical apparatus, autonomy of use, skill, availability of social support and variation in use. For example, Hargittai (2002) conceptualized a second-level digital divide in terms of online skills, as there is considerable variance in the ways in which individuals access and find information online. Consequently, a chasm differentiating between highly-skilled and low-skilled internet users exists, which has implications for users' productivity and the potential benefits that can be reaped from the internet.

Rather than mere physical access to computing resources, internet skills, usage patterns and problem-solving behaviors are the focus of this article, since today's internet users (even seasoned and highly-educated ones) face a growing number of digital opportunities and challenges. Digital opportunities encompass the multiple interactive and creative uses of the internet to obtain, produce and disseminate information. A corpus of research tracking internet use in the USA has shown how the internet has been used by many youths for educational, entertainment and social activities (Lenhart et al., 2005). Similarly, cross-sectional survey research among Singaporean youths also shows that they are using the internet for information-seeking, chatting, playing games, reading news and listening and downloading music (Kuo et al., 2002).

Besides functional or activity-based internet use, part of contemporary internet use also necessitates dealing with computer and internet-related problems on a regular, if not daily, basis. Digital challenges include the perennial threat of computer viruses, chain emails, intrusive cookies and pop-ups, spam and images, with the accompanying need to fix glitches, upload patches and upgrade one's computing systems and software systems to maintain system compatibility with others. Annually, computer parasitic, boot, link and macros viruses, worms and Trojan horse programs cost individuals and organizations millions of dollars. Traditional responses have involved the use of anti-virus software, which remove infections or restrict the transmission of infected communications across firewalls. Yet unprotected systems can become infected and rapidly propagate that infection to other systems, giving rise to multiple social and ethical problems (Neubauer and Harris, 2002). Compounding the effects of actual threats are perceived threats and difficulties in dealing with hoax virus alerts, misinformation, urban legends and even cyberterrorism. As Treese explains, 'in the specialized news of the computer security world, it's nearly impossible to keep up with the flood of detail about new viruses, worms, spam, spyware, or other attacks against computers on the network' (Treese, 2004: 13). Furthermore, research on web-based information systems illustrates how internet use may be fraught with technical glitches, frustrating internet use. For example, Hara and Kling (2000) documented college students' frustrations in a distance education course due to communication breakdowns and technical complexities, noting the importance of peer assistance in the efficacious use of web-based learning systems. Dutton et al. (2004) also illustrated in a case study how hardware and software problems, combined with the pressure of continual system upgrades, represented a more substantial barrier than anticipated for internet users, thereby slowing the processes of technological diffusion and innovative uses of an online e-learning system among college instructors and students.

Hence, technical skills and internet-related social support become important resources for internet use, especially in the light of potential hardware and software problems and global information flows carrying with it rapidly circulating worms and viruses. In their report, the researchers at the Centre for Competence for Informal Education in Germany highlighted the importance of peer structures as influential social capital for youth internet use and access to new online activities (Otto et al., 2003). 'Social support networks', or the availability of others to whom one can turn to seek social support and enhance one's online skills, may play a crucial role in internet use (DiMaggio et al., 2004). More specifically, the results from a study on internet use among Asian youths found that the availability of internet-related help from others affected the scope of their online activities (Jung et al., 2005). Yet youths' perceptions and actual internet problem-solving behaviors remain under-studied, in part due to the assumption of 'problem-free' access perpetuated by popular culture, which is aimed often at portraying today's internet users to be autonomous, technological experts.

The Internet and the Information Society in Singapore

Located in East Asia, Singapore is a country that aims to maintain its role as a financial hub by establishing an advanced telecommunications infrastructure (Tan, 1998). The commitment of policymakers to technological development has made this country one of the most wired nations in the world. In 1997, the Ministry of Education laid out a 'Masterplan for IT' to provide an IT-enriched school environment for every student (see www.moe.edu.sg). In 1999, the Infocommunications Development Authority of Singapore was formed to function as a policymaker, regulator and developer of Singapore's infocommunications industry, in order to create an e-government and promote e-commerce (see www.ida.gov.sg). Some programmes that the Authority has implemented include the Singapore One broadband network and 'eCelebrations Singapore', an annual outreach programme designed to increase the accessibility of the internet. Hence, the case of Singapore illustrates the power of top-down government initiatives to diffuse information technology (Tang and Ang, 2002), thereby socially re-engineering, in some ways, the stratification processes that divide individuals at the micro-level. The results from a national survey of 432 young adults aged 15 to 29 reported that nine out of 10 have personal computers (Wei, 2000). More recently, data from Singapore, Seoul and Taipei show that more than 95 percent of youths are internet users (Jung et al., 2005).

However, it is the contention of this article that high adoption rates do not equate to the demise of the digital divide, since variation of use exists among those who possess the requisite skills and problem-solving behaviors to take advantage of digital opportunities and manage digital challenges. Indeed, past studies in the Asian context have highlighted that cleavages in internet use exist, even among and within Japan, South Korea and Singapore, which have been classified as 'high-access countries' by the International Telecommunications Union (Ono, 2005). According to an analysis of national data collected in 2000 in South Korea, almost half of the population had low digital skills and about one-third cited 'complex and hard to use' as a reason for not using a computer. These findings highlight the presence of new and widening gaps in digital skills among Koreans despite the aggressive promotion of information technology (IT) and broadband access in the last decade (Park, 2002). Building upon prior research, this study takes the view that stratification processes cut down even to the level of internet skills and internet problem-solving behaviors. Therefore, this study investigates the following questions.

> In what ways and to what extent do young adults experience computer and internet problems?
>
> In what ways and what resources do young adults use to solve their computer and internet problems?
>
> How do demographics, internet skills, internet problem-solving behaviors, individually and collectively, affect internet use?

Method

Data

The data for this article were collected as part of a larger project on the internet among young adults (aged 15 to 29 years), supported by the National Youth Council of Singapore. The study employed a multi-method approach where surveys were administered, followed by a series of interviews with a subset of youths. In compliance with the interests of the National Youth Council, data collection targeted young adults who were current students in local institutions of higher education (including universities, polytechnics and institutes of technical education). The data collection process took place between July and September 2005. For the survey, cooperation was sought from various lecturers to email invitations to their students to solicit their participation in a web-based questionnaire (taking about 15 minutes to administer), asking for information such as their overall media usage, problem-solving behaviors and social relationships. The youths were directed to the questionnaire hosted at the National Youth Council website. In compliance with research regulations, the respondents were asked to read an instruction page explaining the study and had to click a box verifying that they were Singapore citizens, understood the instructions and were eligible to participate. The respondents were offered a movie ticket as an incentive to participate. In all, 767 responses were collected. Surveys were complemented by interpersonal interviews with a subset of 38 youths from the survey population. Semi-structured interviews were constructed to provide in-depth information from the youths' point of view. Researchers arranged the interviews at the site of the respondents' convenience and the interviews typically lasted an hour.

Quantitative Measurement

For the survey, the following variables were used to measure internet use, internet skills, internet-related problem-solving behaviors and demographic information.

Internet use was measured by two forms of online behaviors: interactive and creative uses of the internet. Interactive use of the internet was a composite variable created by a summation of 10 online activities adapted from a report on youths and internet use (Livingstone et al., 2004). Responses were derived from the question, 'Here are some of things people do on websites. Do you ever do any of these things?' Activities included: do a quiz, send an email or SMS to a site, vote, send pictures or stories to a site, contribute to a message board, offer advice to others, fill in a form, sign a petition, seek information, seek advice. Creative use of the internet was measured by asking the respondents, 'Have you ever created a website by yourself?' The respondents who were website creators were coded as '1', non-website creators were coded as '0'.

Perceived internet skills were measured by responses to the question, 'On a scale of 1 to 10, where 10 represents extremely good computing and internet skills and 1 represents very poor computer and internet skills, how would you rate yourself?'

Self-efficacy and social support in problem-solving behaviors were measured by two types of problem-solving behaviors: by oneself and social support from one's significant others and friends. First, self-efficacy in problem-solving was measured by responses to the following statement: 'I am very confident with solving computer problems by myself'. Responses were measured on a Likert scale of agreement with the statement (where 1 = 'strongly disagree' and 5 = 'strongly agree'). Second, computer and internet-related social support was measured by the average of the summed response to four statements adapted from a general social support scale measuring the perceived availability of help, that a respondent feels that they have someone to turn to in times of need to fit the context of computing and internet use (Sarason et al., 1987). The statements were: 'I have a special someone on whom I rely on when I need to solve computing problems', 'I can count on my friends when things go wrong with my computer', 'I have a special person who is a real source of help to me when I encounter problems with my internet', and 'I have friends with whom I can share my internet related problems'. Responses were measured on a Likert scale of agreement with the statement (where 1 = 'strongly disagree' and 5 = 'strongly agree'). A principal components analysis with Varimax rotation yielded a single factor with an Eigenvalue of 2.3, explaining 57 percent of the variance. Cronbach's alpha for the scale was acceptably high at .75.

Computer and internet problems The respondents were asked whether and how frequently they experience computer and internet problems. The question involving computer problems was phrased as follows:

> *Computer problems cover hardware and software problems. Examples of computer problems include instances when your computer stalls, your programs will not boot, occurrence of beep codes and bio error messages, when you cannot load programs, open programs, or upgrade a software, etc. How often do you experience computer problems?*

The question involving internet problems was phrased as: 'Internet-related problems refer to virus and email issues. How often do you experience internet-related problems?' Responses were grouped into eight categories, ranging from 'never', to 'once in five years', to 'several times a week for both questions'.

Resources for problem-solving behaviors Following each of the above questions on computer and internet problems, the respondents were asked about the resources they utilize to solve their internet and computer problems. The responses included family members, friends from school, co-workers, customer service, books and references, technical support, online friends or virtual communities, newsgroups or online forums, no one or solve oneself, and website or online documentation. Multiple responses were accepted.

Demographics The respondents were asked their age on their last birthday. Education was measured by a single item, 'highest level of education attained', with six response categories from 'GCE 'N' level' (equivalent to middle school and below) to 'Masters' degree'. Income (total gross family income per month) was measured by eight response categories from '$1,000 or less' (equivalent to US$615) to 'above $7,000' (equivalent to US$4,300). Male gender was coded as '1' and female was coded '0'.

Qualitative Research

The respondents were asked if they had ever encountered any computer and internet-related problems and to describe the nature and consequences of their most recent problems. They were asked what they did to solve their internet-related problems, the resources that they used or the people that they approached for help (if any), and how long it took to solve their most recent problem. Also, they were asked if anyone else has ever asked them for help to solve their computer and internet problems.

Analysis

Quantitative and qualitative data were triangulated to allow comparison of information sources and the verification of the validity of information received (Tashakkori and Teddlie, 1998). Statistical analyses were performed using the Statistical Package for the Social Sciences (SPSS) 12.0 and employed a .05 level of significance. Descriptive statistics (frequency counts) were used to examine the frequency and nature of computer and internet problems and resources used to solve the problems. Descriptive statistics (means, standard deviations) and Pearson product-moment correlations were used to explore the rest of the study factors. Regression analyses were tested using hierarchical analyses to control for their intercorrelations and to identify the variance explained by each block. Block 1 contained the demographic variables age, gender, income and education. Block 2 contained the perceived internet skills factor and Block 3 contained the self-efficacy and social support in problem-solving variables. The interviews were transcribed in full for textual analyses as well as the use of verbatim quotations for reporting the research (Barone and Switzer, 1995). For the purposes of this article, quotes are used to provide descriptive data to complement the survey analysis, in order to elucidate young adults' variation in computing and internet problem-solving behaviors.

Results

Sample

Table 1 summarizes the characteristics of the sample with regard to gender, age, education, internet skills, problem-solving behaviors and internet use. Of the total 767 respondents who were internet users, 470 (61%) were male and 297 (39%) were female. The average age of respondents was 22 years. Of the sample, 15 percent received less than a high school education, 45 percent attained a high school education and 20 percent attained at least a college degree. The respondents' household income ranged from fewer than $1,000 to more than $7,000 per month, with a median household income in the range of $2,001 to $3,000 per month.

Relationships between internet skills, problem-solving behaviors and internet use were moderately positive. As shown in Table 2, correlations between the variables were small, ranging from .04 to .54. The respondents who indicated that they had good computing and internet skills were more likely to be males, older, of a higher education and more confident of solving computing problems by themselves. The respondents who rated themselves as being confident in independently solving computer problems were more likely to be males, older and had attained a higher education.

Frequency of Computer and Internet Problems

The first research question concerned the ways in, and the extent to, which young adults experienced computer and internet problems in their everyday lives. The survey responses showed that almost all of the respondents had experienced computer problems involving hardware and software problems. More than 90 percent of the respondents reported experiencing computer problems at least once a year, about 30 percent of them experienced problems once in three months and about one-fifth experienced problems several times a month. With regard to internet problems, the survey responses showed that more than 85 percent of the respondents reported experiencing internet problems at least once a year. About one-quarter of them experienced virus and email problems once in three months, and about 20 percent experienced internet problems once a month. About nine percent of the respondents reported experiencing internet problems several times a week. Therefore, the survey responses reveal that computing and internet problems were a familiar, if not common, phenomenon for almost all respondents.

TABLE 1 Means and standard deviations for internet use, internet skills, internet problem-solving behaviors, frequency of internet and computer problems and demographic variables ($N = 767$).

Variables	Mean	SD
Interactive internet use	7.0	2.2
Creative internet use	.54	.5
Perceived internet skills	6.6	1.6
Self-efficacy in problem-solving	2.9	1.0
Social support in problem-solving	3.3	.8
Age	21.8	3.2
Gender	.4	.49
Education	3.5	1.0
Income	3.7	2.2

TABLE 2 Pearson correlation coefficients for internet skills, internet problem-solving behaviors and demographic variables ($N = 767$).

Variable	1	2	3	4	5	6	7	8	9
1. Age	—	.64**	.24**	.10**	.13**	.20**	–.08*	.03	–.04
2. Education		—	.04	.14**	.19**	.17**	.00	.04	.01
3. Gender			—	–.13**	.14**	.36**	–.08*	.03	–.05
4. Income				—	.06	.04	.05	.02	.06
5. Perceived internet skills					—	.54**	.04	.18**	.30**
6. Self efficacy in problem solving						—	.04	.21**	.28**
7. Social support in problem solving							—	.01	.06
8. Internet use (interactive)								—	.21**
9. Internet use (creative)									—

Note: *$p < .05$, **$p < .01$ (2-tailed).

Resources for Computer and Internet Problem-Solving Behaviors

The second research question sought to investigate the ways and resources that young adults utilize to solve their computer problems. As summarized in Table 3, more than half of the respondents relied on their friends, 37 percent relied on their family members and 19 percent relied on technical support. Of the respondents, 22 percent did not seek help from others for their computing problems. Fewer than 10 percent used print resources such as books, references and online documentation. With regard to internet problems, about half of the respondents relied on their friends, 35 percent on their family members and 17 percent utilized technical support. About 18 percent of the respondents solved their internet problems by themselves. Hence, interpersonal resources including one's friends and family members were the most common ways in which young adults sought help. Fewer than one in four of the respondents reported to be independent problem-solvers when they encountered problems.

The responses from the respondents illustrate in more detail the ways and the extent to which youths perceive, experience and solve their problems. Most reported having faced computing hardware and software problems and being familiar with computer crashes due to malfunctioning and incompatible hardware and software. Contrary to research in the USA among adult internet users, who said that they are receiving more spam than before but are minding it less (Fallows, 2005), a large proportion of the young adults interviewed said that they were constantly frustrated by junk mail, pop-ups, viruses and spyware. A number of them said that they had enduring problems in dealing with spam and junk mail in their email accounts and were troubled that they had to monitor their inbox and clear mail daily to avoid their inboxes being jammed. In addition, several detailed the consequences of their problems, including comments such as: 'I travel around with this device—if this device doesn't work, it is a big part of myself lost', internet problems 'wasted a lot of my time and effort' and 'To me, it's losing information that is important'.

Thus, most respondents could relate to experiencing internet-related hardware and software problems. However, when asked to describe their most recent computer and internet problem, the majority of the respondents gave vague and unclear descriptions. Some of them sounded hesitant and several were unaware of the reasons and processes leading to their computer breakdowns. For example, some could not define their computing problems in specific ways. The most common phrases appeared to contain elements of surprise and confusion, for example:

> *I don't know what went wrong.*
>
> *The computer just crashed or jammed.*
>
> *Sometimes I go onto some certain websites and something like that, and suddenly I don't know what buttons I press, and I get an error message.*

TABLE 3 Sources of help for internet and computer problems* (*N* = 767).

Sources of Help	Computer Problems (%)	Internet Problems (%)
1. Friends from school	58.5	54.2
2. Other family members	37.4	34.9
3. No one (solve by myself)	21.9	17.7
4. Technical support	19.3	16.8
5. Customer service	17.5	13.3
6. Books and references	8.2	5.3
7. Co-workers	8.6	7.0
8. Online friends/Virtual communities	5.6	5.3
9. Website/online documentation	2.9	4.7
10. Newsgroups/online forums	1.2	1.7

Note: *Multiple answers accepted

A few respondents even expressed bewilderment at the 'weird beeping sounds' emitted by their computers. Those who gave imprecise descriptions of their computing problems also tended to be puzzled about the ways and resources that they could employ to solve their problems.

For example, Adelyn is a 20-year-old college student who uses her computer for schoolwork and utilizes her email regularly to communicate with her friends abroad. When asked about her last computing problem, she said: 'When the hardware gets too overloaded . . . to be honest, I have no idea, I just know the computer is very slow.' She said that her first reaction to a computing problem is to 'whack the computer' and then 'shout' to her brother 'to ask him to solve the problem'. She usually approaches her 18-year-old brother, whom she regards as being 'more well-versed' in the computer than herself, but she does not have any idea what her brother does to her computer. A self-professed 'computer idiot' who 'knows nothing about computers', Adelyn struggled with describing her internet problems, although she shared her frustration in trying to deal with 'weird viruses and those visible worms all the time'.

According to Tapscott (1998), young people are perceived to be more technically savvy than their elders. Yet ironically, some youths said that they looked to older adults and parents to solve their computing and internet problems. For example, Jasmine recalled that she had to reformat her computer last year but was not sure why she had to do it. She said, with an embarrassed look: 'I'm not very tech-savvy person—I just leave it [all computer issues] to my father.' At the point of the interview, she had an ongoing problem with her home computer. She said:

> *For my home desktop, there are a lot of pop-ups and there's this debugging thing, but I always click 'No' as in they'll tell me that my computer is bugged or something like that. But I'll just click 'No', like once or twice, that it goes away [sic]. Yeah, but I don't know what it is.*

She is perplexed that this problem has lasted for 'quite long, a couple of months' and is still unresolved. She said:

> *I don't really know how to solve the problem. Because when I don't, I think my father tried fixing it before, but I think it just came back. So in the end we just left it. Because in a sense, okay, you just click 'No' twice, it'll go away.*

In line with the survey results, most of the young adults interviewed relied on their friends from school for help. Most said that their peer networks were very important resources and many pinpointed specific friends to whom they could turn for help. One particular example came from the accounts of Sylvia, a 17-year-old biomedical science student. Her last computer problem occurred a week before the interview, when she bought a firewire to connect her video camera to her laptop, but the wire did not work. She described how she tried to resolve the problem:

> *I was at the computer, like eight hours long, continuously trying to solve it, but it just didn't work. I don't know, it just cannot, like, [it's] not responsive to the thing I inserted into the laptop.*

She then relied on her friend to help her, she said:

> *My friend is very good at computers because he scored a perfect score at a computer programming module. However, my friend was trying to help me detect the software but it still cannot work.*

To solve her problem, Sylvia related that she had to bring her friend along with her to the shop where she purchased her firewire because she was unsure of how to explain her predicament and needed her friend's assistance to explain the problem to the electronics shop owner.

The above responses illustrated the perceptions and actions of young adults who were less technically aware and skilled at problem-solving. These responses showed how some youths are unaware of technical operations and rather confused, especially when they encounter computer and internet problems in their everyday lives. On the other hand, there were a few respondents who appeared to be more familiar with the hardware and software configurations of their computing equipment and were able to articulate their problems and the ways in which they sought help. For example, Jacqueline, a 19-year-old college student, said that she encountered her last problem when her computer slowed down considerably due to viruses and Trojan horses that entered her computer via a peer-to-peer network system. With regard to email, Jacqueline displayed an awareness of her university's internet development policies. She said she was concerned about her university's email system, as

> *it does not have a spam filter. They just recently, like, last week, introduced the spam filter, but they're still, like, working on it, so we can be exposed to spam and viruses through the school emails.*

She said that people around her usually ask her to solve their problems, 'simple IT questions like how to set up a blog, how to write the HTML code for your blog schemes, like that, simple IT questions'. She occasionally relies on a close friend to help her with any computing and internet issues.

There were also a few others who expressed confidence in their technical and troubleshooting skills. For example, Marcus is a 25-year-old student who also works in a music production company. He said that he is usually 'independent' and did not approach anyone for help with his last computer problem. He also stressed that he is careful in his personal internet use to avoid future problems:

> *Most of the time, I take care of viruses by not opening attachments . . . unlike some people who happily open attachments. I don't anyhow download stuff, have a lot of precautionary measures.*

Marcus explained that he maintains several email accounts 'with different inland and offshore webhosting servers, in the event should one of my inboxes become flooded with spam or becomes inaccessible' so that he can continue with his work. When asked if anyone has asked him to help with their internet problems, he said:

> *Most of the time, I am the computer guy in the office, although for some strange reason I'm not trained to do computers. I'm the default wireless administrator because I am the only one who knows how to configure the wireless modem . . . Other than that, a lot of problems that people come to me have to do with software issues like, 'I don't know how to use this software', 'How do you do this?', 'What's the problem, I can't seem to print', and its just configuration, nothing that the user cannot handle.*

In sum, qualitative evidence yielded several interesting insights about young adults and their computing and internet behaviors. There appeared to be a marked difference in the ways in which the more and less technically-savvy youths defined and approached their problems. Those who were less technically savvy tended to be surprised and caught off-guard by computing problems. Some of them relied on temporary methods such as 'pressing the control–alt–delete with my fingers', or expressed their helplessness over an 'indefinite problem that has gone on for years'. After encounters of computing and internet problems, their efficacious use of the internet was dependent on the assistance and actions of others. On the one hand, there were a few other respondents who exhibited the tendency to give up after first experiences of failure and to desist from engaging the apparently incomprehensible internet hardware and software issues. On the other hand, there were youths who appeared to be more cognizant of troubleshooting strategies and were more confident of their internet use. Young people from the latter group showed greater awareness of, and concern about, IT security issues (e.g. viruses and hackers). Consequently, they reported to have anticipated problems actively and practised preventive strategies to protect their computing systems from breakdowns.

Demographic, Internet Skills and Internet Problem-Solving Factors Affecting Internet Use

The second part of the study sought to examine the relationships between demographics, internet skills, internet problem-solving behaviors and internet use among youths. As indicated in Table 4, the results show that demographic characteristics did not make a significant difference in scope of interactive internet activities online. The findings show that, after controlling for the effects of individual demographics, perceived internet skills was associated significantly with individuals' breadth of online activities ($R^2 = .03$, $F(5,762) = 5.15$, $p < .01$). Supporting the findings from existing research in the realm of secondary digital divides in terms of internet expertise (Hargittai, 2002), the results show that those who considered themselves to have good computing and internet skills were more likely to use the internet more effectively—in this case, for a larger scope of activities ($\beta = .18$, $p < .01$). Results from the third regression showed that individuals' self-efficacy in problem-solving significantly predicted online activity scope ($\beta = .17$, $p < .01$), while computing and internet-related social support from others was not a significant predictor ($R^2 = .05$, $F(7,760) = 5.9$, $p < .01$).

The study also set out to test the variations underlying a creative aspect of internet use. Logistic regression analyses were conducted to test the associations between demographics, internet skills and internet-related social support for those who have and have not used the internet to create their own website. The results show that demographic characteristics did not make a significant difference in the ways in which respondents used the internet for website building purposes. As illustrated in Table 5, the findings show that perceived internet skills were significantly associated with creative internet use, after controlling for the effects of individual demographics. In addition, the third regression analysis showed that self-efficacy in problem-solving and computing and internet-related social support from others were associated significantly with creative internet use, after controlling for the effects of internet skills and individual demographics. The logistic regression analysis of the predictors of creative internet use revealed that about 15 percent of the variance was accounted for by the three predictor blocks.

TABLE 4 Hierarchical regression analysis of predictors of internet use (scope of interactive activities).

	Step 1	**Step 2**	**Step 3**
Demographics			
Gender (Male)	.03	.00	–.05
Age	–.01	–.01	–.01
Education	.05	.01	.01
Income	.02	.01	.00
Perceived skill	—	.18**	.10**
Self-efficacy in problem-solving	—	—	.17**
Social support in problem-solving	—	—	.01
R^2		.03	.05

$N = 767$, *$p < .05$, **$p < .01$

TABLE 5 Logistic regression results for predicting internet use (website creation).

Individual Characteristics	**β**	**SE β**	**WALD**	**OR(95% CI)**
Gender (Male)	–.06	.19	15.33**	.48 (.33–.69)
Education	–.05	.10	.23	.95 (.78–1.61)
Age	–.06	.03	2.93	.94 (.88–1.00)
Income	.04	.04	1.03	1.04 (.97–1.11)
Perceived skill	.28	.06	21.81**	1.33 (1.18–1.50)
Self efficacy in problem-solving	.56	.10	31.02**	1.75 (1.44–2.12)
Social support in problem-solving	.12	.05	5.62*	.78 (.70–.88)
R^2				.15

β = Standardized Coefficient, $N = 767$

*$p < .05$, **$p < .01$

Discussion

This article set out to explore the digital divide issues underlying young adults' internet use and problem-solving behaviors in Singapore, a context where a high proportion of the population have attained internet access. Accordingly, it investigated the relationships between sociodemographics, internet skills and the internet problem-solving behaviors of youths. The survey results showed that almost all young adult internet users have experienced computing and internet problems and a fair proportion of them face these problems rather frequently, i.e. on a monthly or weekly basis. The most popular forms of help for their problems came in the form of interpersonal resources and expertise from their friends and other family members. About one-fifth of the youths said that they usually solved their problems by themselves. Few utilized published resources such as books, references and online documentation.

Contrary to popular conceptualizations of youths as being a cohort of technically-savvy experts, qualitative evidence from the interviews showed considerable variance in youths' internet expertise and problem-solving behaviors. A significant proportion of the young adults interviewed were unfamiliar with the ways in which computing technologies work and were unable to troubleshoot or solve their computing problems. Some young adults even professed to be 'computer idiots' and 'not savvy when it comes to technology'. On the other hand, there were several young adults interviewed who expressed greater knowledge and expertise about their daily internet use. They also expressed greater awareness of computing and internet problem-solving strategies. Some of them said that they usually managed to diagnose and solve the problems, or approach technical support staff on their own.

In this way, the findings support recent research evidence which highlights how a proportion of young people are challenged by various complexities of internet access and most use the internet for a narrow range of activities (Livingstone et al., 2004). The findings of this study further contribute to deconstructing the image of the empowered internet user who is autonomous and technologically competent. More specifically, they show that only about one-fifth of young adults solved computing and internet problems by themselves. These findings challenge the free-agent, self-sufficient image of young technical experts. Specifically, the results of this study illustrate how some young adults are still embedded in their peer and family networks and are highly reliant on their existing social support networks for technological assistance and know-how.

The results from regression analyses showed significant variations in youths' internet use for interactive and creative purposes. However, the results of this study did not show the salience of individual demographics on internet use. The findings here suggest that age, gender and socio-economic status may not be as relevant as other factors such as internet skills for post-internet adoption behaviors in wired contexts such as Singapore, where the internet is reaching maturity and growth rates are stabilizing. Instead, the analyses showed that factors such as the level of internet skills and internet-related problem-solving efficacy to be associated positively with internet use. Hence, although computer ownership is pervasive in Singapore and most of the young adults have adopted the internet, differences exist in levels of technological proficiency and social resources, which in turn may frustrate or facilitate internet use.

Taken together, the findings of this study highlight the presence of secondary digital divides pertaining to the internet experiences of young adults. Digital inequalities in internet skills, problem-solving behaviors and internet usage patterns exist after initial access is obtained. In this way, contrary to popular belief, the digital divide may not be completely bridged in digital societies, as secondary digital divides may be extenuated in these nuclei.

Conclusion

Limitations of the Study

It is acknowledged that limitations to this study exist. First, part of the data were obtained from a web-based survey with a non-probability-based sample of internet users who are students in local tertiary institutions. This sample is unrepresentative of the general young adult population in Singapore as it excludes non-students (14% of the population between 15 and 29 years old; Statistics Singapore, 2000). In addition, coverage is a concern for web-based surveys, although Couper (2000) noted that there are some communities such as college campuses where connectivity is almost universal, thereby mitigating the sampling bias of web surveys. As noted earlier, the internet penetration rate among Singaporean youths is extremely high, making online surveys a low-cost and efficient method of data collection. Moreover, given the exploratory nature of this study, a mix of quantitative and qualitative approaches was employed to gain a textured and more complete understanding of internet use and problem-solving behaviors from the youths' point of view. In light of the dearth of information on youth internet use and their problem-solving behaviors, the ability to generalize is not sought with this sample. Rather, it is hoped that the results of this study will highlight trends in the secondary digital divides underlying youth internet use in wired societies such as Singapore. Second, the survey data is a single-shot, cross-sectional snapshot of youth internet use, thus limiting any ability to draw causal inferences and over-time observations on the relationships among the study factors. Third, measurement of internet expertise is based on a single-item measure and self-reports. Fourth, the model of internet activity scope was statistically significant as a whole but explained a limited amount of variance. Future research could attempt to increase the validity of the study by examining a wider range of attitudinal or motivational variables as antecedents of internet use.

Future Research and Policy Recommendations

In conclusion, this study provides a critical study of internet use and related digital divide issues by examining internet problem-solving behaviors. By discussing the stratification processes underlying internet expertise and problem-solving resources, attention is directed to the epilogue after a society's chapter of rapid internet penetration. The results here point to another dimension of internet use in everyday life and highlight the need to critically (re)examine and enlarge notions of internet 'use' and 'access', in order to take into account internet-related troubleshooting and problem-solving behaviors. In terms of policy implications, the results from this study suggest that future efforts should focus on building knowledge-sharing networks among youths to contribute to a virtuous cycle of internet help, in order to enhance the quality of youths' internet connections. For example, the National Youth Council could start a mentoring system to pair off youth leaders who are able and willing to help other youths in troubleshooting and utilizing the internet to achieve their educational and social goals. To the extent that young adults regard some of their peers as being more technically savvy, peer-to-peer knowledge-sharing would prove to be an economically viable yet constructive way of closing inequalities in internet usage.

About the Author

Pauline Hope Cheong is Associate Professor in the Hugh Downs School of Human Communication, Arizona State University. She researches the social implications of information and communication technologies, particularly social capital building and social cohesion issues among minority and marginalized populations.

Address: Hugh Downs School of Human Communication, Arizona State University, PO Box 871205, Tempe, AZ 85287–1205, USA. [email: pauline.cheong@asu.edu]

Acknowledgments

The author wishes to thank the editors and two anonymous reviewers for their comments and the National Youth Council for its support of this project. A version of this article was presented at the 56th International Communication Association conference, Germany, in 2006.

References

Barone, J.T. and J.Y. Switzer (1995) *Interviewing: Art and Skills*. Needham Heights, MA: Simon and Schuster.

Cheong, P.H. (2007) 'Gender and Perceived Internet Efficacy: Examining Secondary Digital Divides Issues in Singapore', *Women's Studies in Communication* 30(2): 205–29.

Cheong, P.H. and H.A. Wilkin (2005) 'The Internet for Hispanic Immigrants: Health Communication and the Digital Divides', in M. Allen and M. Consalvo (eds) *Internet Research Annual,* Vol. 2, pp. 101–114. New York: Peter Lang.

Compaine, B.M. (2001) *The Digital Divide: Facing a Crisis or Creating a Myth?* Cambridge, MA: MIT Press.

Couper, M.P. (2000) 'Web Surveys: A Review of Issues and Approaches', *Public Opinion Quarterly* 64(4): 464–81.

DiMaggio, P.D., E. Hargittai, C. Celeste and S. Shafer (2004) 'Digital Inequality: from Unequal Access to Differentiated Use', in K. Neckerman (ed.) *Social Inequality,* pp. 355–400. New York: Russell Sage Foundation.

Dutton, W.H., P.H. Cheong and N. Park (2004) 'An Ecology of Constraints on e-Learning in Higher Education: the Case of a Virtual Learning Environment', *Prometheus* 22(2): 131–49.

Fallows, D. (2005) 'CAN-SPAM a Year Later', URL (consulted 15 June 2005): http://www.pewinternet.org/pdfs/PIP_Spam_Ap05.pdf

Graham, S. (2004) '"Beyond the 'Dazzling Light". From Dreams of Transcendence to the "Remediation" of Urban Life: A Research Manifesto', *New Media & Society* 6(1): 16–25.

Hara, N. and R. Kling (2000) 'Student Distress in a Web-based Distance Education Course', *Information, Communication and Society* 3(4): 557–79.

Hargittai, E. (2002) 'Second-level Digital Divide: Differences in People's Online Skills', *First Monday* 7(4), URL (consulted 15 January 2003): http://firstmondayorg/issues/issue7_4/hargittai/index.html

Internet World Statistics (2005) 'The Big Picture. World Internet Users and Population Statistics', URL (consulted 27 October 2005): http://www.Internetworldstats.com/stats3.htm#asia

Internet World Statistics (2008) 'Internet Users and Population Statistics for 35 Countries and Regions in Asia', URL (consulted 30 May 2008): http://www.internetworldstats.com/stats3.htm#asia

Jung, J., L.J. Qiu and Y. Kim (2001) 'Internet Connectedness and Inequality: Beyond the "Divide"', *Communication Research* 28(4): 507–35.

Jung, J.Y., Y.C. Kim, W.Y. Lin and P.H. Cheong (2005) 'The Influence of Social Environment on Internet Connectedness of Adolescents in Seoul, Singapore and Taipei', *New Media & Society* 7(1): 64–88.

Kling, R. (2000) 'Learning about Information Technologies and Social Change: the Contribution of Social Informatics', *The Information Society* 16(3): 217–32.

Kuo, E., A. Choi, A. Mahizhnan, W.P. Lee and C. Soh (2002) *Internet in Singapore. A Study on Usage and Impact*. Singapore: Times Academy Press.

Lenhart, A., C. Marsden and P. Hitlin (2005) 'Teens and Technology. Youths Are Leading the Transition to a Fully Wired and Mobile Nation', URL (consulted 20 October 2005): http://www.pewInternet.org/pdfs/PIP_Teens_Tech_July2005web.pdf

Lenhart, A., S. Arafeh, A. Smith and A.R. Macgill (2008) 'Writing, Technology and Teens', URL (consulted 30 May 2008): http://www.pewinternet.org/pdfs/PIP_ Writing_Report_FINAL3.pdf

Livingstone, S., M. Bober and E. Helsper (2004) 'UK Children Go Online. Surveying the Experiences of Young People and Their Parents', LSE report, URL (consulted 10 December 2004): http://www.children-go-online.net

Livingstone, S., M., Bober and E. Helsper (2005) 'Inequalities and the Digital Divide in Children and Young People's Internet Use', LSE report, URL (consulted 10 October 2005): http://www.children-go-online.net

Lynch, B.P. (2001) 'The Digital Divide or the Digital Connection', *First Monday* 7(10), URL (consulted 1 February 2002): http://firstmonday.org/issues/issue7_10/lynch/index.html

Neubauer, B.J. and J.D. Harris (2002) 'Protection of Computer Systems from Computer Viruses: Ethical and Practical Issues', *Journal of Computing Sciences in Colleges* 18(1): 270–9.

Norris, P. (2001) *Digital Divide: Civic Engagement, Information Poverty and the Internet Worldwide*. Cambridge: Cambridge University Press.

Ono, H. (2005) 'Digital Inequality in East Asia: Evidence from Japan, South Korea and Singapore', *Asian Economic Papers* 4(3): 16–139.

Otto, H.-U., N. Kutscher, A. Klein and S. Iske (2003) *Social Inequality in the Virtual Space: How Do Young People Use the Internet? First Results*. Bielefel: Center for Competence for Informal Education, University of Bielefel.

Park, H.W. (2002) 'The Digital Divide in South Korea: Closing and Widening Divides in the 1990s', *Electronic Journal of Communication* 12(1–2): URL (consulted 4 December 2006): http://www.cios.org.gate.lib.buffalo.edu/getfile/Park_v12n102

Sarason, I.G., B.R. Sarason, E.N. Shearin and G.R. Pierce (1987) 'A Brief Measure of Social Support: Practical and Theoretical Implications', *Journal of Social and Personal Relationships* 4(4): 497–510.

Sng, S. (2001) 'Can't Get Enough of Britney and Ziyi? Take Your PIC', *The Straits Times* (Singapore), 14 August, p. L1.

Statistics Singapore (2000) 'Census of Population', URL (consulted 30 November 2006): http://www.singstat.gov.sg/pdtsvc/pubn/cop2000r2.html

Tan, M. (1998) 'Plugging into the Wired World. Perspective from Singapore', *Information Communication and Society* 1(4): 217–45.

Tang, P.S. and P.H. Ang (2002) 'The Diffusion of Information Technology in Singapore Schools: A Process Framework', *New Media & Society* 4(4): 457–78.

Tapscott, D. (1998) *Growing Digital: The Rise of the Net Generation*. New York: McGraw-Hill.

Tashakkori, A. and C. Teddlie (1998) *Mixed Methodology: Combining Qualitative and Quantitative Approaches*. Thousand Oaks, CA: Sage.

Treese, W. (2004) 'Putting It Together: The State of Security on the Internet', *NetWorker* 8(3): 13–15.

Van Dijk, J. and J. Hacker (2003) 'The Digital Divide as a Complex and Dynamic Phenomenon', *The Information Society* 19(4): 315–26.

Wei, Y.P. (2000) 'Youths Shun Politics, Involvement', *The Straits Times* (Singapore), Home section, p. 2.

Puerto Rican Language Use on MySpace.com

Kevin S. Carroll

The social networking site MySpace.com has quickly become a venue in which Puerto Ricans are able to communicate among themselves within a larger global community, resulting in an interesting range of language use. After examining the amount of Puerto Rican users, the researcher uses five aspects of MySpace.com profiles to analyze language use in fifty profiles of Puerto Ricans ages 18 to 22. The final portion of the paper is a case study of three profiles highlighting current use of Puerto Rican Spanish and netspeak. The paper concludes that many Puerto Rican users of MySpace.com live in a bilingual linguistic reality floating between Spanish and English.

Keywords: Puerto Rican netspeak, MySpace.com, Internet language usage, social networking, Puerto Rican Spanish

Throughout the industrialized world, technology, and specifically the Internet, has overwhelmingly changed the way we think and talk about language. Crystal (2004) argues that the Internet has had the biggest impact on language change since the advent of the written word. As language has changed via its use on the world wide web, so too have the ways in which people create and maintain their identity through language. This paper looks to document language use among Puerto Rican MySpace.com users and identify social networking sites as vibrant and legitimate sites for continued research regarding the use of language. This research will look specifically at the language(s) that Puerto Ricans age 18 to 22 use to create their profiles and how these linguistic features signify in-group identity.

With inexpensive computers and Internet access, Puerto Ricans are taking to the Internet in droves. Currently, approximately 25 percent of Puerto Ricans have Internet access (Miniwatts Marketing Group 2006). While this number is substantially lower than percentage of users in the U.S., Puerto Rico ranks highest in Internet usage among islands in the Caribbean, with more than a half a million people. Although access to the Internet is tied to socioeconomic status, younger generations of Puerto Ricans are often able to connect to the Web at school, and those studying in various universities are required to frequently access the Internet. As a result of younger generations having a better knowledge of computers and the Internet, it is no surprise that the average MySpace.com user in Puerto Rico is under 25 years of age. Through MySpace.com, these young adult users have found a space where they can communicate with the larger global community and still maintain their own local identity, which they signal by their use of language.

Functions of English and Spanish in Puerto Rico

Puerto Rico, a Caribbean island populated by approximately four million people (World-Gazetteer 2006), was originally colonized by the Spanish and currently holds a unique commonwealth status with the United States. Puerto Ricans are given U.S. citizenship, are eligible for federal aid, and are granted a majority of the other liberties afforded by being part of the United States. Puerto Rico's citizens do not pay U.S. federal income tax, however, nor do they have voting rights in U.S. federal elections. Since the U.S. invaded Puerto Rico in 1898, English education, to different extents, has been a mandatory part of public education. Despite U.S. occupation for the past 110 years, 21 percent of Spanish speaking Puerto Ricans report that they do not speak English well, and 45 percent report that they do not speak English at all (U.S. Census Bureau 2003).

Despite overt movements on the part of the U.S. to create an English-speaking island, Puerto Ricans have been very successful in maintaining and bolstering their own dialect of Spanish known as Puerto Rican Spanish (PRS) (Algren de Gutiérrez 1987; Morris 1995). Morris (1995) describes the use of Spanish as a deeply rooted aspect of Puerto Rican identity and one that Puerto Ricans cling dearly to. Furthermore, Morris explains how within the context of politics language "serves as a rallying point for Puerto Rican identity, in direct opposition to the English-speaking United States" (1995: 162). It is difficult, if not impossible, for Puerto Ricans to detach politics from the debate over language instruction because one's opinions on the importance of English education often align with political parties and their support or animosity towards statehood. Hence, any language discussion in Puerto Rico is inherently political (Morris 1996). Consequently, it is often assumed that promoting bilingual education or educational programs that place a significant importance on English education is in some way moving away from "true" Puerto Ricanness and in the direction of statehood (Algren de Gutiérrez 1987; Morris 1995; Morris 1996). While language policies in Puerto Rico are politically charged, many Puerto Ricans acknowledge the importance that English has as a gatekeeper to upward social mobility and therefore continually stress the importance of English education for their children. This is exemplified by the many parents who have the resources to send their children to private schools, where there is often a greater emphasis put on English education (Schweers and Hudders 2000).

While there is not currently a wide body of literature examining Puerto Rican language use, hence the publication of this edition of *CENTRO Journal,* there are publications in the area of applied linguistics that examine different functions between English and Spanish on the island. Fayer (2000) documented the various places where English is used on the island and compiled results from three studies spanning from 1976 to 1996 to analyze the different functions of English on the island. Not surprisingly, there was a high correlation between knowing English and a person's income. Furthermore, the study highlighted an increased access to and use of media among Puerto Ricans, primarily movies and television. Fayer's findings point to an increased exposure to media, which has continued to grow and now undoubtedly includes the Internet. Similarly, Vélez (1993) compared English service encounters in San Juan with those he collected in Austin, Texas, and found that the English used by Spanish speakers in San Juan was more direct and followed particular Spanish discourse patterns, which could be interpreted by a non-Puerto Rican as rude or too informal. Other less formal work on Puerto Rican language use may be found in master's and doctoral theses and dissertations (Nickels 2005).

While Spanish is the language of choice by the vast majority of Puerto Ricans living on the island, islanders have their own unique dialect of Spanish (PRS), which has been well documented. PRS is often categorized with other Caribbean Spanish-speaking countries such as Cuba and the Dominican Republic (Morales 2000). As is customary with different dialects, there are a number of different linguistic differences between PRS and other varieties of Spanish. Some of the more commonly cited and noticed aspects of PRS are phonological, such as: minimal distinction between /l/ and /r/, final /s/ deletion, and using a uvular /x/ in the place of a traditionally trilled r sound in a word like *arroz,* which could be pronounced as [a:xo] (Nash 1977; Hochberg 1986). In addition to the unique aspects of PRS, English in the United States has influenced Spanish spoken in Puerto Rico. As a result of the *vaiven*[1] (García, Morín, and Rivera 2001) amongst Puerto Ricans to and from the United States, as well as influences from U.S. popular culture, Spanglish (Zentella 1997) is commonly used, especially among youth. Spanglish is often marked by consistent code switching and insertion of words that were once English but have been adopted into Puerto Rican Spanish and are pronounced accordingly. Venues such as MySpace.com, where Puerto Ricans are part of both a local and international community provide (and in some instances necessitate) the use of different registers to create and maintain their diverse linguistic identities.

Puerto Ricans are known throughout the Spanish-speaking world as having their own unique dialect. The data in this study have been analyzed using a theoretical framework couched in linguistic anthropology, where language is viewed as playing a major role in the development of personal as well as cultural identity (Bailey 2000; Bucholtz 1999; Eckert 2000). Language thus equally affects the development of society and the particular culture(s) within which it has contact. The field of linguistic anthropology is founded on the work of scholars like Franz Boas, Edward Sapir and Benjamin Whorf, whose work led the field to view language use as a determiner of how one sees and acts in the world (Duranti 1997). As language and culture are inextricably intertwined, the two obviously have an impact on the creation and maintenance of one's identity. The purpose of the remainder of this research is to highlight a new space where Puerto Ricans are using language as an identity marker. This new space is known as MySpace.com.

The MySpace.com Phenomenon

As the Internet has grown in accessibility and speed over the years, so too have programs of computer-mediated communication (CMC). In general terms, CMC stands for the broad type of social networking website that allows Internet users from diverse geographic areas to communicate in virtual worlds. The fourth most-visited English language site in the world, with over 106 million users, MySpace.com is currently the most popular social networking website (Wikipedia 2006). MySpace.com reports that, on a daily basis, more than 230,000 new users join (Wikipedia 2006). Typically a phenomenon for youth and college students, MySpace.com allows its users to create their own profiles using a myriad of different semiotic aspects. Other sites allowing users to do similar things are: Facebook.com, Friendster.com, and Hi5.com. These virtual communities allow users to reconnect and keep in touch with old friends and meet new people, as well as share pictures and music with anyone who is interested enough to open a profile. Started in 2003, MySpace.com is now a space where millions of people around the world spend hours a day creating, maintaining and browsing profiles.

MySpace.com has not operated without controversy. Some secondary schools and even universities, primarily in the United Stares, have banned access to the website in order to protect younger users from pedophiles and to keep students on task while studying. MySpace.com also seems to permeate many social aspects of current school life, as exemplified at a school located in a suburb of St. Louis, where 12 students were expelled for a fight that was organized via MySpace.com (Suhr 2006). While there have been many attempts to curb the website's use by parents, school administrators and record labels (because you can listen to music for free), many parents, school administrators and companies have decided to join the MySpace.com phenomenon by creating their own profiles and managing it that way (Wikipedia 2006). Many famous celebrities, including actors, athletes and musicians, have used the website to their advantage and are able to provide free advertisements for new releases, upcoming shows and new songs, all of which can instantly be viewed by the website's millions of users.

Puerto Rican profiles on MySpace.com range greatly in appearance and theme, and its users' ages, occupations, interests and purposes are equally diverse. The interface also makes it easy to find people through a variety of query options such as the person's name, email, country and many more. User profiles themselves often deviate from the standard white and blue template provided when they first create their profiles. Among the profiles I examined, most were rich with both visual and sound effects. As far as language use is concerned, I have defined a standard MySpace.com profile to have all five of the following aspects that use language: a personal quote (usually a few words, a greeting or quote from a friend or famous person), a personal description (all profiles examined were at least a paragraph long), an interest section (a list of movies, TV shows and different things of interest to the user), at least one song or music video and a comments section where friends can asynchronously[2] leave postings. Collectively, these five aspects make up the linguistic criteria analyzed in this study. Non-language components of the profiles differ in their complexity, but often include elaborate slideshows of pictures, wallpapered pictures, Top Friends pictures and even custom pointers for one's cursor (e.g. the user has changed the standard cursor or pointer into a Puerto Rican flag).

MySpace.com's popularity is immense and growing, but it has just recently caught on as one of the primary social networking websites in Puerto Rico. On 31 October 2006, when I started looking specifically at Puerto Rican MySpace.com, there were approximately 26,047 users between the ages of 18 and 60 who registered as being located in Puerto Rico. As of 11 December 2006, there were 53,262, a 104 percent increase in such users in roughly a month and a half (see Table 1 and 2).

TABLE 1 MySpace Usage in PR as of 11:00 PM 31 Oct. 2006.

Males	Users	Females	Users	
18	1460	18	2641	
19	1380	19	2380	
20	1144	20	1907	
21	934	21	1541	
22	787	22	1209	
23	723	23	1040	
24	587	24	835	
25	518	25	710	
26	493	26	600	
27	329	27	427	
28	334	28	405	
29	272	29	310	
30	196	30	272	
31	188	31	220	
32	139	32	174	
33	97	33	127	
34	93	34	127	
35	103	35	119	
36	67	36	103	
37	62	37	75	
38	46	38	70	
39	40	39	64	
40	29	40	59	
41	28	41	53	
42	16	42	42	
43	19	43	39	
44	23	44	30	
45	17	45	32	
46	12	46	32	
47	8	47	28	
48	12	48	23	
49	7	49	21	
50	9	50	21	
51	11	51	15	
52	7	52	18	
53	3	53	15	
54	2	54	8	
55	4	55	14	
56	2	56	8	
57	6	57	3	
58	1	58	3	
59	3	59	6	
60	5	60	5	
	10216		15831	26047

TABLE 2 MySpace Usage in PR as of 6:00 PM 11 Dec. 2006.

Males	Users	Females	Users	
18	3000	18	3000	
19	2951	19	3000	
20	2498	20	3000	
21	2152	21	3000	
22	1736	22	2743	
23	1651	23	2348	
24	1306	24	1731	
25	1223	25	1728	
26	1109	26	1522	
27	824	27	1068	
28	777	28	1011	
29	691	29	782	
30	452	30	657	
31	437	31	532	
32	333	32	459	
33	310	33	341	
34	257	34	356	
35	241	35	303	
36	197	36	307	
37	162	37	208	
38	128	38	210	
39	113	39	173	
40	96	40	194	
41	87	41	168	
42	62	42	138	
43	65	43	121	
44	56	44	87	
45	48	45	103	
46	40	46	84	
47	29	47	83	
48	34	48	63	
49	39	49	73	
50	35	50	73	
51	28	51	60	
52	23	52	43	
53	13	53	43	
54	10	54	35	
55	13	55	35	
56	12	56	36	
57	13	57	17	
58	9	58	17	
59	12	59	13	
60	11	60	14	
	23283		29979	53262

I estimate there would be at least a few thousand more if it were possible to browse profiles of users under 18, which is restricted for reasons of safety and privacy. As social networking sites like MySpace.com have become more popular, Puerto Ricans have been able to display their cultural and individualistic identities for the world to see.

Methodology

In order to gauge language use among Puerto Rican MySpace.com users, I conducted this study in three parts. The first portion estimates the amount of Puerto Ricans using MySpace.com; this was made easy by the browse function on MySpace.com, which allowed me to search for profiles from Puerto Rico. In order to do so, I set my search criteria in the browse function of MySpace.com for male and then female profiles for each age individually, from eighteen to sixty. The maximum amount of users that MySpace.com will display is 3,000 so breaking my query up into females and then males for each age allowed me to get a more accurate idea of how many users there are. I first collected data on 31 October 2006, and repeated this process on 11 December 2006, to show how fast the site is growing among Puerto Ricans.

The purpose of the second part of the study was to conduct a general survey of language use among fifty Puerto Rican MySpace.com pages. It was not my intent to describe every specific linguistic nuance of MySpace.com users, but to promote the site as a legitimate place where more research can and should be done. In this section, I was interested in the language that Puerto Ricans were using on their profiles and whether or not they were using Spanish, English or a mixture of the two. For this paper, I chose only to look at the language criteria on the profiles, i.e. the user's quote, his or her personal description, the interests section, the song/video and the comments section. From the first part of the study I realized that the bulk of MySpace.com users in Puerto Rico were between the ages of eighteen and twenty-two. Consequently, I looked at five male and five female profiles for each age from eighteen up to twenty-two. In order to have a random sample, I selected every sixth profile. If the profile that I opened did not have all of the five language components of the page, I would return to the browser, count six profiles and try again.

For the final part of the data analysis, I selected three profiles, two of which were primarily in English and one solely in Spanish. I chose the two profiles that are primarily in English because both represented the many profiles that were in English but had comment sections primarily in Spanish. The third profile, which is all in Spanish, was chosen to represent the non-English Puerto Rican users of MySpace.com. These three profiles were not chosen randomly. Instead, I browsed different MySpace.com profiles and chose three profiles that I thought were complete and representative of the various linguistic aspects that were attested on other profiles. The analysis of these three profiles provides specific examples of language use among Puerto Rican MySpace.com users, while simultaneously prompting ideas for future research on this fascinating website.

Puerto Ricans Using MySpace.com

As the MySpace.com phenomenon has hit the United States, it has also attracted millions of people from around the world, including Puerto Rico. At the time of my research, I estimate that there were more than 53,000 users of MySpace.com on the island (Tables 1 and 2). It is impossible to get an exact count because the profiles of users who register as under 18 are not accessible to the general public. As noted above, however, from 31 October 2006 to 11 December 2006 almost all of the categories of users had grown exponentially (Tables 1 and 2). What is not clear is whether the MySpace.com phenomenon will continue to thrive or whether another website (such as Facebook.com, on which it is easier to share pictures but which does not yet have a music component) will become more dominant. I think it is safe to say, however, that Puerto Ricans (and Internet users in general) are not going to stop spending time on social networking websites in the near future.

In order to get a better idea of who uses MySpace.com in Puerto Rico, I compiled the number of female and male Puerto Rican users from age 18 to age 60. After counting only the first six years (age 18–23), it was apparent that the majority of Puerto Rican MySpace.com users are part of the "technology generation," who more than likely have grown up with computers or at least have had access to them. MySpace.com limits searches to 3,000 users. When I did my search on 31 October 2006, neither the male nor female categories were more than 3,000 users for any particular age. However, when I repeated the search on 11 December 2006, I found the 18-year-old male and 18- to 22-year-old female categories showing the maximum of 3,000 users. One would assume that by then there were many more users than I was able to count due to the limitations of the browser function (see Table 2 for details).

MySpace.com is a site where the users are in charge of self-reporting. Thus, it is inevitable that there are users who do not report their real ages. The number of users claiming to be 18 to 20 years of age, however, is indicative that this is a youth and young adult phenomenon (see Table 1). It is hard to tell whether or not there are a substantial number of youth under age eighteen who use the web page. In my research at a local middle school in Tucson, Arizona, I have found that MySpace.com consumes the better part of many of the students' after-school activities. I am not clear whether MySpace.com in Puerto Rico is used primarily by college students, who have faster and more accessible Internet access, or whether MySpace.com usage is similar to the United States, where there is a significant group of users under eighteen.

Another interesting aspect of this study was that there were many more female users than male users. Similarly, while I did not keep an exact tally of those pages I deemed to be incomplete (i.e. they lacked one or more of the five language components being analyzed), it was much easier to find female user profiles that did meet the five component criteria than their male counterparts. Moreover, the female user pages were much more detailed and had substantially more text than their male counterparts' pages.

Language Division

When examining the fifty selected MySpace.com pages, I was surprised by the number of aspects of each profile that were categorized in English. For example, twenty-eight out of fifty pages (56 percent) had a quotation and/or a song in English (see Table 3), and all of the five different language components (except the comments section) were in English for at least seventeen out of fifty pages (35 percent). Moreover, the four areas where I saw the most English were generally the least changed aspects of a person's profile. In other words, when a user creates his or her profile, the user fills in the different components, but rarely goes back to change them. If the user does go back and change them, it is generally to edit or add something new. Therefore, the quotation, personal description, interests and music/video categories are all relatively stagnant aspects of the typical MySpace.com page.[3] The only way I have been able to explain the amount of English language use on the stagnant aspects of Puerto Rican MySpace.com profiles is that Puerto Rican users are doing so in an attempt to appeal to the larger, English-using MySpace.com community.

> *I believe that the amount of profiles in English stems from Puerto Ricans' understanding of English and its role as a lingua franca on the Internet. In using English in their profiles, especially in the description portion, users can automatically appeal to the site's worldwide network of users. Moreover, having portions of one's profile in English gives family members and friends living in the United States, who may or may not speak Spanish, a better opportunity to interact with users on the island. Another reason Puerto Ricans would use English in their quotations or their music is a result of the strong American influence on popular culture in Puerto Rico (Negrón-Muntaner 2004). Moreover, the interests sections of the majority of the profiles I examined used the English titles for movies. In other Spanish-speaking countries, such movies would often be dubbed, and consequently, the titles would be in Spanish, which is rarely the case in Puerto Rico.*

TABLE 3

9 October 2006
Data Results: N = 50
18- to 22-year-olds
The percent of a profile was calculated as the researcher read through the text and determined whether it was primarily in Spanish, English or had at least 30 percent of either language to be considered Mix.

Quote:	**Spanish**	**English**	**Mix**	**NA**
	19	28	2	1
	38%	56%	4%	2%
Desc:	**Spanish**	**English**	**Mix**	
	22	18	10	0
	44%	36%	20%	0%
Inter:	**Spanish**	**English**	**Mix**	
	12	23	15	0
	24%	46%	30%	0%
Song:	**Spanish**	**English**	**Mix**	
	19	28	2	1
	38%	56%	4%	2%
Comment Section:	**Spanish**	**English**	**Mix**	
	23	7	20	0
	46%	14%	40%	0%
Overall Page:	**Spanish**	**English**	**Mix**	
	15	13	22	0
	30%	26%	44%	0%

The language component that exhibited the most amount of Spanish was the comment section. Twenty-three out of fifty (46 percent) of Puerto Rican users had comment sections that were entirely in Spanish, and another forty percent were mixed Spanish and English (mostly with more Spanish than English, but with enough English to be categorized as Mixed and not purely Spanish). In contrast to the other language usage criteria areas, the comments portion of a general profile is the aspect of the page that changes the most frequently. The comment board serves as an asynchronous chat site where friends can say hello to friends, post messages about parties, post pictures or videos and publicly remind a friend of shared experiences.

It should be noted that it was very difficult for me to quickly and systematically distinguish between Spanish and Mixed pages. The primary reason for this difficulty was the amount of copied and pasted postings in English. Postings such as "Happy Halloween," "Happy Thanksgiving," "thanks for the add," and "I love you," among others, were all very common postings that were obviously copied and pasted from other profiles. When a user's comment section was full of commonly used phrases, I tallied them as a "Mix." While it could be argued that the posting of these commonly used phrases and proper nouns are not true examples of the users' (or those of their friends, who are actually posting the comments) language use, I believe that the user's exposure to this text is indeed significant. The use of Spanish on the comments portion of the majority of Puerto Rican MySpace.com profiles is significant because this is the part of the site that is subject to the most change and, consequently, the area that returning users view most.

Language Use on Specific Pages

The next part of this paper will take an in-depth look at three Puerto Rican MySpace.com pages viewed on 28 November 2006. I chose these specific pages because I thought they offered a glimpse of many of the linguistic aspects used on many Puerto Rican profiles I viewed. In the interest of space, I have decided to discuss some of the similarities in language use, specifically: the division of language use (in the two profiles with English descriptions), Puerto Rican netspeak, spelling and lexical borrowings.

The profiles examined come from two males, Roberto (age 20 from Ponce) and Angel (age 19 from Camuy), and a 19-year-old female from Mayagüez named Veronica. The profiles for Veronica and Angel were predominantly in English with both of their quotations, descriptions, interests and music being in English. Roberto's profile, on the other hand, was almost 100 percent in PRS (with the exception of a few common English expressions like "I love you"). Roberto and Veronica study at different campuses in the University of Puerto Rico system, and it is unclear whether Angel is in school or not.

Division of English and Puerto Rican Spanish

Puerto Rican MySpace.com pages vary in the way their users use both English and PRS. However, even those whose pages are primarily in English consistently have comment sections that are primarily, if not 100 percent, in PRS. As noted above, Veronica's and Angel's quotations, descriptions, interests and song sections were all in English. Veronica's description reads:

> *HEY! whats up? im Veronica. im from Mayaguez, P.R., im a college student @ UPRM and mayoring (for now) in Sociology, but want to mayor in Psychology. im a pretty fun, loving, friendly, liberal, and cool person to chill with. im into fantasy, music, reading, jokes, STAR WARS, and having tons of fun with my ppl. anything else u wanna know just ask ;)! ciao!!! :)*

Angel's description reads:

> *ohh well my name is angel but my friends call me junito, juno, june or like i call my self U_KNow hehehe i am from a litte island call Puerto Rico (remember we beat the dream team in basket ball at the olympics?) i am very shy person if u want to talk to me u have to come to me because i am not going to walk to u. when i am whit my friends i want to be center of entertainment the funny guy of the group (i dont have idd) i just like too see my friends laugh of me and have a good time whit them. i also play soccer is my new luv <3 i play midfilder and i luv it my favorite player is snoop dog.................................... (i am not good whit jokes) is zinadine zidane he is harry potter whit the ball aaaaaaaand i dont know what more to say just ask me...*

One might think, after reading the description portions of Veronica's and Angel's pages, as well as viewing their English use in the interests, quote and music sections, that their comments sections would also be in English, but this is not the case. While the comments on Veronica's and Angel's pages are not their own (users do not generally post on their own comment section), the fact that their friends write to them in Spanish, despite the fact that the rest of their profiles are in English, is telling of how Puerto Ricans actually communicate.

In order to confirm that Veronica uses Spanish with her friends, I viewed one of her friend's profiles to see how she was posting with her friends. This is what I found:

> *heyyyy nikole todo bn chika aki bn clava cn el test d manana . . . pero pssss . . . . sabes hay q janguiar tomorrow night! me tengo q desquitar por la mierda d semestre q tuve . . . . ahhh!! jejeje . . dale babe me dejas saber a dnd vas k??? muax cdt mucho chika. . . . ciao (retrieved from: http://profile.myspace.com/index.cfm?fuseaction =user.viewprofile&friendid=16857524 on 7 December 2006)*

The only English words that Veronica uses in her response to her friend's posting are "heyyy," "test," "tomorrow night," and "babe." When writing to her friends, she deviates from the English-speaking persona that she presents in her description section and uses PRS as a gesture of solidarity with her friend. Her use of this form of netspeak is evidence that Spanish is the daily language of Puerto Ricans regardless of their knowledge of English. Furthermore, Veronica's comments also highlight an emerging form of communication that the three users all appear to be well versed in: Puerto Rican netspeak.

Puerto Rican Netspeak

Netspeak, as explained in Crystal (2004), is a complex form of language usage that is vastly different than the traditional forms of communicating, i.e., writing and speaking. Crystal highlights how the Internet has taken away many aspects of traditional communication, such as facial expressions, tone, intonation and other semiotic aspects of language. Furthermore, as a person types to communicate, speed is also a factor resulting in many localized abbreviations.

Both Veronica and Angel use many features of netspeak, including the lack of capitalization and apostrophization in words like "*im*" and the shortening of "you" to "*u*" or "people" to "*ppl.*" This manner of communication undoubtedly allows them to type faster, but it simultaneously deviates from standard written Spanish and English text. As a result, this creates an in-group who understands the language being used and an out-group who does not. Also researching in-group language use are Bailey (2000), Kiesling (2004) and Mendoza-Denton (1999), all of whom discuss the importance of language variation that sets particular subcultures apart from those who do not belong. Thus, Veronica and Angel's use of Puerto Rican netspeak and the mixing of Spanish and English sets them as well as other similar Puerto Ricans apart from MySpace.com users who are not Puerto Rican.

Specific aspects of Spanish, and Puerto Rican netspeak, were seen in the comment sections of all three profiles. The use of "*tambn*" for the word "también" and "*CDT M*" for "cuidate mucho," in Roberto's profile are examples of what I would label Puerto Rican netspeak, which is so new that I am not sure if these usages are common in other Spanish-speaking countries.

Similar to English netspeak, users often signify a change in subject or a pause by writing three dots such as ". . .". For example, the posting on Roberto's profile reads: "Gracias por el comment . . . Espero q estes bn . . . Te cuidas . . . bye." Traditional aspects of U.S. or English netspeak are also exhibited on these profiles, for example, smiley faces written as :) and using "lol" to mean *laughing out loud* (Internet Slang & Translator 2006). Again, this is just a small glimpse at the complexity of netspeak in Puerto Rico. As an active user of MySpace.com, one would consistently be in contact with a variety of both U.S.-coined netspeaks as well as less documented PRS netspeaks.

Puerto Rican netspeak is a fascinating aspect of everyday Puerto Rican language use on the Internet. In my experience taking Spanish classes at universities in Puerto Rico as well as my experiences in other Spanish-speaking countries, there is great importance and attention placed on proper spelling (especially accents), pronunciation and the use of proper grammar. Reproducing the ideals of Spain's Royal Academy of Spanish, many Spanish-speaking countries and people for that matter work against the "pollution" or loss of what is deemed to be correct or traditional Spanish. However, the usage of Puerto Rican netspeak appears to be a venue in which traditional language use is consciously and consistently ignored and in a sense frowned upon.

Spelling

Netspeak also pertains to the spelling of particular words. All three profiles examined in this section used unconventional forms of traditional Spanish spelling. All three of the pages, for the most part, are absent of accents and tildes, which are customary in standard varieties of Spanish, but can take additional time to type in on a standard English keyboard (the primary keyboard used in Puerto Rico).

Additionally, all three users consistently substitute a "k" when the word requires a "qu," as in Roberto's usage of "Cualkier" and "aunke." This is an aspect inherent to netspeak, where users substitute letters that make the same sound in order to cut down on the time spent writing. The use of unconventional spelling is evidence of in-group language usage because it is very difficult to read and understand what a person is saying if one is not aware of these consistent spelling changes. While I personally speak a very non-formal "street variety" of PRS, I have not spent an equal amount of time chatting online with friends in Puerto Rico. As a result, I initially had a difficult time understanding the different spellings and abbreviations. My feelings as an outsider reading these profiles is additional evidence as to how language creates and bolsters in-group identity or keeps others out.

As is evident in Veronica's posting to her friend, she has refused to spell words in their conventional manner by inserting a "k" for all words with a /k/ sound, which would traditionally be spelled with a "c" or "qu" in Spanish. Furthermore, her abbreviations of Spanish words such as *donde* as "dnd" and *cuidate* as "cdt" provide more examples of her sophisticated use of Puerto Rican netspeak.

English use on the three pages would not be considered standard English either. Roberto spells a variety of words incorrectly, such as "basket ball," "litte," and "whit." Similarly, Veronica spells some of the English words as if she was writing in Spanish, for example, "mayoring," where she inserted the y (which in Puerto Rican Spanish is pronounced as a "j" in *juror*) or in the writing of *wanna* for "want to."

Lexical Borrowings

Some of the examples of words tallied as English words were originally English words now incorporated into PRS. Such words include: el shot; el turkey; or the word "break" in "T vere Hoy si me da BREAK." The borrowing of English words in PRS is commonly associated with pop culture and can be seen to varying extents throughout the three profiles. The two profiles with descriptions in English had many more words borrowed from English. On the other hand, Roberto's profile (which was all in Spanish) only had a few English words in the comment section, such as "cheque" used at the end of a posting meaning "check you later" or "see you later." It is important to remember that while I am examining a particular person's profile, the comment section of the profile is composed of many comments from other MySpace.com users, so the language on this portion of the profile is representative of many more users.

One aspect of lexical borrowing that all of the profiles had was technical MySpace.com interface words. Because the interface of MySpace.com is in English, much like most Microsoft Windows interfaces on Puerto Rican computers, users refer to the different aspects of the page in English. So a user might write "gracias por el pic comment" (Angel's page) or "gracias por add" where the person commenting is thanking the profile owner for adding the person to their list of friends. Another example of using technical terms borrowed from English came from Veronica's comments section where her friend wrote "lokilla hiciste ese slideshow . . ." (referring to a slideshow of rotating pictures that is common to many profiles). Thus, the MySpace.com interface itself has changed the way users use their language, just as Microsoft Windows has influenced the way many people talk about computers. Puerto Ricans generally refer to aspects of the Windows interface in English, despite the fact that the interface can be changed to Spanish.

Puerto Rican Identity in Language

Puerto Ricans on MySpace.com are part of a larger community of users, and thus many choose to present themselves as people who are able to communicate in a language of wider communication, i.e., English, and do so by creating a variety of different aspects of their profile in English. One specific example came from Angel's profile, when he wrote "i am from a litte island call Puerto Rico." This is an example of how Angel appeals to MySpace.com users living outside of Puerto Rico. At the same time, Angel's comments section is almost entirely in PRS, exemplifying his everyday language use and his identity as a Puerto Rican. Thus, Angel uses two different languages in two parts of his profile in order to appeal to different audiences. Angel's shift to appeal to different audiences draws resemblance to Wilson, the Dominican-American participant in Bailey (2000), who was able to jump from English to Dominican Spanish in order to foster the identity of a Dominican-American or an African-American, depending on which identity suited him at the time. Wilson's ability to play both identities to his advantage is similar to what Puerto Ricans are doing on their MySpace.com profiles when they write their descriptions and quotations in English. I would argue that they are trying to appeal to and attract other off-island English users with English in their descriptions, while simultaneously keeping their ties to locals by exchanging PRS messages in comment sections.

The division of Spanish and English language use on these pages suggests that English is being used by more Puerto Ricans than was reported in the 2000 census. Or because one must have access to the Internet in order to create and maintain a MySpace.com page, the amount of English we see on these pages could instead be evidence of the division of the haves and have-nots as described in Schweers and Hudders (2000). In other words, those with higher socioeconomic status are able to attend private schools where more emphasis is put on English, and they also have increased access to the Internet and the millions of other users who are also connected. Similarly the language use on these MySpace.com profiles indicates a change in language use among the younger ages of the general Puerto Rican population.

I believe language maintenance of PRS has been highly successful because of Puerto Ricans' acceptance and creativity in embracing new forms of expression. While Puerto Rican Spanish is inherently "Spanish," it has its own linguistic nuances, which have developed over the past five hundred years and are much of the reason why Puerto Rican identity is so entrenched in the dialect of Spanish spoken on the island. Puerto Ricans' understanding of the relationship between their dialect of Spanish and their identity as Puerto Ricans is one of the principal reasons for such successful language maintenance (Morris 2003: 155). The Internet and emerging PRS netspeak is a current venue which will continue to reinforce language maintenance on the island. Hence, PRS and particularly PRS netspeak, creates a community of users that is different to other users, e.g., those in the United States, who can only limitedly tap into the identity structure at play due to the rampant use of PRS netspeak. Thus, the use of Spanish in the comment sections, as well as the code switching and borrowing of English words, forms an aspect of Puerto Rican identity among MySpace.com users that ultimately aids in the maintenance of PRS as the primary language on the island. Given the vastness of the Internet, the use of Spanish could have potential implications for language maintenance among Puerto Ricans living off the island as well.

Despite the amount of English use in the profiles, only seven out of fifty (14 percent) comment sections were primarily in English, and only one out of fifty (2 percent) was entirely in English. This is evidence that Puerto Ricans continue to identify as PRS speakers and users. Furthermore, the Internet and specifically, websites like MySpace.com have provided a forum in which PRS can continue to change and evolve into a language that is representative of the people and culture that use it.

In addition to the use of both Spanish and English, the pages exhibited a plethora of examples of an emerging area in Puerto Rican netspeak, namely non-traditional ways of spelling and lexical borrowings. The complex uses of these features constrain what it means to be a "true" Puerto Rican user of MySpace.com and, in doing so, continue to construct Puerto Rican identity for this group of 18- to 22-year-olds.

Conclusion

English language use in Puerto Rico has been a politically charged topic since the United States' takeover in 1898. Throughout the past century, Puerto Ricans have been successful in maintaining and bolstering Puerto Rican Spanish at the expense of English. As mass media has taken a stronger hold on everyday life on the island, Puerto Ricans have more access to English than ever before. As is evidenced on MySpace.com, Puerto Ricans are cognizant of their place in a multilingual world where English is the lingua franca. However, despite the use of English, Puerto Rican MySpace.com users prefer to interact with their friends and family members in PRS and Puerto Rican netspeak. As more and more Puerto Ricans start using the Internet, the creativity and ingenuity of PRS will become even clearer as it is used on different social networking sites. Likewise, it appears that MySpace.com profiles are a site where traditional Spanish prescriptivist language ideals are blissfully ignored and innovative, and where contemporary Puerto Rican netspeak is substituted for those conservative ideals. While the language use documented in this research deviates from standard varieties of Spanish, it is representative of how a particular group of Puerto Ricans actually use language and therefore can be understood as an in-group marker of Puerto Rican identity.

Acknowledgments

I would like to acknowledge my family and friends who have helped me throughout this process, as well as those at the University of Puerto Rico Mayagüez and the University of Arizona, without whom this would not have been possible.

Notes

1. The continual movement between Puerto Ricans between Puerto Rico and the United States.
2. Friends post comments in the comments section at different times. Often times a user will post a question or comment on a person's comment section and the receiving friend will later respond on their same friend's profile.
3. An argument could be made that the music/video sections change often, but in my experience users rarely change the genre and language their songs are in and thus this change is not as relevant for this study.

References

Algren de Gutiérrez, Edith. 1987. *The Movement Against Teaching English in the Scbools of Puerto Rico.* Lanham, MD: University Press of America.

Bailey, Benjamin. 2000. Language and Negotiation of Ethnic/racial Identity among Dominican Americans. *Language in Society* 29(4): 555–82.

Buchhotz, M. 1999. "Why Be Normal?": Language and Identity Practices in a Community of Nerd Girls. *Language in Society* 28(2): 203–23.

Crystal, David. 2004. *The Language Revolution,* Maldin, MA: Polity Press.

Duranti, Alessandro. 1997. *Linguistic Anthropology.* Cambridge, UK: Cambridge University Press.

Fayer, Joan M. 2000. Functions of English in Puerto Rico. *International Journal of the Sociology of Language* 142: 89–102.

García, Ofelia, José Luis Morín, and K. M. Rivera. 2001. How Threatened is the Spanish of New York Puerto Ricans?: Language Shift with Vaivén. In *Can Threatened Languages be Saved?,* ed. Joshua A. Fishman, 44–73. Toronto: Multilingual Matters.

Hochberg, Judith G. 1986. Functional Compensation for /s/ Deletion in Puerto Rican Spanish. *Language* 62(3): 609–21.

Internet Slang Dictionary & Translator. 2006 [cited 12 November 2006]. Available from http://www.noslang.com/top20.php.

Kiesling, Scott F. 2004. Dude. *American Speech* 79(3): 281–305.

Mendoza-Denton, Norma. 1999. Turn-initial *No:* Collaborative Opposition among Latina Adolescents. In *Reinventing Identities: The Gendered Self in Discourse,* eds. Liang Bucholtz, A. C. Liang and Laurel A. Sutton, 273–92. New York: Oxford University Press.

Miniwatts Marketing Group. Internet World Stats: Usage and Population Statistics. 2006 [cited 3 December 2006]. Available from http://www.internetworldstats.com/car/pr.htm.

Morales, Amparo. 2000. ¿Simplificación o interferencia?: El español de Puerto Rico. *International Journal of the Sociology of Language* 142: 35–62.

Morris, Michael A. 2003. The Effects of North American Integration on Linguistic Diversity. In *Languages in a Globalising World,* eds. Jacques Maurais and Michael A. Morris, 143–56. New York: Cambridge University Press.

Morris, Nancy. 1995. *Puerto Rico: Culture. Politics, and Identity.* Westport, CT: Praeger.

———. 1996. Language and Identity in Twentieth Century Puerto Rico. *Journal of Multilingual and Multicultural Development* 17(1): 17–32.

Nash, Rose. 1977. *Comparing Spanish and English: Patterns in Phonology and Orthography.* San Juan: Parallexicon Publishing Company.

Negrón-Muntaner, Frances. 2004. *Boricua Pop: Puerto Ricans and the Latinization of American Culture.* New York: NYU Press.

Nickels, Edelmira L. 2005. English in Puerto Rico. *World Englishes* 24(2): 227–37.

Schweers Jr., C. W., and M. Hudders. 2000. The Reformation and Democratization of English Education in Puerto Rico. *International Journal of the Sociology of Language* 142: 63–87.

Suhr, Jim. 2006. Dozen Students Expelled over MySpace Fight. Canada.com, 30 November. http://www.canada.com/topics/technology/story.html?id=02e61206-07ca-43d1-98c6-a67797941c&k=30824.

U.S. Census Bureau. 2003. Census 2000 data for Puerto Rico. [cited 5 September 2007]. Available from http://www.census.gov/census2000/states/pr.html.

Vélez, Jorge. 1993. Contrasts in Language Use: Office Service Encounters in San Juan and Austin. In *Puerto Rican Communication Studies,* ed. J. Fayer, 63–78. Puerto Rico: Fundación Arqueológica, Antropológica, e Histórica de Puerto Rico.

Wikipedia. MySpace. 2006 [cited 10 December 2006]. Available from http://en.wikipedia.org/wiki/Myspace.

World-Gazetteer. Puerto Rico: Largest Cities and Towns and Statistics of Their Population. 2006 [cited 4 December 2006]. Available from http://www.world-gazetteer.com/.

Zentella, Ana Celia. 1997. *Growing Up Bilingual: Puerto Rican Children in New York.* Malden, MA: Blackwell Publishers.

'Access Denied': Arguments about Equality and Access to New Media in the Information Society

Stephen Lax

One of the claims about new information and communications technologies (ICTs) is that they can help to bring about a fairer, more equal society. Studies of technology suggest that assumptions about their impact are often unfulfilled. Nevertheless, arguments about a new information society have again elevated the importance of technology as a social force. Here we will examine the ways in which public policy making is bound up with assumptions about the kind of society we live in and the role of technology within it.

Keywords: poverty and inequality, information society, meritocracy, technological determinism

Issues:

- *Can new media technologies create a more equal society?*
- *Is the increase in information creating a new kind of society?*
- *What is the role of governments?*

Introduction

There can be no doubt that the world exists in a state of staggering inequality. The United Nations reports some sobering statistics: the world's richest 500 individuals have a combined income greater than that of the poorest 416 million; the 2.5 billion people living on less than $2 a day account for 5 per cent of global income while the richest 10 per cent account for 54 per cent. Within individual countries, whether rich or poor, similar inequalities persist; inequality in both the USA and the UK is significantly greater than in Albania or Ethiopia (UNDP, 2005, pp. 4, 55).

Not surprisingly, these general material inequalities are reflected in levels of access to technologies. While in Sweden and the UK in 2004, for example, there were more mobile phones than people, in Mozambique just 3.7 per cent of the population had one; in North America, 62 per cent of people had Internet access compared with 2.6 per cent in Africa (ITU, 2005). The World Summit on the Information Society (WSIS), convened under the auspices of the United Nations, noted at its first meeting in 2003 that:

> *The benefits of the information technology revolution are today unevenly distributed between the developed and developing countries and within societies. We are*

> *fully committed to turning this digital divide into a digital opportunity for all, particularly for those who risk being left behind and being further marginalized.*
>
> *The rapid progress of these technologies opens completely new opportunities to attain higher levels of development. The capacity of these technologies to reduce many traditional obstacles, especially those of time and distance, for the first time in history makes it possible to use the potential of these technologies for the benefit of millions of people in all corners of the world. (WSIS, 2003, p. 2)*

In many countries governments seek to overcome this division by providing access to, for example, Internet-connected computers in public libraries.

Think About It

Should ICTs be considered a basic skill, alongside numeracy and literacy? What are the criteria by which we should decree an area of human activity an essential or 'basic' skill? Are there any other skills which should similarly be regarded as essential? How about driving? Are those who can't drive socially disadvantaged any less than those who can't do ICT?

While it is hard to imagine anyone arguing against opening up ICT access to all, the suggestion that a more equitable exposure to new communications technologies will lead to a reduction in inequalities more generally depends on a number of assumptions, firstly about an understanding of the term 'equality' and secondly about the role of technology in society. At an instinctive level, we may feel we know what an egalitarian society would look like, but in fact we probably know better what an *unequal* society looks like. Gross disparities in material wealth and possessions have been documented above; further, most would acknowledge the unequal nature of the barriers to *acquiring* wealth faced between those born into the poorest families in even the wealthiest nations and those growing up in families that are readily able to afford to pay for the best schools and healthcare. The intractability of social position is underscored by research revealing static or even reducing levels of social mobility. For example, Blanden and Gibbons's (2006) study of UK adults found that those who had grown up in poor families in the 1980s were *more* likely to remain poor themselves in later life in comparison with children growing up ten years earlier. In other words, levels of **poverty** were increasingly persistent across generations. Equivalent studies in the USA also find low or reducing levels of intergenerational mobility (Corak, 2004; Hertz, 2006). Thus we have the compound inequalities of the absolute differences in material wealth and the obstacles, evidently increasing, which prevent those at the bottom easing their way out of poverty and closing the gap with the better off.

Poverty and inequality Poverty can be absolute or relative. How thresholds of poverty are defined is contentious—what are basic minimum living standards? Relative poverty is defined in terms of inequalities: the poor in a wealthy country are likely to be better off than the poor in a developing country.

The Digital Divide

There are two possible objectives in tackling these aspects of **inequality**: *equality of outcome* or *equality of opportunity*. The former approach seeks to ensure, through social policies targeted at both disadvantaged *and* advantaged groups, that levels of inequality between rich and poor are reduced; the latter meanwhile is more concerned with targeting the poorest in society through highly focused policies to support them in getting on the first rung of the social ladder, to set a basic minimum level of entitlement that therefore doesn't directly affect those already free from social deprivation. The outcome-focused approach tends to be associated with currently unfashionable, interventionist politics (based on redistribution of wealth through tax and welfare policies) while the equality of opportunity approach is popular with liberal (or neoliberal) politicians who favour governments playing merely an enabling role that seeks to impact minimally on the free market. While this characterisation is necessarily crude, its significance lies in the approach to new media technologies taken by governments and non-governmental organisations across the globe. Policy makers argue that we are in danger of developing, or deepening, a 'digital divide', a schism between those who are able to access and make use of new media technologies and those who, for a variety of reasons, cannot. Governments are increasingly migrating everyday services such as tax returns to the World Wide Web and conducting online consultations as part of the policy making process. Further, campaigning and political organisations outside government are using the Web to publicise and debate civic and political issues. The Internet is extensively used too for commercial transactions with all kinds of goods and services being traded. It is thus feared that those who do not have access to these communications technologies risk being marginalised, having restricted choice as consumers and also being left out of an important part of the democratic process (Lax, 2004).

In most industrialised nations, therefore, initiatives have taken place to secure public access to ICTs through public provision of Internet-connected computers. Examples include national projects, such as 'Networking the Nation' in Australia, in which rural communities were given funding to establish public access points, or the installation of 'the People's Network' of Internet-connected computers in the UK's public libraries. These are supplemented by countless smaller-scale projects in which public terminals are placed in schools or community centres and so in such places few people can be said to be completely 'disconnected' in the sense of having no *material* access to the Internet. For non-industrialised nations, as we saw earlier in the WSIS declaration, similar goals are defined for universal access through public provision of Internet terminals.

However, there is no guarantee that this provision will lead to more equitable use of the networks. The ways in which people use the Internet, and indeed whether they do so at all, are far more complex than simply a question of access to the technology itself. For example, countless market research studies show that in advanced countries like the USA and the UK, household access to the Internet via home computer is reaching a plateau of around 60 per cent—while computers and the cost of Internet access itself have both become cheaper, more than one in three households remain without Internet access (ITU, 2005; Ofcom, 2006, p. 157). Evidence about the ways in which public Internet terminals are used is variable. Hardy and Johansen (2003) studied users of public terminals placed in libraries in Victoria, Australia, and concluded that users tended to be drawn more from lower-income groups or the unemployed than in the population at large, although the educational background of users was much higher than typical. Simpson et al. (2004) and Strover et al. (2004), in their respective Australian and US studies, both concluded that in many cases community provision of Internet terminals was characterised by a lack of clarity as to their purpose. The notion of 'access' is nebulous—is access about increasing the total number of users, or increasing usage among a particular social group? Such questions affect the choice of location for Internet terminals, and in the absence of an answer it is difficult to interpret, in terms of success or failure, the consequent usage.

Think About It

Leadbeater reckons our children will work with their brains, not with their hands. Can all manual labour be replaced with machines? If not, whose kids will do the dirty work?

Beyond questions about who actually uses these public terminals, evidence about the wider social consequences of Internet usage is variable. Studies of Internet users' political engagement suggests that those who are politically active online (for instance emailing representatives, taking part in consultations or political organisations) are generally the same people who were active in the absence of the Internet—the evidence that a new political engagement is engendered by access to the Internet is insubstantial (Gibson et al., 2004). Another area where the use of the Internet was anticipated to result in social gain was in education, in particular lifelong learning for adults. The UK government's view has been consistently positive about the effects of ICTs on educational provision—the education secretary has declared, 'sometimes people can't access the learning they need because courses simply don't fit around their daily lives. New technologies open up huge opportunities to give people access to the learning they want in a way that suits their needs' (Morris, 2002). While this may be an almost universal position among policy makers, again evidence suggests that ICTs in public places make almost no difference to the take up of educational provision in later life. Family, geography and other social factors are much more important and patterns of 'lifelong learning' in the new technological era are almost unchanged from the 'non-technological' past (Selwyn et al., 2006).

An Information Society

So how do we begin to explain the enthusiasm with which new technologies like the Internet are proposed as furthering the cause of equality and fulfilment of individuals? The explanation lies in a particular view of a new technological society. Influential thinkers have been prominent in the past few decades in describing far-reaching social changes, in part wrought by new technologies, heralding a new stage of social evolution following the agricultural and industrial phases (Toffler, 1980). While there is no coherent account of this new era, it is generally labelled the **information society**. This notion identifies information (and knowledge) as the central feature of society: your place in that society is bound up with your relationship with information, your access to it and what you do with it (May, 2002).

Glossary

Information society A society where access to and manipulation of information becomes the key determinant of one's place in society.

Industrial society A society where position depends on social relations under capitalism.

Agrarian society A society where position depends on feudal relations such as land ownership.

The information society idea can be traced back to Daniel Bell. He began writing in the late 1960s of the 'post-industrial society', one in which the reliance on manufacturing as the main source of wealth was being replaced by the service industries. These depend far more for their success on access to information rather than the raw materials of industrial manufacturing. For Bell this heralded a new kind of social order, with shifts in the balances of wealth and power. These transformations would be every bit as important as the change from an **agrarian society** to an **industrial society** in the seventeenth and eighteenth centuries, in which the absolute rule of feudal landlord (the aristocracy, who assumed power solely by birth) was replaced with the rise of a capitalist class (where power was conferred by monetary wealth) and its own form of rule (parliamentary democracy). In the 'post-industrial society', power no longer depends upon access to wealth and physical labour, but instead on access to knowledge and the skills to select and manipulate that knowledge (mental labour).

> *It becomes clear that the 'old' industrial order is passing and that a 'new society' is indeed in the making. To speak rashly: if the dominant figures of the past hundred years have been the entrepreneur, the businessman, and the industrial executive, the 'new men' are the scientists, the mathematicians, the economists, and*

> *the engineers of the new computer technology. And the dominant institutions of the new society—in the sense that they will provide the most creative challenges and enlist the richest talents—will be the intellectual institutions. The leadership of the new society will rest, not with businessmen or corporations as we know them (for a good deal of production will have been routinized), but with the research corporation, the industrial laboratories, the experimental stations, and the universities (Bell, 1970, p. 394).*

The 'post-industrial society' became translated during the 1980s and 1990s into the 'information society', a term which Bell himself also came to use, though it should be noted that Bell's explanation of modern society was not as exclamatory as many of the more recent 'information society' advocates. While the notion of 'post-industrial' society is clear and precise in that it implies a break with the previous industrial society, the term 'information society' is more vague, and this is part of the problem in dealing with it.

However vague we may find the idea, and Webster (2005) points out clearly the lack of precision in most attempts at definition, it has nevertheless captured the hearts and minds of political leaders and advisers. Shortly before becoming the UK's Chancellor of the Exchequer, Gordon Brown wrote about equality in the 'fast-changing information-based economy dominated by the importance of knowledge'. He continued, 'the defining characteristic of economy is less an individual's ability to gain access to capital and far more his or her ability to gain access to knowledge and to use it creatively' (Brown, 1996). Here was a new kind of economy that offered a new kind of politics, the 'Third Way'. Keenly adopted by US President Clinton and the UK Prime Minister Tony Blair, it was articulated by Antony Giddens as a rejection of both 'old left', social democratic politics and the neoliberalism of the 'new right' (Giddens, 1998). In fact, Third Way ideas embrace much of the neoliberal economic agenda (Callinicos, 2001). A fascinating insight is given by Charles Leadbeater, a former adviser to the Blair Government and member of the think tank Demos, which was highly influential on UK government thinking in the late 1990s. Indeed, Leadbeater's book, *Living on Thin Air* (2000), carried an endorsement by Prime Minister Tony Blair on the front cover: '. . . an extraordinarily interesting thinker. His book raises critical questions for Britain's future'. Two passages give a flavour of the ideas:

> *Most of us earn our livings providing service, judgement, information and analysis, whether in a telephone call centre, a lawyer's office, a government department or a scientific laboratory. We are all in the thin-air business. That should allow our economies, in principle at least, to become more humane; they should be organized around people and the knowledge capital they produce. Our children will not have to toil in dark factories, descend into pits or suffocate in mills, to hew raw materials and turn them into manufactured products. They will make their livings through their creativity, ingenuity, and imagination (p. ix).*
>
> *An economy which becomes more knowledge-intensive has the potential to become more inclusive and meritocratic. Everyone with an education can have a go. . . . In an economy which trades know-how and ideas, everyone seems to have a chance to make it, working from a garage, their kitchen or their bedroom. Twenty-five-year-old drop-outs can create best-selling computer games; a nerd fresh out of college can create the Internet's best browser; a boy with no formal education can become Europe's most precocious fashion designer (Leadbeater, 2000, p. 33).*

A Knowledge Economy

This new knowledge economy then is one which is overhauling social institutions, and in the process, becoming more equal, 'in principle at least'. It is a seductive argument. Leadbeater outlines how he himself deliberated long and hard before deciding to become a 'portfolio worker', but having made the leap is now reaping the rewards. No longer having to work for an organisation where 'some ambitious manager points the spotlight at you', he can organise his life around his family and earn his living 'by finding people who will pay me to do things I am interested in' (2000, p. 2). While acknowledging that he is helped by having a good degree and a list of useful contacts through his earlier career as a journalist, for Leadbeater this simply helps make his argument: there are no barriers to this kind of life provided a good education is available to all.

Again we see the importance of new technology to the claims made about the new economy. When wealth is derived from manipulating and trading information, access to technology becomes vital.

Activity 1

Think of some more examples of 'thin air' kinds of employment. Try and envisage what living in a society that relies only on these categories of work would actually be like. Will any other workers be required? What will those people be doing? Develop a short description of such a society.

Whereas in an industrial society 'access to technology' (in the form of factories and machinery) could equally be argued as key to wealth, access was also required to raw materials (including labour) and markets—and access here meant physical transportation of goods. In an industrial society, then, this kind of access was inevitably privileged and uneven. But the technologies of the new economy are different. Leadbeater chastens those who are pessimistic:

> *Yet this ought to be an age of excitement and optimism. Consider, first, the breathtaking possibilities of technology. By 2007, the hard disk in the average television set top box should have enough memory to store all the songs ever recorded. By 2010, it should be able to take every film. Telecommunications bandwidth is doubling every 12 months. The capacity of fibre to transmit information has increased by a factor of 16,000 in less than five years; it will soon carry everything we can say, write, compose, play, record, film, draw, paint or design. The entire contents of the US Library of Congress could be passed through an optical switch in less than three seconds (Leadbeater, 2004, p. 30).*

Not only is this an impressive list of technological achievements, for the new economy the point is that these are relatively cheap and therefore accessible technologies. Leadbeater's college dropouts can be on a similar footing with long-established corporate giants when it comes to creating some best-selling software. It was no surprise when the **meritocracy** that Leadbeater alluded to in 2000 became a topic of earnest debate in Third Way economies. In the lead-up to the 2001 general election, Tony Blair declared that the UK would become a truly meritocratic society, a natural consequence of the emergence from industrial society into an information society. The contrast with old industrial societies seemed clear. Inequalities and unfairness are endemic to industrial society, where prosperity resulted from ownership of large capital assets. These assets were often acquired not through any mechanism that rewarded skill, capability or effort, but more often were inherited through accident of birth, however poorly suited the recipient might be to the role. Now, thanks to the accessibility of the tools of the information economy, the new ICTs, it would be possible to open up its wealth-creating potential to all. This logic underpins the policy objective of providing public access to ICTs for those who do not have access elsewhere.

Meritocracy A *meritocracy* is a social order in which each individual acquires social status on the basis of merit: skill, ability and effort, rather than gender or class. Superficially a fairer society, the notion is in reality hypothetical due to complexities of defining merit and tends to cement social position, consigning 'non-elite' members of society to unfulfilling manual labour.

A Meritocratic Society?

The same logic of a meritocratic society explains the changing role of education in many industrialised nations. In the UK, information and communications technology has been placed at the centre of educational policy, not merely as a tool for learning but as a subject in its own right as one of the only three subjects (alongside maths and English) required to be taught to all throughout compulsory schooling. Amid much publicity in 1997, a National Grid for Learning was established and plans put in place to connect all schools to the Internet—the importance of ICT was in no doubt (Lax, 2001). The government could claim that it had set in place the basis of an egalitarian society—one in which all would have access not only to the technologies of the information society but also to the basic skills needed to exploit them. Having established a level playing field, whatever anyone chose to do with those skills and technologies was their own choosing. If they applied their skills successfully, then they deserved whatever reward came their way; should they be unsuccessful, then really they could not complain. In Leadbeater's words, 'everyone with an education can have a go'. Michael Young, who coined the term in 1958 in *The Rise of the Meritocracy 1870–2033,* was concerned that the point of his satire had been missed: the meritocracy would be a divided society. When those at the top were able to believe (wrongly) they had truly acquired their status on their own merits, then equally those beneath them deserved their lot and certainly no sympathy nor, logically, any assistance (Young, 2001). Yet here we can see why a belief in the possibility of meritocracy means that calls for equality are in terms of opportunity rather than outcome. If one believes in the possibilities of an information society, then the role of government is limited to providing equality of opportunity, and that is delivered by equalising access to technology and the requisite basic skills. One can readily extend the argument to accept the logic of 'laptop welfare' where, rather than cash benefits, welfare provision might be more valuably conferred by giving the poor incentives to buy laptop computers; such schemes were proposed in 1995 by US Senate leader Newt Gingrich and in 1999 by UK Chancellor Gordon Brown.

The foregoing has suggested that to believe we live in an information society can imply that the conditions exist in which a true meritocracy can be achieved. In such circumstances, governments can justify intervening only minimally in the workings of society, ensuring that there is a basic provision of access to ICTs and skills. As the entry costs to ICTs are so low, equality is achieved as everyone has the same *opportunity* to acquire wealth. But such beliefs are readily challenged on two grounds: firstly on the singling out of technology as the main driver in this social change; and secondly on interpreting general shifts in economies from manufacturing to service industries as signifying more fundamental social changes.

The first objection focuses on the assumption that the technologies of communication (such as those listed by Leadbeater) are themselves independent entities in this process of social change. In other words, the technologies merely exist, evolving according to some internal logic of technical change. This view, labelled **technological determinism**, sees technology as autonomous, largely outside of social forces, but nevertheless having far-reaching social consequences. This claim seems intuitively true; in some instances it is even enshrined in dictums such as 'Moore's Law' (which states that computer processing power doubles every eighteen months). No doubt computers and networks will get faster and generally more capable, and this can seem like some natural process. However, it doesn't require much investigation to see that technological change is most certainly impacted upon by social forces. Economic forces selectively impel some technologies forward, such as ICTs today, just as different economic needs were crucial to developing and establishing other technologies such as radio communication 100 years ago, and the prime funders of scientific and technical research are governments and large industrial enterprises. Thus instead of seeing the technologies of the information society as somehow *naturally* progressing along the particular trajectory they currently follow, we should recognise that this path is socially shaped. The progress of a particular technology and its consequences are reflections of the society from which they emerge as much as any inherent capabilities of the technology.

Technological determinism A belief that technology develops independently of society and in so doing is the central cause of consequent social 'impacts'. Widely challenged, nevertheless the idea is found in many historical accounts of social change and lies behind countless predictions of future social trends.

The second problem with the claims about the information society is that it is assumed that the changes that have undeniably taken place over the past few decades, in particular the increase in the quantity and speed of information in its many guises, imply fundamentally new social forms. For example, the new information workers identified by Leadbeater above include those in 'a telephone call centre, a lawyer's office, a government department or a scientific laboratory': all are lumped together in the 'thin air' business. Meanwhile, the factory worker, the weaver and the miner are all from a past age. Leadbeater believes that the divisions of old, between manual worker and manager, are swept away in the new age of information. Yet working conditions in today's call centres share far more with twentieth-century factories or nineteenth-century weaving sheds than a lawyer's office or government department (Poynter & Miranda, 2000). That a call centre worker should see herself as aligned with lawyers and scientists in an information economy is readily dismissed.

More fundamentally, we have seen already that despite the growth of service industries over several decades, traditional inequalities persist doggedly. Whether we study social mobility, participation in political activity or the take-up of educational opportunities, we find no convincing evidence of novel trends that might suggest some social transformation. The alternative conclusion is that we are simply witnessing a particular phase of the continuing social order known as capitalism. Webster makes the connections explicit. He suggests that, rather than today's society requiring redefinition, it is enough to look at the historical nature of capitalist society to explain the present. For if we look at what characterises capitalism, we see a number of tendencies: access to goods and services being dependent on ability to pay; private rather than public provision; market criteria (profitability) as the dominant factors in deciding whether something is available; wage labour as the norm in the workplace (Webster, 2000). There is little here that does not apply when we include the growth of information availability and the growth of information work. For example, the growth in the numbers employed in the service industries or information work has not substantially (if at all) altered the relationship between employer and employee. The increasing numbers entering (or staying in) education is part of a long-term trend of professionalisation of work roles rather than a feature just of the past few years (and it is significant that many of these professionals, teachers and doctors for instance, now find themselves in conditions of less professional autonomy and more subject to metric and prescriptive practices historically associated with manual workers). Hence it is quite possible to explain today's information-rich age as fully part of a continuing but ever-transforming capitalist system.

Think About It

- The information society debate obviously keys into a discussion of globalisation. Although globalisation is a big area, we haven't had time to consider it here, but it is worth thinking a little about the connection between the two.
- When the telephone was introduced over 100 years ago, the telephone companies issued advice and instructions in its usage (e.g. how to speak into the microphone: 'distinctly and directly'). The idea seems laughable now. Is it just a matter of time then before everyone can use the Internet without giving it a moment's thought?

Conclusion

If we are to seek a society that is more equal, in which more people not only have greater *opportunities* to improve their own lives and those around them, but also are actually able to *achieve* those objectives, then we cannot rely simply on the provision of access to technology. Technology here is a distraction. Despite the ubiquitous presence (in industrialised nations) of ICTs and the information they process, there is no reason to see this as changing our understanding of society. Instead, as has been debated (and fought over) for decades or even centuries, the route to equality and fairness requires us to look more fundamentally at the ways in which society is divided, and to seek to overcome the assumptions and activities that give rise to those divisions.

- This chapter has described how communications technologies are claimed to deliver social change.
- In particular we have seen that public provision of Internet access is seen as a public policy goal.
- It has explained that such policies are based on assumptions about the role of technology in society and about the kind of society in which technologies develop.
- We have explored issues such as the information society and technological determinism.

Tutorial Questions

- Can access to new media technologies create a more equal society?
- Which characteristics were argued in this chapter to be indicative of a new kind of society and economy?
- How can the notions of an 'information society' and a 'meritocratic society' work together to justify limited government intervention in the provision of access to ICTs and skills?

Recommended Reading

MacKenzie, D. & Wajcman, J. (eds), 1999, *The Social Shaping of Technology,* 2nd edn, Open University Press, Buckingham.

Warschauer, M., 2004, *Technology and Social Inclusion: Rethinking the Digital Divide,* MIT Press, Cambridge, Massachusetts.

Webster, F. (ed.), 2004, *The Information Society Reader,* Routledge, London.

Williams, R., 1990 [1974], *Television: Technology and Cultural Form,* Routledge, London.

Winston, B., 1998, *Media Technology and Society,* Routledge, London.

Websites

Digital Divide: see www.benton.org and www.digitaldividenetwork.org

European Union's Information Society Portal:
www.europa.eu.int/information_society/text_en.htm

Wikipedia: www.en.wikipedia.org/wiki/Technological_determinism

World Summit on the Information Society (WSIS): www.itu.int/wsis/index.html

References

Bell, D., 1970, 'Notes on the Post-industrial Society' in Olsen, M. (ed.), *Power in Societies,* Collier Macmillan, London.

Blanden, J. & Gibbons S., 2006, *The Persistence of Poverty Across Generations: A View from Two British Cohorts,* Policy Press, Bristol.

Brown, G., 1996, 'In the Real World', *Guardian,* 2 August, p. 13.

Callinicos, A., 2001, *Against the Third Way,* Polity, Cambridge.

Corak, M. (ed.), 2004, *Generational Income Mobility in North America and Europe,* Cambridge University Press, Cambridge.

Gibson, R., Römmele, A. & Ward, S., 2004, *Electronic Democracy: Mobilisation, Organisation and Participation via New ICTs,* Routledge, London.

Giddens, A., 1998, *The Third Way: The Renewal of Social Democracy,* Polity, Cambridge.

Hardy, G. & Johansen, G., 2003, 'Characteristics and Choices of Public Access Internet Users in Victorian Public Libraries', *Online Information Review* 27 (5), pp. 344–58.

Hertz, T., 2006, *Understanding Mobility in America,* Center for American Progress, Washington.

ITU, 2005, *ICT Statistics,* available at www.itu.int/ITU-D/ict/statistics/ (accessed 30 November 2005).

Lax, S., 2004, 'The Internet and Democracy' in Gauntlett, D. & Horsley, R. (eds) *Web Studies* (2nd edn), Arnold, London.

Lax, S., 2001, 'Information, Education and Inequality. Is New Technology the Solution?' in Lax, S. (ed.) *Access Denied in the Information Age,* Palgrave, Basingstoke.

Leadbeater, C., 2004, 'Globalisation: Now the Good News.' *New Statesman* 1 July, pp. 29–31.

Leadbeater, C., 2000, *Living on Thin Air: the New Economy,* Penguin, London.

May, C., 2002, *The Information Society: A Sceptical View,* Polity, Cambridge.

Morris, E., 2002, 'Estelle Morris welcomes report on learning with new technology,' press notice 0161, Department for Education and Skills.

Ofcom, 2006, *The Communications Market 2006,* Office of Communications, London.

Poynter, G. & de Miranda, A., 2000, 'Inequality, Work and Technology in the Services Sector' in Wyatt, S., Henwood, F., Miller, N. & Senker, P. (eds), *Technology and Inequality: Questioning the Information Society,* Routledge, London.

Selwyn, N., Gorard, S. & Furlong, J., 2006, *Adult Learning in the Digital Age: Information Technology and the Learning Society,* Routledge, Abingdon.

Simpson, L., Daws, L. & Pini, B., 2004, 'Public Internet Access Revisited', *Telecommunications Policy* 28(3–4), pp. 323–37.

Strover, S., Chapman, G. & Waters, J., 2004, 'Beyond Community Networking and CTCs: Access, Development, and Public Policy,' *Telecommunications Policy* 28(7–8), pp. 465–85.

Toffler, A., 1980, *The Third Wave,* Collins, London.

UNDP, 2005, *Human Development Report 2005,* UNDP, New York.

Webster, F., 2005, 'The Information Society Revisited' in Lievrouw, L. & Livingstone, S. (eds), *The Handbook of New Media: Social Shaping and Consequences of ICTs,* Sage, London.

Webster, F., 2000, 'Information, Capitalism and Uncertainty', *Information, Communication & Society* 3(1) 69–90.

WSIS, 2003, *Declaration of Principles,* WSIS, Geneva.

Young, M., 2001, 'Down with Meritocracy', *Guardian,* 29 June.

MODULE SEVEN

Communities and the Hyperlinks that Bind

Grooming, Gossip, Facebook and Myspace: What Can We Learn about These Sites from Those Who Won't Assimilate?

Zeynep Tufekci

This paper explores the rapid adoption of online social network sites (also known as social networking sites) (SNSs) by students on a US college campus. Using quantitative (n = 713) and qualitative (n = 51) data based on a diverse sample of college students, demographic and other characteristics of SNS users and non-users are compared. Starting with the theoretical frameworks of Robin Dunbar and Erving Goffman, this paper situates SNS activity under two rubrics: (1) social grooming; and (2) presentation of the self. This study locates these sites within the emergence of social computing and makes a conceptual distinction between the expressive Internet, the Internet of social interactions, and the instrumental Internet, the Internet of airline tickets and weather forecasts. This paper compares and contrasts the user and non-user populations in terms of expressive and instrumental Internet use, social ties and attitudes toward social-grooming, privacy and efficiency. Two clusters are found to influence SNS adoption: attitudes towards social grooming and privacy concerns. It is especially found that non-users display an attitude towards social grooming (gossip, small-talk and generalized, non-functional people-curiosity) that ranges from incredulous to hostile. Contrary to expectations, non-users do not report a smaller number of close friends compared with users, but they do keep in touch with fewer people. Users of SNS are also heavier users of the expressive Internet, while there is no difference in use of instrumental Internet. Gender also emerges as an important predictor. These findings highlight the need to differentiate between the different modalities of Internet use.

Keywords: social network sites, Dunbar, Goffman, presentation of self; social grooming, Internet, Facebook, Myspace

Social computing applications, led by social network(ing) sites (SNSs) like Facebook and Myspace, have burgeoned in the past few years; scholarly research on the social consequences of social computing has not caught up. A focus of past research has been possible inequalities rising from the Internet's capacity to provide access to information, jobs and economic mobility, education, access to government services and similar benefits (Dimaggio et al. 2004). As Internet access has become near-universal in rich nations, earlier concerns regarding a digital divide have mostly faded from public and policy discussions. Yet looking at Internet use in a fine-grained manner, disaggregating specific modalities of practice, does reveal persistent divisions and differences. The specific character of Internet use and socio-psychological disposition of users can also influence social outcomes, even after controlling for demographics and total Internet use (Zhao 2006; Livingstone 2007). It is not just the Internet but what you do with it—as well as who and what kind of person you are.

The rise of social computing opens a new dimension of benefits (and harms) stemming from differential use. These applications have the potential to create gaps in social capital (Putnam 2000), transform the role of weak ties (Granovetter 1973) and shift the boundaries between public and private.

The rapid diffusion of SNSs within a few years invites two connected questions about their adoption process. Why have these applications, which initially had little corporate backing, no paid advertisements and significant negative media attention, attracted so many users so quickly? Conversely, even on college campuses, where access is rarely the bottleneck, why has a small but seemingly persistent minority chosen not to adopt these technologies—and what can we learn from an analysis of these non-users? How do these applications relate to the larger context of Internet use? Finally, what are the possible social implications of use and non-use of these sites?

Social Network(ing) Sites (SNSs): A Brief Overview

These sites center on the *profile,* which for users is 'a representation of their sel[ves] (and, often, of their own social networks)—to others to peruse, with the intention of contacting or being contacted by others' (Gross 2005). Boyd offers the following definition: 'A "social network site" is a category of websites with profiles, semi-persistent public commentary on the profile, and a traversable publicly articulated social network displayed in relation to the profile' (Boyd 2006a).

The most prevalent examples of SNSs are Facebook, which started as a college site and is still dominated by college users, and Myspace, which has always been open to the general public. Multiple studies show that somewhere between 80 and 90 per cent of all college students have a profile on an SNS (Gross et al. 2005; Lampe et al. 2006; Stutzman 2006). All SNS allow users to articulate their social network via links between their profile page and other profiles. Profiles linked to each other in this manner are called *friends*. Profile owners also express an online persona through pictures, words and page composition, as well as through data fields where information ranging from favourite books and movies to sexual orientation and relationship status (single, in a relationship, etc.) is indicated.

Partly due to the high level of offline–online integration (Ellison et al. 2007), students tend to use their real names and engage in high levels of self-disclosure, especially on Facebook (Tufekci 2008). Facebook allows users to 'tag' individuals on photographs uploaded to the site, which means identifying the person in the photograph and thereby linking the picture to that person's profile, and thus creating a searchable digital trail of a person's social activities. A 'news feed' feature shows what one's 'friends' have been doing on the site: a typical entry might read 'Sally has Left a Message on Jim's Wall', or 'Alice and Bob are now friends'. Users also provide status reports—'Mary has a headache from studying for the organic chemistry exam!' All of this activity is framed by semi-public comments people leave on each other's profiles—short salutations, humorous repartee and more. A profile on an SNS is not a static entity; rather, it is a locus of social interaction that evolves and changes to reflect various dynamics within social networks and communities.

Conceptualizing SNS

I draw on the work of Robin Dunbar (1998), who proposed that gossip, people-curiosity and small talk, all of which are seemingly non-functional and are often popularly understood as mere distraction or deviation, are in essence the human version of social grooming in primates: an activity that is essential to forging bonds, affirming relationships, displaying bonds, and asserting and learning about hierarchies and alliances. Dunbar suggests that our seemingly insatiable appetite for gossip is neither a random, irrelevant fact, nor simply a construction of a singular culture. While the particular forms of gossip are entangled in culturally shaped constructions, ranging from celebrity gossip in our mediated mass culture to daily chatting around the village well in a peasant society, gossip in general can be seen as a corollary of our disposition towards sociality, which integrally involves figuring out where we and all others stand in relation to each other (Dunbar 1998).

Social grooming should be seen as both a bonding activity and a *competitive* activity: it is a means to improve one's reputation and status as well as access to resources and social and practical solidarity. An inability, unwillingness or lack of talent in social grooming activities may be of real detriment to a person's interests through a disadvantage in accumulation of *social capital,* which can be understood as resources that accrue to an individual through 'more or less institutionalized relationships of mutual acquaintance and recognition' (Bourdieu & Wacquant 1992, p. 119; see also Putnam 2000).

SNSs replicate many of the functions of gossip or social grooming: users display their own bonds and observe those of others through profile 'friends', leave semi-public messages for each other (which serve mainly as acknowledgement), present a public self for their community, and watch and participate as all others also engage in these activities in an interlocked dance of community formation.[1] At their core, these sites are about mutual acknowledgement, status verification and relationship confirmation (Boyd 2006b).

Much of the activity on an SNS can also be conceptualized as a form of presentation of the self, in the sense of Goffman (1959). Users engage in impression management by adjusting their profiles, linking to their friends, displaying their likes and dislikes, joining groups, and otherwise adjusting the situated appearance of their profiles (Boyd & Heer 2006; Lampe et al. 2007; Tufekci 2008).

Dunbar's notion of social grooming and Goffman's concepts of the presentation of the self and impression management are complementary aspects of the construction of the social self. As Goffman articulated, 'for a complete man to be expressed, individuals must hold hands in a chain of ceremony' (Goffman 1956, p. 493). It is through social interaction and socially embedded public or semi-public action that we affirm our relations, construct our status and ultimately produce the social 'me' in the sense proposed by Mead (1934).

Factors in SNS Adoption

Given the recent rise of these sites, it is not surprising that there are relatively few large-scale, quantitative analyses of SNS use that incorporate details about users and non-users, types of use and possible consequences. However, existing research already finds important social consequences as well as differences in access and usage. Due to the limited amount of existing research on SNS, this study was developed in a two-step approach which involved incorporating early qualitative findings into the research design in the latter stages.

Expressive and Instrumental Internet

Previously, scholars have generally categorized Internet use under three broad headings: commercial, informational and communicative uses (Kraut et al. 1999; Weiser 2001). Following Petric (2006) and Zhao (2006), I suggest a related but somewhat different division: between social and non-social uses, between the *expressive Internet* and the *instrumental Internet*. By the *expressive Internet,* I mean the practice and performance of *technologically mediated sociality:* using the Internet to perform and realize social interactions, self-presentation, public performance, social capital management, social monitoring, and the production, maintenance and furthering of social ties. The expressive Internet should be recognized as 'a social ecology involving other people, values, norms and social contexts' (Petric 2006). The *instrumental Internet,* on the other hand, refers to information seeking, knowledge gathering and commercial transactions on the Internet, and non-social communication involved in such transactions. This is the Internet of online banking, shopping and checking the weather.[2]

The expressive Internet has been expanding rapidly, a process often described in the popular press as the rise of social computing. Studies show that these tools have been assimilated as a means of social interaction and social integration for increasing numbers of people and communities (Haythornthwaite 2005; Quan-Haase 2007), and that people are increasingly using the expressive Internet in ways that complement or further their offline sociality (Wellman et al. 2001; Hampton & Wellman 2003; Hampton 2007). Given the social grooming conceptualization of SNS proposed in this paper, users of SNS can be expected to be interested in social uses of the Internet.

RQ1: How does social network site usage fit with different types of Internet use as well as total amount of use?

Friendship and Social Ties

While these applications are designed in order to facilitate social interaction, there is fairly little known about the strength of their impact on existing or potential social relations. Ellison et al. (2007) found some differences in social capital formation among college students depending on their intensity of Facebook use. The strongest association they found was between Facebook use and *bridging social capital,* i.e. looser ties between people situated in different, non-overlapping social groups. Such ties, also referred to as weak ties, are crucially important in providing a window of access to opportunities outside one's own immediate network (Granovetter 1973). Due to the high levels of offline integration (Ellison et al. 2007) SNSs could also promote strong ties.

RQ2: How does SNS use relate to number of friends kept in touch with as well as to weak or strong ties?

People may differ in their inclination to view the Internet as a place for engaging in social activities and friendship management. While some researchers have argued that the Internet has serious shortcomings for community and friendship building (Blanchard & Horan 1998), others find that it may facilitate interaction (Walther 1996).

RQ3: Do people's attitudes toward online sociality in general impact on their SNS adoption?

Demographics and Place of Residence

Hargittai (2007), in one of the few academic publications concentrating on non-users of SNS, provides a unique level of detail regarding the demographic and socioeconomic background of college level users and non-users of SNSs. Her findings were that demographic and place of residence matter and that being female, not living with parents, having ready Internet access and spending more time on the Internet were associated with higher rates of SNS adoption. Gender remains a strong pivotal point. Although access and total amounts of use have converged between genders, strong divergences in modalities of use, in self-perceived and real skill levels, and in attitudes remain (Boneva 2001; Jackson et al. 2001; Hargittai & Shafer 2006).

RQ4: How do demographics and place of residence influence SNS adoption?

Privacy

Another important consequence of SNS adoption concerns privacy. Participation in these sites involves extensive self-disclosure (Tufekci 2008) and may discourage those who are concerned about privacy (Acquisti and Gross 2006). While privacy controls exist for most sites, a significant number of users do not employ them. This makes sense if, as I argue, the purpose of using these sites is to be seen. Further, the privacy controls often simply mean that only SNS 'friends', who may number in the hundreds and thousands, can access the profile. Most SNS users, especially college students, freely add most people who ask to be 'friended'. There have already been many cases reported in the media about negative consequences stemming from SNS posts, ranging from denied diplomas to lost jobs.

RQ5: How do students' privacy concerns impact on their decision to join SNS?

Social Grooming

While social grooming through language may well be an important human activity (Dunbar 1998) there is no reason to presuppose that everyone will be equally disposed to such activity. Interest in exchanging and browsing social information about friends and acquaintances, and curiosity about people, is likely to be related to interest in how an application specifically facilitates such activity.

RQ6: How does general interest in social grooming relate to adoption of SNS?

Methods

Sample and Data Collection

This study reports results based on a sample of college students in a diverse, midsized public research university. The quantitative results (n = 713) were collected at three points in time (spring and fall of 2006 and spring of 2007) and qualitative results (n = 51) at two (spring and fall of 2006).

The qualitative phase consisted of focus groups. Undergraduate respondents were recruited through announcements in classes specifically excluding those in which the survey was administered. The focus groups were divided into non-users (no profile), light users (rarely checks the profile), medium users (checks the profile a couple of times a week) and heavy users (checks and updates almost everyday) in order to better gauge the consensus beliefs of these subgroups (citation) and to understand the differences between users and non-users. The interviews were transcribed and coded.

The survey instrument was developed in light of the findings of the qualitative phase and pretested with a group of non-users and users. The survey was then administered to students enrolled in multiple sections of an introductory social science course, a popular elective for fulfilling general education requirements. The survey was administered in class, in a paper-and-pencil format.

Questions were added to the survey after the first round of data collection, based on a preliminary analysis of the first round results and further qualitative interviews. The advantage of this approach was that this allowed for further probing based on data in an area for which there was virtually no published research during the time of the study. However, this incremental approach also introduces a weakness in that not all questions were asked across all the samples. The data from the three rounds of survey data collection were compared with regard to key variables, and there were no statistically significant differences in racial or gender composition or in average age of the students. The sample was thus combined for analysis for questions that were asked in all three rounds. Analysis was performed separately for questions that were asked only in later rounds. While I do not have any reason to suspect that the six-month apart samples differed in any key aspects, it must be kept in mind that some analyses are in effect from a different, later sample.

Demographic characteristics of the sample are described in Table 1. The combined sample was generally representative of the undergraduate population of the university, as the targeted class was a popular choice to fulfill mandatory requirements. Women were somewhat overrepresented (52.6 per cent, as opposed to 44.0 per cent in the university as a whole). The school reports minority enrollment of 41 per cent, arrived at by aggregating African-American, Hispanic and Asian-American students, leaving 59 per cent as combined sum of white and 'other'—the similar total in the sample was 53.8. Thus, white students might potentially be somewhat underrepresented, although it is not possible to ascertain this for certain. The sample had a significant portion of first-year students (44.0 per cent), who tend to be heavy users of social network sites. Students, to the degree

TABLE 1 Demographics of the combined sample.

	percentage of the sample (n = 713)
gender	
female	52.6
male	47.4
race/ethnicity	
White	45.2
African-American	12.7
Hispanic	2.3
Asian-American	31.2
other	8.6
grade	
freshman	44.7
sophomore	28.9
junior	19.2
senior	9.2
social network site (SNS)	
user	85.4
non-user	14.6

they had declared majors, ranged from humanities and social sciences to engineering and physical sciences and were not concentrated in any particular major.

Measures and Variables

To address the relationship between SNS usage and maintenance of social ties, the students were asked how many friends they kept in touch with at least once a week (*Weekly Friends*). (The response options were 0–5, 6–10, 11–16 friends a week, etc.) A subsample was about their ties. We distinguished the strength of the tie (Marsden & Campbell 1984) by asking about the number of people to whom they felt very close (*Very Close Ppl*) and the number of people to whom they felt somewhat close (*Somewhat Close Ppl*). Following previous research on social networks, *Somewhat Close* was defined to include those who are more than acquaintances but less than those who are very close.

The instrument also measured the amount of daily Internet use (*Daily Internet*) and whether students used Instant Messaging at all (*Uses IM*). Daily Internet was measured in intervals of 30 minutes, topping at 3 hours or more. Similar to previous research on SNS (Ellison et al. 2007; Hargittai 2007) place of residence was controlled by asking students if they lived in dormitories (*Lives Dorm*).

Students were asked whether they thought that a person could have a close friendship with someone they know only through online methods in order to gauge the impact of a general disposition toward online sociality. Students were also asked about their level of concern with online privacy (*Privacy*) with options ranging from 1 = not concerned at all to 4 = very concerned.

Questions used to probe the types of Internet activities were adapted largely from the Pew Internet & American Life Project which has been ongoing for many years. The selected questions were used in multiple Pew surveys, including the 'Parents and Teens' survey of 2004 which used these questions for both teenagers and adults (Lenhart & Madden 2005). However, unlike the Pew survey, which asked whether the activity was performed *at all,* this instrument asked about the frequency, coded from very often to never (4 to 1). Table 2 lists the wording of these questions as well as means and standard deviations for responses from the total sample.

Items that corresponded to social interaction, self-expression, communication or entertainment were aggregated into a variable called *Expressive,* and those that corresponded to informa-

TABLE 2 Internet practices among college students.

expressive			***instrumental***		
question	***mean***	***SD***	***question***	***mean***	***SD***
send or read email	3.805	0.486	look for health, dieting, or physical fitness information online	2.202	0.895
send or receive instant messages	3.391	0.946	go online to get news or information about current events	3.079	0.846
send or receive text messages using a cell phone	3.128	10.058	look for news or information about politics	2.429	10.016
go online for no reason at all, just for fun or to pass the time	3.632	0.687	look for religious or spiritual information online	2.127	0.875
read blogs of other people	2.386	10.083	look for information about a job online	2.321	0.946
go online to create or work on your own web page	1.761	10.003	go online to get information about a college, university or other school you were/are thinking about attending	2.890	0.911
			go to web sites about movies, TV shows, music groups, or sports stars you are interested in	2.431	10.085
			look for information online about a health topic that's hard to talk about, like drug use, sexual health, or depression	1.751	0.905
			buy things online, such as books, clothing or music	2.213	0.724
			go online to do school work or research	3.511	0.608

Scale: very often (4) = everyday or almost everyday; sometimes (3) = once a week or so; rarely (2) = once a month or so; and never (1).

tional and commercial uses to a variable called *Instrumental*. *Expressive* thus comprised the use of email, instant messaging, sending or receiving text messages, going online just for fun, reading blogs, and working on a personal web page. *Instrumental* included purchasing things online, doing school work, looking for health and fitness information, news, politics, job searching, going online to look up a college, looking up websites of movies and seeking information about health issues. Online games were excluded because interviews revealed the complicated nature of games, ranging from very social to very insulating (Steinkuehler & Williams 2006). The *Instrumental* scale, summing nine variables with a maximum value of 36, had a mean of 22.4 and a standard deviation of 4.4 (Cronbach's alpha = 0.74). The *Expressive* scale, summing five variables with a maximum value of 20, had a mean of 14.2 and a standard deviation of 3 (Cronbach's alpha = 0.66).

Social Grooming

Since there were no existing scales that measure social grooming, eight questions that closely followed social grooming themes of people curiosity, social interaction and keeping in touch (Dunbar 1998) were created.

EFFICIENCY

Also, as non-users strongly suggested in interviews that they believed efficiency was the main motivation for SNS use, five questions measuring interest in efficiency on and off the Internet were added. Since these questions were derived using preliminary results, they were asked only of a subsample of the students in the later waves (listwise N = 498). These questions are shown and analysed in Table 4.

Analysis and Findings

Predictors of SNS Use

To model the odds that a person would become an SNS user, I ran three logistic regressions (Table 3). In our first model, our predictors were female (dummy coded), age (ranging from 18 to 25+), dorm residence (dummy coded), amount of time spent on the Internet, number of friends kept in touch with every week, Instant Message use (dummy coded user or not), Expressive Internet Use scale and Instrumental Internet Use scale. In the second model, I added number of close friends and somewhat close friends, and in the third model, I added a variable indicating whether the student believed friendship through only online methods was possible (dummy coded as possible or not).

Amount and Type of Internet Use

The total amount of time a student used the Internet in general was not linked to the likelihood of SNS use. In the interviews, non-users of SNS indicated that they felt comfortable using the Internet. Similarly, use of instant messaging, another popular method of communication, was not associated with a change in the odds of becoming an SNS user.

TABLE 3 Logistic regression results: odds ratio (e^b) of using a social network site.

	model one		***model two***		***model three***	
	uses SNS	***p value***	***uses SNS***	***p value***	***uses SNS***	***p value***
female	4.191***	0.000	5.021***	0.000	4.987***	0.000
age	0.867†	0.086	0.836†	0.077	0.837†	0.081
lives dorm	1.725	0.105	1.245	0.573	1.235	0.589
internet per day	1.022	0.886	1.108	0.543	1.130	0.477
weekly friends	1.257*	0.031	1.340*	0.050	1.333*	0.054
uses IM	0.680	0.357	0.474	0.153	0.464	0.143
online privacy concern	0.709*	0.027	0.667*	0.029	.666*	0.029
expressive	1.464***	0.000	1.486***	0.000	1.493***	0.000
instrumental	0.950	0.176	0.928	0.110	0.923	0.094
no. of very close people			1.031	0.878	1.036	0.860
somewhat close people			1.049	0.734	1.051	0.727
online friendship					0.787	0.512
baseline odds (constant)	1.570	0.828	2.945	0.683	3.093	0.669
Cox and Snell R2	0.19		0.21		0.22	
N	506		369		363	

*Significant at 0.05; **significant at 0.01; ***significant at 0.001; †marginally significant (0.10 > p > 0.5). Since not all questions were asked in all samples, and since not all subjects answered all the questions, the N for the models varies. Specifically, number of close friends was asked to a subsample of subjects. Running model one specifically for the subsample used in models two and three revealed no substantive differences in the results.

However, while the total amount of Internet use was similar between non-users and users, the manner of Internet use mattered significantly. In the interviews, non-users of SNS also reported Internet use concentrated around practical needs: online banking, shopping, researching and such. Quantitative results confirmed that the use of the Internet for expressive purposes was highly significant in predicting SNS use, while use of the Internet for instrumental purposes was not.

Our interviews revealed that the differences in disposition towards Internet use persisted even when non-users of SNS were employing expressive applications. To the degree they used social network sites, they had specific objectives—looking up musical groups, finding ideas on fashion and makeup, and so on. The interviews revealed that the non-SNS users could understand looking something up, but not necessarily looking around just for fun. One non-user confessed that the constant pressure to join these sites had her seriously considering whether she should: 'I'm not going to do it because what's in it for me? Nothing.' This pattern was also observed in other mediated methods: one student expressed, with the understanding that she knew this was out of the ordinary, that she checked email only when she knew someone was going to email her.

Social Ties and Online Sociality

In interviews, SNS non-users emphatically stated that they had friends in their lives and were involved in social interactions with people around them. As can be seen in all the models in Table 3, the number of friends students kept in touch with weekly was significantly associated with SNS use. However, in our third model, I added the number of close or somewhat close friends and found no association between SNS usage and numbers of close friends. Also, in model three, I added a dichotomous variable which indicated whether or not the student believed close friendship could be achieved through online interaction alone. This variable was also not statistically significant.

Privacy Concerns

In the interviews, the non-users were often concerned with online privacy, but they did not see online SNS as dangerous. Many pointed out that disclosure on these sites was voluntary: 'But if you don't put it on there, no one can find it.' Another compared it to threats in real life: 'Some random person from the mall could follow you home, that's way more dangerous than someone tracking you down on the Internet.' In line with previously reported results (Tufekci 2008), the logistic regressions did show that higher online privacy concerns somewhat lowered the odds that a student would use SNS.

Demographics and Other Characteristics

Gender was the strongest predictor in the model and the only demographic variable that was statistically significant. In all the models, the odds of a woman using SNSs were four to five times the odds of a man.

Living in dorms was not associated with an increased likelihood of using SNS. While age appeared to be marginally significant, I found that the significance of age declined with each round of data collection, suggesting that this effect may have been an artefact of the spread of SNS initially among a younger population.

Social Grooming, Efficiency, and SNS Users and Non-Users

In the interviews, a majority of the SNS users talked about how much they enjoyed learning about their friends' and even strangers' lives. The heavy users in particular expressed that their use of these sites was partially driven by their curiosity about how people from their pasts were doing or whether they had changed. Some students also reported 'getting lost' in social networking, continually checking profile after profile, leaving message after message. One student talked about how receiving messages made her feel good, so she tried to leave messages to extend the good feeling to her friends. The important element in this interaction was not the content of the message, but the act of leaving the message as a means of acknowledging the other person.

Among non-users of SNSs, a very different response emerged. The idea of SNSs as fun did not seem comprehensible to the non-users. They were all familiar with these sites, and all had been asked by their friends to join. But why they should *want* to use them was just not clear. One non-user remarked how her friends would check out other people's profiles and sighed, 'I don't understand what people get out of looking at other people's profile. Live your life.'

Non-users generally reported that that they did not find it interesting to keep up with friends from their past. One student was somewhat exasperated: 'People from my high school would try to find me. . . . I had 39 pending friendship requests. I looked at the list and I knew five of these people. Five. Who the hell are you? Why are you bothering me? I haven't seen you in seven years, if I've seen you at all.'

Non-users expressed that they also did not understand why people are so curious about people they do not know, commenting not just about social networking sites, but life in general. There was a spirited discussion about how confusing, and stupid, they found it that many of their friends followed the lives of celebrities, or gossiped about people they hardly knew. One said in a mocking voice, to approving nods: 'Look at what Katie did this weekend, she's with who now? [sighs] You don't even know this person.'

These differences are clearly visible in Table 4, where the responses to questions clustered around social grooming, especially regarding curiosity about other people, people from one's past and

TABLE 4 T-tests comparing social grooming and efficiency disposition between users and non-users of SNS.

		N	*mean*	*p value*
social grooming				
I am curious about other people's lives**	user	443	3.12	0.037
	non-user	86	2.93	
I like keeping in touch with friends**	user	443	3.46	0.006
	non-user	86	3.24	
I am curious about people from my past***	user	446	3.13	0.001
	non-user		2.82	
I am outgoing*	user	449	3.19	0.023
	non-user	86	2.98	
I like to follow trends***	user	445	2.43	0.001
	non-user	85	2.11	
I do not enjoy social events*	user	439	1.81	0.013
	non-user	85	2.08	
I am shy	user	446	2.43	0.819
	non-user	86	2.45	
I like meeting new people	user	446	3.32	0.116
	non-user	86	3.17	
efficiency				
I am worried about wasting time on the Internet	user	445	2.29	0.130
	non-user	86	2.10	
I value efficiency highly	user	441	3.44	0.421
	non-user	86	3.50	
I am a very busy person	user	442	3.13	0.405
	non-user	86	3.21	
I am usually bored	user	442	2.32	0.398
	non-user	86	2.23	
I am always in a hurry	user	443	2.50	0.594
	non-user	85	2.55	

*Significant at 0.05; **significant at 0.01; ***significant at 0.001.

enjoyment of keeping in touch with friends, were significantly different between users and non-users. SNS users were also more likely to enjoy social events and reported being more outgoing. However, shyness or enjoyment at meeting *new* people was not significantly different between the two groups. The differences between the groups were more about how and how much to keep in touch with existing friends—be they weak or strong ties.

Efficiency

When asked about what they thought drew people to SNSs, the consensus theory among non-users was efficiency—people must be using these sites as a timesaver. One non-user suggested that, as college students, they were all very busy, and SNSs were 'more and more taking the work out of meeting people. If you are a college student you have many things to do and it's cutting one element out and also giving you the benefit you desire.' While SNS users did mention the efficiencies of SNSs, such as being able to find classmates to get notes for missed classes, or updating large numbers of people quickly on developments in one's life, for most users of these sites, the main theme was one of satisfaction derived from social interaction and observation itself.

I tested the efficiency/time-crunch hypothesis with multiple questions and found that this was not the difference between the users and non-users, no matter how the question was phrased. Questions such as 'I value efficiency highly', 'I am always in a hurry', 'I am worried about wasting time on the Internet', 'I am a very busy person', and 'I am usually bored' produced similar responses across both sub-groups (see Table 4).

Self-Presentation

The non-users also did not like the idea of engaging in presentation of self through these sites. One remarked, shaking her head: 'America is so self-obsessed. They are so about "Look at me. Look how I cool I look in this dress I'm wearing right now drinking with my friends. Check out how long this beer-bong cable is." I don't understand it.' Another complained that people were just fishing for affirmation: 'People will out-and-out lie on their profile. "My favourite movie is this movie; my favourite band is this band." I've been friends with you for ten years and I've never in my life heard you listen to that band. You are just saying so that people will look at your profile and will say, "Oh man, that is so cool. That's cool".'

One non-user remarked: 'They are wasting their talents. They should be paid for that kind of work.' Once again, engaging in this kind of activity without a direct goal, such as making money, seemed hard to comprehend for this group.

Are the Non-Users Disappearing?

While the number of non-users decreased from 17.1 per cent to 12.5 per cent during the sampling time frame, the differences were not statistically significant (Table 5). The per cent of female users remained steady across the three time periods (9 per cent of women were non-users in all three waves) while men steadily increased their participation (26 per cent of the men in the first wave, 21.7 in the second wave, and 15.4 in the last wave were non-users).

TABLE 5 SNS site usage over time.

	wave one (%)	*wave two (%)*	*wave three (%)*	*total (%)*
user	17.1	14.5	12.5	15.1
non-user	82.9	85.5	87.5	84.9

(χ = 1.896, asymp. two-sided p = 0.388).

Limitations

This study examined a purposive sample of undergraduate college students. Since the data in this paper are from a purposive sample, this limits the generalizability of the results. The study also represents data collection at more than one point in time and some of the questions were only asked of subsamples. In effect, this can be seen as related to subsequent studies. Causal assertions cannot be made, as the data are cross-sectional in nature.

Discussion and Conclusions

Contrary to the perceptions of non-users, SNS adoption does not seem to be about efficiencies, or yet another shortcut in the students' busy lives. The students who do not use SNSs are neither hermits, nor socially isolated, nor fearful of the Internet. However, the non-users are less interested in activities that can be conceptualized as social grooming.

Also contrary to what one might expect, the non-users reported similar numbers of very close and somewhat close friends as compared with SNS users. However, the number of friends kept in touch with weekly was significantly higher among SNS users. Keeping in touch may be conceptualized partially as a form of social grooming. Even if social grooming is essential to the functioning of complex societies, it may be that not engaging in it as much as other people do is not harmful to a small minority of individuals' capacity for maintaining friendship.

In fact, the lack of overlap between instant message non-use and SNS non-use shows that SNS non-users are not reticent toward communicating via the Internet; nor are they closed to the possibility of genuine social interaction through online methods. It is the social browsing (Lampe et al. 2006) and social grooming functions of SNSs that they are less interested in. Conversely, activities which can be grouped under these headings were often mentioned as among the most attractive features of SNSs in interviews with users.

The most significant predictor of SNS use in the logistic models, besides gender, was the tendency to use the Internet for *expressive* purposes: reading blogs, creating web pages, emailing, etc. Importantly, there was no relationship with *instrumental* uses of the Internet, such as information seeking and commercial transactions. This result highlights the need to differentiate between the different modalities of Internet practices.

In the interviews, non-users confessed to disinterest in and bafflement by social grooming activities and this was demonstrated in the quantitative analysis. Non-users might understand why one might sit in a sidewalk café with a friend and chat, but not comprehend why one would spend hours there simply to watch people go by, see who is sitting with whom, and observe how others interact among themselves. It was as if the non-users were people without a sense of smell, wondering why others buy expensive water with which to squirt themselves. Why waste so much money? People must like the shape of the bottle, they might imagine.

The combined analysis of quantitative and qualitative results suggests two principal clusters that influence SNS adoption among undergraduate students: disposition towards social grooming and privacy concerns. Our sample of non-users was too small to perform factor analysis to model these clusters. They may represent different groups of non-users, or they may overlap partially or totally.

About the Author

Zeynep Tufekci is an assistant professor in the Department of Sociology and Anthropology at the University of Maryland, Baltimore County. Her research interests include social and cultural impacts of technology, gender, inequality, social networks, surveillance and privacy. *Address:* University of Maryland, Baltimore County, Sociology, 1000 Hilltop Circle, Baltimore MD, 21250, USA. [email: zeynep@umbc.edu]

Acknowledgments

The author would like to thank the editors of this issue, Keith Hampton and Barry Wellman, as well as the anonymous reviewers for their insightful and constructive comments on an earlier draft of this paper. Portions of this research were supported by the Department of Sociology and Anthropology at the University of Maryland, Baltimore County.

Notes

1. Donath (2007) has made a similar argument in a paper published while this article was under review.
2. As with any ideal typology, there will be activities that reside on the boundaries. People do chat about products on Amazon and blog about the weather.

References

Acquisti, A. & Gross, R. (2006) *Imagined Communities: Awareness, Information Sharing, and Privacy on the Facebook,* 6th Workshop on Privacy Enhancing Technologies, Robinson College, Cambridge University, UK.

Blanchard, A. & Horan, T. (1998) 'Virtual communities and social capital', *Social Science Computer Review,* vol. 16, no. 3, p.293.

Boneva, B. (2001) 'Using e-mail for personal relationships: the difference gender makes', *American Behavioral Scientist,* vol. 45, no. 3, pp. 530–549.

Bourdieu, P. & Wacquant, L. (1992) *An Invitation to Reflexive Sociology,* University of Chicago Press, Chicago.

Boyd, D. (2006a) 'Social network sites: my definition', [Online] Available at: http://www.zephoria.org/thoughts/archives/2006/11/10/social_network_1.html (1 September 2007).

Boyd, D. (2006b) 'Friends, friendsters, and MySpace Top 8: writing community into being on social network sites', *First Monday*. 11(12)

Boyd, D. (2007) 'Social network sites: public, private, or what', *Knowledge Tree,* 13, [Online] Available at: http://kt.flexiblelearning.net.au/tkt2007/edition-13/social-network-sites-public-private-or-what/ (1 September 2007).

Boyd, D. & Ellison, N. (2007) 'Social network sites: definition and conception', *Journal of Computer-Mediated Communication,* vol. 13, no. 1, pp. 210–230.

Boyd, D. & Heer, J. (2006) 'Profiles as conversation: networked identity performance on friendster', in *Proceedings of the Hawai'i International Conference on System Sciences, Persistent Conversation Track,* IEEE Computer Society, 4–7 January, Kauai, HI.

Boyd, D. & Jenkins, H. (2006) 'MySpace and Deleting Online Predators Act (DOPA)', *MIT Tech Talk,* 26 May. [Online Available at: http://www.danah.org/papers/MySpaceDOPA.html

DiMaggio, P., Hargittai, E., Celeste, C. & Shafer, S. (2004) 'Digital inequality: From unequal access to differentiated use', in *Social Inequality,* ed. K. Neckerman, Russell Sage Foundation, New York, pp. 355–400.

Donath, J. (2007) 'Signals in social supernets', *Journal of Computer-Mediated Communication,* vol. 13, no. 1, [Online] Available at: http://jcmc.indiana.edu/vol13/issue1/donath.html (January 2008).

Dunbar, R. (1998) *Grooming, Gossip, and the Evolution of Language,* Harvard University Press, Cambridge, MA.

Ellison, N. B., Steinfield, C. & Lampe, C. (2007) 'The benefits of Facebook "friends:" Social capital and college students' use of online social network sites', *Journal of Computer-Mediated Communication,* vol. 12, no. 4, pp. 1143–1168.

Goffman, E. (1956) 'The nature of deference and demeanor', *American Anthropologist* vol. 58, no. 3, pp. 475–499.

Goffman, E. (1959) *The Presentation of Self in Everyday Life,* Doubleday Anchor Books, Garden City, New York.

Granovetter, M. (1973) 'The strength of weak ties', *American Journal of Sociology,* vol. 78, no. 6, pp. 1360–1380.

Gross, R., Acquisti, A. & Heinz, H. J. III (2005) 'Information revelation and privacy in online social networks', in *Proceedings of the 2005 ACM workshop on Privacy in the Electronic Society,* New York, pp. 71–80.

Hampton, K. & Wellman, B. (2003) 'Neighboring in netville: how the internet supports community and social capital in a wired suburb', *City and Community,* vol. 2, no. 4, pp. 277–311.

Hampton, K. N. (2007) 'Neighborhoods in the network society: the e-Neighbors study', *Information, Communication & Society,* vol. 10, no. 5, pp. 714–748.

Hargittai, E. (2007) 'Whose space? Differences among users and non-users of social network sites', *Journal of Computer-Mediated Communication,* vol. 13, no. 5, pp. 276–297.

Hargittai, E. & Shafer, S. (2006) 'Differences in actual and perceived online skills: the role of gender', *Social Science Quarterly,* vol. 87, no. 2, pp. 432–448.

Haythornthwaite, C. (2005) 'Social networks and Internet connectivity effects', *Information, Communication & Society,* vol. 8, no. 2, pp. 125–147.

Jackson, L. A., Ervin, K. S., Gardner, P. D. & Schmitt, N. (2001) 'Gender and the Internet: Women communicating and men searching', *Sex Roles,* vol. 44, nos 5/6, pp. 363–379.

Kraut, R., Mukhopadhyay, T., Szczypula, J., Kiesler, S. & Scherlis, W. (1999) 'Communication and information: alternative uses of the internet in households', *Information Systems Research,* vol. 10, no. 4, pp. 287–303.

Lampe, C., Ellison, N. & Steinfield, C. (2006) 'A Face(book) in the crowd: Social searching vs. social browsing', in *Proceedings of the 2006 20th Anniversary Conference on Computer Supported Cooperative Work,* ACM Press, New York, pp. 167–170.

Lampe, C., Ellison, N. & Steinfield, C. (2007) 'A familiar Face(book): profile elements as signals in an online social network', in *Proceedings of the SIGCHI Conference on Human Factors in Computing Systems,* ACM Press, New York, pp. 435–444.

Lenhart, A. & Madden, M. (2005) *Teen Content Creators and Consumers.* [Online] Available at: http://www.pewInternet.org/pdfs/PIP_Teens_Content_Creation.pdf (28 January 2008).

Lenhart, A. & Madden, M. (2007) *Social Networking Websites and Teens: An Overview.* [Online] Available at: http://www.pewInternet.org/PPF/r/198/report_display.asp (10 October 2007).

Livingstone, S. & Helsper, E. (2007) 'Gradations in digital inclusion: children, young people, and the digital divide', *New Media and Society,* vol. 9, no. 4, pp. 671–696.

Marsden, P. & Campbell, K. E. (1984) 'Measuring tie strength', *Social Forces,* vol. 63, no, 2, pp. 482–501.

McKenna, K. & Bargh, J. (2000) 'Plan 9 from cyberspace: the implications of the internet for personality and social psychology', *Personality & Social Psychology Review,* vol. 4, no. 1, pp. 57–75.

Mead, G. H. (1934) *Mind, Self, and Society from the Standpoint of a Social Behaviorist,* ed. C. W. Morris, University of Chicago Press, Chicago.

Petric, G. (2006) 'Conceptualizing and measuring the social uses of the internet: the case of personal web sites', *The Information Society,* vol. 22, no. 5, pp. 291–301.

Putnam, R. (2000) *Bowling Alone: The Collapse and Revival of American Community,* Simon & Schuster, New York.

Quan-Haase, A. (2007) 'University students' local and distant social ties: using and integrating modes of communication on campus', *Information, Communication & Society,* vol. 10, no. 5, pp. 671–693.

Steinkuehler, C. & Williams, D. (2006) 'Where everybody knows your (screen) name: online games as "third places"', *Journal of Computer-Mediated Communication,* vol. 11, no. 4, pp. 885–909.

Stutzman, F. (2006) 'An evaluation of identity-sharing behavior in social network communities', *iDMAa Journal,* vol. 3, no. 1, pp. 10–18.

Tufekci, Z. (2008) 'Can you see me now? Audience and disclosure regulation in online social network sites', *Bulletin of Science, Technology and Society,* vol. 28, no. 1, pp. 20–36.

Walther, J. (1996) 'Computer-mediated communication: impersonal, interpersonal, and hyperpersonal interaction', *Communication Research,* vol. 23, no. 1, p. 3.

Weiser, E. B. (2001) 'The functions of Internet use and their social and psychological consequences', *CyberPsychology & Behavior,* vol. 4, no. 6, pp. 723–743.

Wellman, B., Quan-Haase, A., Witte, J. & Hampton, K. (2001) 'Does the internet increase, decrease, or supplement social capital?: social networks, participation, and community commitment', *American Behavioral Scientist,* vol. 45, no. 3, pp. 436–455.

Zhao, S. (2006) 'Do internet users have more social ties? A call for differentiated analyses of internet use', *Journal of Computer-Mediated Communication,* vol. 11, no 3, pp. 844–862.

Weaving Webs of Faith: Examining Internet Use and Religious Communication among Chinese Protestant Transmigrants

Pauline Hope Cheong and Jessie Poon

This paper examines the relationship between new media use and international communication that addresses religiosity and affirms users' standpoints occupied by transmigrants that are marginalized in dominant societal structures. Drawing from focus group interviews among recent Chinese Protestant immigrants in Toronto, we argue that new media 'use' is broadened by users' cultural appropriation in situational contexts to include proxy Internet access as accommodative communication given the political and legal constraints in their home country. Chinese transmigrants not only re-interpret and alter semantic associations that spiritualize the Internet, they also engage in innovative strategies that involve the intertwining of offline and online communicative modes. These include deploying complementary media forms or communicating in codes that are mutually understood among participating members to facilitate intragroup networking among Chinese religious communities. Implications are discussed with regard to the importance of cultural norms and situational context in shaping mediated international communication.

In recent years, the analytical ante has been raised in the areas of international and religious communication as nation-states respond to civil disturbances and heightened sensitivities concerning inter-religious relations, and social cohesion. Such a revival prompts needed academic inquiry into the study of religion beyond sacred texts, as religion is critical in identity and community development in many parts of the world (Golden, Niles & Hecht, 2002; Hecht, Warren, Jung & Krieger, 2005; Kong, 2001). In addition, in light of web-based technology developments, expectations regarding the role of the Internet for advancing globalization and transnational communication have been raised (Chen, 2006). As a new avenue for religious communication, the Internet provides an international platform weaving new immigrants to their countries of origin, local places of worship, as well as to transnational religious movements. Given the increasing Internet adoption rates around the world, Internet use may facilitate a "diasporic public sphere" (Parham, 2004), a "safe" discursive space for identity-construction (Mitra, 2002) and debate among religious bloggers (Cheong, Halavis & Kwon, 2008), and a "cultural sphere" for ethnically affiliated immigrants (Yang, 2003). Indeed the Internet may play a more critical role than older media in global religious networks, or in what Levitt (2001) refers to as religiously relevant "social remittances" including religious ideas, beliefs, and social capital that is circulated from host to home countries. Although the Internet can facilitate geographical dispersion of religious groups (Helland, 2007; McAlister, 2005), scarce empirical research has focused on the communicative dimensions of transnational religious flows (Khagram & Levitt, 2008), rendering the mediated interrelationships between local and global religious connections a fertile ground for study.

Extant research on international migration and religious communication has seen two separate but related literatures. The first literature increasingly emphasizes the transnational dimension of migration, yet there are few studies on the forms and nature of transnational religious ties maintained by a growing number of "transmigrants" (Ebaugh & Chafetz, 2002). The second recognizes the webs of religious communication in transnational flows (e.g. Levitt, 2007), but this literature has paid scant attention to computer-mediated communication. This paper therefore seeks to integrate and expand both literatures by critically examining the role of the Internet in religiously-related communication and analytically assessing the expectation that Internet use will enhance and expedite cross-border interactions between religious actors. Following calls for research in co-cultural communication strategies to explicate ways in which minority and marginalized members communicate within dominant societal structures (Orbe, 1998; Orbe & Spellers, 2005), we extend the traditional foci on face-to-face communication to show that examining online transnational religious communication can unveil innovative media practices, thereby broadening conceptions of Internet 'use' to include immigrants' creative weaving of online and offline communication, including proxy Internet access. Concomitantly, we investigate and discuss the extent to which the Internet is appropriated for transnational religious communication, particularly how users tactically enact their mediated communicative practices as resistance (de Certeau, 1984), under conditions of heightened religious sensitivity and potential conditions of censorship.

The foci and context of investigation here are the Chinese Protestant immigrants in Toronto, Canada. Recent studies point to the surge in spirituality in China including the rise in self-identified Christians reaching up to 40 million (Cody, 02/08/07, *The Washington Post Foreign Service*). Interest in spirituality in China is in part attributed to the quest for new value systems to replace the instability engendered by the abandonment of communist doctrine for market economies (Aikman, 2003). Moreover, issues of immigrant integration and management of religious diversity have long attracted academic and policy interest in Canada and the United States where significant shares of residents in cities like Toronto, Vancouver, Los Angeles, and New York are recent immigrants. Investigating the nature of transnational religious communication of Chinese immigrants should help provide a barometer of community building where ties to the (universal) Church family and home churches are valued, even as nation-state boundaries are being challenged by popular images of globalization and intensified transnational communication.

Religious Transnationalism: Whither the Internet?

Historically, religious believers, especially those affiliated with world religions like Christians, Buddhists and Muslims, have utilized a variety of older media including newspapers, radio and television, to propagate and fulfill their faith missions (Soukup, 2002). It is thus unsurprising that the introduction of newer media applications online has been heralded to advance dramatic changes to religious communities and leadership. Yet, such expectations have not been verified by studies lodged in particular locales and conditions of religious individuals' everyday Internet use. In the case of religious transnationalism, previous studies on Chinese migration highlighted the presence of transnational religious communication but mostly did not examine the role of newer media. For example, Yang's (2002) case study of Chinese transnationalism of a church in Texas identified multiple formal and informal ties between members, ministers, and missionaries with Christian organizations in Asia as members contribute knowledge, time, and organizational skills to overseas churches. Chinese migrants maintained their transnational relationships via letters, phone calls, and the sharing of religiously-related magazines and tapes through visiting pastors or laypersons. However, findings of this study conducted between 1997 and 1999, during the nascence of the Internet's popularization, did not explicate the role of the Internet in facilitating the flows of religious communication. Similarly, Guest's (2003) ethnographic research of migrants in New York's Chinatown illustrated how local clergy and church members served as bridges when Chinese churches in Fuzhou received books, pamphlets, and tapes produced in New York from immigrants returning to visit or to serve as missionaries. Perhaps because the population studied comprised primarily of monolingual and low-income migrants who were presumably not Internet savvy, the role of the Internet was also not investigated.

More recently, in *God Needs No Passport,* Levitt's (2007) findings from four transnational communities in Boston, highlighted how many first generation immigrants live transnational lives through religion. Levitt (2007) contends that transmigrants structure a "new religious architecture" that operates flexibly via transnational supply chains of religious artifacts and decentralized ritual practices. However, although technological innovations like television, videos, and a few websites were mentioned in her book, an explication of the role of online religious communication in supporting the proposed "global religiosity" was absent in her analysis. Yet recent research evidences rising numbers of Internet users who practice "faith online" (Hoover, Clark & Rainie, 2004), highlighting the potential of the Internet to act as a global platform for the exchange of religiously relevant information, goods and services, with consequences for transnational communication (Cheong & Poon, 2008). However, as the Internet is a meta-medium (Agre, 1998), the notion of 'Internet use' tends to be conflated and thus needs to be better understood with regard to new migrant standpoints (Hiller & Frank, 2004). More recently, Chen (2006) explored how the Internet affords transnational entrepreneurship among Chinese immigrants in Canada, and stressed the importance of investigating both online and offline contexts for international communication. Thus, we argue that the notion of 'Internet use' and (non)-use needs to be unpacked and critically considered in light of distinct contexts in which media is embedded.

Cultured Technology and Transnational Religious Appropriation of the Internet

The Internet plays a significant role in facilitating globalization and transnational ties as newer media support a higher intensity of exchanges, new modes of transactions, and increase in transnational travel and contact (Portes et al., 1999). According to Krueger (2004), the spread of religious ritual and dogmatic knowledge is increasingly reliant on Internet-based discourse in religious newsgroups and forums. Furthermore, people who find themselves in a religious minority are often enthusiastic about using the Internet to foster ties with other marginalized group members (O'Leary & Brashers, 1996). Therefore, for some migrant communities, such as the Chinese immigrants being examined here, the Internet appears to be offering a new opportunity for re-establishing the spatial bonds of fledgling religious community. Helland's (2007) analysis of the early Usenet system, for instance, highlighted how Jewish and Hindu online forum users connected with others in their home countries as well as practiced long-distance rituals and cyber pilgrimages. However, what is less understood is how users shape Internet use in sync with certain social obligations and local conditions that promote or impede online religious connections, with implications for their transnational lives.

Indeed a corpus of literature has argued that computer-mediated communication does not operate in a social vacuum; rather newer media is shaped by users, embedded in local contexts and social relations (e.g. Ess, Kawabata & Kurosaki, 2007; Kling, 2000). This is relevant for transmigrants because the Internet may be subjectively appropriated by religious users in various ways. The level of technology appropriation, according to Eglash (2004), exists along a continuum of consumption and production. On one side, "reinterpretation" occurs on the consumer end of the spectrum, with changes only in language associated with the technological artifact. In the middle of the spectrum is "adaptation" which modifies both the semantic association and use. Adaptation is based on the flexibility of the technology to be modified by discovering a latent function or using it in a different way as originally conceived. The strongest changes are "reinvention" at the production end of the spectrum: where the technological artifact is significantly altered and where new functions are created through structural change.

Broadly speaking, technology appropriation is an extension of the "social shaping of technology" perspective, which emerged as a critique to "technology determinism" (Mackay & Gilispie, 1992). According to Kling (2000), social context is influential in how people use online technologies, and thus affects the consequences of the technology for the maintenance of social relationships. Technology is pliant, bearing imprints of its domestication within specific cultural and ethnic enclaves (Leonardi, 2003) and reinvention in ways that embody critique, resistance, and even revolt (Eglash, 2004). For instance, fringe group members' experimentation may reconfigure media and facilitate technological innovations outside of established corporate institutions (Sawhney & Lee, 2005). According to Mackay & Gilispie (1992), technology appropriation is integral for its cultural shaping, as people "may reject technologies, redefine their functional purpose, customize, or even invest idiosyncratic symbolic meanings in them" (p.699). As such, 'Internet use' or non-use may be attributed to the extent by which religious users reinterpret, adapt, and reinvent web-enabled applications within their daily locales. In the case of religious communication, these levels of appropriation may intepretatively constitute tactics of resistance (de Certeau, 1984), or assimilation, accommodation, and separation co-cultural communication strategies (Orbe, 1998) among minority ethnic, religious individuals on the margins of dominant society.

According to Barzilai-Nahon & Barzilai (2005), the Internet represents a form of "cultured technology" as their analysis of an ultra-orthodox Jewish community highlights how believers selectively use the Internet to maintain, even reinforce the fundamental values of their community. As such, a "spiritual shaping of technology" perspective examines the extent to which religious adherents frame technology and create online narratives within a spiritually rich worldview (Campbell, 2005). More recently, Kluver & Cheong (2007) maintain that instead of secularization, technological modernization and religion mutually reinforce and co-exist in Singapore, a highly wired city, as the Internet is shaped for community building by Asian religious leaders of diverse faiths.

In the context of transnational religious communication, in addition to local influences, spiritual shaping of the Internet may be further complicated by interacting conditions in both the host and home countries. As Levitt puts it (2001), migrants create "transnational social fields" with core and extended activities that vary in complexity and scope, depending on migrant involvement as well as how the State and religious organizations interact within specific geographical contexts. It follows that for transmigrants, online and offline religious flows need to be considered within the social and political contexts in which they occur. Such transnational contexts can be quite significant in influencing communicative behavior, as subsequent sections will show below, because the patterns of social interactions among Chinese immigrants appear to be contingent upon the political and institutional environments of their home country, in particular, constraints associated with the regulation of the Internet in China. Traditionally, diaspora is seen as antithetical to government control, and migrants may use the Internet to "narrowcast" to other members, thereby overcoming border restrictions and national regulations (Karim, 2003). The Internet may function as a site of resistance for activism, political and religious discourse in highly regulated societies (Ho, Baber & Khondker, 2002). For example, Robinson (2004)'s analysis of websites devoted to the spread of religious fundamentalist beliefs suggests that the Internet is important for the creation of transnational religious identities among the Hindu middle class in the United States. Yet, the ability of religious actors to maintain rich and varied transnational communication may be adversely affected by uneven transnational social fields.

Yang (2002) highlights that transnational religious ties between the Chinese in the U.S. and China remain mostly informal primarily because of the Chinese government's control of religious groups. The absence of a strong denominational infrastructure also encourages loose associations between congregants, with many believers in underground 'house churches'. China however regulates both religious and Internet practices. While the post-Mao government has allowed limited religious freedom, the latter is subjected to legal restrictions (Potter, 2003). Internet use is governed by laws prohibiting "controversial" online content, protecting "state secrets", sanctioning the employment of filters on websites and emails, as well as controls on Internet service and content providers (Opennet initiative, 2005). These constraints raise questions regarding if and how Chinese migrants negotiate transnational religious practices, online and offline.

In sum, this paper examines the interface of international communication and the cultural shaping of new media as manifested in the relationships between transnationalism and Internet use for religious communication. In the following sections, we seek to investigate how recent Chinese immigrants approach the Internet and negotiate their religious ties with implications for their transnational lives. Specifically, the literature reviewed above raises the following research questions. In what ways do recent migrants view and utilize the Internet and other media, for the exchange of transnational religious relevant remittances and the construction of transnational religiously-related ties? Given that recent migrant groups can create a new international religious architecture to facilitate transnational flows, how do they appropriate the Internet to participate in the rising forms of global religiosity?

Method

This study draws upon focus group interviews that were conducted as part of a larger project on the Internet and religion in Toronto, located in the province of Ontario, Canada, from July 2005 to May 2007. This included interviews with religious leaders, ethnographic observations of religious services, as well as a website and hyperlink analysis of religious organizations to provide rich contextual data from different perspectives (Yin, 1994). Historically and demographically, Toronto is ideal for studying transnationalism as it is a cosmopolitan city of approximately 2.5 million people, including new ethnic minority immigrants of varying religions (Census 2001, Statistics Canada). According to the latest Census, Chinese immigrants accounted for more than a quarter of the ethnic minority population in Canada. Toronto, together with Vancouver and Montreal, is amongst the top three cities of residence for more than 80 per cent of immigrants from Mainland China (Census 2001, Statistics Canada). In addition, Toronto affords the opportunity to examine the role of the Internet in religious practice: the latest Canadian Internet usage survey report (Zamaria & Fletcher, 2008) shows that a large proportion of urban residents in Ontario use the Internet (75% in 2004, and 81% in 2007).

This study draws from three focus group interviews conducted among 24 Chinese Protestant immigrants who were recruited after one year of fieldwork in collaboration with local religious organizations, in line with recommendations from prior ethnographic research among Protestant Chinese immigrants advocating vigilance in building trust with local official and lay religious leaders over time (Guest, 2004). Given the research interests of this study, recruitment was targeted at current adult users of the Internet. All interviewees were first generation immigrants from China (16 females, 8 males, with an average age of 33 years). The majority of them had been residents of Canada for less than five years. Generalizability is not sought with this sample, instead given the dearth of information on religious transnationalism, and to respond to the mandate for further grounded research to investigate "transnationalism from below" (Guarnizo & Smith, 1998), focus group interviews provide rich, in-depth data that are sensitive to the social context in which the data are produced (Krueger, 1997).

The discussion sessions typically lasted about one and half hours and the groups met in a local residence near a large church identified by religious leaders as a regular gathering place for discussions and prayers. The authors, who are effectively bilingual in Chinese and English, led the interviews in Chinese. Participants were given the choice to respond in Chinese or English. Sessions began with the participants introducing themselves and participants were then asked to discuss their perceptions of the Church and the Internet, and Internet use including transnational religious communication, information seeking, and community building. The project arranged for a fifty-dollar honorarium and provided refreshments to all participants. The interviews were transcribed in full for textual analyses as well as the use of verbatim quotations for reporting the research (Barone & Switzer, 1995). They were then subjected to thematic codification using The Ethnograph software, version 5.0 for qualitative analysis and analytic interpretation in conjunction with memos constructed from field research (Glaser & Strauss, 1967; Charmez, 2007). To standardize the presentation of the Chinese and English quotations from interview transcriptions, we have translated the Chinese responses into grammatically correct English. To protect the anonymity of the respondents, and in light of the sensitive information shared, personal names and identifiers are not mentioned, and acronyms, instead of names of the actual places, are used in this article.

Analysis

Our interviews revealed that the Internet carries various affirmative and liberatory meanings for Chinese migrants. They have a largely positive approach to the Internet and use it for various religiously-related purposes. Respondents described how the Internet was appropriated for transnational communication as web-enabled applications were reinterpreted and adapted to meet faith goals within the social milieu undergirding their local religious and transnational relationships.

Internet Perception and Appropriation for Religious Information Seeking

For all respondents, the Internet was used regularly to communicate with others and for most to search for information, including religious information and instruction. Our respondents indicated that religion online presents several functional advantages over older media connections in terms of religious information seeking. The Internet is perceived to be "full of religious and cultural resources" and "valuable for nurturing the spiritual life." For some, the Internet was seen as "critical" and "exceptionally helpful", for others "a matter of convenience" for connecting to textual and audio forms of religious information, including devotional materials, scripture, bibliographies of saints, and Chinese translations of short stories and commentaries, sermons and hymns. Their comments included the following:

> *"I use the Internet mainly to locate more news and information about Christianity . . . when I need to clarify certain bible verses, look up songs for praise and worship and retrieve sermons from English and Chinese speaking pastors . . . It is really convenient, like the library, and it's easier to go online for the information that you need."*

> *"I will log on daily to read the daily devotionals on the homepage. I have also subscribed to get short stories updates."*

> *"You can see the biographies of famous saints, bible movies, and articles about their lives online. The biographies are very interesting and beneficial for my spiritual development."*

In particular, several respondents mentioned how their Internet use enables them to acquire sermons by certain pastors, to whom they previously had limited access. Here, some respondents indicated that the increased connections to a number of religious leaders online were helpful for spiritual growth, although a few were less enthusiastic about listening to unfamiliar pastors. Several comments touched on the positive aspects of accessing religious teachings online:

> *"In the past, when you want to listen to the preaching of a certain pastor, you have to wait . . . Even if he comes for a day or two, he cannot say very much. On the Internet, I can listen to sermons whenever I want. It is very convenient."*

> *"I am more focused when I listen to sermons online as I feel that the pastor is just preaching to me. I wear my headphones and stay focused on what he is preaching. But in church, it is very tough to concentrate, as someone may cough, and another may sneeze during the service. If something comes up, I can stop the*

> *recording and listen to the sermon later . . . if there are certain parts that are harder to understand, I can listen to it again to gain a deeper understanding."*
>
> *"Every pastor and their main areas of emphasis in religious teachings are different. For example, one may emphasize the power of prayer; another may emphasize personal spiritual experiences. Therefore, I feel that listening to various sermons online helps my faith and increases my Bible knowledge."*

According to Leonardi (2003), minority ethnic populations align their perceptions of the Internet with their cultural values regarding 'good communication'. Specifically, in the case of religious communication, some emerging research points to how Christian email-based communities frame the Internet within a sacramental worldview (Campbell, 2005). Interestingly, some respondents interviewed here reinterpreted the Internet by altering its semantic associations to spiritualize this communication medium. For example, one respondent stressed that the Internet is "not a technology" but a "weapon of spiritual warfare" for believers to use as "a sword to bring truth to the world." A few others referred to the Internet as an "investment" where believers are to use their time online wisely to "reap optimal spiritual growth." In line with this reinterpretation, these respondents expressed that they do not participate or will limit their participation in electronic chatrooms and forums, as they deem these applications to be "very time-consuming" and thus a "bad stewardship" of time. One respondent said that electronic chatrooms are an "inappropriate" technology as it "generates too much confusion" which may lead to heresy and doubts about one's faith. Consequently, some of the respondents will "read postings", "criticisms and comments about their religion", but will "not join the forum", or "reason with non-believers" as online interactions are perceived to be less helpful than weekly church meetings for spiritual conversion. As one respondent explained, "I feel that it is very difficult to use the Internet to spread the gospel to a non-believer, especially in a textual form. It is very difficult to convince the other party through online discussions."

With regard to the global exchange of religiously-related remittances, the Internet is conceived as particularly significant as it is "an international place that allows a huge amount of information flow and exchange", "connecting people from all over the world." Many indicated that they relied upon religious information online in their daily lives and have connected to multiple sources and organizations in various countries including the United States, Taiwan, and Hong Kong. In addition, a number of our respondents agreed that the Internet is well suited for the search and exchange of religious information as it is relatively more difficult to censor information online as compared to other traditional media. Several respondents maintain that they could find ways to "get around" imposed controls and restricted informational access. For example, two interviewees said:

> *"It is not easy to restrict it, because the Internet is open and big. Even if there are restrictions, you can find a way to go around it, and find the information that you need."*
>
> *"The best part about the Internet is that it provides you with a free source. However, it can never replace the pastor's preaching. But during your spare time when you wish to obtain some information, for example in China where some information from overseas are censored, you can go online. In the past, it is very difficult to locate spiritual books in China but now you can easily find some of them online."*

Given the rising adoption rates of the Internet among different groups in the world, transmigrants are perceived to be increasingly active and enthusiastic agents in using and shaping the Internet for international communication. Yet our study indicates that the appropriation of the Internet is socially-situated, and in this case, religious information seeking is related to both local and transnational contexts in certain aspects of Internet use. In some instances, religious information seeking online exists in the context of the local church in which meaning and consumption of religious information takes place. Several participants mentioned their dependency on interpersonal relationships to locate religious websites, although the Internet enables autonomous quests for religious knowledge. We found that some responses expressed a deep trust in their pastoral leadership and reliance on other local church members to help pinpoint and locate relevant online resources, with some interviewees in turn recommending resources to others. For example, two interviewees said:

> *"In our church group, our brothers and sisters share their thoughts of websites which they think are good and they recommend the good ones to us."*

> *"I feel that in church, when I tell people that I visited a particular website, and if the people around me feel that the website is not good, they will tell me. I am sure if they know that the website is bad, they will not tell me, yes, that is good."*

These responses reveal the intertwining relationships between online and offline contexts as religious Internet use is socially informed and shaped by interpersonal influences that the interviewees believe to be reliable.

Furthermore, besides local influences, online religious information seeking is also influenced by transnationalism, particularly a change in the context of information seeking. Many interviewees agreed that they were keenly aware of the difference in transnational fields in terms of religious diversity and tolerance, although to a certain extent, the change in context may be more acutely perceived by recent immigrants. Freedom of religious belief is constitutionally protected in both Canada and China, yet several reports point to strong regulation, surveillance, fines and even imprisonment of independent Christians in China who are not affiliated with state approved associations that are supervised by an atheistic communist party (Madsen, 2003; Marshall, 2000). On the other hand, it has been recognized that the treatment of religious freedom in Canada and United States specifically privileges Christianity, in particular Protestantism (Beaman, 2003). Many interviewees said that they experienced a significant difference in their faith practices and Internet connections when they relocated from China to Canada; a country which they perceived to be more open to the expression and practice of personal and religious freedom:

> *"I became a Christian when I was still in China. However, I did not lead a life that was different from non-believers. For example, I did not have the courage to say grace openly when I was eating in public . . . After coming here, there is more information to help me understand about my Lord and Savior. This is a huge turning point of my life. Now it's very convenient to search for information online. For example, I can quickly look for information about church history."*

> *"Occasionally, we would listen to Christian radio stations from Hong Kong and United States and realize that we are in a different world. We felt alienated when they talked about God, prayer or about their daily lifestyles. That's why we hardly tuned in to them. In addition, we rarely came across such information. In fact, the first time that I read the Bible was when I came to the United States. You cannot compare China to the United States or Canada. Here, we have a wide range of information on Christianity unlike China because of the government's restrictive policies."*

In this way, our responses reveal that Internet adoption, particularly in this case, accessing religious information is shaped by respondents' previous experiences as lived in their past locales which allow for less scope for personal access and flexible appropriation of information online. Protestantism in contemporary China is growing but some house churches continue to experience intense persecution and regulation of their activities (Madsen, 2003; U.S. Commission on international religious freedom, 2008). In the absence of political and media regulations outside of China and the officially sanctioned 'Three Self Patriotic church' (Aikman, 2003), some Chinese transmigrants are actively connecting to religious websites to obtain and in turn disseminate previously forbidden didactic instruction and publications.

Internet Appropriation for Transnational Religious Communication and Ties

Focus group respondents were asked about their perception of the word 'church'. The majority saw the church as "jia", which translates into family. The Church is more than just a "jiaotang" (a physical religious meeting place), "it is a jiaohui", which means networks of members that fellowship with one another, implying the need to create and maintain transnational religious ties with their "fellow brothers and sisters" at home and worldwide. Our interviews reveal that Chinese immigrants use web-enabled applications quite intensively for a variety of relational goals, including email, Skype (Internet telephony), Instant Messenger, and the Webcam, to connect to their families and friends. Yet, with regard to religiously-related ties to their home country, Internet use was less frequent and constrained as the participants were mindful of considerable Internet restrictions in the context of transnational religious communication.

As noted above, prior literature suggests that the Internet may expand international religious flows as distance and time for communication are significantly reduced. However, more frequent transnational ties are sometimes deliberately curtailed to avert communication away from potentially controversial religious issues. For example, a number of interviewees believe that conventional email use is ineffective and even reject its use under some circumstances as it may destroy existing relationships or "guanxi" (Gao & Ting-Toomey, 1998) with other counterparts in China. One interviewee explained that in spite of his personal enthusiasm to proselytize, evangelism online is difficult, if not futile with home audiences who are either not interested, or are fearful of being involved in transnational religious networks, (e.g. the banned 'Falungong' in China). He said:

> *"Sometime ago, I sent an email to my friend in China about my religion here. I told him that I now trust Jesus Christ, but after hearing about my new belief, he has disappeared and does not respond to my emails. We are no longer in contact with each other."*

Other interviewees pointed out that although it is generally believed that regulation in China is not as strict as before, with the publication of the Bible and other religious materials, it is not the technological availability of information that matters but the social conditions surrounding media access. In this way, "situational context" is central to the co-cultural communication practices of these minority ethnic Christians (Orbe & Spellers, 2005). Two interviewees referred to how "phobia" and the "stigma of fear" restrain their online access to religious information:

> *"If I am in China, and someone tells me that I can go online to this website and find a lot of information, I will go and look at it. But I would have some reservations about downloading the information there because of fear. When I download things here, I am not afraid that people will come and check what I have downloaded."*
>
> *"The information is all online, but you do not dare to go look at it, in case it becomes evidence against you . . . If the police wants to check on you, they will check and take everything. They can say that they found this and that in your computer and where it came from, and the accusations against you will get longer." [sic]*

Given the difference in the transnational contexts of communication, several interviewees stressed the need to be circumspect about maintaining loose ties. A few said it was imperative to be "very careful" about online religious communication so they "have contacts, but not very close contacts . . . our contacts are much alerted" [sic]. In this sense, although the Internet has been popularly touted to facilitate and strengthen the density of global online ties, a low frequency of exchanges here is perceived by our interviewees to be beneficial under certain circumstances. They are cautious to practice secrecy, and even exercise self-censorship so as not to jeopardize their relationships with other members living in a context that is less than tolerant of religious communication outside the officially sanctioned grounds. They said:

> *"We have more freedom here. You can do or say anything you want here. The people in China do not want to have any trouble, and you do not want to cause them any trouble."*
>
> *"My job is related to the Internet so I know about [China's] regulations. As long as you are online, they [the police] can check on you. They know everything that you say or do. So, for their safety, we have to be careful."*

One participant stated, "Your words should not spoil the relationship". In other words, a premium is placed on mutual trust (Gao & Ting-Toomey, 1998) between transnational members to engage in religious communication that enhances flows of religious information throughout the network and at the same time minimizes negative consequences on members in China.

Hence, in lieu of direct email communication, a number of interviewees explained that transnational religious ties are more efficaciously sustained by either indirect communication exchanges involving the Internet and other media and/or coded communication online. The above tactics are adaptations that take into account the surveillance and strict regulatory conditions in their home country. For instance, in the face of potential censorship, some interviewees alter the conventional notion of 'Internet use' by practicing a tactic of 'indirect' use where they "figure out how to send online religious content over" to their mainland Chinese counterparts. This is done by first obtaining the needed information online and then converting online data into other forms. Correspondingly, they bridge religious information between Canada and China by bringing other media such as CDs back when they visit their friends and families, or through lay persons visiting China, mainly because certain media forms are less subjected to rigorous customs inspection. Two interviewees explained:

> *"People who can really help are the overseas Chinese who save religious information in their laptops or CDs then bring it over and show it to them . . . People in China can check some information online but they cannot open certain religious websites on their personal computers or on public computers in cyber cafes."*
>
> *"For example, some brothers or sisters will tell me that there is this mp3 now, and they want to hear the particular preaching of someone . . . Then I can go find it, and make it into a CD, and either send it to them or bring it to them when I travel to China. That is very convenient."*

In this way, various immigrants providing a wider range of religiously-related remittances that traditionally included books, artifacts, and pamphlets may act as new informational mediators in transnational religious networks. They serve as valuable transnational 'proxy Internet users' on behalf of their counterparts in China who require religious materials that are either unavailable or cannot be freely accessed or downloaded. As such, restricted access experienced by friends of our participants to religious information is alleviated somewhat through alternative but complementary media use.

In other cases, transnational religious communication operates clandestinely, or as one interviewee puts it, like "underground networks", in terms of having to guilefully deploy codes in religious interactions online. A number of interviewees said that it is "very common for emails to China to be rejected", thus necessitating innovative forms of indirect communication customized for religious purposes. For example, a few respondents regularly engage in the practice of not putting any content on emails, but rather sending information only in the form of attached files, as one interviewee explained:

> *"A lot of emails with words like god, lord, or prayer get rejected. So I always send emails as an attachment and nothing happens."*

Another adaptation of the Internet identified was the circulation of online content that has been coded to prevent the interception of direct communication in religious terms. One tactic is to specifically substitute religious terminology for other terms that have been agreed upon in advance by both parties, as one interviewee elaborated:

> *"We have some friends in China that work in religious organizations, and when we send email to them, it is very special. Like when we say prayer, it is not prayer, and god is not god, but you use the word X or something like that. The email will not make sense [to non-members], but both parties can understand it."*

In addition, certain religious communicative exchanges are deliberately composed as incomplete for recipients to decode or fill in between the lines, as one respondent said, "for example, if you send an email, you do not disclose your whereabouts." Another respondent related an incident that illustrated the need to flexibly react and creatively interpret information exchanges in order to protect religious communities meeting in private 'house churches':

> *"One religious group in SH requested for me to meet them . . . They then gave me a place to go . . . After I had reached the place, they sent someone to tell me that their house was actually at SM. I know that they are very secretive, so I did not initially ask them the number or road name of their meeting place. They said that as long as someone goes to look for them at their home, a government officer will visit them the next day."*

In the above instances, responses highlight how individual believers reinvent the Internet in multiple ways that embody resistance and "everyday creativity" to the dominant "discipline" restrictions (de Certeau, 1984) in transnational religious communication, particularly in response to China's regulation of the Internet and religion. Responses illustrate both the salience of inventive Internet appropriation by religious actors as well as the significance of transnational contexts for the exchange of religious-related remittances and maintenance of transnational relationships.

Discussion

In her study of the history of 19th century 'new media' inventions like the telephone and electric light, Marvin (1988) noted that media are "not fixed, natural objects". "They are constructed complexes of habits, beliefs, and procedures embedded in elaborate codes of communication", and thereby carry "the seeds of subversion" (p.8). In this way, the identification and explication of the behaviors of marginalized persons surrounding contemporary newer media access are valuable for understanding communicative responses to dominant societal structures that render some as outsiders (Orbe, 1998). This paper has sought to critically explore and assess how recent Chinese Protestant transmigrants in Toronto culturally shape and appropriate the Internet for transnational religious communication. Given the popular belief that Internet may facilitate, even revive world religiosity beyond state borders and local houses of worship, an analytic examination of transnational 'Internet use' helps to clarify the varied processes of new media appropriation contextualized within transmigrants daily experiences of local and transnational media spaces. Our study reveals an acceptance and enthusiasm for the utilization of the Internet for religious information seeking, but a more restrained and cautious approach of the Internet toward the building of religiously-related ties among Chinese transmigrants operating outside officially sanctioned religious organizations in China. The Internet is highly valued and used to connect to local and international websites for a range of textual and audio resources for personal spiritual growth but mediated religious communication is also adapted to political contexts and socially-situated needs.

In particular, with regard to transnational religious ties, our research points to interesting findings which are contrarian to conventional expectations that the Internet will accelerate international communication and increase cross border interactions between transmigrants. The presence and adoption of the Internet does not automatically translate into an increase in international religious ties online, in the face of political and legal restrictions that constrain aspects of religious expression and Internet use. If the Internet is used, it is appropriated in varying levels, in line with social norms operant in local and transnational contexts. Several respondents explained how they reinterpreted and adapted the Internet, and tactics deployed included the translation of online religious information into different technological forms, like portable CDs, sending 'non' emails with attachments and exchanging coded messages negotiating face-to-face meetings. In this way, critical examination of the religious webs here contributes to the clarification of the function of religious beliefs in shaping international communication, extending research utilizing a spiritual shaping of technology framework to account for the non-use and domestication of the Internet.

In addition, findings here evince how contextual factors related to the accommodative management (Orbe & Spellers, 2005) of geo-political sensitivities and non-state legal orders between host and sending nations become significant for a critical understanding of Internet use, at least among the Chinese immigrants in Toronto. As de Certeau (1984) notes, resistance in the forms of media appropriation evinces how the "weak" may subvert laws imposed on them by the "strong" and create for themselves a sphere of communicative action as a mechanism to redistribute power. Our findings identified ways whereby the informal manipulation of old and new media forms of communication may disrupt the 'logic' of institutionalized Internet regulation, consequently reconstituting new flows of communication across international borders previously constructed as impenetrable by formal laws.

More specifically, our interest on the social appropriation of the Internet suggests that the communicative behavior of Chinese religious transnational communities is influenced by artful bricolage tactics of "making do" (de Certeau, 1984), dependent on a network of relationships or what we may suggest as a form of 'religious guanxi'. Guanxi (which translates into 'relationships') interaction is described by affective rather than strictly market and contractual exchanges between social and economic actors (Kipnis, 1997). Interacting members respond to community rewards and sanctions that operate through informal social mechanisms of trust and reciprocal obligations. Members are rewarded when they engage in reciprocal exchanges that in turn enhances social harmony and community stability. When interpersonal exchanges are non-reciprocal, this invokes the loss of trust among injured parties (Gao & Ting-Toomey, 1998). Punishment is swift in the form of social sanctions through a decrease in interactions and thereby loss in community standing. Furthermore, guanxi relations can operate across uneven social ranks since it is sustained by behavioral conducts, which are not embedded in any written or institutionalized codes.

The notion that guanxi relations underscore both community and transnational interactions among Chinese is popularly documented in the literature on Chinese capitalism (e.g. Xin & Pearce, 1996; Seung & Luo, 2001). This literature suggests that guanxi bonds drive economic decisions in Chinese business networks. To the extent that this study points to the influence of social bonds, and the premium placed on maintaining social and community harmony in transnational religious activities, there is reason to expect that the transnational religious communication of Chinese transmigrants examined here is also influenced by *guanxi* where governance of social relations based on mutual trust and reciprocal understanding is not confined only to the economic sphere, but extends to the religious sphere as well. This is because religious communicative flows constituted from *guanxi* networks may be argued to be particularly effective in the context of legal deficits in China; in this case, the lack of legal protection for religious communication in China. The relevance of *guanxi* practices is that under conditions of uncertainty brought on by the absence of legal protection, personal relationships become an important community tool that substitutes for the lack of institutional support. Our interviewees stressed the need to maintain social harmony in their spiritual family or jia, both locally and transnationally, and care is taken to ensure that interacting members in China are not adversely affected by their Internet use for religious communication. As such, the Internet is often used in combination with other media, and online religious communication is augmented by offline religious communication especially when our interviewees visit their home country. This suggests that transnational religious communication also needs to be understood within the larger cultural fabric that social relations are embedded. Indeed, in some cases, for the transnational religious web to function at all, mediated religious communication involves direct interpersonal encounters, including unwritten codes of communication, to facilitate the exchange of religious remittances between agents in socially-situated circumstances. In this sense, our study indicates that Chinese transmigrants' appropriation of the Internet involves both consumptive and productive elements where Internet functions (e.g. email) are not only adapted but also reinvented for transnational religious communication.

Finally, a discussion of the limitations of the research and recommendations for future research is in order. In response to recent injunctions for studies to investigate "transnationalism from below" embedded in immigrants' daily milieu, our grounded research explored the transnational religious communication from the standpoint of recent Chinese immigrants who connect to the Internet in their everyday lives in Toronto. The interviews conducted come from a small sample of middle class, Internet users, and the results may not be applicable to other Chinese migrant groups. This may warrant more research to examine transnational communication in other contexts, and within other religious traditions. In addition, the findings here have uncovered how transmigrants are appropriating the Internet in varying levels of reinterpreting and adaptation, although results here did not point to instances of reinvention. Future research could extend the investigation to ferret out drastic alterations or new functions created by structural changes in religious communities. Furthermore, our paper highlights how religious *Guanxi* may facilitate communication and points the way for added research to examine how widespread the influence of religious *Guanxi* has on Chinese religious communication, both locally and transnationally. Beyond the domain of religious communication, these investigations should help contribute to a broadened conceptualization of 'Internet use' and a deepened understanding of the role of cultural norms in the social shaping of newer and older media, in contemporary international communication increasingly mediated by communication technologies.

References

Agre, P. (1998). The Internet and public discourse. *First Monday* 3(3). Retrieved March 20, 2005 from http://www.firstmonday.org/issues/issue3_3/agre/

Aikman, D. (2003). *Jesus in Beijing: How Christianity is transforming China and changing the global balance of power*. Washington, DC: Regnery.

Barone, J.T., & Switzer, J.Y. (1995). *Interviewing: Art and skills*. Needham Heights, MA: Simon & Schuster.

Barzilai-Nahon, K., & Barzilai, G. (2005). Cultured technology: The Internet and religious fundamentalism. *The Information Society, 21*(1), 25–40.

Beaman, L.G. (2003) The myth of plurality, diversity and vigour: Constitutional privilege of Protestantism in the United States and Canada. *Journal for the Scientific Study of Religion,* 42(3), 311–325.

Campbell, H. (2005). *Exploring religious community online: We are one in the network*. New York: Peter Lang.

Charmaz, K. (2006) *Constructing grounded theory: A practical guide through qualitative analysis*. Thousand Oaks, CA: Sage Publications.

Chen, W. (2006). The impact of Internet use on transnational entrepreneurship: The case of Chinese immigrants. In P Luo, L Fortunati & S Yang (Eds.), *New technologies in global societies* (pp. 197–220). New Jersey: World Scientific Press.

Cheong, P.H., Halavais, A. & Kwon, K. (2008). The chronicles of me: Understanding blogging as a religious practice. *Journal of Media and Religion,* 7(3), 107–131.

Cheong, P.H., & Poon, J.P.H. (2008). 'WWW.Faith.Org': (Re)structuring communication and social capital building among religious organizations. *Information, Communication and Society,* 11(1), 89–110.

Cody, E. (2007). Poll finds surge in religion among Chinese. *The Washington Post Foreign Service,* p. A15

de Certeau, M. (1984). *The practice of everyday life*. Berkeley: University of California Press.

Ebaugh, R. & Chafetz, S. (2002). *Religion across borders: Transnational religious networks*. Walnut Creek, CA: AltaMira Press.

Eglash, R. (2004). Appropriating technology: An introduction. In R. Eglash, J. Croissant, G. Di Chiro, & R. Fouche (Eds.), *Appropriating technology: Vernacular science and social power* (p. vii–xxi). Minneapolis: University of Minnesota Press.

Ess, C., Kawabata, A., & Kurosaki, H. (2007). Cross-cultural perspectives on religion and computer-mediated communication. *Journal of Computer-Mediated Communication,* 12(3), 939–955.

Gao, G., & Ting-Toomey, S. (1998). *Communicating effectively with the Chinese*. Thousand Oaks, CA: Sage.

Glaser, B., & Strauss, A. (1967). *The discovery of grounded theory: Strategies for qualitative research*. Chicago: Aldine De Gruyter.

Golden, D.R., Niles, T. A., & Hecht, M. L. (2002). Jewish American identity. In J. N. Martin, T. K. Nakayama & L. A. Flores (Eds.), *Readings in intercultural communication* (pp. 44–52). Boston: McGraw-Hill.

Guarnizo, L.E., & Smith, M.P. (1998). The locations of transnationalism. In M.P. Smith & L. E. Guranizo (Eds.). *Transnationalism from below* (pp. 3–27). New Brunswick: Transaction Publishers.

Guobin, Y. (2003). The Internet and the rise of a transnational Chinese cultural sphere. *Media, Culture & Society,* 25(4), 469–490.

Guest, K.J. (2003). *God in Chinatown: Religion and survival in New York's evolving immigrant community*. New York: New York University Press.

Hecht, M. L., Warren, J. R., Jung, E., & Krieger, J. L. (2005). A communication theory of identity: Development, theoretical perspective and future directions. In W. B. Gudykunst (Ed.), *Theorizing about intercultural communication* (pp. 257–278). Thousand Oaks, CA: Sage.

Helland, C. (2007). Diaspora on the electronic frontier: Developing virtual connections with sacred homelands. *Journal of Computer Mediated Communication,* 12(3), 956–976.

Hiller, H.H., & Franz, T.N. (2004). New ties, old ties and lost ties: The use of the Internet in diaspora. *New Media & Society,* 6(6), 731–752.

Ho, K.C., Baber, Z., & Khondker, H. (2002). Sites of resistance: alternative websites and state-society relations. *British Journal of Sociology,* 53(1), 127–148.

Hoover, S., Clark, L., & Rainie, L. (2004). Faith online. Retrieved February 17, 2006 from http://www.pewInternet.org/pdfs/PIP_Faith_Online_2004.pdf

Karim, K.H. (2003) Mapping diasporic mediascapes. In K.H. Karim (Ed.) *The Media of Diaspora* (pp. 1–17). London: Routledge.

Kipnis, A. (1997). *Producing Guanxi: Sentiment, self, and subculture in a north China village*. Durham, NC: Duke University Press.

Kling, R. (2000). Learning about information technologies and social change: The contribution of social informatics. *The Information Society,* 16(3), 217–232.

Kluver, R. & Cheong, P.H. (2007). Technological modernization, the Internet, and religion in Singapore. *Journal of Computer Mediated Communication,* 12(3), 1122–1142.

Kong, L. (2001). Mapping 'new' geographies of religion: politics and poetics in modernity, *Progress in Human Geography,* 25(2), 211–233.

Krueger, O. (2004). The Internet as distributor and mirror of religious and ritual Knowledge, *Asian Journal of Social Science,* 32(2), 183–197.

Krueger, R.A. (1997). *Analyzing and reporting focus group results*. Thousand Oaks, CA: Sage.

Leonardi, P.M. (2003). Problematizing "New Media": Culturally based perceptions of cell phones, computers and the Internet among United States Latinos. *Critical Studies in Media Communication,* 20 (2), 160–179.

Levitt, P. (2007). *God needs no passport: How immigrants are changing the American religious landscape*. New York: The New Press.

Levitt, P. (2001). Transnational migration: taking stock and future directions. *Global Networks,* 1(3), 195–216.

Mackay, H., & Gillespie, G. (1992). Extending the social shaping of technology: Ideology and appropriation. *Social Studies of Science, 22*(4), 685–716.

Madsen, R. (2003) Chinese Christianity: indigenization and conflict. In E. Perry & M. Selden (Eds.) (2nd ed), *Chinese Society: Change, conflict and resistance* (pp. 271–287). London: Routledge.

Marshall, P. (Ed) (2000 *Religious freedom in the world: A global report on freedom and persecution*. Nashville: Broadman & Holman.

Marvin, C. (1988). *When old technologies were new. Thinking about electric communication in the late nineteenth century*. New York: Oxford University Press.

McAlister, E. (2005). Globalization and religious production of space. *Journal for the Scientific Study of Religion, 44*(3), 249–255.

Mitra, A. (2005). Creating immigrant identities in cybernetic space: Examples from a non-resident Indian website. *Media, Culture & Society, 27*(3), 371–390.

Orbe, M.P. (1998). *Constructing co-cultural theory: An explication of culture, power, and communication*. Thousand Oaks, CA: Sage.

Orbe, M. P., & Spellers, R. E. (2005). From the margins to the center: Utilizing co-cultural theory in diverse contexts. In W. B. Gudykunst (Ed.) *Theorizing about intercultural communication* (pp. 173–191). Thousand Oaks, CA: Sage

Opennet Initiative. (2005). Internet filtering in China: A country study. Retrieved April 3, 2006 from http://www.opennetinitiative.net/studies/china/

Parham, A. (2004). Diaspora, community and communication: Internet use in transnational Haiti. *Global Networks,* 4(2), 199–217.

Park, S., & Luo, Y. (2001). *Guanxi* and organizational dynamics: organizational networks in Chinese firms. *Strategic Management Journal, 22*(5), 455–77.

Portes, A., Guarnizo L.E., & Landolt, P. (1999). The study of transnationalism: pitfalls and promise of an emergent research field, *Ethnic and Racial Studies, 22*(2), 217–237.

Potter, P.B. (2003). Belief in control: Regulation of religion in China, *The China Quarterly, 174*(June), 317–337.

Robinson, R. (2004). Virtual warfare: The Internet as the new site for global religious conflict. *Asian Journal of Social Science, 32*(2), 198–215.

Sawhney, H., & Lee, S. (2005). Arenas of innovation: Understanding new configurational potentialities of communication technologies. *Media, Culture & Society,* 27(3), 391–414.

Soukup, P. (2002). Media and religion. *Communication Research Trends, 21*(2), 1–44.

Statistics Canada. (2001) Census of Canada. Retrieved March 3, 2004 from http://www12.statcan.ca/english/census01/home/index.cfm.

U.S. Commission on International Religious Freedom (2008). Annual report. Retrieved September 9, 2008 from http://www.uscirf.gov/images/AR2008/china.pdf

Xin, K.R., & Pearce, J.L (1996). Guanxi: Connections as a substitute for formal institutional support. *Academy of Management Journal, 39*(6), 1641–58.

Yin, R. (1994). *Case study research*. Thousand Oaks, CA: Sage.

Yang, F. (2002). Chinese Christian transnationalism: Diverse networks of a Houston Church. In R. Ebaugh & J.S. Chafetz (Eds.) *Religions Across Borders: Transnational Immigrant Networks* (pp. 129–148). Walnut Creek, CA: AltaMira Press.

Zamaria, C., & Fletcher, F. (2008). Canada online! The Internet, media and emerging technologies: Uses, attitudes, trends and international comparisons, 2007. Toronto: Canadian Internet Project, 2008. Retrieved October 20, 2008 from http://www12.statcan.ca/english/census01/home/index.cfm.

Community and Social Interaction in the Wireless City: Wi-Fi Use in Public and Semi-Public Spaces

Keith N. Hampton and Neeti Gupta

A significant body of research has addressed whether fixed internet use increases, decreases or supplements the ways in which people engage in residential and workplace settings, but few studies have addressed how wireless internet use in public and semi-public spaces influences social life. Ubiquitous wi-fi adds a new dimension to the debate over how the internet may influence the structure of community. Will wireless internet use facilitate greater engagement with co-located others or encourage a form of 'public privatism'? This article reports the findings of an exploratory ethnographic study of how wi-fi was used and influenced social interactions in four different settings: paid and free wi-fi cafes in Boston, MA and Seattle, WA. This study found contrasting uses for wireless internet and competing implications for community. Two types of practices, typified in the behaviors of 'true mobiles' and 'placemakers', offer divergent futures for how wireless internet use may influence social relationships.

Keywords: community network, cafes, coffee shops, network, mobile computing, Muni wi-fi, parochial realm, privatism, social networks, third places

Introduction Emailed

Recent years have seen rapid growth in the availability of wireless broadband internet access in public spaces. Providers and points of access take the form of municipal wi-fi networks (Muni wi-fi), such as those operating in Philadelphia and Toronto, community wireless networks, such as New York Wireless or Île Sans Fil in Montreal, advanced mobile phone networks (e.g. 3G), and wi-fi cafes, restaurants, bookstores and related spaces (hereafter abbreviated as 'wi-fi'). While there is a significant body of research addressing whether fixed internet use increases, decreases or supplements the ways in which people engage in residential (Hampton, 2007; Hampton and Wellman, 2003) and workplace settings (Quan-Haase and Wellman, 2006), few studies have addressed how the use of wireless broadband in public and semi-public spaces influences social life. Ubiquitous wi-fi adds a new dimension to the debate over how the internet may influence the structure of community—the network of supportive ties that exist between individuals. It is unclear whether wireless internet use in public spaces will facilitate greater engagement with people in public spaces or encourage a form of 'public privatism'. Will wi-fi use support public disengagement, with people withdrawing from the public realm in exchange for private spheres of influence, or will it facilitate new interactions and contribute to the development of a new public sphere?

This article reports the findings of an exploratory study which examined how wi-fi was used and influenced social interactions in a series of wi-fi coffee shops. Observations were drawn from four different settings: paid and free wi-fi cafes in Boston and Seattle. The goal of this article is to provide an initial framework for understanding how wi-fi influences the interactions and structure of personal networks in a wireless city.

Privatism

In the past, the 'wired' nature of desktop computing limited the potential for internet use to blend into urban public spaces. With a few exceptions, such as libraries, internet cafes and community technology centers, internet use was confined to the home and workplace. The connection between internet use and home-centeredness generated concern that new media use was increasing privatism (Graham and Marvin, 1996). Indeed, personal networks increasingly have become privatised, consisting of densely-knit networks of interactions centered around the home, rather than diverse, loosely-coupled interactions in more public settings. For example, a study of the size and composition of people's core 'discussion networks' in 1985 and 2004 identified a shift from ties formed through voluntary associations, neighbors and interactions in the public realm, towards networks increasingly dominated by kin and based around the home (McPherson et al., 2006). While the authors of that study did not link internet use directly with changes in the structure of social networks, it is notable that the time period observed by McPherson et al. (2006) corresponds with the rise of the 'network society' (Castells, 1996). Their findings are consistent with observations of other home-based media, including television and the telephone, which have been linked to increased privatism (Fischer, 1992; Putnam, 2000).

The concern with privatism is the sacrifice of 'bridging social capital' for 'bonding social capital' (Putnam, 2000). Bonding social capital is formed through the interaction of tightly-knit networks of similar others, often close friends and kin. Personal communities high in this form of social capital tend to provide generalized social support and to be high in reciprocity (Wellman and Wortley, 1990), but they can be repressive and tend to be racially, culturally, behaviorally and ideologically homogeneous (McPherson et al., 2001). Bridging social capital exists through access to diverse and relatively 'weak' social ties that provide specialized social support and access to novel information and resources (Burt, 1992; Granovetter, 1973). Individuals who have more bridging social capital, which can come only from participation in diverse social milieus, are more trusting, demonstrate greater social tolerance, cope with daily troubles and trauma more effectively, tend to be physically healthier (Cohen et al., 2000) and have access to more diverse information and resources, which has been shown to assist in search processes (such as finding a job; Granovetter, 1974).

The earliest evidence of the role that internet use plays in personal networks appeared to verify that the internet amplifies the existing trend toward privatism. The work of Kraut et al. (1998) and Nie et al. (2002) found that internet use contributed to a decrease in the size of people's social circles, a reduction in public participation and an increase in homecenteredness. However, later research has found that the internet does not significantly influence the allocation of day-to-day activities (Robinson et al., 2002), and supplements rather than replaces traditional modes of communication (Quan Haase et al., 2002). Those who use the internet to communicate with their closest and most significant social ties are also in frequent contact in person and through other media (Baym et al., 2004; Boase et al., 2006). Similarly, email users tend to have more social ties than nonusers and email appears to be a particularly useful medium for maintaining contact with a larger number of relatively weak social ties (Boase et al., 2006; Zhao, 2006). Face-to-face and telephone contact remain the dominant modes of connectivity when people communicate with their closest ties (Boase et al., 2006). The general conclusion, that internet use increases overall communication and possibly leads to larger networks, suggests that it is a possible counter-force to privatism. However, the evidence on frequency of communication and network size alone does not address directly the underlying concern of privatism: that networks are increasingly home-centered and homogeneous as a result of new media.

In an attempt to examine more closely the circumstances under which the internet does or does not encourage privatism, a series of studies have examined the role that internet use plays in the formation and maintenance of neighborhood ties (Hampton, 2007; Hampton and Wellman, 2003). Early internet adopters were found to have smaller neighborhood networks, but experience using the internet was found to inoculate them from increased privatism. Over time, experienced internet users increased the number of ties that they had from the parochial realm. The neighborhood networks of noninternet users and those with less internet experience lost ties over time; they become increasingly privatized (Hampton, 2007). These studies also found that the introduction of a neighborhood email list increased the number of weak social ties at the local level and facilitated public participation (Hampton, 2007; Hampton and Wellman, 2003; Mesch and Levanon, 2003). The observation that internet use affords both global and local connectivity has been termed 'glocalization' (Hampton, 2001). While these studies provide some promising evidence that home-based internet use does not encourage privatism and may even help to reverse the trend, they are not conclusive, especially in light of the findings of McPherson et al. (2006). What is conclusive is that the internet has become integrated increasingly into everyday life (Haythornthwaite and Wellman, 2002).

The Public

With the launch of Muni wi-fi, for the first time it is possible to integrate intensive internet use with the use of urban public space. Public spaces and public life play a unique role in the formation of social networks, opinions and democracy. When referencing public space, urbanists typically refer to a 'city's street, its parks, its places of public accommodation' such that 'public space may be distinguished from private space in that access to the latter may be legally restricted' (Lofland, 1973: 19). Semi-public spaces—those spaces that are not completely 'a world of strangers' (Lofland, 1973) or domesticated—are recognized for the role that they play in public life (sometimes these spaces are termed the 'parochial realm'). Habermas (1989) noted the role of such places, in the form of London coffee houses and French salons, in the development of a public sphere for cultural and political debate. While Habermas (1989) argued that the growth of capitalism diminished the public sphere, Ray Oldenburg (1989) suggested that while such 'third places' (differentiated from work and home) have declined, they continued to play an important role in the social life of Americans well into the 20th century. Oldenburg (1989) noted that these semi-public spaces provide exposure to diverse social ties, they create a sense of place and community, and provide both serendipity and companionship.

As with the history of 'community' (Hampton and Wellman, 2003;Wellman, 1999), the 'public' has an extensive literature that documents its birth, transformation, death and rebirth at the hands of societal (e. g. capitalism, industrialism, bureaucratization, etc.) and technological change (e. g. electricity, telephone, automobile, etc.) (Fischer, 1992; Habermas, 1989; Marvin, 1988; Sennett, 1977). The dominant interpretation of the relationship between public space and social interaction suggests that the modern urban environment is responsible for increasing social segregation, isolation and non-involvement. Public spaces are seen to afford bystander apathy (Latané and Darley, 1976), to generate stimulus overload (Milgram, 1970) and to be increasingly sanitized (Hannigan, 1998; Zukin, 1995). Yet a considerable literature exists to suggest that street life is far from anonymous: it is full of symbolic interaction (Goffman, 1959, 1963, 1971); contains planned and fleeting encounters (Berkowitz, 1971; Lofland, 1973;Whyte, 1980); is a source of serendipity (Merton and Barber, 2004); and is the setting for a range of informal interactions that contribute to social norms and public safety (Jacobs, 1961).

Public Privatism

It is unclear how wireless internet access, which penetrates public spaces as the internet already has penetrated private spaces, will influence the structure of people's networks and social interactions. If people use wi-fi in the same way as they use mobile phones, it is likely that wi-fi use will exasperate a trend toward 'public privatism'.

Mobile phones make community instantly accessible, social ties are reachable anywhere at any time; a form of community that Wellman et al. (2003) termed 'networked individualism'. Most people use mobile phones to call a small set of mostly strong ties (Ito and Okabe, 2006; Ling and Yttri, 2006). Email or Short Message Service (SMS, 'texting') is used when voice communication is perceived to be inappropriate (Ito and Okabe, 2006) and to keep in touch with a larger, more diverse set of contacts (Matsuda, 2005). Whether used for voice or SMS, mobile phones create a private sphere of interaction within public spaces. When people engage with their mobile phone, they create a private 'cocoon' that reduces the likelihood of serendipitous public encounters (Harris, 2003), contradicts common expectations of public behavior (Ling, 2004) and diverts attention away from co-present others (including existing social ties) (Humphries, 2005). The mobile phone has made it less necessary to rely on anyone other than those who are already highly familiar; 'those who have come into our sphere of friendship are always available' (Ling, 2000: 83). As argued by Goldberger (2003):

> *The great offense of the cellphone in public is not the intrusion of its ring, although that can be infuriating when it interrupts a tranquil moment. It is the fact that even when the phone does not ring at all and is being used quietly and discretely, it renders a public place less public. It turns the boulevardier into a figure of privacy. And suddenly the meaning of the street as a public place has been hugely diminished.*

The 'public privatism' of interactions as a result of the mobile phone mirrors the findings from studies of how fixed internet access has been used in semi-public spaces: 'Private uses in Public Spaces' (Lee, 1999).

The trend of people socializing in small, intimate groups in private homes rather than with large, diverse groups in public spaces, might be augmented through the use of new mobile media by a tendency to socialize remotely with small, intimate groups in any space, at any time.

Ubiquitous Wi-Fi

Public spaces play a unique role in shaping and maintaining personal networks. Unlike the close, homogeneous, densely-connected nature of social relationships that are likely to dominate private spaces (McPherson et al., 2006; Putnam, 2000), of which the private home is the best example, public and semi-public spaces are more likely to be the setting for diverse social interactions. It is unclear how the penetration of the wireless internet will influence interactions in these spaces or the broader structure of people's social networks. There may even be variation based on local culture, climate, the built environment and how wi-fi is deployed (e. g. free vs. paid). At the most basic level, the growth of wireless internet access suggests two competing, although not mutually exclusive, possibilities:

1. The ubiquitous availability of wireless internet access will encourage greater participation in public spaces, lead to increased public interaction and possibly diversify the composition of people's social networks; and
2. Public wi-fi use will consist of private cocoons of interaction that benefit existing close ties, distract from interactions with copresent others and ultimately reinforce the existing trend toward privatism.

In the absence of Muni wi-fi projects fully operational and active for any extended period at the time of this research, to uncover initial evidence of how wi-fi use will influence the structure of community interaction, this article relies on observations from those examples where wi-fi has already penetrated public and semi-public spaces: wi-fi coffee shops. The observations of wi-fi use are exploratory; this study did not attempt to test specific hypotheses in advance of the observations. Instead, based on what was observed of how wi-fi was used in coffee shops, this article provides a first in-depth view of public wi-fi use and sketches a theoretical framework for how interactions and networks may be augmented in the context of wi-fi.

Method

Selection of Participants

The observations were limited to four wi-fi cafes in two cities, Seattle,WA and Boston, MA. One cafe in each city offered paid wi-fi, the other offered free access. The selection of cafes was not random: all four of the cafes were familiar to at least one of the authors in advance of the study. However, neither of the authors was a regular at these coffee houses. The cafes were selected to help control for variables that were exogenous to the influence of wi-fi. This included the potential for bias in the observations as a result of the culture of any one coffee house, the characteristics of surrounding neighborhoods and the social qualities of different cities. It was recognized that the contrast between paid and free wi-fi might be a source of variation in users' experiences. Given the dominance of the Starbucks coffee chain (more than 7200 stores in 30 countries), its early adoption of wi-fi service (starting in 2001) and a deal with T-Mobile (Deutsche Telekom) and AT&T to offer paid wi-fi in the majority of its cafes, Starbucks seemed to be a natural choice for the observations in order to maximize the ability to generalize the findings. Given that Starbucks provides paid wi-fi use, this necessitated that independent coffee houses were selected for the free wi-fi comparison.

Initially we had hoped to observe wi-fi use in diverse urban environments, central business districts and suburbs. However, after spending many hours in suburban coffee houses, these plans were abandoned after making very few observations of wi-fi use. Instead, the study limited itself to commercial areas in or near the downtown core and the areas bordering the University of Washington and Massachusetts Institute of Technology (MIT).

It was important to insure that the cafes selected were not unusual in a way that would raise questions as to whether the observations were as mundane as desired. For example, initially the plan was to make observations at Starbucks' first location, in Seattle's Pike Place Market, but it became obvious very quickly that its role as a tourist destination made it atypical. In the end, time was spent at eight different Seattle Starbucks locations before settling on the Starbucks at 6th and Union as a 'typical' Starbucks setting: steady foot traffic, a small number of large, stuffed, comfortable purple chairs and many less comfortable steel-framed chairs with matching small tables. After a similar process, a similarly pedestrian Starbucks located in Central Square (Cambridge/Greater Boston) between Harvard and MIT was selected.

It was surprisingly difficult to find completely free wi-fi cafes. Many had hidden price tags in the form of required purchases or time-limits on use. Others advertised wi-fi but in practice served up such unreliable service that there were few takers. In Seattle the Chaco Canyon Cafe (near the University of Washington) was selected for observation. The owner was a local community activist. The cafe offered an extensive selection of raw foods, organic juices and fair trade coffees, and had an 'at-home feeling', with ample tables made from golden oak. Wi-fi access was installed and operated by the owner and his brother as an experiment to draw in new customers. In Boston, Trident Booksellers & Cafe (located on Newbury Street) was selected. Twenty years ago, Trident's owner was an early pioneer of the bookstore–cafe combination that is now a standard configuration for the big bookstore chains. The cafe has an open look with large windows that open out to a trendy commercial street. The store provides wi-fi as part of NewburyOpen.net, a free community wi-fi provider.

Data Collection

The methods of observation were primarily qualitative. A total of 120 hours were spent in the four cafes between December 2003 and March 2004. A total of 30 hours were spent on direct observations in each cafe. Observations were made in two-hour time blocks systematically distributed across hours of operation, roughly one-third of the observations were made on weekends, the rest on weekdays.

Observations consisted of extended visits to each cafe with laptop in hand. The time in each coffee house was spent making detailed notes of how patrons with mobile devices interacted with each other and cafe staff. Careful notes documented each interaction, including the gender and approximate age of those involved, how the exchange was initiated and the duration of the exchange. In addition to unobtrusive observations, a short web-based survey was created with questions on basic demographics, social networks, technology and prior experiences in wi-fi cafes. As the patrons left the cafe, every fifth person who had used a laptop was given the web address to our online survey and a letter explaining the study. However, a low response rate led us to abandon this survey. However, when patrons were approached with the survey, many spontaneously stopped to talk about their experiences. In total 20 unstructured exit interviews were completed, representing about eight percent of the total number of people observed using laptops. Most interviews were conducted on the spot; when necessary they were scheduled for a later time and on rare occasions they were completed through email correspondence.

Results

The selection of four field sites was intended as a source of differentiation with the expectation of observing variation based on city, individual location and free vs. paid wi-fi services. However, the observations did not support even this simple expectation. The most significant distinctions in social interactions were observed, based on different practices of wi-fi use. The settings observed attracted users with two distinct activities. These practices are presented as a typology with two ideal types: 'true mobiles' and 'placemakers'. It is important to recognize that while these activities are presented as ideal types (Weber, 1946), there was some variation in the practices observed and such variation is noted where appropriate.

True Mobiles

For 'true mobiles', wi-fi coffee shops functioned as a backdrop for activities focused on the completion of 'work' (studying, paid work, etc.). True mobiles identified the cafe as a 'space of productivity'. They typically would suggest that the store offered a change of setting that helped them to focus or provided a source of creativity. One participant offered: 'It is nice to get out of the office, if I don't have a specific reason to be there. The change of pace seems to be good for my productivity.' Another noted that 'background noise helps me focus and I know other people who think so too.' The limited number of true mobiles that did not refer directly to the coffee shop as a productive space told us that the cafe provided an 'escape' that, in turn, would aid productivity when they returned to their place of work:

> *I do a lot more writing at home, actually, but sitting in a coffee house is a temporary break. At home, this usually means that I walk a mile down to town, work there a while and then walk back, so it's a matter of changing the dynamic. (Joey, male, 35, Boston Starbucks)*

Early in the observations, the initial assumption was that semi-public spaces were not ideal for work productivity and that the participants were offering a cover to justify 'having no purpose' (Goffman, 1963: 58), possibly as an excuse for disengaging from a space which had a norm of social involvement, or as a reason for taking a break where there was a norm of work, possibly resulting from the presence of wi-fi. However, direct observations ultimately were inconsistent with that initial assumption: true mobiles were not providing a cover for disengagement, true mobiles were truly in the cafe with the sole purpose of work.

Whether a true mobile described their visit to the wi-fi cafe as an 'escape' or a 'space of productivity', there were no actual observable differences in their activities. All true mobiles spent their time almost completely engrossed in wi-fi and laptop use. The observations and interviews indicated that they were primarily engaged in sending email and surfing the web. Their laptops were not mere props, they were a means to a specific ends: productivity. Despite their own internal differentiation, there were no observable clues that those on an 'escape' were immediately more or less 'productive' in semi-public spaces than those specifically seeking efficiency.

True mobiles included all the 'mobile workers' that were observed (those who were currently traveling 'out of town' as part of their occupation), but did not specifically exclude those who did not travel or were not traveling currently for paid employment. True mobiles who were mobile workers often sought out wi-fi coffee shops to serve as primary, although temporary, locations for employment activities. This contrasted with the majority of true mobiles that were observed: most true mobiles lived or work within a short distance of the cafes studied. They tended to report a need to work from a fixed office location for most of their day, and used the cafe as a secondary, occasional extension of local workplaces. It was not uncommon for true mobiles to report that they scheduled a specific day each week to spend at the cafe: 'I work Monday to Thursday at the office and every Friday from Starbucks'.

Local-based true mobiles identified with the wi-fi cafe as a space of productivity, but they were likely to cite the coffee houses as the location for another type of 'escape': from the physical presence of their co-workers. This included home-based employees who by choice or other arrangement worked from home on specific days of the week, but sought out the wi-fi cafe as a refuge from distractions at home—escape from partners, children and television. When compared to true mobiles of the more local variety, mobile workers differed only in whether they designated the wi-fi coffee shop as their primary workplace, not in task; using wi-fi and their laptops to check email and surf the web.

As with most of the cafe customers that were observed, true mobiles participated in a minimal level of overt interaction with co-present others (both patrons and coffee shop employees). Much of their interaction was subtle and non-verbal. Their base level of engagement could be characterized by what Goffman called 'civil inattention':

> *One gives to another enough visual notice to demonstrate that one appreciates that the other is present (and admits openly to have seen him), while the next moment withdrawing one's attention from him so as to express that he does not constitute a target of special curiosity or design. (1963: 84)*

The majority of true mobile's interactions were observed in the form of near-constant keyboard use and, when interviewed, the communication they reported in the form of email and instant messages with colleagues, friends and other existing members of their social network. The activities and interactions of the coffee shop were peripheral.

True mobiles both reported and were observed repeatedly avoiding the gaze of staff and other store inhabitants. Like the majority of other coffee shop patrons, true mobiles employed 'portable involvement shields' (Goffman, 1963). Goffman described portable shields as fans, masks and the use of people's hands to conceal facial expressions, used literally to shield oneself from others and to signal unavailability for more overt interactions. When seated at a table, the technology 'have nots' used portable shields in the form of newspapers, magazines and books and the 'haves'—true mobiles—used laptops. Cellphones were too small to make for good shields, but they did play a unique role for cafe patrons of all types. After ordering coffee at the counter, but before being seated and before coffee was poured, there was a particularly strong tendency for those with cellphones to use their mobile phones as a 'legitimate momentary diversion' (Goffman, 1963: 59). Customers would take out their phone, stare at the screen, possibly move a dial or push a few buttons—presumably reviewing some content—but almost never initiated a new phone call. This was most frequent at our two Starbucks locations, where there was a norm that customers wait next to the counter for their coffee to be made before taking a seat.

In the use of involvement shields, what differentiated true mobiles from other patrons was their persistence in their use of shields as barriers to interaction. Activities related to 'work' were paramount; the extent that true mobiles could be distracted from these activities depended on the tempo and atmosphere in the cafe. When other patrons ignored the subtle (or not so subtle) signals of a shield and attempted to initiate verbal communication with a true mobile, they were unlikely to be met with eye contact and were more likely to be met with no response (completely ignoring the other) or an abrupt one to two-word retort, than when verbal contact was initiated with non-wi-fi users. For example, one wi-fi user approached a true mobile at a neighboring table, asking, 'Do you know how to get this working on my machine?' Without looking up, the true mobile replied 'No' and continued surfing the web. In another situation which was repeated often, where a customer would enter a cafe and ask, 'Is this seat taken?', true mobiles were unlikely to do more than shake their heads; other patrons would make at least fleeting eye contact and provide a verbal response. When we pressed true mobiles about these encounters, they offered an explanation consistent with their attempt to remain focused on activities associated with work. At the same time, the same true mobiles were interacting with online contacts through email and instant messaging, although it was unclear whether the exchanges were completely work-related. The few times that true mobiles were observed in more extended, unplanned, in-person interactions, were exchanges that involved meeting clients or other true mobiles. Usually, true mobiles who participated in extended co-present interactions were interacting with co-workers who arrived together, but on one occasion a true mobile was observed talking to another patron who had the same new model of laptop.

Placemakers

In contrast with true mobiles, the primary activity of 'placemakers' was 'not to engage in paid work'. They came to wi-fi coffee shops to 'hang out'. The coffee house was not intended as a direct or indirect place of productivity. For the placemaker, the cafe was center stage, not peripheral. They were drawn by what one subject described as the 'inherently casual sociability' of the physical setting. Placemakers used their laptops as a premise to enter and engage in the 'social hubbub' of the space. This could mean direct co-present participation with existing members of their social network, unplanned encounters or the pleasure that Lofland (1998) ascribes to 'public solitude' and 'people watching'. As with true mobiles, placemakers were observed regularly using wireless connectivity for email, web surfing and instant messaging. However, the laptop was never their primary focus; the availability or potential for co-present sociability was their primary activity.

While placemakers were more likely than true mobiles to engage in unplanned interactions, they were also more likely than true mobiles to enter a cafe alone. True mobiles occasionally arrived in pairs and settled in to complete some sort of 'business', such as meeting with a client, but placemakers almost never arrived in pairs or larger groups.

A typical placemaker arrived alone, bought a coffee and took a seat at a table for two next to a window. They would pull out their laptop, set it on the table and become masters of the 'momentary diversion' (Goffman, 1963). A great deal of time would be spent gazing out the window, looking around the cafe, adjusting personal belongings, slowly sipping coffee, searching for a power outlet, powering up and then casually surfing the web and checking email, with prolonged intermittent pauses to glance around and outside the cafe.

In stark contrast to true mobiles, placemakers did not actively avoid the gaze of other patrons within the shared space. While placemakers participated in the rituals of civil inattention, a casual glance from another customer was more likely to be met with a fleeting smile than a quick look away. Once mutual awareness had been established, there was a higher probability that additional encounters would take place with a placemaker than with a true mobile. As one placemaker described his experience over the previous week:

> *Met people face to face. Spoken to people several times. People ask me about laptops frequently and about wireless services. Helped several people learn what they need to buy. Also, while in line about to order, meet people sometimes. Religious people sometimes use [the cafe] to make connections and invite me to their church. (Nancy, female, 28, Seattle Chaco Canyon)*

While placemakers resembled true mobiles in their use of laptops and other devices as portable interaction shields, with true mobiles these same props were also the most likely observable sources for new interactions. For example, it was common, if not routine, for coffee shop patrons to glance at wi-fi users' computer screens. True mobiles typically would ignore such a glance or reposition their device to indicate unavailability; this behavior contrasted with placemakers, who were less likely to signal unavailability. With surprising regularity, glances towards placemakers developed into discussions based on shared interest. People were observed engaging over products on an auction website, a site devoted to a local art show and an online news site. This behavior also worked in reverse: wi-fi users were observed glancing at the books and activities of other patrons, they engaged when there was shared affinity and on occasion used their laptops to find and share information. Serendipitous interactions were observed that were as brief as a few seconds to those that extended more than 20 minutes. When interviewed, the placemakers described many encounters of this nature, as often with coffee shop employees as with customers.

Overall, the total number of observed conversations between previously unknown customers was small, in each hour maybe a couple of interactions that were more than the most fleeting: requesting a chair, moving a bag, passing a condiment. However, more serendipitous verbal interactions were recorded between placemakers and other patrons than between patrons in general. When interviewed, almost 50 percent of placemakers reported meeting someone new at a wi-fi cafe, very few true mobiles reported an unscheduled meeting. When true mobiles did report a serendipitous encounter, they tended to be instrumental (such as a commonly observed exchange associated with negotiating access to a power plug), and were more fleeting (such as a glance or small gesture exchanged with other 'wi-fi regulars'—which were observed most frequently with a group of true mobiles who shared a morning routine: Starbucks, email, a latte and a smile).

While placemakers almost always arrived alone and more often than not left alone, both placemakers and true mobiles experienced scheduled and unplanned encounters with existing social ties. About one-third of placemakers, but only a small fraction of true mobiles, would be met by an acquaintance at some point. In all the situations observed the arriving party would not engage in their own private wi-fi use. Instead, the laptop was likely to become a shared focus of attention. For example, one person would read off the screen to another, or two people would watch a video together.

Unlike true mobiles, it was unusual to find a placemaker who was not local. Placemakers were almost always regular customers at the same cafe and lived or worked in close proximity. Placemakers were more frequent visitors to wi-fi cafes than true mobiles. Most true mobiles reported one to two visits per week, whereas placemakers visited almost daily. All wi-fi users spent in excess of 30 minutes in the coffee shop on a single visit and three in 10 stayed more than four to five hours.

While the observations were of a continuum that ranged from true mobiles to placemakers, not an absolute dichotomy, for the most part the distinctions were obvious: users did not move back and forth between types within the same setting. Placemakers did not immerse themselves in shielded, private cocoons of interaction with the goal of completing work. Placemakers were openly, if not actively interested, in communicating with co-located others: it was their primary activity.

Conclusion

The Future of Community

In the observations of the semi-public space offered by wi-fi cafes, contrasting uses were found for wireless internet and competing implications for community. Two types of practices, typified in the behaviors of true mobiles and placemakers, offer divergent futures for the way in which the deployment of ubiquitous wi-fi may influence the structure of social networks and social relationships in public spaces.

A shift in internet use away from the home and workplace and into the semi-public environment of a cafe is, by definition, a shift away from privatism. True mobiles advanced this trend in their use of wi-fi spaces as an 'escape', but in this case the substitution of private for public space did not address the underlying implications of privatism. While true mobiles escaped the confines of private space, they did not embrace new opportunities for public interaction, at least not unplanned, serendipitous encounters with co-present others. Instead, true mobiles actively resisted the public: they attempted to erect barriers, physically in the form of interaction shields, and sociologically through their avoidance of gaze and verbal contact. True mobiles used public space for very private activities, those they associated with the productivity of work and likely for computer-mediated maintenance of their existing social network. As with mobile phone users, true mobiles embraced wireless connectivity for 'public privatism'. True mobiles used wireless internet access to help them transform the network structure of work and community to enable connections to people in any place, rather than to people in place: 'networked individualism' (Wellman et al., 2003). In addition, the true mobiles who also spent part of their time doing paid work from home consciously used wi-fi-enabled spaces as an escape from home-based ties (partner and children). It is not clear if this escape from the nuclear family is a reaction to home-centric personal networks, an escape from the psychological overload of intensive kinship relations, or an additional trend toward individualism and social isolation which has the potential to narrow further the size and composition of social networks.

In contrast with true mobiles, placemakers embraced the wireless internet precisely for its ability to connect to the activities afforded by public space. The primary activity of placemakers was 'not work', rather, it was interactions with co-present social ties, serendipitous exchanges and availability for interactions with strangers. Placemakers used the public setting of wireless internet connectivity as a means for local, place-based interactions, what Hampton and Wellman (2003) have previously referred to as 'glocalization'. This is not to say that placemakers did not use the environment of wi-fi cafes to maintain their existing networks, as with true mobiles many of their wi-fi enabled activities were likely to be computer-mediated exchanges with established social ties; but consistent with the traditional activities of coffee shops, placemakers also had many planned and unplanned face-to-face encounters. What fundamentally differentiated placemakers from true mobiles was their use of public space for interactions that were not private.

'Community' is not a normative concept. It is the structure of supportive relations that exists between individuals. Large-scale social change can affect how individuals structure their networks and in turn constrain behavior, influence how information is channeled and affect the allocation of resources. The underlying question that this article has attempted to address is whether or not the introduction of ubiquitous wireless internet connectivity into the urban environment will alter the prevailing trend in how personal networks are structured: the tendency toward privatism. The answer is that it is too early to tell, but there are signs leading in divergent directions.

Networked individualism and glocalization are parallel paths: each involves a transformation to the structure of community that is a result of the affordances of the internet. It is possible that the tendency toward networked individualism or glocalization will vary by individual, and possibly vary for that individual at different stages in the life cycle (Hampton, 2007), but it is not only a matter of personal choice. In a situation where the activities of 'public privatism' dominate within even a marginal segment of public spaces, a 'neighborhood' or 'contextual' effect may be the result. Typically, contextual effects are used to explain the role of community-level characteristics in social tie formation in the context of neighborhood communities (Sampson et al., 2008). Specifically, individuals who are highly motivated to form social ties, but who live in a neighborhood where few others are available or interested in forming relationships, are structurally disadvantaged relative to similarly motivated (or even lesser motivated) persons in a neighborhood where people are open to tie formation. The network constraints are very different. The same effect applies to the likelihood of serendipity and other encounters in public spaces; it takes at least two for interaction, and if no one else is interested or available, ultimately you will remain alone. The higher the number of people engrossed in public privatism within any space, the less opportunity for new tie formation.

The Future of Wireless Internet in Urban Public Spaces

The deployment of wireless internet access in public spaces will initiate a path dependence that ultimately will lead to the domination of either a new public privatism or increased public participation which will help to counter the existing trend of privatism. However, wi-fi is not in and of itself deterministic, but decisions related to the deployment of the technology afford different types of social interactions. Although it has not been stated in this article until now, it is true that more true mobiles were found in the paid wi-fi Starbucks locations and a greater number of placemakers where free wi-fi could be found. However, a simple one-to-one relationship, where the presence of free wi-fi simply created public interactions, was not observed. The environment of paid wi-fi was simply more conducive to the activities of true mobiles and less so to placemakers. As one participant described:

> *[People] go to Starbucks because people don't want to experiment with their coffee. They usually like to get coffee they like and Starbucks promises them exactly that. Similarly, people who are looking for wi-fi connections will go to places where they know they will get a good connection and where they can sit for some time and work. (Sandy, male, 35, Seattle Chaco Canyon Cafe)*

Even in those environments that were more favorable to placemakers, coffee houses often adjusted their environment to limit their presence. Free wi-fi cafes employed strategies to discourage people from gathering or feeling overly welcome: for example, the Trident Cafe's management removed all power outlets from customer areas. While visiting other free cafes, this was found to be a common practice, along with actively limiting wi-fi access during peak hours, reducing table space and established rules controlling access to toilet facilities. Employees would use the guise of customer service to ask patrons if they needed anything, but many admitted that it was a strategy to encourage 'wireless squatters' to buy something or leave. The perception of free wi-fi providers was that their customers had a tendency to loiter and stay around socializing for long periods of time, taking space away from 'legitimate' customers. Even in situations where wi-fi cafes had the potential to afford the broad community interactions of a 'third place' (Oldenburg, 1989), private control over semi-public spaces actively worked to reduce the potential for wireless internet to afford social interactions.

As Muni wi-fi blossoms, it is reasonable to assume that the strategies employed by the owners of semi-public spaces to limit the activities of true mobiles will be carried over by governments and private organizations (such as business improvement districts and internet service providers) in their attempt to regulate public spaces to make them most welcoming to paying consumers. Yet, commercial interests that drive away placemakers may find themselves driving away the majority of public wi-fi uses. Even true mobiles are drawn to public wi-fi by the social characteristics of the setting: as William H. Whyte (1980) noted in *The Social Life of Small Urban Spaces,* what attracts people the most is other people. As cities plan for the deployment of Muni wi-fi—whether purely public initiatives, partnerships with private service providers or driven solely by private investment—if public spaces are to support diverse opportunities for social interaction and the benefits to democracy and public safety that such interactions afford, local governments must re-evaluate existing policy on the access and design of public space. Conventional considerations related to the design of public space to maximize for social uses (Whyte, 1980) must be reconsidered in light of the unique requirements of new media use, such as the provision of power outlets, flat surfaces for laptops and shade to view digital displays. It is also important to encourage additional research on wi-fi use, in public settings such as parks and plazas (eg Forlano, 2008; Hampton et al., 2008; Powell, 2008). Observations of these spaces will help us to understand how the built environment can be used to help wi-fi users balance privacy, mutual surveillance, public safety, the opportunity for serendipitous encounters and other social behaviors. In addition, observation of wi-fi use must be coupled with longitudinal studies of wireless users to clarify what ethnographic studies are unlikely to be able to examine: that is, the extent to which the networks of placemakers and true mobiles are otherwise home-centered, the overall diversity of their social ties and whether they are actually exchanging time that otherwise would have been spent in the sanctity of the private sphere for a new public life.

About the Authors

Keith N. Hampton is an assistant professor at the Annenberg School for Communication, University of Pennsylvania. His research interests focus on the relationship between new information and communication technologies, social networks, and the urban environment (for papers related to this work, see http://www.mysocialnetwork.net). Address: Annenberg School for Communication,

University of Pennsylvania, 3620 Walnut Street, Philadelphia, PA 19104, USA. [email: khampton@asc.upenn.edu]

Neeti Gupta is a product manager for the Unified Communications Group at Microsoft. Her research and professional work has allowed her to straddle the worlds of technology and culture, production and consumption, design and user experience and exploring experience design in online and offline community.

Acknowledgments

We would like to thank the Department of Urban Studies and Planning and the Comparative Media Studies Program at MIT. This manuscript benefited from the comments of Pablo Boczkowski, Henry Jenkins, Marc Smith, Barry Wellman and the editors and anonymous reviewers of New Media & Society. We would also like to thank the anonymous reviewers of the 2007 International Conference on Communities and Technologies, where a preliminary version of this article was presented.

References

Baym, N.,Y.B. Zhang and M.-C. Lin (2004) 'Social Interactions across Media: Interpersonal Communication on the Internet, Telephone and Face-to-Face', *New Media & Society 6*(3): 299–318.

Berkowitz, S.D. (1971) 'A Cross-national Comparison of Some Social Patterns of Urban Pedestrians', *Journal of Cross-cultural Psychology 2*(2): 129–44.

Boase, J., J. Horrigan, B.Wellman and L. Rainie (2006) *The Strength of Internet Ties*. Washington, DC: Pew Internet & American Life Project.

Burt, R. (1992) *Structural Holes*. Chicago, IL: University of Chicago Press.

Castells, M. (1996) *The Rise of the Network Society*. Oxford: Blackwell.

Cohen, S., I. Brissette,W. Doyle and D. Skoner (2000) 'Social Integration and Health: The Case of the Common Cold', *Journal of Social Structure 1*(3): 1–7.

Fischer, C. (1992) *America Calling: A Social History of the Telephone to 1940*. Berkeley, CA: University of California Press.

Forlano, L. (2008) 'Anytime? Anywhere? Reframing Debates Around Municipal Wireless Networking', *The Journal of Community Informatics 4*(1) URL (consulted 16 September 2008): http://ci-journal.net/index.php/ciej/article/view/438/401

Goffman, E. (1959) *The Presentation of Self in Everyday Life*. London: Penguin.

Goffman, E. (1963) *Behavior in Public Places: Notes on the Social Order of Gatherings.* New York: Free Press.

Goffman, E. (1971) *Relations in Public: Micro-studies of the Public Order*. New York: Basic Books.

Goldberger, P. (2003) 'Disconnected Urbanism: The Cellphone Has Changed Our Sense of Place More Than Faxes, Computers and E-Mail', *Metropolis Magazine,* November, URL (consulted 27 March 2007): http://www.metropolismag.com/cda/story.php?artid_254

Graham, S. and S. Marvin (1996) *Telecommunications and the City: Electronic Spaces, Urban Places.* London: Routledge.

Granovetter, M. (1973) 'The Strength of Weak Ties', *American Journal of Sociology 78*(6): 1360–80.

Granovetter, M. (1974) *Getting a Job*. Cambridge, MA: Harvard University Press.

Habermas, J. (1989) *The Structural Transformation of the Public Sphere*. Cambridge, MA: MIT Press.

Hampton, K. (2001) 'Living the Wired Life in the Wired Suburb: Netville, Glocalization and Civic Society', PhD thesis, Department of Sociology, University of Toronto.

Hampton, K. (2007) 'Neighborhoods in the Network Society: The e-Neighbors Study', *Information, Communication and Society* 10(5): 714–48.

Hampton, K., O. Livio and C. Trachtenberg (2008) 'Is WiFi Use Compatible with Public Space? Wireless Internet Use, Sociability, and Social Networks', presented at the annual meeting of the International Communication Association, Montreal.

Hampton, K. and Wellman, B. (2003) 'Neighboring in Netville: How the Internet Supports Community and Social Capital in a Wired Suburb', *City and Community 2*(3): 277–311.

Hannigan, J. (1998) *Fantasy City: Pleasure and Profit in the Postmodern Metropolis*. New York: Routledge.

Harris, K. (2003) 'Keep Your Distance: Remote Connection', *Journal of Community Work and Development 4*(4): 6–28.

Haythornthwaite, C. and B. Wellman (2002) 'The Internet in Everyday Life: An Introduction', in B. Wellman and C. Haythornthwaite (eds) *The Internet in Everyday Life,* pp. 3–44. Oxford: Blackwell.

Humphries, L. (2005) 'Cellphones in Public: Social Interactions in a Wireless Era', *New Media & Society 7*(6): 810–33.

Ito, M. and D. Okabe (2006) 'Intimate Connections', in R. Kraut, M. Brynin and S. Keisler (eds) *Computers, Phones and the Internet: Domesticating Information Technology,* pp. 235–47. New York: Oxford University Press.

Jacobs, J. (1961) *The Death and Life of Great American Cities*. New York: Random House.

Kraut, R., M. Patterson, V. Lundmark, S. Kiesler, T. Mukhopadhyay and W. Scherlis (1998) 'Internet Paradox: A Social Technology that Reduces Social Involvement and Psychological Well-being?', *American Psychologist* 53(9): 1017–31.

Latané, B. and J. Darley (1976) *Help in a Crisis: Bystander Response to an Emergency*. Morristown, NJ: General Learning Press.

Lee, S. (1999) 'Private Uses in Public Spaces: A Study of Internet Cafes', *New Media & Society 1*(3): 331–50.

Ling, R. (2000) 'Direct and Mediated Interaction in the Maintenance of Social Relationships', in A. Sloane and F. van Rijn (eds) *Home Informatics and Telematics: Information, Technology and Society,* pp. 61–86. Boston, MA: Kluwer.

Ling, R. (2004) *The Mobile Connection*. San Francisco, CA: Morgan Kaufmann.

Ling, R. and B.Yttri (2006) 'Control, Emancipation and Status', in R. Kraut, M. Brynin and S. Keisler (eds) *Computers, Phones and the Internet: Domesticating Information Technology,* pp. 219–34. New York: Oxford University Press.

Lofland, L. (1973) *A World of Strangers*. New York: Basic Books.

Lofland, L.H. (1998) *The Public Realm: Exploring the City's Quintessential Social Territory*. New York: Aldine de Gruyter.

McPherson, M., L. Smith-Lovin and M.E. Brashears (2006) 'Social Isolation in America: Changes in Core Discussion Networks over Two decades', *American Sociological Review 71*(3): 353–75.

McPherson, M., L. Smith-Lovin and J.M. Cook (2001) 'Birds of a Feather: Homophily in Social Networks', *Annual Review of Sociology 27*: 415–44.

Marvin, C. (1988) *When Old Technologies Were New: Thinking about Electric Communication in the Late Nineteenth Century*. New York: Oxford University Press.

Matsuda, M. (2005) 'Mobile Communications and Selective Sociality', in M. Ito, M. Okabe and M. Matsuda (eds) *Personal, Portable, Pedestrian: Mobile Phones in Japanese Life,* pp. 123–42. Cambridge, MA: MIT Press.

Merton, R.K. and E. Barber (2004) *The Travels and Adventures of Serendipity: A Study in Sociological Semantics and the Sociology of Science*. Princeton, NJ: Princeton University Press.

Mesch, G.S. and Y. Levanon (2003) 'Community Networking and Locally-based Social Ties in Two Suburban Localities', *City & Community 2*(4): 335–51.

Milgram, S. (1970) 'The Experience of Living in Cities', *Science 167*(3924): 1461–8.

Nie, N., S. Hillygus and L. Erbring (2002) 'Internet Use, Interpersonal Relations and Sociability: A Time Diary Study', in B.Wellman and C. Haythornthwaite (eds) *The Internet in Everyday Life,* pp. 215–43. Oxford: Blackwell.

Oldenburg, R. (1989) *The Great Good Place: Cafes, Coffee Shops, Community Centers, Beauty Parlors, General Stores, Bars, Hangout and How They Get You Through the Day*. New York: Paragon House.

Powell, A. (2008) 'Co-productions of Technology, Culture and Policy in North America's Community Wireless Networking Movement', PhD thesis, Communication Studies, Concordia University.

Putnam, R. (2000) *Bowling Alone*. New York: Simon & Schuster.

Quan-Haase, A. and B. Wellman (2006) 'Hyperconnected Net Work', in C. Heckscher, and P. Adler (eds) *The Firm as a Collaborative Community: Reconstructing Trust in the Knowledge Economy,* pp. 281–333. New York: Oxford University Press.

Quan Haase, A., B. Wellman, K. Hampton and J. Witte (2002) 'Internet, Social Capital and Information Seeking', in B. Wellman and C. Haythornthwaite (eds) *The Internet in Everyday Life,* pp. 291–324. Oxford: Blackwell.

Robinson, J.P.,M. Kestnbaum, A. Neustadl and A. Alvarez (2002) 'The Internet and Other Uses of Time', in B.Wellman and C. Haythornthwaite (eds) *The Internet in Everyday Life,* pp. 244–62. Oxford: Blackwell.

Sampson, R., J. Morenoff and T. Gannon-Rowley, (2002) 'Assessing "Neighborhood Effects": Social Processes and New Directions in Research', *Annual Review of Sociology 28*: 443–78.

Sennett, R. (1977) *The Fall of Public Man*. New York: Knopf.

Weber,M. (1946) *From Max Weber: Essays in Sociology*. New York: Oxford University Press.

Wellman, B. (1999) 'The Network Community: An Introduction', in B.Wellman (ed.) *Networks in the Global Village,* pp. 1–47. Boulder, CO:Westview Press.

Wellman, B. and S.Wortley (1990) 'Different Strokes from Different Folks: Community Ties and Social Support', *American Journal of Sociology 96*: 558–88.

Wellman, B.,A. Quan-Haase, J. Boase,W. Chen, K. Hampton, I.I. de Diaz and K. Miyata (2003) 'The Social Affordances of the Internet for Networked Individualism', *Journal of Computer-Mediated Communication 8*(3), URL (consulted 2 April 2007): http://jcmc.indiana.edu/vol8/issue3/wellman.html

Whyte,W.H. (1980) *The Social Life of Small Urban Spaces*. Washington, DC: Conservation Foundation.

Zhao, S. (2006) 'Do Internet Users Have More Social Ties? A Call for Differentiated Analyses of Internet Use', *Journal of Computer-Mediated Communication,* 11(3), URL (consulted 2 April 2007): http://jcmc.indiana.edu/vol11/issue3/zhao.html

Zukin, S. (1995) *The Cultures of Cities*. Cambridge, MA: Blackwell.

The Rise and Fall of a Facebook Hate Group

Andre Oboler

One Facebook group has repeatedly caught the media's attention. The group is called "Israel" is not a country! . . . . . . Delist it from Facebook as a country! *Despite the opinions of experts who highlighted the racist nature of the group, Facebook refused to take action. After unsuccessfully lobbying Facebook for intervention, a organization known as the Jewish Internet Defense Force took control of the Facebook group in late July 2008 and began to manually dismantle it from the inside. The rise, fall and eventual deletion of this group open questions on the correct response to online hate in user-generated content.*

Meet Facebook

With over 115 million users Facebook is the largest social networking site on the Internet (Holahan, 2008). It was initially limited to colleges in the United States, though later expanded to educational institutions in other countries and then to the general public (Arrington, 2006).

Facebook allows users to share personal information, join groups, send private messages and leave public notes on either a group or an individual's "wall" (a section of screen real estate designated for this purpose). Members can also share photographs and multimedia and install and use third-party applications based on the Facebook platform.

Facebook's founder and chief executive officer is Mark Zuckerberg, a Jewish self-made billionaire, aged 24. The valuation is based largely on Microsoft's investment of $US240 million to buy a 1.6 percent stake in Facebook in 2007 (Stone, 2007). Rather than going public or selling out, Zuckerberg has expressed a desire to develop the technology and push the boundaries. He believes in an "intense focus on openness, sharing information, as both an ideal and a practical strategy to get things done" (McGirt, 2007). The things that get done could however be for better or for worse. Asked if Facebook would take proactive measures to fight against anti-semitism, Zuckerberg stated that Facebook does not need to be proactive about it and that Facebook users should use the platform to generate more worldly perspectives (O'Neill, 2008).

Left to their own devices in a wild west–like atmosphere, those using Facebook have also been pushing the boundaries, including the boundary of acceptable content. The rules governing the use of Facebook are created and modified by Facebook, published in their "terms of use" (Facebook, 2008a). They include a prohibition on "content that, in the sole judgment of Company, is objectionable or which restricts or inhibits any other person from using or enjoying the Site". Facebook also has a "code of conduct", stating that "certain kinds of speech simply do not belong in a community like Facebook" and enumerated examples including material that is "derogatory, demeaning, malicious, defamatory, abusive, offensive or hateful" (Facebook, 2007). Despite these conditions of use, Facebook has, in some cases, failed to act.

A Facebook Hate Group

The group *"Israel" is not a country! . . . . . . Delist it from Facebook as a country!* (Facebook, 2008b) is a Facebook group established around January 2007. A group with an almost identical name and description was created in the Hi5 social networking site on 9 January 2007 (Hi5, 2008). That group today has only 1,076 members, while the Facebook group at its height had over 48,000 members, over 120,850 posts, over 150 videos and over 100 photographs. While some of this content was in disagreement with the nature of the group, the majority was decidedly anti-Israel and often anti-semitic.

The long running controversy was mostly likely started by Facebook itself when, in October 2006, they removed Palestine from the list of countries individuals could join. It had been included in Facebook's original list but was replaced by an entity known as "Gaza and the West Bank". Some Palestinians objected to this alteration, claiming it advocated a position that denied Palestinians ownership of the old city of Jerusalem as well as East Jerusalem. Facebook quietly restored Palestine to the country listing in early 2007 (Jacobs, 2007). Shortly thereafter a group was established in protest at the re–inclusion, and another group, the now infamous "Israel is not a country" group was created in response (Zerbisias, 2007). The growth of the group from October 2007 is shown in Figure 1.

While they may have started off in a similar manner, there is something very different in the nature of the "Israel is not a country" and "Palestine is not a country" groups. The difference is not about international law or the legitimacy of countries. In the electronic world there is no requirement to mirror reality, comply with international standards, or even include all countries in any given list. To give one example, Israel and about four other countries are missing from the list of countries in the causes application (in this case it is in the donations section and the missing listing means people in Israel cannot donate money via Israeli credit cards). These things happen, people complain, and usually they get fixed.

The "Israel is not a country" and "Palestine is not a country" groups differ in size, but this is not the fundamental difference, though it does have an effect on the impact the groups can have. The most popular "Palestine is not a country" group has 3,353 members at the time of writing. This is less than seven percent of the size of the "Israel is not a country" group—prior to it being taken over and members being booted out by new administrators. Size matters, but it is not what makes this group so different.

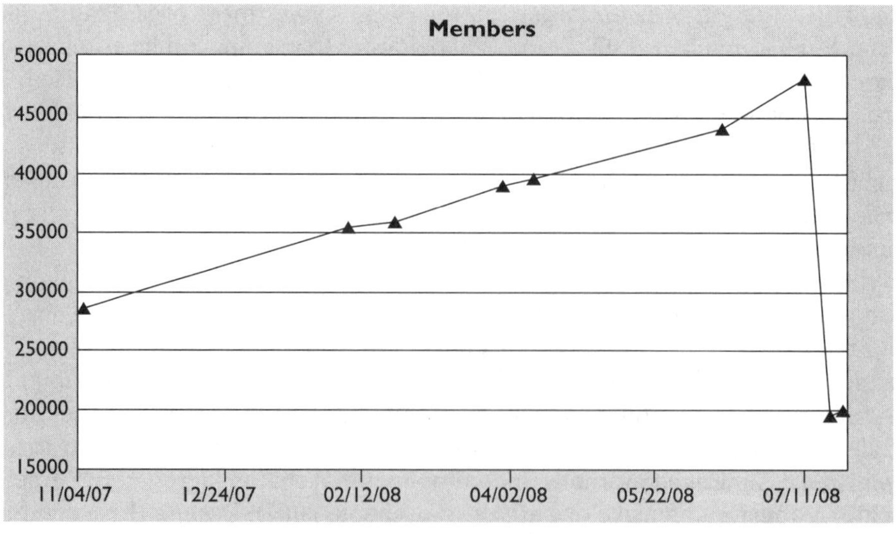

FIGURE 1 Growth and decline of a Facebook hate group.

What Makes This Group Different?

What makes "Israel is not a country" different from other groups in Facebook is the way it is used. The "Israel is not a country" group was not designed to seriously protest the listing of Israel in Facebook, it was simply part of Facebook politics. That however changed as more radical elements joined the group—something Facebook themselves seems to have missed.

The earliest press report about the group appeared in the Toronto Star on 3 May 2007 (Zerbisias, 2007). The author, Antonia Zerbisias, notes a proliferation of groups and counter–groups noting that "not unlike college campuses in the real world, when it comes to Israel, it's an all-out war . . . of words." Zerbisias also contacted Matt Hicks, senior manager of corporate communications at Facebook, who had until then been unaware of the problem. This newspaper story establishes that Facebook knew about the group for over a year before the takeover, and for over year Facebook decided to do nothing. If Zerbisias' analysis of this as a college campus-style war of words remained accurate the group would not have grown as large as it did, or would it have attracted the attention it received.

This protest group eventually altered, becoming decidedly anti-Israel and anti-semitic, attracting more like-minded Facebook members and changing administrators a number of times. Having Palestine re-instated, the group's initial goal was met, and the collective membership began to form new goals. The group's logo, a map with Israel entirely replaced by "Palestine", promoted replacement geography in a similar manner to the anti-Israel campaign in Google Earth (Oboler, 2008a). The group became highly linked with anti-Israel and anti-semitic Web sites and campaigns. It was used as a base to promote general anti-Israel sentiment and to promote an ideology that advocated the destruction of the Jewish state. This group, with its large membership base, was used as a recruitment site where new campaigns against Israel could prosper.

The role of this group was to promote online hate, and specifically online anti-semitism. As a result it drew significant attention in early 2008. Addressing the Global Forum for Combat Anti-semitism in February, I described this group and highlighted its use of apartheid rhetoric and clever use of denial. Half the statements in the group's introduction were anti-semitic, the other half were designed to fend off charges of anti-semitism. The New York Jewish Week carried a front–page story on Antisemitism 2.0 on 20 February 2008, noting the growth of the "Israel is not a country" group (Snyder, 2008). That report was partly based on a draft of the Jerusalem Center for Public Affairs' Anti-semitism 2.0 report (Oboler, 2008b) which provided the first extensive coverage of anti-semitism on the social Web and coined the phrase "Antisemitism 2.0." The "Israel is not a country" group was one of two detailed case studies examined in the report. Particular attention was paid to the group's clever introduction of itself as a group against racism, while simultaneously attacking both Israel and the Jewish people. Israel, it claimed, has no right to exist.

The introduction was reused in 75 different groups which anti-Israel elements took over in a very similar manner to that used five months later by the JIDF. The anti-semitic takeovers were originally reported in the press in February 2008. Many of the groups taken over had names completely unrelated to the conflict. They included local sports or film appreciation groups. The common thread is simply that the groups had their location set to Israel or otherwise indicated Israeli or Jewish interests. Some of these groups are still up with the original message still visible. The message is a diatribe familiar to any who deal with anti-semitism; Israel is "an apartheid regime", Israel "has no right to exist", "Israelis accuse people of anti-Semitism every time someone criticizes Israel", "Arabs are Semites . . . unlike most Jews", "80–90 percent of Jews today are Ashkenazi (which is the term for European Jews that are descendants of converts)", they "use the Holocaust to silence critics of their own crime", "Israel never met the conditions for its entry into the UN", the lies go on in some of the remaining clones despite the efforts of the JIDF.

Not all the attention showed the group in a negative light. Ala'a Ghosha, the hostess of a television show aimed at teenagers, discussed the "Israel is not a country" group on official Palestinian TV in March 2008. She praised the group, describing Facebook as "a new battleground between the Arabs and Israel" and encouraged viewers to join it (Marcus and Crook, 2008).

The JIDF Response

On 27 July 2008, Arutz Sheva published a short report titled "Jewish Activists Hack Anti-Semitic Facebook Group". The report notes how the Jewish Internet Defense Force (JIDF) had taken control of the "Israel is not a country" group. The JIDF confirmed that the takeover occurred earlier on 27 July.

Following the takeover, the JIDF worked around the clock to empty the group of members. Once a group is empty it can be deleted. As a first step, the problematic description, multimedia and the wall were removed, significantly mitigating the hate that this group was promoting.

The *Jerusalem Post* followed up with a story in its 29 July issue (Rubenstein, 2008). This corrected the misrepresentation of JIDF's activities as a hack. It was certainly a takeover, but nothing illegal or indeed outside the rules of Facebook had occurred. Next in the U.K., the *Telegraph* picked up the story with a headline on 31 July declaring "Facebook: 'Anti-Semitic' group hijacked by Jewish force" (Moore, 2008).

Hate groups picked up the story too. The Neo-Nazi site Stormfront had the story almost from the start as did forums at Al-Jazeera. On Facebook itself two groups were immediately setup up against JIDF.

On 1 August, only a few days after the takeover, the JIDF lost control of the group. The group's size had been reduced, with its multimedia and wall gone, but the group's problematic description reappeared. Even if the JIDF has maintained control of the group, setting up a new group takes only a few clicks. JIDF's action, however, was effective.

During the few days that the JIDF controlled the group they expelled 59 percent of the group's membership. Based on the membership increase in the days immediately following JIDF's loss control, it looked like the group has been set back 12 and 18 months in terms of its membership numbers. A month later Facebook, however, took decisive action. They deleted the group in response to continued press coverage and the emergence of a consensus in Wikipedia that Facebook was indeed publicly hosting an anti-semitic group.

Responding to Hate in User-Generated Content

Facebook is not the only place where user-generated content promotes hate. Google Earth has come under fire for allowing replacement geography (Oboler, 2008a), eBay came under fire for its role in selling Nazi memorabilia and other goods that encourage hate (ADL, 2001), and Craigslist has had issues with racist accommodation postings (Beck, 2008). In all these cases there was an effort by corporations in control of the sites to acknowledge the existence of this sort of material and they sought ways to resolve the issue. In this case, after a year of inaction by Facebook, users stepped in to respond to the problem. This reaction seems in keeping with Zuckerberg's hands-off approach on the topic of anti-semitism and his desire for users to resolve these matters within the Facebook platform. Facebook in the end acted to protect its public image, and only under increasing public pressure.

The actions of the JIDF, however, raise interesting questions. If they acted within the law, and within Facebook's terms of service . . . did they do anything wrong? Should similar attempts be encouraged or avoided? One might interpret the closure of specific groups as acts of censorship, but is censorship always wrong? The most common approach to anti-semitic Web sites in the past, before Web 2.0, was to push for the groups to be shut down by their specific Internet service providers. Even in this case, many were from the start urging Facebook to step in and remove the group. There are only really two differences between what occurred through the JIDF's action and what later occurred when Facebook themselves intervened. The differences are as much about values as they are about technical details.

The first difference is that the action taken by Facebook required far less effort. Facebook staff, with a few clicks, can remove material that it would take the JIDF hundreds of people hours to accomplish. As the owner of the platform, Facebook also has the legal and moral authority to determine when a breach of the rules has occurred and to take whatever action it deems as appropriate. Had Facebook taken action, the result would have been new protest groups, added publicity and accusations that Facebook was taking a "political" position. This is not so different from what occurred back in October 2006, and again with another controversy in March 2008 (Reuters, 2008). Facebook, however, as a large corporation with a huge amount of power—absolute power, in fact, within Facebook—needs to realize it can't avoid these issues. By simply stepping back and trying to give everyone whatever they ask for, they become a platform open to misuse and abuse. Despite their terms and conditions, Facebook seems ill-prepared to deal with anti-semitism 2.0. Such online hate is designed to appear socially acceptable. As such, it aims to have itself classed as legitimate political discourse. Allowing such hate speech to be published is indeed taking a position by default.

Facebook itself is falling victim to those pressures that spread anti-semitism 2.0, the social acceptability of anti-Jewish racism online. Given their responsibly to Facebook users, not to mention to society at large, Facebook's current hands-off approach is legitimizing online hate and allowing it to spread. Facebook's senior management team needs to take another look at the implementation of their terms of service and need to consider more carefully what counts as legitimate discussion and what counts as the promotion of hate.

The problem can only really be solved by Facebook, either voluntarily or if left alone, through external pressure and even legislation. The Facebook platform is too powerful and widespread to be manipulated by those promoting hate. The only question is who realizes this first—Facebook, the public, or legislators.

As for the JIDF, it will continue to take action against anti-semitism in all of its digital forms. Some will agree, some will disagree with their strategies and solutions. Unfortunately, Facebook will probably continue to hope in vain that problems will be resolved without specific intervention by Facebook itself.

About the Author

Andre Oboler is a Post-Doctoral Fellow in Political Science at Bar-Ilan University and Legacy Heritage Fellow at NGO Monitor in Jerusalem. Web: http://www.ZionismOnTheWeb.org. E-mail: oboler@zionismontheweb.org

References

Anti-Defamation League (ADL), 2001. "ADL applauds eBay for expanding guidelines to prohibit the sale of items that glorify hate," press release (4 May), at http://www.adl.org/PresRele/Internet_75/3820_75.htm, last accessed 29 October 2008.

Michael Arrington, 2006. "Facebook just launched open registrations," *TechCrunch* (26 September), at http://www.techcrunch.com/2006/09/26/facebook-just-launched-open-registrations/, last accessed 29 October 2008.

Amanda Beck, 2008. "Craigslist cleared over racist and sexist ads," *News.com.au* (17 March), at http://www.news.com.au/technology/story/0,25642,23385890-5014239,00.html, last accessed 29 October 2008.

Facebook, 2008a. "Terms of Use" (23 September), at http://www.new.facebook.com/terms.php?ref=pf, last accessed 29 October 2008.

Facebook, 2008b. "Group: 'Israel' is not a country! . . . . . . Delist it from Facebook as a country!" at http://www.facebook.com/, accessed 12 August 2008.

Facebook, 2007. "Content Code of Conduct" (24 May), at http://www.new.facebook.com/codeofconduct.php, last accessed 12 August 2008

Hi5, 2008. "Group: Israel is not a country! . . . . . . Delist it from hi5 as a country!" at http://hi5.com/, accessed 12 August 2008.

Catherine Holahan, 2008. "Facebook's new friends abroad," *Business Week* (14 May), at http://www.businessweek.com/technology/content/may2008/tc20080513_217183.htm, last accessed 29 October 2008.

Megan Jacobs, 2007. "Facebook sparks 'Palestine' debate," *Jerusalem Post* (10 October), at http://www.jpost.com/, last accessed 29 October 2008.

Itamar Marcus and Barbara Crook, 2008. "PATV lauds Internet site calling for Israel's elimination," *Palestinian Media Watch* (14 March), at http://www.pmw.org.il/Bulletins_apr2008.html, last accessed 29 October 2008.

Ellen McGirt, 2007. "Facebook's Mark Zuckerberg: Hacker. Dropout. CEO," *Fast Company,* issue 115 (May), at http://www.fastcompany.com/magazine/115/, last accessed 29 October 2008.

Matthew Moore, 2008. "Facebook: 'Anti-Semitic' group hijacked by Jewish force," *Telegraph* (31 July), at http://www.telegraph.co.uk/news/2478773/Facebook-Anti-semitic-group-destroyed-by-Israeli-hackers.html, last accessed 29 October 2008.

Andre Oboler, 2008a. "Google Earth: A new platform for anti–Israel propaganda and replacement geography," *Jerusalem Center for Public Affairs,* volume 8, number 5 (26 June), at http://www.jcpa.org/, last accessed 29 October 2008.

Andre Oboler, 2008b. "Online antisemitism 2.0. 'Social antisemitism' on the 'Social Web'," *Jerusalem Center for Public Affairs,* number 67 (1 April), at http://www.jcpa.org/, last accessed 29 October 2008.

Nick O'Neill, 2008. "My interview With Mark Zuckerberg," *AllFacebook* (11 March), at http://www.allfacebook.com/2008/03/my-interview-with-mark-zuckerberg/, last accessed 29 October 2008.

Reuters, 2008. "Facebook face-off: Settlers win right to list country as Israel, not Palestine," *Haaretz.com* (17 March), at http://www.haaretz.co.il/hasen/spages/965215.html, last accessed 29 October 2008.

Stephanie Rubenstein, 2008. "Jewish Internet Defense Force 'seizes control' of anti-Israel Facebook group," *Jerusalem Post* (29 July), at http://www.jpost.com/, last accessed 29 October 2008.

Tamar Snyder, 2008. "Anti-semitism 2.0 going largely unchallenged," *New York Jewish Week* (20 February), p. 1, and at http://www.zionismontheweb.org/antisemitism/antisemitism_2.0_going_largely_unchallenged.htm, last accessed 29 October 2008.

Brad Stone, 2007. "Microsoft to pay $240 million for stake in Facebook," *New York Times* (25 October), at http://www.nytimes.com/2007/10/25/technology/24cnd-facebook.html, last accessed 29 October 2008.

Antonia Zerbisias, 2007. "Playing politics on Facebook," *Toronto Star* (3 May), at http://www.thestar.com/News/article/209925, last accessed 29 October 2008.

MODULE EIGHT

Civic Engagement and Democracy

Smart Mobs: The Power of the Mobile Many

Howard Rheingold

Bypassing the complex of broadcasting media, cell phone users themselves became broadcasters, receiving and transmitting both news and gossip and often confounding the two. Indeed, one could imagine each user becoming a broadcasting station unto him or herself, a node in a wider network of communication that the state could not possibly even begin to monitor, much less control. Hence, once the call was made for people to mass at Edsa, cell phone users readily forwarded messages they received, even as they followed what was asked of them.

Cell phones then were invested not only with the power to surpass crowded conditions and congested surroundings brought about by the state's inability to order everyday life. They were also seen to bring a new kind of crowd about, one that was thoroughly conscious of itself as a movement headed towards a common goal.

—*Vicente Rafael, "The Cell Phone and the Crowd: Messianic Politics in Recent Philippine History"*

Netwar—Dark and Light

On January 20, 2001, President Joseph Estrada of the Philippines became the first head of state in history to lose power to a smart mob. More than 1 million Manila residents, mobilized and coordinated by waves of text messages, assembled at the site of the 1986 "People Power" peaceful demonstrations that had toppled the Marcos regime.[1] Tens of thousands of Filipinos converged on Epifanio de los Santas Avenue, known as "Edsa," within an hour of the first text message volleys: "Go 2EDSA, Wear blck."[2] Over four days, more than a million citizens showed up, mostly dressed in black. Estrada fell. The legend of "Generation Txt" was born.

Bringing down a government without firing a shot was a momentous early eruption of smart mob behavior. It wasn't, however, the only one.

- On November 30,1999, autonomous but internetworked squads of demonstrators protesting the meeting of the World Trade Organization used "swarming" tactics, mobile phones, Web sites, laptops, and handheld computers to win the "Battle of Seattle."[3]
- In September 2000, thousands of citizens in Britain, outraged by a sudden rise in gasoline prices, used mobile phones, SMS, email from laptop PCs, and CB radios in taxicabs to coordinate dispersed groups that blocked fuel delivery at selected service stations in a wildcat political protest.[4]
- A violent political demonstration in Toronto in the spring of 2000 was chronicled by a group of roving journalist-researchers who webcast digital video of everything they saw.[5]

- Since 1992, thousands of bicycle activists have assembled monthly for "Critical Mass" moving demonstrations, weaving through San Francisco streets en masse. Critical Mass operates through loosely linked networks, alerted by mobile phone and email trees, and breaks up into smaller, tele-coordinated groups when appropriate.[6]

Filipinos were veteran texters long before they toppled Estrada. Short Message Service (SMS) messaging was introduced in 1995 as a promotional gimmick.[7] SMS messaging, free at first, remained inexpensive. Wireline telephone service is more costly than mobile service, and in a country where 40 percent of the population lives on one dollar a day, the fact that text messages are one-tenth the price of a voice call is significant.[8] A personal computer costs twenty times as much as a mobile telephone; only 1 percent of the Philippines' population own PCs, although many more use them in Internet cafés.[9] By 2001, however, 5 million Filipinos owned cell phones out of a total population of 70 million.[10]

Filipinos took to SMS messaging with a uniquely intense fervor. By 2001, more than 70 million text messages were being transmitted among Filipinos every day.[11] The word "mania" was used in the Manila press. The *New York Times* reported in 2001:

> *Malls are infested with shoppers who appear to be navigating by cellular compass. Groups of diners sit ignoring one another, staring down at their phones as if fumbling with rosaries. Commuters, jaywalkers, even mourners—everyone in the Philippines seems to be texting over the phone.... Faye Slytangco, a 23-year-old airline sales representative, was not surprised when at the wake for a friend's father she saw people bowing their heads and gazing toward folded hands. But when their hands started beeping and their thumbs began to move, she realized to her astonishment that they were not in fact praying. "People were actually sitting there and texting," Slytangco said. "Filipinos don't see it as rude any more."*[12]

Like the thumb tribes of Tokyo and youth cultures in Scandinavia, Filipino texters took advantage of one of the unique features of texting technology—the ease of forwarding jokes, rumors, and chain letters. Although it requires effort to compose messages on mobile telephone keypads, only a few thumb strokes are required to forward a message to four friends or everybody in your telephone's address book. Filipino texting culture led to a national panic when a false rumor claimed that Pope John Paul II had died.[13]

Many Filipino text message jokes and rumors were political. Vicente Rafael, professor at the University of California, San Diego, sees Filipino texting culture as inherently subversive:

> *Like many third world countries recently opened to more liberal trade policies, the Philippines shares in the paradox of being awash in the latest technologies of communication such as the cell phone while mired in deteriorating infrastructures such as roads, postal services, railroads, power generators and land lines. With the cell phone, one appears to be able to pass beyond these obstacles. And inasmuch as such infrastructures are state run so that their breakdown and inefficiencies are a direct function of governmental ineptitude, passing beyond them also feels like overcoming the state, which to begin with is already overcome by corruption. It is small wonder then that cell phones could prove literally handy in spreading rumors, jokes, and information that steadily eroded whatever legitimacy President Estrada still had.*[14]

The "People Power II" demonstrations of 2001 broke out when the impeachment trial of President Estrada was suddenly ended by senators linked to Estrada. Opposition leaders broadcast text messages, and within seventy-five minutes of the abrupt halt of the impeachment proceedings, 20,000 people converged on Edsa.[15] Over four days, more than a million people showed up. The military withdrew support from the regime; the Estrada government fell, as the Marcos regime had fallen a decade previously, largely as a result of massive nonviolent demonstrations.[16] The rapid assembly of the anti-Estrada crowd was a hallmark of early smart mob technology, and the millions of text messages exchanged by the demonstrators in 2001 was, by all accounts, a key to the crowd's esprit de corps.

Professor Rafael sees the SMS-linked crowd that assembled in Manila as the manifestation of a phenomenon that was enabled by a technical infrastructure but that is best understood as a social instrument:

> *The power of the crowd thus comes across in its capacity to overwhelm the physical constraints of urban planning in the same way that it tends to blur social distinctions by provoking a sense of estrangement. Its authority rests on its ability to promote restlessness and movement, thereby undermining the pressure from state technocrats, church authorities and corporate interests to regulate and contain such movements. In this sense, the crowd is a sort of medium if by that word one means the means for gathering and transforming elements, objects, people and things. As a medium, the crowd is also the site for the generation of expectations and the circulation of messages. It is in this sense that we might also think of the crowd not merely as an effect of technological devices, but as a kind of technology itself. . . . Centralized urban planning and technologies of policing seek to routinize the sense of contingency generated in crowding. But at moments and in areas where such planning chronically fails, routine can at times give way to the epochal. At such moments, the crowd . . . takes on a kind of telecommunicative power, serving up channels for sending messages at a distance and bringing distances up close. Enmeshed in a crowd, one feels the potential for reaching out across social space and temporal divides.*[17]

The Battle of Seattle saw a more deliberate and tactically focused use of wireless communications and mobile social networks in urban political conflict, more than a year before texting mobs assembled in Manila. A broad coalition of demonstrators who represented different interests but were united in opposition to the views of the World Trade Organization planned to disrupt the WTO's 1999 meeting in Seattle. The demonstrators included a wide range of different "affinity groups" who loosely coordinated their actions around their shared objective. The Direct Action Network enabled autonomous groups to choose which levels of action to participate in, from nonviolent support to civil disobedience to joining mass arrests—a kind of dynamic ad hoc alliance that wouldn't have been possible without a mobile, many-to-many, real-time communication network. According to a report dramatically titled, "Black Flag Over Seattle," by Paul de Armond:

> *The cohesion of the Direct Action Network was partly due to their improvised communications network assembled out of cell phones, radios, police scanners and portable computers. Protesters in the street with wireless Palm Pilots were able to link into continuously updated web pages giving reports from the streets. Police scanners monitored transmissions and provided some warning of changing police tactics. Cell phones were widely used.*
>
> *Kelly Quirke, Executive Director of the Rainforest Action Network, reports that early Tuesday, "the authorities had successfully squashed DAN's communications system." The solution to the infrastructure attack was quickly resolved by purchasing new Nextel cell phones. According to Han Shan, the Ruckus Society's WTO action coordinator, his organization and other protest groups that formed the Direct Action Network used the Nextel system to create a cellular grid over the city. They broke into talk groups of eight people each. One of the eight overlapped with another talk group, helping to quickly communicate through the ranks.*
>
> *In addition to the organizers' all-points network, protest communications were leavened with individual protesters using cell phones, direct transmissions from roving independent media feeding directly onto the Internet, personal computers with wireless modems broadcasting live video, and a variety of other networked communications. Floating above the tear gas was a pulsing infosphere of enormous bandwidth, reaching around the planet via the Internet.*[18]

From Seattle to Manila, the first "netwars" have already broken out. The term "netwar" was coined by John Arquilla and David Ronfeldt, two analysts for the RAND corporation (birthplace of game theory and experimental economics), who noticed that the same combination of social networks, sophisticated communication technologies, and decentralized organizational structure was surfacing as an effective force in very different kinds of political conflict:

> *Netwar is an emerging mode of conflict in which the protagonists—ranging from terrorist and criminal organizations on the dark side, to militant social activists on the bright side—use network forms of organization, doctrine, strategy, and technology attuned to the information age. The practice of netwar is well ahead of theory, as both civil and uncivil society actors are increasingly engaging in this new way of fighting.*
>
> *From the Battle of Seattle to the "attack on America," these networks are proving very hard to deal with; some are winning. What all have in common is that they operate in small, dispersed units that can deploy nimbly—anywhere, anytime. All feature network forms of organization, doctrine, strategy, and technology attuned to the information age. They know how to swarm and disperse, penetrate and disrupt, as well as elude and evade. The tactics they use range from battles of ideas to acts of sabotage—and many tactics involve the Internet.*[19]

The "swarming" strategies noted by Arquilla and Ronfeldt rely on many small units like the affinity groups in the Battle of Seattle. Individual members of each group remained dispersed until mobile communications drew them to converge on a specific location from all directions simultaneously, in coordination with other groups. Manila, Seattle, San Francisco, Senegal, and Britain were sites of nonviolent political swarming. Arquilla and Ronfeldt cited the nongovernmental organizations associated with the Zapatista movement in Mexico, which mobilized world opinion in support of Indian peasants, and the Nobel Prize-winning effort to enact an anti-landmine treaty as examples of nonviolent netwar actions. Armed and violent swarms are another matter.

The Chechen rebels in Russia, soccer hooligans in Britain, and the FARC guerrillas in Colombia also have used netwar strategy and swarming tactics.[20] The U.S. military is in the forefront of smart mob technology development. The Land Warrior experiment is scheduled to field-test wearable computers with GPS and wireless communications by 2003.[21] The Joint Expeditionary Digital Information (JEDI) program links troops on the ground directly to satellite communications. JEDI handheld devices combine laser range-finding, GPS location awareness, direct satellite telephone, and encrypted text messaging.[22] Remember the DARPA-funded startup MeshNetworks from Chapter 6, the company whose technology enables military swarms to parachute onto a battlefield and self-organize an ad hoc peer-to-peer wireless network? Small teams of special forces, wirelessly networked and capable of calling in aircraft or missile strikes with increasing accuracy, were introduced by the United States and its allies in Afghanistan: netwar.

Examples later in this chapter demonstrate that smart mobs engaging in either violent or nonviolent netwar represent only a few of the many possible varieties of smart mob. Netwars do share similar technical infrastructure with other smart mobs. More importantly, however, they are both animated by a new form of social organization, the network. Networks include nodes and links, use many possible paths to distribute information from any link to any other, and are self-regulated through flat governance hierarchies and distributed power. Arquilla and Ronfeldt are among many who believe networks constitute the newest major social organizational form, after tribes, hierarchies, and markets. Although network-structured communications hold real potential for enabling democratic forms of decision-making and beneficial instances of collective action, that doesn't mean that the transition to networked forms of social organization will be a pleasant one with uniformly benevolent outcomes. Arquilla and Ronfeldt note the potential for cooperation in examples like the nongovernmental organizations that use netwar tactics for public benefit, but they also articulated a strong caution, worth keeping in mind when contemplating the future of smart mobs:

> *Most people might hope for the emergence of a new form of organization to be led by "good guys" who do "the right thing" and grow stronger because of it. But history does not support this contention. The cutting edge in the early rise of a new form may be found equally among malcontents, ne'er-do-wells, and clever opportunists eager to take advantage of new ways to maneuver, exploit, and dominate. Many centuries ago, for example, the rise of hierarchical forms of organization, which displaced traditional, consultative, tribal forms, was initially attended, in parts of the world, by the appearance of ferocious chieftains bent on military conquest and of violent secret societies run according to rank—long before the hierarchical form matured through the institutionalization of states, empires, and professional administrative and bureaucratic systems. In like manner, the early spread of the market form, only a few centuries ago, was accompanied by a spawn of usurers, pirates, smugglers, and monopolists, all seeking to elude state controls over their earnings and enterprises.*[23]

In light of the military applications of netwar tactics, it would be foolish to presume that only benign outcomes should be expected from smart mobs. But any observer who focuses exclusively on the potential for violence would miss evidence of perhaps an even more profoundly disruptive potential—for beneficial as well as malign purposes—of smart mob technologies and techniques. Could cooperation epidemics break out if smart mob media spread beyond warriors—to citizens, journalists, scientists, people looking for fun, friends, mates, customers, or trading partners?

Substitute the word "computers" for the words "smart mobs" in the previous paragraph, and you'll recapitulate the history of computation since its birth in World War II.

Lovegety and p2p Journalism

Organized conflict is undoubtedly a site of intensive cooperation. Humans enjoy cooperating to each other's benefit, as well, given the right conditions and payoff. Alexis de Tocqueville made an important observation in regard to early-nineteenth-century America:

> *The best-informed inhabitants of each district constantly use their information to discover new truths which may augment the general prosperity; and, if they have made any such discoveries, they eagerly surrender them to the mass of the people. . . . Men attend to the interests of the public, first by necessity, afterwards by choice: what was intentional becomes an instinct; and by dint of working for the good of one's fellow-citizens, the habit and the taste for serving them is at length acquired.*[24]

Elinor Ostrom and other students of common pool resource management (discussed in Chapter 2) have detailed the ways farmers, fishers, and foresters around the world devise ingenious social arrangements to balance cooperation and self-interest.[25]

Consider a few experiments on the fringes of mobile communications that might point toward a wide variety of nonviolent smart mobs in the future:

- "Interpersonal awareness devices" have been evolving for several years.[26] Since 1998, hundreds of thousands of Japanese have used Lovegety keychain devices, which signal when another Lovegety owner of the opposite sex and a compatible profile is within fifteen feet.[27] In 2000, a similar technology for same-sex seekers, the "Gaydar" device, was marketed in North America.[28] Hong Kong's "Mobile Cupid service" (www.sunday.com) sends a text description of potential matches who are nearby at the moment.[29]

- ImaHima ("are you free now?") enables hundreds of thousands of Tokyo i-mode users to alert buddies who are in their vicinity at the moment.[30]
- Upoc ("universal point of contact") in Manhattan sponsors mobile communities of interest; any member of "manhattan celebrity watch," "nyc terrorism alert," "prayer of the day," or "The Resistance," for example, can broadcast text messages to and receive messages from all the other members.[31]
- Phones that make it easy to send digital video directly to the Web make it possible for "peer-to-peer journalism" networks to emerge;[32] Steve Mann's students in Toronto have chronicled newsworthy events by webcasting everything their wearable cameras and microphones capture.[33]
- Researchers in Oregon have constructed "social middleware," which enables wearable computer users to form ad-hoc communities, using distributed reputation systems, privacy and knowledge-sharing agents, and wireless networks.[34]

In the fall of 2001, I visited the office of ImaHima in Tokyo's ultra-modern Ebisu Garden Place Tower. ImaHima founder, Neeraj Jhanji, was the only person in the office on a Saturday morning. The DoCoMo skyscraper I had visited the day before was visible through the window. Neeraj, twenty-nine, a native of India, remained in Tokyo after a stint with an international consulting firm. One sunny Saturday, walking alone in one of Tokyo's most popular and crowded districts, he wondered if any of his friends were nearby. "I looked at my phone and the answer seemed obvious," he told me.[35] Even without GPS location awareness, it would be possible to use the Internet to coordinate locations. At the time I spoke to Jhanji, ImaHima had won the prestigious Prix Ars Electronica, had been adopted as an official i-mode site, and had gained 250,000 users with a median age of twenty-five. ImaHima was planning to launch in European markets by 2002.[36]

Jhanji showed me how the service works. When you join, you fill out a profile and set up a buddy list similar to the kind used with Internet instant messaging; each person must give permission before someone else can know automatically where they are. You also list your favorite places. When you select the "update" link on your mobile's ImaHima menu, everyone on your buddy list knows, for example, that you are within a few blocks of Shibuya station and are free for lunch.

The just-in-time, just-in-place matchmaking service for strangers, an intriguing aspect of ImaHima, is also where the most caution is required. With young women in Tokyo being targeted on the street by solicitors for "hostess bars," no service could hope to attract any females without strict controls—nor would DoCoMo's strict policies allow a service to become a lucrative official i-mode site. "You can search through the list of your friends," Jhanji told me, "or you can ask permission to contact a stranger whose profile matches your request and who is nearby. But if you request permission to communicate and the other person denies your request, the system blocks you from communicating with that person again."

The ambience of the Manhattan location of Upoc differs sharply from the milieu of Ebisu Garden Tower. Upoc's building on lower Broadway is close enough to ground zero for the lingering stench to have been strong outside the building when I visited Upoc in November 2001. Upoc had used its own service as a virtual office in the days after the September attack. Upoc employee Alex LeVine sent a group SMS message to three dozen others employees immediately after he saw the second plane crash: "Do not go to work. Stand by for more directions."[37] Then he messaged nine employees already at work, telling them to evacuate. Although wireline telephone, cellular telephone, and email were all down, Upoc employees discovered that their text messaging service, based out of a server safely in New Jersey, stayed up and enabled them to regroup.

I met LeVine, Andrew Pimentel, and Upoc founder Gordon Gould in their office, a standard open-plan geek farm. It was heartening to see that at least a few rooms full of twenty-somethings in Aeron chairs still existed. Gould had been an enthusiastic participant in virtual communities. He knew the power of online social networks and noticed how today's teens have taken to mobile phones and pagers the way his generation had taken to computer keyboards. Upoc provides instant infrastructure for a smart mob, whether it is a group of shopping buddies, fans, families, political street theater, or affinities as yet undefined. The confusing clash of standards and services that has slowed the adoption of SMS services in the United States created an opportunity to provide a platform for mobile communities among users of different services. Register for Upoc on the Web, join an existing group, or start one of your own and invite your friends and family, and suddenly you can receive and broadcast text messages to your group, no matter what mobile telephone service they use or where they are located. Link up to your roving tribe from your desktop email and vice versa. More than 100,000 users have registered for hundreds of groups.[38]

I registered for an Upoc account and observed from afar for two weeks before I visited New York. I joined "nyc celebrity sightings," a mobile community of celebrity stalkers, and "nyc terror alert," which promised immediate messages in the case of terrorist attack. I also joined the "channel" for a youth entertainer named "lil bowwow" and received offers of tickets and opportunities to download lil bowwow's latest ringtone. After an afternoon of buzzing around California, feeling my phone buzz, and noting that Julia Roberts had been spotted in midtown Manhattan or that a fifteen-year-old lil bowwow fan in Brooklyn just got out of school, I switched to receiving my messages as email until I went to New York in person. Scanning the scores of messages exchanged every day in just a few groups made it clear that some kind of community ferment was underway.

"Communities started forming from the week we started testing the service," Gould told me.[39] Andrew Pimentel, who had conducted Upoc's marketing research, added, "There are virtual cliques, groups of friends, enemies, grudges, gangs, fights, and double agents with multiple handles who spy in groups to make sure nobody is badmouthing them or their clique." Upoc members can set up groups in any of three ways—secret, private, or public. Anyone can join a public group. Private groups are listed in the directory, but people join by applying and can be expelled by the founder. Secret groups are not listed and are known only to their members.

I unexpectedly experienced the "nyc terror alert" in action. The next to last day of my stay in Manhattan, walking up Fifth Avenue toward a morning meeting, my pocket started buzzing. I looked at the screen of my mobile and learned that two minutes earlier, American Airlines flight 587 had crashed after takeoff from JFK. I immediately reserved a train ticket to Boston for the next day, in case the airports remained closed. It was another one of those living-in-the-future moments. I had become one of those people I had first observed at Shibuya Crossing a year and a half earlier. My pocket buzzed again. Another plane crash? No. A celebrity spotted in an upscale deli downtown.

What if smart mobs could empower entire populations to engage in peer-to-peer journalism? Imagine the impact of the Rodney King video multiplied by the people power of Napster. What if people beamed WearComp video to the Web, offering continuous views of breaking events that hitherto have been available only from Newscorp, AOL-Time-Wamer, and Disney? Would it be possible to turn the table on the surveillance society and counter the media monopolies? What would be the effect on public opinion if thousands of WearComp-equipped citizens webcast all they saw and heard? Wild as it sounds, mobile squads of citizen telejournalists have already surfaced. Whether today's experiments will even make it onto the radar of the media giants remains to be seen, but the first stirrings of p2p journalism have already been reported in Toronto and Tokyo.

In 2000, WearComp researcher, innovator, and evangelist Steve Mann launched "ENGwear, an experiment in wearable news-gathering systems conducted by students and researchers at the Humanistic Intelligence Lab at the University of Toronto."[40] In the spring of 2000, Mann and a group of his students, all wearing computers equipped with "EyeTaps," which broadcast everything they saw and heard to the Web, showed up at a demonstration in Toronto called by the Ontario Coalition Against Poverty (OCAP). Violence broke out. Mann reported, "We, along with the journalists and various television crews, ran for cover. However, unlike the reporters, my students and I were still broadcasting, capturing almost by accident the entire event. Whatever we saw before us was captured and sent instantly in real time to the World Wide Web, without our conscious thought or effort."[41]

Mann claims that the WearComp journalist-researchers who made their first appearance at the OCAP demonstration could be a model for a wider movement, which could influence as well as chronicle events:

> *WearComp represents a solution to this legacy of suppressed creativity and confining imagination in an age where ever-fewer sources of information seem to reach us, even as the conduits of information grow exponentially. What my students and I undertook in deciding to "cover" the OCAP protest was an experiment in media diversification. This is the process by which we merge our cyborg narratives with the demands of a growing cyberspace that we should, and one day will, be able to interact with and control. Facilitating the individual's creation and broadcast of their own narratives and perspectives is an important part of wearable computing technology. . . .*
>
> *What my students and I did—and continue to do—is something far more important than just providing "home movies" and "alternative" images for viewing on the Internet. We are also engaging in a process of cultural reclamation, where the individual is put back into the loop of information production and dispensation.*[42]

Justin Hall, the journalist who helped me interview Shibuya youth, recently reported that Tokyo's G3 videophones, like the one I carried around Tokyo, make it possible to send video to a Web site in real time: "With the technology in place," wrote Hall, "it's only a matter of time before an important amateur news video is directly distributed to the web, or to ten friends with video-mail in a news chain letter. When that happens, this new form of news distribution will become the news, and then ultimately, it will be no big deal."[43]

Hall reported that some of the videophones offered digital editing capabilities and that a new service in Japan made it possible to post photos and text on the Web directly from a mobile phone. People already use weblogging software to "blog" in real-time from conferences and conventions, continuously updating their Web pages through 802.11b connections.[44] Putting cameras and high-speed Net connections into telephones, however, moves blogging to the streets. By the time this book is published, I'm confident that street bloggers will have constructed a worldwide culture.

Mobile Ad Hoc Social Networks

Imagine my excitement, many months into my smart mob odyssey, when I came across a research report titled "When Peer-to-Peer Comes Face-to-Face: Collaborative Peer-to-Peer Computing in Mobile Ad Hoc Networks," from the "Wearable Computing Group" at the University of Oregon.[45] The Oregon group, assembled by Professor Zary Segall and led by Gerd Kortuem, had designed a test bed for smart mobs around the same time I began to believe such a development was possible.

"Mobile ad hoc social network" is a longer, more technical term than "smart mob." Both terms describe the new social form made possible by the combination of computation, communication, reputation, and location awareness. The *mobile* aspect is already self-evident to urbanites who see the early effects of mobile phones and SMS. *Ad hoc* means that the organizing among people and their devices is done informally and on the fly, the way texting youth everywhere coordinate meetings after school. *Social network* means that every individual in a smart mob is a "node" in the jargon of social network analysis, with social "links" (channels of communication and social bonds) to other individuals. Nodes and links, the elements of social networks made by humans, are also the fundamental elements of communication networks constructed from optical cables and wireless devices—one reason why new communication technologies make possible profound social changes.

The Wearable Computing Group specializes in exploring the community aspects of wireless, wearable, and peer-to-peer technologies. Kortuem agreed with my assessment when I called him to talk about the research at the University of Oregon. "When I talk about community," he told me, "I mean both the users who form social networks when they interact personally and communities of developers, like the open source community, where each member shares ideas and contributes to building something larger."[46] In Oregon, Toronto, Pittsburgh, Atlanta, Palo Alto, and Tokyo, small bands of researchers are beginning to walk around the same geographic neighborhoods while wearing intercommunicating computers.

Kortuem and colleagues realized that p2p computing and wireless networking technologies made it possible to design ad hoc networks of mobile devices to support the ad hoc social networks of the people who wear them. The fundamental technical unit cited by Kortuem and other wearable computing researchers has come to be known as the "personal area network," an interconnected network of devices worn or carried by the user. The concept was first described by Tom Zimmerman, now at IBM's Almaden Research Center, who had invented the VR "dataglove" while he was an MIT student.[47]

Kortuem and colleagues treat the personal area networks as building blocks of a dynamic community of networks with emergent capabilities of its, own. The research is as much behavioral as it is computational, beginning with simple experiments matching properties of mobile computing networks with the needs of social networks. The community of personal area network users within geographic proximity, for example, could serve as a wireless mesh network, dynamically self-organizing a cloud of broadband connectivity as nodes came in and out of physical proximity, providing always-on Internet connections to members. Using Bluetooth and other short-range wireless technologies such as very-low-power wideband radio, individual members of the community could engage in more intimate and timely information exchanges when face to face, whereas WiFi technologies could provide the infrastructure for neighborhood-wide and Internet-wide communication:

> *Mobile ad hoc systems provide opportunities for ad hoc meetings, mobile patient monitoring, distributed command and control systems and ubiquitous computing. In particular, personal area networks enable the creation of proximity-aware applications in support of face-to-face collaboration.*
>
> *Mobile devices like cell phones, PDAs and wearable computers have become our constant companions and are available wherever we go. . . . Personal area networks open the opportunity for these devices to take part in our everyday social interactions with people. Their ability to establish communication links among devices during face-to-face encounters can be used to facilitate, augment or even promote human social interactions.*
>
> *In some sense, an ad hoc mobile information system is the ultimate peer-to-peer system. It is self-organizing, fully decentralized, and highly dynamic.*[48]

Short-range radio frequency links such as those used by Bluetooth chips and wearable computers create a sphere of connectivity within the immediate vicinity of the wearer. Paul Rankin, at Philips Research laboratory in England, wrote about the need for intermediary agents to negotiate transactions between the "aura" of one person and radio beacons in the environment, or another person's aura.[49] "Auranet" is what Jay Schneider, Kortuem, and colleagues named their "framework for structuring encounters in social space based on reputations and trust."[50] The wireless instantiation of a 12-foot information bubble around wearable computer users is a physical model of what sociologist Erving Coffman calls the "Interaction Order," the part of social life where face-to-face and spoken interactions occur.[51] Goffman claimed that the mundane world of everyday interactions involves complex symbolic exchanges, visible but rarely consciously noticed, which enable groups to negotiate movement through public spaces. Although people use the ways they present themselves to "give" information they want others to believe about themselves, Goffman noted that people also "give off" information, leaking true but uncontrolled information along with their more deliberate performance.

One form of information that people give off, called "stigma" by Goffman, is markings or behaviors that locate individuals in a particular social status. Although many stigma can have negative connotations, stigma can also mark positive social status. The information we give off by the way we behave and dress helps us coordinate social interaction and identify likely interaction partners. When the Interaction Order is formalized and modeled automatically in an Auranet, the social network and the technological network meet in a way that makes possible new capabilities such as automated webs of trust for ad hoc interactions—for example, assembling a carpool of trustworthy strangers when you drive downtown or seek a ride.

Kortuem et al., noting the lack of fully embodied "human moments" in purely virtual worlds, concentrated on ways to enhance the most basic sphere of human social behavior, the face-to-face encounters of everyday life. Indeed, the primary question asked by the Oregon researchers is the primary question regarding smart mobs: What can communities of wearable computer users do in their face-to-face encounters? At a technical level, the wearable devices can share bandwidth by acting as nodes in an ad hoc wireless network. The devices could exchange media and messages, similar to the way Napster and Usenet use links between individual nodes to pass data around. However, as soon as the members of the community allow their computers to exchange data automatically, without human intervention, complex issues of trust and privacy intervene—the unspoken norms of the interaction order. Kortuem et al. explored the social and technical implications of personal agent software, which filters, shields, and acts as a go-between for their users.

A number of social and technical barriers must be overcome in order for mobile ad hoc communities to self-organize cooperatively. Nobody is going to contribute their personal area network to a community internetwork unless they feel secure about privacy and trust—who snoops whom, and who can be counted on to deal honestly? Privacy requires data security, and security is complicated by wireless communications. Encryption techniques make secure wearable community infrastructure possible, but someone has to figure out how to build them. Trust means a distributed reputation system, which the Oregon group has prototyped. When you break down the interesting idea of mobile, ad hoc social networks into the elements needed to make it work in practice, a rich and largely undeveloped field for research opens.

Another experiment by the Eugene group mediates social encounters by comparing personal profiles automatically and alerting participants in a face-to-face encounter of mutual interests or common friends that they might not know about (a recommendation system for strangers).[52] Each social encounter of wearable computer users involving automatic exchanges of personal data, sharing of bandwidth, or passing of messages from others would necessarily involve individual computations of where each participant's self-interest lies in relation to a computation of the other party's trustworthiness. Kortuem et al. recognized this complex weighing of trust versus self-interest as an example of our old friend, the Prisoner's Dilemma, and designed an experimental system called WALID to test some of these issues, taking advantage of the fact that the Oregon wearable computing researchers lived and worked in the same general neighborhood in Eugene, Oregon:

> *WALID implements a digitized version of the timeworn tradition of borrowing butter from your neighbor. You do a favor for others because you know that one day they will do it for you.*
>
> *With WALID two individuals use their mobile devices to negotiate about and to exchange real world tasks: dropping off someone's dry cleaning, buying a book of stamps at the post office, or returning a book to the local library.*
>
> *WALID employs personal agent software to find close-by community members and to negotiate the exchange of tasks. The agents maintain a user's task list, become fully aware of the locations and activities involved. When an encounter occurs, the agents produce a negotiation. If both users approve, a deal is struck.*
>
> *The role of the agent in a negotiation is to evaluate the value of favors and to keep scores. Having to run across town just to drop off someone's mail compares unfavorably with buying milk for someone if the grocery store is just a block away. Agents employ ideas from game theory to ensure that results of negotiations are mutually beneficial; they cooperate only if there is the opportunity to enhance the user's goals.*[53]

In our telephone conversation, Kortuem noted that at the beginning of wearable computing research, the main goals involved either creating tools for professionals, such as maintenance and repair specialists, or creating tools to augment individuals, in the manner promoted by Steve Mann. "I came to realize," Kortuem told me, "that what is really interesting is not the technology of a specialized application at a job site, but what happens if ordinary people are empowered to use this technology and what effects might emerge when technology penetrates society."[54] These words will be worth remembering when millions of people carry devices that invisibly probe and cloak, reach out, evaluate, interconnect, negotiate, exchange, and coordinate invisible acts of ad hoc cooperation that create wealth, democracy, education, surveillance, and weaponry from pure mind-stuff, the way the alchemy of inscribing ever-tinier patterns on purified sand invokes the same forces from the same place.

Swarm Intelligence and the Social Mind

Massive outbreaks of cooperation precipitated the collapse of communism. In city after city, huge crowds assembled in nonviolent street demonstrations, despite decades of well-founded fear of political assembly. Although common sense leads to the conclusion that unanimity of opinion among the demonstrators explained the change of behavior, Natalie Glance and Bernardo Huberman, Xerox PARC researchers who have studied the dynamics of social systems, noted that a *diversity* of cooperation thresholds among the individuals can tip a crowd into a sudden epidemic of cooperation. Glance and Huberman pointed out that a minority of extremists can choose to act first, and if the conditions are right, their actions can trigger actions by others who needed to see somebody make the first move before acting themselves—at which point the bandwagon-jumpers follow the early adopters who followed the first actors:

> *Those transitions can trigger a cascade of further cooperation until the whole group is cooperating.*
>
> *The events that led to the mass protests in Leipzig and Berlin and to the subsequent downfall of the East German government in November 1989 vividly illustrate the impact of such diversity on the resolution of social dilemmas. . . . The citizens of Leipzig who desired a change of government faced a dilemma. They could stay home in safety or demonstrate against the government and risk arrest—knowing that as the number of demonstrators rose, the risk declined and the potential for overthrowing the regime increased.*
>
> *A conservative person would demonstrate against the government only if thousands were already committed; a revolutionary might join at the slightest sign of unrest. That variation in threshold is one form of diversity. People also differed in their estimates of the duration of a demonstration as well as in the amount of risk they were willing to take. Bernhardt Prosch and Martin Abram, two sociologists from Erlangen University who studied the Leipzig demonstrations, claim that the diversity in thresholds was important in triggering the mass demonstrations.*[55]

Sudden epidemics of cooperation aren't necessarily pleasant experiences. Lynch mobs and entire nations cooperate to perpetrate atrocities. Decades before the fall of communism, sociologist Mark Granovetter examined radical collective behavior of both positive and negative kinds and proposed a "threshold model of collective behavior." I recognized Granovetter's model as a crucial conceptual bridge that connects intelligent (smart mob) cooperation with "emergent" behaviors of unintelligent actors, such as hives, flocks, and swarms.

Granovetter studied situations in which individuals were faced with either-or decisions regarding their relationship to a group—whether or not to join a riot or strike, adopt an innovation, spread a rumor, sell a stock, leave a social gathering, migrate to a different country. He identified the pivotal statistic as the proportion of *other* people who have to act before an individual decides to join them. Thresholds appear to be an individual reaction to the dynamics of a group.

One of Granovetter's statements yielded a clue to smart mob dynamics: "By explaining paradoxical outcomes as the result of aggregation processes, threshold models take the 'strangeness' often associated with collective behavior out of the heads of actors and put it into the dynamics of situations."[56] Smart mobs might also involve yet-unknown properties deriving from the dynamics of situations, not the heads of actors. Goffman's Interaction Order, the social sphere in which complex verbal and nonverbal communications are exchanged among individuals in real time, is precisely where individual actions can influence the action thresholds of crowds. Mobile media that can augment the informal, mostly unconscious information exchanges that take place within the Interaction Order, or affect the size or location of the audience for these exchanges, have the potential to change the threshold for collective action.

I started looking for ways to connect these congruent ideas operationally. How would they map onto an ad hoc social network of wearable computer users, for example? When my idea hunting brought me to "the coordination problem," a social dilemma that is *not* a Prisoner's Dilemma, separate ideas began to fit together into a larger pattern.

A coordination problem does not involve the Prisoner's Dilemma zero-sum game between self-interest and common resources but instead represents the quandary that confronts individuals who are ready to cooperate, but whose cooperation is contingent on the prior cooperation of others. Monitoring and sanctioning are important not simply as a way of punishing rule breakers but also as a way of assuring members that others are using common resources wisely. That is, many people are contingent cooperators, willing to cooperate as long as most others do (what Ostrom referred to as a "commitment problem"). Thus, monitoring and sanctioning serve the important function of providing information about others' actions and levels of commitment.

In *Rational Ritual: Culture, Coordination, and Common Knowledge,* Michael Suk-Young Chwe claims that public rituals are "social practices that generate common knowledge," which enables groups to solve coordination problems. Suk-Young Chwe writes: "A public ritual is not just about the transmission of meaning from a central source to each member of an audience; it is also about letting audience members know what other audience members know."[57] Everyone in a group has to know who else is contributing, free riding, and sanctioning in order to solve both free rider and coordination problems on the fly with maximum trust and minimum friction. This is the key to the group-cooperation leverage bestowed by reputation systems and many-to-many communications media.

Threshold models of collective action and the role of the Interaction Order are both about media for exchange of coordinating knowledge. Understanding this made it possible to see something I had not noticed clearly enough before—a possible connection between computer-wearing social networks of thinking, communicating humans and the swarm intelligence of unthinking (but also communicating) ants, bees, fish, and birds. Individual ants leave chemical trail markers, and the entire nest calculates the most efficient route to a food source from a hundred aggregated trails without direction from any central brain. Individual fish and birds (and tight-formation fighter pilots) school and flock simply by paying attention to what their nearest neighbors do. The coordinated movements of schools and flocks is a dynamically shifting aggregation of individual decisions. Even if there were a central tuna or pigeon who could issue orders, no system of propagating orders from a central source can operate swiftly enough to avoid being eaten by sharks or slamming into trees. When it comes to hives and swarms, the emergent capabilities of decentralized self-organization can be surprisingly intelligent.

What happens when the individuals in a tightly coordinated group are more highly intelligent creatures rather than simpler organisms like insects or birds? How do humans exhibit emergent behavior? As soon as this question occurred to me, I immediately recalled the story Kevin Kelly told at the beginning of *Out of Control,* his 1994 book about the emergent behaviors in biology, machinery, and human affairs.[58] He described an event at an annual film show for computer graphics professionals. A small paddle was attached to each seat in the auditorium, with reflective material of contrasting colors on each side of the paddle. The screen in the auditorium displayed a high-contrast, real-time video view of the audience. The person leading the exercise, computer graphics wizard Loren Carpenter, asked those on one side of the auditorium aisle to hold the paddles with one color showing and asked the other half of the audience to hold up the opposite color. Then, following Carpenter's suggestions, the audience self-organized a dot that moved around the screen, added a couple of paddles on the screen, and began to play a giant game of self-organized video Pong, finally creating a graphical representation of an airplane and flying it around the screen. Like flocks, there was no central control of the exercise after Carpenter made a suggestion. Members of the audience paid attention to what their neighbors were doing and what was happening on the screen. Kelly used this as an example of a self-conscious version of flocking behavior.[59]

Musician and cognitive scientist William Benzon believes that the graphical coordination exercise led by Carpenter and described by Kelly is similar to what happens when musicians "jam" and that it involves a yet unexplored synchronization of brain processes among the people involved:[60]

> *The group in Carpenter's story is controlling what appears on the screen. Everyone can see it all, but each can directly affect only the part of the display they control with his or her paddle. In jamming, everyone hears everything but can affect only that part of the collective sound that they create (or withhold).*
>
> *Now consider a different example. One of the standard scenes in prison movies goes like this: We're in a cell block or in the mess hall. One prisoner starts banging his cup on the table (or on one of the bars to his cell). Another joins in, then another, and another, until everyone's banging away and shouting some slogan in unison. This is a simple example of emergent behavior. But it's one that you won't find in chimpanzees. Yes, you will find them involved in group displays where they're all hooting and hollering and stomping. But the synchrony isn't as precise as it is in the human case.*
>
> *And that precision is critical to my argument. That precision allows me to treat the human group as a collection of coupled oscillators. Oscillation is one of the standard and simplest emergent phenomena. Once a group has become coupled in oscillation, we can treat the group as a single entity. To be sure, there's more to music than simple oscillation. But oscillation is the foundation, the starting point, and all the elaboration and complexities take place within this framework.*
>
> *In effect, in musical performance (and in dance), communication between individuals is pretty much the same as communication between components of a single nervous system. It's continuous and two-way, and it does not involve symbolic mediation. Think of Goffman's interaction order, but drop verbal communication from it. It is a public space that is physically external to the brains of participating individuals, but it is functionally internal to those brains.*[61]

Kevin Kelly traced back the new theories regarding emergent properties to William Morton Wheeler, an expert in the behavior of ants.[62] Wheeler called insect colonies "superorganisms" and defined the ability of the hive to accomplish tasks that no individual ant or bee is intelligent enough to do on its own as "emergent properties" of the superorganism. Kelly drew parallels between the ways both biological and artificial "vivisystems" exhibit the same four characteristics of what he called "swarm systems":

- The absence of imposed centralized control
- The autonomous nature of subunits
- The high connectivity between the subunits
- The webby nonlinear causality of peers influencing peers[63]

Steven Johnson's 2001 book, *Emergence,* shows how the principles that Kelly extrapolated from biological to technological networks also apply to cities and Amazon.com's recommendation system: "In these systems, agents residing on one scale start producing behavior that lies on one scale above them: ants create colonies; urbanites create neighborhoods; simple pattern-recognition software learns how to recommend new books. The movement from low-level rules to higher level sophistication is what we call emergence."[64] In the case of cities, although the emergent intelligence resembles the ant-mind, the individual units, humans, possess extraordinary onboard intelligence—or at least the capacity for it.

At this point, connections between the behavior of smart mobs and the behavior of swarm systems must be tentative, yet several of the earliest investigations have shown that the right kinds of online social networks know more than the sum of their parts: Connected and communicating in the right ways, populations of humans can exhibit a kind of "collective intelligence." In the summer between my smart mob inquiries in Scandinavia and my expedition to Tokyo, my inquiries brought me to a fellow who seems to have discovered the underpinnings of group intelligence. Bernardo Huberman, formerly at Xerox PARC, now scientific director of Hewlett-Packard's Information Dynamics research laboratory, was doing intriguing research on the emergence of primitive forms of collective intelligence.

I visited Huberman in his office, located in the same Palo Alto complex as the CoolTown laboratory. Huberman is a master of thinking of new ways of looking at familiar phenomena, seeing computer networks as ecologies, markets as social computers, and online communities as social minds. Originally a physicist, Huberman presents his findings in pages of mathematical equations. When I visited him in his office, he seriously agreed that "the Internet enables us to building collective intelligence."[65] At PARC, he had directed investigations of "the ecology of computation." As soon as I told him about smart mobs, he jumped up and exclaimed, "The social mind!" And he dug out a chapter on "The Social Mind" that he had published in 1995. Huberman thought it useful to think of emergent intelligence as a social computation:

> *Intelligence is not restricted to single brains; it also appears in groups, such as insect colonies, social and economic behavior in human societies, and scientific and professional communities. In all these cases, large numbers of agents capable of local tasks that can be conceived of as computations, engage in collective behavior which successfully deals with a number of problems that transcend the capacity of any individual to solve. . . . When large numbers of agents capable of symbolic-processing interact with each other, new universal regularities in their overall behavior appear. Furthermore, these regularities are quantifiable and can be experimentally tested.*[66]

The interesting statement is the last one. There have been varieties of theories about the Internet as the nervous system of a global brain, but Huberman and colleagues have made clever use of markets and game simulations as computational test beds for experiments with emergent group intelligence. The fall that I visited Huberman, he and his colleagues had used "information markets" to perform experiments in emergent social intelligence and found that group forecasts were more accurate than those of any of the individual participants' forecasts.[67] In information markets, members trade symbolic currency representing predictions of public information. The Hollywood Stock Exchange, for example, uses the market that emerges from the trading of symbolic shares to predict box office revenues and Oscar winners. The HP research team makes the extraordinary claim that they have created a mathematically verifiable methodology for extracting emergent intelligence from a group and using the group's knowledge to predict the future in a limited but useful realm: "One can take past predictive performance of participants in information markets and create weighting schemes that will predict future events, even if they are not the same event on which the performance was measured."[68]

Decades ago, computer scientists thought that someday there would be forms of "artificial intelligence," but with the exception of a few visionaries, they never thought in terms of computer-equipped humans as a kind of social intelligence. Although everyone who understands the use of statistical techniques to make predictions hastens to add the disclaimer that surprises are inevitable, and one of the fundamental characteristics of complex adaptive systems is their unpredictability, the initial findings that internetworked groups of humans can exhibit emergent prediction capabilities are potentially profound.

Another research group that takes emergent group intelligence seriously is the laboratory at Los Alamos, where a group of "artificial life" researchers issued a report in 1998, "Symbiotic Intelligence: Self-Organizing Knowledge on Distributed Networks, Driven by Human Interaction."[69] The premise of this interdisciplinary team is based on the view proposed by some in recent years that human society is an adaptive collective organism and that social evolution parallels and unfolds according to the same dynamics as biological evolution.[70] According to this theory, which I will revisit in the next chapter, new knowledge and new technologies have made possible the evolution of the maximum size of the functioning social group from tribes to nations to global coalitions. The knowledge and technologies that triggered the jump from clan to tribe to nation to market to network all shared one characteristic: They each amplified the way individual humans think and communicate, and magnified their ability to share what they know.

The Los Alamos team, looking at some of the same characteristics of the Internet that Huberman and his colleagues investigated and citing a range of research that has only recently begun to emerge as a discipline, claim that "self-organizing social dynamics has been an unappreciated positive force in our social development and has been significantly extended, at least in scope, by new technologies."[71] The Los Alamos group cited evidence for their hypothesis that the self-organizing social systems that have driven human social evolution will be enhanced by self-organized, distributed, information and communication systems. The research conducted directly by the Los Alamos researchers reinforced Huberman et al.'s claim that groups of humans, linked through online networks, can make collective decisions that prove more accurate than the performance of the best individual predictors in the group. If it isn't a dead end, the lines of research opened by Huberman's team, the Los Alamos researchers, and others could amplify the powers of smart mobs into entirely new dimensions of possibility, the way Moore's Law amplified the powers of computer users.

Will self-organized, ad hoc networks of computer wearers, mediated by privacy-protecting agents, blossom into a renaissance of new wealth, knowledge, and revitalized civil society, or will the same technological-social regime provide nothing more than yet another revenue stream for Disinfotainment, Inc?

Or is that the wrong question? Given the direction of the technological, economic, and political changes I have touched on so far, I propose the following questions:

- What do we know now about the emergent properties of ad hoc mobile computing networks, and what do we need to know in the future?
- What are the central issues for individuals in a world pervaded by surveillance devices—in terms of what we can *do* about it?
- What are the long-term consequences of near-term political decisions on the way we'll use and be affected by mobile, pervasive, always-on media?

I hope that the understandings I've shared from my investigations of the past two years make it clear that smart mobs aren't a "thing" that you can point to with one finger or describe with two words, any more than "the Internet" was a "thing" you could point to. The Internet is what happened when a lot of computers started communicating. The computer and the Internet were designed, but the ways people used them were not designed into either technology, nor were the most world-shifting uses of these tools anticipated by their designers or vendors. Word processing and virtual communities, eBay and e-commerce, Google and weblogs and reputation systems *emerged.* Smart mobs are an unpredictable but at least partially describable emergent property that I see surfacing as more people use mobile telephones, more chips communicate with each other, more computers know where they are located, more technology becomes wearable, more people start using these new media to invent new forms of sex, commerce, entertainment, communion, and, as always, conflict.

Notes

Epigraph: Vincente Rafael, "The Cell Phone and the Crowd: Messianic Politics in Recent Phillipine History," 13 June 2001. http://communication.ucsd.edu/people/f_rafael.cellphone.html.

1. Michael Bociurkiw, "Revolution by Cell Phone," *Forbes,* 10 September 2001, <http://www.forbes.com/asap/2001/0910/028.html> (1 March 2002).
2. Ibid.
3. Paul de Armond, "Black Flag Over Seattle," *Albion Monitor* 72, March 2000, <http://www.monitor.net/monitor/seattlewto/index.html> (1 March 2002).
4. Alexander MacLeod, "Call to Picket Finds New Ring in Britain's Fuel Crisis," *Christian Science Monitor,* 19 September 2000. See also: Chris Marsden, "Britain's Labour Government and Trade Union Leaders Unite to Crush Fuel Tax Protest," *World Socialist Web Site,* 15 September 2000, <http://www.wsws.org/articles/2000/sep2000/fuel-s15.shtml> (1 March 2002).
5. Steve Mann and Hal Niedzviecki, *Cyborg: Digital Destiny and Human Possibility in the Age of the Wearable Computer* (Mississauga: Doubleday Canada, 2001), 177–178.
6. Critical Mass, <http://www.critical-mass.org/> (6 March 2002).
7. Anne Torres, "4 SME, Txtng is Lyf," *TheFeature.com,* 18 April 2001, <http://www.thefeature.com> (1 March 2002).
8. Bociurkiw, "Revolution by Cell Phone."
9. Rafael, "The Cell Phone and the Crowd."
10. Ibid.
11. Arturo Bariuad, "Text Messaging Becomes a Menace in the Philippines," *Straits Times,* 3 March 2001.
12. Wayne Arnold, "Manila's Talk of the Town Is Text Messaging," *New York Times,* 5 July 2000, C1.
13. Bariuad, "Text Messaging Becomes a Menace."
14. Rafael, "The Cell Phone and the Crowd."
15. Ibid.
16. Richard Lloyd Parry, "The TXT MSG Revolution," *Independent Digital,* 23 January 2001, <http://www.independent.co.uk/story.jsp?story=51748> (1 March 2002).
17. Rafael, "The Cell Phone and the Crowd."
18. de Armond, "Black Flag Over Seattle."
19. David Ronfeldt and John Arquilla, "Networks, Netwars, and the Fight for the Future," *First Monday* 6, 10 (October 2001), <http://firstmonday.org/issues/issue6_10/ronfeldt/index.html> (1 March 2002).
20. John Arquilla and David Ronfeldt, eds., *Networks and Netwars: The Future of Terror, Crime, and Militancy* (Santa Monica, Calif.: RAND, 2001).
21. Jim Lai, "The Future of Infantry," *Mindjack* 28, January 2002, <http://www.mindjack.com/feature/landwarrior.html> (1 March 2002).
22. Ian Sample, "Military Palmtop to Cut Collateral Damage," *New Scientist,* 9 March 2002, <http://www.newscientist.com/news/news.jsp?id=ns99992005> (29 March 2002).
23. Arquilla and Ronfeldt, eds., *Networks and Netwars,* 310–313.
24. Alexis de Tocqueville, *Democracy in America,* ed. J. P. Mayer and Max Lerner, trans. George Lawrence (New York: Harper and Row, 1966), 483–484.
25. Elinor Ostrom, *Governing the Commons: The Evolution of Institutions for Collective Action* (Cambridge: Cambridge University Press, 1990).
26. Nick Montfort, "My Pager, My Matchmaker," *Ziff Davis Smart Business,* 7 July 2000, <http://techupdate.zdnet.com/techupdate/stories/main/0,14179,2577889,00.html> (1 March 2002).
27. "Bleep at First Sight," *Reuters,* 15 May 1998, <http://www.wired.com/news/news/culture/story/12342.html> (1 March 2002).

28. Craig Wilson, "'Gaydar' Device Clears Up Mixed Signals," *USA Today,* 25 February 2000, <http://www.usatoday.com/life/cyber/tech/review/crg942.htm> (1 March 2002).
29. Montfort, "My Pager, My Matchmaker."
30. ImaHima, <http://www.imahima.com/index.jsp> (1 March 2002).
31. Diego Ibarguen, "Tracking Celebrities Via Cell Phones, Web Sites," *San Francisco Chronicle,* 21 January 2001.
32. Justin Hall, "Mobile Reporting: Peer-to-Peer News," *TheFeature.com,* 20 February 2002, <http://www.thefeature.com/printable.jsp?pageid=14274> (1 March 2002).
33. Mann and Niedzviecki, *Cyborg.*
34. Gerd Kortuem et al., "Close Encounters: Supporting Mobile Collaboration through Interchange of User Profiles," in *Proceedings of the First International Symposium on Handheld and Ubiquitous Computing (HUC99),* 1999, Karlsruhe, Germany, <http://www.cs.uoregon.edu/research/wearables/Papers/HUC99-kortuem.pdf> (6 March 2002).
35. Neeraj Jhanji, interview by author, October 2001, Tokyo.
36. ImaHima Press Releases, <shttp://www.imahima.com/en/news3.html> (1 March 2002).
37. Deborah Mendez-Wilson, "Users Turned to Pagers When Phone Lines Went Down in Crisis," *Wireless Week,* 24 September 2001, <http://www.wirelessweek.com/index.asp?layout=story&articleId=CA160205> (1 March 2002).
38. John Geirland, "Mobile Community," *TheFeature.com,* 24 September 2001, <http://thefeature.com/article.jsp?pageid=12836> (1 March 2002).
39. Gordon Gould, Alex Levine, and Andrew Pimentel, interview by author, November 2001, New York.
40. ENGwear: Wearable Wireless Systems for Electronic News Gathering, <http://www.eyetap.org/hi/ENGwear/> (1 March 2002).
41. Mann and Niedzviecki, *Cyborg,* 175–176.
42. Ibid., 177–178.
43. Hall, "Mobile Reporting."
44. Henry Jenkins, "Digital Renaissance," *Technology Review,* March 2002, <http://www.technologyreview.com/articles/jenkins0302.asp> (24 February 2002).
45. Gerd Kortuem et al., "When Peer-to-Peer Comes Face-to-Face: Collaborative Peer-to-Peer Computing in Mobile Ad Hoc Networks," 2001 International Conference on Peer-to-Peer Computing (P2P2001), 27–29 August 2001, Linköping, Sweden, <http://www.cs.uoregon.edu/research/wearables/Papers/p2p2001.pdf> (6 March 2002).
46. Gerd Kortuem, telephone interview by author, 27 February 2002.
47. T. G. Zimmerman, "Personal Area Networks: Near-Field Intrabody Communication," *IBM Systems Journal* 35, 3&4 (1996), <http://www.research.ibm.com/journal/sj/mit/sectione/zimmerman.html> (6 March 2002).
48. Kortuem et al., "When Peer-to-Peer Comes."
49. Paul Rankin, "Context-Aware Mobile Phones: The Difference Between Pull and Push, Restoring the Importance of Place," Philips Research Laboratories, Redhill, Surrey, U.K.
50. Jay Schneider et al., "Auranet: Trust and Face-to-Face Interactions in a Wearable Community," Technical Report WCL-TR15, July 2001, <http://www.cs.uoregon.edu/research/wearables/Papers/auranet.pdf> (6 March 2002).
51. Erving Goffman, *The Presentation of Self in Everyday Life* (Garden City, N.Y.: Doubleday, 1959).
52. Jay Schneider et al., "Disseminating Trust Information in Wearable Communities," 2nd International Symposium on Handheld and Ubiquitous Computing (HUC2K), 25–27 September

2000, Bristol, England, <http://www.cs.uoregon.edu/research/wearables/Papers/HUC2K.pdf> (6 March 2002).

53. Gerd Kortuem et al., "When Cyborgs Meet: Building Communities of Cooperating Wearable Agents," *Proceedings Third International Symposium on Wearable Computers,* 18–19 October 1999, San Francisco, California, <http://www.computer.org/proceedings/iswc/0428/04280124abs.htm> (6 March 2002).
54. Kortuem, telephone interview by author.
55. Natalie S. Glance and Bernardo A. Huberman, "The Dynamics of Social Dilemmas," *Scientific American,* March 1994, 76–81.
56. Mark Granovetter, "Threshold Models of Collective Behavior," *American Journal of Sociology* 83, 6 (1978): 1420–1443.
57. Michael Suk-Young Chwe, *Rational Ritual: Culture, Coordination, and Common Knowledge* (Princeton: Princeton University Press, 2001), <http://www.chwe.net/michael/r.pdf> (6 March 2002).
58. Kevin Kelly, *Out of Control* (Reading, Mass.: Addison-Wesley, 1994), <http://www.kk.org/outofcontrol/index.html> (6 March 2002).
59. Ibid.
60. William Benzon, *Beethoven's Anvil: Music in Mind and Culture* (New York: Basic Books, 2001).
61. William Benzon, email interview by author.
62. William Morton Wheeler, *Emergent Evolution and the Development of Societies* (New York: W. W. Norton, 1928).
63. Ibid.
64. Steven Johnson, Emergence: The Connected Lives of Ants, Brains, Cities, and Software (New York: Scribner, 2001).
65. Bernardo Huberman, interview by author, October 2001, Palo Alto, California.
66. Bernardo A. Huberman, "The Social Mind," in *Origins of the Human Brain,* ed. Jean-Pierre Changeux and Jean Chavaillon (Oxford: Clarendon Press, 1995), 250.
67. Kay-Yut Chen, Leslie R. Fine, and Bernardo Huberman, "Forecasting Uncertain Events with Small Groups," HP Laboratories, Palo Alto, California, 3 August 2001, <http://papers.ssrn.com/sol3/papers.cfm?abstract_id=278601> (6 March 2002).
68. Ibid.
69. Norman Johnson et al., "Symbiotic Intelligence: Self-Organizing Knowledge on Distributed Networks, Driven by Human Interaction," in *Artificial Life VI: Proceedings of the Sixth International Conference on Artificial Life* (*Complex Adaptive Systems,* No. 6), ed. C. Adami, R. Belew, H. Kitano, and C. Taylor (Cambridge: Bradford Books/MIT Press, 1998).
70. George Dyson, *Darwin Among the Machines: The Evolution of Global Intelligence* (Reading, Mass.: Addison-Wesley, 1997).
71. Johnson et al., "Symbiotic Intelligence."

Interactive Audiences? The "Collective Intelligence" of Media Fans

Henry Jenkins

If "Star Trek *Rerun, Reread, Rewritten" represented my first public airing of the ideas in* Textual Poachers, *"Interactive Audiences" was my first attempt to layout the reconceptualization of fandom that would shape* Convergence Culture. *The goal I set for myself with "Interactive Audiences" was to write about fans without once mentioning Michel de Certeau. We should change our theory every five thousand miles just like we change oil in our cars. New injections improve performance and keep us from clogging up the system. I am frustrated that despite a growing number of younger scholars writing about fans, many still operate primarily in relation to the paradigms from the late 1980s and early 1990s. There are so many other potential ways of looking at the topic.*

When my friend Christopher Weaver handed me a copy of Pierre Levy's Collective Intelligence, *I realized that this approach addressed many of the questions I had trouble talking about in* Textual Poachers*—specifically the social dimensions of fan communities. Levy gave us a way of thinking about fandom not in terms of resistance but as a prototype or dress rehearsal for the way culture might operate in the future. Levy describes his vision of "collective intelligence" as an "achievable utopia"—not something that grows inevitably from the new configuration of technologies but rather something we must work toward and fight to achieve. Fandom is one of those spaces where people are learning how to live and collaborate within a knowledge community. We are trying out through play patterns of interaction that will soon penetrate every other aspect of our lives. Levy, in short, gives us a model for a fan-based politics.*

"Interactive Audiences?" first appeared in Dan Harries, ed., The New Media Book *(London: British Film Institute, 2002). Some dimensions of the convergence culture argument emerged in two other essays, "The Stormtroopers and the Poachers: Cultural Convergence in a Digital Age," in Phillipe Le Guern, ed.,* Les cultes médiatiques: culture fan et oeuvres cultes *(Rennes, France: Presses universitaires de Rennes, 2002), and "Quentin Tarantino's Star Wars? Digital Cinema, Media Convergence and Participatory Culture," in David Thorburn and Henry Jenkins, eds.,* Rethinking Media Change: The Aesthetics of Transition *(Cambridge, MA: MIT Press, 2003). (I should acknowledge that there are significant terminological shifts and rethinking between these three essays.)*

"You've Got Three Seconds. Impress Me."

An advertisement for Applebox Productions depicts the new youth consumer: his scraggly dishwater blonde hair hangs down into his glaring eyes, his chin is thrust out, his mouth is turned down into a challenging sneer, and his finger posed over the remote. One false move and he'll zap us. He's young, male, and in control. No longer a couch potato, he determines what, when, and how he watches media. He is a media consumer, perhaps even a media fan, bur he is also a media producer, distributor, publicist, and critic. He's the poster child for the new interactive audience.

The advertisement takes for granted what cultural studies researchers struggled to establish throughout the 1980s and 1990s—that audiences were active, critically aware, and discriminating. Yet, this advertisement promises that Applebox Productions has developed new ways to overcome his resistance and bring advertising messages to this scowling teen's attention. The interactive audience is not autonomous; it still operates alongside powerful media industries.

If the current media environment makes visible the once invisible work of media spectatorship, it is wrong to assume that we are somehow being liberated through improved media technologies. Rather than talking about interactive technologies, we should document the interactions that occur among media consumers, between media consumers and media texts, and between media consumers and media producers. The new participatory culture is taking shape at the intersection between three trends:

1. New tools and technologies enable consumers to archive, annotate, appropriate, and recirculate media content;
2. A range of subcultures promote Do-It-Yourself (DIY) media production, a discourse that shapes how consumers have deployed those technologies; and
3. Economic trends favoring the horizontally integrated media conglomerates encourage the flow of images, ideas, and narratives across multiple media channels and demand more active modes of spectatorship.

In this essay, I will try to describe how these three trends have altered the way media consumers relate to each other, to media texts, and to media producers. In doing so, I hope to move beyond the either-or logic of traditional audience research—refusing to see media consumers as either totally autonomous from or totally vulnerable to the culture industries. It would be naive to assume that powerful conglomerates will not protect their own interests as they enter this new media marketplace, but at the same time, audiences are gaining greater power and autonomy as they enter into the new knowledge culture. The interactive audience is more than a marketing concept and less than "semiotic democracy."

Collective Intelligence

In *Collective Intelligence,* Pierre Levy offers a compelling vision of the new "knowledge space," or what he calls "the cosmopedia," that might emerge as citizens more fully realize the potentials of the new media environment. Rejecting technological or economic determinism, Levy sees contemporary society as caught in a transitional moment, the outcome of which is still unknown, but which has enormous potentials for transforming existing structures of knowledge and power. His book might best be read as a form of critical utopianism framing a vision for the future ("an achievable utopia"), offering an ethical yardstick for contemporary developments. Levy explores how the "deterritorialization" of knowledge, brought about by the ability of the net and the Web to facilitate rapid many-to-many communication, might enable broader participation in decision-making, new modes of citizenship and community, and the reciprocal exchange of information. Levy draws a productive distinction between organic social groups (families, clans, tribes), organized social groups (nations, institutions, religions, and corporations), and self-organized groups (such as the virtual communities of the Web). He links the emergence of the new knowledge space to the breakdown of geographic constraints on communication, of the declining loyalty of individuals to organized groups, and of the diminished power of nation-states to command the exclusive loyalty of their citizens. The new knowledge communities will be voluntary, temporary, and tactical affiliations, defined through common intellectual enterprises and emotional investments. Members may shift from one community to another as their interests and needs change, and they may belong to more than one community at the same time. Yet, they are held together through the mutual production and reciprocal exchange of knowledge. As Levy explains,

> *the members of a thinking community search, inscribe, connect, consult, explore. . . . Not only does the cosmopedia make available to the collective intellect all of the pertinent knowledge available to it at a given moment, but it also serves as a site of collective discussion, negotiation, and development. . . . Unanswered questions will create tension within cosmopedic space, indicating regions where invention and innovation are required.*[1]

Online fan communities might well be some of the most fully realized versions of Levy's cosmopedia, expansive self-organizing groups focused around the collective production, debate, and circulation of meanings, interpretations, and fantasies in response to various artifacts of contemporary popular culture. Fan communities have long defined their memberships through affinities rather than localities. Fandoms were virtual communities, "imagined" and "imagining" communities, long before the introduction of networked computers.[2] The history of science fiction fandom might illustrate how knowledge communities emerged. Hugo Gernsbeck, the pulp magazine editor who has been credited with helping to define science fiction as a distinctive genre in the 1920s and 1930s, was also a major advocate of radio as a participatory medium. Gernsbeck saw science fiction as a means of fostering popular awareness of contemporary scientific breakthroughs at a moment of accelerating technological development.[3] The letter column of Gernsbeck's *Astounding Stories* became a forum where laypeople could debate scientific theories and assess new technologies. Using the published addresses, early science fiction fans formed an informal postal network, circulating letters and amateur publications. Later, conventions facilitated the face-to-face contact between fans from across the country and around the world. Many of the most significant science fiction writers emerged from fandom. Given this history, every reader was understood to be a potential writer, and many fans aspired to break into professional publication; fan ideas influenced commercially distributed works at a time when science fiction was still understood predominantly as a micro-genre aimed at a small but passionate niche market. The fan-issued Hugo Award (named after Gernsbeck) remains the most valued recognition a science fiction writer can receive. This reciprocity among readers, writers, and editors set expectations as science fiction spread into film and television. *Star Trek* fans were, from the start, an activist audience, lobbying to keep its series on the air and later advocating specific changes in the program content to better reflect its own agendas. Yet, if fans were the primary readers for literary science fiction, they were only a small fraction of the audience for network television. Fans became, in John Tulloch's words, a "powerless elite," unable to alter the series content but actively reshaping the reception context through grassroots media production.[4] *Star Trek* fandom, in turn, was a model for other fan communities to create forums for debating interpretations, networks for circulating creative works, and channels for lobbying the producers.

Fans were early adopters of digital technologies. Within the scientific and military institutions where the Internet was first introduced, science fiction has long been a literature of choice.[5] Consequently, the slang and social practices employed on the early bulletin boards were often directly modeled on science fiction fandom. Mailing lists that focused on fan topics took their place alongside discussions of technological or scientific issues. In many ways, cyberspace is fandom writ large.

The reconstitution of these fandoms as digital enclaves did not come without strenuous efforts to overcome the often overtly hostile reception fan women received from the early Internet's predominantly male population. Operating outside of those technical institutions, many female fans lacked computer access and technical literacy. Heated debates erupted at conventions as fans were angered at being left behind when old fan friends moved online. At the same time, as Sue Clerc notes, fan communities helped many women make the transition to cyberspace; the group insured that valued members learned to use the new technologies, since "for them, there is little benefit to net access unless many of their friends have it."[6] Fan women routed around male hostility, developing Web communities "that combine the intimacy of small groups with a support network similar to the kind fan women create off-line." Discussion lists, mailing groups, Web rings, and chatrooms each enabled fan communication.

Nancy Baym has discussed the important functions of talk within online soap fandom: "Fans share knowledge of the show's history, in part, because the genre demands it. Any soap has broadcast more material than any single fan can remember."[7] Fans inform each other about program history or recent developments they may have missed. The fan community pools its knowledge because no single fan can know everything necessary to fully appreciate the series. Levy distinguishes between shared knowledge (which would refer to information known by all members of a community) and collective intelligence (which describes knowledge available to all members of a community). Collective intelligence expands a community's productive capacity because it frees individual members from the limitations of their memory and enables the group to act upon a broader range of expertise. As Levy writes, within a knowledge community, "no one knows everything, everyone knows something, all knowledge resides in humanity."[8] Baym argues:

> *A large group of fans can do what even the most committed single fan cannot: accumulate, retain, and continually recirculate unprecedented amounts of relevant information. . . . [Net list] participants collaboratively provide all with the resources to get more story from the material, enhancing many members' soap readings and pleasures.*[9]

Soap talk, Baym notes, allows people to "show off for one another" their various competencies while making individual expertise more broadly available. Fans are motivated by epistemaphilia—not simply a pleasure in knowing but a pleasure in exchanging knowledge. Baym argues that fans see the exchange of speculations and evaluations of soaps as a means of "comparing, refining, and negotiating understandings of their socioemotional environment."[10] Matthew Hills has criticized audience researchers for their preoccupation with fans' meaning production at the expense of consideration of their affective investments and emotional alliances.[11] Yet, as Baym's term "socio-emotional" suggests, meanings are not some abstracted form of knowledge, separated from our pleasures and desires, isolated from fandom's social bonds. When fans talk about meaningful encounters with texts, they are describing what they feel as much as what they think. . . . Fan speculations may, on the surface, seem to be simply a deciphering of the aired material, but increasingly speculation involves fans in the production of new fantasies, broadening the field of meanings that circulate around the primary text. . . .

Levy contrasts his ideal of "collective intelligence" with the dystopian image of the "hive mind," where individual voices are suppressed. Far from demanding conformity, the new knowledge culture is enlivened by multiple ways of knowing. This collective exchange of knowledge cannot be fully contained by previous sources of power—"bureaucratic hierarchies (based on static forms of writing), media monarchies (surfing the television and media systems), and international economic networks (based on the telephone and real-time technologies"—that depended on maintaining tight control over the flow of information. The dynamic, collective, and reciprocal nature of these exchanges undermines traditional forms of expertise and destabilizes attempts to establish a scriptural economy in which some meanings are more valuable than others.[12]

The old commodity space was defined through various forms of decontextualization, including the alienation of labor, the uprooting of images from larger cultural traditions so that they can circulate as commodities, the demographic fragmentation of the audience, the disciplining of knowledge, and the disconnect between media producers and consumers. The new information space involves multiple and unstable forms of recontextualization. The value of any bit of information increases through social interaction. Commodities are a limited good and their exchange necessarily creates or enacts inequalities. But meaning is a shared and constantly renewable resource and its circulation can create and revitalize social ties. If old forms of expertise operated through isolated disciplines, the new collective intelligence is a patchwork woven together from many sources as members pool what they know, creating something much more powerful than the sum of its parts.

How Computers Changed Fandom

For Levy, the introduction of high-speed networked computing constituted an epistemological turning point in the development of collective intelligence. If fandom was already a knowledge culture well before the Internet, then how did transplanting its practices into the digital environment alter the fan community? The new digital environment increases the speed of fan communication, resulting in what Matthew Hills calls "just in time fandom."[13] If fans once traded ideas through the mails, they now see the postal service as too slow—"snail mail"—to satisfy their expectations of immediate response. Hills explains, "The practices of fandom have become increasingly enmeshed with the rhythms and temporalities of broadcasting, so that fans now go online to discuss new episodes immediately after the episode's transmission time or even during ad-breaks perhaps in order to demonstrate the 'timeliness' and responsiveness of their devotion."[14] Where fans might have raced to the phone to talk to a close friend, they can now access a much broader range of perspectives by going online.

This expectation of timeliness complicates the global expansion of the fan community, with time lags in the distribution of cultural goods across national markets hampering full participation from fans that will receive the same program months or even years later. International fans often complain that they are additionally disadvantaged because their first-time experience of the episodes is spoiled by learning too much from the online discussions.

The digital media also alters the scope of communication. Fandoms centering on Asian popular culture, such as Japanese anime or Hong Kong action films, powerfully exploit the Internet's global reach. Japanese fans collaborate with American consumers to insure the underground circulation of these cultural products and to explain cultural references, genre traditions, and production histories.[15] Anime fans regularly translate and post the schedule of Japanese television so that international fans can identify and negotiate access to interesting programs. American fans have learned Japanese, often teaching each other outside of a formal educational context, in order to participate in grassroots projects to subtitle anime films or to translate manga (comics). Concerned about different national expectations regarding what kinds of animation are appropriate for children, anime fans have organized their own ratings groups. This is a new cosmopolitanism—knowledge sharing on a global scale.

As the community enlarges and reaction time shortens, fandom becomes much more effective as a platform for consumer activism. Fans can quickly mobilize grassroots efforts to save programs or protest unpopular developments. New fandoms emerge rapidly on the Web—in some cases before media products actually reach the market. As early participants spread news about emergent fandoms, supporters quickly develop the infrastructure for supporting critical dialogue, producing annotated program guides, providing regular production updates, and creating original fan stories and artwork. The result has been an enormous proliferation of fan Web sites and discussion lists. . . . As fandom diversifies, it moves from cult status toward the cultural mainstream, with more Internet users engaged in some form of fan activity.

This increased visibility and cultural centrality has been a mixed blessing for a community used to speaking from the margins. The speed and frequency of communication may intensify the social bonds within the fan community. In the past, fans inhabited a "week-end only world," seeing each other in large numbers only a few times a year at conventions.[16] Now, fans may interact daily, if not hourly, online. Geographically isolated fans can feel much more connected to the fan community and home-ridden fans enjoy a new level of acceptance. Yet, fandom's expanded scope can leave fans feeling alienated from the expanding numbers of strangers entering their community. This rapid expansion outraces any effort to socialize new members. For example, fandom has long maintained an ethical norm against producing erotica about real people rather than fictional characters. As newer fans have discovered fan fiction online, they have not always known or accepted this prohibition, and so there is a growing body of fan erotica dealing with celebrities. Such stories become a dividing point between older fans committed to traditional norms and the newer online fans who have asserted their rights to redefine fandom on their own terms.

Online fan discussion lists often bring together groups who functioned more or less autonomously offline and have radically different responses to the aired material. Flame wars erupt as their taken-for-granted interpretive and evaluative norms rub against each other. In some cases, fans can negotiate these conflicts by pulling to a metalevel and exploring the basis for the different interpretations. More often, the groups splinter into narrower interests, pushing some participants from public debates into smaller and more private mailing lists.

Levy describes a pedagogical process through which a knowledge community develops a set of ethical standards and articulates mutual goals. Even on a scale much smaller than Levy's global village, fandoms often have difficulty arriving at such a consensus. While early accounts of fandom stressed its communitarian ideals, more recent studies have stressed recurring conflicts. Andre MacDonald has described fandom in terms of various disputes—between male and female fans, between fans with different assumptions about the desired degree of closeness of the producers and stars, between fans who seek to police the production of certain fantasies and fans who assert their freedom from such constraints, between different generations of fans, and so forth.[17] MacDonald depicts a community whose utopian aspirations are constantly being tested against unequal experiences, levels of expertise, access to performers and community resources, control over community institutions, and degrees of investment in fan traditions and norms. Moreover, as Nancy Baym suggests, the desire to avoid such conflicts can result in an artificial consensus that shuts down the desired play with alternative meanings.[18] Levy seemingly assumes a perfect balance between mechanisms for producing knowledge and for sustaining affiliations. Yet, MacDonald and Baym suggest a constant tension between these two goals, which can reach a crisis as list memberships have expanded alongside the exponential growth of net subscribers. . . .

Networked computing has also transformed fan production. Web publication of fan fiction, for example, has almost entirely displaced printed zines. Fanzines arose as the most efficient means of circulating fan writing.[19] Fan editors charged only the costs of reproduction, seeing zines as a vehicle for distributing stories and not as a source of income. In some fandoms, circuits developed for loaning individually photocopied stories. In other cases, readers and editors came to see zines as aesthetic artifacts, insisting on high-quality reproduction and glossy color covers. Fans have increasingly turned to the Web to lower the costs of production and to expand their reading public. Fans are also developing archives of older zine stories, helping to connect newer fans with their history. . . .

Digital technologies have also enabled new forms of fan cultural production. Photoshop collage has become popular as a means of illustrating fan fiction, and now digital art may go to auction at cons (conventions) alongside illustrations done in pen and ink, colored pencil, or oil. For a time, mp3s of fan-generated music (filk) could be readily downloaded alongside commercial favorites through Napster. . . . Fan artists have been part of the much larger history of amateur film and video production. George Lucas and Steven Spielberg were themselves amateur filmmakers as teenagers, producing low-budget horror or science fiction movies. *Star Wars,* in turn, has inspired Super 8 filmmakers since its release in the early 1970s. Some British fan clubs produced original episodes of *Doctor Who,* sometimes filming in the same gravel quarries as the original series. As the videocassette recorder became more widely available, fans re-edited series footage into music videos, using popular music to encapsulate the often-unarticulated emotions of favorite characters.[20] As fan video makers have become more sophisticated, some fan artists have produced whole new storylines by patching together original dialogue.

The World Wide Web is a powerful distribution channel, giving what were once home movies a surprising degree of public visibility. Publicity materials surface while these amateur films arc still in production, most of the films boast lavish movie posters, and many of them include downloadable trailers to attract would-be viewers impatient with download times. *Star Wars* fans were among the first to embrace these new technologies, producing at last count more than three hundred Web movies.[21] These fans exploited the various merchandise surrounding this blockbuster film franchise for raw materials to their homegrown movies. . . . These fan filmmakers have used home computers to duplicate effects Lucasfilm had spent a fortune to achieve several decades earlier; many fan films create their own light saber or space battles. . . .

Knowledge Culture Meets Commodity Culture

Levy distinguishes between four potential sources of power—nomadic mobility, control over territory, ownership over commodities, and mastery over knowledge—and suggests a complex set of interactions and negotiations between them. The emergent knowledge cultures never fully escape the influence of the commodity culture, any more than commodity culture can totally function outside the constraints of territoriality. But knowledge cultures will, he predicts, gradually alter the ways that commodity culture operates. Nowhere is that transition clearer than within the culture: industries, where the commodities that circulate become resources for the production of meaning: "The distinctions between authors and readers, producers and spectators, creators and interpretations will blend to form a reading-writing continuum, which will extend from the machine and network designers to the ultimate recipient, each helping to sustain the activities of the others."[22]

Creative activity, he suggests, will shift from the production of texts or the regulation of meanings toward the development of a dynamic environment, "a collective event that implies the recipients, transforms interpreters into actors, enables interpretation to enter the loop with collective action."[23] Room for participation and improvisation are being built into new media franchises. Kurt Lancaster, for example, has examined how commercial works (including computer, role-playing, and card games) surrounding the cult science fiction series *Babylon 5* facilitate a diverse range of fan performances, allowing fans to immerse themselves in the fantasy universe.[24] . . . Cult works were once discovered; now they are being consciously produced, designed to provoke fan interactions. The producers of *Xena: Warrior Princess,* for example, were fully aware that some fans wanted to read Xena and Gabrielle as lesbian lovers and thus began to consciously weave "subtext" into the episodes. As Levy explains, "The recipients of the open work are invited to fill in the blanks, choose among possible meanings, confront the divergences among their interpretations."[25]

To be marketable the new cultural works will have to provoke and reward collective meaning production through elaborate back stories, unresolved enigmas, excess information, and extratextual expansions of the program universe.[26] The past decade has seen a marked increase in the serialization of American television, the emergence of more complex appeals to program history, and the development of more intricate story arcs and cliffhangers. To some degree, these aesthetic shifts can be linked to new reception practices enabled by the home archiving of videos, net discussion lists, and Web program guides. These new technologies provide the information infrastructure necessary to sustain a richer form of television content, while these programs reward the enhanced competencies of fan communities.

Television producers are increasingly knowledgeable about their fan communities often soliciting their support through networked computing. *Babylon* 5 producer J. Michael Straczinski actively courted the science fiction fan community long before his proposed series was approved for production. He cited the fan buzz to demonstrate its market potential, and the fans lobbied local stations to purchase the syndicated series. The series producer, known affectionately by his user name, JMS, went online daily, responding to questions about his complex and richly developed narrative. Kurt Lancaster estimates that JMS may have made more than 1,700 posts to the fan community, sometimes actively engaging in flame wars with individual fans as well as conducting what he saw as a continuing seminar on the production of genre television.[27] While JMS sought to be more accessible to fans, he found it difficult to shed his authority or escape a legal and economic system designed, in part, to protect corporate interests from audience appropriation. His lawyers warned him that he would have to leave the group if there was danger that he would be exposed to fan speculations that might hold him hostage to potential plagiarism suits. Such restrictions reimpose the hierarchy of commodity culture over the informal reciprocality of the knowledge culture.

While JMS is perhaps unique in the degree of his exposure to fans, other producers have shown a similar awareness of online fan discourse. For example, when the WB Network postponed the season finale of *Buffy the Vampire Slayer* in the wake of the Columbine shootings, producer Joss Whedon made a notorious public call for Canadian fans to "bootleg that puppy" and distribute it via the Web to American viewers. Fans, in turn, rallied to Whedon's defense when the religious right launched a letter-writing campaign against the introduction of a lesbian relationship involving series regulars.[28] By contrast, *Survivor* producer Mark Burnett engaged in an active disinformation campaign to thwart audience efforts to predict the winner of its million-dollar competition, burying false leads in the official Web site awaiting discovery by fan hackers. When longtime World Wrestling Federation announcer Jerry Lawler was fired, he brought his side of his disputes with Vince McMahon directly to online fans. Some of these producers sought to deceive, others to inform the fan community, but each showed an awareness of how online discourse reframed the reception context for television programs.

For many media producers, who still operate within the old logic of the commodity culture, fandom represents a potential loss of control over their intellectual property. The efforts of the recording industry to dismantle Napster demonstrated that the traditional media companies were prepared to spend massive sums in legal action against new forms of grassroots distribution. The recording industry explicitly framed the case as a chance to "educate" the public about corporate intellectual property rights and thus avoid future "piracy."[29] Television producers, film studios, and book publishers have been equally aggressive in issuing "cease and desist" letters to fan Web sites that transcribe program dialogue or reproduce unauthorized images. If new media has made visible various forms of fan participation and production, then these legal battles demonstrate the power still vested in media ownership.

The horizontal integration of the entertainment industry—and the emergent logic of synergy—depends on the circulation of intellectual properties across media outlets.[30] Transmedia promotion presumes a more active spectator who can and will follow these media flows. Such marketing strategies promote a sense of affiliation with and immersion in fictional worlds. The media industry exploits these intense feelings through the marketing of ancillary goods, from T-shirts to games, with promises of enabling a deeper level of involvement with the program content. However, attempts to regulate intellectual property undercut the economic logic of media convergence, sending fans contradictory messages about how they are supposed to respond to commercial culture.[31] . . . Often, the conflict boils down to an issue of who is authorized to speak for a series, as when a Fox television executive justified the closing of *Simpsons* fan sites by saying: "We have an official Web site with network approved content and these people don't work for us." It is perhaps symptomatic of this highly charged legal culture that fandom.com, a company created to support fan community activities and thwart "cyberbullying," almost immediately began issuing "cease and desist" letters to other sites that used the term "fandom." Ultimately, fandom.com was forced to back down, but only after it had totally undercut its claims to be "by and for fans."

Levy sees industry panic over interactive audiences as short-sighted: "By preventing the knowledge space from becoming autonomous, they deprive the circuits of commodity space . . . of an extraordinary source of energy." The knowledge culture, he suggests, serves as the "invisible and intangible engine" for the circulation and exchange of commodities.[32] The online book dealer Amazon.com has linked bookselling to the fostering of online book culture. Readers are encouraged to post critical responses to specific works or to compile lists of their favorite books. Their associates program creates a powerful niche marketing system: Amazon patrons are offered royalties for every sale made on the basis of links from their sites. Similarly, the sports network ESPN sponsors a fantasy baseball league, a role-playing activity in which sports fans form teams, trade players, and score points based on the real-world performance of various athletes. Such activities give an incentive for viewers to tune into ESPN for up-to-the-minute statistics.[33]

Attempts to link consumers directly into the production and marketing of media content are variously described as "permission-based marketing," "relationship marketing," or "viral marketing" and are increasingly promoted as the model for how to sell goods, cultural and otherwise, in an interactive environment. Jupiter Communications notes that 57 percent of consumers visit a new site based on word of mouth.[34] As one noted industry guide explains, "Marketing in an interactive world is a collaborative process with the marketer helping the consumer to buy and the consumer helping the marketer to sell."[35] Researchers are finding that fandom and other knowledge communities foster a sense of passionate affiliation or brand loyalty that insures the longevity of particular product lines.[36] In viral marketing, such affiliations become self-replicating as marketers create content that consumers want to circulate actively among their friends. Even unauthorized and vaguely subversive appropriations can spread advertising messages, as occurred through Internet spoofs of the Budweiser "whazzup" commercials.

Building brand loyalty requires more than simply coopting grassroots activities back into the commodity culture. Successful media producers are becoming more adept at monitoring and serving audience interests. The games industry, which sees itself as marketing interactive experiences rather than commodities, has been eager to broaden consumer participation and strengthen the sense of affiliation players feel towards their games.[37] LucasArts has integrated would-be *Star Wars* garners into the design team for the development of their massively multiplayer online game. A Web page was created early in the design process and ideas under consideration were posted for fan feedback. Kurt Squire describes the benefits of this "participatory design" process: "Ordinary users, who are ordinarily left out of the design process, can bring their expertise using products to the conversation, and help ensure more usable products. This ends up saving money for the designers, who can spend less energy in user/customer support. And, of course, this process results in more usable products, which benefits everyone."[38] Game companies often circulate their game engines as shareware, seeking to unleash the creative potential of their consumers. In some cases, fan-designed "mods" or game worlds (such as *Counterstrike*) have been integrated into the commercial releases. Maxis, the company that manages the *Sims* franchise, encourages the grassroots production and trading of "skins" (new character identities), props, and architectural structures, even programming code. *Sims* creator Will Wright refers to his product as a "sandbox" or "doll house," viewing it more as an authoring environment where consumers can play out their own stories than as a "hard-rails" game. Ultimately, Wright predicts, two-thirds of Sims content will come from consumers.[39]

It remains to be seen, however, whether these new corporate strategies of collaboration and consultation with the emerging knowledge communities will displace the legal structures of the old commodity culture. How far will media companies be willing to go to remain in charge of their content or to surf the information flow? In an age of broadband delivery, will television producers see fans less as copyright infringers and more as active associates and niche marketers? Will global media moguls collaborate with grassroots communities, such as the anime fans, to insure that their products get visible in the lucrative American market?

From Jammers to Bloggers

In his 1993 essay "Culture Jamming: Hacking, Slashing and Sniping in the Empire of Signs," Mark Dery documented emerging tactics of grassroots resistance ("media hacking, informational warfare, terror-art and guerilla semiotics") to "an ever more intrusive, instrumental techno culture whose operant mode is the manufacture of consent through the manipulation of symbols."[40] In citizens band (CB) radio slang, the term "jamming" refers to efforts to "introduce noises into the signal as it passes from transmitter to receiver." Culture jammers refused to be "passive shoppers" and insisted on their right to insert alternative ideas into the meme-stream. . . .

Dery's essay records an important juncture in the history of DIY media. Over the past several decades, emerging technologies—ranging from the photocopier to the home computer and the video cassette recorder—have granted viewers greater control over media flows, enabled activists to reshape and recirculate media content, lowered the costs of production, and paved the way for new grassroots networks. Recognizing that their revolution would not be televised, the 1960s counterculture created an alternative media culture, using everything from rock to underground newspapers, from poster art to people's radio, to communicate outside the corporately controlled media, and in the process, student leaders proposed theories of participatory culture that would influence subsequent activists. The DIY aesthetic got a second wind in the 1980s as punk rockers, queer activists, and third-wave feminists, among others, embraced photocopied zines, stickers, buttons, and T-shirts as vehicles for cultural and political expression.[41] These groups soon recognized the radical potential of videotape for countersurveillance and embraced the "digital revolution" as an extension of earlier movements toward media democracy.[42]

Many of the groups Dery describes, such as Adbusters, ACT UP, Negativeland, The Barbie Liberation Army, Paper Tiger Television, and the Electronic Disturbance Community, would happily embrace his "culture jammer" banner. Yet, Dery overreached in describing all forms of DIY media as "jamming." These new technologies would support and sustain a range of different cultural and political projects, some overtly oppositional, others more celebratory, yet all reflecting a public desire to participate within, rather than simply consume, media. Culture jammers want to opt out of media consumption and promote a purely negative and reactive conception of popular culture. Fans, on the other hand, see unrealized potentials in popular culture and want to broaden audience participation. Fan culture is dialogic rather than disruptive, affective more than ideological, and collaborative rather than confrontational. Culture jammers want to "jam" the dominant media, while poachers want to appropriate their content, imagining a more democratic, responsive, and diverse style of popular culture. Jammers want to destroy media power, while poachers want a share of it.

"The territory mapped by this essay ends at the edge of the electronic frontier," Derry wrote, expressing optimism about the emerging political and cultural power grassroots media activists might enjoy in a context where media flows are multidirectional.[43] Yet, he also cautions that the media industries will find alternative means of marginalizing and disenfranchising citizen participation. . . . Returning to this same terrain at the end of the decade, it is clear that new media technologies have profoundly altered the relations between media producers and consumers. Both culture jammers and fans have gained greater visibility as they have deployed the Web for community building, intellectual exchange, cultural distribution, and media activism. Some sectors of the media industries have embraced active audiences as an extension of their marketing power, have sought greater feedback from their fans, and have incorporated viewer-generated content into their design processes. Other sectors have sought to contain or silence the emerging knowledge culture. . . . The old rhetoric of opposition and cooptation assumed a world where consumers had little direct power to shape media content and where there were enormous barriers to entry into the marketplace, whereas the new digital environment expands their power to archive, annotate, appropriate, and recirculate media products. . . . Levy describes a world where grassroots communication is not a momentary disruption of the corporate signal but the routine way that the new media system operates: "Until now we have only reappropriated speech in the service of revolutionary movements, crises, cures, exceptional acts of creation. What would a normal, calm, established appropriation of speech be like?"[44]

Perhaps, rather than talking about culture jammers, we might speak of bloggers. The term "blog" is short for "Web log," a new form of personal and subcultural expression involving summarizing and linking to other sites. In some cases, bloggers actively deconstruct pernicious claims or poke fun at other sites; in other cases, they form temporary tactical alliances with other bloggers or with media producers to insure that important messages get more widely circulated. These bloggers have become important grassroots intermediaries—facilitators, not jammers, of the signal flow. Blogging describes a communication process, not an ideological position.

As Levy writes:

> *The new proletariat will only free itself by uniting, by decategorizing itself, by forming alliances with those whose work is similar to its own (once again, nearly everyone), by bringing to the foreground the activities they have been practicing in shadow, by assuming responsibility—globally, centrally, explicitly—for the production of collective intelligence.*[45]

Bloggers take knowledge in their own hands, enabling successful navigation within and between these emerging knowledge cultures. One can see such behavior as cooptation into commodity culture insofar as it sometimes collaborates with corporate interests, but one can also see it as increasing the diversity of media culture, providing opportunities for greater inclusiveness, and making commodity culture more responsive to consumers. In an era marked both by the expanded corporate reach of the commodity culture and the emerging importance of grassroots knowledge cultures, consumer power may now be best exercised by blogging rather than jamming media signals.

Notes

1. Pierre Levy, *Collective Intelligence: Mankind's Emerging World in Cyberspace* (Cambridge: Perseus, 1997), 217.
2. The phrase "imagined community" comes from Benedict Anderson, *Imagined Communities: Reflections on the Origin and Spread of Nationalism* (New York: Verso, 1991). Anderson argues that we feel strong affiliations with nation-states even though they are too large for us to have personal contacts with all of the other citizens and cites the role media plays in providing the social cement between these scattered populations. Levy (*Collective Intelligence,* 125) introduces the concept of an "imaging community" to describe how a sense of affiliation emerges from an active process of self-definition and reciprocal knowledge transfer.
3. A fuller account of Gernsbeck's role in the development of science fiction fandom can be found in Andrew Ross, *Strange Weather: Culture, Science and Technology in the Age of Limits* (New York: Verso, 1991). For a fuller account of contemporary literary SF fandom, see Camille Bacon-Smith, *Science Fiction Culture* (Philadelphia: University of Pennsylvania Press, 2000).
4. John Tulloch and Henry Jenkins, *Science Fiction Audiences: Watching Doctor Who and Star Trek* (London: Routledge, 1995).
5. Sherry Turkle, *The Second Self: Computers and the Human Spirit* (New York: Touchstone, 1984), provides some glimpse of the centrality of science fiction in that early hacker culture, as does my study of *Star Trek* fans at MIT in Tulloch and Jenkins, *Science Fiction Audiences.*
6. Susan J. Clerc, "Estrogen Brigades and 'Big Tits' Threads: Media Fandom Online and Off," in *Wired Women: Gender and New Realities in Cyberspace,* ed. Lynn Cherney and Elizabeth Reba Weise (Seattle: Seal, 1996).
7. Nancy Baym, "Talking about Soaps: Communication Practices in a Computer-Mediated Culture," in *Theorizing Fandom: Fans, Subculture, and Identity,* ed. Cheryl Harris and Alison Alexander (New York: Hampton Press, 1998).
8. Levy, *Collective Intelligence,* 20.
9. Baym, "Talking about Soaps," 115–16.
10. Ibid., 127.
11. Matthew Hills, *Fan Cultures* (London: Routledge, 2002).
12. For a useful discussion of the ways that the net is challenging traditional forms of expertise, see Peter Walsh, "That Withered Paradigm: The Web, the Expert and the Information Hegemony," http://media-in-transition.mit.edu.
13. Hills, *Fan Cultures,* 78–79.
14. Ibid.
15. For an overview of anime and its fans, see Susan J. Napier, *Anime from Akira to Princess Mononoke: Experiencing Contemporary Japanese Animation* (New York: Palgrave, 2001).
16. The phrase "week-end only world" is discussed in the concluding chapter of Henry Jenkins, *Textual Poachers: Television Fans and Participatory Culture* (New York: Routledge, 1992).
17. Andre McDonald, "Uncertain Utopia: Science Fiction Media Fandom and Computer-Mediated Communication," in *Theorizing Fandom: Fans, Subculture, and Identity,* ed. Cheryl Harris and Alison Alexander (New York: Hampton Press, 1998).
18. Nancy Baym, *Tune In, Log On: Soaps, Fandom and Online Community* (New York: Corwin, 1999).
19. Stephen Duncombe, *Notes from Underground: Zines and the Politics of Alternative Culture* (New York: Verso, 1997).
20. For a fuller discussion of fan video practices, see *Textual Poachers.* For a larger context on amateur media production, see Patricia R. Zimmermann, *Reel Families: A Social History of Amateur Film* (Indianapolis: Indiana University Press, 1995).

21. Henry Jenkins, "Quentin Tarantino's Star Wars? Digital Cinema, Media Convergence and Participatory Culture," in *Dfilm,* ed. Bart Cheever and Nick Constant (Cambridge, MA: MIT Press, 2001).
22. Levy, *Collective Intelligence,* 121.
23. Ibid., 123.
24. Kurt Lancaster, *Interacting with Babylon 5: Fan Performances in a Media Universe* (Austin: University of Texas Press, 2001).
25. Levy, *Collective Intelligence,* 125.
26. Amelie Hastie, "Proliferating Television in the Market and in the Know," paper presented at the "Console-ing Passions" conference, Bristol, UK, July 6, 2001.
27. Lancaster, *Interacting with Babylon 5,* 26. See also Alan Wexelblat, "An Auteur in the Age of the Internet," in *Hop on Pop: The Politics and Pleasures of Popular Culture,* ed. Henry Jenkins, Tara McPherson, and Jane Shattuc (Durham, NC: Duke University Press, 2002).
28. Allison McCracken, "Bronzers for a Smut-filled Environment: Reading Fans Reading Sexual Identity at Buffy.com," paper presented at the "Consoleing Passions" conference, Bristol, UK, July 6, 2001.
29. David Spitz, "Contested Codes: Toward a Social History of Napster" (Masters thesis, Comparative Media Studies Program, MIT, June 2001).
30. See, for example, Eileen Meehan, "Holy Commodity Fetish, Batman!": The Political Economy of a Political Intertext" in *The Many Lives of the Batman: Critical Approaches to a Superhero and His Media,* ed. Roberta Pearson and William Uricchio (New York: Routledge, 1991).
31. This formulation of the issue was inspired by Sara Gwenllian Jones, "Conflicts of Interest? The Folkloric and Legal Status of Cult TV Characters in Online Fan Culture," paper presented at the Society for Cinema Studies Conference, Washington, DC, May 26, 2001.
32. Levy, Collective Intelligence, 237.
33. For example, see Amy Jo Kim, *Community Building on the Web: Secret Strategies for Successful Online Communities* (Berkeley: Peachpit Press, 2000).
34. Jupiter Communication, as cited in "Just Exactly What Is Viral Marketing?" http://marketsherpa.co.uk.
35. Don Peppers, "Introduction," in Seth Godon, *Permission Marketing: Turning Strangers into Friends, and Friends into Customers* (New York: Simon and Schuster, 1999), 12.
36. Robert V. Kozinets, "Utopian Enterprise: Articulating the Meanings of *Star Trek*'s Culture of Consumption," *Journal of Consumer Research* (June 2001): http://www.journals.uchicago.edu/JCR/journal/.
37. See, for example, Elizabeth Kolbert, "Pimps and Dragons: How an Online World Survived a Social Breakdown," *New Yorker,* May 28, 2001.
38. Kurt Squire, "Wars Galaxies: A Case Study in Participatory Design," *Joystick 101,* www.joystick101.org, forthcoming.
39. Personal interview, April 2001.
40. Mark Dery, *Culture Jamming: Hacking, Slashing and Sniping in the Empire of Signs* (Open Magazine Pamphlet Series, 1993), http://web.nwe.ufl.edu/~mlaffey/cultjam1.html.

    For elaboration on the concept of culture jamming, see also Gareth Branwyn, *Jamming the Media: A Citizens Guide for Reclaiming the Tools of Communication* (San Francisco: Chronicle, 1997); and David Cox, "Notes on Culture Jamming," http:www.syntac.net/hoax/manifesti/notes.php.
41. For a useful overview of media activism in this period, see Douglas Rushkoff, *Media Virus! Hidden Agendas in Popular Culture* (New York: Ballantine, 1996).

42. Philip Hayward, "Situating Cyberspace: The Popularization of Virtual Reality," in *Future Visions: New Technologies of the Screen,* ed. Philip Haywood and Tana Wollen (London: British Film Institute).
43. Dery, *Culture Jamming.*
44. Levy, *Collective Intelligence,* 171.
45. Ibid., 36–37.

The Practices of Internet Networking—A Resource for Alternative Political Movements

Tobias Olsson

Over the past ten years, the political significance of the Internet has become a recurrent theme within the social science literature, with questions centering on the Internet's part in the emergence of political movements and in the reshaping of people's roles as citizens. These discussions have been predominantly theoretical and deterministic in nature, with the political significance of the Internet being viewed as a consequence of its features; for example, the Internet's network structure is addressed within this literature as a forerunner to a politics following network logic. For the most part, extant research has not told us much about the Internet's significance in individuals' everyday political engagement. This article aims to modestly compensate for this shortcoming by presenting empirical results from a study on the perception and use of the Internet among young people in four alternative political movements. Specifically, by drawing on semi-structured interviews with twenty-one activists and viewing the interview data through the lens of the 'network metaphor', the article illustrates and discusses three different networking practices in which the Internet is an important resource: (1) the activists' use of the Internet to maintain their organizations' network-like character, (2) their use of the Internet for networking within their organizations, and (3) their use of the Internet for networking in between different, alternative political organizations. The article concludes by discussing the potential significance of these networking practices for politics, and presenting a comparative outlook towards a similar, previous study of young members within established political parties' youth organizations.

Keywords: political, network, Internet, alternative movements, Sweden

Introduction

It is no exaggeration to claim that the concept 'network' has become a temporal metaphor—especially when it is attached to the word 'society'. In terms of capturing the essence of contemporary society in one phrase, 'network society' is close to becoming as widespread as that of 'information society' or 'postindustrial society'.

As we have seen, Manuel Castells is one of the scholars who have worked diligently to move this metaphor into its prominent position in contemporary social science. In his work of three volumes,

The Information Age: Economy, Society and Culture, he describes networks of all kinds as the most important material and social entities of our age. He writes, for instance:

> *Our exploration of emergent social structures across domains of human activity and experience leads to an overarching conclusion: as a historical trend, dominant functions and processes in the information age are increasingly organized around networks. Networks constitute the new social morphology of our societies, and the diffusion of networking logic substantially modifies the operation and outcomes in processes of production, experience, power, and culture. While the networking form of social organization has existed in other times and spaces, the new information technology paradigm provides the material basis for its pervasive expansion throughout the entire social structure. (Castells 1996, p. 469)*

From this position it follows that all parts of society—from the stock market, to politics, to our everyday working lives—are subject to and/or dominated by network logic. Castells is also careful to explain that network logic is dominant to the extent that it even asserts itself over people and practices situated outside of the networks, and which are essentially offline (Castells 1996, p. 376).

Castells and other contemporary social scientists (e.g. Barney 2004; Hassan 2004; van Dijk 2006) have ascribed substantial weight to network configurations, and in doing so have devoted much attention to understanding emerging digital media. The similarities between the digital media's technical features (nodes, networks, flows, etc.) and the networks that are identified in society at large emerge quite clearly within their work. In taking this a step further, we can even posit that digital media developments have, to some extent, triggered and inspired conceptualizations about society as being dependent on, and shaped by, various network configurations (see also Castells 2001).

One of the fields within research on new, digital media that is commonly intersected by reflections on network logic's significance is that which centers on the study of the Internet's political dimensions. Manuel Castells—as an obviously central scholar within this arena—pays heed to how network-based technologies, i.e. the Internet, open up new possibilities for the emergence of new social movements (Castells 1997). Other scholars have also contributed to this discussion, including Jenny Pickerill (2004), with her case studies on Internet activism, as well as Neil Stammers and Catherine Eschle (2005), who have contributed with theoretical reflections on networking within social movements (see also Loader et al. 2004).

This article situates itself among these works. It presents and analyzes empirical data from a study on the perception and use of the Internet among young people in alternative political movements. Within the study, semistructured interviews with some 20 activists serve as the primary source of data. By drawing o the network metaphor as a theoretical frame for understanding significant parts of these young, 'alternative' people's use of the Internet, the article illustrates and discusses three different networking practices in which the Internet is an important resource—answering how they 'network' when it comes to their political engagement, and shedding light on the ways that the Internet is a resource for their practices.

Mapping the Research Field

This article seeks to contribute to an understanding of the Internet's political dimensions. In doing so it draws its primary source of inspiration from two theoretical horizons in contemporary research—namely, research on the Internet's contribution to the establishment of the so-called network society, and studies that discuss the Internet's political dimensions and implications. With regards to the former, the article attempts to add empirical weight to existing works, which have been predominantly theoretical in nature. In terms of the latter, it aims to make further use of the network metaphor in understanding the Internet's political implications.

As indicated in the introduction to this article, the theorization of contemporary society as a network society has been both encompassing and suggestive. It has addressed a range of questions about the social significance of new information and communication technologies—from their role in organizational change (Bouwman et al. 2005), to the creation of communities on-line (Jankowski 2002), to flows in the global economy (Castells 1996), to changes within social structure (van Dijk 2005).

Of relevance to the article at hand are those works that apply the metaphor in analyzing how new information and communication technologies (ICTs) in general, and the Internet in particular, can reshape the political landscape. For instance, Jan van Dijk (2006) has theorized about how the Internet creates new threats to personal integrity, how it opens up the political processes for increasing civic participation, and how it can help bring about new political communities (see also Smith and Kollock 1999; Feenberg & Barney 2004). And Manuel Castells has conceptualized how new network configurations can lead to new political movements by allowing previously disconnected, undeveloped political identities to take shape and rise to a prominent position. Identity politics, Castells writes, is helped by the Internet by way of the internal coordination between small nodes within political movements as well as in the movement's external opinion formation (Castells 1997).

The literature is thus replete with discussions on how networks inform political action and the overall political landscape; however, most are primarily theoretical in nature and lack empirical substantiation or concrete exemplification. While they have done much to elucidate our understanding of what might happen to politics in the emerging network society, they have done less in the way of informing our knowledge on how networking actually occurs within alternative political movements and on how members of various social movements make use of new technology in their networking practices.

In turning to research into the Internet's political dimensions, on the other hand, we find an increasing number of empirical studies within the literature. These studies have, however, often started from an analysis of the new information and communication technology per se, and then tried to estimate its political significance (Hill & Hughes 1998; Hague & Loader 1999; Margolis & Resnick 1999; Bentivegna 2002)—thus reading the new ICT's political significance off the technology in and of itself (Holmes 1997; Poster 1997).

Of late, however, a somewhat new direction within this research has begun to surface, which aims to empirically substantiate existing theoretical reflections on the Internet's political significance. Examples include a study of web initiatives aimed at cultivating political participation among young people (Montgomery et al. 2004), statistical studies mapping youths' interest in participating in politics on-line (Livingstone & Bober 2005), and studies centering on how young, active citizens interpret and make use of the Internet as a political resource (Dahlgren 2007).

Among these efforts to map the Internet's political significance in a more empirical manner, little attention has been paid to making use of the network metaphor. There are of course exceptions, such as Anna Kavada's (2005) studies of Internet use within civil society organizations, and Jenny Pickerill's (2004) studies of Internet activism. This article, however, aims to call attention and add weight to this nascent strand of empirical study, by making further use of the network metaphor in analyzing and understanding the Internet's significance for political movements.

Short Notes on Network

Of the numerous definitions of 'network' that exist within the social science literature, that which is offered by David Barney in *The Network Society* serves as a good starting point for the purpose of our empirical endeavour. Barney defines a network as follows:

> *The word 'network' describes a structural condition whereby distinct points (often called 'nodes') are related to one another by connections (often called 'ties') that are typically multiple, intersecting and often redundant. A network exists when many nodes (people, firms, computers) are linked to many other nodes, usually by many ties which cross the ties connecting other node. (Barney 2004, p. 2)*

Barney's definition does however, require further qualification here. That is, in the context of this study, networks and network connections are defined as being primarily confined to informal connections between persons, persons and organizations and between organizations (Stammers & Eschle

2005). Hence, formal connections and communications, such as those encompassing the employee-employer relation or the distribution of invoices from one company to another, are not considered as networks. This qualification, of course, comes with the recognition that the distinction between formal and informal connections is not always obvious—as informal network connections can surface within and around formal organizational relations, as the empirical data of this study will show.

In thus defining networks in terms of informal, abundant and redundant ties between people and the organizations they make up, this article aims to analyze and describe the ways in which the Internet serves as a resource for different kinds of network connections among activists within alternative political movements. As we shall see, applying the network metaphor to the analysis will inform our understanding of how the Internet substantiates—or even encourages—various forms of networking within the sphere of alternative political movements.

The Empirical Study

The empirical data for this study was derived from twenty-one semi-structured interviews with young, active members (ages 15 to 20) from four alternative political organizations in Sweden. The interviews focused mainly on the youths' use and perception of the media and the Internet as resources for their identities as active citizens (cf. Olsson 2005a; Olsson 2005b; Dahlgren & Olsson 2007).

The four political organizations under analysis have been assigned pseudonyms, and include *Green Globalization, Friends of the Animals, Young Feminists* and *Just Say No* (to racism). Green Globalization is an environmental, alter-globalization movement. Although the organization has international connections, it is organized on a national level and most of its activities are initialized within small, local groups. Friends of the Animals is a large organization within the animal rights movement, which also initializes its activities at the small, local group level. The Young Feminists group is made up young activists from various feminist interest groups, which have been combined under this single organization for the purpose of this study. Finally, Just Say No is a small group of activists focusing on anti-racism. It is loosely connected to a larger organization with similar interests; however, Just Say No is new in and of itself—for at the time of this study it had only been active for six months. For the study, three respondents were recruited from each of the organizations, with the exception of the Young Feminists, from which twelve activists were recruited (10 women and two men).

A more in depth explanation of these respondents as well as their organizations will be presented in the sections that follow; however, it is important to note here that through the study, the four organizations did not always surface as 'organizations' in the everyday sense of the term. As the empirical analysis will show, they instead sometimes revealed themselves as loose clusters of people working toward their political aims at the outskirts of organized political action.

In analyzing the data from the semi-structured interviews, a comparative approach has been used. That is, the interviews with the young activists were compared to another set of interviews—also conducted within the frame of this study—which focused on members within the established Swedish political parties' youth organizations (cf. Olsson 2004; Olsson 2006a). Using the empirical material from that study as a point of reference facilitated the identification of differences between the two groups, thus pointing to ways in which the Internet practices of youth activists in the alternative political organizations are unique.

Internet in Network-Like Organizations

From the point of view of the empirical data it is quite easy to associate the respondents' descriptions of their organizations, as well as the activities taking place within them, with the network metaphor. All four of the organizations, or associations, reveal qualities that are similar to those commonly affiliated with network-like organizations, and furthermore, all four of them draw on the Internet as a central resource in their networking practices.

To begin with, the empirical data points to the presence of loose, informal affiliations between members, as opposed to structured organizational relations. The interviews with the respondents in the Young Feminists and with those from Just Say No indicate that the members do not always understand themselves as parts of large organizations. Instead, they conduct their political activities within small groups of young people, who float around the internal and external boarders of organized political action.

The Young Feminists are organized in small groups of a few activists each. These include, for example, equality groups in school or feminist student associations. Their activities take place within and around these small groups, and their engagement in feminist issues is not subordinate to any larger organization. The interviews reveal, however, that the activists do often participate in—or loosely affiliate themselves with—other groups of activists. Some of the activists are connected to other feminist groups, while others are involved with different activist groups, like those within the animal rights movement.

In turning to the group Just Say No, which the respondents themselves describe as anti-racist, the empirical data reveal that it comprises less then 10 members. The group is independent, even though it is partially connected to a large, well-known human rights organization. Thus, like the Young Feminists, Just Say No is a small, flat organization with only loose connections between members in different groups, and wherein all activities occur at the level of small groups of activists.

For the respondents from Just Say No, the Internet surface as an important source of information on issues of anti-racism. Due to its small size and inability to serve as an encompassing source of knowledge, the members are asked to look outside of the group to gain information and perspective on their issues. The interviews with young respondents bring forth several examples of this occurrence, but an extract from an interview with Lena, a 17-year-old high school student, serves as especially depictive:

Interviewer:	When it comes to your engagement within Just Say No: How do you use the computer and the Internet?
Lena:	Quick Response [a website] is very good, we're supposed to visit it every now and then within Just Say No. They've got quite a lot there, good stuff that you can read.
Interviewer:	What do you usually read about on Quick Response?
Lena:	It's a lot of, you know, anti-racist things, like discrimination and stuff like that and . . . well . . . It's real good.
Interviewer:	You indicated that you're supposed to visit it. What do you mean by that?
Lena:	Ehh . . . well, no, but it's more as if . . . When we started [the group] they said that it was a good site for us to visit. You can . . . There are questions, for example, that you can get when you visit schools. For instance like: 'How many per cent of the people conducting violent crimes are immigrants?' Then it [the website] offers information that makes you prepared for such questions, how to answer them. It also offers a lot of useful statistics.

The website Quick Response is not in any formal sense related to Just Say No. It is instead an independent source of information on the representation of immigrants in the Swedish media, and a watchdog on immigrant politics. But even though the website has no formal connection to Just Say No, all three respondents refer to the site as an important frame of reference in their political activities. It is considered somewhat compulsory for the members in Just Say No to visit it.

It might therefore be suggested that Just Say No is basically outsourcing its knowledge production. The organization is too small and too loose in itself to manage to produce and store knowledge useful for its area of political interest. Instead, it must ask its members to become independent information seekers, or consultants of other sources. Just Say No thus relies on knowledge produced by other nodes in the network of antiracist organizations.

A similar pattern emerges among the Young Feminists. Due to its loose, small-group organizational structure, and the fact that some of the respondents are involved in several different feminist organizations, and in several different activities (some of which the respondents had difficulty in specifying which organization these were associated with), the Internet serve as an important source of knowledge. Maria, a 20-year-old Young Feminist member who is involved in various feminist activities, provides an obvious example of the Internet's importance to gaining knowledge:

Interviewer: What kind of websites do you visit then? [. . .]

Maria: Well, there is one—I think it's called 'feminist dot org', I think, or perhaps 'dot com', I'm not quite sure. I used to consult it quite a lot. It suggests a lot of books to read. That's how I find books to read. It's a good thing to consult a website that . . . you know, you can find hundreds of books there and you can read something about all of them. Then you can go to the library [to borrow them]. Other websites . . . Well, I guess . . . I can't quite recall their names, but I visit some regularly.

Interviewer: Bang's [an intellectual feministic website] website for instance?

Maria: Yes, I go there every now and then, when there's something interesting to read.

Maria is not one of the most frequent Internet users among the respondents. Nevertheless, the Internet has become an important resource to her identity as feminist. The Internet allows her to keep up-to-date on feminist issues—she regularly visits feminist.org, and occasionally visits the feminist journal 'Bang'. Thus, similar to the members in Just Say No, she is shaping the new information and communication technology into a tool that compensates for the lack of local, 'in house' feminist knowledge within the networks she is a part of.

In turning to the other two organizations under study, which are both larger and somewhat more organized, the Internet likewise seems to serve as a central tool in facilitating networks, although in a different manner. The alter-globalization organization, Green Globalization, is essentially a national organization with global connections. On the surface, Green Globalization is organized in terms of a formal structure covering all levels of organization, from the local to the national. However, almost all activities are determined within local groups that are independent and only loosely affiliated with the regional and national levels.

To a large extent, the (limited) coordination within and between the local groups takes place on the Internet; the network-like organization coordinates its activities through the Internet on both local and, to a lesser extent, regional and national levels. Asta, who holds an important position on the local level, details the ways in which the Internet is a resource for her organization:

Interviewer: You also said Green Globalization . . .

Asta: Mailing lists . . . There is one for [the local club], you can subscribe to it, and you can send . . . if you want to send to [all local members in Green Globalization] . . . I use that quite a lot to distribute schedules and calls for meetings. [. . .]

Interviewer: Then you also have a website . . .

Asta: Yes, the region has one and we also have one on a national level.

Interviewer: What do you put on the regional one?

Asta: There we've got regional activities as well as activities on a local level, in the clubs that send us their schedule. Then we've got some links and . . . phone numbers and things like that.

Interviewer: Do you get mail related to Green Globalization everyday?

Asta: Yes, I usually do. It's usually at least two or three mails a day.

In the extract Asta describes rather carefully how both she and Green Globalization make use of the Internet. Firstly, she mentions the local mailing lists, though which she receives and distributes information about meetings and schedules. Then she also describes the regional mailing lists, which are for more general information. A website also exists at the regional level. It is used both for internal issues and as a tool for external information about Green Globalization's ideas and activities. Taken together, these activities indicate that a somewhat loose organization (Green Globalization) has managed to shape the Internet into a resource that at least partially ties its disparate groups and activities together.

The Internet is used in a similar manner within Friends of the Animals. The members call for meetings, make plans for action and discuss ethical guidelines for their activities on the Internet. Most of the interaction between the members takes place over the Internet, except for the interactions among members of local groups—as they primarily interact face-to-face.

It is rather obvious that all four organizations bring the network metaphor to the fore, however in different ways. The organizations vary in size and the extent to which they are organized, from loosely organized to very loosely organized. Nevertheless, all of them have shaped the Internet into a resource for their political activities. Within the Young Feminists and Just Say No, the Internet seems to have been shaped into a resource compensating the loose, network organizations' lack of knowledge. The organizations per se are not large enough to contain all knowledge of relevance to their political interests. Therefore, the Internet comes in as a useful tool, making it possible for the organizations—so to speak– to outsource the production and storage of knowledge of relevance for its political interests. In the cases of Green Globalization and Friends of the Animals, where the activities are slightly more organized, the Internet has instead been shaped into glue holding the organizations' disparate, network-like parts together.

Networking within the Organization through the Internet

This study's two larger organizations, Green Globalization and Friends of the Animals, are large enough to contain different kinds of networking within themselves. Within these different networking practices—of which the Internet appears to be a central resource—there are two aspects of importance. The first is that of making connections between members within the organizations that crosscut the organizations' loose but still existing structures. The second aspect is that the actual political activities within the organizations take place only within small groups of people, and that the members tend to mingle between these various small groups.

The intra-organizational networking within Green Globalization becomes evident through the interviews. Most of the organization's activities are accomplished on the local level. The other levels, the regional and the national levels, have little (if any) say in the planning and carrying out of activities. Interestingly, though, the individual members themselves seem to have many internal network connections with other people within the organization, regionally as well as nationally. Hence, there are substantial links between various geographical levels within the organization, but these connections typically transgress the existing organizational structures and instead take on an informal, network like character.

The respondents themselves refer to these connections with other people within the organization as 'network connections'. The respondents describe themselves as nodes in, for instance, 'the wood network', or 'the feminist network' in order to emphasize that these networks bring together members with similar special interests. These networks are virtual in the sense that they bring together members who basically only meet through the Internet, but also in the sense that they would not exist within any formal description of the organization.

From the interviews with the members of Friends of the Animals, we find several examples of the second intra-organizational networking practice—that of shifting between various small groups when conducting political activities. An extract from the interview with Emma, who is 18 years old and has been a member of Friends of the Animals since she was 12, illustrates this quite well:

> Emma: And then we've got different groups with different responsibilities, like the fur group and the 'experiment on animals'-group. To begin with I was in a youth group, but we were a bit too few so we went into other groups. I ended up helping out a little here and a little there. [. . .]

Interviewer:	What do you do within Friends of the Animals here, locally?
Emma:	We've got different activities. We have a tent on the festival where we sell food and things like that. Then we write petitions against the fur industry and . . . And we have actions, against the fur industry, for instance.

In the extract Emma first explains what part of the organization she formally belongs to. Technically speaking she still belongs to the youth section, but it has too few members. Emma has instead chosen to move into other groups within Friends of the Animals. Today she helps out 'a little here and a little there'. In this network-like gliding both into and in between various groups, the Internet serves as an important resource; it is mainly through the Internet that Emma keeps in touch with the different groups in which she participates. The Internet has basically been shaped into an infrastructure holding together the loosely organized members in the rather informal groups within Friends of the Animals.

Networking Between Different Organizations

The network-like character of political engagement also stretches outside the organizations from which the study's respondents have been recruited. The interviews offer several examples of how the young, active citizens channel their engagements into several different organizations at the same time. For some respondents it is a matter of being engaged in several different alternative groups with a similar profile, while others activate themselves in different alternative groups focusing on different political issues. In all of these network-like connections, the Internet holds an important position. For instance, mailing lists are in several cases the main means of connection between the individual activists and their networks of colleagues.

The interview with Gina, who was recruited to the study as a member of Friends of the Animals, illustrates this point well. On the question of what she specifically does within Friends of the Animals, she gives a long answer that, among other things, points to how some of her activities within Friends of the Animals also affiliate her with other organizations with similar political interests:

Gina:	At the moment I'm not that active [within Friends of the Animals]. But the first thing I attended was the fourth of October, the Animals' Day. We celebrate that . . . Not really celebrate . . . We usually have demonstrations . . . But we were handing out brochures about how animals are treated within the fur industry.
Interviewer:	What have you been doing since then?
Gina:	Well, I've been . . . ehhh . . . I've been handing out brochures. I'm also into the Animal Rights Association. It's also a legal organization . . . But it's not really an organization. It's more like a network. I've been affiliated with it since last November . . . I went to this meeting with them and became a member. Well, not really a member, you're not a member. They put you on their mailing list and you get their e-mails. It's not like a real . . . organization with membership cards and all that, it's built on active members . . . If everybody ignores a meeting, there won't be any meeting. We don't have anyone who's responsible or anything. But . . . Yes, I've been with them in town. We use to block the entrance to fur shops. That's legal, at least as long as you aren't standing right outside . . . I've done that quite a few times now. And we've also been selling books and pins and stuff like that, and I've also been serving vegetarian bread . . . But I guess that's pretty much it.

The most interesting part of the extract is not the fact that Gina is politically involved in several activities, but that her engagement with Friends of the Animals is connected to the activities of other organizations with similar political profiles. Gina does not solely participate in activities within

Friends of the Animals. Instead, her engagement is also channeled into a loosely organized network of animal rights activists; she even refers to it as a 'network' herself. The only thing holding this network together is a mailing list that connects the activists, discussing and informing them about upcoming events and meetings.

Rita, who was recruited to the study through her connection with Green Globalization, brings another example of how the young activists participate in inter-organizational networking. Rita is 17 years old and divides her engagement between Green Globalization, a feminist network and an animal rights movement. She was also formerly engaged in party politics, but has abandoned this involvement, explaining that the political party was, for her, too influential in framing her political opinions.

The networking between various organizations is also evident among the young feminists. Bengt, one of the study's two male feminist activists, is not only a feminist but also describes himself as 'radical activist' and channels his engagement through several different organizations, feminist as well as others. Also Maria—who is 20 years old—is affiliated with at least three different feminist organizations, which she stays connected with mainly through the Internet.

Basically, the above serves to reveal how the Internet has been shaped into a resource enabling the young activists to move into and out of various kinds of organizations at will. Through the new ICT, these loose connections are easily turned both on and off, which, at first glance, makes the young activists' political engagement appear ephemeral. However, it is not the engagement itself that is transient—for it is usually rather robust among the respondents; instead it is their way of channeling that engagement into various, volatile constellations.

Discussion and Conclusions

The network concept has been beneficial in the context of this study. It has helped to bring together what otherwise would have been disparate impressions of activities within various alternative organizations. We have been able to see that the networking practices within these organizations differ, but that they also have a number of common threads, including the fact that the Internet is a fundamentally valuable resource to their networking.

Some could argue, however, that networking practices similar to those identified in this study can be found in a range of organizations, from corporate entities to football clubs. The question thus becomes whether there is something unique about this study's networking practices. I would argue that they are. The basis for this argument is rooted in an examination of comparative empirical data—which I also conducted—from a study of youth activities within established political parties (Olsson 2004; Olsson 2006a). Specifically, comparing this later data with the data from the alternative political activists reveals that their networking practices differ in that they follow their organizations' already-established structures. This difference did not surface as evident in terms of the types of networking taking place, but instead surfaced in terms of the extent and nature of these practices. That is, for the youths within alternative organizations, networking through the Internet appears to be fundamental to their political activities, while to the respondents within the political parties' youth organizations it appears to be merely supplemental.

This is also one of the major differences found between the two groups when it comes to Internet practices. Another important difference between the groups has been presented and analyzed elsewhere (Dahlgren & Olsson 2007): the Internet's importance for construction of alternative political identity. Among the young respondents from alternative political movements the Internet has been shaped into a support for their alternative political identity. This is made specifically obvious in comparison to how they perceive traditional news media. Traditional news media hardly ever pay interest in their issues, and when they finally do they always seem to get the issues wrong. On the other hand, all respondents agree on the Internet's value to them as a resource for their construction of alternative political identity. Among other things, the Internet provides them with alternative information, opportunities for internal coordination, contacts with likeminded people, access to public spheres beyond their respective organizations, as well as a resource for everyday activism.

The significance of this study's results is at least twofold. Firstly, they are useful as an example of the social shaping of technologies; that is, of how the Internet—like other ICTs—is shaped and reshaped by the contexts in which it is appropriated (cf. Miller & Slater 2000; Bakardjieva 2005; Olsson 2006b). This study has shown how the Internet is shaped into a technology supporting the alternative organizations' already emerging network-like structures. As a medium, the Internet seems to correspond very well with the alternative organizations' network-like structures, and possibly even encourages further organizational development along networking lines.

Secondly and conclusively, knowledge of the networking practices' significance, and the Internet's position within them, is valuable to our understanding of the 'political' in the digital era. Our understanding of political organizing becomes limited if we neglect the extent to which alternative organizations and their members make use of networking in striving for their political aims. This article has brought several illustrations of various networking practices related to the Internet, examples that hopefully can inspire research in the area to further develop the understanding of the political landscape in the Internet era.

About the Author

Tobias Olsson is senior lecturer and associate professor in Media and Communication Studies at Växjö University, Sweden. He is also currently a researcher at Lund University, working within the European project 'Young People, The Internet and Civic Participation'. He has published widely on issues of new media, citizenship and politics. Address: Växjö University, School of Social Sciences, 35195 Växjö, Sweden. [email: tobias.olsson@vxu.se]

Acknowledgment

A previous version of this text was presented as a paper at the IAMCR conference in Cairo in July 2006. I am grateful to the colleagues who attended my presentation, and especially to those whose comments and recommendations have inspired this new version.

The empirical data in this article are from the research project 'Young Citizens, ICTs and Learning', which was funded by the Swedish Knowledge Foundation's research programmer, Learn-IT (2003–2007).

References

Bakardjieva, M. (2005) Internet Society: *The Internet in Everyday Life,* Sage, London.

Barney, D. (2004) *The Network Society,* Polity, Oxford.

Bentivegna, D. (2002) 'Politics and new media', in *The Handbook of New Media,* eds L. Lievrouw & S. Livingstone, Sage, London, pp. 50–61.

Bouwman, H., van den Hooff, B., van de Wijngaert, L. & van Dijk, J. (2005) *Information and Communication Technology in Organizations,* Sage, London.

Castells, M. (1996) *The Information Age: Economy, Society and Culture*. Vol. 1: The Rise of the Network Society, Blackwell Publishers, London.

Castells, M. (1997) *The Information Age: Economy, Society and Culture*. Vol. 2: The Power of Identity, Blackwell Publishers, London.

Castells, M. (2001) *The Internet Galaxy: Reflections on the Internet, Business and Society,* Oxford University Press, Oxford.

Dahlgren, P. & Olsson, T. (2007) 'From public sphere to civic culture: young citizens' internet use", in *Media and the Public Sphere,* ed. R. Butsch, Palgrave Macmillan, Basingstoke, pp. 198–209.

Dahlgren, P. (ed.) (2007) *Young Citizens and New Media: Learning Democratic Engagement,* Routledge, London.

Feenberg, A. & Barney, D. (eds) (2004) *Community in the Digital Age,* Rowman & Littlefield, Lanham, MD.

Hague, B. & Loader, B. D. (eds) (1999) *Digital Democracy: Discourse and Decision Making in the Information Age,* Routledge, London.

Hassan, R. (2004) *Media, Politics and the Network Society,* Open University Press, Maidenhead.

Hill, K. & Hughes, J. (1998) *Cyberpolitics: Citizen Activism in the Age of the Internet,* Rowman & Littlefield, Lanham, MD.

Holmes, D. (ed.) (1997) *Virtual Politics: Identity and Community in Cyberspace,* Sage, London.

Jankowski, N. (2002) 'Creating community with media: history, theories and scientific investigations', in *The Handbook of New Media,* eds L. Lievrouw & S. Livingstone, Sage, London, pp. 34–49.

Kavada, A. (2005) 'Civil society organizations and the internet: the case of Amnesty International, Oxfam and the World Development Movement', in *Global Activism,* Global Media, eds W. de Jong, M. Shaw & N. Stammers, Pluto Press, London, pp. 208–225.

Livingstone, S. & Bober, M. (2005) *UK Children Go Online: Final Report of Key Project Findings,* Department of Media and Communications, London School of Economics and Political Science, London.

Loader, B., Nixon, P., Rucht, D. & van de Donk, W. (eds) (2004) *Cyberprotest: New Media, Citizens and Social Movements,* Routledge, London.

Margolis, M. & Resnick, D. (2000) *Politics as Usual: The Cyberspace 'Revolution',* Sage, London.

Miller, D. & Slater, D. (2000) *The Internet: An Ethnographic Approach,* Berg, Oxford.

Montgomery, C., Gottlieb-Robles, B. & Larson, G. (2004) *Youth as E-Citizens: Engaging the Digital Generation,* Center for Social Media, School of Communication, American University, Washington, DC.

Olsson, T. (2004) *Oundgängliga resurser: Om medier, IKT och lärande bland partipolitiskt aktiva ungdomar* [Indispensable resources: On media, ICTs and learning among young, politically active people], Media and Communication Studies Research Reports, Lund.

Olsson, T. (2005a) 'Young citizens, ICTs and learning: a design for a study of the media and political activity', *Nordicom Review,* vol. 26, no. 1, pp. 131–140.

Olsson, T. (2005b) *Alternativa resurser: Om medier, IKT och lärande bland ungdomar i alternativa rörelser* [Alternative resources: On media, ICTs and learning among young people in alternative movements], Media and Communication Studies Research Reports, Lund.

Olsson, T. (2006a) 'Active and calculated media use among young citizens: empirical examples from a Swedish study', in *Digital Generations: Children, Young People and New Media,* eds D. Buckingham & R. Willet, Lawrence Erlbaum, Mahwah, NJ, pp. 115–130.

Olsson, T. (2006b) 'Appropriating civic information and communication technology: a critical study of Swedish ICT-policy visions', *New Media and Society,* vol. 8, no. 4, pp. 611–627.

Pickerill, J. (2004) 'Rethinking political participation: experiments in internet activism in Australia and Britain', in *Electronic Democracy: Mobilisation, Organization and Participation via New ICTs,* eds R. Gibson, A. Römmele & S. Ward, Routledge/ ECPR Studies in European Political Science, London, pp. 170–193.

Poster, M. (1997) 'Cyberdemocracy: internet and the public sphere', in *Internet Culture,* ed. D. Porter, Routledge, New York, pp. 201–218.

Smith, M. & Kollock, P. (eds) (1999) *Communities in Cyberspace,* Routledge, London.

Stammers, N. & Eschle, C. (2005) 'Social movements and global activism', in *Global Activism,* Global Media, eds W. de Jong, M. Shaw & N. Stammers, Pluto Press, London, pp. 50–67.

Van Dijk, J. (2005) *The Deepening Divide: Inequality in the Information Society,* Sage, London.

Van Dijk, J. (2006) *The Network Society: Social Aspects of New Media,* 2nd edn, Sage, London.

Net Gains in Political Participation: Secondary Effects of Internet on Community

Andrea Kavanaugh, B. Joon Kim,
Manuel A. Pérez-Quiñones, Joseph Schmitz
and Philip Isenhour

Broad and diverse civic participation is essential to a democratic society. Studies of opinion leadership show that politically active citizens report that Internet information and communication helped increase civic involvement by enabling them to keep up more easily with news, interact with fellow citizens or engage in collective action. Yet information about less active citizens remains scant. Does the Internet influence the politically passive majority of citizens to become more involved in political talk or other forms of participation? Do they report that the Internet has been helpful in increasing their involvement in political issues, interactions with other citizens, or with local government? These kinds of impacts that follow the primary effect of gaining access to information are considered secondary effects of the Internet upon political participation (Sproull & Kiesler 1991). This article presents quantitative and qualitative findings from a case study of local political participation from the Blacksburg, Virginia region, within the context of a mature community computer network (the Blacksburg Electronic Village). Not only do politically active, but some politically passive citizens also report increased Internet use to communicate with other citizens and with fellow members of local groups about local or national issues. Further, in addition to politically active citizens, some politically passive citizens report that web logs (blogs) have fostered greater online exchange with other citizens through ad hoc political talk and knowledge sharing. These research findings help us to understand the secondary effects of the Internet on political participation in local communities.

Keywords: political participation, survey research, community computer networks, social impact

Introduction

Whether called 'strong democracy' (Barber 1984), 'discursive democracy' (Dryzek 1990) or 'deliberative democracy' (Guttman & Thompson 1996), public discussion and deliberation are intended to help citizens increase their awareness and knowledge about civic issues and shared problems—to foster solutions that benefit a wider range of stakeholders. Policy decisions without broad citizen involvement more often reflect minority views and the limited interests of power elites. Although most of the population is not politically active, it does stay informed to some extent. Non-involved citizens hold opinions on issues, although they may not necessarily be well informed (Delli Carpini & Keeter 1996).

Numerous studies have investigated why a 'silent majority' of citizens does not engage in political activities to a greater extent (Bimber 1999; Gastil & Levine 2005; Lupia & McCubbins 1998; Norris 1999; Verba et al. 1995). These studies suggest that lesser political participation may stem from lower political efficacy, interest, and motivation—conditions often associated with lower socioeconomic status. Further, even more interested passive citizens have found it difficult to access background information and context regarding news about particular issues, even at local levels. But presently the Internet makes information more accessible to a much broader population, just as it offers greater opportunities for discussion and political discourse through online bulletin boards, forums and presently, web logs (or blogs).

Evidence clearly demonstrates increasing electronic political participation (Barabas 2004; Bimber 2003; Chadwick 2006; Dahl 1989; Dahlberg 2001; Hague & Loader 1999; Kavanaugh et al. 2005a). Studies also show that cost reductions to obtain and share political information coupled with more easily accessible information stem from the Internet and foster greater political participation among politically active citizens (Dutton 1999; Johnson & Kaye 1998; Kavanaugh & Patterson 2001; Norris 2000; Putnam 2000; Verba et al. 1995). E-mail and listservs can increase interested citizens' knowledge of events and provide a means to coordinate and mobilize collective action. But much of this political communication transpires among politically active citizens—a small minority in US society, roughly 8 percent to 12 percent (Bimber 2003; Shane 2004). Moreover, announcements by e-mail or listservs are predominantly in broadcast, rather than discussion, mode (e.g. 'come to the 7:00 p.m. meeting at this location').

Bimber's (2001) large study of US data demonstrated that information cost reductions and increased access to information via many media, including the Internet, have not led to aggregate increases in political participation throughout the general population. Consequently, Bimber argues that social and psychological factors shape participation of less-interested or motivated citizens. Social networks can affect participation of passive citizens through affective, cognitive and social pathways (such as, local groups and clubs). Putnam (2000) also emphasized the importance of social gatherings (formal and informal, e.g. bowling leagues and car pools) where casual conversation and political talk raise awareness and may spur collective action. These mundane gatherings have potential to foster political participation.

Some emerging technologies transcend information broadcasting and encourage greater interaction and discussion among users. For example, blogs and wikis can foster more interactive and collaborative participation and information production by creating forums for discussion and knowledge sharing (Kavanaugh et al. 2005b). Easy set-up and content authoring coupled with free availability make blogs ideal Internet-based media for both inexperienced and veteran Internet users. Moreover, while activists and some group leaders use blogs and wikis to express opinions, share information and raise awareness about political issues—most people who write, read and comment on blogs are not politically active (Coleman 2005; Drezner & Farrell 2004; Gruhl et al. 2004; Herring et al. 2004; Johnson & Kaye 2004; Kavanaugh et al. 2006a; Kumar et al. 2003; Lenhart & Fox 2006; Nardi 2004; Rainie 2005).

Our prior research demonstrated that politically active citizens used the Internet to increase involvement in issues of interest to them and to connect both like-minded *and* diverse citizens within the community (Kavanaugh & Patterson 2001; Kavanaugh et al. 2005a). Still, important questions remain: Can Internet use increase political or collective efficacy among politically passive citizens? Do these citizens find that Internet use increases their involvement in issues of interest to them, or that it helps to connect them with similar and/or diverse residents of their local community?

The present article examines how the Internet may be used to spur political discussion and participation among politically passive citizens. We examine offline and online local political participation in the context of the Blacksburg Electronic Village (BEV). We present findings from household survey, interview and focus group data from residents of the town of Blacksburg and surrounding Montgomery County, Virginia.

Prior Research

Political Participation

Political participation can include many diverse activities and processes; we acknowledge that some political acts or apathy may serve to resist democratic norms or demonstrate alienation from the body politic. Because present research envisions political participation *within* traditional American norms and values and our research is set in the context of the BEV, a community network designed to foster local ties and political enfranchisement, the present research seeks to understand and thereby extend participative democracy within local communities. Therefore we use validated instruments that capture such political participation as voting, keeping up with news, attending public meetings, contacting elected officials, contributing campaign donations, engaging in formal discussions and informal political talk (Almond & Verba 1963; Verba et al. 1995).

Levels of Political Participation

Many studies categorizing political participation levels rest on the work of Almond and Verba (1963). While subsequent researchers made refinements, most studies employ three broad categories of participation: active, passive and apathetic. Active political participation includes the most active community (or national) leaders who demand responsible government and may engage in protests, meetings and demonstrations. This category also encompasses less active citizens who exhibit more conventional behaviour, such as working with others on local problems, maintaining membership in community organizations and contacting officials on social issues. Nonetheless, these less active citizens keep informed about politics, engage in political discussions and may write letters (or e-mails) to newspaper editors. These persons may also contact officials about unique problems related to personal needs or agendas.

Politically passive citizens are less likely to engage in political activity individually or within the groups to which they belong. Even so, they may have important and unique knowledge about their neighbourhood, business or recreational activities. They represent latent political participation especially when they become personally affected (e.g. issues of 'not in my backyard'), in response to important local issues (e.g. controversial zoning) or important national issues (9/11 and Hurricane Katrina). The blog use of politically passive citizens (reading, posting comments or authoring) sometimes includes political opinions as part of the larger mix of online musings about their lives.

The final category, politically apathetic, refers to citizens who have virtually no interest in politics. These persons do not vote, do not engage in political activity or join local community groups. They rarely discuss politics. This segment typically comprises about one-fifth of the total US population (Verba et al. 1995).

People elect to participate in politics for many different reasons (Almond & Verba 1963; Putnam 2000; Verba & Nie 1972). Those with higher education and income more often own property and may feel they have more at stake during political decision-making. They typically have a greater sense of civic duty—an obligation to contribute to the common good. Increased political participation by socio-economic elites has also been associated with greater political participation by parents and attributed to their upbringing. Such individuals may have learned to care about community, in addition to their own well-being. Higher socio-economic status may foster greater generosity towards those who are less fortunate. Politically less active citizens more typically become involved, especially at local levels, when issues directly affect them or their immediate neighbourhoods (e.g. land development, proposed zoning changes or the local schools).

Political Efficacy

People with higher levels of education and income more often believe that they can influence what happens to them in their daily lives. This construct of self-efficacy (Bandura 1997, 2000, 2001) has been well established by many studies that found that people with higher levels of education and income more often strongly believed they could influence political outcomes. These beliefs have been called subjective political competence (Almond & Verba 1963) or internal political efficacy (Robinson et al. 1999; Verba et al. 1995). In contrast, persons who believe that government is not interested in what people like them think have low levels of *internal* political efficacy while individuals who believe that a political system is responsive to the populace are said to have a high sense of *external* political efficacy. At the local community level, Carroll and Reese (2003) developed a measure they call 'community collective efficacy' which measures a person's belief that the local community, as a whole, can work together to solve problems despite such obstacles as delays, shortage of resources or other setbacks.

Local Groups

Typically, citizens find it difficult to engage individually and directly in national (or even local) debates, although highly motivated individuals sometimes do. More often, citizens engage in political processes through local and national groups. The importance of local groups to civil society and democratic participation has been firmly established. Local groups and voluntary associations provide common ways for individuals to learn about and participate in civic life (Norris 2000; Putnam 2000; Verba & Nie 1972). Political activity is thereby embedded in non-political *and* political groups or institutions (Verba et al. 1995). Even churches have taken stands on political issues such as abortion and war. Recreational groups also engage in politics when they campaign for resources, such as bike trails, soccer fields or land preservation. In these ways, groups collectively influence political outcomes. These mundane face-to-face interactions can be supplemented, reinforced and extended by the additional online sharing of ideas and information (Kavanaugh & Schmitz 2004; Kavanaugh et al. 2006b; Wellman et al. 2001). Writing well before the Internet, Mancur Olson (1965) stressed that new technology revolutionizes the ways in which groups work just as it enhances the potential to work in groups. Noveck (2005) agreed, claiming that Internet-based technologies in particular could give rise to collective institutions that engage more people in new forms of participation. Our previous household surveys of Blacksburg and Montgomery County found that the computer-mediated communication use by local groups (religious, social, professional and civic) had increased over time, especially among instrumental (goal-oriented) groups, such as civic and political organizations (Kavanaugh et al. 2006b).

Internet Use

The design of ICT shapes its capacity to support and foster citizen interaction, information production, information consumption and content manipulation (Dearden & Walker 2003). While many persons use electronic mail, group mailing lists (such as listservs) and websites to disseminate political information, these technologies are optimally designed for one-to-many communication. Important differences between listservs and blogs shape each technology's capacity to enhance citizen participation and political communication. Listservs provide communication for people in which membership in a group (virtual or 'real') and subscription to listserv mailings are prerequisites for persons to get messages. In contrast, blogs (and to some extent wikis) are much more open systems that typically extend participation to a universe of potential contributors. The 'subscription' bars for participation are usually low, although password and login requirements may sometimes restrict access. Blog writers often describe their daily lives and experiences; they stress their own insights about the world around them while subscribers to a particular listserv usually share commonality with others on that listserv—homophily that brought listserv members together in the first place (e.g. church, interest group, academic researchers).

Notably, the informal conversational style of blog writing, which usually includes comments from readers, provides a personal exchange of political views (typically interspersed with musings about daily life, photos and hyperlinks) that might better engage politically passive citizens in political discussion. Most blog writers are politically passive and focus their blog content on their own lives, experiences and perspectives (Gruhl et al. 2004; Lenhart & Fox 2006; Nardi et al. 2004; Rainie 2005). Even so, most of these 'politically passive' blog writers sometimes express concerns about local or national political problems (Coleman 2005; Kavanaugh et al. 2006a). Therefore, the pool of blog readers, which may also be politically passive, will read these political writings just as they read other content on the blogs that they follow. This process offers a silent majority segment of the population an opportunity to become more aware of and subsequently become drawn into discussions of local issues.

Thus, we pose the following research questions: Do politically passive citizens report increases in political participation, including discussion about local issues, as a consequence of Internet use? Do they experience increased political or collective efficacy that may be attributed to Internet use?

Methodology

The present research is part of a three-year-period longitudinal case study that investigated citizens' and local governments' use of the Internet, especially citizen-to-citizen deliberation and governmental integration of civic deliberation into decision-making. It also assessed the impacts of innovative, collaborative tools designed to support these participative civic governance. This research used both quantitative and qualitative methods. It employed methodological triangulation and thus included two waves of random sample household surveys in addition to interviews with government representatives and citizens. We also used homogenous focus groups designed to elaborate on findings from the survey data about different levels of political activity and Internet use (Morgan 1998). A primary goal of the larger research project is to modify existing online tools and to develop new tools that support citizen-to-citizen online discussion and knowledge sharing.

The larger research project employed a random sample of adult participants from Blacksburg and Montgomery County community members. While focus group results represent the views of a few individuals, they are consistent with quantitative survey data and qualitative interview data. They offer rich insights into how innovative tools for citizen interaction and collaboration can enable citizens, especially 'politically passive' individuals, to collaboratively address local neighbourhood and community issues. We recruited focus group participants from household telephone survey respondents (Spring 2005; N = 717; 40 per cent response rate). Upon survey completion, we asked respondents if they would later participate in a focus group. More than 70 per cent of respondents agreed and comprised a database of focus group candidates.

Exploratory analyses of the Spring 2005 survey data included developing constructs and testing relationships among variables based on our prior research (National Science Foundation IIS-0080864). One year later, we conducted a second round with the same telephone survey using the same respondents. From the original 717 first wave respondents, we obtained 430 respondents who completed the second wave (response rate: 76 per cent). This second survey wave provides confirmatory analyses for the present results. Except for changes over time, this article reports findings from the second wave of data (N = 430).

Household Survey Constructs

Demographics

We investigated demographics, such as age, gender, education and socio-economic status. We also examined psychological factors such as trust, extroversion, measures of efficacy; we measured behavioural factors such as political participation, local group affiliation and Internet uses and consequences of that use.

Political Participation

The composite measure of political participation that we called Political Discussion Network consisted of 11 items on the survey questionnaire (see Table 1). This composite construct is composed of two smaller constructs: Political Talk and Willingness to Talk About an Important Local Issue. The Political Talk construct (Cronbach alpha = 0.83) measured the frequency of respondent's talk with family and non-family members about local, national or global issues while the Local Issue construct measured respondents' willingness to speak about an important (to them) local issue at a public forum. The Political Discussion Network construct had good internal reliability (Cronbach alpha = 0.74) based on 337 survey respondents. The distribution of responses on the Political Talk construct mapped very closely with the distribution of responses on Political Discussion Network and it provided more focus group candidates. We divided the respondent population into quartiles according to the respondents' scores on Political Discussion Network based on the findings of prior studies

TABLE 1 Political discussion network and political talk constructs.

construct name (range) mean	alpha value (N)	survey questions
Political discussion network (Standardized z scores) (−1.37–1.94) −0.045	0.74 (337) (Round 1)	In the last six months, on average how frequently have you: 1. Attended a local political talk or meeting? 2. Attended a public meeting? 3. Discussed politics? 4. Talked with family members about local issues or concerns? 5. Talked with family members about national or global issues? 6. Talked with people outside your family about local issues or concerns? 7. Talked with people outside your family about national or global issues? 8. How many people (outside your family) do you talk with about an issue that you consider to be the most important issue facing the area? 9. What is likelihood of your attending a public meeting on this issue? 10. What is likelihood of your speaking at a public meeting on this issue? 11. What is likelihood of your expressing a different opinion on this issue at a public meeting?
Political talk (1–6) 3.60	0.83 (708)	In the last six months, on average how frequently have you: 1. Talked to family members about local issues or concerns? 2. Talked to family members about national or global issues or problems? 3. Talked to people outside your family about local issues or problems? 4. Talked to people outside your family about national or global issues or problems?

Note: Standardized z scores are used when the scales are not identical for different questions in the construct. Maximum and minimum scores are reported.

that the 'active' population comprised only 8 per cent to 12 per cent of the total US population while the 'apathetic' segment constituted about 15 per cent to 20 per cent of this populace. We standardized the responses with z scores because our constructs used questions with different scales.

We created and tested constructs that measured levels of civic and political interest and activities both offline and online (see Tables 2 and 3).

Local Group Affiliation

We obtained the number of formal local groups that respondents belonged to and we asked them detailed questions about each of their groups. Respondents provided the different modes of communication (offline and online) that their group used to communicate, their level of involvement in each group (leader, member, financial contributor, attendee, volunteer). Finally, we asked three questions about political discussion in the group: (1) Are political topics discussed formally?, (2) Do group members chat informally about politics? and (3) Does this group take a stand on local or national issues?

Political Efficacy, Collective Efficacy, Trust in Government

Political efficacy has two aspects: internal (subjective) and external. We measured internal political efficacy using the level of agreement with: 'Sometimes local politics and government seem so complicated that persons like me can't truly understand what's going on.' We measured external political efficacy with the level of agreement with: (1) 'I don't think local public officials care much what people like me think'; and (2) 'There are plenty of ways for people like me to have a say in what our local government does.' We measured collective efficacy with the level of agreement with: 'I am convinced that we can improve the quality of life in the local community, even when resources are limited'. We

TABLE 2 Civic and political interest and behavior offline.

construct name (range) mean	alpha value (N)	survey questions
Political interest offline (1–4) 1.39	0.525 (430) (Round 2)	In the last six months, how frequently have you: 1. Attended a local political talk or meeting? 2. Wrote or called a local government official?
Political activity offline (1–3.43) 1.38	0.533 (422) (Round 2)	In the last two years, how frequently have you: 1. Attended a neighborhood meeting? 2. Wrote a letter or email to a local area newspaper editor? 3. Called in or emailed a local radio station? 4. Circulated or signed a petition for a local candidate or issue? 5. Watched a town council or board of supervisors meeting on cable television? 6. Worked locally for a political campaign? 7. Contacted a local public school official about an issue of concern to you? 8. Protested about a local issue?
Civic activity offline (1–5.4) 3.22	0.57 (429) (Round 2)	In the last six months, how frequently have you: 1. Read local news in the paper? 2. Read national or global news in the newspaper? 3. Attended a religious service? 4. Attended a public meeting? 5. Did volunteer work?

TABLE 3 Political attitudes: efficacy and trust.

construct name (range) mean	alpha value (N)	survey questions
Political efficacy (1–4) 2.73	0.6 (417) (Round 2)	Level of agreement with: 1. Sometimes local politics and government seem so complicated that persons like me can't truly understand what's going on (reverse coded). 2. I don't think local public officials care much what people like me think (reverse coded). 3. There are plenty of ways for people like me to have a say in what our local government does.
Collective efficacy (1–4) 3.34	NA (423) (Round 2)	I am convinced that we can improve the quality of life in the local community, even when resources are limited (level of agreement).
Trust government (1–5) 3.18	NA (423) (Round 2)	In general, do you trust your local government in handling local problems a great deal, a good deal, a fair amount, not very much, or not at all?

measured Trust in Local Government with: 'In general, do you trust your local government in handling local problems?' (see Table 4).

INTERNET USE

We measured respondents' Internet use with their yes or no reply to the question: 'Do you use the Internet from any location?' We then asked respondents where they used the Internet: home, work, school and other? Many questions determined how often respondents used the Internet for a plethora of political and civic purposes (Political Activity Online and Civic Activity Online constructs); see Table 3. We measured respondents' use of blogs with three frequency-level questions about blog reading, commenting, and writing (0 = never, 1 = less than monthly, 2 = more than monthly). Using recoded response categories—the sum of scores across these three questions, plus the question 'heard of blogs' (yes = 1, no = 0)—yielded a scale that ranged from 0 to 7.

Focus Group Interviews

We used the focus group discussions to understand more completely how civic participation might be linked to Internet use. We sought to examine the synergies among citizens' use of various communication media (newspaper, radio, telephone, postal mail and Internet). We wanted to discover how these media helped participants stay informed and interact with other citizens about local community issues. We also explored citizens' relationships to and their interactions with local government regarding their personal interests, problems and concerns. We selected focus group participants in order to create homogeneous groups that maximized group variance on political participation and Internet (including blog) use (see the following section).

We conducted focus group interviews with adult residents of the town of Blacksburg and Montgomery County (population 83,000 divided evenly between the towns of Blacksburg, Christiansburg and the surrounding rural county). Two hour focus groups were held in each town's public library; sessions were audio taped and transcribed using digital tapes. Participants were assured of confidentiality and were free to *not* take part in focus groups at any time for any reason. They were compensated $50. We report findings from four of the six groups. Because this article focuses on Internet

TABLE 4 Political and civic activity online.

construct name (range) mean	alpha value (N)	survey questions
Political activity online (1–3.3) 1.20	0.57 (336) (Round 2)	In the last six months, how frequently have you: 1. Worked online for a political party or candidate? 2. Tried to influence a government policy or affect a politician's view? 3. How often in the past six months did you send email to a local government official?
Civic activity online (1–3.8) 1.85	0.68 (336) (Round 2)	Internet use to: 1. Get national or global news. 2. Get local news. 3. Look for information on BEV website. 4. Look for information on the Montgomery website. 5. Look for information on the Town of Blacksburg website. 6. Communicate with other residents about local concerns or issues that interest you. 7. Post factual information for other citizens. 8. Express your opinion in online forums or group discussions.
Blog use (round 1 total *N* = 717) sum of blog activity: heard/read/post/write (0–7) 1.15	SUM (four variables) (555) (Round 1)	1. Have you ever heard of web logs or blogs? (yes/no) How often in the past six months have you used the Internet to (never, less than monthly/more than monthly): 2. Read a web log or blog, or an online journal (0–2)? 3. Post a comment on a web log, blog, or online journal (0–2)? 4. Write in your own web log, blog, or online journal (0–2)?

Note: Blog use scale (0–7) derived from sum of four variables shown: yes/no = 0/1; 0 = never; 1 = less than monthly; 2 = more than monthly.

impacts we do not report findings from a politically active, non-Internet group. We also planned for a politically passive, low income, cell phone only group. Because only two participants showed up for that planned focus group, we were not able to conduct the cell phone focus group.

Participant Selection Constructs

We selected focus group participants from the master list of candidates based on their (1) level of political participation and (2) Internet use. We used the construct Political Talk to stratify participants according to their level of political participation in the community. The Political Talk construct was correlated highly with the Political Discussion Network construct (0.89) and had high internal reliability (alpha = 0.83). The response distribution of the Political Talk construct mapped very closely with Political Discussion Network response distribution while it provided more eligible focus group candidates. We determined whether an interview candidate was a politically *active* blog user or politically *passive* blog user by their score on the Political Talk construct (see Table 1). Table 5 compares

TABLE 5 Demographics of focus group participants.

focus group	age (range)	percent male	percent Caucasian (N)	political talk score (range)
1. Politically passive; blog user	40.7 (24–76)	50% (5/10)	80% (8/10)	3.5 (2.3–5)
2. Politically active; blog user	43.25 (31–65)	33.3% (4/12)	91.7% (11/12)	5.4 (4–6)
4. Politically active; Internet user	47.4 (23–72)	36.4% (4/11)	100% (11/11)	5.7 (5.25–6)
5. Passive–active; Internet user	43.4 (24–74)	45.5% (5/11)	81.8% (9/11)	4.7 (4.25–5)

Note: There is no focus group 3 reported here, as that was a non-Internet user group.

demographic and Political Talk scores of participants in the four focus groups. Selection of some participants into blog user focus groups did *not* mean that all of them were experienced Internet users. Some blog users described themselves as computer or Internet novices; often these users had grown children who helped them set up and bookmark favourite blogs.

Focus Group Discussion Protocol

The three overarching focus group discussion topics were: (1) What information technology was being used, for what purposes, and how did its use differ among users? (2) How did information technology foster local political participation by different types of users? and (3) What Internet capabilities would spur use of information technology for local political participation? During focus groups, we probed about participants' community interests and activities, their involvement with local community groups, and asked about their online and offline interactions with government and other citizens. We asked how and why they used different communication media (newspapers, radio, TV and the Internet) for civic purposes. We also sought to understand different persons' motivations for local political participation both offline and online, their sense of political and collective efficacy and their level of trust. The focus group protocol included probes adapted from questions we had developed for prior BEV research supported by the National Science Foundation, and by incorporating questions developed by other researchers, such as Horrigan et al. (2004), Putnam (2000) and Verba et al. (1995).

Interview Data Analyses

We analysed the focus group discussions using NVIVO software. We coded focus group data (words, phrases) as closely as possible to match the comparable variables and constructs that we first used in the survey data. Coding categories that we used included: (1) length and types of Internet use for various purposes (e-mail, Web, listservs), (2) political involvement offline and online, (3) communication with government, citizens and local groups and (4) political or collective efficacy.

Results

Due to space limitations, we do not include all details for the statistical tests. Please see the project website: http://java.cs.vt.edu/public/projects/digitalgov/Data for more details on project statistics.

Demographics

As expected, political participation was correlated with demographic factors such as education and income. It was also associated with sociability (e.g. extroversion), affiliation with formal and informal groups, political and collective efficacy and trust in government. Consistent with other studies, Internet users differed from non-users: Internet users were more likely to be affluent, have higher education, to be younger, married and have children living at home. They were also more likely

TABLE 6 Correlates of Internet use variables [VTCHCI12].

variable name	mean (SD)	valid N	Spearman's ρ correlations	valid N
Age	44.4 (15.37)	552	−0.393**	709
[a]Household income	1.58 (0.49)	518	0.298**	661
[b]Education	4 (1.4)	553	0.467**	712
[c]Children at home	0.43 (0.49)	495	0.111**	602
Formal group affiliations	1.15 (1.26)	554	0.179**	713
Informal Group Affiliations	0.70 (1.02)	553	0.209**	703

** $p < 0.01$

Note: Of 717 (Round 1) respondents, 557 or 77.7% reported they used the Internet. Spearman's ρ Correlations calculated between responses to the question 'Do you use the Internet from any location' and selected demographic characteristics show that Internet use is correlated with age, estimated household income, education, marital status, living with children or not, extroversion, number of formal and informal group affiliations. Similar correlations were found for Round 2. Notes on Response Codes. [a]1 = less than $50,000, 2 = $50,000 or more. [b]1 = eighth grade or less, 2 = some high school, 3 = high school grad GED, 4 = some college/certificate program, 5 = graduated from college or certificate program, 6 = some graduate level work, 7 = completed graduate school/professional school. [c]1 = some of the HH members are younger than 18 years, 0 = none of the HH members are younger than 18 years.

than non-users to belong to formal and informal local groups (see Table 6). In our sample, Internet users were more likely to live in Blacksburg than in Christiansburg or the surrounding (rural) Montgomery County. Focus group participant demographics were not representative of the local population, since we selected participants to contrast their level of political participation and Internet use, and sought a diverse mix of men and women of different ages.

Household Survey

The Political Discussion Network construct responses (total Round 2 sample, N = 430) were normally distributed. We divided respondents into quartiles *based on their score (standardized z score) on the Political Discussion Network construct* (see Table 7); standardized z scores ranged from a minimum of −1.42 to a maximum of 1.65 with a mean of −0.042. Each quartile for the Political Discussion Network construct has a range of 77 points. Respondents in the lowest quartile of political activity had scores in the range of −1.42 to −0.65 range (standardized z score); respondents in the highest quartile of political activity had scores in the 0.88 to 1.65 range (standardized z score) as shown in Table 7. Thus, the percentage of respondents that fell into each 'political activity quartile' is roughly what would be expected based on most prior studies noted earlier (i.e. that political activists make up only about 8–12% of the US population). For example, as shown in Table 7, our 'politically active quartile' is 7.9% of the total respondent population; our 'politically apathetic quartile' is 15.8% of the population. It is important to bear in mind that the quartiles refer to respondents' *political activity scores,* not the percentage of the population. Thus, the most active political quartile has only a small percentage of the population.

The broad, sweeping pattern within the data showed that Passive–Active citizens responded quite similarly to Active citizens on the primary survey constructs; further they significantly differed from Passive–Apathetic and Apathetic citizens. Important differences included: (1) civic and political interests and activities offline and online, (2) political and collective efficacy, (3) discussing politics formally or informally at meetings of local groups with which they were affiliated, and (4) whether these group(s) took stands on local or national issues. Further, similar patterns emerged regarding the impacts (secondary effects) of Internet use on: (1) respondents' involvement in local issues of interest, (2) connections and interaction with other citizens (both homophilous and diverse individuals) and (3) interactions with local government. We next examine these findings in detail.

TABLE 7 Quartiles for the level of political participation construct.

level of participation	N	percent	range: z score	mean z score
Apathetic	68	15.8	−1.42 to −0.65	−0.994
Passive–apathetic	189	44.0	−0.65 to 0.12	−0.253
Passive–active	143	33.2	0.12 to 0.88	0.442
Active	30	7.0	0.88 to 1.65	1.135

One-way ANOVA tests indicated significant differences in Political Efficacy across the four levels (quartiles) of political participation. A substantially higher level of political efficacy was reported in the Politically Active group, followed by Passive–Active, Passive–Apathetic and Apathetic. Politically Active respondents were significantly different from Passive–Apathetic and Apathetic respondents while the responses of Passive–Active citizens were similar to (not significantly different from) the other quartiles. For Collective Efficacy, one-way ANOVA tests indicated significant differences across the four levels of political participation. The highest level of Collective Efficacy was reported in the Passive–Active group, followed by Active, Passive–Apathetic and Apathetic. Politically Active and Passive–Active citizens were not significantly different from each other, but Passive–Active respondents reported significantly higher Collective Efficacy than did Passive–Apathetic and Apathetic respondents. None of the quartiles differed significantly on measures of Trust in Government. One-way ANOVA tests of the three constructs for Offline Political and Civic Activity and Interest, each showed significant differences across the quartiles. In all three constructs, Active and Passive–Active respondents differed significantly from each other and from the lower quartiles. Passive–Apathetic and Apathetic respondents were not significantly different from each other on all three constructs for Offline Political and Civic Interest and Civic Activity (see Table 8).

One-way ANOVA tests of Political Discussion in Local Groups, again broken down by quartile, showed significant differences across the four levels of political participation. Active citizens reported the highest level of political discussion in their groups ($M = 4.5$), followed by Passive–Active citizens, then Passive–Apathetic ($M = 2.4$); finally politically Apathetic citizens reported the lowest levels of political discussion within their local groups. Politically Active respondents were significantly higher on measures of discussion with local groups than Passive–Apathetic and Apathetic citizens. But the politically Passive–Active were not significantly lower on these measures than the politically Active (see Table 9).

Paired t-tests combined the groups to determine if all respondents showed significant increases in Political Discussion within their local groups across time (rounds one and two), for example, secondary, longitudinal Internet effects. When we controlled for different types of groups (Civic/Political, Religious/Charitable, Educational/Professional and Social/Recreational), we found that significant increases in Political Discussion within local groups had occurred in Civic/Political, Religious/Charitable and Social Recreational groups; the magnitude of the increase in Educational/Professional groups was substantial but the effect was not statistically significant, most likely because of the small number of these groups (see Table 10).

One-way ANOVA tests of the Politics Online and Civic Online constructs showed significant differences across the four quartiles. Active and Passive–Active respondents were significantly different from all three quartiles for Online Politics. Although Active and Passive–Active respondents did not significantly differ for Civic Online, both groups differed significantly on this construct from Passive–Apathetic and Apathetic respondents (see Table 11).

One way to examine 'secondary' Internet effects (across the spectrum of political activity) was to assess differential respondents' attributions (in the Political Participation quartiles) of Internet effects on political and collaborative activity. We conducted one-way ANOVA tests on variables that measured how much the Internet helped to increase respondents' local involvement in issues of interest and their feeling more connected with diverse and like minded citizens. Regarding the helpfulness of the Internet for (1) becoming more involved in issues of interest, (2) improving interaction with local

TABLE 8 Political attitudes and offline behavior.

	apathetic (AP)	passive–apathetic (PAP)	passive–active (PAC)	active (AC)
Political efficacy* $F = 5.540$, DF = 3/426, $p < 0.001$, $\eta^2 = 0.038$	2.60 [AC]	2.65 [AC]	2.87	3.71 [AP, PAP]
Collective efficacy* $F = 5.366$, DF = 3/409, $p < 0.001$, $\eta^2 = 0.038$	3.20 [PAC]	3.23 [PAC]	3.52 [AP, PAP]	3.47
Trust local in government NS	3.12	3.14	3.27	3.07
Political interest offline* $F = 30.739$, DF = 3/335, $p < 0.001$, $\eta^2 = 0.216$	1.07 [PAC, AC]	1.20 [PAC, AC]	1.56 [AP, PAP, AC]	2.43 [AP, PAP, PAC]
Political activity offline* $F = 93.865$, DF = 3/426, $p < 0.001$, $\eta^2 = 0.398$	1.12 [PAP, PAC, AC]	1.25 [AP, PAC, AC]	1.51 [AP, PAP, AC]	2.12 [AP, PAP, PAC]
Civic offline* $F = 16.901$, DF = 3/426, $p < 0.001$, $\eta^2 = 0.106$	2.83 [PAC, AC]	3.09 [PAC, AC]	3.43 [AP, PAP, AC]	3.99 [AP, PAP, PAC]

*$p < 0.01$

Note: Eta-squared (η^2) is defined as the proportion of total variation attributable to the factor, and it ranges in value from 0 to 1. Eta-squared values typically range from 0.01 to 0.09 in the social sciences (see, for example, Cohen 1988).

Post hoc Multiple Comparisons (Bonferroni) test was used on all variables; equal variances were assumed. Group superscripts indicate significant differences between the respective groups (Bonferroni test).

TABLE 9 Political discussion in local groups.

	apathetic (AP)	passive–apathetic (PAP)	passive–active (PAC)	active (AC)
Political discussion in local groups** $F = 7.441$, DF = 3/280, $p < 0.001$, $\eta^2 = 0.072$	2.18 [PAC, AC]	2.37 [PAC, AC]	3.57 [AP, PAP]	4.50 [AP, PAP]

**$p < 0.01$.

Note: Post hoc Multiple Comparisons (Bonferroni) test was used on all variables; equal variances were assumed. Group superscripts indicate significant differences between the respective groups (Bonferroni test).

TABLE 10 Political discussion in local groups by type of group (paired *t*-test).

local groups:	mean round 1	mean round 2	*p*-values (*t*)	*n*
All groups combined	2.25	3.00	0.000** (–4.52)	245
Civic/political groups	2.76	3.56	0.013* (–2.55)	84
Religious/charitable	2.18	2.83	0.001** (–3.30)	173
Educational/professional	3.53	4.37	NS	38
Social/recreational	2.8	3.61	0.008* (–2.72)	75

$^{**}p < 0.01$, $^{*}p < 0.05$.

TABLE 11 Online political and civic behavior.

	apathetic (AP)	passive–apathetic (PAP)	passive–active (PAC)	active (AC)
Politics online** $F = 30.739$, DF = 3/335, $p < 0.001$, $\eta^2 = 0.216$	1.06 ^{PAC,AC}	1.10 ^{PAC,AC}	1.26 ^{AP,PAP,AC}	1.73 ^{AP,PAP,PAC}
Civic online** $F = 9.934$, DF = 3/335, $p < 0.001$, $\eta^2 = 0.082$	1.62 ^{PAC,AC}	1.73 ^{PAC,AC}	1.98 ^{AP,PAP}	2.28 ^{AP,PAP}

$^{**}p < 0.01$.

Note: Post hoc multiple comparisons (Bonferroni) test was used on all variables; equal variances were assumed. Group superscripts indicate significant differences between the respective groups (Bonferroni test).

government and (3) improving interaction with other citizens, there were significant differences across the four quartiles of political participation. Active citizens reported the highest levels followed by Passive–Active, followed by Passive–Apathetic and Apathetic. Active respondents were not significantly different from Passive–Active respondents; both differed significantly from Passive–Apathetic and Apathetic citizens. Similar results emerged regarding the extent to which the Internet aided connections with like-minded citizens. Active respondents were not significantly different from Passive–Active respondents, but they were significantly different from Apathetic citizens. These data show clear and important trends: that as offline political participation increases, the Internet becomes increasingly important to sustain both online political activity and interaction with other community members—effects which seem to increase over time (see Table 12).

Overview of Focus Group Participants

We selected focus group participants for variance in the level of political participation, and Internet and blog use. We used the Political Talk construct, shown in Table 1, as our measure of political participation. Beyond these selection priorities, we also sought diversity in gender, age and ethnicity. Because almost all survey respondents were Caucasian (reflecting the local population), most focus group members were also Caucasian. Just as female respondents predominated in household surveys, they outnumbered males in focus groups. Focus group participants' ages ranged from 23 to 76. Although we invited 10 participants to each focus group (and reminded them several days before scheduled interviews) we did not always have 10 attendees. Thus, the final focus group make-up (and participant number) was determined by who actually showed up (see Table 6).

We invited the participants on a priority basis (i.e. we first called people who were highest on the selection measures of political participation, Internet use and blog use). Therefore, focus group participants are not representative of the larger survey sample or the local population. The Political

TABLE 12 Secondary effects of Internet.

net helpful for:	apathetic (AP)	passive–apathetic (PAP)	passive–active (PAC)	active (AC)
... becoming more** involved in issues $F = 7.186$, $DF = 3/330$, $p < 0.001$, $\eta^2 = 0.061$	1.58 [PAC,AC]	1.94 [PAC,AC]	2.33 [AP,PAP]	2.76 [AP,PAP]
... feeling more connected to like-minded citizens** $F = 4.653$, $DF = 3/323$, $p < 0.001$, $\eta^2 = 0.041$	2.00 [PAC,AC]	2.32	2.60 [AP]	2.68 [AP]
... feeling more connected to diverse citizens* $F = 2.808$, $DF = 3/323$, $p < 0.05$, $\eta^2 = 0.025$	1.87	2.16	2.37	2.42
Improving interaction with local government** $F = 16.441$, $DF = 3/279$, $p < 0.001$, $\eta^2 = 0.150$	1.43 [PAC,AC]	1.73 [PAC,AC]	2.30 [AP,PAP]	2.74 [AP,PAP]
Improving interaction with other citizens** $F = 15.100$, $DF = 3/326$, $p < 0.001$, $\eta^2 = .122$	1.41 [PAC,AC]	1.78 [PAC,AC]	2.27 [AP,PAP]	2.76 [AP,PAP]

$^{**}p < 0.01$, $^{*}p < 0.05$.

Note: Post hoc multiple comparisons (Bonferroni) test was used on all variables; equal variances were assumed. Group superscripts indicate significant differences between the respective groups (Bonferroni test).

Talk scores for focus group participants in the Politically *Passive* Blog User category (focus group 1) ranged from 2.3 to 5 (on a scale of 6) with a mean of 3.5 which is higher than the mean for the total respondent pool in the survey data. Similarly, the Political Talk scores for focus group participants in the Politically *Active* Blog User category (focus group 2) is higher than that of the total survey respondent pool; the focus group scores on the Political Talk construct ranged from 4 to 6 (on a scale of 6) with a mean of 5.4. With regard to *Blog Use,* the average and range of scores among focus groups 1 and 2 was typically higher than the total survey respondent population, as we had hoped. Blog Use scores for Politically *Passive* Blog Users (focus group 1) ranged from 3 to 5 on a scale of 7, with a mean of 3.7; for the Politically Active Blog Users (focus group 2), blog use scores ranged from 2 to 6 (on a scale of 7), with a mean of 3.4.

Focus groups 4 and 5 comprised survey respondents who reported using the Internet but had not heard of blogs. (There is no focus group 3 reported here, as that was a non-Internet user group.) Internet use was measured by a simple yes response to the question, 'Do you use the Internet from any location?' The two groups differed on measures of Political Talk. Specifically, the Political Talk scores of Politically *Active* Internet Users (Group 4) ranged from 5.3 to 6 (on a scale of 6) with an average of 5.7. The Political Talk scores of Politically *Passive* Internet Users (Group 5) ranged from 4.3 to 5 (on a scale of 6), with an average of 4.7. Thus, Politically *Active* Internet users had higher range and average scores on Political Talk by precisely one point than Politically *Passive* Internet users.

Political Discussion, Internet, and Blog Use: Evidence from Focus Group Interviews

We were struck by the emphasis of focus group participants' statements about how their use of various media (e-mail, listservs, websites) greatly helped them understand and navigate their involvement in their neighbourhoods, schools or communities—particularly among participants with lower levels of political participation (Politically passive participants with moderate Political Talk scores). Politically passive participants said that they usually acted when their interests were at stake, as exemplified by these statements:

> *I do like reading both sides of the argument on things that matter to me. The problem is, the Internet, I mean it reaches the whole world and so many of the things they are discussing don't have anything to do with me. And so I don't know why I have, I need to have an opinion on this or that. The things that I do have an opinion on I've listened to and thought about and um, have based my decision on listening to both sides. You do need to listen to both sides. But I don't see why you have to have an opinion on everything.*
>
> *I am somewhat active in the community because, I bike a lot. I commute to work and I'm a firm believer that we should get an infrastructure for bicyclers biking to work, um, for many reasons . . . So I am pretty active in like 'rails to trails' and stuff like that. Bicycle advocacy groups. But most of that is through mailings . . . like the postal mailings.*
>
> *Our entire community and neighborhood gets together and shows up at the forums or contacts the city manager or whoever the developer is that is going to do whatever and we are the proprietors to that large 3.5 acre lot . . . they keep on wanting to build up there and we keep on telling them ain't going to happen. So we've got John who sends out the listserv and they go to you, you get some sort of information or at least you get some word to what's going on and we all show up.*
>
> *I had various things I was trying to find. Maybe um, just community events, thing to do with my children you know . . . and I did a lot of searching just through the parks and rec[creation] site.*
>
> *I also just recently because of a situation at my son's school, have finally found my voice and found a cause I'm willing to fight for and I have been using it a lot to research where Montgomery County stands on what I'm interested in and what I want to know, communicating with his teachers, the principal, with the superintendent. Um, and also for research for my son's medical problem.*

Moreover, Passive–Active Internet users were not necessarily sophisticated technologically. Often they were not very savvy, but through their social networks, by other informal means, or through self-learning, they eventually discovered how to do what interests them with regard to blogs, as these statements illustrate:

> Now, how did you find those blogs?
>
> My wife found them. I have to be honest with you, I don't know how to get into any of this stuff. She found them and she earmarked them or whatever you call it and I push the button.
>
> Yeah, my daughter wrote a cheat for me to get into the email and I do read some of the blogs that I have a very clear map to get to. You know, push this button now type of thing. We started getting computers where I work about three years before I retired. I didn't use them at work and didn't have them when I was a kid.

Both Politically Active and Politically Passive–Active focus group members reported using the Internet to extend their interpersonal relationships with other community members, especially with local groups they're involved in; they used these technologies within their neighbourhoods to complement offline activity:

> *[F]or the community we have a community bulletin paper that goes out I guess about six times a year and so I'm in contact with that group 'The Linkletter' and I for that and, you know, I get, they send information. A couple of ladies at my church are always looking for controversial issues and so they will send me*

> *whatever's the latest thing that Proctor and Gamble has done wrong or something, you know.*
>
> *[J]ust as a family, soccer team communications, the Internet is essential. Un, PTA communications. I'm on the board of my church. We all communicate by email, you know, there any organization that I'm either involved with uh, professionally or as a volunteer. I mean that's just the way we communicate and I think if you don't have the Internet, you're at a serious disadvantage now, I think, because it's real hard for to [sic] people who are part of an organization that have to get the phone call instead of email.'*

Diverse Political Information Sources

Both Politically Active and Politically Passive–Active focus group members stated that the Internet, and blogs in particular, helped them to access political information that they did not encounter in mainstream media (e.g. newspapers, radio and TV). In this way, websites and blogs provided good alternative sources for information and news:

> *Um, national blogs, they are political mostly and I try to get sources of news other than the Roanoke Times because it is very biased and you only get one side of the news. Um, and so I try to read a couple of other newspapers on the Internet to try to get different view points and some of the blogs from both sides of the spectrum. Because I figure half way between their stands is going to be the truth. Um, and some of the things that are brought out . . . are never in the Roanoke Times or the Washington Times or the Washington Post. You will see it in the blogs is the only place you're going to ever see it. Uh, it's like some of the interviews and truth there following the hurricane.*
>
> *[O]ne of the advantages of a blog is you actually do get eye witnesses. Now you can't tell whose the eye witness and whose making it up, but some of the information is the source, and some of it's not . . . and some of it, the political blogs, generally are opinion.*
>
> *[T]he first place you saw the fact that the people in the nursing home didn't die after four days [when Katrina hit] was in the blogs. And so, it's a matter of getting information from more than one view point.*

However, diverse focus group participants also complained about misinformation online and especially within blogs:

> *I followed the New Orleans situation with great detail. I have tons of stuff on it . . . I put out about three newsletters on the New Orleans timeline and then, there was too much, uh, misinformation. I usually, when I get the material together I like to get at least two different sources so that I don't have to go by one source.*
>
> *I find that so many things, well especially whenever you're just talking about blogging out there. Nothing is sourced, there is no way to validate, other than in somebody's opinion.*
>
> *To me that's part of the problem with blogs, is a lot of people just have more time than sense and, just go on and on and on and on.*

So, while the Internet and blogs offer alternative news sources and eyewitness accounts, they also offer a sea of opinions and suspect remarks that are difficult to verify. Nonetheless, the diversity of information and knowledge represented by the many different websites and blogs can far exceed the fare presented by mainstream media.

Regarding intrinsic political efficacy, that is, citizens' belief that they can be effective in trying to influence government, is typically stronger among Internet users, especially discussion list users (i.e. notably listservs); they used the Internet to raise awareness, educate others and organize collective action:

> *Um, I also use it [the Internet] to whip up the neighbors when we, we had to stop a subdivision in our neighborhood last year and our homeowner's association all talk by email and so we showed up at town council meetings in large numbers.*
>
> *I go to Blacksburg.gov a lot [the town government website]. And I found that they're awfully responsive. You can go and tell them you've got a streetlight out and they're there. Uh, I find it to be a wonderful website.*

Some Passive–Apathetic participants, however, especially those with lower socio-economic status, show low political efficacy and strongly believe that important decisions have already been made by power elites:

> *Well, you get so many minutes to speak and no matter where you are, 'you're done' and they wave, you know, at that time, I don't know if they still do it but we were there until 1 am because they put anybody who wants to speak at the end in hopes that everybody will go. But it didn't matter what anybody had to say, they did what they wanted anyway. So, I felt that was pretty useless. A waste of our time . . .'*

This final point resonates with Bandura's (2001) warning, that citizens will use the Internet for political purposes, but only to the extent that they believe it is efficacious to do so. Thus, offline circumstances must be favourable to political participation coupled with online deliberative potential in order to support citizen-to-citizen interaction and information production.

Conclusion

Clearly citizens engage in local political life that ranges from minimal to very active and involved; they use communication media that ranged from traditional face-to-face interaction to an intensive use of information and communication technology. The present data from household surveys and focus groups provide a glimpse into current technology uses and potential benefits for political engagement via emerging technologies. We find that blogs are becoming integrated into the mix of news sources and political discussion outlets even for relatively unsophisticated Internet users.

Notably, Politically Passive–Active citizens (the passive quartile closest to the Politically Active) were more similar to Politically Active citizens than their 'other half'—the Passive–Apathetic (the passive quartile close to the Apathetic citizens) on most measures in our study. Passive–Active citizens are not as similar to Active citizens in offline political interest and activities or in formal online politics, but they are quite similar to them in online civic activities. These online civic activities include, (1) using the Internet to get local, national and global news; (2) looking for information on the Blacksburg Electronic Village website, and the local town and county websites; (3) communicating with other residents about local concerns or issues that interest them; (4) expressing their opinion in online forums or group discussions; and (5) in reading, posting or writing blogs.

With respect to secondary effects of Internet use, Passive–Active respondents also often report similar Internet effects as Active citizens. They report increased involvement in issues of interest to them, improved interaction with other citizens and with local government. Focus group discussions help us to understand how this occurs in peoples' mundane lives—through everyday conversations during formal and informal political discussion within local neighbourhood and community groups.

Thus, we found evidence of overall increases in Internet benefits that extend beyond politically active citizens but to less politically active respondents. Notably focus group discussions indicated that blogs, in particular, drew not only the politically active but *also* helped to engage less politically active citizens into greater online exchange with other citizens through ad hoc political talk and knowledge sharing.

Thus, the differences among various online tools for communication (such as, listservs, blogs and wikis) seem often more marked within the respective social system they support rather than with respect to the technology itself. In this way, listservs are better suited to maintain communication and awareness within a group of like-minded individuals, while blogs seem more effective for individuals to express their opinions towards a broader, more heterogeneous audience, and thereby introduce more diverse sources of information and knowledge, even without knowledge of who reads their content or opinions.

Still, in stark contrast, for passive-apathetic and apathetic citizens we find little evidence yet that the Internet (including blogs) helps bring these individuals into community or political decision-making spheres. These persons may still feel disconnected from 'power games' and that they lack influence in political outcomes. This finding differs considerably from popular press reports that suggest that blogs effectively bring more voices to civic discourse. Although we find that these communication technologies add voices from engaged segments of the population, voices from passive-apathetic and apathetic groups largely remain silent. The multi-method, longitudinal nature of the present study lends confidence to these findings; if we are to broaden enfranchisement, it seems powerful social and technological interventions remain needed.

About the Authors

Andrea Kavanaugh, a Fulbright scholar and Cunningham Fellow, is a Senior Research Scientist in the Department of Computer Science at Virginia Tech. She is also the Associate Director of the Center for Human-Computer Interaction (HCI). Her research lies in the areas of social and educational computing, communication behaviour and effects, and development communication. Prior to joining the HCI Center in 2002, she served as Director of Research for the community computer network known as the Blacksburg Electronic Village (BEV) from its inception in 1993. Kavanaugh holds an MA in Communication from the Annenberg School for Communication, University of Pennsylvania, and a PhD in Planning from Virginia Tech. *Address:* Virginia Tech, Department of Computer Science, 2202 Kraft Drive, Blacksburg 24060, US. [e-mail: kavan@vt.edu]

B. Joon Kim is a PhD candidate in Public Administration and Policy at Virginia Tech, and a Graduate Research Assistant with the Digital Government project at the Center for Human Computer Interaction, Virginia Tech. His research interest is the use and social impact of information and communication technology for citizen interaction and collaboration on local governance. *Address:* Virginia Tech, Department of Computer Science, 2202 Kraft Drive, Blacksburg 24060, US. [e-mail: bjkim@vt.edu]

Manuel A. Pérez-Quiñones is an associate professor in the Department of Computer Science at Virginia Polytechnic Institute and State University. His research interests include human-computer interaction, personal information management, multiplatform user interfaces, user interface software, and educational uses of computers. Pérez-Quiñones received a DSc in computer science from The George Washington University. He is a member of the ACM, the IEEE Computer Society, and Computer Professionals for Social Responsibility. *Address:* Virginia Tech, Department of Computer Science, 2202 Kraft Drive, Blacksburg 24060, US. [e-mail: perez@vt.edu]

Joseph Schmitz is an Assistant Professor in the Communication Department at Western Illinois University. His doctoral research at the University of Southern California developed and tested the Social Influence Model of Technology. Schmitz helped the City of Santa Monica create its Public Electronic Network, the first local government sponsored community network in the United States. His research concentrates on the uses and consequences of communication technology within organizations, local communities and universities. Schmitz chaired the Communication and Technology Division of the International Communication Association and is presently a member of the

Board of Governors of the American Communication Association. *Address:* Western Illinois University, Communication Department, Macomb, US. [e-mail: J-Schmitz@wiu.edu]

Philip Isenhour is Research Associate with the Center for Human Computer Interaction at Virginia Tech. His research areas include distributed collaborative systems, tools development and interface design. His recent research has been published in *International Journal of Human-Computer Studies, Internet Research, Educational Technology and Society, Interacting with Computers, Educational Technology Research and Development, and Computer Supported Cooperative Work*. His research has been supported primarily by the National Science Foundation. *Address:* Virginia Tech, Department of Computer Science, 2202 Kraft Drive, Blacksburg 24060, US. [e-mail: isenhour@vt.edu]

Acknowledgments

We gratefully acknowledge support from the US National Science Foundation, Digital Government Program (IIS-0429274) and Information Technology Research Program (REC-0080864). We also thank our outside advisers, John M. Carroll and Mary Beth Rosson. We thank Jaideep Godara, Alain Fabian, Than Than Zin, Matthew Stover and Andrew Mike for focus group assistance; Tamara Smith and Tabitha Kim provided extensive help with focus group transcription. Carol Bailey provided expert advice and leadership in planning and facilitating all focus group sessions. We thank Susan Willis-Walton and the Virginia Tech Center for Survey Research for conducting the telephone survey interviews and coordinating subsequent focus group invitations. We would also like to thank our anonymous reviewers for many helpful suggestions.

References

Almond, G. & Verba, S. (1963) *The Civic Culture: Political Attitudes and Democracy in Five Nations,* Princeton University Press, Princeton, NJ.

Bandura, A. (1997) *Self-efficacy: The Exercise of Control,* W.H. Freeman, New York.

Bandura, A. (2000) 'Exercise of human agency through collective agency', *Current Directions in Psychological Science,* vol. 9, no. 3, pp. 75–78.

Bandura, A. (2001) 'Social cognitive theory: an agentic perspective', *Annual Review of Psychology,* vol. 53, pp. 1–26.

Barabas, J. (2004) 'Virtual deliberation: knowledge from online interaction versus ordinary discussion', in *Democracy Online: The Prospects for Political Renewal through the Internet,* ed. P. Shane, Routledge, New York, pp. 239–251.

Barber, B. (1984) *Strong Democracy: Participatory Politics for a New Age,* University of California Press, Berkeley, CA.

Bimber, B. (1999) 'The Internet and citizen communication with government: does the medium matter?' *Political Communication,* vol. 16, pp. 409–428.

Bimber, B. (2001) 'Information and political engagement in America: the search for effects of information technology at the individual level', *Political Research Quarterly,* vol. 54, no. 1, pp. 53–67.

Bimber, B. (2003) *Information and American Democracy: Technology in the Evolution of Political Power,* Cambridge University Press, New York.

Carroll, J. M. & Reese, D. D. (2003) 'Community collective efficacy: structure and consequences of perceived capacities in the Blacksburg Electronic Village', *Proceedings of HICSS-36, Hawaii International Conference on System Sciences,* IEEE Computer Society.

Chadwick, A. (2006) *Internet Politics: States, Citizens, and New Communication Technologies,* Oxford University Press, Oxford.

Cohen, J. (1988) *Statistical Power Analysis for the Behavioral Sciences,* 2nd edn, Lawrence Erlbaum, Mahwah, NJ.

Coleman, S. (2005) 'Blogs and the new politics of listening', *The Political Quarterly,* vol. 76, pp. 273–280.

Dahl, R. (1989) *Democracy and its Critics,* Yale University Press, New Haven, CT.

Dahlberg, L. (2001) 'The Internet and democratic discourse: exploring the prospects of online deliberative forums extending the public sphere', *Information, Communication and Society,* vol. 4, no. 4, pp. 615–633.

Dearden, A. & Walker, S. (2003) 'Designing for civil society', *Proceedings of HCI 2003: Designing for Society,* vol. 2, pp. 157–158.

Delli Carpini, M. & Keeter, S. (1996) *What Americans Know About Politics and Why It Matters,* Yale University Press, New Haven.

Drezner, D. W. & Farrell, H. (2004) 'The power and politics of blogs', paper presented at the *2004 American Political Science Association,* April 14–18, 2004, Chicago, IL, [Online] Available at: http://www.utsc.utoronto.ca/~farrell/blogpaperfinal.pdf (15 April 2007).

Dryzek, J. (1990) *Discursive Democracy: Politics, Policy and Political Science,* Cambridge University Press, New York.

Dutton, W. (1999) *Society on the Line: Information Politics in the Digital Age,* Oxford University Press, Oxford.

Gastil, J. & Levine, P. (2005) *Deliberative Democracy Handbook: Strategies for Effective Civic Engagement in the Twenty first Century,* Jossey-Bass, San Francisco, CA.

Guttman, A. & Thompson, D. (1996) *Democracy and Disagreement,* Belknap Press, Cambridge, MA.

Gruhl, D., Guha, R., Liben-Nowell, D. & Tomkins, A. (2004) 'Information diffusion through blogspace', *Proceedings of the Thirteenth International World Wide Web Conference,* May 17–22, 2004, New York.

Hague, B. & Loader, B. (1999) *Digital Democracy: Discourse and Decision Making in the Information Age,* Routledge, London.

Herring, S. C., Scheidt, L. A., Bonus, S. & Wright, E. (2004) 'Bridging the gap: a genre analysis of weblogs', *Proceedings of the Thirty Seventh Hawaii International Conference on System Sciences (HICSS-37),* IEEE Press, Los Alamitos.

Johnson, T. & Kaye, B. (1998) 'A vehicle for engagement or a haven for the disaffected? Internet use, political alienation, and voter participation', in *Engaging the Public: How Government and the Media Can Reinvigorate American Democracy,* eds T. Johns, C. Hays & S. Hays, Rowman & Littlefield, New York, pp. 123–135.

Johnson, T. J. & Kaye, B. K. (2004) 'Was the blog: how reliance on traditional media and the internet influence credibility perceptions of weblogs among blog users', *Journalism and Mass Communication Quarterly,* vol. 81, pp. 622–642.

Kavanaugh, A. & Patterson, S. (2001) 'The impact of computer networks on social capital and community involvement', *American Behavioral Scientist,* vol. 45, no. 3, pp. 496–509.

Kavanaugh, A. & Schmitz, J. (2004) 'Talking in lists: the consequences of computer mediated communication on communities', in *Internet Research Annual: Selected papers from The Association of Internet Researchers Conferences 2000–2002,* Vol. 1, eds M. Consalvo, N. Baym, J. Hunsinger, K. Jensen, J. Logie, M. Murero & L. Regan Shade, Peter Lange, New York, pp. 250–259.

Kavanaugh, A., Carroll, J. M., Rosson, M. B. & Zin, T. T. (2005a) 'Participating in civil society: the case of networked communities', *Interacting with Computers,* vol. 17, pp. 9–33.

Kavanaugh, A., Isenhour, P., Cooper, M., Carroll, J. M., Rosson, M. B. & Schmitz, J. (2005b) 'Information technology in support of public deliberation', in *Communities and Technologies,* eds P. Besselaar, G. de Michelis, J. Preece & C. Simone, Springer, The Netherlands, pp. 19–40.

Kavanaugh, A., Zin, T. T., Carroll, J. M., Schmitz, J., Perez-Quinones, M. & Isenhour, P. (2006a) 'When opinion leaders blog: new forms of citizen interaction', *Proceedings of the Seventh Annual Digital Government Conference,* 20–22 May 2006, San Diego, ACM Press, New York.

Kavanaugh, A., Zin, T. T., Rosson, M. B., Carroll, J. M. & Schmitz, J. (2006b) 'Local groups online: Political learning and participation', *Journal of Computer Supported Cooperative Work,* vol. 15, no. 5 (November), [Online] Available at: http://www.springerlink.com/content/tk770mu72w4v6483/ (6 December 2007).

Kumar, R., Novak, P., Raghavan, S. & Tomkins, A. (2003) 'On the bursty evolution of Blogspace', *Proceedings of the Twelfth International World Wide Web Conference,* May 2003, Budapest, Hungary.

Lenhart, A. & Fox, S. (2006) 'Bloggers: a portrait of the internet's new storytellers', Pew Internet & American Life Project, [Online] Available at: http://www.pewinternet.org/pdfs/PIP%20Bloggers%20Report%20July%2019%202006.pdf (20 July 2006).

Lupia, A. & McCubbins, M. (1998) *The Democratic Dilemma,* Belknap Press, Cambridge, MA.

Morgan, D. L. (1998) *Planning Focus Groups: Focus Group Kit,* SAGE Publications, Thousand Oaks, CA.

Nardi, B. (2004) 'Why we blog', *Communications of the ACM,* vol. 47, no. 12, pp. 41–46.

Norris, P. A. (1999) 'Who surfs? New technology, old voters and virtual democracy in America', in *Democracy.com? Governance in a Networked World,* eds E. Kamarck & J. S. Nye, Jr., Hollis Publishing NH.

Norris, P. A. (2000) Virtuous Circle: Political Communications in Postindustrial Societies, Cambridge University Press, New York.

Noveck, B. S. (2005) 'A democracy of groups', *Firstmonday,* vol. 10, no. 11, [Online] Available at: http://www.firstmonday.org/issues/issue10_11/noveck.html (20 January 2006).

Olson, M. (1965) *The Logic of Collective Action: Public Goods and the Theory of Groups,* Harvard University Press, Cambridge, MA.

Rainie, L. (2005) 'The State of Blogging, Pew Internet & American Life Project', [Online] Available at: http://www.pewinternet.org (30 January 2005).

Robinson, J., Shaver, P. & Wrightsman, L. (1999) *Measures of Political Attitudes,* Academic Press, New York.

Putnam, R. (2000) *Bowling Alone: The Collapse and Revival of American Community,* Simon and Schuster, New York.

Shane, P. (2004) *Democracy Online: The Prospects for Political Renewal through the Internet,* Routledge, New York.

Sproull, L. & Kiesler, S. (1991) *Connections: New Ways of Working in the Networked Organization,* MIT Press, Cambridge, MA.

Verba, S. & Nie, N. (1972) *Participation in America: Political Democracy and Social Equality,* Harper and Rowe, New York.

Verba, S., Schlozman, K. & Brady, H. (1995) *Voice and Equality: Civic Voluntarism in American Politics,* Harvard University Press, Cambridge, MA.

Wellman, B., Haase, A. Q., Witte, J. & Hampton, K. (2001) 'Does the Internet increase, decrease or supplement social capital?' *American Behavioral Scientist,* vol. 45, no. 3, pp. 436–455.

MODULE NINE

Work and Commerce

Ecash

David D. Friedman

I pay for things in one of three different ways—credit card, check, or cash. The first two let me make large payments without having to carry large amounts of money. What are the advantages of the third?

One is that a seller does not have to know anything about me in order to accept cash. That makes money a better medium for transactions with strangers, especially strangers from far away. It also makes it. a better medium for small transactions, since using cash avoids the fixed costs of checking up on someone to make sure that there is really money in his checking account or that his credit is good. It also means that money leaves no paper trail, which is useful not only for criminals but for anyone who wants to protect his privacy—an increasingly important issue in a world where data processing threatens to make every detail of our lives public.

The advantage of money is greater in cyberspace, since transactions with strangers, including strangers far away, are more likely on the Internet than in my realspace neighborhood. The disadvantage is less, since my ecash would be stored inside my computer, which is usually inside my house, and hence less vulnerable to theft than my wallet.

Despite its potential usefulness, there is as yet no equivalent of cash available online, although there have been unsuccessful attempts to create one and successful attempts to create something close. The reason is not technological; those problems have been solved. The reason is in part the hostility of governments to competition in the money business, and in part the difficulty of getting standards, in this case private monetary standards, established. I expect both problems to be solved sometime in the next decade or two.

Before discussing how a system of electronic currency, private or governmental, might work, it is worth first giving at least one example of why it would be useful—for something more important than allowing men to look at pornography online without their wives or employers finding out.

Slicing Spam

My email contains much of interest. It also contains READY FOR A SMOOTH WAY OUT OF DEBT?, A Personal Invitation from make_real_money@BIGFOOT.COM, You've Been Selected . . . from friend@localhost.net, and a variety of similar messages, of which my favorite offers "the answer to all your questions." The Internet has brought many things of value, but for most of us unsolicited commercial email, better known as *spam,* is not one of them.

There is a simple solution to this problem—so simple that I am surprised it is not yet in common use. The solution is to put a price on your mailbox. Give your email program a list of the people you wish to receive mail from. Mail from anyone not on the list is returned with a note explaining that you charge five cents to read mail from strangers—and the URL of the stamp machine. Five cents is a trivial cost to anyone with something to say that you are likely to want to read, but five cents times ten million recipients is quite a substantial cost to someone sending out bulk email on the chance that one recipient in ten thousand may respond.

The stamp machine is located on a web page. The stamps are digital cash. Pay $10 from your credit card and you get in exchange 200 five-cent stamps—each a morsel of encrypted information that you can transfer to someone else who can in turn transfer it.

A virtual stamp, unlike a real stamp, can be reused; it is paying not for the cost of transmitting my mail but for my time and trouble reading it, so the payment goes to me, not the post office. I can use it the next time I want to send a message to a stranger. If lots of strangers choose to send me messages, I can accumulate a surplus of stamps to be eventually changed back into cash.

How much I charge is up to me. If I hate reading messages from strangers, I can make the price $1, or $10, or $100—and get very few of them. If I enjoy junk email, I can set a low price. Once such a system is established, the same people who currently create and rent out the mailing lists used to send spam will add another service—a database keeping track of what each potential target charges to receive it.

What is in it for the stamp machine—why would someone maintain such a system? Part of the answer is seigniorage—the profit from coining money. After selling a hundred million five-cent stamps, you have five million dollars of money. If your stamps are popular, many of them may stay in circulation for a long time—leaving the money that bought them in your bank account accumulating interest.

In addition to the free use of other people's money, there is a second advantage. If you own the stamp machine, you also own the wall behind it—the web page people visit to buy stamps. Advertisements on that wall will be seen by a lot of people.

One reason this solution to spam requires ecash is that it involves a large number of very small payments. It would be a great deal clumsier if we used credit cards—every time you received a message with a five-cent stamp, you would have to check with the sender's bank before reading it to make sure the payment was good. A second reason is privacy. Many of us would prefer not to leave a complete record of our correspondence with a third party—which we would be doing if we used credit cards or something similar. What we want is not merely ecash but anonymous ecash—some way of making payments that provides no information to third parties about who has paid what to whom.

Constructing Ecash

Suppose a bank wants to create a system of ecash. The first and easiest problem is how to provide people with virtual banknotes that cannot be counterfeited.

The solution is a digital signature. The bank creates a banknote that says "First Bank of Cyberspace: Pay the bearer one dollar in U.S. currency." It digitally signs the note, using its private key. It makes the matching public key widely available. When you come into the bank with a dollar, it gives you a banknote in the form of a file on a flash drive. You transfer the file to your hard disk, which now has a one-dollar bill with which to buy something from someone else online. When he receives the file he checks the digital signature against the bank's public key.

The Double Spending Problem

There is a problem—a big problem. What you have gotten for your dollar is not a single dollar bill but an unlimited number of them. Sending a copy of the file in payment for one transaction does not erase it from your computer, so you can send it again to someone else to buy something else. And again. That is going to be a problem for the bank, when twenty people come in to claim your original dollar bill.

One solution is for the bank to give each dollar its own identification number and keep track of which ones have been spent. When a merchant receives your file he sends it to the bank, which deposits the corresponding dollar in his account and adds its number to a list of banknotes that are no longer valid. When you try to spend a second copy of the note, the merchant who receives it tries to deposit it, is informed that it is no longer, valid, and doesn't send you your goods.

This solves the problem of double spending, but it also eliminates most of the advantages of ecash over credit cards. The bank knows that it issued banknote 94602 . . . to Alice, and it knows that it came back from Bill, so it knows that Alice bought something from Bill, just as it would if she had used a credit card.

The solution to this problem uses what David Chaum, the Dutch cryptographer who is responsible for many of the ideas underlying ecash, calls *blind signatures*. It is a way in which Alice, having rolled up a random identification number for a dollar bill, can get the bank to sign that number (in exchange for paying the bank a dollar) without having to tell the bank what the number they are signing is. Even though the bank does not know the serial number it signed, both it and the merchant who receives the note can check that the signature is valid. Once the dollar bill is spent, the merchant has the serial number, which he reports to the bank, which can add it to the list of serial numbers that are now invalid. The bank knows it provided a dollar to Alice, and it knows it received back a dollar from Bill, but it does not know that they are the same dollar. So it does not know that Alice bought something from Bill. The seller has to check with the bank and know that the bank is trustworthy, but it does not have to know anything about the purchaser.

Curious readers will want to know how it is possible for a bank to sign a serial number without knowing what it is. I cannot tell them without first explaining the mathematics of public key encryption, which requires more math than I am willing to assume my average reader has. Those who are curious can find the answers via the virtual footnotes (www.daviddfriedman.com/Future_Imperfect.html), which point to webbed explanations of both public key encryption and blind signatures.

So far I have been assuming that people who receive digital cash can communicate with the bank that issues it while the transaction is taking place—that they and the bank are connected to the Internet or something similar. That is not a serious constraint if the transaction is occurring online. But digital cash could also be useful for realspace transactions, and the cabby or hotdog vendor may not yet have an Internet connection.

The solution is another clever trick (Chaum specializes in clever tricks). It is a form of ecash that contains information about the person it was issued to but only reveals that information if the same dollar bill is spent twice. For an explanation of how it works, you must again go to the virtual footnotes.

Skeptical readers should at this point be growing increasingly unhappy at being told that everything about ecash is done by mathematics that I am unwilling to explain—which they may reasonably enough translate as "smoke and mirrors." For their benefit I have invented my own form of ecash—one that has all of the features of the real thing and can be understood with no mathematics beyond the ability to recognize numbers. It is a good deal less convenient than Chaum's version but a lot easier to explain, and so provides at least a possibility proof for the real thing.

Low-Tech Ecash

I randomly create a very long number. I put the number and a dollar bill in an envelope and mail it to the First Bank of Cybercash. The FBC agrees—in a public statement—to do two things with money it receives in this way:

1. If anyone walks into the FBC and presents the number, he gets the dollar bill associated with that number.
2. If the FBC receives a message that includes the number associated with a dollar bill it has on deposit, instructing the FBC to change it to a new number, it will make the change and post the fact of the transaction on a publicly observable bulletin board. The dollar bill will now be associated with the new number.

Let's see how this works:

Alice has sent the FBC a dollar, accompanied by the number 59372. She now wants to buy a dollar's worth of digital images from Bill, so she emails the number to him in payment. Bill emails the FBC, sending them three numbers: 59372, 21754, and 46629.

The FBC checks to see if it has a dollar on deposit with number 59372; it does. It changes the number associated with that dollar bill to 21754, Bill's second number. Simultaneously, it posts on a publicly observable bulletin board the statement "the transaction identified by 46629 has gone through." Bill reads that message, which tells him that Alice really had a dollar bill on deposit and it is now his, so he emails her a dollar's worth of digital images.

Alice no longer has a dollar, since if she tries to spend it again the bank will report that it is not there to be spent—the FBC no longer has a dollar associated with the number she knows. Bill now has a dollar, since the dollar that Alice originally sent in is now associated with a new number and only he and the bank know what it is. He is in precisely the same situation that Alice was in before the transaction, so he can now spend the dollar to buy something from someone else. Like an ordinary paper dollar, the dollar of ecash in my system passes from hand to hand. Eventually someone who has it decides he wants a dollar of ordinary cash instead; he takes his number, the number that Alice's original dollar is now associated with, to the FBC and exchanges it for a dollar bill.

My ecash may be low-tech, but it meets all of the requirements. Payment is made by sending a message. Payer and payee need know nothing about the other's identity beyond the address to send the message to. The bank need know nothing about either party. When the dollar bill originally came in, the letter had no name on it, only an identifying number. Each time it changed hands, the bank received an email but had no information about who sent it. When the chain of transactions ends and someone comes into the bank to collect the dollar bill he need not identify himself; even if the bank can somehow identify him he has no way of tracing the dollar bill back up the chain. The virtual dollar in my system is just as anonymous as the paper dollars in my wallet.

With lots of dollar bills in the bank there is a risk that two might by chance have the same number, or that someone might make up numbers and pay with them in the hope that the numbers he invents will, by chance, match numbers associated with dollar bills in the bank. But both problems become insignificant if instead of using 5-digit numbers we use 100-digit numbers. The chance that two random 100-digit numbers will turn out to be the same is a good deal less than the chance that payer, payee, and bank will all be struck by lightning at the same time.

Robot Mechanics

It may have occurred to you that if you have to roll up a 100-digit random number every time you want to buy a dollar of ecash from the bank and two more every time you receive one from anyone else, not to mention sending off one anonymous email to the bank for every dollar you receive, ecash may be more trouble than it is worth. Don't worry—that's your computer's job, not yours. With a competently designed ecash system, the program takes care of all mathematical details; all you have to worry about is having enough money to pay your (virtual) bills. You tell your computer what to pay to whom; it tells you what other people have paid to you and how much money you have. Random numbers, checks of digital signatures, blind signing, and all the rest are done in the background. If you find that hard to believe, consider how little most of us know about how the tools we routinely use, such as cars, computers, or radios, actually work.

Ecash and Privacy

When Chaum came up with the idea of ecash, email was not yet sufficiently popular to make spam an issue. What motivated him was the loss of privacy created by the ability of modern information processing to combine publicly available information into a detailed portrait of each individual.

Consider an application of ecash that Chaum has actually worked on—automated toll collection. It would be very convenient if, instead of stopping at a toll booth when getting on or off the interstate, we could simply drive past, making the payment automatically in the form of a wireless communication between the (unmanned) tollbooth and the car. The technology to do this exists and has long been used to provide automated toll collection for busses on some roads.

One problem is privacy. If the payment is made with a credit card, or if the toll agency adds up each month's tolls and sends you a bill, someone has a complete record of every trip you have taken on the toll road, every time you have crossed a toll bridge. If we deal with auto pollution by measuring pollutants in the exhaust plumes of passing automobiles and billing their owners, someone ends up with detailed, if somewhat fragmentary, records of where you were when.

Ecash solves that problem. As you whiz past the tollbooth, your car pays it fifty cents in anonymous ecash. By the time you are thirty feet down the road, the (online) tollbooth has checked that the money is good; if it isn't an alarm goes off, a camera triggers, and if you do not stop a traffic cop eventually appears on your tail. But if your money is good you go quietly about your business—and there is no record of your passing the tollbooth. The information never came into existence, save in your head. Similarly for an automated system of pollution charges.

It works for shopping as well. Ecash—this time encoded in a smart card in your wallet, a palmtop computer in your pocket, or perhaps even a tiny chip embedded under your skin—could provide much of the convenience of a credit card with the anonymity of cash. If you want the seller to know who you are, you are free to tell him. But if you prefer to keep your transactions private, you can.

Private Money: A New Old Story

My examples so far assume that ecash will be produced and redeemed by private banks but denominated in government money. Both are likely, at least in the short run. Neither is necessary.

Private money denominated in dollars is already common. My money market fund is denominated in dollars, although Merrill Lynch does not actually have a stack of dollar bills in a vault somewhere that corresponds to the amount of money "in" my account. My university ID card doubles as a money card, with some number of dollars stored on its magnetic strip—a number that decreases every time I use the card to buy lunch on campus. A bank could issue ecash on the same basis. Each dollar of ecash represents a claim to be paid a dollar bill. The actual assets backing that claim consist not of a stack of dollar bills but of stocks, bonds, and the like—which have the advantage of paying the bank interest for as long as the dollar of ecash is out there circulating.

While I do not have to know anything about you in order to accept your ecash, I do have to know something about the bank that issues it—enough to be sure that the money will eventually be redeemed. That means that any ecash expected to circulate widely will be issued by organizations with reputations. In a world of almost instantaneous information transmission, those organizations will have a strong incentive to maintain their reputations, since a loss of confidence will result in money holders bringing in virtual banknotes to be redeemed, eliminating the source of income that the assets backing those banknotes provided.

Some economists, in rejecting the idea of private money, have argued that such an institution is inherently inflationary. Since issuing money costs a bank nothing and gives it the interest on the assets it buys with the money, it is always in the bank's interest to issue more. The rebuttal to this particular argument was published in 1776. When Adam Smith wrote *The Wealth of Nations,* the money of Scotland consisted largely of banknotes issued by private banks, redeemable in silver.[1] As Smith pointed out, while a bank could print as many notes as it wished, it could not persuade other people to hold an unlimited number of its notes. A customer who holds $1,000 in virtual cash—or Scottish banknotes—when he only needs $100 is giving up the interest he could have been earning if he had held the other $900 in some interest-earning asset instead. That is a good reason to limit his cash holdings to the amount he actually needs for day-to-day transactions.

What happens if a bank tries to issue more of its money than people wish to hold? The excess comes back to be redeemed. The bank is wasting its resources printing money, trying to put it into circulation, only to have each extra banknote promptly returned for cash—in Smith's case, silver. The obligation of the bank to redeem its money guarantees its value, and at that value there is a fixed amount of its money that people will choose to hold.

> *Let us suppose that all the paper of a particular bank, which the circulation of the country can easily absorb and employ, amounts exactly to forty thousand pounds; and that for answering occasional demands, this bank is obliged to keep at all times in its coffers ten thousand pounds in gold and silver. Should this bank attempt to circulate forty-four thousand pounds, the four thousand pounds which are over and above what the circulation can easily absorb and employ, will return upon it almost as fast as they are issued.* (Wealth of Nations, *Bk I, chapter 2)*

So far I have assumed that future ecash will be denominated in dollars. Dollars have one great advantage—they provide a common unit already in widespread use. They also have one great disadvantage—they are produced by a government, and it may not always be in the interest of that government to maintain their value in a stable, or even predictable, way. On past evidence, governments sometimes increase or decrease the value of their currency, inadvertently or for any of a variety of political purposes. In the extreme case of hyperinflation, a government tries to fund its activities with the printing press, rapidly increasing the amount of money and decreasing its value. In less extreme cases, a government might inflate in order to benefit debtors by inflating away the real value of their debts—governments themselves are often debtors, hence potential beneficiaries of such a policy—or it might inflate or deflate in the process of trying to manipulate its economy for political ends.

Dollars have a second disadvantage, although perhaps a less serious one. Because they are issued by a particular government, citizens of other governments may prefer not to use them. This has not prevented dollars from becoming a *de facto* world currency, but it is one reason why a national currency might not be the best standard to base ecash on. The simplest alternative would be a commodity standard, making the unit of ecash a gram of silver or gold or some other widely traded commodity.

Under such a commodity standard the monetary unit, while no longer under the control of a government, is subject instead to the forces that affect the value of the particular commodity it is based on. If large amounts of gold are discovered or if someone invents new and better techniques for extracting gold from low-grade ore, the value of gold, and of gold-based money, will decline.[2] If, on the other hand, important new uses for gold are found but little new gold is mined, the value of gold will rise and prices fall. Thus commodity money carries with it at least some risk of unpredictable fluctuations in its value, and hence in prices measured in it.

That problem is solved by replacing a simple commodity standard with a commodity bundle. Bring in a million Friedman dollars and I agree to give you in exchange 10 ounces of gold, 40 ounces of silver, ownership of 1,000 bushels each of grade A wheat and grade B soybeans, a ton of grade S30040 stainless steel . . . If the purchasing power of a million of my dollars is less than the value of the bundle, it is profitable for people to assemble a million Friedman dollars, exchange them for the bundle, and sell the contents of the bundle—forcing me to make good on my promise and, in the process, reducing the amount of my money in circulation. If the purchasing power of my money is more than the worth of the commodities it trades for, it is in my interest to issue more money. Since the bundle contains lots of different commodities, random changes in commodity prices can be expected to roughly average out, giving us a stable standard of value.

A commodity bundle is a good theoretical solution to the problem of monetary standards, but implementing it has a serious practical difficulty—getting all the firms issuing ecash to agree on the same bundle. If they fail to establish a common standard, we end up with a cyberspace in which different people use different currencies and the exchange rates between them vary randomly.

That is not an unworkable situation—Europeans lived with it for a very long time—but it is a nuisance. Life is easier if the money I use is the same as the money used by the people I do business with. On that fact our present world system—multiple government moneys, each with a near monopoly within the territory of the issuing government—is built. It works because most transactions are with people near you and people near you probably live in the same country you do. It works less well in Europe than in North America because the countries are smaller, which is why the European countries have largely moved from national currencies to the euro.

A system of multiple monopoly government moneys works less well in cyberspace because in cyberspace national borders are transparent. For information transactions, geography is irrelevant—I can download software or digital images from London as easily as from New York. For online purchases of physical objects geography is not entirely irrelevant, since the goods have to be delivered, but less relevant than in realspace shopping. With a system of national currencies, everyone in cyberspace has to juggle multiple currencies in the process of figuring out who has the best price and paying it. The obvious solution is to establish a single standard of value, either by adopting one national currency, probably the dollar, possibly the euro, or by establishing a private standard such as the sort of commodity bundle described earlier.

That may not be the only solution. The reason that everyone wants to use the same currency as his neighbors is that currency conversion is a nuisance. But currency conversion is arithmetic and computers do arithmetic fast and cheap. Perhaps, with some minor improvements in the interfaces on which we do online business, we could make the choice of currency irrelevant, permitting multiple standards to coexist.

I live in the United States; you live in India. You have goods to sell, displayed on a web page, with prices in rupees. I view that page through my brand new browser—Firefox v 9.0. One feature of the new browser is that it is currency transparent. You post your prices in rupees but I see them in dollars. The browser does the conversion on the fly, using exchange rates read, minute by minute, from my bank's web page. If I want to buy your goods, I pay in dollar-denominated ecash; my browser sends it to my bank, which sends rupee-denominated ecash to you. I neither know nor care what country you are in or what money you use—it's all dollars to me.

Currency transparency will be easiest online, where everything filters through browsers anyway. One can imagine, with a little more effort, realspace equivalents. An unobtrusive tag on my lapel gives my preferred currency; an automated price label on the store shelf reads my tag and displays the price accordingly. Alternatively, the price is displayed by a dumb price tag, read by a smart video camera set into the frame of my glasses, converted to my preferred currency by my pocket computer, and written in the air by the heads-up display generated by the eyeglass lenses.

As I write, the countries of Europe are in the final stages of replacing their multiple national currencies with the euro. If the picture I have just painted turns out to be correct, they may have finally achieved a common currency just as it was becoming unnecessary.

We now have three possibilities for ecash. It might be produced by multiple issuers but denominated in dollars or some other widely used national money. It might be denominated in some common nongovernmental standard of value—gold, silver, or a commodity bundle. It might be denominated in a variety of different standards, perhaps including both national monies and commodities, with conversion handled transparently, so that each individual sees a world where everyone is using his money. Any of these forms of ecash might be produced by private firms, probably banks, or by governments.[3]

Will It Happen?

During World War II, George Orwell wrote regular articles for *Partisan Review,* an American magazine. Near the end of the war, he wrote a retrospective in which he discussed what he had gotten right and what wrong.[4] One of his conclusions was that he was generally right about the way the world was moving, wrong about how fast it would get there. He correctly saw the logical pattern but failed to allow for the enormous inertia of human society.

Similarly here. David Chaum's articles laying out the groundwork for fully anonymous electronic money were published in technical journals in the 1980s and summarized in a 1992 article in *Scientific American*. Ever since then various people, myself among them, have been predicting the rise of ecash along the lines he sketched. While pieces of his vision have become real in other contexts, there is as yet nothing close to a fully anonymous ecash available for general use. Chaum himself, working with the Mark Twain Bank of Saint Louis, attempted to get a semi-anonymous ecash into circulation—one that permitted one party to a transaction to be identified by joint action of the other party and the bank. The effort failed and was abandoned.

One reason it has not happened is that online commerce has only very recently become large enough to justify it. A second reason, I suspect but cannot prove, is that national governments are unhappy with the idea of a widely used money that they cannot control and so are reluctant to permit (heavily regulated) private banks to create such a money. A third and closely related reason is that a truly anonymous ecash would eliminate a profitable form of law enforcement. There is no practical way to enforce money-laundering laws once it is possible to move arbitrarily large amounts of money anywhere in the world, untraceably, with the click of a mouse. A final reason is that ecash is only useful to me if many other people are using it, which raises a problem in getting it started.

These factors have slowed the introduction of ecash. I do not think they will stop it. It only takes one country willing to permit it and one issuing institution in that country willing to issue it, to bring ecash into existence. Once it exists, it will be politically difficult for other countries to forbid their citizens from using it and practically difficult, if it is forbidden, to enforce the ban. There are a lot of countries in the world, even if we limit ourselves to ones with sufficiently stable institutions so that people elsewhere will trust their money. Hence my best guess is that some version of one of the moneys I have described in this chapter will come into existence sometime in the next decade or so.

Notes

1. White, 1995.
2. The example is not wholly imaginary. At the end of the nineteenth century, the combined effect of the discovery of the South African gold fields and the invention of the cyanide process for extracting gold from low-grade ore led to a gold inflation, although a mild one compared to fiat money inflations. An earlier gold (and silver) inflation was produced in the sixteenth century by the inflow of both metals from the New World.
3. For a current product along these lines, see http://www.e-gold.com. So far as I know, their money is not anonymous ecash, but it is set up so that your holding is defined in gold but payments can be converted into a variety of currencies.
4. "London Letter to *Partisan Review,*" in Orwell, 1968.

Consuming Code: Use-Value, Exchange-Value, and the Role of Virtual Goods in Second Life

Jennifer Martin

In recent years, there has been significant growth in consumption of commodities in virtual social worlds, such as Second Life, and in the economies that arise from this practice. While these economic systems have been acknowledged and studied, there remains relatively little understanding of the reasons why individuals choose to purchase such goods, despite the fact that reasons for consumption are strong enough to drive a virtual goods industry with annual profits in the millions of dollars. Virtual goods, the author argues, meet no immediate needs for avatars or individuals and, as such, are purchased based exclusively on their exchange- and symbolic-values. Due to the graphical nature of Second Life and the consequent visibility of commodities within the environment, these reasons for purchasing virtual goods are explored in terms of their roles for users, and especially in terms of their potential for expressing wealth, power, status, individuality, and belonging. As such, this paper considers the roles of consumption in a way that relies on and further illuminates theories of consumption and value with respect to virtual environments and commodities.

Keywords: consumption, Second Life, use-value, exchange-value, symbolic-value.

Every day, thousands of people spend millions of dollars on goods that exist only within a virtual environment and the computer servers on which they are housed. This world is the online social world known as Second Life, and these people are many of its millions of residents. With regards to Second Life, there is no question that virtual goods are being purchased, but there are questions as to why they are being consumed. Within this virtual commodity system goods are not needed or used in a conventional way. Although they could be coded otherwise, avatars within Second Life are not programmed to have any requirements for their survival—they do not get cold, hungry, or thirsty, and therefore have no immediate need for goods that will slake these conditions. As such, the immaterial goods within this world have no capacity to fulfill physical needs for either the user of their virtual representation. However, despite these limitations millions of dollars are still being spent on the purchase of virtual goods (Dibbell, 2006).

Modern theories of consumption suggest that use-value, or the ability of a good to fulfill a need, has been overridden by exchange-value (Jhally, 1987) as well as sign- and symbolic-value (Baudrillard, 2000). Since they lack any use-value in terms of physical needs, but still appear to be compelling commodities, an examination of virtual goods highlights and further develops ongoing claims that conventional use-value is no longer a characteristic that must be present for goods to be consumed. By considering the meanings of virtual goods for their users, it is possible to elaborate on current understandings of consumption in virtual worlds beyond the purely economic elements of such environments. An analysis of virtual goods within Second Life reveals that despite the fact that they fulfill no material needs for participants or avatars, the role of virtual commodities as symbols of status, individuality, and belonging are significant enough to subsume use-value promote consumption as seen in a substantial number of sales transactions and significant economic activity.

Consumption in Second Life

In recent years, there has been increasing development of graphically rendered online social environments. These social worlds provide an environment where individuals can create an avatar as a visual representation of themselves and, using this virtual body, interact with the world and other individuals in a variety of different ways. Created by Linden Lab, Second Life is one virtual world that focuses on social interaction while also offering a variety of different forms of entertainment, including socializing, playing, and shopping. As with many other such worlds, Second Life is graphical, and provides users with a visible world and a body through which to experience and interact within it. The environment was made public in October of 2003, and as of August 17, 2008 housed 14,760,237 "residents" (Linden Lab, 2003–2008b). Currently, Second Life offers a highly developed online community and market as a result of the five years that it has had to develop and build not only a solid base of residents, but also of commerce.

Despite their focus on social interaction, virtual social worlds tend to involve consumption as an element of the world's environment. Where once online spaces, and especially those involved in socializing, were largely free of such practices, consumption is rapidly moving into online worlds through the presence of virtual goods (Poster, 2004). On the "What is Second Life?" webpage, three main features of the environment are presented to potential users: community, creativity, and commerce (Linden Lab, 2003–2008b). While it is listed on its own, both community and creativity also feed into and further commerce. In terms of virtual goods, creativity is required in their development, while community is required in their consumption. As a result, the two remaining features in Second Life can also be associated with virtual consumerism.

As in the offline world, virtual worlds are populated with a wide variety of goods that are available to outfit the avatar in their virtual life. Virtual goods are especially prevalent in graphical worlds, since in these environments goods can be visually rendered onscreen and therefore become visible to both the immediate user and any other users who happen to be present within the environment (Bartle, 2003). As a result, there is no dearth of goods to be bought and sold within the world, and commodities for sale include clothes, electronics, property, houses, furniture, vehicles, and almost anything else that a developer can imagine and for which they can create the computer code.

Prices for these goods differ based on a number of factors, including rarity, custom work, time and effort required to create the item, and the whims of the developer, but ultimately all virtual goods are graphically rendered code. In many cases, these goods are developed by users who have the technical skill to create not only the commodities themselves, but ways of customizing, animating, and making functional their goods, especially those that are intended to be interactive or animated in nature. These goods should not only be rendered visually, but they must also function within the world—clothes must cover the bodies of those who buy them, and vehicles must respond to steering and braking commands—which requires a certain level of knowledge and skill to make them functional and attractive enough for purchase (Bartle, 2003).

While not everyone has the skills to generate a full income, by selling their goods users accumulate in-world currency, known as the Linden dollar (LD), that they can use to purchase commodities. Conversely, they also have the option to "cash out" and exchange their Lindens for offline currencies, such as the US dollar. Conversely, these links between offline and online currency also allow for Linden dollars to be purchased with other currencies. Therefore, while currency can be earned in-world by creating and selling goods, or by purchasing a membership with an allowance, Linden dollars (LD) may also be easily obtained through a standard currency exchange between online and offline economies. The Linden dollar is valued at approximately $240 LD to $1 USD and rises and falls with the market for virtual goods (Walsh, 2006). No matter how users acquire their money, though, with cash in (virtual) hand they are able to begin to purchase virtual goods with which to outfit their virtual lives.

To date, a significant amount of work has been generated on the economic implications of consumption in virtual worlds. Edward Castronova has produced extensive examinations of the costs of virtual goods (2001, 2002, 2003), how money circulates through virtual and real markets (2001, 2002, 2005), and the role of the avatar within these systems and in virtual worlds in general (2003, 2005). For Castronova, "Virtual worlds may be the future of ecommerce, and perhaps of the internet itself" (2001: 3), suggesting that worlds like Second Life may be heralds of larger economic systems yet to come. Within this analysis, desire for virtual goods is shown through the fact that individuals actively seek and are willing to work for and even spend offline money on virtual goods. Similarly, F. Gregory Lastowka and Daniel Hunter claim that virtual goods are subject not only to conventional forms of supply and demand, but also to unconventional sales tactics, such as migration to online auction sites (2007). These tactics, which require extra time, effort, and money to purchase commodities also indicate increased desire for such goods. As such, there is little doubt that there is desire furthering the sale of virtual goods.

Although the desire to consume virtual goods is frequently acknowledged as a driving force of this economy, in many cases the factors underlying and driving this consumption are largely absent from analyses of virtual capital. There are multiple signs pointing to a rising desire for virtual goods, but in order to understand this desire in more than general terms, it is necessary to interrogate reasons for virtual consumerism in more specific terms.

Use-Value and Exchange-Value: The Debate

Despite customary readings of use-value and exchange-value through the lenses of materiality and abstraction (Marx, 2000), these terms have been contested and debated in ways that are further illuminated by the consumption of virtual goods. Recent debates surrounding consumption have primarily centered on the links between use-value and exchange-value, and especially on the ways in which exchange-value is implicated in use-value. Traditional readings of the work of Karl Marx consider use-value to be distinct from exchange-value. In this context, use-value is a material reality inherent in a physical object while, in contrast, exchange-value is a socially assigned system of value determining an object's worth in relation to other goods, and not by virtue of any inherent characteristics within the object itself (Marx, 2000: 10).

However, this divide between use-value and exchange-value has not gone unproblematized. For many theorists there are faults in Marx's argument, especially in his failure to acknowledge that the symbolic elements of exchange-value can affect use-value (Baudrillard, 2000; Jhally, 1987). Here, as suggested by Sut Jhally, the notion that exchange-value can supplant use-value is especially relevant. Virtual goods offer an exemplary case of this subsumption because they lack use-value and yet continue to be bought and sold at a rapid rate. In July of 2008, Second Life users made 19,937,851 user-to-user transactions (Linden Lab, 2008a) and collectively spent close to $84 million USD in one quarter (*Ibid,* 2008b). Despite their inability to meet any material needs, trade in these goods is reaching millions of dollars per month. As a result, exchange-value has subsumed a use-value that never was, not only because virtual goods are incapable of meeting physical needs, but also because virtual bodies in Second Life *are* not programmed to have them.

In Marx's account of the valuation of goods, use-value is positioned as the ability of a good to fulfill a material but not necessarily a social need (2000). Although Marx does not explicitly make clear that use-value is purely practical and not linked to social needs, it is this perspective that is generally taken up by later theorists dealing with consumption (Baudrillard, 2000; Debord, 1994). This is not to say, however, that commodities do not have a use beyond their application to material needs. While use-value is commonly associated with physical needs, such as shelter and food, through their symbolic application commodities can meet less immediately material but equally important needs such as belonging and individuality.

Given that these needs exist even in the absence of use-value, exchange-value may also be determined in the context of sign- and symbolic-value, or the social values that are given to commodities (Baudrillard, 2000). For Baudrillard, "that which precisely the commodity was for Marx—is no longer today properly either commodity or sign, but indissolubly both" (1981, p. 148). Here, the symbolic-value of an object comes into play and works to value commodities within their social context. This movement towards the value of the symbolic can be seen in Second Life, where exchange-values associated with virtual goods can be read as a product of what they signify for individuals as well as what they represent for the community at large. Given the absence of use-value within Second Life, it is the symbolic role of virtual goods that will be seen to determine the roles of virtual goods within Second Life, their exchange-values and the rates at which they are sold.

The Roles of Virtual Goods

The roles of virtual goods and the drive to purchase them can be tied to recent debates around the ways in which use-value and exchange-value function in a world increasingly based on capitalism, consumption, and consumerism. Since virtual goods do not meet physical needs, but are still purchased in abundance, it stands that there must be forces behind this consumption that are not rooted in use-value. As a result of the lack of use-value inherent in virtual goods, this driving force is linked to exchange-value in multiple ways, including profitability and affordability, social belonging, status, conspicuous consumption, identity and selfhood, individuality, and social lubrication.

Although they are intrinsically linked, the roles of virtual consumption in Second Life can be divided between the focus of the virtual economy and the focus of the virtual consumer. The focus of the virtual economy is to sell goods and, for many residents, to generate profit. Here, producers seek to create goods and position them in such a way that they are attractive enough to residents that they will be purchased. Conversely, the focus of the virtual consumer is to use those goods in a way that augments their virtual life. Consequently, they seek out the goods that are best going to fill their need and desire for items that will allow for the appearance of status and uniqueness within the virtual world. By positioning goods in a way that plays into what consumers wish to gain from their purchases, the virtual economy is able to use only the symbolic and social needs and desires of virtual consumers to sell their products.

Virtual Goods and the Economy

Profitability

In general, the virtual economy is concerned with selling its goods to generate profits. Since users are able to exchange their Linden dollars for offline currencies, profitability can be a significant goal for producers. This opportunity to make money is positioned as an attractive feature of the world, and Second Life assures its residents that, "Thousands of residents are making part or all of their real life income from their Second Life Businesses" (2008). Making a profit can be difficult. Since there is an abundance of virtual goods within Second Life, developers must balance the cost of their goods with what people are willing to pay. However, generating income is not impossible, and a number of Second Life users are able to create goods that are attractive enough to other users that they can live off the proceeds of their sales (Bridge, 2007; Craig, 2006).

The creation of virtual goods in order to generate profits can be a complicated endeavour, especially in an environment that boasts thousands of items for sale and where no commodity will meet a pressing physical need. Given the availability of an extensive selection of well-designed items in Second Life, basic goods do not tend to sell as well as those created by people with more advanced skill sets and design abilities (Lavalee, 2006). Those able to make money from their virtual creations are generally seen to exhibit exceptional devotion to their work in addition to creativity, coding, business, and design skills (Bridge, 2006). Anyone can learn to code virtual goods, and many thousands of commodities are available for purchase in-world, as well as through online auction sites and personal web pages.

As a result of these conditions, producers must find ways to appeal to potential customers. This appeal is especially important given that they are competing for individuals who have a wealth of both offline and virtual goods on which to potentially spend their money. Therefore, those individuals who create and produce virtual goods must do so in a way that makes them especially appealing to other individuals. Producers seem to create appeal in their virtual goods in two main ways. First, appeal can be created by the fact that virtual goods—even those goods that are high in quality, status, or other positive attributes—tend to be vastly cheaper than their offline counterparts. Second, and the strategy that is most directly linked to virtual consumers, is that producers can create a sense of status and belonging for the consumer that arises from the purchase and possession of virtual goods. These elements, combined with the desire to generate profits, define the focus of the virtual economy.

Affordability

In addition to the Linden to dollars exchange rate, an analysis of Second Life commodities reveals that virtual goods are much less costly than their offline equivalents. An in-world Hummer sells for $500 LD, an amount that is far more affordable than the offline version on which it is modeled. A live-aboard yacht sells for $2,600 LD or $9.97 USD, a significantly lower price than even an offline rowboat, and a pitbull dog can be had for $1,800 LD, or $6.87 USD. Furthermore, none of these goods have any associated fees or costs that will arise in the future. In Second Life, vehicles do not need maintenance, upgrades, or gas and dogs do not require food, veterinarian visits, or toys. While the initial outlay of money for these commodities is more affordable than it would be offline, virtual goods are also cheaper because most do not necessitate spending after their initial purchase.

This affordability is a trait of most virtual goods, not just those within Second Life. Dibbell's account of consumption in *Ultima Online* offers the information that Dugger's house cost him $750 USD which, while not an insignificant amount at more than a single week's wages (Dibbell, 2003, 2006), is a small fraction of the cost of buying an offline house. While the online house does not fill a need for shelter, it is a relatively affordable purchase within the virtual world. More importantly, it is a worthwhile purchase for Dugger, who sees the house as a way of better establishing himself within the environment (*Ibid*).

Juliet Schor writes that Americans are driving themselves into serious debt as they purchase (offline) status in an attempt to appear more affluent. In contrast, there are few, if any, accounts of individuals who have gone into debt purchasing virtual goods. This is not to say that people are not going into debt or overworking themselves in order to be able to afford virtual commodities. However, when money is tight for some individuals, online virtual world accounts are one of the first things sold to generate money (Dibbell, 2003, 2006). Given the possibilities for gaining status in the virtual world, these considerations suggest that despite the fact the people are making frequent purchases, virtual commodities remain more affordable than their offline counterparts and are therefore even more attractive to users than they otherwise might be.

Social Belonging

In order to sell their wares and sustain the virtual economy, producers of virtual goods have an abiding interest in imbuing their virtual goods with meanings that make them attractive to consumers. Jhally writes that, "given the central role of objects in the constitution of human societies, human culture and human meaning, one can provide an answer as to where the power of advertising comes from: it derives not from the ingenuity of advertisers but from the need for meaning" (1987, p. 197). Jhally's point regarding the "need for meaning" is well taken in this context. There is no inherent material use-value in virtual goods, both by virtue of their immateriality and the fact that they are simply not necessary for the basic requirements of offline or virtual life. Again, they do not feed us, keep us warm, shelter us, or offer any tangible benefits that fulfill basic physical needs. They do, however, sell at an impressive rate for reasons that have, as Jhally asserts, everything to do with meaning, and especially meaning that producers are able to position in terms of status, belonging, and individuality.

This sentiment is echoed in the work of Schor, who argues that the purchase of goods is rooted in the meanings and social status that they offer. For Schor, increasing consumption is driven by a desire to enjoy the benefits associated with a higher status group by purchasing the trappings and appearance of such a lifestyle. By engaging in consumption and allowing other people to see these goods, consumers are able to appear as though they belong in a particular status group within society even when their income would indicate otherwise. Therefore, through purchasing virtual goods, individuals can increase their apparent status within the virtual world. At the same time, those who produce virtual goods can use their ability to offer status as means to position their goods attractively within the economic system.

This focus on status also arises in the Second Life economy in the pricing and discourse of selling virtual goods. For virtual objects, as with their offline equivalents, "the value character of the products of labour become firmly established only when they act as magnitudes of value" (Marx, 2000, p. 13). These goods are still the product of labour, for it takes time and effort to develop and code even the simplest of virtual goods. Here, different objects require different amounts of time and effort to design and code (Lavalee, 2006). However, these differences can affect the price of a commodity and their success within the economy, with those requiring more time and effort being more expensive but also more desirable.

These differences in prices are important since virtual goods are computer code and, beyond labour, offer no intrinsic reason for variable prices. Despite the fact that virtual goods are all constructed of code, there is a tendency for producers to value goods differently within the economy depending on what is being represented. For example, while clothing is common to most avatars, art requires a place for display and is therefore more likely to be a status item and less likely to be widespread within Second Life. Consequently, one dress for sale in the Plaza Mall costs $150 LD while directly across the aisle a painting sells for $800 LD. As such, the pricing structure in Second Life can be seen to be as much of a product of exchange- and symbolic-value as any offline commodity.

This desire for status and social belonging is not lost on developers and appears frequently in the discourse around the sale of virtual goods. To denote higher status goods, sellers frequently advertise items in a way that makes them sound special or rare in a bid to attract buyers. One designer selling a piece of art writes, "only 50 of these will be released, this is your chance to own something unique" (SLExchange, 2006c). In turn, the creator of a pre-fab rainforest home writes, "This is a truly one-of-a-kind structure" (*Ibid,* 2006), which could be outfitted with "The Candarian Dweller", a "unique 7 person sitting room custom" (Maelstrom, 2006, May 13). This advertising discourse—involving terms such as "unique," "one-of-a-kind," "exclusive," and "rare"—appeals to the desire for status through individuality (Mueller, 1986). By positioning commodities in terms of the status and prestige they convey, developers are able to sell their goods to residents despite the absence of use-value by playing off social and symbolic needs and desires.

Aura and Limited Editions

Status associated with rarity can be traced to Walter Benjamin's notions of the aura of an original piece of work (1968). Aura is the sense of uniqueness based on authority granted through history and presence that emanates from an original work of art. This aura is lost when the work is disseminated through reproduction. Benjamin asserts that, "For the first time in world history mechanical reproduction emancipates the work of art from its parasitical dependence on ritual" (*Ibid,* p. 96), suggesting that there is value to be found in the process of reproduction, despite the loss of the aura. However, it remains that the original piece of work is subject to a loss of its sense of history, presence, and uniqueness when it is rapidly reproduced and disseminated. This loss becomes an issue since, in a world with millions of participants and virtual goods, status can be conveyed through that which is rare, expensive, historical, and possessing an aura.

As with the art world, high-status virtual goods can be protected by ensuring that they are singular, or at least not widely distributed or reproduced. In order to protect its authority from reproduction, Buskirk claims that art "requires a fully elaborated external structure of support, which includes . . . the adherence to external conventions that limit and control the reproduction of otherwise inherently reproducible works" (2005, p. 4), especially around the convention of the limited edition. Similarly, by only releasing single units or limited edition runs of a particular item, developers have tried to ensure that their goods retain their status and value in Second Life.

However, the protection offered to virtual goods through the use of limited editions has not gone unchallenged. Conflict within Second Life over the program CopyBot—a tool designed to backup in-world content that is capable of copying and reproducing the code for any item—have highlighted this importance of individuality, rarity, and status associated with commodities by threatening these qualities. Protests against the use of this software came largely from developers of virtual goods, who feared that the ability to replicate any commodity would allow them to be undersold and render their profit margins non-existent. By dealing with the issue quickly and declaring the use of CopyBot in violation of the Terms of Service, Linden Lab reaffirmed the value of virtual goods.

Given this response, it appears that the development and availability of such tools may also have implications for the meanings associated with virtual goods. Since CopyBot can recreate any item, it could also affect the loss of the individuality, rarity, and status associated with in-world commodities that were for sale or had already been purchased. To manage this threat, one resident suggests, "I'd also put up a store in Second Life and sell signed and numbered digital copies of the player so copybot wouldn't be an issue" (Nathan, 2006). By acknowledging the need to reaffirm the authenticity of the work through the virtual equivalent of numbered editions of art, the threat to its aura is simultaneously addressed. When any commodity can be copied, including rare or custom-made items, the potential for losing the aura of individuality and status is very real, and is therefore a threat to those who create and sell virtual goods.

Virtual Goods and the Consumer

Status and Conspicuous Consumption

Despite their importance to the Second Life economy, the role of virtual goods as a driving force in the economy also relies on consumers' desire for commodities. While the creation and purchase of virtual goods is common within Second Life, there would be no sale of virtual goods—and quite likely less development of the same—without individuals who desire to purchase those commodities. Since virtual goods do not fulfill physical needs, sellers and buyers must have a means of determining an appropriate exchange-value for commodities, while buyers must also have reasons for purchasing goods. In the absence of use-value, the sale of virtual goods in Second Life depends on symbolic-value to drive their purchase by consumers.

Since many virtual goods are graphically rendered, and therefore visible to other users, they can convey status through their display. Due to the visibility of virtual goods in Second Life, in-world status can be linked to theories of conspicuous consumption. Thorstein Veblen writes, "the basis on which good repute in any highly organized industrial community ultimately rests is pecuniary strength; and the means of showing pecuniary strength, and so of gaining or retaining a good name, are leisure and a conspicuous consumption of goods" (2000, p. 39). By making consumption visual, users and developers are working within a world in which status can be expressed and recognized through commodities, which leads to conspicuous consumption.

The tendency towards conspicuous consumption is seen in the drive to acquire noticeably high-status virtual goods. This drive is not exclusive to Second Life, and has been documented in other online environments. In writing about a player named Troy Stolle from *Ultima Online,* another virtual world, Dibbell describes not only the process of buying a house—a rare, expensive, and high-status good in and of itself—but also what the player does with that property. Once the dwelling has been purchased, "The first floor, once the austere workshop of a hardcore craftsman, has become a bright, busy public gallery for Stolle's collection of rares and semi-rares" (Dibbell, 2003). As with Second Life, other users are free to come and go within the space, and presumably take an admiring look at the display.

Key words in Dibbell's account of what he terms "The Unreal Estate Boom" are "public gallery" and "rares and semi-rares" (*Ibid*). The purpose of this space is to show off Stolle's uncommon goods as a sign of his status within the *Ultima Online* community. Stolle's goods suggest status on their own. Both the dwelling and the goods that it contains are rare, and therefore signal difficulty or expense in their acquisition. However, by making this space public, Stolle is engaging conspicuous consumption in an overt way. Not only does he possess a house—a very visible sign of status, given that it leaves a mark on the very landscape of the virtual world—but he wishes to display his other virtual goods in a way that makes them visible to others. It is not enough that he owns them. His consumption must be made conspicuous to reveal the status associated with his goods in a similar manner to that seen in Second Life.

Similarly, consumption is conspicuous in Second Life in multiple ways. First, even the smallest of purchases of clothing or jewelry can be seen on the avatar, since it is graphically rendered. Second, users can more easily see significant purchases, such as cars and boats, because they are larger than the avatar, and therefore more immediately visible. Third, people tend to be inclined to show off their goods, and private dwellings are frequently made public to all Second Life residents so that others can see both the dwelling and the goods that it houses. All three forms of consumption allow almost any commodity that can be purchased within the world to be put on display for others to see.

While these first two expressions of conspicuous consumption are readily apparent in offline life (Schor, 1999), the third one is somewhat unusual. It is rare that individuals open their offline homes to anyone who wishes to enter, and yet virtual homes are frequently accessible to other users, if not the entire environment's population. Within the virtual world there is no dirty virtual laundry lying on the floor, or virtual medicine chests with embarrassing contents for visitors to peek into. As a result, privacy is not as much of an issue as it would be offline. Therefore, it is easier for users to open up their virtual abodes to show off their goods without worrying about many of the small details that could keep an offline home off limits to other people.

More importantly, those dwellings that are open to the public tend to be set up nicely with furniture, art, electronics, and a significant amount of space for users to move around in. In some cases users are so proud of their spaces that they post virtual web sites specifically dedicated to this kind of display (Samiam, 2005). By and large, virtual homes can be kept open to the public more easily than offline dwellings because there is less upkeep, there are fewer concerns about privacy, and theft is unlikely, if not impossible. However, there is also a very real sense of display that goes into the construction of the space and the decision to open it up to the public, either in-world or on the web. Given this tendency towards display, consumption within Second Life can be read as conspicuous in its attempts to make purchases visible within the world.

Identity and Selfhood

In addition to conspicuous consumption, the graphical nature of virtual goods also lends itself more specifically to augmenting personal appearance and in-world identity through consumption. Despite the importance of rare, custom, and high-status goods within the Second Life economy, it is also important to consider other ways through which residents can use the wide variety of more widespread goods to seek status and establish their identity and place within the world. A more common and, in many cases, more affordable way to mark identity and generate status is through developing the avatar's appearance using the thousands of items that are readily available for purchase. In virtual worlds, many people simply want to create an avatar that looks distinctive and is representative of who they wish to be (Castronova, 2005). The easiest way to accomplish this task is through the purchase of smaller but no less significant virtual goods.

Erving Goffman's theories of identity speak to the idea that appearance is a critical way of providing personal information to other people, and that props are a means by which we can convey personal information to others about who we are and want to be (1973). Graphical virtual goods associated with the avatar can be seen and appreciated by other players both for their aesthetics and for the fact that their owner has chosen them (Castronova, 2004). Consequently, by obtaining and displaying virtual goods, residents are able to customize the appearance of their avatars, performing their virtual self to others in particular ways. Although these goods can be relatively small, they can also be a sign to other players of how much money, status, time, or effort has been invested in the avatar.

Goffman's ideas of performativity can be also linked to multiple conceptions of the self, including ideal, multiple, and extended selves. Second Life offers its users the opportunity to create a virtual self with which to experience the environment. However, due to the anonymity associated with online interaction, the virtual self can be different from the offline self in terms of how it is constructed and defined as well as how many selves are created. Rather than being limited to one representation of the offline self, online selfhood can be multiple, variable, idealized, and changeable. In the context of these many options, consumption functions as a way to expand on and delineate between different expressions of selfhood within Second Life.

In addition to the possibility of creating only one self within the virtual world, residents are able to create multiple selves through which to interact with the virtual world (Turkle, 1995; Taylor, 2002; Boellstorff, 2008). While it is possible to simply create multiple avatars, this process can be augmented through consumption. As Goffman suggests, consuming particular types of virtual goods allows individuals to relatively quickly and easily position themselves in a particular way (1973). Because different items designate different aspects of identity, the consumption of a variety of goods may be used to delineate different identities from each other, especially within the virtual context (Kafai, Fields, and Cook, 2007). Since these items can be and often are made visible to other members of the community, they serve as a visible marker to distinguish between and reinforce the different characteristics of multiple selves.

Within Second Life, consumption can also be linked to the idea of extended selves. The idea of the extended self refers to the way in which individuals' possessions help to define who they are, expanding the sense of self beyond the immediacy of the body and its associated personality and persona (Belk, 1988; Ahuvia, 2005). Rather than focusing on augmenting the performativity of selfhood for others, the extended self relies more on possessions as a way to define who they are and, in some cases, who they wish to be (*Ibid*). Within Second Life, virtual goods offer ways to extend the self through a wide variety of possessions, from clothes defining a particular group membership to houses that indicate a particular class of property ownership. This extension of the self is especially interesting in terms of virtual goods that can be scripted or coded so that they function in particular ways within the world. While buying a sailboat in offline life may allow individuals to associate themselves with sailing, it does not guarantee the skills necessary to safely take a boat out on open water. In contrast, buying a virtual scripted sailboat in Second Life allows the resident not only the feeling of being a sailor, but the ability to be one as well. As a result of their meanings and capabilities, the associations made with virtual goods can be a powerful way of extending the self within Second Life.

The influence of virtual goods on providing an extended sense of self is perhaps most clearly seen in reactions to the loss of virtual goods. Russel W. Belk asserts that, "If possessions are viewed as part of self, it follows that an unintentional loss of possessions should be regarded as a loss or lessening of self (1988: 142). Within Second Life, it is not uncommon for bugs in the code to delete the inventories of residents, who respond negatively to such losses. On the official Second Life blog, one resident who lost the contents of their inventory writes, "My inventory is back but I'm a shadow of my former self" (Seaton, 2007), alluding to both the literal loss of items that defined her visual appearance and the loss of feelings of self generated by such an experience. Although it is not solely defined by the individual's possessions, the loss of such goods can have a profound effect on a sense of self.

Finally, since individuals are free to construct their Second Life avatar as they please, virtual consumption can be further expanded through the development of an ideal version of the self. Research into consumption practices has suggested that purchases can be used as a way for individuals to work towards the development of their ideal self (Landon, 1974). In online worlds, rather than simply purchasing virtual goods to define and distinguish online identity, goods can be purchased so as to create an idealized persona with which to engage with the virtual world. In Dibbell's account of *Ultima Online,* the acquisition of an in-world tower and creation of a public gallery within it helps to create an ideal version of the self for the purchaser (2003, 2006). While the offline user is positioned as rather average, the in-world self is a property-owning, visible, well-outfitted member of the virtual community. By consuming virtual goods that construct the self in a particular way it is possible for individuals to not only create alternate selves online, but also to construct an ideal self through which to interact with the world and other residents within it.

Personal Appearance

Within Second Life, personal appearance can be strongly linked to the construction of virtual selves and identity. By offering visible cues about avatar style and preferences, virtual goods can help to establish a particular identity through the appearance of the avatar. While larger goods, such as dwellings and vehicles, have the advantage of being more easily noticeable, commodities associated with appearance are attractive to users because they are always present with the avatar, and therefore always provide information about status and identity. While a virtual house cannot be taken into an in-world nightclub, a distinctive outfit can. Therefore, it is these smaller, portable goods that come to signify status immediately. In this way, appearance is the mark of status and a designator of identity that is always present with the avatar.

Due largely to this focus on appearance, clothing is the most popular commodity within Second Life. Of the 2,353,390 goods currently available on the Second Life exchange, 79,422 are clothing—a category larger than any other by almost 38,000 items—33,565 are accessories, and 16,998 are otherwise related to avatar appearance. It is also possible to buy upgrades to the avatar body, and designers offer a multitude of eye colors, hairstyles, hair colors, piercings, and facial features that can be used to change the body of the avatar and achieve a desired appearance.

Another expression of this focus on using goods to develop personal appearance and identity in Second Life is the production and sale of "skins," or full-body tattoos that cover the original body and can change a variety of the avatar's features at once. One newspaper reports that, "a rocked-out, tattoo-covered sex god skin can sell for upwards of $3000 LD, or about $15 in U.S. currency. People can't buy them fast enough" (Stafford, 2003). Skins allow users a way not just to customize their appearance through clothing and other goods, but also a means of altering the entire body. In this way, Second Life offers residents a variety of options for customizing their avatar's appearance. The wide range of goods available for customizing the avatar body and the rate at which they sell suggests that appearance and the commodities that can be used to change it are important markers of status within the Second Life environment.

The prevalence of clothing and other appearance-related goods in Second Life is also a phenomenon found in other virtual worlds. One analysis of There.com, a virtual world similar to Second Life, states that, "as with the real world, much attention is given to personal appearance, and one of the main activities in *There* is designing virtual clothes and selling them through in-game auctions" (Brown and Bell, 2004, April 24–29). It is not possible to definitively suggest that this focus on appearance is an integral part of all virtual social worlds. However, the fact that appearance is such an important element of two virtual worlds does suggest a common focus on goods as a sign of status and as a means of customizing the virtual body. Although the goods associated with an avatar do not fulfill physical needs, they are created, sold, and bought at high rates. As such, it follows that their ability to symbolically fulfill needs and desires for status and a pleasing appearance within the virtual world is driving their popularity.

Uniqueness and Individuality

While the almost unlimited combinations of skins, clothing, and features available in-world are a way to ensure avatar uniqueness, numerous advertisements in the Second Life forums offer custom designs, which can further help users to maintain their individuality. These custom orders can cost significantly more than their conventional counterparts—in one shop, a regular dress is $100 LD, while a custom dress is $500—but they ensure that the avatar can be singular in its appearance. Custom work ensures that no other avatar will be wearing a particular dress, tattoo, or skin, and also provides a way of creating an appearance or item that is not only unique, but representative of the user. Out of 6,329 advertisements for goods in Second Life's "New Product" forum, 351 vendors, or 5.5 per cent of vendors refer to their work as "custom," while in Second Life *Exchange,* 2,704 advertisements, or 14 per cent claim to offer custom goods. These numbers suggest that despite higher costs, there is enough demand for custom work that it is an important factor behind purchases within Second Life.

Although custom designs are commonly found in clothing and appearance-related goods, similar development can be completed for almost any virtual commodity, providing that there is a willing developer. Custom cars and motorcycles are available to those willing to pay, and a customized castle currently costs $20,000 LD. However, it remains that these virtual goods still do not fulfill material needs. What they do offer is a fulfillment of desires. Custom goods are closely linked with appearance and status, and offer users a way to establish themselves within the virtual community through the goods that they possess and the things that they suggest about their bearer. Moreover, these goods can be seen as representations of status at a financial level. While custom goods ensure the individuality of the avatar, they are also a visual sign that the user has the means to pay for unique goods rather than simply purchasing items that are available to other residents. By purchasing virtual goods, and especially those that are custom made, users are expressing their status, individuality, and position within the virtual world to others.

Social Lubrication

In addition to their potential to indicate status, virtual goods can also function as a social lubricant to bring people together through common interests and shared activities. Despite the fact that goods in Second Life lack conventional use-value, this does not mean that they are not still useful within the world. In some instances, commodities can bring people together into shared activities out of which social networks and communities develop. Some of the clearest examples of the use of virtual goods as social lubricant surround larger commodities and the events that surround them. For owners of cars and dune buggies, it is not uncommon to have owners come together for a race and enjoy time together socializing. Although individuals do come together of their own accord and form in-world communities, virtual goods can serve as a common bond and a reason to get together with other users.

To this end, one boat designer and owner writes, "I was thinking of having a race, or series of races, some time in the next couple of weeks. I know nothing about real sailboat races, so it would probably be a simple affair. Everyone would race the same class of boat, though—i.e., a Flying Tako" (Massiel, 2006, June 22). Membership in a particular group of individuals is signaled by ownership of specific goods—in this case, specific to the point of owning the same class of boat. More importantly, it not only signals membership but allows for participation in activities that are associated with or facilitated by the goods in question. Without owning these goods, membership may not be possible and individuals can be excluded. Conversely, while there is status associated with these goods, they are also wrapped in intangible notions and feelings of belonging that can be employed in their sale. Brown and Bell write, "By choosing to do leisure activities with certain people they become our friends" (2004, p. 353). Although virtual goods may not be necessary, through their purchase they are able to bring people together into communities based on common interests and activities, and can be sold as such.

Conclusions

Despite the necessity of moving away from conceptions of use-value and need fulfillment when considering virtual goods, it cannot be said that there is not a sense of need attached to consumption within Second Life. However, this sense of need is constructed not on conventional notions of physical and material needs and use-values, but on symbolic meanings around individuality, power, status, community, and belonging. Baudrillard asserts that individuals, "in the rupture of symbolic exchange, autonomize themselves and rationalize their desire, their relation to others and to objects, in terms of needs, utility, satisfaction, and use value" (2000, p. 24). Although residents may justify their purchases in terms of their use-value, it remains that they are purchasing goods whose meaning is located in social- and symbolic-value, rather than the capacity to meet physical needs.

Although there remains a lack of use-value in virtual goods, it is important to note that their consumption should not be read as irrational or meaningless. A significant number of individuals who enter into forms of online engagement highly value their interactions (Turkle, 1994, 1995). As Thomas Malaby suggests, even within spaces that are virtual, "People may, in any specific context, imbue them with deep, normatively charged meanings" (2007: 96).

As such, it can be argued that it makes sense for them to focus their efforts in terms of time and money on developing reputation and status within the online world through virtual goods, if this is the environment and community that matters most to them. The roles that these commodities play for individuals is still meaningful, and it is these representations of status, individuality, and belonging that have driven the sale of virtual goods even as they have supplanted use-value.

The continued purchase of millions of dollars worth of virtual goods suggests that their consumption is being driven by a quality apart from that of use-value. An examination of virtual goods reveals that use-value is not a characteristic that is necessary for acquisition, and that exchange- and symbolic-value can be the driving force behind consumption. As a result of their existence as visually rendered code stored on a computer server, none of the virtual goods available for purchase in online worlds can be linked to a tangible expression of use-value. Yet, walking through a virtual world, these goods are everywhere—on avatars, in shops, advertised on the street, and placed in homes and buildings—and despite their ephemeral and immaterial qualities, they affect Second Life users, residents, communities, and interactions in significant ways that continue to justify their purchase.

Bibliography

Ahuvia, A. C. (2005). Beyond the Extended Self: Loved Objects and Consumer's Identity Narratives. *Journal of Consumer Research* (32), 171–184.

Bartle, R. (2003). *Designing Virtual Worlds*. Indianapolis: New Riders.

Baudrillard, J. (1981). *For a Critique of the Political Economy of the Sign*. St. Louis: Telos Press.

———. (2000). Beyond Use Value. In M. J. Lee (Ed.), *The Consumer Society Reader*. Malden, MA: Blackwell Publishing.

Belk, R. W. (1988). Possessions and the Extended Self. *Journal of Consumer Research, 15,* 139–168.

Benjamin, W. (1968). The Work of Art in the Age of Mechanical Reproduction (H. Zohn, Trans.). In H. Arendt (Ed.), *Illuminations*. New York: Harcourt, Brace & World.

Boellstorff, T. (2008). *Coming of Age in Second Life: An Anthropologist Explores the Virtually Human*. Princeton, NJ: Princeton University Press.

Bridge, M. (2007). Second Life, Second Income. *The Times* Online. Retrieved 2007, February 17, from http://business.timesonline.co.uk/article/0,,9558-2479694,00.html.

Brown, B., & Bell, M. (2004a). *CSCW at play: 'There' as a collaborative virtual environment.* Paper presented at the CSCW.

———. (2004b, April 24–29). *Social Interaction in 'There'.* Paper presented at the Conference on Human Factors in Computing Systems, Vienna, Austria.

Buskirk, M. (2005). *The Contingent Object of Contemporary Art*. Cambridge, MA: MIT Press.

Castronova, E. (2001). Virtual Worlds: A First-Hand Account of Market and Society of the Cyberian Frontier. *CESifo Working Papers*.

———. (2002). On Virtual Economies. *CESifo Working Papers*.

———. (2003). Theory of the Avatar. *CESifo Working Papers*.

———. (2004). The Price of Bodies: A Hedonic Pricing Model of Avatar Attributes in a Synthetic World. *KYKLOS, 57*(2), 173–196.

———. (2005). *Synthetic Worlds: The Business and Culture of Online Games*. Chicago: University of Chicago Press.

Craig, K. (2006, February 8). Making a Living in Second Life. *Wired* Online. Retrieved April 2, 2008, from http://www.wired.com/news/technology/0,70153-0.html.

Debord, G. (2004 (1967)). *The Society of the Spectacle* (D. Nicholson-Smith, Trans.). New York, NY: Zone Books.

Dibbell, J. (2003, January 2003). The Unreal Estate Boom. *Wired,* 11.

———. (2006). *Play Money: Or, How I Quit My Day Job and Made Millions Selling Virtual Loot*. Cambridge, MA: Basic Books.

Goffman, E. (1959). *The Presentation of Self in Everyday Life*. New York, NY: Anchor Books.

Gwinn, E. (2003, December 4). What is There? More Options for More Players. *The Chicago Tribune*.

Jhally, S. (1987). *The Codes of Advertising: Fetishism and the Political Economy of Meaning in the Consumer Society*. London: F. Pinter.

Kafai, Y. B., Fields, D. A., & Cook, M. (2007, September 24–28). *Your Second Selves: Avatar Designs and Identity Play in a Teen Virtual World.* Paper presented at the Situated Play: DiGRA 2007, Tokyo.

Landon, E. L. J. (1974). Self Concept, Ideal Self Concept, and Consumer Purchase Intentions. *Journal of Consumer Research, 1*(2), 44–51.

Lastowka, F. G. H., Dan. (2003). *The Laws of Virtual Worlds.* Institute for Law and Economics.

LaVallee, A. (2006, September 22). Now, Virtual Fashion—Second Life Designers Make Real Money Creating Clothes for Simulation Game's Players. *The Wall Street Journal*.

Linden Lab. (2003–2008a). Second Life, from www.secondlife.com.

———. (2003–2008b). Second Life: Your World. Your Imagination. *Second Life.* Retrieved August 15, 2008, from http://SecondLife.com.

———. (2003–2008c). What is Second Life? *Second Life.* Retrieved August 15, 2008, from http://SecondLife.com/whatis/.

———. (2006, August 14). Free C&C Hum-vee for sale. *SL Forums.* Retrieved April 14, 2008, from http://forums.SecondLife.com/showthread.php?t=10025814.

———. (2008a). Economic Statistics. *Second Life.* Retrieved August 15, 2008, from http://secondlife.com/whatis/economy_stats.php.

———. (2008b). Economy Graphs. *Second Life.* Retrieved August 15, 2008, from http://secondlife.com/whatis/economy-graphs.php.

Maelstrom, C. (2006, May 13). "Candarian Dweller" NEW @Bleed Designs. *SL Forums*. Retrieved April 14, 2008, from http://forums.SecondLife.com/showthread.php?t=100063&highlight =unique.

Malaby, T. M. (2007). Beyond Play: A New Approach to Games. *Games and Culture, 2*(2), 95–113.

Marx, K. (2000). The Fetishism of the Commodity and its Secret. In M. J. Lee (Ed.), *The Consumer Society Reader*. Malden, MA: Blackwell Publishing.

Massiel, M. (2006, June 22). Sailboat Races. *SL Forums*. Retrieved April 14, 2008, from http://forums.SecondLife.com/showthread.php?t=51075&highlight=unique

Metros, S. E. (1999). Making Connections: A Model for Online Interaction. *Leonardo: Journal of the International Society for the Arts Sciences and Technology, 32*(4), 281–291.

Mueller, B. (1986, August 3–6). *Reflections of Culture: An Analysis of Japanese and American Advertising Appeals*. Paper presented at the Annual Meeting of the Association for Education in Journalism and Mass Communication, Normal, OK.

Nathan, M. (2006, December 4). RCWcast Hardware Update. *Bulldog Financial's Bullblog*. Retrieved February 10, 2007, from bulldogfinancial.com/bullblog/?p=30.

Poster, M. (2004). Consumption and Digital Commodities in the Everyday. *Cultural Studies, 18*, 409–423.

Reuters. (2006, March 18). Video Game Therapy—A New Frontier. *CNet News*. Retrieved February 18, 2008, from http://news.com.com/2100-1043_3-6051269.html.

Samiam, C. (2005, July 1). My House. Retrieved April 16, 2006, from http://www.sluniverse.com/pics/pic.aspx?id=17022&Type=Album&RelatedID=122&sorce =ViewAlbum.aspx%3fid%3d122&sourcename=My+House.

Schor, J. (1998). *The Overspent American: Why We Want What We Don't Need*. New York, NY: Basic Books.

Seaton, A. (2007, March 7). Comment in, "Inventory Database is Feeling Better." *Official Second Life Blog*. Retrieved October 21, 2008, from http://blog.secondlife.com/2007/03/07/inventory-database-isunhappy/.

SLExchange. (2006a). Inspirations LE. Retrieved April 2, 2008, from http://www.slexchange.com/modules.php?name=Marketplace&file=item&ItemID=76203

———. (2006b). Rainforest Prefab Home. Retrieved April 3, 2008, from http://www.slexchange.com/modules.php?name=Marketplace&file=item&ItemID=50400

———. (2006c). THE COMPLETE GUIDE TO MAKING CLOTHES, v2.0! Retrieved April 16, 2006, from http://www.slexchange.com/modules.php?name=Marketplace&file=item&ItemID=59953

Stafford, J. (2003, January 23). Your second life; An online world where you can try on another identity, explore your fantasies, choose mates, shape a digital city—even make real money. *Austin American Statesman*.

Taylor, T. L. (2002). Living Digitally: Embodiment in Virtual Worlds. In R. Schroeder (Ed.), *The Social Life of Avatars: Presence and Interaction in Shared Virtual Environments*. London: Springer-Verlag.

Turkle, S. (1994). Constructions and Reconstructions of Self in Virtual Reality: Playing in the MUDs. *Mind, Culture and Activity, 1*, 158–167.

———. (1995). *Life on the Screen: Identity in the Age of the Internet*. New York, NY: Simon and Schuster.

Veblen, T. (2000). Conspicuous Consumption. In M. J. Lee (Ed.), *The Consumer Society Reader*. Malden, MA: Blackwell Publishing.

Walsh, T. (2006, April 16). 'Second Life' Economy Tweaked. *Clickable Culture* Retrieved April 2, 2008, from http://www.secretlair.com/index.php?/clickableculture/.

My Obsession: I Thought I Was Immune to the Net. Then I Got Bitten by eBay.

William Gibson

When I was a young man, traversing the '70s in whatever post-hippie, pre-slacker mode I could manage, I made a substantial part of my living, such as it was, in a myriad of minuscule supply-and-demand gaps that have now largely closed. I was what antique dealers call a "picker," a semi-savvy haunter of Salvation Army thrift shops, from which I would extract objects of obscure desire that I knew were upmarketable to specialist dealers, who sold in turn to collectors. To this day I am often unable to resist a professionally quick, carefully dispassionate scan over the contents of any thrift shop, though I almost never buy anything there. Mainly because the cut-rate treasures, the "scores" of legend, are long gone. The market has been rationalized. We have become a nation, a world, of pickers. There are several reasons for this. One has to do with boomer demographics and the cult of nostalgia. There are now more fifty-somethings than there are primo childhood artifacts of a similar vintage. Most of our toys, unlike the wood and pot-metal of yore, were extrusion-molded ephemera, fragile styrene simulacra, highly unlikely to survive the random insults of time. A great deal of the boomers' remembered world has been melted down, or crushed into unreadable fragments in forgotten strata of landfill. What remains, particularly if it's "mint in box," becomes increasingly rarefied.

Another reason, and this one is more mysterious, has to do with an ongoing democratization of connoisseurship, in which curatorial privilege is available at every level of society. Whether one collects Warhol prints or Beanie Babies becomes, well, a matter of taste.

The idea of the Collectible is everywhere today, and sometimes strikes me as some desperate instinctive reconfiguring of the postindustrial flow, some basic mammalian response to the bewildering flood of sheer *stuff* we produce.

But the main driving force in the tidying of the world's attic, the drying up of random, "innocent" sources of rarities, is information technology. We are mapping literally *everything,* from the human genome to Jaeger two-register chronographs, and our search engines grind increasingly fine.

"Surely *you* haven't been bitten by the eBay bug," said my publishing friend Patrick. We were in the lobby of a particularly bland hotel somewhere within the confines of a New England technology park, and I was in fact feeling twinges of withdrawal. eBay, which bills itself as Your Personal Trading Community™, is a site that hosts well over 800,000 online auctions per day, in 1,086 categories. eBay gets around 140 million hits per week, and, for the previous few months, a certain number of those hits had been from me.

And, in the process of adding to eBay's gargantuan hit-pile, several days before, I had gotten myself in trouble. In Uruguay. How this happened: I'm home in Vancouver, midway through that first cup of morning coffee, in front of the computer, ready to work straight from the dream-state.

I am deep into eBay, half-awake, staring at a scan of this huge-ass Zenith diver's watch. And I am, mind you, a practicing ectomorph. I have wrists like pipestems. I am not going to get too much wear out of a watch that's actually wider than my wrist.

But a little knowledge is a dangerous thing, and I know, having already become a habitué of eBay's Clocks, Timepieces, and Wrist Watches, that the movement in this particular Zenith is the very one Rolex installs in the big-ticket Daytona. Making this both a precision timepiece and possibly an undervalued one—the identical thing having sold on eBay, the week before, albeit in better cosmetic condition, for around two grand.

"I didn't even know you had Web access," Patrick said. "You mean you've overcome your infamous resistance to using the Net, but only in order to service an eBay addiction?"

Well, yes. Sort of. Not exactly.

eBay is simply the only thing I've found on the Web that keeps me coming back. It is, for me anyway, the first "real" virtual place.

In Patrick's hotel room, we used his laptop to get onto eBay, where I discovered that, yes, I was *still* high bidder for the damned Zenith: $500 American. This bid, you see, was the result of Fiddling Around. I'd sat there in my office, not quite awake yet, and had poked around with modest but increasingly higher bids, assuming that the seller's hidden "reserve," the lowest bid he'd accept, would be quite high. But no, at $500 I hit it, and suddenly I was listed there as high bidder. This had happened before, and I had always been outbid later. So I didn't worry.

But I didn't really want to have to buy this very large watch. Which was in Uruguay. And now I was still high bidder, and the auction would be run off before I got back to Vancouver. I thought about having to resell the Zenith.

When I did get back, though, I discovered, to mixed emotions, that I'd been "sniped." Someone, or rather their automated bidding software, had swooped in, in the last few seconds, and scooped the Zenith for only the least allowable increment over my bid.

How did I get into this, anyway?

I went happily along for years, smugly avoiding anything that involved a modem. Email address? Sorry. Don't have one.

And then I got a Web site. Had one foisted upon me, actually, and quite brilliantly, by Christopher Halcrow, who created William Gibson's Yardshow, an "official" home page. So I kept having to go into my kids' bedrooms and beg for Web access to look at it, which bugged them.

Then Chris, who knows a bargain when he sees one, happened to buy this Performa 5200CD from someone who was leaving town. He passed the Performa on to me for what he'd paid for it, and suddenly I had this video-ready machine I could look at my site on, and the video-ready part brought cable into the office, so I got a cable modem, because it was faster, and I already had a hole drilled in the wall, and then I discovered that, damn, I had an email address. It was part of the deal. So email, over the course of about 15 minutes, replaced the faxes I'd been using to keep in touch with certain people.

In the way of things, very shortly, I no longer had a Web site, Chris having found it necessary to get a life. But there was the rest of the Web, waiting to be explored. And I did, and promptly got bored. It was fun, at first, but then gradually I found that there wasn't really anywhere in particular I wanted to go. I went a lot of places, but I seldom went back.

Then I found eBay. And I wanted to go back. Because eBay is, basically, just a whole bunch of *stuff.* Stuff you can look at and wonder if you want—or let yourself want and then bid on.

Mechanical watches are so brilliantly unnecessary.

Any Swatch or Casio keeps better time, and high-end contemporary Swiss watches are priced like small cars. But mechanical watches partake of what my friend John Clute calls the Tamagotchi Gesture. They're pointless in a peculiarly needful way; they're comforting precisely because they require tending. And vintage mechanical watches are among the very finest fossils of the predigital age. Each one is a miniature world unto itself, a tiny functioning *mechanism,* a congeries of minute and mysterious *moving parts.* Moving parts! And consequently these watches are, in a sense, alive. They have heartbeats. They seem to respond, Tamagotchi-like, to "love," in the form, usually, of the expensive ministrations of specialist technicians. Like ancient steam-tractors or Vincent motorcycles, they can be painstakingly restored from virtually any stage of ruin.

And, as with the rest of the contents of the world's attic, most of the really good ones are already in someone's collection. But the best of what's still available, below the spookily expensive level of a Sotheby's watch auction, can still be had for a few thousand dollars at most. At the time of this writing, the most desirable vintage Rolex on one New York dealer's website, a stainless steel "bubble back" automatic, is priced at $3,800, a fraction of the cost of many contemporary watches by the same maker. (And it's so much cooler, possesses so much more *virtu,* than one of those gold-and-diamond Pimpomatic numbers!)

My father bought a stainless steel Rolex Oyster with a stainless band at a duty-free in Bermuda in the early '50s.

After his death, not very long after, my mother put it away in a bank vault, from whence I wheedled it when I was 18 or so. I had a Rolex dealer in Tucson replace its white dial with a black one, so that it would be more like the one James Bond wore in Fleming's novels. I loved it, and, one very sad night a few years later, I sold it for very little to a classmate of mine, in order to pay for a hotel room in which to enjoy, if that's the word, a final bitter tryst with my high school sweetheart. It was one of those dumb-ass, basically self-destructive gestures, and I actually don't regret it. I needed that hotel room. But I've always missed that watch, that Rolex Oyster Precision, and have always had it in the back of my mind to replace it one day with another of similar vintage. I had never done anything about it, though, and made do quite happily with quartz. My last quartz watch was a French faux-military job I bought at the airport in Cannes, on my way home from the film festival. Cost about a hundred dollars. Perfectly adequate for everything—everything except the Tamagotchi Gesture.

Last year, for some reason, I was struck by an ad, one that ran repeatedly in the British men's fashion magazines, for the Oris "Big Crown Commander." It was just a very good-looking watch, I thought, and it was Swiss, and mechanical, and not terribly expensive as such things go. Driven in part by my then brand-new Web access, I used a search engine to determine that Oris had no Canadian distributor. This made the watch seem even cooler, so I went on, via the Web, to locate a Seattle retailer who carried what a sarcastic friend had taken to calling the Big Dick Commando. (The crown, the bit you twist to set it, see, is rather more than usually prominent, so that you can do it without removing your whacking great RAF pilot's gloves.)

And I was and am quite happy with it.

Except that, though I didn't know it at the time, my search for the Big Crown Commander had inadvertently exposed me to the eBay bug.

I got a little compulsive, eventually.

I found myself coming down to my basement office every morning and going straight to that one particular bookmark. New auctions are posted daily on eBay, so there was always something new to look at.

The first watch I bought was a Croton "Aquamedico," a rare—or obscure, depending on how you look at it—Swiss manual-wind from the late '40s or early '50s, black dial with a white medical chapter ring. (Getting the terminology down was a big part of the kick, for me; a medical chapter ring is an outer, 60-second set of graduations that facilitate taking a patient's pulse.) It had been listed by a seller who didn't seem to be particularly into watches; the language of the listing was casual, nonspecialist, and not much mention was made of the watch's condition. Email to the seller elicited the opinion that the watch looked as though "it hadn't been worn very often," which I liked. The scan was intriguingly low-res, but I really liked the design of the numerals. And I really liked its name, "Aquamedico," which for some reason evoked for me the back pages, circa 1956, of *Field & Stream* and *True.*

Tentatively (but compulsively), I placed a low bid and waited to see if the Aquamedico attracted much attention from the eBay watch buffs. It didn't.

In the meantime, I determined that Croton was a long-established Swiss maker whose watches had been a lot more prominent in the United States in the '40s and '50s. Full-page ads in wartime *Fortune.*

I decided to go for it. To try and buy this thing. To import a unique object, physically, out of cyberspace. Well, out of Pennsylvania, actually, but I did experience this peculiar yearning to turn the not-so-clear scan on my screen into a physical object on my desk. And for all I knew, it might be the only Croton Aquamedico left, anywhere. (And in fact I've only ever seen one other Aquamedico since on eBay, and it was gold-filled with a white dial, neither of which I liked as much.) On the night of the auction, after having carefully considered bidding strategy (and this with no prior experience of bidding in any kind of auction), I placed a bid for considerably less than the $200 limit I'd set for myself.

That put me in high-bidder position. And then I sat there. What if, it occurred to me, someone noticed my Croton in the auction's last few minutes and decided to grab it? eBay's system of proxy bidding encourages buyers to offer the most they're willing to pay for an item—their "maximum" bid. My maximum bid was $140. But on eBay you don't necessarily end up having to pay your maximum bid. In an auction house, if you raised your hand to bid $200 on a watch, you'd be on the hook for that amount. But on eBay, each auction has a set "bid increment"—with some as little as 5 cents. With a $2 set bid increment, a rival could bid $200 on my watch, beat me out, and end up having to pay only $142, or $2 over my maximum.

I started to get nervous. (And this, mind you, was before I even knew about sniping software and autobid bots.) What if someone else got this watch, this watch I'd never seen but which I now, somehow, was emotionally invested in winning? I began to have some sense of the power of the psychology of auctions, something I hadn't really experienced before. I'm not a gambler. I've never put money on a horse, bought a lottery ticket, or bet on a hand of cards. Just doesn't do it for me. I've engaged in compulsive risk-taking behavior, certainly, but not the kind involving games of chance. But here, I recognized, I was starting to experience a buzz that I suspected was very much like a gambling buzz.

And what if, I asked myself, the Croton was really not all that desirable an object, a lemon, something a serious watch-nerd would find laughable?

What if the seller simply cashed my money order and did a runner? But I'd already checked his profile in the Feedback Forum, and there were lots of people on record there as saying he was honest, prompt, goods as described, and pleasant to deal with. (All of which turned out to be true.)

Meanwhile, with less than an hour to go before the auction closed, I was robotically punching the Netscape Reload button like a bandit-cranking Vegas granny, in case somebody outbid me. I knew how long it would take me to counterbid (not long), but I didn't know how quickly I could expect the server to process my bid.

Into the final countdown, nobody else showing up, when one more click on the Reload button produced . . . a new high bidder! Galvanized, I scrambled frantically through the bid process, and hit Bid. Real heart-in-mouth stuff, this. And, I must say, really fun.

Reload. And I was back, reinstated.

The auction closed.

The Aquamedico was mine.

I examined the address of the buyer who'd tried to outbid me at the last minute. An "hk" suggested that he was out of Hong Kong, which I already knew to be a hotbed of serious vintage-watch action. (The day before, I had found a wonderfully bizarre site in Taiwan, a sort of micro wrecker's yard, exclusively devoted to selling parts of Rolex watches: cases, dials, hands, et cetera.) I loved it that this Hong Kong bidder had popped in at the last minute, hoping to scoop what he, with his no doubt very considerable watch-savvy, knew to be an extremely desirable piece. But I had been there, ready, and I had prevailed.

I emailed the seller, sending my physical address and asking for his.

In the morning, I went out to buy a postal money order, the standard medium of exchange on eBay.

When the Aquamedico arrived, however, I was dismayed to find that it was peculiarly small, probably a "boy's" watch. I went back to its page on eBay and noted that, yes, it was indeed described as being a 30-mm watch. But the scan was larger than the watch itself, and I had assumed that 30 mm was standard (36 mm is actually closer to the contemporary men's standard). And while the steel case was very nearly mint, even better than the description, the crystal was in such rough shape that it was impossible to get a clear idea of the condition of the dial and hands. It had arrived from cyberspace, but it didn't really look like the scan. It looked as though it had been sitting in a sock drawer, somewhere in Pennsylvania, for 40-some years. Which it probably had.

But the seller's performance had been excellent, so I added my own note of positive feedback to his profile, and he gave me one in turn.

I took the Aquamedico to Otto Friedl, élite specialist in the care of vintage Swiss Tamagotchis, down in the lower lobby of the Hotel Vancouver, and asked to have it cleaned, lubricated, and the crystal replaced. When I went back for it, I discovered that it was a beautiful object indeed, the black dial immaculate, *virtu* intact.

But it wasn't *the* watch.

I told myself that there wasn't any *the* watch, and that I had simply found my own way, after avoiding it for years, of compulsively wasting time on the Net.

But I kept doing it. Opening that same bookmark and clicking down through pages and pages of watches. Learning to read a restricted code. And there was everything there, really: Swatches (which are collected like Barbies), the same battered Gruens you would see gathering dust in a Kansas City pawnshop, every sort and vintage of Rolex, wartime Omegas with the British broad arrow stamped into the caseback, German Sinn chronographs that you aren't really supposed to be able to buy here, Spiro Agnew campaign watches. . . .

And bidding. I'd bid a few times per week and was usually content to let myself be outbid. But I did buy another watch, from London, an oddly named Tweka with a two-tone copper dial. It went for around $150 and had been listed as "NOS," which means new old stock, something that supposedly has sat in the back of a jeweler's drawer since 1952. Very nice, after a trip to Otto, but still not *the* watch. . . .

eBay is a cross between a swap meet in cyberspace and a country auction with computer-driven proxy bidding. The auctioneer is one of eBay's servers.

Buyers don't pay anything to eBay; they just pay sellers for the items they buy. Sellers, however, pay a fee on each item they list, and another fee if it sells. You can set this up so that your eBay seller's account comes off your credit card. I doubt if anyone's seller account amounts to much in a given month, but eBay moves a lot of items.

There's a sense of taking part in an evolving system, here. I suspect that eBay is evolving in much the way the Net did. I started visiting eBay just as user IDs were coming in. You can opt to do business on eBay under a handle. I think that this was introduced in order to foil spam-miners, who were sending bots into eBay to scoop up email addresses. And I actually did get spam, my very first, after my initial foray into eBay. But then I got a user ID and the spam stopped.

Looking over the Announcement Board recently, I saw that eBay now requires credit card information before allowing users into such categories as firearms and X-rated adult material: an age-checking strategy.

One thing I can imagine changing on eBay is the current requirement that sellers who want to display scans of their items find an off-site page on which to host their HTML. eBay links to tutorials on how to do this, but it's just enough of a learning curve to discourage some people. Myself included. If it were possible to send a scan directly to eBay, I think selling would take a major step toward becoming a ubiquitous activity.

I find clutter, in my personal environment, oppressive. But crazed environments of dead tech and poignant rubbish turn up in my fiction on a regular basis, where they are usually presented as being at once comforting, evocative, and somehow magical. The future as flea market. I really do tend to see the future that way, though not exclusively. My first impulse, when presented with any spanking-new piece of computer hardware, is to imagine how it will look in 10 years' time, gathering dust under a card table in a thrift shop. And it probably will.

The pleasure afforded by browsing eBay is the pleasure afforded by any flea market or garage sale. Something ruminative, but with an underlying acuity, as though some old hunter-gatherer module were activated. It's a lot like beachcombing.

Where eBay departs from the traditional pleasure of a flea market, though, is in its sheer scale and its searchability. If you can think of a thing, you can search it on eBay. And, very probably, you can find it.

If randomness is what you're after, though, there are ways to surf eBay, rather than search it. Modes of sheer drift. Every item offers you a chance to peruse Seller's Other Auctions, which can take you off into categories of merchandise you wouldn't have thought of. A search for Hopi silver, for instance, brought up other kinds of Native American artifacts, much older ones, so that a series of clicks through stone adzes and Clovis points led to an obscure monograph on mound-excavation in Florida in the 1930s.

But it was the watches that kept me coming back.

And I started to get sniped.

I'd find a watch I wanted, work my way up to high-bidder position, check my position regularly (eBay regularly informs you of your bidding status, and outbid notices arrive promptly, but it's still fun to check), and find, as the auction ran off, that I'd been zapped, in the last five minutes of bidding, by someone offering just one increment more than I had bid. I began to smell a rat.

The nature of the rat became apparent when I started checking out "Dutch" (multiple buyer) auctions of eBay-specific software, and discovered that one could buy plug-ins that automated the bidding process.

This bothered me. I thought about it. It bothered me more. The idea of this software ran entirely counter to the peculiar psychology of bidding at auction. The software-driven sniper isn't really bidding; he's shopping. Skimming an existing situation. The sniper (or his software package) is able to look at the final minutes of any auction as a done deal, then decide whether or not to purchase that item at the fixed price, plus one bid increment. Which pissed me off, and took some of the fun away.

A friend's hacker boyfriend, in Chicago, offered to write me a piece of software that would outsnipe anything on the market. Tempting, but not very. Instead, I sent eBay a message to the effect that allowing auto-bid software detracted from the eBay experience. That it spoiled the chemistry of the thing, which in my view was a large part of what they offered as a venue. I also suspected, though I couldn't think of a convincing way to put it, that sufficient proliferation of sniping software could eventually, theoretically, bring the whole community to a halt.

I got no reply, and I hadn't expected to, but the problem seems in the meantime to have been resolved. Entirely to my satisfaction, and in a way that illustrates exactly how things have a way of finding their own uses for the street.

Text of a message sent to all vendors of third-party bidding software at eBay, 8/13/98:

> *eBay bid system change: Yesterday, through the help of an eBay user, we detected and disarmed a "bid bot" which had placed bids on hundreds of items. A bid bot is a program which bids on many items or the same item over and over again. Our SafeHarbour team is tracking down the source of the bot, and will be working with our lawyers and the authorities to take appropriate action. In an effort to prevent this type of system attack in the future, eBay plans to make an internal change to the bidding process. Most of you will not notice this change. It will NOT affect the interface you use at all. All bidding processes will remain the same as they were before. Unfortunately, the change may disable most, if not all "automated bidding programs" (aka sniping programs). We apologize for this, but it's important that we make eBay safe from robots of this kind.*

I'd love to know what that bot was bidding on. Beanie Babies, probably. (A follow-up message partially reversed course: eBay would not outlaw bid bots but would require that they conform to sign-on procedures.)

With a level playing field restored, I decided to kick this eBay watch-buying habit in the head.

Addictive personality that I am, I decided that the best way to do that was to binge: to do a whole bunch of it at one time and get it out of my system. To that end, I decided to buy a couple of fairly serious watches. Keepers.

I bid on, and won, a late-'40s Jaeger two-register chronograph in Hong Kong. The idea of sending a check off to Causeway Bay for more than a thousand dollars to someone I'd never heard of, let alone met, seemed to be stretching it a little. But Eric So, a B Tech (Mech) at the Hong Kong Water Supplies Department and an avid watch fancier, was so evidently honest, so helpful, and responded to email so readily, that I soon had no reservations whatever. Once the check had cleared, the Jaeger arrived with blinding speed and was even nicer than described.

And I did have one authentic auction-frisson over the Jaeger when, very near the end of the auction, someone bidding "by hand" topped me. This gentleman, when I checked his profile, appeared to be a European collector of some seriousness. After I bid again, I waited nervously, but he never came back. My other binge watch was a Vulcain Cricket, an alarm-watch introduced in the late '40s, which sounds like a very large, very mechanical *cricket.* I wanted one of these because the older ones look terrific, and because "Vulcain Cricket" is one of the finest pieces of found poetry I've ever stumbled across.

I found the best one I'd ever seen, offered by Vince and Laura of Good Timing, who, by virtue of tagging all their items "(GOOD TIMING)," have built themselves the equivalent of a stall in cyberspace. Most sellers' goods on eBay are spread, as it were, on the same huge blanket, but Vince and Laura's tag allows them an edge in rep-building.

I think it worked, the binge cure. Possibly because getting serious about choosing serious watches made the shuffling of pages a chore rather than a pleasure. Whereas before I'd been able to veg out, in the style of watching some version of the Shopping Channel that actually interested me, I now felt as though I were buying real estate. Investing. Collecting.

I'd always hoped that I wouldn't turn into the sort of person who collected anything.

I no longer open to watches on eBay first thing in the morning. Days go by without my contributing so much as a single hit.

Or maybe I just have enough wristwatches.

I wonder, though, at the extent to which eBay facilitated my passage through this particular consumer obsession. Into it and out the other side in a little under a year. How long would it have taken me to get up to speed on vintage watches without eBay? Would I have started attending watch shows? Would I have had to travel? Would it have taken years? Would I have gotten into it at all?

Probably not.

In Istanbul, one chill misty morning in 1970, I stood in Kapali Carsi, the grand bazaar, under a Sony sign bristling with alien futurity, and stared deep into a cube of plate glass filled with tiny, ancient, fascinating things.

Hanging in that ancient venue, a place whose on-site café, I was told, had been open 24 hours a day, 365 days a year, literally for centuries, the Sony sign—very large, very proto-*Blade Runner,* illuminated in some way I hadn't seen before—made a deep impression. I'd been living on a Greek island, an archaeological protectorate where cars were prohibited, vacationing in the past.

The glass cube was one man's shop. He was a dealer in curios, and from within it he would reluctantly fetch, like the human equivalent of those robotic cranes in amusement arcades, objects I indicated that I wished to examine. He used a long pair of spring-loaded faux-ivory chopsticks, antiques themselves, their warped tips lent traction by wrappings of rubber bands.

And with these he plucked up, and I purchased, a single stone bead of great beauty, the color of apricot, with bright mineral blood at its core, to make a necklace for the girl I'd later marry, and an excessively mechanical Swiss cigarette lighter, circa 1911 or so, broken, its hallmarked silver case crudely soldered with strange, Eastern, aftermarket sigils.

And in that moment, I think, were all the elements of a real futurity: all the elements of the world toward which we were heading—an emerging technology, a map that was about to evert, to swallow the territory it represented. The technology that sign foreshadowed would become the venue, the city itself. And the bazaar within it.

But I'm glad we still have a place for things to change hands.

Even here, in this territory the map became.

Fortune-Telling on eBay: Early African American Textual Artifacts and the Marketplace

Eric Gardner

Fortune—To make a sudden fortune in your dream is a bad omen.
Paper—To dream of white paper, denotes respect; if scribbled or printed, you will do some mean or unjust action.

—*Chloe Russel, The Complete Fortune Teller (1824), 8, 10; eBay item #1982259838, ended on December 22, 2002*

In late 2002, after perhaps three years of using eBay to collect early editions of texts by nineteenth-century women and African Americans, I discovered one of only four extant copies of a hitherto unknown piece of Afro-Americana: Chloe Russel's twenty-page booklet (called a "chapbook"), *The Complete Fortune Teller and Dream Book*. The "dream book" portion, a list of items paired with explanations of what the presence of those items in a dream portends for the dreamer, is the source of this chapter's epigraph—a chilling prediction for any book collector. The volume was produced in the 1820s by Exeter, New Hampshire, publisher Abel Brown and notes on its title page that Russel was "A Woman of Colour in the State of Massachusetts."[1] Subsequent research has documented a free African American woman named Chloe Russel (c.1770–c.1835), who worked as a laundress and then a cook in Boston, owned property, and seems to have been a seer of some repute.[2] While *The Complete Fortune Teller* is certainly *about* Russel—and uses her exoticism to promote its sales—it may or may not be *by* Russel.[3] Nonetheless, as one of the earliest American texts to take an African American woman and African American divination as subjects, it is a find of some importance to the scholarly community—as evinced by its recent republication, with my introduction, in the *New England Quarterly*.[4]

The recovery of *The Complete Fortune Teller* through eBay foregrounds a number of questions that swirl around not only the collection and dissemination of rare Afro-Americana in an electronic age, but also, perhaps more pointedly, the ways in which race and "recovery" function in light of the socioeconomic, cultural, and political practices engendered by eBay's structure and user base. After briefly examining the "high-toned" world of serious book collecting and the ways in which Afro-Americana and African American collectors have begun to challenge this world, this chapter describes functional approaches to collecting Afro-Americana through eBay, and then steps back to critically examine the language of any such "how to" guide. While it recognizes democratizing tendencies in such work, this chapter also suggests that this very language, which is full of racist undertones as well as metaphors of hunting (and so also of hegemonic power over racial artifacts), causes the over-determined space of the marketplace (and specifically of eBay) to enact, given the place of the slave market in African American history, a set of curious revisions of those early sales of human beings. In this, the act of collecting Afro-Americana on eBay functions in complex dialogue with the language surrounding the entry of Afro-Americana into the traditionally "genteel" world of upper-class book dealers and collectors, academics, and librarians.

For decades, the representative book collector looked much like the massively wealthy Percy Gryce of Edith Wharton's *The House of Mirth,* whose sole purpose in life seemed to be a conservation of both his family name and a famous collection of rare Americana. Far from a figment of Wharton's imagination, this scion of the upper-class patriarchy of book collecting—the rich man who deigns to indulge in a passion for books (as opposed to more ostentatious modes of consumption)—is still celebrated in a range of contemporary texts on book collecting. For such collectors, fortunes and printed paper would seem far from the bad omens promised by Chloe Russel's fortune-telling guide and dream book. Consider, for example, the posh coffee table book *At Home with Books: How Booklovers Live with and Care for Their Libraries,* which retailed for 50 USD when released in 1995. Loaded with lush illustrations of private book collections, it features informative sections on "Library Lighting" and "The Art of the Bookshelf" and could stand alongside most contemporary design magazines.[5] The collectors profiled here are the financial elite: Paul Getty talks about his "thermostatically controlled" library, built because "[b]ooks, like wine, need to be kept at a regular, unfluctuating temperature," and Andrew, eleventh duke of Devonshire, whose library is "a mixture between an illustrated manuscript and a French restaurant."[6]

The entry of Afro-Americana into this (very white) world is a much more recent phenomenon: many books about books say little—when anything—about black books, and few have nonwhite faces among their sometimes posh illustrations. Arthur Schomburg and Charles Blockson, for example, *combined* get ten pages in Nicholas Basbanes's 638-page memoir-like study of book collecting and collectors entitled *A Gentle Madness.* Both entries focus on "the need to document black history"—rather than, say, Lincoln collector Louise Taper's claim that "I enjoy it. I love it, actually; I am driven by it."[7] Basbanes is correct that most major early black bibliophiles collected for reasons beyond personal obsession—and so also often left the genteel world of high-end dealers to "pick" among society's refuse in hopes of preserving a rare piece of an oppressed past.[8]

The end of the twentieth century saw some melding of these two discursive fields—passion and preservation—embodied in some ways in Henry Louis Gates's discussion of finding Hannah Crafts's manuscript of *The Bondwoman's Narrative* in 2001:

> *Each year, Swann Galleries conducts an auction of "Printed & Manuscript African Americana". . . . I have the pleasure of receiving Swann's annual mailing of the catalog . . . [which] consists of descriptions of . . . artifacts that have managed, somehow, to surface from the depths of the black past. . . . [T]o me, there is a certain poignancy to the fact that these artifacts, created by the disenfranchised, have managed to survive . . . and have found their way . . . to a place where they can be preserved and made available to scholars, students, researchers, and passionate readers. . . .*
>
> *I made my way through [the catalog] leisurely, keeping my precious copy on the reading stand next to my bed, turning to it each evening to fall asleep in wonder at the astonishing myriad array of artifacts that surface, so very mysteriously, from the discarded depths of the black past.*[9]

The thickness of the language of gentility here—Gates's comments on both poignancy and passion—masks some of the marks of socioeconomic power: most Swann catalogs cost about 30 USD, and items are priced at a premium; Gates has the leisure to read the catalog and find the manuscript—as well as the somewhat rarefied knowledge necessary to mark its importance; and Gates speaks of his "modest cap" for bidding on the manuscript, which was actually 8,500 USD, a figure not discussed in most treatments of Crafts's novel.[10]

Still if Gates's financial flexing seems akin to that of the collectors in *At Home with Books,* it actually signifies quite differently. Rhetorically, Gates is demonstrating that some African Americans are now able both to, in essence, join the country club and change it. His rhetoric and his actions suggest the less consumeristic goals of collectors like Schomburg and Blockson. Indeed, instead of keeping the rare manuscript after he got it, he engineered the publication of Crafts's novel (which brought it into the popular sphere—and so rescued it from obscurity) and then donated the manuscript, which had increased exponentially in monetary value, to a university library where it could be both preserved and ensured a place in American cultural remembrance.[11]

If Gates's work within the formerly white world of high-end book collectors offers these radical elements, then, because eBay in essence removes the country club walls surrounding serious book collecting, it would be easy to say that eBay democratizes such collecting much more fully. And it does. Both book collectors and booksellers took to eBay early, and eBay has long had a thriving market in Afro-Americana that extends far beyond the reach of the Swann Gallery catalogs.[12] Early editions of important texts such as Frederick Douglass's *Narrative* and even Olaudah Equiano's *Interesting Narrative* come up for auction with some regularity, and occasionally much rarer texts also appear—including two first editions (of perhaps only one hundred extant) of Harriet Jacobs's *Incidents in the Life of a Slave Girl.*[13]

The immediate surface difference between eBay and the more rarefied sense of collecting embodied in the Swann catalogs and, even more, Percy Gryce, is hierarchical: while eBay certainly features some of the most respected high-end dealers, it also brings in wares from original sources and "pickers" who canvas estate sales, flea markets, and library discard sales.[14] For decades among the more genteel collectors, the book dealer served as a skilled middle link in the chain of consumption. He (and occasionally she) purchased books from other dealers and from pickers, and then presented the artifacts (often cleaned up and prefaced by the obligatory brief textual description of content, context, condition, and price) to the collector in the "safety" of the bookshop (often open only by appointment), the general catalog, or the auction catalog—or even a private showing.

Within this realm, "finding" on eBay is not—at least not primarily—simply finding the obvious, complete with a detailed description written by an expert and trustworthy dealer who articulates the object's significance and value. Rather, "finding" is about getting as close as possible to the estate sales and attics where objects have been "lost" and then about being able to recognize and articulate hidden value. As such, both finding and acquiring objects are, by nature, competitive—and dependent on some sellers listing objects the value of which they do not fully understand (and thus cannot articulate).[15]

In part for these reasons, by necessity eBay searches combine features of contemporary information technology (using Boolean logic and truncation to find variations of key words) and echoes of the nineteenth-century language of race and racism. Because Afro-Americana is not a separate eBay category (and not all sellers would list merchandise in that category even if it were), narrowing searches to a small set of categories limits the scope of the buyer's task. In addition to the obvious "Books: Antiquarian and Collectible" and, less so, "Antiques: Books, Manuscripts: American" subcategories, objects can occasionally be found in "Collectibles: Ethnic and Cultural" among racist memorabilia such as Mammy cookie jars and Little Black Sambo dolls. Searching both titles and descriptions is essential, as many sellers unfamiliar with Afro-Americana (and race studies in general) use idiosyncratic terminology in listing titles. The obvious search terms, "Black" and "African American," have notable problems on eBay. "Black" comes up much more often as a color—e.g., "black and white illustrations"—than as a racial signifier. ("Color," "colour," "colored," and "coloured," while certainly the terms of choice for many antebellum African Americans, are ineffective for the same reasons as "Black.") Adding "American"—for example, searching for "Black American"—improves searching with such terms, but, much like "African American," is usually only effective in finding items with descriptions by sellers who know what they have—and ask a premium for it.[16]

This means that terms common in early titles and book texts are often more usable for searching. Entering the truncated search term "slave*," which also hits "slavery," and the truncated "abol*," which brings up "abolitionism," "abolitionist," and so on, is often somewhat effective, though it can bring up a range of twentieth-century historical texts as well as other kinds of collectibles. The term "negro" finds texts dating from the late nineteenth century and early twentieth century; in addition to appearing in early titles, the term is used by sellers out of touch with the development of terminology like "African American." Combining search terms with truncated dates often narrows the search field further; for example, "slave* 185*," though not foolproof, generally brings up a number of items tied to slave narratives, antislavery texts, and proslavery texts because it hits a common term in early texts and a set of publication dates (which is common even in the most basic of book descriptions).

Still, to find the truly "lost" artifact, collectors look for texts without clear racial markers—texts not labeled by any term that obviously suggests a piece of Afro-Americana. Again, searching for terms common to nineteenth-century authors—yet unfamiliar to many twenty-first-century readers—offers some potential; the truncated "coloniz*," for example, brings up variants of texts by and about colonization societies. Searching by gender using nineteenth-century constructions of authorial names also offers possibilities: narrowing the focus to the category "Books: Antiquarian and Collectible" and searching for "mrs* 187*," for example, often finds a range of texts by women writers of the 1870s (including, once, an 1879 edition of Julia Foote's slave narrative *A Brand Plucked from the Fire,* the seller of which was almost certainly unaware of Foote's race, as the description was silent on the issue).

Finally, searching by eBay user ID can be useful to those buyers familiar with the relatively small circle of bidders most skilled at finding such texts; often, though, such bidders attempt to keep other buyers from doing this kind of searching by resorting to "sniper bidding" or, at least, holding their bids until late in auctions. Still, the fact that eBay maintains a sixty-day history of items bid on by each user, as well as complete lists of current bids and bidders for most items, means that users can track bidding patterns, examine tides and descriptions for items initially missed and so learn more effective search strategies, and get a rough sense of the amounts a given bidder is willing to pay for an item.

Within this framework, *The Complete Fortune Teller* was not an ideal find; its seller was aware of the byline and so titled the listing "1824 Black Fortune Teller Freemasons, Irish"—with the last two words noting other chapbooks included in the bound volume. Further, he both quoted the byline and noted, "If Chloe Russel was real, she must have been one of the earliest black American women to be published." Still, the identifying terms here—"Black" and "color"—are not those most friendly to (or used by) most collectors of Afro-Americana, and no other commonly searched and used terms—abolition, slavery, negro, and so on—appeared. This must have limited the number of bidders. A counter placed in the description—marking the total number of hits to the listing rather than different users—showed only a few hundred by the end of the auction, perhaps forty of which (including refreshes) were made by myself.[17]

Nonetheless, it was clear, when the item turned up in a search for "Black America* 18*," that the seller was aware of at least the potential that the text was black-authored and was fully aware that the text had black subject matter, and so put something of a premium (100 USD) as an opening bid. The seller did not, though, have the knowledge of Russel's identity or the potential importance and rarity of the text. Between finding the item and placing a bid close to the auction's end, I was able to check for other copies using WorldCat, to use an extensive range of genealogical and historical sources including Ancestry.com to find the initial proof that there was an African American named Chloe Russel, and to do initial research on the links between African Americans and divination in early national New England. Probably few potential bidders were able to do such work, and few bid on the text. And thus a piece of rare Afro-Americana came to a middle-class professor who can't justify spending the 30 USD for the luxury of a Swann catalog—much less actually bidding on something in it.

The tone in the above—like that in much collecting literature (and even some of the high-end books on books)—is, of course, mingled pride in a successful hunt and excitement about a win in a competitive process.[18] Again, however, it was not the ideal find; it might have been had it simply been advertised as an early fortune-telling text with an opening bid of, say, 20 USD, and some clue that only I had caught.

The problems with such a tone, though, are myriad. Of course, my narrative is full of class markers—albeit not the level of markers in *At Home with Books.* My eBay acquisition might have been more "democratic" than the Swann auction, but. . . .

In some ways, the quality of gentility found in the Swann catalog is almost lost amidst the racist language that marks the discursive field of Afro-Americana on eBay. Indeed, racist items and/or descriptions turn up much more often in the searches described above (which are themselves dependent upon using racist terminology). Consider the early 2005 sales of the 1896 and 1897 picture book *Ten Little Niggers* by eBay ID rebel_soldier, who specializes in Civil War material; the book's backward counting in racist rhyme ("Seven Little Niggers chopping up sticks. One chopped himself in halves, and then there were Six") is accompanied by gruesome illustrations—several of which were reproduced in the item description.[19] In the first auction, which offered the 1896 edition, the seller quoted eBay's policy on "Offensive Items," which allows sellers to use such language "provided that the offensive words are actually part of the title of the listed item" and that their own language "shows appropriate sensitivity to those in the Community that might view it." In the second auction, in which the 1897 edition was presented in similarly great detail, the seller expanded the quotation from the policy to note, "Artists occasionally use offensive words and phrases such as 'Nigger' and 'Jap' in the titles of media items" and that, nonetheless, "eBay urges all of its community members to treat others as they themselves would like to be treated." The second auction was also a "private listing"; that is, bidder identification was visible only to the seller. Thus, though its policy statements suggest otherwise, eBay not only allows but even authenticates racist discourse.[20] Finding such texts, only a step from finding *The Complete Fortune Teller,* thus involves support of eBay's mixed message and direct use of racist language.

Perhaps more troubling still, all aspects of auctions of early Afro-Americana call to mind the most common auctions in African American history. The flipness of my narrative of acquisition—for example, wishing I could have bought the text for 20 USD instead of 200—frankly, insults: while we are not bidding on bodies, we are bidding on memories.[21] While many of the attributes of this narrative—the ascribing of monetary value to the invaluable, the sense of "knowledgeable" bidders, the contest between seller and bidder to get the "best price," and the bragging surrounding a "good find" and a "good buy"—echo most capitalist exchanges, the use of racist language to identify and mark racial objects and the subject matter and purposes of these texts call to mind a specific nexus of capitalism and racism, the antebellum slave market.[22]

While purchasing an antebellum text and purchasing a slave are obviously very different activities, the collector of Afro-Americana may certainly be buying the text—as slaves were bought—for both its use value and exchange value in all their complex shadings. As a collector, or, in Walter Benjamin's words, "a real collector, a collector as he ought to be," "ownership is the most intimate relationship that one can have to objects. Not that they come alive in him; it is he who lives in them."[23] I may want ownership of the text—as so many slave buyers seem to have wanted ownership of slaves—to articulate myself. Further, while the discovery of Chloe Russel certainly adds to our sense of African American history and literature, discovering and sharing that text will likely advance my academic career (Gates's discovery of *Our Nig* certainly advanced his).[24] I am also, frankly, using Chloe Russel's text in different ways than it was intended—including the making of specific arguments about the landscape of early African American literature. Such appropriation is complicated by the fact that *The Complete Fortune Teller* may well have been written by someone other than Chloe Russel, someone who hoped to capitalize on her name and race to sell books. If I am not fully akin to a buyer at a slave market, I may have more in common with the tradition of (sometimes well-meaning) whites—one thinks of the editors with whom African Americans such as Frederick Douglass and Harriet Jacobs struggled over control of their stories—who have appropriated the texts and the stories of African Americans.

It would be easier—and, in the end, hopefully more true—to say that my own purposes were not pecuniary or oppressive, but better, higher, like those I ascribe to Gates and others who hope to rescue a piece of the black past. In explaining why he collects racist memorabilia, for example, David Pilgrim, founder of the Jim Crow Museum, says,

> *I am a garbage collector, racist garbage. For three decades I have collected items that defame and belittle Africans and their American descendants . . . 4,000 . . . items that portray blacks as Coons, Toms, Sambos, Mammies, Picaninnies, and other dehumanizing racial caricatures. I collect this garbage because I believe, and know to be true, that items of intolerance can be used to teach tolerance.*[25]

If even *this* material—and the approaches demanded by collecting it—can be redeemed, then perhaps my ends can in part justify the means I employed—the eBay requirement to use racist language in searching, the form of the auction, and the reenactment of scenes of racism. Were it not for these means, I submit, we would not today be able to remember Harriet Wilson, Hannah Crafts, Chloe Russel. . . .

But collectors of Afro-Americana cannot simply "dream" of clean and easy resolution, of "white paper," as Russel notes, and assume that such dreams portend respect. Just as slave narrators from Frederick Douglass to Harriet Jacobs, in speaking of how their friends purchased them with the express intent of freeing them, mingle gratitude with painful questions about participating in the slave economy, so too must we recognize that the quality of democratic openness fostered by eBay often opens the barely healing scars of racism. In short, the paper is already scribbled upon. And so when we dream of great fortunes based on printed pages, when we engage with Jim Crow language and the auctioning of memories, the cautionary fortune-telling of Chloe Russel's dream guide—embodied in this chapter's epigraph—and the cautionary tale of how Chloe Russel has come back to us via eBay may remind us of the complex terrain and the ironic series of trade-offs necessary to study slavery's remembrance in a capitalist society.

Notes

The author wishes to thank Jodie Gardner for her comments.

1. Chloe Russel, *The Complete Fortune Teller and Dream Book* (Exeter, NH: Abel Brown, 1824), in *The Amusing Bud[get]* (Exeter, NH: Abel Brown, 1827). The other copies of Russel's text include an 1824 chapbook not bound with *The Amusing Budget,* a partial copy that is owned by the American Antiquarian Society; a circa 1800 edition by Boston printer Tom Hazard, owned by the Boston Athenaeum; and an undated partial copy of another (perhaps earlier) edition owned by the Library Company of Philadelphia. The Library Company edition contains what purports to be an autobiography of Russel—which narrates a birth in Africa, years as a slave in Virginia, and the gaining of the "gift" of the powers of divination.
2. On Russel, see my introduction to "The Complete Fortune Teller and Dream Book: An Antebellum Text 'By Chloe Russel, a Woman of Colour,'" *New England Quarterly* 78, no. 2 (June 2005): 259–88.
3. Errors in the narrative included in the copy at the Library Company of Philadelphia, paired with what may well be incidents of plagiarism in the text, yield serious questions about its authenticity as a black-authored text. I posit three possible scenarios: that Russel wrote the text, that a white publisher (possibly Hazard) wrote it and hoped to capitalize on her name, or that they collaborated. Research following the republication of the text hopefully will help us make better guesses at just which scenario is true. It should also be noted that the seeming plagiarism in *The Complete Fortune Teller* echoes issues raised about *The Bondwoman's*

Narrative, which borrows liberally from Charles Dickens even as it constructs a unique tale around such borrowings. See, for example, Hollis Robbins, "Blackening Bleak House: Hannah Crafts' The Bondwoman's Narrative," in *In Search of Hannah Crafts: Critical Essays,* ed. Henry Louis Gates Jr. and Hollis Robbins (New York: Basic Civitas, 2004).

4. Scholars have found only a small bookshelf full of texts written by nineteenth-century black women; few of these date to before 1830. The number of texts about African American women and divination is similarly small.
5. Estelle Ellis et al., *At Home with Books: How Booklovers Live with and Care for Their Libraries* (New York: Carol Southern Books, 1995), 132–35, 136–37. All of the coauthors have design backgrounds—including Ellis's tenure at *House and Garden.*
6. Ibid., 37 and 21. The books of Nicholas A. Basbanes, the dean of writers about bibliophilia (which he terms "a gentle madness"), often parallel such treatment, though at times they have less of the country club feel. See Basbanes, *A Gentle Madness: Bibliophiles, Bibliomanes, and the Eternal Passion for Books* (New York: Henry Holt, 1995). See also Nicholas A. Basbanes, *Among the Gently Mad: Perspectives and Strategies for the Book Hunter in the Twenty-First Century* (New York: Henry Holt, 2002); Lawrence and Nancy Goldstone, *Slightly Chipped: Footnotes in Booklore* (New York: St. Martin's Griffin, 2000); and Lawrence and Nancy Goldstone, *Used and Rare: Travels in the Book World* (New York: St. Martin's Griffin, 1997).
7. Basbanes, *A Gentle Madness,* 394–404, 427.
8. Robert Adger's collecting (which resulted in one of the most important collections of early Afro-Americana, now at Wellesley)—as well as the book collections created for the early World's Fairs and the now-famous efforts of librarians at historically black colleges (Howard's Dorothy Porter Wesley stands most notable here)—focused on preserving cultural memory because, as librarian and historian Tony Martin notes, "In the book-collecting world, books dealing with blacks have been treated with casual regard." See Wendy Ball and Tony Martin, *Rare Afro-Americana: A Reconstruction of the Adger Library* (Boston: G. K. Hall, 1981).
9. Henry Louis Gates Jr., "Introduction," in Hannah Crafts, *The Bondwoman's Narrative,* ed. Henry Louis Gates Jr. (New York: Warner Books, 2003), xxi–xxii.
10. The purchase reportedly totaled 9,775 USD (8,500 USD plus the commission), which was actually below the 10,000 USD price set by Swann. Gates's was the only bid, and Swann accepted it; see Alexander J. Blenkinsopp, "Gates Acquires Slave's Novel," *Harvard Crimson,* November 14, 2001. For Swann Gallery catalog information, see their website, www.swanngalleries.com.
11. On the donation, see "Henry Louis Gates Donates Slave Novel to Yale University," *Black Issues in Higher Education* 20, no. 9 (June 19, 2003): 12. The manuscript's monetary value is now reportedly assessed at well over 100,000 USD, thanks in part to the place of Gates's edition on the *New York Times* Bestseller List. Thus, Gates is far different from the curiously conservative sense articulated by Walter Benjamin (and, one would assume, espoused by the collectors in *At Home with Books*) that "[t]o a book collector, you see, the true freedom of all books is somewhere on his shelves"; see Walter Benjamin, "Unpacking My Library: A Talk about Book Collecting" in *Illuminations,* ed. Hannah Arendt (London: Fontana Collings, 1973), 64.
12. On most days in early 2005, at least seventy-five items came up on a search of the "Books: Antiquarian and Collectible" category using the very specific phrase "African American"; given what is said below about searching—as well as about other categories—the volume of Afro-Americana available on eBay in a month is probably larger than that of many small auction catalogs.
13. The most recent auction of a first edition of *Incidents* ended with a last-second bid of over 2,000 USD on June 26, 2005 (item number 6540838524).
14. My discussion of collecting here and throughout the essay owes much to Russell W. Belk, *Collecting in a Consumer Society* (New York: Routledge, 1995), especially his sense of late-twentieth-century collecting as a partially democratic phenomenon enabled by the rise of the industrial

and postindustrial capitalist state. See also Werner Muensterberger, *Collecting: An Unruly Passion* (Princeton, NJ: Princeton University Press, 1994). Books on books privilege the traditional sense of the book dealer. Basbanes, for example, quotes historian and bibliographer Katharine Kyes Leab: "Bookselling is theater, and that is something the Internet cannot duplicate. It cannot create the drama that the dealer of the auction fires up in a collector's mind" (*Among the Gently Mad,* 89). Basbanes also notes, "Another difficulty some professionals have with online auctions, especially those mounted on www.eBay.com, is the total absence of accountability on the part of the auction house" (ibid., 89).

15. In fascinating ways, Swann Galleries clearly knew what it had yet did not, at all, know what it had. Central to Gates's decision to bid on the Crafts manuscript, for example, was the Swann catalog's notation that it came from the collection of black bibliophile Dorothy Porter Wesley (Gates, "Introduction," xxiv). Still, the gallery was unable to pursue authentication as far as Gates eventually did—including conversations with several rare book librarians and, with the backing of Lawrence Kirshbaum, the CEO of the AOL Time Warner Book Group (*The Bondwoman's Narrative* was published by Warner), an extensive examination by historical documents dealer Kenneth Rendell and manuscripts expert Joe Nickell.
16. Such sellers may not be dealers—or even knowledgeable about the field of Afro-Americana, per se—but may simply recognize that an object's exoticism or connections to the well-publicized collecting of "Black Americana" artifacts should call for a premium price. See David Pilgrim, "The Garbage Man: Why I Collect Racist Objects," *The Jim Crow Museum of Racist Memorabilia,* www.ferris.edu/news/jimcrow/collect/.
17. Though I hopefully never became the "bandit-cranking Vegas granny" continually hitting the Reload button who William Gibson describes in chapter 1 of this book, I was also far from the "reflective" and "beachcombing" browser Gibson comparatively notes. In essence, the moment when the "hunt"—the capitalist struggle—begins is marked by such obsessive behavior.
18. Belk, 93, as well as 61–101 generally.
19. eBay item numbers 6954160399, ended April 9, 2005, and 6950967178, ended March 19, 2005, respectively.
20. For the full policy, see http://pages.ebay.com/help/policies/offensive.html. A full study of eBay's policy is beyond the scope of this essay—though such certainly needs to be done. Several users play on eBay's policy to specifically sell items; rebel_soldier, for example, in an auction for the collection of supposedly "true stories about 'Hill-billies' and 'Negroes,'" Carl Carmer's 1934 *Stars Fell on Alabama,* noted that the sale was "ALMOST BANNED FROM eBAY!!!" See eBay item number 6950874224, ending on March 19, 2005.
21. On the conflicts surrounding racialized memories in the marketplace, consider, for example, the recent coverage of the discovery by Holly Jackson that Emma Dunham Kelley-Hawkins, long thought to be one of the earliest African American woman novelists, may have been white—including her summary dismissal from inclusion in future editions of the Schomburg Library of Nineteenth Century Black Women Writers. These changes to the construction of the scholarly memory of her race will undoubtedly shape the prices demanded for original editions of her books, which, in the hands of high-end, specialized dealers, were, just prior to the announcement of the discovery, upward of 1,000 USD—with the rare *Four Girls at Cottage City* at 8,500 USD on abebooks.com by Between the Covers Rare Books, Inc. See Holly Jackson, "A Case of Mistaken Identity," *Boston Globe,* February 20, 2005; and David Mehegan, "Correcting a Case of Mistaken Identity," *Boston Globe,* March 5, 2005. There has even been some internet debate on just who has claim to this discovery—and how memory should be presented. Katherine Flynn, a well-known genealogist who had been researching Kelley and was ready to publish her findings in a peer-reviewed journal when Jackson's newspaper accounts came out, is quoted on one site as saying, "I think that with all that is at stake here it is time to stop publishing in the *Boston Globe* as if it was now the new premiere journal in English lit or AA lit studies without any references to census records or any systematic study of the whole record"; see www.plastic.com/article.html;sid=05/03/02/10155368;cmt=11.

22. The best work on the rhetoric—and the history—of the slave market is Walter Johnson, *Soul by Soul: Life inside the Antebellum Slave Market* (Cambridge, MA: Harvard University Press, 1999). Johnson's discussion on "Marking Race," 136–61, particularly influences my comments here.
23. Benjamin, 69.
24. In part hoping to balance the mercenary with the community oriented, I am in the process of donating my copy of *The Complete Fortune Teller* to a university library.
25. www.ferris.edu/news/jimcrow/collect/.

MODULE TEN

Privacy and Surveillance

Situating Privacy Online: Complex Perceptions and Everyday Practices

Ana Viseu, Andrew Clement and Jane Aspinall

Media and research reports point to the issue of privacy as the key to understanding online behaviour and experience. Yet it is well recognized within privacy-advocacy circles that 'privacy' is a loose concept encompassing a variety of meanings. In this article we view privacy as mediating between individuals and their online activities, not standing above them, and as being constantly redefined in actual practice. It is necessary to examine, therefore, what individuals are reacting to when asked about online privacy and how it affects their online experience. This article is based on data generated in the Everyday Internet study, a neighbourhood-based, ethnographic project being conducted in Toronto, Canada, that investigates how people integrate online services in their daily lives. We propose that there are three organizing 'moments' of online privacy: the moment of sitting in front of the computer, the moment of interaction with it, and the moment after the data has been released.

Keywords: privacy, ethnographic study, Internet, practices, perceptions

Privacy is a buzzword in the media these days. Accounts of the loss or invasion of privacy, threats to and worries about privacy, and even the death of privacy appear daily. Similarly, there are reports of privacy-enhancing technologies and calls for legislation to defend privacy. At the same time, however, privacy is a slippery concept. It has been categorized as a countervailing force to the development of the information society and e-commerce, denounced as an outdated and defunct concept, and called the civil rights issue of the twenty-first century.

The one constant in the privacy debate are public opinion surveys indicating that the overwhelming majority of people are 'very concerned' about their privacy. Yet empirical evidence suggests that there is a significant discrepancy between privacy principles and privacy practices. Furthermore, given the many definitions and interpretations of privacy, it is important to investigate what meanings of privacy are involved in the practice of being online.

This study[1] explores the perceptions and practices of online privacy. It draws upon qualitative, empirical data generated in the Everyday Internet study, conducted in Toronto, Canada, the goal of which is to investigate how regular Internet users experience online services in the context of their daily lives. Our objective is not to draw final conclusions about the state of privacy online; instead—and given the limited fieldwork research conducted so far—this study illustrates patterns and themes that are useful in understanding the broader privacy and policy-setting discourses.

Background

Privacy is a far-reaching concept that encompasses and can be applied to a variety of core rights and values such as freedom of thought, freedom of expression, freedom from surveillance, personhood, human dignity, identity and anonymity, secrecy, and so on.[2]

So hard is it to define that Alan Westin, a noted privacy scholar, observes that 'no definition of privacy is possible, because privacy issues are fundamentally matters of values, interests and power' (Westin 1995, quoted in Gellman 1998, p. 194). Others have argued differently, saying that defining privacy is unnecessary since the concept itself serves no function, merely standing in for other primary interests (Thompson 1984, quoted in Solove 2002).

Within the policy discourse, however, privacy has been consistently discussed in terms of 'fair information practices' (FIPs) based on the principle of 'informational self-determination'. The concept of FIPs, which constitutes an attempt to regulate the collection, usage, security and dissemination of personal data (Cavoukian & Tapscott 1995; Gellman 1998), originated almost simultaneously in the USA and the UK in the early 1970s (Flaherty 1999).[3] Thereafter, the need for transnational data protection policies was quickly identified (Bennett 1998), and in 1980 the Organization for Economic Co-operation and Development (OECD) issued its *Guidelines on the Protection of Privacy and Transborder Flows of Personal Data,* the first attempt to 'harmonize national privacy legislation without interrupting the free flow of information between borders' (Center for Democracy and Techology (CDT) n.d.). Since then, a number of jurisdictions have passed data protection legislation in an effort to regulate the collection of personal information, including the European Union's Data Protection Directive,[4] introduced in 1995, and Canada's Personal Information Protection and Electronic Documents Act (PIPEDA), passed in 2000 (Privacy Commissioner Canada n.d.). Finally, the term 'informational self-determination' was used in 1983 by the German Federal Constitutional Court to assert privacy protection as a fundamental human right. The expression points to the right of every individual—not the government—to determine and control what happens to his/her personal information (Cavoukian & Tapscott 1995) and to know who knows what about him/her at what time (Virtual Privacy Office n.d.).

Nonetheless, it is commonly acknowledged that the privacy landscape is riddled with ambiguities. For instance, surveys routinely show that an overwhelming number of people are 'very concerned' about their privacy: the 2001 Internet Report of the University of California at Los Angeles (UCLA) puts this figure at 95 per cent, while America Online (OL)/Roper Starch's (2000) Worldwide Adult Cyberstudy gives a figure of 94 per cent.[5] Yet these surveys are accompanied by what could be called counter-surveys, i.e. factual demonstrations of a disconnection between privacy perceptions and practices. Daniel Solove (2001) presents the example of the survey conducted, in 1998, by Condé Nast Publications:

> *[Condé Nast] sent out a booklet of 700 questions concerning detailed information about an individual's hobbies, health (including drugs used, acne problems, vaginal/yeast infections, etc.), shopping preferences, etc. . . . In return for the data, the survey said: 'Just answer the questions below to start the conversation and become part of this select group of subscribers to whom marketers listen first.' (p. 1428)*

Almost 400,000 people responded. Another example of the privacy gap is described by Lyon (1994, p. 140):

> *In the mid-1980s passers-by on a New York street were asked one day their opinion concerning the 'invasion of privacy' by 'modern technology'. Ninety per cent expressed concern. But the next day, when offered a credit card with a favourable interest rate, ninety per cent of passers-by on the same street filled their entire application forms requiring Social Security number, bank account numbers and information about other credit cards.*

Given the varied interpretations of privacy, it is important to investigate what individuals are reacting to when asked about privacy online, as well as how they act upon their perceptions and how these perceptions affect their online practices. The need for a contextualized approach to privacy is well documented in the literature. In 1996, Alan Westin found that 50 per cent of consumers were 'pragmatists' in their approach to privacy (Federal Trade Commission (FTC) 1996). Other researchers, such as Sheehan (2002), Hine & Eve (1998), and Regan (1995), have also argued in favour of such an approach. In this article we base our analysis on the framework set out by Daniel Solove (2002). Solove criticizes contemporary privacy theories and proposes new methods for analysing the issue. He argues that, rather than trying to find a one-size-fits-all definition of privacy, we must search for a pragmatic approach that allows us to emphasize and explore the situated and dynamic nature of privacy. Privacy must be viewed within the practices that constitute it and give it meaning, and studying threats to privacy means studying the types of disruptions to these particular practices (Solove 2002). Privacy, then, should be viewed as a concept that stands, not above individuals and their activities, but between them, and as one that is constantly being (re)defined in actual practice.[6]

Research Design

The Everyday Internet project,[7] an ongoing research study initiated in the summer of 2001, is informed by a variety of contemporary information-policy issues, of which privacy is an example. By not focusing specifically on privacy, we attempt to diminish the risks of biasing our sample towards individuals who are not concerned about privacy, and are thus following a strategy that has been previously adopted by researchers such as Hine and Eve (1998). This approach also fits our objective of discussing privacy from a grounded perspective, i.e. seeing if, and how, privacy concerns shape online experience and vice versa. In addition, because we aim to obtain a better understanding of the integration of networked services in everyday life, we take a 'holistic' approach towards our participants. Privacy studies frequently sectoralize individuals as consumers (e.g. Phelps *et al.* 2000; Miyazaki & Fernandez 2001; Earp & Baumer 2003), but in this study we treat them as multifaceted, active citizens.

In our recruiting strategy, we attempt to include participants from those socio-demographic categories that are repeatedly identified (e.g. National Telecommunications and Information Administration (NTIA) 2000; Dryburgh 2001; Sciadas 2002; Pew Internet and American Life Project 2003) as lagging behind in Internet access: low income, low education, English as a second language, and seniors. Our goal is not to hypothesize on or extrapolate the differences between these variables but rather to give voice to those who constitute them. Taking advantage of Toronto's cultural and ethnic diversity, we utilize a 'geography-based' recruiting strategy similar to that used by Miller and Slater (2000). Through census data and Internet access data, we have settled on a downtown multicultural neighbourhood as the focus of our investigation.[8] We define the regular Internet user as someone who logs on to the Internet on average at least once per week.

In order to be consistent with the situational dimensions of our ethnographic study, we have recruited and interviewed participants at their regular access site, whether that be at home or at a public facility. Relying on in-depth, semi-structured interviewing and participation-observation methods to collect data, we have invited participants to give us a demonstration of their online activities and to show us their familiar websites, while also structuring our interviews as informal conversations based on the interviewee's online practices (see Clement *et al.* 2002 for a more detailed description of the Everyday Internet project). Most participants have been interviewed twice, with an interval of two to four months between interviews.[9]

Participants have been recruited via two methods. To find home Internet users, we used a door-to-door strategy. If nobody was home, we left a brochure with information about the project and a questionnaire, which the person can send to us if he/she is interested in becoming involved.

Since we wanted to talk to both home Internet users and individuals reliant on public facilities for their access, we approached St Christopher House, a community centre located in our study area, to help us with the recruitment process. We spoke to the person in charge of their Internet services and left some questionnaires in the computer room. One day later we returned to find that more than twenty-five individuals had volunteered.

TABLE 1 Overview of participants.

pseudonym	access type (interview location)	language	age‡	education§
Barbara	Home	Portuguese†/English	40–59	Low
Camila	PF*	Portuguese	Over 60	High
Danielle	Home	English	20–39	High
Ignacio	Home	Spanish†/English	Over 60	High
Julia	Home	German/English†	20–39	High
Marianna	PF	English	20–39	High
Nicholas	Home	English	40–59	High
Tammy	PF	English	20–39	High
Ursula	PF	German/English†	Over 60	Low
Xavier	PF	Portuguese†/English†	40–59	High

*Public facility
†Denotes the language in which the interview was conducted
‡Each of the age groups presented here is the result of the merging of four categories from the 1996 census data.
§Low education: High school diploma or less; High education: BA degree and over.

We have so far interviewed a total of fourteen participants, ten of whom are mentioned in this article. Of those ten, half were interviewed in public facilities and the other half in their homes. Their socio-demographic characteristics are summarized in table 1.[10] To protect the participants' anonymity and privacy, pseudonyms are used throughout this article.

The small size of our participant sample is countered by the potential for depth offered in our interactions with participants. The contacts with participants were designed to be as 'naturalistic' as possible, offering them the opportunity to express their opinions and demonstrate their abilities in familiar settings. We recognize, however, the inherent and insurmountable 'artificiality' of our interactions with participants and acknowledge that their opinions were expressed in a specific context, that of a conversation with a stranger, often perceived as an expert.

The participant pool is rich in experience—from professionals to novices—and diverse in character—immigrants, youth, and seniors. In fact, a noteworthy characteristic of this sample is the number of participants for whom English is a second language.[11] The scope of Internet experience among our participants ranges from six months to more than five years. All participants use the Internet regularly, ranging from several times per day to two or three times a week.

All interviews conducted during this project have been transcribed and then carefully read and revised by several members of the research team. The interviews done in languages other than English have been transcribed and then selectively translated. For the analysis presented here, the research team reviewed and coded the interview transcripts. A cluster of privacy-related concepts—for instance, security, surveillance, identity, control, and seclusion—were then employed, in our first round of analysis, to flag and code all comments regarding privacy within the context of each individual transcript. Through this method, we selected a total of ten respondents—out of the fourteen who had been interviewed—for further analysis, on the basis of the pertinence of their observations. After this initial selection, topics and patterns among transcripts were identified and quotes were organized thematically.

Privacy Perceptions and Online Practices

Privacy-related conversations were held with all participants in the study, making it the only topic about which all participants had an opinion. This resulted not only from our interest in the issue but also from the extremely varied meanings of the term 'privacy' itself. With the exception of one or two participants, the topic of privacy surfaced 'naturally' in the course of our conversations. For instance, if discussing issues of financial security, we would probe for connections with the safety of personal information. Hence, at no point did we provide a guiding privacy definition—nor were we asked to do so—although at times we did guide the discussion in a specific direction.

Considering the broader goals of the study, and the ambiguous nature of the term 'privacy', we found it significant that none of our participants seemed surprised to be asked privacy-related questions and that no one requested clarification about our understanding of the concept. In fact, participants seemed to view our interest in privacy issues as a natural extension of our interest in their Internet use, even when they objected to the concept itself. This indicates that the debate has moved beyond the scholarly, activist and legal realms to reach the larger public.

When the interview transcripts were coded by different research members and the quotes were organized thematically, three interwoven 'moments' of online privacy stood out: sitting in front of the computer, interacting with it, and releasing data. In the present study, these moments are not organized sequentially or temporally, nor are they mutually exclusive; for instance, the moment of sitting in front of the access point extends throughout the entire interaction. However, taken together, the three moments are constitutive of the experience of being 'online' and are helpful in understanding the different factors at play during this experience.

The Places and Spaces of Going Online

Characteristics of physical space, associated with life stories and experiences, affected participants' perceptions of privacy, suggesting that the first 'moment' of privacy online is shaped not only by the immediate surroundings of the individual but also by his/her overall experience. For example, Barbara and Marianna access the Internet from both home and public facilities.[12] Although their life stories are very different, they share a concern over the privacy issues involved in the use of public computers for financial transactions. Barbara is in her forties and is barely literate. Three years ago, she was pressured by her teenage children to buy a computer and get high-speed Internet access. However, until recently, Barbara did not know how to use the computer. Six months before our first interview, she had an accident that still prevents her from working, and, with the free time available to her, she started taking computer classes at St Christopher House. Barbara does not explore the web—she mainly visits websites of entities that are already familiar to her. She readily admits to not knowing much about the Internet, and her main online activity consists of financial transactions—shopping and banking—which are traditionally associated with high privacy and security concerns. Barbara speaks of the Internet as a twenty-four-hour mall, a great improvement over catalogue and telephone shopping. When probed further about her views of online banking, Barbara praises the service's convenience but says that, if she did not have home access, she would not be willing to use it. The reason for this is that it would not be safe to bank using a computer that is not hers. Barbara's feeling of security does not come from an understanding of the Internet's workings—she knows nothing of 'cookies' left in the computer's hard drive or network hackers—but rather from a sense of ownership: the terminal used for access is hers and is located in her private space, and therefore it is safe. Previous studies (e.g. Gronroos 1994; Pitkow & Kehoe 1996; Rogers 1996; Earp & Baumer 2003) have shown that privacy concerns decrease when individuals are familiar with the collecting entity. What our study seems to show is that familiarity, or indeed ownership, of the instruments used for access can be just as important.

Marianna is the youngest of our participants. She is doing a BA degree and started using the Internet five years ago. She describes herself as someone who does not like computers but has a fair understanding of the language of the web. She says that this is because she grew up with it. She started using the Internet for chatting and has now 'moved on' to using it for research and email. Marianna explains that her father is trying to push her to do her banking online, but she is reluctant to do so. 'I think it's because I don't have access at home,' she says, 'and I don't really feel like banking in a public place, like, more so than I already do.'

The amount of privacy (here associated with security) that is granted by the access site—public or private—is one of the factors that shape online activities.[13] Online banking, owing to its sensitive nature, is assumed to be a personal activity, one that should be conducted in a private space or at least with the support of a private access point.

The issue of the intersection of private and public spaces and activities was reflected in areas other than financial transactions. Several participants spoke about the need to keep their public and private personas separate, to establish boundaries between personal and work life. Tammy, a PhD student at the University of Toronto, relies on public facilities for her Internet access. She has a modem-less laptop that she refuses to take home:

> *I don't want to bring either my work and all those outside concerns into a space which, for me, is a bit of a refuge.... Somehow [the laptop] speaks to me of work, it speaks to me of obligation, it speaks to me of so many things I don't want to have when I come home.*

Personal life stories also affect perceptions of online privacy. Camila and Xavier—recent Brazilian immigrants to Canada—describe the existence of two Internets, one Canadian and one Brazilian. Camila, a retired nutritionist who started using the Internet three years ago while she was still in Brazil, discusses the difference between the Brazilian Internet and the Canadian one, and how in Canada it is possible to do a variety of things that would be unthinkable in Brazil:

> *In Brazil, it's hard... many people are robbed, things get stolen, they steal your personal data, in Brazil it is not advisable.... There they take advantage of anything to be able to clone your personal data.... Now, here, the little I heard from people here is that, bank accounts included, everything, everything, payments and everything, is done through the web and that there is... I mean, if there are problems they are minor.*

This point of view, whether valid or not, points to the importance of place and cultural practices in the understanding of online privacy, or at least to the perception that the physical location of online access matters.

Releasing Personal Information

The second privacy 'moment' has to do with how the study's participants understand the process of releasing their personal information online. It refers to the dynamic interaction between the goal of the individual's online activity, his/her perception of the medium, and the strategies used in his/her self-presentation.

In August 2000, the Pew Internet and American Life Project released 'Trust and privacy online: why Americans want to rewrite the rules', which described the results of a telephone survey. Researchers found that about a quarter of 'internet users have provided a fake name or personal information in order to avoid giving a Web site real information' (PEW 2000). Lying came ahead of the two other privacy-protection strategies mentioned in the report, namely email encryption software (9 per cent) and anonymizing software (5 per cent). This report caused quite a stir in the media (e.g. Charny 2000) and even prompted some zealous business executives to argue that lying when releasing personal information should be made illegal and punishable by law.

When we started our study, we assumed that the number of users employing deception strategies when releasing personal information would have grown since the Pew report and that using pseudonyms to protect one's identity would be common practice. We were therefore surprised when none of our participants reported using fake identities online, not even when specifically questioned about it. (Marianna did tell us that in her old chat room days she used a nickname, but now she no longer chats.) We found that respondents were selective when releasing their personal information, but when they did so they always gave their real identity.

The reasons given by our participants for releasing their real identity in their online interactions ranged from a certain innocence that comes from lack of experience in a new environment—their not realizing that, in Internet communications, lying is not only possible but also extremely easy—to a desire for customization and minimization of the information flow and a need to establish trust.

Barbara, our shopping enthusiast, falls into some of these categories. She is selective in her release of personal information in the sense that her online experiences are circumscribed to a limited number of businesses that she already knows and trusts. Since we often give away our identity when paying in 'real life', this is business as usual for her. But there is more to it: Barbara enjoys the 'value-added' touch that comes with personalization, i.e. not having to introduce her credit card number again and having her preferences displayed on the screen. Because she does not know much about the technicalities of the Internet, she is not aware of where her personal information is stored, or to what uses her information is put. Furthermore, she is not affected by spam because she does not have an email address. To her, the usefulness of personalization is clear in that it simplifies her Internet interactions.

Danielle, a human resources consultant and writer who works from her home, uses the Internet both in dealing with people and as a research tool. She explains that she has only one email account (with real identity) as a strategy to try to reduce the incoming data flows. She uses it in her Internet exchanges but is selective when registering at a site because she does not want to be inundated with spam; that is why she does not have a hotmail account—'those [kinds of email accounts],' she says, 'are the true spam'. In the second interview, Danielle is even more explicit. She is becoming irritated by our questions about privacy, a concept that she deems useless. She says:

> *[Being reluctant to release personal information is] more a concern about viruses [than a privacy issue]. . . I don't sign up for lots of services because: I don't want lots of email, and I don't want lots of email even if I ask for it, and I don't want spam—period. So that's just to not be annoyed rather than not to have my privacy invaded.*

Although openly resisting the association between privacy and personal information, Danielle is embracing another facet of privacy, that of solitude and control of access to one's personal space. She recognizes that her desire for control of her personal space may have to do with privacy, but she is also adamant in her belief that she does not care for privacy per se. This refusal to accept the *idea* of privacy has not been studied enough by researchers.

Xavier arrived in Canada in late 2000 and, aside from the odd job, has since been unemployed. He relies on public access for his Internet use and goes to St Christopher House every day, spending three or four hours doing nothing but looking for jobs. He estimates that, in the last year and a half, he has sent out about a hundred curricula vitae (CVs). In our conversations, Xavier described some of the constraints he is subjected to when releasing personal information:

> Interviewer: [Let me ask you a couple of questions about privacy] Are you concerned about information you gave to others about yourself?
>
> Xavier: Yes. Of course. I give good information, real information about me, and I would like if people on the other side believe my information.

Xavier goes on to argue that establishing a solid and truthful identity in his online search for work is crucial. Finding a job depends on, among other things, his ability to convince others that he is who he is claiming to be, and so his release of personal information is vital if he is to be able to participate fully in the job market and, thus, in society at large.

These constraints are not much different from those that Xavier would have in 'real life', but they do point to the flaws of resting a privacy argument on 'freedom of choice', which is made explicit in the recurrent 'opt-in/opt-out' debate. This debate involves at least three unproven assumptions: that one is indeed free to choose and that acting upon one's preferences is easy; that one is aware of the dangers; and that the context in which one chooses will remain stable over time (Viseu 2001).

Until now we have discussed matters of conscious presentation of self; however, not all personal information disclosure is conscious and/or active. In fact, one of the often cited online privacy threats is the monitoring ability of the Internet, specifically of cookies.[14] Cookies run in the background, seamlessly part of the 'normal' functioning of the web, and are operated by a party other than the user. Contrary to what we expected, after years of public controversy—the DoubleClick story[15] that started in late 1999 is an example—only half of our participants were aware of cookies, a proportion strikingly similar to the results obtained by the Pew report. Furthermore, only two participants ever acted upon their fears of cookies and they now say that these fears are behind them. Our designer participant, Nicholas, emphatically states: 'I used to really worry about cookies, but I don't really care anymore.'

What Happens to Data in 'Cyberspace' and Does It Really Matter?

The third online privacy 'moment'—when personal data is released, whether voluntarily or not, into 'cyberspace'—is at the core of policy discussions on privacy. In the policy-setting discourse, privacy is generally understood as the rights of individuals vis-à-vis those who collect their personal data, and the responsibilities of the collector entities towards these individuals (Agre & Rotenberg 1998; Gellman 1998). These rights and responsibilities are assembled under codes of FIPs.

Although they are a favourite subject of scholars and activists, FIPs have not made their way into the public consciousness. In our study, participants seemed oblivious to them. In their view, once personal data has been released, there are no expectations that it will be treated with their interests in mind. Moreover, as we describe below, participants' privacy perceptions seem to be associated with an attitude of resignation and even of total disregard for the whole subject of privacy. The dismissal of privacy issues and annoyance at our questions were a surprise to us. The conviction with which some participants dismissed privacy concerns altogether is an issue that has not been sufficiently explored in privacy studies.

One of the clearest features of our subjects' privacy perceptions and practices was their passivity towards the issue. Of the participant sample discussed in this article, only one participant said that he used privacy-enhancing technologies, and he was not sure if the software was still working. This passivity seems to derive, to a large extent, from the argument that, if you have nothing to hide, you have nothing to fear. It is an argument that has been well documented in the privacy literature (e.g. Agre 1996; Marx 2003, n.d.; Lyon 2001b) as a stumbling block to the development of pragmatic privacy-protection strategies, and it, too, is related to the ambiguous and symbolic nature of the term 'privacy' itself. Privacy, as a value, becomes a concern only once it is lost (Marx n.d.). This abstract, yet real, concept is best understood with an ecological analogy. While most people are aware of the problems of global warming and ozone depletion, the immediate gains of driving the car to work or putting on hairspray outweigh the often invisible losses of polluting the environment.

Nicholas summarizes the 'nothing to hide, nothing to fear' argument when saying:

> *I'm not worried about [privacy] and I'm not doing anything about it, and no issues have come up about it either. You don't worry about being broken into until someone breaks into your house . . . or some people do things. So, I think if I ever did get hacked I would start worrying about it, but at this point I don't have anything to hide, and don't know why anyone would want to hack into my system anyway.*

But the abstract character of privacy does not fully account for the range of responses participants had towards the issue. Other frequently mentioned—and often related—attitudes were resignation, dismissal, and annoyance.

Resignation was the most common feeling among the participants. This is not surprising given the reasons set out above, but what is noteworthy is the inclination of the participants to associate their resignation with a lack of knowledge. Camila says that she has never read a privacy policy because she is a novice with computers. Ignacio, an immigrant from Latin America who is now unemployed, makes similar remarks. He is a self-instructed computer aficionado who spends most of his time assembling and repairing old computers; yet he, too, believes that when it comes to privacy he does not know enough about it to protect it.

Xavier also displays an attitude of self-blame and resignation. He comments that the personal information he discloses on his CV has been sold to third parties and that he receives a lot of junk mail and marketing phone calls. When asked about how he feels about this, Xavier says, 'I don't know. I didn't give my information to be sold but it's a risk.' We then ask him if he is aware of the existence of Canadian federal data protection legislation. Xavier says that he is not, but he believes that his mishaps are his own fault. After all, when signing on to a new service he never reads the privacy policy: 'I have never read this information, so I'm wrong to . . . I'm not the kind of person to complain because I didn't read this.'

Julia is working towards her PhD in social and political thought. She has resigned herself to using the Internet in general, and in particular to the risks involved in online banking. She discusses her resignation in terms of lack of knowledge and lack of options:

> *I guess I didn't know how secure [online shopping] was. I don't really know how the Internet works exactly, or I don't really know much about computers. . . . And I'm not particularly interested in finding out in detail. I have other things I want to spend my energy on.*

Ursula is in her mid-sixties and is a self-described obsessive compulsive with a tumultuous personal life, financial difficulties and a daughter with serious drug addiction problems. For her, the Internet (she has a home computer and regularly uses public facilities) is a tool for personal growth and empowerment. It allows her to look for information on topics that concern her, which in many cases are rather delicate ones, such as sites on drugs or gay and lesbian issues. Given the nature of her online explorations, we expected Ursula to be worried about privacy in general, and specifically about the monitoring of her activities and the information that could be drawn from it.

In fact, however, Ursula dismisses privacy concerns, equating them with hiding, with a lack of the openness and transparency that she believes is an essential value in society:

> *You know something, it wouldn't upset me if someone made up their mind whether I'm a drug user or whether I ever used drugs, or whether I was gay or heterosexual, it doesn't bother me . . . I mean the idea is that it is not to bother us, that's why those [social and community] programs are available, the idea is to be comfortable with ourselves, certainly concealing drug use is not making it better, and certainly not acknowledging or not coming out is not making their life any better, so the whole idea is to be out there, period.*

Furthermore, for Ursula, privacy concerns are nothing short of arrogance, an unjustified feeling of self-importance:

> *And really, when you think about it, why would anybody be interested [in what I do online]? I mean what does somebody want to know what you think of your mother, or what your argument was? . . . I think there's something arrogant or something wrong with thinking that everybody's interested in what I'm doing, for starters.*

Danielle and Nicholas—the two most Internet-literate participants in our study—also associate privacy concerns with a paranoid mindset. Danielle is our most conspicuous example of the rejection of privacy concerns. She is annoyed at our questions on the subject and defends her position by saying that what she does online is simply not interesting enough to justify paranoia about privacy. Her dislike for the concept of privacy is such that she mentions that she once 'dated a programmer who was obsessed with Internet privacy and had firewalls and all sorts of *crazy crap* because he's one of those people that can go into people's computers' (emphasis added).

What we may be observing here is an effect similar to that documented by Schneier (2000) in *Secrets and Lies: Digital Security in a Networked World*. Schneier points out that a classical problem in the field of security systems is that the more alerts are generated, the less valuable they become. This may well be what is happening with our participants, since, after years of constant media exposure, we have yet to develop concrete solutions to concerns about privacy. As a result, it is possible that people are becoming insensitive to the problem. Also contributing to this is the belief of most of our participants (aside from Xavier, who thinks that his personal data have been sold to third parties) that their online experiences have been free of privacy hazards, at least as far as they can tell.

Conclusions

Given the small size of our sample and the nature of our research evidence, we are unable to draw final conclusions about the state of privacy online. We also acknowledge that, owing to the sensitive nature of the topic, it is possible that our sample is biased towards those who are not concerned about privacy.[16] This type of bias is difficult to surmount and will always be a problem in studies of privacy. By focusing on the experience of using networked services, rather than on privacy per se—a technique previously used by Hine and Eve (1998)—and by offering anonymity to participants, we have attempted to minimize any potential bias.

With these caveats in mind, the patterns and themes identified here are useful to understanding the broader issues involved in the field of privacy. In the first place, some of this study's findings confirm what is frequently asserted in popular and scholarly literature: for instance, the significant reach of the privacy debate, the ambiguity of the concept itself, and the gap between people's worries about privacy and the concrete actions they take, or do not take, to defend their privacy. Others, however, run counter to the prevalent assumptions. We found, for example, that privacy had a minimal impact in shaping our participants' online activities, and that their understanding of fair information codes is close to nil. This indicates that, at best, the privacy 'movement' is failing to meet its objectives, and that, at worst, it is misguided.

In our study we found that all participants interviewed understood the term 'privacy' without a need for further explanation. This shows that the privacy debate has reached beyond policy and advocacy circles and has become an inseparable part of the online experience. It is equally evident that individuals approach privacy from the context of their own actual practices, associating it with their individual experiences and concerns, such as place of access, spam, security and personal boundaries. This confirms the need for empirical, situated analysis of the meanings of privacy, rather than overarching definitions (cf. Schoeman 1984; Regan 1995; Hine & Eve 1998; Sheehan 2002; Solove 2002).

We believe that one of the reasons for the slippery character of privacy is that the word itself points in the direction of the individual rather than the social. The individual character of privacy is revealed in our participants' predominant mode of approach to privacy, a 'nothing to hide, nothing to fear' attitude. This posture emphasizes a short-term vision of individual interests and preferences rather than a broader societal perspective.[17] This bias towards the individual makes it hard, if not impossible, to make visible its value as a social good, to find a common approach, and to devise coordinated, global policy actions.[18] In a sense, the paradox of privacy is that, in order to be effectively studied, it is approached from a situated, often singular, context, thus reinforcing the policy-inhibiting bias.

'Nothing to hide, nothing to fear' behaviour is best understood through the analogy between privacy and environmental concerns. Both entail risks that are abstract, dispersed and distant, and benefits that are immediate and personal. In the same way that the benefits of driving one's car to work are immediately felt, so the gains of releasing personal information are immediately realized on the Internet.

What is lacking, from a policy standpoint, is knowing where to go next. To continue with our ecological analogy, just as once it was very difficult for an individual to make progress towards more ecologically sound practices with household waste, it is now very difficult, and perhaps frustrating for some, to make progress towards a privacy ideal that is neither well defined nor perceived as attainable. But in the same way that the communal ideal of a cleaner environment has been 'inscribed' in the daily practice of separating recyclables with thoroughgoing education and an enabling mechanism (e.g. recycling bins and fortnightly pick-up), so the ideal of more socially sound privacy practices may be inscribed through education, legislation and practical mechanisms by which people can take steps for their own and the communal good. It is necessary to create a 'cyberspace' where an individual knows, and can verify, that when he/she releases personal information, it is processed in a privacy-protective way by the appropriate entities.

As things stand so far, however, this constitutes a critical shortcoming of the privacy movement. Efforts to frame privacy discussions as an issue of FIPs have not been very successful in establishing accountability mechanisms that convey to individuals the knowledge of how, when and where to release personal information, and how to enforce their rights to their own data.

Among the websites that our participants visit, what little reference there is to the site owners policies regarding the handling of personal information is confined mainly to generic 'privacy statements', which go largely unread. This, in part, could be because these statements appear designed to provide vague reassurance and are often hard to find within the website. Rather like with the 'fine print' of licensing agreements, which people routinely click through on the way to their main business, most users have neither the time nor the expertise to understand these privacy statements, let alone the ability to compare them systematically with the statements of other sites or standard privacy norms.[19]

An alternative approach would be for sites to display their compliance with established FIP norms prominently and, in particular, at the point where people enter personal information, to make visible in standard ways some of the key FIP ingredients—notably the purpose for collection, whether there will be third-party use, how to inspect and correct one's own record, who the responsible data manager is and how to contact her, how to make a complaint, and how to view other complaints along with the compliance record of the organization. Not only would such measures help tilt the balance in favour of individuals in holding data-collecting organizations more accountable, it would also help educate people about their rights and how to exercise them in ways that are currently unavailable. Of course, without external pressure, web hosts will not voluntarily reform their activities in such a manner, but this suggests a role for privacy advocates. To complement efforts on the legislative front, attention should also be paid to the contingencies of everyday online practices through which people enact apparently insignificant acts of personal data disposition, but which cumulatively constitute a data-handling regime potentially dangerous to collective rights and well-being.

In this article we describe online privacy as being constituted by three 'moments', each with its own meanings. However, online-privacy discussions are hardly ever directed at the first two moments—that of sitting in front of the computer and that of the actual interaction with it.[20] Instead, attention is focused almost exclusively on the third moment, when the personal information has already been released into 'cyberspace'. The emphasis is not on the actions that can lead to a better understanding of privacy, but on the consequences of not doing so. Privacy becomes an effect rather than a process, a result rather than a continuum. The physical context and the specific characteristics of the individual who holds the personal information are lost.

Focusing on this third moment has succeeded in keeping the media's attention on the topic of online privacy, a fact that explains why most individuals are aware of the issue. Yet it has also fostered the recurrent association of privacy with the creation of a Big Brother society, and thus with a catastrophic all-or-nothing game. In this approach, individuals have privacy if they do not go online, and lose it once their personal data are in 'cyberspace'. Privacy becomes a risky choice: being online or not, opting in or opting out. Activists and policy developers have not been able—or perhaps willing[21]—to counter this view, which is best represented by the (in)famous statement of Scott McNealy, Sun Microsystem's chief executive officer: 'You have zero privacy anyway. Get over it'. The black-or-white approach to privacy has merely dramatized the issue, and in a society that values sound bites over substance, this is often seen as more than half the battle. However, the victory, if such it is, might come at the cost of losing the privacy war.

We speculate that another unexpected consequence of the emphasis on the third moment—the attempt to maintain a high state of alert—is the privacy backlash we observed in our participants. It is possible that, after years of constant media exposure, in combination with minimal experience of personal privacy problems, people are reaching a saturation stage and becoming inured to the issue. For privacy advocates, the next great challenge will be to counter this backlash.

Our study suggests that online privacy is best understood as an environment that simultaneously contains and is contained in people's activities. Borrowing from Francisco Varela's (1999) explanation of the nature of reality, we could say that privacy is neither pre-given nor constructed but rather is interpreted through one's personal experience. We must re-evaluate the privacy discourse to comprise all the moments that are involved in the practice of being online, while also creating better accountability mechanisms that make clear the options of those releasing the information and that expose the responsibilities of those involved in the process of data collection.

About the Authors

Ana Viseu (http://fcis.oise.utoronto.ca/~aviseu) is a researcher currently working at the University of Toronto on her PhD. Her thesis work combines a field survey of the visions and desires that drive the development of wearable computers, with an ethnographic study of their implementation in the workplace. Ana's research interests also include questions of privacy, social dimensions of technology and everyday use of digital media. From 2000 to 2002, she founded and directed the *Privacy Lecture Series* at the University of Toronto. Ana is also a Marshall McLuhan Fellow and a Knowledge Media Design Institute (KMDI) Graduate Fellow. *Address*: Human Development and Applied Psychology, University of Toronto/OISE, 252 Bloor Street West, 9th Floor, Toronto, Ontario M5S 1V6, Canada. [email: ana.viseu@utoronto.ca]

Andrew Clement (http://www.fis.utoronto.ca/faculty/clement/) is a professor in the Faculty of Information Studies at the University of Toronto, where he coordinates the Information Policy Research Program; he is also the Director of the Collaborative Graduate Program in Knowledge Media Design. He has had longstanding research and teaching interests in the social implications of information/communication technology and human-centred systems development. His recent research has focused on public information policy, Internet use in everyday life, digital identity constructions, public participation in information/communication infrastructures development, and community networking. Address: Information Policy Research Program, Faculty of Information Studies, University of Toronto, 140 St George Street, Toronto, Ontario M5S 3G6, Canada. [email: clement@fis.utoronto.ca]

Jane Aspinall is a Masters student at the Faculty of Information Studies at the University of Toronto. *Address*: Information Policy Research Program, Faculty of Information Studies, University of Toronto, 140 St George Street, Toronto, Ontario M5S 3G6, Canada. [email: aspinal@fis.utoronto.ca]

Acknowledgments

We appreciate the willingness of our informants to let us into their homes and lives, spending hours with us answering our many questions. Randall Terada of St Christopher House helped us greatly in recruiting many of our informants. We found the suggestions of the anonymous reviewers and journal editors useful in improving the paper. The Social Sciences and Humanities Research Council has supported this research financially.

Notes

1. Previously presented at the Association of Internet Researchers (AoIR) conference in October 2002.
2. There is a large literature offering diverse perspectives on privacy. For a legal treatment, see Gellman (1998) or Solove (2002); for a philosophical inquiry, see Schoeman (1984); for a policy-grounded analysis, see Bennett (1992); for a sociological perspective, see Lyon (2001a); and, finally, for a historical view, see (Flaherty 1999).

3. For more on the subject of fair information practices see, for instance, the Center for Democracy and Technology (http://www.cdt.org/privacy/guide/basic/fips.html), Junkbusters (http://www.junkbusters.com/fip.html) or OECD's report (http://www1.oecd.org/publications/e-book/9302011E.pdf).
4. Directive on the Protection of Personal Data and on the Free Movement of Such Data (Directive 95/46/EC).
5. For the latest privacy surveys, see: http://www.privacyexchange.org/iss/surveys/surveys.html.
6. As Solove (2002) notes, some privacy expectations can also be instigated by the law. For instance, letters became private means of communication only when, in 1877, the US Supreme Court held that letters were protected by the Fourth Amendment. Until that time, letters were seen as insecure and there was no expectation of privacy in personal correspondence.
7. See: http://www.fis.utoronto.ca/everydayinternet/.
8. Census data, from 1996, report that, of the 735 people in our selected neighbourhood, there is a fairly even spread in ages: from young children to seniors over seventy years old. More than half have immigrated to Canada, with the largest such group being Portuguese. Demographic data on education and income also support our location choice: of the 620 adults, over a third (225) have a less than grade nine education, and 150 more have completed between nine and thirteen grades. Only seventy-five have a bachelor's degree or higher. The number earning less than $24,000 per annum (130), which is below the Canadian national average, exceeds the number earning over $65,000 (120).
9. The first interview consisted of a general discussion of the participant's online activities. In the second interview, participants were asked to keep a record of their online activities and these were discussed in detail.
10. We did not specifically ask participants about income, and thus cannot present figures. However, it is clear from talking to participants and, in certain cases from taking note of their houses, that most participants come from middle- and low-income strata, which is consistent with the neighbourhood being studied.
11. The presence of a fluent speaker of Portuguese and Spanish (AV) on the research team allowed us to accommodate the linguistic needs of participants.
12. Mainly for financial reasons, Marianna has recently started to rely on public facilities only.
13. For more on the issue of the blurring of public and private spaces see, for example, Shapiro (1998) and Marx (2001).
14. Cookies are small text files sent out to the user's computer when visiting a website. These text files are stored on the computer's hard drive and help track that user's browsing activities. Cookies are composed of a unique identifier that identifies that computer to the website that is sending out the cookie.
15. In November 1999, DoubleClick, an Internet advertisement company, and Abacus Direct, a direct-marketing agency, announced that they were going to merge. Public concern emerged when DoubleClick announced (in February 2000) that it was going to change its privacy policy and start correlating personal information with its cookie identifiers. This information would be gathered in a database of consumer profiles, which would include each user's name and address, online surfing habits, online purchase history and demographic data. Consumers would be able to 'opt out' by going to the company's website.
16. Our main concern at the start of the study was that the number of people who claim to be 'very concerned' about privacy would be inflated. We expected participants to want to please us by seeming concerned about privacy.
17. This is not to imply that individual responses to privacy are not important. In 'A Tack in the shoe: neutralizing and resisting the new surveillance', Marx (2003) details the significance of individual responses to surveillance.

18. The individualistic character of privacy has been widely debated in the privacy literature (e.g. Schoeman 1992). Among the attempts to shift the discourse towards a more social approach are that of Raab and Bennett (1998), who argue for a 'sectoral approach' to the subject, i.e. the search for identifying patterns among certain social groups.
19. In Canada, the principal established privacy norms are the Canadian Standards Association (CSA) Model Code (1994) and, since 2001, the Personal Information Protection and Electronic Documents Acts (PIPEDA), which is based explicitly on the CSA Model Code. The authors have not seen any site, including those that the research participants visited, that makes prominent reference to its compliance with these standards.
20. A notable exception to this statement is the work of the World Wide Web Consortium (W3C) in developing the Platform for Privacy Preferences (P3P). However, this effort so far has had little practical influence on everyday online experiences and faces formidable obstacles in becoming an established part of them.
21. A look at the privacy literature shows that a number of privacy advocates frequently use the Big Brother metaphor to justify privacy fears in an online world. See, for instance, Davies (1992).

References

Agre, P. E. (1996) 'Arguments against privacy and what's wrong with them', [Online] Available at: http://dlis.gseis.ucla.edu/people/pagre/arguments.html (30 Aug. 2002).

Agre, P. E. (1998) 'Introduction' in *Technology and Privacy: The New Landscape,* ed. P. E. Agre & M. Rotenberg, MIT Press, Cambridge, MA, pp. 1–28.

Agre, P. E. & Rotenberg, M. (1998) *Technology and Privacy: The New Landscape,* MIT Press, Cambridge, MA.

America Online (AOL)/Roper Starch (2000) 'The America Online/Roper Starch Worldwide Adult Cyberstudy 2000', [Online] Available at: http://www.corp.aol.com/PDF/Cyberstudy2000.pdf (30 Aug. 2002).

Bennett, C. J. (1992) *Regulating Privacy: Data Protection and Public Policy in Europe and the United States,* Cornell University Press, Ithaca, NY.

Bennett, C. J. (1998) 'Convergence revisited: toward a global policy for the protection of personal data?', in *Technology and Privacy: The New Landscape,* ed. P. E. Agre & M. Rotenberg, MIT Press, Cambridge, MA, pp. 99–123.

Cavoukian, A. & Tapscott, D. (1995) *Who Knows: Safeguarding Your Privacy in a Networked World,* Random House, Toronto.

Center for Democracy and Technology (CDT) (n.d.) 'Privacy basics: fair information practices', [Online] Available at: http://www.cdt.org/privacy/guide/basic/fips.html (26 Sept. 2002).

Charny, B. (2000) 'Protect your Internet privacy . . . by lying', *ZDNet News,* [Online] Available at: http://dailynew.yahoo.com/h/zd/20000822/tc/protect_your_internet_privacy_by_lying_1.html (22 Aug. 2000).

Clement, A., Aspinall, J., Viseu, A. & Suchman, L. (2002) 'The *Everyday Internet* project: a 'neighbourhood ethnography' informed by current Canadian policy concerns', paper presented at a conference on *Ethnographies of the Internet: Grounding Regulation in Lived Experience,* Oxford University, UK, 8 Mar., [Online] Available at: http://www.fis.utoronto.ca/EverdayInternet/oxford.html

Davies, S. (1992) *Big Brother: Australia's Growing Web of Surveillance,* Simon & Schuster, Sydney.

Dryburgh, H. (2001) 'Changing our ways: why and how Canadians use the Internet', *Statistics Canada,* Catalogue No. 56F0006XIE, [Online] Available at: http://www.statcan.ca/english/research/56F0006XIE/56F0006XIE.pdf (11 Aug. 2003).

Earp, J. B. & Baumer, D. (2003) 'Innovative web use to learn about consumer privacy and online behavior', *Communications of the ACM,* vol. 46, no. 4, pp. 81–83.

Federal Trade Commission (FTC) (1996) 'Consumer information privacy hearings', [Online] Available at: http://www.ftc.gov

Flaherty, D. H. (1999) 'Visions of privacy: past, present, and future', in *Visions of Privacy: Policy Choices for the Digital Age,* ed. C. J. Bennett & R. Grant, University of Toronto Press, Toronto, pp. 19–38.

Gellman, R. (1998) 'Does privacy law work?', in *Technology and Privacy: The New Landscape,* ed. P. E. Agre & M. Rotenberg, MIT Press, Cambridge, MA, pp. 193–218.

Gronroos, C. (1994) 'From marketing mix to relationship marketing: towards a paradigm shift in marketing', *Management Decisions,* vol. 32, no. 2, pp. 4–20.

Hine, C. & Eve, J. (1998) 'Privacy in the marketplace', *The Information Society,* vol. 14, no. 4, pp. 253–262.

Lyon, D. (1994) *The Electronic Eye: The Rise of Surveillance Society,* University of Minnesota Press, Minneapolis, MN.

Lyon, D. (2001a) *Surveillance Society: Monitoring Everyday Life,* Open University Press, Buckingham, and Philadelphia, PA.

Lyon, D. (2001b) 'Terrorism and Surveillance: Security, Freedom, and Justice after September 11 2001', Paper presented at the *Privacy Lecture Series,* University of Toronto, 12 Nov., [Online] Available at: http://privacy.openflows.org/pdf/lyon_paper.pdf (12 Nov. 2001).

Marx, G. T. (2001) 'Murky conceptual waters: the public and the private', *Ethics and Information Technology,* vol. 3, no. 3, pp. 157–169.

Marx, G. T. (2003) 'A tack in the shoe: neutralizing and resisting the new surveillance', *Journal of Social Issues,* vol. 59, no. 2, pp. 369–390.

Marx, G. T. (n.d.). 'Privacy and technology', revision of material in *The World and I,* Sept. 1990; *Telektronik,* Jan. 1996, [Online] Available at: http://web.mit.edu/gtmarx/www/privantt.html (30 Aug. 2002).

Miller, D. & Slater, D. (2000) *The Internet: An Ethnographic Approach,* Berg, Oxford, and New York, NY.

Miyazaki, A. D. & Fernandez, A. (2001) 'Consumer perceptions of privacy and security risks for online shopping', *Journal of Consumer Affairs,* vol. 35, no. 1, pp. 27–44.

National Telecommunications and Information Administration (NTIA) (2000) 'Falling through the net, toward digital inclusion', US Department of Commerce, Washington, DC, [Online] Available at: http://search.ntia.doc.gov/pdf.fttn00.pdf

Pew Internet and American Life Project (PEW) (2000) 'Trust and privacy online: why Americans want to rewrite the rules', [Online] Available at: http://www.pewinternet.org/reports/pdfs/PIP_Trust_Privacy_Report.pdf (30 Aug. 2002).

Pew Internet & American Life Project (2003) 'The ever-shifting Internet population: a new look at Internet access and the digital divide', [Online] Available at: http://pewinternet.org/reports/pdfs/PIP_Shifting_Net_Pop_Report.pdf

Phelps, J., Nowak, J. & Ferrell, N. (2000) 'Privacy concerns and consumer willingness to provide personal information', *Journal of Public Policy & Marketing,* vol. 19, no. 1, pp. 27–41.

Pitkow, J. E. & Kehoe, C. (1996) 'Sixth WWW User Survey', [Online] Available: http://www/cc/gatech/edu/gvu/user_surveys/survey-04–1996

Privacy Commissioner of Canada (n.d.) *The Personal Information Protection and Electronic Documents Act* (official version), [Online] Available at: http://www.par.gc.ca/36/2/parlbus/chambus/house/bills/government/C-6/C-6_4/C-6_cover-E.html

Raab, C. D. & Bennett, C. J. (1998) 'The distribution of privacy risks: who needs protection?', *The Information Society,* vol. 14, no. 4, pp. 263–274.

Regan, P. (1995) *Legislating Privacy: Technology, Social Values and Public Policy,* University of North Carolina Press, Chapel Hill, NC.

Rogers, J. L. (1996) 'Mail advertising and consumer behavior', *Psychology and Marketing,* vol. 13, no. 2, pp. 211–233.

Schneier, B. (2000) *Secrets and Lies: Digital Security in a Networked World,* John Wiley & Sons, New York.

Schoeman, F. D. (ed.) (1984) *Philosophical Dimensions of Privacy: An Anthology,* Cambridge University Press, Cambridge.

Schoeman, F. D. (1992) *Privacy and Social Freedom,* Cambridge University Press, New York.

Sciadas, G. (2002) 'Unveiling the digital divide', Connectedness Series, *Statistics Canada,* Catalogue no. 56F0004MIE, No. 7, [Online] Available at: http://www.statcan.ca/english/research/56F0004MIE/56F0004MIE2002007.pdf (11 Aug. 2003).

Shapiro, S. (1998) 'Places and spaces: the historical interaction of technology, home, and privacy', *The Information Society,* vol. 14, no. 4, pp. 275–284.

Sheehan, K. B. (2002) 'Toward a typology of Internet users and online privacy concerns', *The Information Society,* vol. 18, no. 1, pp. 21–32.

Solove, D. J. (2001) 'Privacy and power: computer databases and metaphors for information privacy', *Stanford Law Review,* vol. 53, no. 6, pp. 1393–1462, [Online] Available at: http://law.shu.edu/faculty/fulltime_faculty/soloveda/Database-Privacy%20FINAL%20VERSION.doc (5 Sept. 2002).

Solove, D. J. (2002) 'Conceptualizing privacy', *California Law Review,* vol. 90, no. 4, pp. 1087–1155.

Thomson, J. J. (1984) 'The right to privacy', in *Philosophical Dimensions of Privacy: An Anthology,* ed. F. D. Schoeman, Cambridge University Press, Cambridge, pp. 272–289.

University of California at Los Angeles (UCLA) Center for Communication Policy (2001) 'The UCLA Internet Report 2001: surveying the digital future', [Online] Available at: http://www.ccp.ucla.edu/pdf/UCLA-Internet-Report-2001.pdf (30 Aug. 2002).

Varela, F. J. (1999) *Ethical Know-How: Action, Wisdom, and Cognition,* Writing Science Series, Stanford University Press, Stanford, CA.

Virtual Privacy Office (n.d.) 'Informational self-determination—what does that mean?', [Online] Available at: http://www.datenschutz.de/recht/grundlagen/ris.xml (26 Sept. 2002).

Viseu, A. (2001) 'Privacy and freedom of choice', posted on *Nettime-l,* republished by *Noema,* [Online] Available at: http://www.noemalab.com/sections/ideas/ideas_articles/viseu_privacy.html

Westin, A. (1995) 'Privacy in America: a historical and socio-political analysis', paper presented at *National Privacy and Public Policy Symposium,* Hartford, CT.

Publicly Private and Privately Public: Social Networking on YouTube

Patricia G. Lange

YouTube is a public video-sharing website where people can experience varying degrees of engagement with videos, ranging from casual viewing to sharing videos in order to maintain social relationships. Based on a one-year ethnographic project, this article analyzes how YouTube participants developed and maintained social networks by manipulating physical and interpretive access to their videos. The analysis reveals how circulating and sharing videos reflects different social relationships among youth. It also identifies varying degrees of "publicness" in video sharing. Some participants exhibited "publicly private" behavior, in which video makers' identities were revealed, but content was relatively private because it was not widely accessed. In contrast, "privately public" behavior involved sharing widely accessible content with many viewers, while limiting access to detailed information about video producers' identities.

Introduction

As social network sites (SNS) gain users and visibility, a wide range of websites have adopted SNS features (boyd & Ellison, 2007). YouTube began as a video sharing platform, but it also offers users a personal profile page—which YouTube calls a "channel page"—and enables "friending." Research on SNSs has shown that the meanings of social network site practices and features differ across sites and individuals (boyd, 2006). This article explores how YouTube users employ the technical and social affordances of the site to calibrate access to their videos by members of their social circle. Specifically, it examines how video sharing can support social networks by facilitating socialization among dispersed friends. An important goal of the article is to document how youth and young adults use YouTube's video sharing and commenting features to project identities that affiliate with particular social groups. The analysis also evaluates to what extent YouTube's video sharing practices are conducted in public view (versus as private video exchanges) and how these choices reflect different kinds of social relationships. For example, how does sharing videos with certain friends but not others reflect different kinds of friendships? More generally, how do young people maintain and delineate distinct social networks through public video sharing?

In order to address these questions and themes, I begin by describing theoretical frameworks that illustrate the relationship between media exchange and the processes whereby social networks are negotiated online and in public. I then discuss how youth and young adults manipulate both technical features and social mechanisms to achieve different levels of visibility in their media sharing activities. Using a combination of ethnographic methods that included semi-structured, formal interviews and analyses of videos, comments, and video responses, I argue that "friendship linkages" are not the only—or even the primary—way in which some participants express social linkages. Furthermore, I suggest that YouTube participation involves more than strictly public or private interaction. Through an examination of two categories that complicate this dichotomy—"publicly private" and "privately public" behavior—this article describes how participants manipulate media to maintain social networks and intimacy amid public scrutiny. It is shown that expectations about what should be shared or withheld on YouTube vary considerably according to individual needs and according to different types of social relationships.

Background

The concept of "social networks" is difficult to define. Wellman (1996) argues that online and offline social networks do not exist as such, but that they are useful analytic constructs for understanding social dynamics. A social network will look different depending upon how one measures it (counting the number of interactions between members versus rating the closeness of relationships, for instance). A social network is defined here as relations among people who deem other network members to be important or relevant to them in some way (Wellman, 1996). Using media to develop and maintain social networks is an established practice (Baym, 2000; Horst & Miller, 2006; Ito & Okabe, 2005; Kendall, 2002).

One way that social networks are articulated and negotiated on social network sites is through linking and viewing profiles (Donath & boyd, 2004; Gross & Acquisti, 2005). The linking of profiles through friendship requests and acceptances and the ability to view the resulting connections on others' profiles are tangible mechanisms that reflect existing social networks. They also reflect change, as when friends are added (Ellison, Steinfield, & Lampe, 2007). By studying social and technical choices that participants make on social network sites, it is possible to analyze how different social networks are created, maintained, and negotiated through media (Ellison et al., 2007; Gross & Acquisti, 2005).

Social network sites are defined as websites that allow participants to construct a public or semi-public profile within the system and that formally articulate their relationship to other users in a way that is visible to anyone who can access their profile (boyd & Ellison, 2007). This definition does not specify the closeness of any given connection, but only that participants are linked in some fashion. Similarly, this article considers social networks in a broad sense and examines their sub-instantiations as they are visibly reflected on YouTube through friending practices, video sharing, and posting comments to videos.

Maintaining Social Networks Through Media Circuits

Writing about transnational migration, Rouse (1991) proposed a useful framework for understanding the role of media in social network maintenance. He analyzed how Mexican immigrants used media to stay in touch, engage in community decision-making, and "participate in familial events even from a considerable distance" (p. 13). The telephone was not used merely to maintain contact; it also helped immigrants engage intensely with distant loved ones and friends. The use of media by members of a social group to stay connected or to interact with other members of the group constitutes a "media circuit." In Rouse's example, the media circuit connected people via the telephone. Media circuits in other forms—such as interconnections through video sharing—may also be examined to understand the social dynamics of a particular group. A media circuit is not a social network itself, but rather it supports social networks by facilitating and technically mediating social interactions among people within a network.

By examining a media circuit, it is possible to understand important social dynamics among the people who use the media. On YouTube, for example, a person may make and share certain kinds of videos with one set of friends, while making and sharing other types of videos with a different set of friends. Although it is not feasible to obtain information about all the media that are exchanged across a social network, Rouse's framework provides a window into particular instantiations of a social network as tangibly reflected in related media use. A media circuit helps some members of a social group to stay connected or interact in qualitatively meaningful ways. Not all members of a social group may participate in a particular media circuit, but examining media use reveals crucial aspects of relationships among participants. In the context of video sharing, posting videos that are shared with certain individuals and not others materially reflects different social configurations.

People who do not regularly participate on YouTube may not understand why people watch seemingly poor quality or odd videos on the site. Yet those videos may serve important social functions, and their utility is not necessarily judged by technical criteria. Media utility is often assessed in terms of the bandwidth or quality of information that a particular medium affords (Nardi, 2005). However, in the context of instant messaging, Nardi (2005) demonstrates how certain exchanges are not focused on information gathering. Rather they strive to establish an "affinity," which Nardi defines as a "feeling of connection" between people who "[experience] an openness to interacting with another person" (p. 92). Evaluating an instant message in terms of novel information ignores how short, similarly worded messages may transmit feelings of openness and affinity to the people to whom they are circulated. How and why media like instant messages or videos are viewed, enjoyed, and forwarded reveal much about the participants and their relationships. How they are shared also provides insight into distinctions between public and private.

Interrogating the Public/Private Dichotomy

Scholars invoking the public/private dichotomy typically use one of two analytically distinct metaphors (Weintraub, 1997). The first is "what is hidden or withdrawn versus what is open, revealed, or accessible" (pp. 4–5). Private things are "things that we are able and/or entitled to keep hidden, sheltered, or withdrawn from others" (p. 6). Similarly, "individuals have privacy to the extent that others have limited access to information about them, to the intimacies of their lives, to their thoughts or bodies" (Sheehan, 2002, p. 22). This includes preventing physical access to materials, as well as using symbols that are not easily understood by people outside a small social group. Intent, however, does not guarantee success, which depends upon the capabilities of interpreting parties, who may detect someone's identity or decipher arcane meanings in videos. The second line of analysis used to distinguish between the public and private concerns "what is individual, or pertains only to an individual, versus what is collective, or affects the interests of a collectivity" (Weintraub, 1997, pp. 4–5). This article analyzes the public and private both in terms of what is physically or symbolically more or less visible, and what counts as important or socially relevant for social network members. How "public" a video is can be evaluated according to both factors; it can be judged in terms of how much information a person yields about their identity and how much information in a video is accessed and meaningful to many people, as opposed to a select few.

Social and legal theorists have grappled with the subtleties of the public/private distinction. Nissenbaum (2004) introduces the concept of contextual integrity, which states that within a context of an interaction, people have expectations about what information is appropriate to collect and whether it should be distributed. For instance, when one friend tells another a secret, expectations within the context of "friendship" mean that a confidant will not widely distribute sensitive information. Nissenbaum stresses that ideas about contextual integrity vary across time, place, and culture. However, she does not account for variations within a particular context. On YouTube, people have different expectations about what information can be shared or what constitutes sensitive information. One useful lens for understanding differences in expectations about identity and content sharing within social networks is the concept of the fractalization of the public and private.

The Fractalization of the Public and Private

Some scholars suggest that communication technologies may be "eroding the boundaries between 'publicity' and 'privacy' in fundamental ways" (Weintraub & Kumar, 1997, p. xi). At the same time, Weintraub and Kumar (1997) rightly assert that "the enormous bodies of discourse that use 'public' and 'private' as organizing categories are not always informed by a careful consideration of the meanings and implications of the concepts themselves" (pp. xi–xii). Recent scholarship on the public/private distinction in mediated contexts confirms this observation. Several recent studies discussing public and private media do not acknowledge nuanced meanings. For instance, they 1) consider either visibility or social relevance to determine public/private boundaries, rather than incorporating both lenses (Barnes, 2006; Sandvig, 2006) or 2) do not consider how contexts other than the one they are analyzing shed light on public/private dimensions (Al-Saggaf, 2006). Such limitations are unsurprising, given that entire volumes grapple with these complex terms (Schoeman, 1984; Weintraub & Kumar, 1997). Yet, nuances and gradations exist in who can access, manipulate, and distribute information. What do these nuances and gradations look like within different contexts? How do social and technical affordances enable or limit particular configurations? If communication technologies are eroding these boundaries, what are the social responses to this erosion?

Gal (2002) provides a useful lens for interrogating the meaning of and response to public/private erosions. According to Gal, "public" and "private" are relative terms and shift according to individual perspectives. The public/private dichotomy is thus more productively visualized as a "fractal distinction." A fractal is defined as "a shape made of parts similar to the whole in some way" (Feder, 1988). Gal explains, "Whatever the local, historically specific content of the dichotomy, the distinction between public and private can be reproduced repeatedly by projecting it onto narrower contexts or broader ones" (p. 81). For instance, a home is private when contrasted with the neighborhood. At the same time, public and private spaces exist within the home. At a higher level, community matters may appear private relative to affairs at the level of state government.

Examining the fractalized patterns of public and private video access and interpretation reveals different social networks and their dynamics within the context of public video sharing. Just as small towns constitute a socio-spatial domain that subdivides when smaller social networks require neighborhood interaction, so, too, do video-sharing practices sub-divide in ways that reflect different social networks and relationship dynamics. Studying the fractalization of video-sharing practices thus provides a valuable lens through which to view how social relations shift and subdivide within the context of visible media sharing.

Method

The analysis that follows is based on an ethnographic investigation into young people's engagement with online video sharing on YouTube. The data include semi-structured interviews, fieldnotes on observations conducted several times a week over the course of one year on YouTube, analyses of posted videos and comments, and examination of subscription and friending practices.[1] The 54 interviewees ranged in age from 9 to 43, although most were in the mid-teens to early 20s range. In keeping with the project's focus, most of the interviewees were from the United States, although two were from Canada and two were from Europe. Formal interviews lasted from one to three hours and included a pre-interview survey on interviewees' demographic background and media use. Twenty-nine interviews were conducted over the phone, 15 interviews were conducted via instant messaging chat, and 10 interviews were conducted face-to-face.

Observations were collected during varied activities, including watching the main YouTube page to see what videos were featured, reading the YouTube video blog, watching videos of interviewees and others, reading video comments, and observing how YouTube participants responded to commentary (such as addressing a comment in a video). Other data include observations gathered at video-themed events that I attended in Los Angeles, San Francisco, and New York, including a YouTube meet-up on July 7, 2007 in New York City.

Interviewees were recruited in several ways, such as asking members of the research team's personal networks for interviews and sending email requests for interviews on the basis of YouTube content. Soliciting interviews from the researchers' social networks, such as acquaintances of project members, was helpful because several of these participants did not promote their videos widely on

YouTube, if they posted videos at all. Such a strategy helped to illuminate a broader participation base than would have been possible by examining only youth who were popular or particularly visible on the site.

The interview protocol included questions about why the individuals participated on YouTube, how they responded to comments, and what they thought of certain issues on YouTube, such as how "haters" (people posting excessively negative commentary) should be dealt with (Lange, 2007). The interview items included questions such as:

- What are the major advantages and disadvantages about participating on YouTube? Are you a "YouTuber"? Why or why not?
- Do you know all the people who have posted comments to your videos? Where do you know them from (YouTube, elsewhere online, offline)? What were your reactions to the comments on your videos?
- Under what circumstances do you "friend" someone? Do you only make YouTube friends with people you know offline? Do you make YouTube friends with people only online or only from YouTube? Why? How do you decide to whom you will subscribe? How is the practice of "friending" someone on YouTube different from subscribing to their videos?

The interview was adapted to the interviewee's particular interests and background. Interviewees were also asked specific questions related to their media choices and interactions. For instance, interviewees who indicated that they had received text comments on their videos were asked if they knew specific people, either from online or offline contexts, who posted comments on their videos.

Interviewees represented a wide range of visibility, as indicated by the numbers of YouTube friends and viewers they had. Some were YouTube celebrities who uploaded dozens of videos, had thousands of friends and subscribers, and were recognized within the YouTube community and beyond. Others were not well known within YouTube, had few subscribers and friends, and posted few or no videos.

Results

Creating and Negotiating Social Networks Through Video Sharing

Posting text comments and video responses to videos enacts Rouse's (1991) concept of media circuits that illustrates particular social network patterns. For example, an aunt of two boys that I interviewed posted a comment on one of their videos. Living several states away, the boys rarely see her. But by viewing her nephews' videos and posting comments, she re-affirms her position in a familial social network. When the boys read her posted comments, they complete a media circuit that began with an experience that was encoded in video, displayed, and commented upon by their aunt, who participated from afar. In this case, the media circuit helps maintain a social network that already exists. Media circuits take different forms and not only help maintain, but can also help create connections and negotiate relationship changes, as described below.

While some media circuits support existing social networks that began in-person, other social networks would not exist as such without online media circulation. For example, one YouTube celebrity expanded his network after posting videos for dispersed, unknown others to see. When a YouTube participant posts a video, viewers may "friend" the celebrity or comment on the video to garner the celebrity's attention. Several interviewees reported that an intelligent comment on their video usually prompted them to examine the commenter's work. In an environment where friending practices can become liberal, in that some participants routinely automatically accept friendship links, participants often post comments to increase their social visibility and connection to a video maker.

YouTube participants can broaden or limit physical access to their videos and thus create larger or smaller media circuits by using technical features such as limited "friends-only" viewing or strategic tagging. Viewers may locate videos using keywords or "tags" that video makers designate for their videos or write into the video's title or description. Given the limited capabilities of YouTube's search features at the time of this research, several interview participants reported that appropriate tagging, titling, and video descriptions were critical for finding relevant videos. Vague or generic tags, such as "cats," can yield a list of over 200,000 videos, which are difficult to sift through. Manipulation of video descriptions and tagging systems thus becomes crucial for video makers to regulate physical access to their videos. For example, one video maker used his YouTube name as his only tag. Unless one were a close enough friend to know this tag, it would be difficult to find his videos using the tagging system.

In addition to supporting social networks, video-sharing practices helped create new connections and develop social networks. Several interviewees said that they were not friends with their fellow video makers until they began making videos together and distinguishing their videos from other YouTube participants' work. For instance, despite living in the same dorm, one group of college students only began interacting in a qualitatively meaningful way after an incident that they considered ideal to record for YouTube. In this incident, two boys wrestled in their dorm's hallway. One was wearing combat gear and the other, a set of medieval armor. Several people filmed the incident, one of whom filmed it to share on YouTube, given its reputation for similar spontaneous and playful videos. The video, referred to here as Ninjas and Knights, was posted on September 30, 2006. Some of the students noted that they did not know each other well prior to the mediated incident. They further said that filming the incident for YouTube encouraged them to meet and interact with each other.

> Brian2: Yeah. We made new friends with the people who filmed us because we'd never met them before. They just came out of their rooms like, "Hey, let's film this." So one—two of the guys that were filming became new friends.

Making the video helped launch connections that did not exist prior to the mediated event. The goal was not to make a high-quality video, but rather to encode an interesting, affective experience that they could share. People who do not regularly use YouTube are often critical of the quality of videos on the site. Some participants were dismissive of the standard of other people's videos. Their objections often related to technical issues (including poor editing, lighting, sound, or some combination) or content (too many videos about people sparring). But critics fail to understand that video quality is not necessarily the determining factor in terms of how videos affect social networks. Rather, creating and circulating video affectively enacts social relationships between those who make and those who view videos. Posting the Ninja video expanded and articulated the network of people interested in participating in the video makers' experience. For students, identifying one's college in the video description and tagging system helps other students and alumni locate the video and participate in a larger community of individuals who have similar affinities.

> Brian1: . . . and we know a lot of other people watch [YouTube] like our friends, and it's—if we wanted to tell our friends, "Hey, come and watch this," it'd be a lot easier if we just put it on YouTube instead of sticking it in an email and waiting for the email to get there and waiting for them to open it which would take forever 'cause the file was so big. So we just put it on YouTube and go the link, sent the link to everyone, and they watched it.

Not all videos target general consumption. Some reaffirm pre-existing social networks by including material intended to appeal to shared affinities between the senders and certain receivers. Similarly, sending so-called viral videos to intimates demonstrates readiness to maintain social associations and bond with recipients. People who post criticisms, expecting a high-bandwidth, high-quality experience, fail to display affinity with the video. They effectively reject a social link to the video maker's social network members, for whom the video represents an emotional connection point.

For example, in the case of the Ninja video, someone posted the comment, "Can I please have my five minutes and seven seconds back?" In an interview, one of the video makers said that he did not know the person who had posted this comment from either his online or offline social networks. Further, he did not appreciate the comment, nor did he feel compelled to interact with the commenter. Thus, by posting a comment that did not display affinity with the video maker, the comment discouraged the video maker from pursuing additional social contact.

Conversely, interviewees reported that people who post comments of affinity or leave thoughtful comments may prompt the video maker to respond to the poster of the comment. Video makers may react to comments by: 1) reading them; 2) posting comments in response to the comments they received on a video; 3) posting a comment or question on the commenter's video or channel page; 4) extending a friend request to the commenter; or 5) viewing or subscribing to the commenter's videos. Although further interaction is not guaranteed through comments, interviewees noted that these interactions were often the initial steps in broadening social connections through media circuits.

Fractalized Public and Private Dynamics Reveal Different Social Networks

Applying the fractal public/private distinction (Gal, 2002) is useful because it reveals social patterns that otherwise remain hidden. The fractal distinction between the public and private is reproduced at various levels on YouTube. At a high level, YouTube is public, especially when contrasted with home video viewing, but within YouTube, fractalized divisions occur. While some video makers heavily promote their videos to connect with many people, others tell only a few friends about their work. Moving down to the level of specific video makers, the fractalized pattern repeats. Within a single video maker's work, some videos are personal, whereas others address issues such as environmental sustainability or racism and are intended for a larger audience.[2] Video makers may also have more than one YouTube account. Some celebrities reported having one account that encourages broad participation and social network linkages with fans and another account that targets circles of intimate friends.

Publicly Private

On YouTube, some participants broadcast extensive information about their identity. Moreover, they craft videos with content that is broadly appealing, and they aggressively promote and disseminate their videos. This can be considered quite public video making and viewing, because all three factors—identity information about the video maker, content relevance, and technical access to videos—are designed to be broadly appealing and widely available. On the private end of the spectrum, video makers may choose to restrict information about their identity in videos, and they may make videos with content that appeals only to a few close friends. They can also restrict access to the videos technically, by using few or cryptic tags or by invoking YouTube's "friends-only" viewing feature, according to which only members designated as the video maker's friends may access the video.

Other important categories of participation exist. Participants may share private experiences through video but do so in a "public way," by revealing personal identity information. This behavior may be characterized as "publicly private," in that they disclose identity information about themselves. At the same time, they use mechanisms to limit physical access to the videos or to limit understanding of their contents. The first part of this compound expression thus refers to the amount of identity information imparted by video makers. The second part refers to the available physical access to the video and to the interpretative content that may be understood only by the video maker and a few viewers.

Mechanisms that limit access may take technical and social forms. In a video referred to here as *Hand Puppets,* both the video's tags and content limit access to it. Theoretically, anyone can view the video and the identity information it contains. A girl's face intermittently appears on camera and the video's channel page contains identifying information. Nevertheless, only people who find it can analyze its contents. It also exhibits occasionally uneven audio, which offers another type of physical access barrier. This video was posted on July 8, 2006, and over the course of a year, it garnered only 88 views. The creator's tagging choices, which include common first names and cryptic abbreviations, complicate casual viewing access. For instance, one abbreviated tag is a reference to the religious convention that the girls attended while making the video. People who attended this convention or who are members of that religious network might use the tag to find other videos in other media circuits with the same tag. In this way, they can try to connect with other members of this social network.

Once they obtain physical access to videos, participants may have numerous interpretations, but these interpretations may not coincide with the girls' feelings about what is happening in the video or their in-jokes about the people they discuss. The video is filled with inside references and stylistics that are not intended for general YouTube viewership.

Interviewer: What was your story idea?

Allison: It was just basically about this guy that I met and myself, and [my friend] kept like messing up our names when we were talking about it. Like she kept calling him James instead of John, which is her brother's name. And so we decided to just change the names in the movie and such. And then we also talked about some random things that he and I had talked about on the phone and such, and yeah, it just kind of random, spur of the moment sort of things.

Anjin, a youth in his late teens, also uses cryptic tagging practices, such as using only his YouTube name. Unless one knows this name, the video's title, or words in the video description, it will be impossible to use the tagging and search systems to search for his videos. Only his close friends and family—whether online or offline—are likely to find his videos using the tagging system as it was implemented at the time of this writing. Unlike some other participants, he does not widely promote his videos (such as by posting comments and directing viewers to his videos). Of his nine videos, six of them received fewer than 200 views over periods ranging from five to 11 months. Even though YouTube enables global video sharing, relatively few people view his videos there. With regard to tagging practices, it should be noted that nuances in public and private access to videos are not only based on the video maker's intent. For instance, one interviewee claimed that she was simply "lazy" about putting up a lot of tags.

Certainly, viewers' and creators' interpretations may not coincide. Anjin's videos contain personal information, including his real name, the places he visits, and his and his friends' faces. In one video, Anjin and a few friends take a hike. At one point, a young man faces the camera and gives a thumbs-up sign in front of some graffiti, which is difficult to see. Anjin enlarges the graffiti in a graphic and places it next to the youth's face. The box contains the image of swastika. For many people, a swastika symbolizes Nazis and mass genocide. But as Anjin explains, the incident had a particular meaning for the small network of friends with whom he participated on the hike and with whom he shared the video.

Anjin: And there was this graffiti there and it had a swastika. And Ben is Jewish, so we're like; ha-ha, Ben. And he's like, what. So then the kind of joke was that Ben is Jewish and he's making this thumbs up at this swastika. There's some sort of irony there. And, once again, that's just kind of our odd inbred sense of humor at work, even though it's actually pretty terrible.

The image has symbolic value to the boys who participated in the nature walk. For the youth, giving a thumbs-up sign to a swastika does not aim to transmit political information but rather creates an affinity among the friends. One of Anjin's friends later posted a comment on another video in Anjin's oeuvre. The comment—No Swastika . . . =(—expresses mock disappointment that the current video contained no swastika. If one has not seen the prior video, the comment's intertextual meaning is likely to be lost. Even if viewers had seen Anjin's prior video, they may not understand that the swastika is an affective symbol that refers to humor shared by a small group of friends, rather than genuine disappointment that Anjin does not place swastikas in his videos.

Why did Anjin post the video publicly, given its inside references and symbolism? Although he is aware of the friends-only viewing option, he declined to restrict physical access to his videos. For Anjin and other people interviewed, it was a matter of convenience. Although many people watch YouTube, a number do not have accounts, which are required to "friend" other participants and enable selective viewing. Notably, just as social network site participants have different interpretations of what constitutes a "friend" (boyd, 2006), YouTube participants apply different definitions of what constitutes a "public" video. Despite the fact that YouTube offers potential connections to a large public, some people choose to create less public videos, some of which are not available or interesting to people beyond a few close friends.

> *Anjin*: . . . And like I said, I don't really feel too bad about [the swastika video] because I don't, like, display it super publicly. I guess people could stumble onto it randomly but nobody has given me a hard time about it yet.

Within Anjin's YouTube oeuvre, fractalized gradations between public and private videos exist. Although he does not consider most of his videos "super public," since they contain inside references and few people actually watch them on YouTube, some of his videos are "more public" than others. For example, on a martial arts video he used tags that included the name of the martial art. Over a seven-month period, the video garnered over 2,000 views, which is far more than Anjin's other videos. The martial art is practiced around the world and thus has potentially broad social relevance to other dispersed practitioners. One person posted the comment, "Very cool . . . if you guys are ever in New York, make sure to stop by our branch. =) Best wishes." This poster simultaneously invites Anjin into his social network and expresses interest in joining Anjin's network based on common interest in the video's content.

Privately Public

YouTube participation may take another hybrid form.[3] Being "privately public" means making connections with many other people, while being relatively private with regard to sharing identity information. Participants in this category conceal certain aspects of their identity, while expanding their friend and subscriber base and making videos with widely accessible content. Privately public individuals who were interviewed included YouTube celebrities and non-famous participants who wished to retain some anonymity. For instance, one person did not want his identity revealed, lest it compromise professional credibility with clients and co-workers. Other people cited concerns about safety and the desire to avoid stalkers.

One privately public video, which is referred to here as *Vlad The Impaler,* showed a mask-wearing monster. "Vlad's" menacing speech echoes the sentiments of other participants who wish to interact with many people, but in a way that conceals identifying information. The video announces a series of videos that he and his friends intend to make. In this video, Vlad outlines the level of identity that people should expect.

> *Vlad*: We have a strict rule [governing] our society that there are some things that should never be seen, should never be known, [which is] more for our protection than yours. If people knew that we were doing such things, [we] would be punished, so you'll never know our real names or our real faces but you will know our entertainment.

In a chat interview, CT, a youth in his mid-teens, said that he created this character with select friends as a secret joke. Although CT is in the costume, another friend provided a narrative voice over, in which his voice was modulated to sound lower. CT made the video after another friend (RJ) "jokingly" told him that he didn't need CT's help to make videos. RJ implied that CT's contribution was technically lacking. In response, CT and some other friends created the "Vlad the Impaler" character.

CT's goal with the "Vlad" films was to become quite popular on YouTube and then reveal himself to his friend RJ, so that RJ would know that he had misjudged CT's talent.

CT: the first vid was to make it seem like we had no clu what we were doing and i wanted it to [build] up. and when the time came if we had a lot more views than RJ we would [unveil] our selves, but we havent been able to make any videos for them

Interviewer: Wow!

CT: we were going to make them in a way where we wouldnt have to show our faces

Interviewer: Why [wouldn't] you show your faces?

CT: well so nobody on youtube, or RJ would [know]. just to prove him wrong

Interviewer: . . . It is rather [surprising] to me that RJ did not notice that video. I scanned through [Vlad's] subscriber list and it is not that long.

CT: well one day he had seen it and then he showed it to me that day because we had subscribed to him (he always checks that) and was like hey [CT] look at this noob[4]. i had to comment on the cool [lighting] effect

CT uses video in a way that articulates two separate social networks to which he belongs. The first includes his friend and video-making partner, RJ. The second includes a group of friends from which RJ is excluded. RJ was not invited to create the "Vlad the Impaler" video, because he had "insulted" CT. Even though the video is "public," in the sense that anyone using YouTube, including RJ, can view it, RJ did not recognize his own friend in the video and did not associate CT with Vlad. Despite being a close friend of CT's, RJ cannot access the symbolic meaning of the video that refers to conflict between CT and RJ. In this sense, although public, the "Vlad" video was not interpretable in the same way by all who viewed it. Again, there appears to be a fractalization of the public and private in terms of how CT presents himself in different media circuits that support social connections in varyingly public and private ways. When he makes videos with RJ, he reveals his face, name, and location. Yet when he made the Vlad film, he withheld that information and sought more "privately public" connections to new fans. It should be noted that CT's intentions do not guarantee identification privacy; RJ's role in failing to recognize CT was critical for CT's ruse to work.

Some well-known YouTubers also exhibit this type of privately public participation. A further fractalization of relative public and private behavior appears across different participants within this category in terms of their levels of identity revelation. Examples include MysteryGuitarMan (MGM) and MadV. MGM discloses more personal information than does MadV. For instance, MGM speaks and shows himself on camera, albeit usually behind glasses and often wearing a hat.[5] In contrast, MadV wears a Guy Fawkes[6] mask, does not disclose much personal information about himself, and never speaks.[7]

Although MadV's participation is public, in that a large number of people view his work and have a connection to him (as a YouTube "friend" or subscriber), the person behind the persona does not disclose identity details such as name, location, occupation, age, and gender.[8] MadV has maintained relative anonymity, although he said that this is becoming "trickier" as he attains greater popularity and participates in a widening social network on YouTube. MadV interacts with fans while maintaining a mysterious persona in a Guy Fawkes mask, a symbol that appears in many art works such as the vigilante character named "V" in the dystopic film, *V for Vendetta*. MadV said that he did not adopt the mask to show alliance with this V character. Indeed, MadV does not allow posted comments that suggest that he wishes to overthrow a government. People attempting to enter his YouTube media circuit using these comments are barred from entry both technically and socially, since MadV also deletes such comments from his YouTube pages. These types of negotiations determine whom MadV will allow into his social network.

People may friend or subscribe to MadV because his videos, such as One World, are broadly appealing and popular. People may assume that MadV is aligned with the "One World" movement,[9]

which promotes a more just global society in which people can access information and make connections worldwide. Instead, MadV said he wished to raise global awareness, a topic that also has wide appeal and interests many. In this video, MadV invites others to "make a statement" and to "make a difference." He holds up his hand to the camera. On it is written the phrase, "One World." Many viewers joined this media circuit by posting videos in which they wrote special messages on their hands (thus becoming video makers as well as viewers). Within an eight-month span, One World garnered more than one million views and over 2,000 video responses. At one point, it also held the honor of being the "most responded to video of all time on YouTube."

MadV encourages broad participation, albeit anonymously, by making videos that potentially concern many people and adopting liberal friending practices. Although MadV connects with many people, he nevertheless allows only certain kinds of comments to be posted on his videos. By refusing to post comments or videos from others who try to guess or knowingly reveal his identity, MadV tries to maintain a degree of privately public participation. He writes people who make such attempts at disclosure out of his YouTube media circuit. By placing a comment of affinity to MadV's work, a supporter can write him- or herself into a social network in which MadV is the center by showing how MadV is relevant to them. The way in which a participant responds to other participants illustrates different relationship patterns. Examining these patterns reveal particular media circuit topologies. For example, MadV said in an interview that he does not subscribe to other people's channels or initiate friend requests. In this sense, the media circuit topology that MadV supports is one in which he is the center of a directed social network in which the fans watch and comment on his videos, but he does not engage in these same practices toward fans. His practices are quite different from those that other interviewees reported, in which small, close groups of friends post and watch videos intended for consumption by a limited group. This and the other examples presented above illustrate how by examining participants' choices with regard to video sharing practices, processes of social network maintenance, delineation, and negotiation are revealed.

Discussion

This research suggests that for many participants, profile linkages are not the only or even the primary way of supporting a social network through YouTube. A more common practice among the interviewees was posting videos that friends and family could see and respond to. Posting comments also enabled people to express feelings of affinity for the video or the video makers. Media circuits are useful for understanding how social networks are created, maintained, and negotiated in public arenas such as on YouTube. Sharing and commenting on videos helped users maintain connections with their friends and relatives at a distance, and such behaviors facilitated membership negotiation within social networks. Interviewees reported that intelligent commentary on a video could stimulate closer social connections, if the video maker continues to communicate with and interact with the poster of the commentary.

Why do networks fractalize in ways that rupture yet maintain relative public and private social relations? Nissenbaum (2004) argues that some privacy is required for individuals to self-actualize. Privacy is arguably necessary for advancing the self and protecting the integrity of relationships. Having insulation against outside scrutiny is important for experimenting with aspects of the self without fear of retribution. For instance, as part of CT's "privately public behavior," he cloaked himself in a character in order to develop video skills, garner a fan base, and then reveal himself, once successful, to his old friend RJ. Different kinds of relationships require distinctive levels of protection from outside parties; these are manifested in varied levels of publicness and privacy in video making and sharing.

The question also arises: Why do people engage in "publicly private" behavior? This study identified two main reasons. First, the technical options for delineating media circuits as implemented at the time of the research had drawbacks. Many interviewees wished to share media with particular friends or family members but did not want to use the friends-only viewing option, because it required viewers to have an account. Nevertheless, YouTube was a convenient way to circulate videos with dispersed friends and relatives whom they knew watched YouTube. Some interviewees responded to the problem of sharing in public by using limited tagging, such that only people in the social network who knew the tags—which included cryptic references such as the video maker's YouTube name—would be able to use to locate participants' videos.

In addition, some interviewees used their view counts and comment systems as a way of gauging viewership, determining that their videos were not "super public" if they did not garner much response. Just as prior research on blogging demonstrated (Nardi, Schiano, & Gumbrecht, 2004; Viégas, 2005), a few interviewees were indifferent to being watched. Further, they did not express awareness about how their material could be accessed in ways outside their control. For instance, after the purchase of YouTube, Google enabled its search engine to return results from general searches that included YouTube videos. Such a feature widens video makers' exposure to being viewed, if they use tags that people searching Google also use to find material. YouTube also lists a number of "related" videos next to the current video being watched. If YouTube places a video maker's video on a "related list," the video may get more exposure than the video maker anticipated. Video source code can also be placed on profiles of other social network sites, allowing people to access videos from YouTube.

This article reveals that what constitutes how "public" a media circuit is varies, depending on participants' display of information about themselves and the content that they produce. For instance, MadV kept his identity guarded and reported that he does not know most people in his YouTube media circuit offline, but he encourages broad online viewership and friendships in a way that is not "mutual" with his fans. Posted videos and related comments create media circuits that take on a variety of different forms according to the relationships that they reflect. In MadV's case, fans watch and comment on his videos, but the social trajectory only moves in one direction, since MadV does not subscribe to other people's videos nor initiate friend requests. Different media circuit types reflect different interaction dynamics.

Future studies could examine different types of media exchange to see what they reveal about underlying social relationships. The social connections in a group in which media are circulating among a small number of people may differ from those in a group in which media are distributed only in one direction, as is the case between certain YouTube celebrities and their fans. A small private network may more interactively exchange media in bi-directional exchanges than a fan-centric media circuit with a YouTube star. Viewing the media exchange trajectories in different media circuits may thus shed light on other types of media-supported social relationships.

The analysis has shown that the public and private fractalize in complex ways in video making and sharing on YouTube. It also supports previous scholarship that claims that watching media is not merely a passive exercise, but rather that film or video viewing in general involves active interpretations that shape reception of media messages (Friedman, 2006). On YouTube, frequent interaction between video makers and viewers is a core component of participation on the site. Viewers and commenters are often themselves video makers, who comment with the strategic intent of forming social relationships with others who will support their work. Some interviewees reported that their attempts to be "friends" with popular YouTube stars were lost in large, mediated, social networks. In the context of liberal friending practices, discourse through comments and video responses provides another way for participants to establish meaningful social connections.

Conclusion

This article demonstrated how YouTube participants used both technical and symbolic mechanisms to attempt to delineate different social networks. It proposed new categories of nuanced behavior types that are neither strictly public nor strictly private. Moreover, it demonstrated that parts of social networks, as supported by media circuits, can be examined to shed light on the dynamics of social network creation, maintenance, and negotiation. Beyond profile-based friendship connections, the analysis showed how video sharing can become an important way for participants to negotiate membership in social networks.

These findings also have design implications. As social groups and professional organizations increasingly supplement their websites with social network site components, it is important to be alert to fractalization touch points, where greater amounts of publicity or privacy may be required to meet different individuals' and groups' social needs. Technical features that provide participants more customization and control in creating public and private interactions could help to optimize social network site usage.

As intentional and unintentional surveillance becomes more commonplace in contemporary society, people will likely continue to seek ways to carve out privacy in highly visible media environments. While corporations and institutions may require additional information sharing and employee monitoring on the company website, the present analysis suggests that some participants will nevertheless manipulate public systems in ways that preserve, in a fractalized pattern, different desired levels of informational and behavioral publicity and privacy. Analyzing micro-processes of networking through video sharing is one way to understand how social groups strive to protect varied levels of privacy amid increasing public scrutiny.

About the Author

Patricia Lange [plange@cinema.usc.edu] is a postdoctoral fellow in the School of Cinematic Arts at the University of Southern California. Her research explores issues of technical identity performance, identity in online environments such as video sharing sites, emotions in online encounters, and bodily experience of place. Address: 746 East Adams Boulevard, Los Angeles, CA 90089, USA

Notes

1. On YouTube channel pages, participants can display friend and subscription links or withhold them from view, even from other YouTube friends. This contrasts to Donath and boyd's (2004) finding that friendship links are assumed by many participants to be "mutual" and "public" on social network sites. Subscriptions mean that when a new video is posted, all the video maker's subscribers are alerted through email about it. Some interviewees had difficulty distinguishing the social difference between accepting a friend request and subscribing to their videos.
2. The claim here is not that all viewers interested in racism or other subjects will actually view a particular video, but rather, as Weintraub (1997) points out, that some subjects are considered more public than others. What scholars have characterized as relatively more public issues are those that concern a larger collective versus, for example, gossip that involves an event that relatively few people know or care about.
3. The two intermediate forms ("publicly private" and "privately public') discussed here are not the only ones. However, these forms emerged as particularly salient in the dataset used in this study.
4. A "noob" (short for "newbie") is a person who is new to a social group, typically an online environment. The term is often used in a derogatory way to characterize someone who is unfamiliar with the technical and social aspects of a particular online community or activity.
5. A sense for his level of participation can be gotten from his channel page: http://www.youtube.com/profile?user=mysteryguitarman. Retrieved October 29, 2007.
6. For more information about Guy Fawkes, see "The Gunpowder Plot" (2006).
7. MadV's channel page can be accessed at: http://www.youtube.com/profile?user=madv. Retrieved October 29, 2007.
8. Here I follow his fans' convention and refer to him as "he."
9. http://us.oneworld.net/. Retrieved October 29, 2007.

References

Al-Saggaf, Y. (2006). The online public sphere in the Arab world: The war in Iraq on the Arabiya Website. *Journal of Computer-Mediated Communication,* 12(1), article 16. Retrieved July 1, 2007 from http://jcmc.indiana.edu/vol12/issue1/al-saggaf.html

Barnes, S. B. (2006). A privacy paradox: Social networking in the United States. *First Monday,* 11(9). Retrieved July 1, 2007 from http://www.firstmonday.org/issues/issue11_9/barnes/index.html

Baym, N. (2000). *Tune in Log On: Soaps, Fandom, and Online Community*. Thousand Oaks, CA and London: Sage.

boyd, d. (2006). Friends, Friendsters, and Top 8: Writing community into being on social network sites. *First Monday,* 11(12). Retrieved February 28, 2007 from http://www.firstmonday.org/issues/issue11_12/boyd/

boyd, d.m. & Ellison, N.B. (2007). Social Network Sites: Definition, History & Scholarship. *Journal of Computer-Mediated Communication.* 13(1).

Donath, J., & boyd, d. (2004). Public displays of connection. *BT Technology Journal,* 22(4), 71–82.

Ellison, N. B., Steinfield, C., & Lampe, C. (2007). The benefits of Facebook "friends:" Social capital and college students' use of online social network sites. *Journal of Computer-Mediated Communication,* 12(4), article 1. Retrieved August 28, 2007 from http://jcmc.indiana.edu/vol12/issue4/ellison.html

Feder, J. (1988). *Fractals*. New York and London: Plenum Press.

Friedman, S. L. (2006). Watching twin bracelets in China: The role of spectatorship and identification in an ethnographic analysis of film reception. *Cultural Anthropology,* 21(4), 603–632.

Gal, S. (2002). A semiotics of the public/private distinction. *Differences: A Journal of Feminist Cultural Studies,* 13(1), 77–95.

Gross, R., & Acquisti, A. (2005). Information revelation and privacy in online social networks. *Proceedings of WPES'05* (pp. 71–80). Alexandria, VA: ACM.

"The gunpowder plot." (2006). Retrieved February 28, 2007 from http://www.parliament.uk/documents/upload/g08.pdf

Horst, H. A., & Miller, D. (2006). *The Cell Phone: An Anthropology of Communication*. Oxford, UK and New York: Berg.

Ito, M., & D. Okabe (2005). Technosocial situations: Emergent structurings of mobile email use. In M. Ito, D., Okabe, & M. Matsuda (Eds.), *Personal, Portable, Pedestrian: Mobile Phones in Japanese Life*. Cambridge, MA: MIT Press.

Kendall, L. (2002). *Hanging Out in the Virtual Pub: Identity, Masculinities, and Relationships Online*. Davis: University of California Press.

Lange, P. G. (2007 March). Commenting on comments: Investigating responses to antagonism on YouTube. Paper presented at the Annual Conference of the Society for Applied Anthropology. Retrieved August 29, 2007 from http://sfaapodcasts.files. wordpress.com/2007/04/update-apr-17-lange-sfaa-paper-2007.pdf

Nardi, B. A. (2005). Beyond bandwidth: Dimensions of connection in interpersonal communication. *Computer-Supported Cooperative Work,* 14, 91–130.

Nardi, B., Schiano, D., & Gumbrecht, M. (2004). Blogging as social activity, or, would you let 900 million people read your diary? *Proceedings of Computer-Supported Cooperative Work 2004*. New York: ACM Press.

Nissenbaum, H. (2004). Privacy as contextual integrity. *Washington Law Review,* 79(1), 101–158.

Rouse, R. (1991). Mexican migration and the social space of postmodernism. *Diaspora,* 1(1), 8–24.

Sandvig, C. (2006). The Internet at play: Child users of public Internet connections. *Journal of Computer-Mediated Communication,* 11(4), article 3. Retrieved July 1, 2007 from http://jcmc.indiana.edu/vol11/issue4/sandvig.html

Schoeman, F. (1984). *Philosophical Dimensions of Privacy*. Cambridge, UK: Cambridge University Press.

Sheehan, K. B. (2002). Toward a typology of Internet users and online privacy concerns. *The Information Society,* 18(1), 21–32.

Viégas, F. B. (2005). Bloggers' expectations of privacy and accountability: An initial survey. *Journal of Computer-Mediated Communication,* 10(3), article 12. Retrieved August 20, 2007 from http://jcmc.indiana.edu/vol10/issue3/viegas.html

Weintraub, J. (1997). The theory and politics of the public/private distinction. In J. Weintraub & K. Kumar (Eds.), *Public and Private in Thought and Practice* (pp. 1–42). Chicago and London: University of Chicago Press.

Weintraub, J., & K. Kumar, (Eds.) (1997). *Public and Private in Thought and Practice*. Chicago and London: University of Chicago Press.

Wellman, B. (1996). Are personal communities local? A Dumptarian reconsideration. *Social Networks,* 18(4), 347–354.

Imagined Communities: Awareness, Information Sharing, and Privacy on the Facebook[1]

Alessandro Acquisti[2] and Ralph Gross[3]

Online social networks such as Friendster, MySpace, or the Facebook have experienced exponential growth in membership in recent years. These networks offer attractive means for interaction and communication, but also raise privacy and security concerns. In this study we survey a representative sample of the members of the Facebook (a social network for colleges and high schools) at a US academic institution, and compare the survey data to information retrieved from the network itself. We look for underlying demographic or behavioral differences between the communities of the network's members and non-members; we analyze the impact of privacy concerns on members' behavior; we compare members' stated attitudes with actual behavior; and we document the changes in behavior subsequent to privacy-related information exposure. We find that an individual's privacy concerns are only a weak predictor of his membership to the network. Also privacy concerned individuals join the network and reveal great amounts of personal information. Some manage their privacy concerns by trusting their ability to control the information they provide and the external access to it. However, we also find evidence of members' misconceptions about the online community's actual size and composition, and about the visibility of members' profiles.

1 Introduction

"Students living in the scholarship halls [of Kansas University] were written up in early February for pictures on facebook.com that indicated a party violating the scholarship halls alcohol policy" [1]. "'Stan Smith' (not his real name) is a sophomore at Norwich University. He is majoring in criminal justice even though he admits to shoplifting on his MySpace page" [2]. "Corporations are investing in text-recognition software from vendors such as SAP and IBM to monitor blogs by employees and job candidates" [3]. Although online social networks are offering novel opportunities for interaction among their users, they seem to attract non-users' attention particularly because of the privacy concerns they raise. Such concerns may be well placed; however, online social networks are no longer niche phenomena: millions of people around the world, young and old, knowingly and willingly use Friendster, MySpace, Match.com, LinkedIn, and a hundred other sites to communicate, find friends, dates, and jobs—and in doing so, they wittingly reveal highly personal information to friends as well as strangers.

Nobody is literally forced to join an online social network, and most networks we know about *encourage,* but do not *force* users to reveal—for instance—their dates of birth, their cell phone numbers, or where they currently live. And yet, one cannot help but marvel at the nature, amount, and detail of the personal information some users provide, and ponder how informed this information sharing is. Changing cultural trends, familiarity and confidence in digital technologies, lack of exposure or memory of egregious misuses of personal data by others may all play a role in this unprecedented phenomenon of information revelation. Yet, online social networks' security and access controls are weak by design—to leverage their value as network goods and enhance their growth by making registration, access, and sharing of information uncomplicated. At the same time, the costs of mining and storing data continue to decline. Combined, the two features imply that information provided even on ostensibly private social networks is, effectively, public data, that could exist for as long as anybody has an incentive to maintain it. Many entities—from marketers to employers to national and foreign security agencies—may have those incentives.

In this paper we combine survey analysis and data mining to study one such network, catered to college and high school communities: the Facebook (FB). We survey a representative sample of FB members at a US campus. We study their privacy concerns, their usage of FB, their attitudes towards it as well as their awareness of the nature of its community and the visibility of their own profiles. In particular, we look for underlying demographic or behavioral differences between the communities of the network's members and non-members; we analyze the impact of privacy concerns on members' behavior; we compare members' stated attitudes with actual behavior; and we document the change in behavior subsequent information exposure: who uses the Facebook? Why? Are there significant differences between users and non-users? Why do people reveal more or less personal information? How well do they know the workings of the network?

Our study is based on a survey instrument, but is complemented by analysis of data mined from the network before and after the survey was administered. We show that there are significant demographic differences between FB members and non-members; that although FB members express, in general, significant concern about their privacy, they are not particularly concerned for their privacy on FB; that a minority yet significant share of the FB population at the campus we surveyed is unaware of the actual exposure and visibility of the information they publish on FB; and we document that priming about FB's information practices can alter some of its members' behavior.

The rest of the paper is organized as follows. In Section 2 we discuss the evolution of online social networks and FB in particular. In Section 3 we highlight the methods of our analysis. In Section 4 we present our results. In Section 5 we compare survey results to network data.

2 Online Social Networks

At the most basic level, an online social network is an Internet community where individuals interact, often through profiles that (re)present their public persona (and their networks of connections) to others. Although the concept of computer-based communities dates back to the early days of computer networks, only after the advent of the commercial Internet did such communities meet public success. Following the SixDegrees.com experience in 1997, hundreds of social networks spurred online (see [4] for an extended discussion), sometimes growing very rapidly, thereby attracting the attention of both media and academia. In particular, [5], [6], and [7] have taken ethnographic and sociological approaches to the study of online self-representation; [8] have focused on the value of online social networks as recommender systems; [4] have discussed information sharing and privacy on online social networks, using FB as a case study; [9] have demonstrated how information revealed in social networks can be exploited for "social" phishing; [10] has studied identity-sharing behavior in online social networks.

2.1 The Facebook

The Facebook is a social network catered to college and high school communities. Among online social networks, FB stands out for three reasons: its success among the college crowd; the amount and the quality of personal information users make available on it; and the fact that, unlike other networks for young users, that information is personally identified. Accordingly, FB is of interest to researchers in two respects: 1) as a mass social phenomenon in itself; 2) as a unique window of observation on the privacy attitudes and the patterns of information revelation among young individuals.

FB has spread to thousands of college campuses (and now also high schools) across the United States, attracting more than 9 million (and counting) users. FB's market penetration is impressive: it can draw more than 80% of the undergraduate population in many colleges. The amount, quality, and value of the information provided is impressive too: not only are FB profiles most often personally and uniquely identified, but by default they show contact information (including personal addresses and cell phone numbers) and additional data rarely available on other networks.

FB requires a college's email account for a participant to be admitted to the online social network of that college. As discussed in [4], this increases the expectations of validity of the personal information therein provided, as well as the perception of the online space as a closed, trusted, and trustworthy community (college-oriented social networking sites are, ostensibly, based "on a shared real space" [11]). However, there are reasons to believe that FB networks more closely resemble *imagined* [12] communities (see also [4]): in most online social networks, security, access controls, and privacy are weak *by design*; the easier it is for people to join and to find points of contact with other users (by providing vast amounts of personal information, and by perusing equally vast amounts of data provided by others), the higher the utility of the network to the users themselves, and the higher its commercial value for the network's owners and managers. FB, unlike other online networks, offers its members very granular and powerful control on the privacy (in terms of searchability and visibility) of their personal information. Yet its privacy default settings are very permeable: at the time of writing, by default participants' profiles are *searchable* by anybody else on the FB network, and actually *readable* by any member at the same college and geographical location. In addition, external access to a college FB community (e.g., by non-students/faculty/staff/alumni, or by non-college-affiliated individuals) is so easy [4], that the network is effectively an open community, and its data effectively public.

3 Methods

Our study aims at casting a light on the patterns and motivations of information revelation of college students on FB. It is based on a survey instrument administered to a sample of students at a North American college Institution, complemented by analysis of data mined from the FB network community of that Institution.

3.1 Recruiting Methods

Participants to the survey were recruited in three ways: through a list of subjects interested in participating in experimental studies maintained at the Institution where the study took place (and containing around 4,000 subscribed subjects); through an electronic billboard dedicated to experiments and studies, with an unknown (to us) number of campus community subscribers; and through fliers posted around campus. The above two lists are populated in majority by undergraduate students. The emails and the fliers sought participants to a survey on "online networks," and offered a compensation of $6, plus the possibility to win a $100 prize in a lottery among all participants.

Around 7,000 profiles were mined from the FB network of the same Institution. In order to automate access to the Facebook we used Perl scripts [13], specifically the Perl LWP library [14], which is designed for downloading and parsing HTML pages. The data was mined before and after the survey was administered.

3.2 Survey Design

The survey questionnaire contained around forty questions: an initial set of screening questions; a consent section; a set of calibration questions (to ascertain the respondents' privacy attitudes without priming them on the subject of our study: privacy questions were interspersed with questions on topics such as economic policy, the threat of terrorism, same-sex marriage, and so on); and, next, FB-related questions. Specifically, we asked respondents to answer questions about their usage, their knowledge, and their attitudes towards FB. Finally, the survey contained a set of demographics questions.

Only respondents currently affiliated with the Institution were allowed to take the survey (students, staff, and faculty). Respondents received somewhat different questions depending on whether they were current FB members, previous members, or never members. The survey is available on request from the authors.

3.3 Statistical Analysis

We analyzed survey results using STATA 8.0 on Windows and other *ad hoc* scripts. The study was performed on dichotomous, categorical (especially 7-point Likert scales), and continuous variables. We performed a number of different tests—including Pearson product-moment correlations to study relations between continuous variables, χ^2 and t tests to study categorical variables and means, Wilcoxon signed-rank test and Wilcoxon/Mann-Whitney test for non-normal distributions, as well as logit, probit, and linear multivariate regressions.

4 Results

A total of 506 respondents accessed the survey. One-hundred-eleven (21.9%) were not currently affiliated with the college Institution where we conducted our study, or did not have an email address within that Institution's domain. They were not allowed to take the rest of the survey. A separate set of 32 (8.2%) participants had taken part in a previous pilot survey and were also not allowed to take the survey. Of the remaining respondents, 318 subjects actually completed the initial calibration questions. Out of this set, 278 (87.4%) had heard about FB, 40 had not. In this group, 225 (70.8%) had a profile on FB, 85 (26.7%) never had one, and 8 (2.5%) had an account but deactivated it. Within those three groups, respectively 209, 81, and 7 participants completed the whole survey. We focus our analysis on that set—from which we further removed 3 observations from the non-members group, since we had reasons to believe that the responses had been created by the same individual. This left us with a total of 294 respondents.

4.1 Participants

In absolute terms, we had exactly the same number of male participants taking the survey as female participants, 147. We classified participants depending on whether they were current members of the FB campus network (we will refer to them as "members"), never members, or no longer members (we will often refer to the last two groups collectively as "non-members").

A slight majority of FB members in our sample (52.63%) are male. Our sample slightly over-represents females when compared to the target FB population, whose data we mined from the network (males represent 63.04% of the Institution's FB network, but it is important to note that the gender distribution at the Institution is itself similarly skewed). However, we know from the information mined from the network that 79.6% of all the Institution's undergraduate males are on the FB (91.92% of our sample of male undergrads are FB members) and 75.5% of all the Institution's undergraduate females are on the FB (94.94% of our sample of female undergrads are FB members). In other words (and expectably), our total sample of respondents slightly over-represents FB members.

The gender distribution of our sample is reversed among respondents who were never or are no longer members of FB: 56.46% are female. This gender difference between current members and current nonmembers is not statistically significant (Pearson $\chi^2(1) = 2.0025$, $Pr = 0.157$). However, when

FIGURE 1 Gender distribution of the survey participants for the three FB member groups.

	What is your gender		
User	Male	Female	Total
Current FB Member	110 52.63 74.83	99 47.37 67.35	209 100.00 71.09
Former FB Member	3 42.86 2.04	4 57.14 2.72	7 100.00 2.38
Never FB Member	34 43.59 23.13	44 56.41 29.93	78 100.00 26.53
Total	147 50.00 100.00	147 50.00 100.00	294 100.00 100.00

we test usage by contrasting actual FB users and non-members plus members who claim to "I never login/use" their profile, the gender difference becomes more radical (54.19% of users are male, but only 40.66% of *non-users* are) and significant (Pearson χ^2 (1) = 4.5995 *Pr* = 0.032). See Figure 1 for the gender distribution in the three FB member groups.

There is no significant difference among the distributions of undergraduate versus graduate students in our sample and in the overall FB population.

Overall, sixty-four percent of our respondents (64.29%) are undergraduate students; 25.17% are graduate students; 1.36% are faculty; and 9.18% are staff. We did not consider alumni in our study. This distribution slightly oversamples undergraduate students when compared to the actual Institution's population (total student population in 2005/06: 10,017. Undergraduate students: 54.8%). This was expected, considering the available recruiting tools and the comparatively higher propensity of undergraduate students to take paid surveys and experiments. However, when checking for current FB membership in our sample, we find that undergraduates dominate the picture (84.21%), followed by graduate students (14.35%) and staff (1.44%). These numbers are comparable to the distribution of the target population discussed in [4] when correcting for alumni (91.21% were undergraduate students on the Facebook network).

Again, the distribution of non-members is reversed: graduate students dominate (51.76%), followed by staff (28.24%). The distributions of user types (undergraduates, graduates, staff, or faculty) by FB membership status are significantly diverse (Pearson χ^2 (3) = 135.3337 *Pr* = 0.000). See Figure 2 for a breakdown of the academic status of survey participants across the three FB groups.

Unsurprisingly, age is a strong predictor of membership (see Figure 3). Non-members tend to be older (a mean of 30 years versus a mean of 21) but their age is also more broadly distributed (*sd* 8.840476 vs. *sd* 2.08514). The difference in the mean age by membership is strongly significant ($t = -14.6175$, $Pr<t = 0.0000$).

4.2 Privacy Attitudes

Age and student status are correlated with FB membership—but what else is? Well, of course, having heard of the network is a precondition for membership. Thirty-four participants had never heard of the FB—nearly half of the staff that took our survey, a little less than 23% of the graduate students, and a negligible portion of the undergraduate students (1.59%).

FIGURE 2 Distribution of survey participant status for FB members, non-members and people who never had a FB account.

	Are you ...				
User	**Undergrad**	**Graduate**	**Faculty**	**Staff**	**Total**
Current FB Member	176 84.21 93.12	30 14.35 40.54	0 0.00 0.00	3 1.44 11.11	209 100.00 71.09
Former FB Member	2 28.57 1.06	4 57.14 5.41	0 0.00 0.00	1 14.29 3.70	7 100.00 2.38
Never FB Member	11 14.10 5.82	40 51.28 54.05	4 5.13 100.00	23 29.49 85.19	78 100.00 26.53
Total	189 64.29 100.00	74 25.17 100.00	4 1.36 100.00	27 9.18 100.00	294 100.00 100.00

FIGURE 3 Distribution of age for FB members and non-members.

	Group_age				
usercomb	**17–24**	**25–34**	**35–44**	**45+**	**Total**
Current member	204 97.61 88.31	4 1.91 9.30	1 0.48 9.09	0 0.00 0.00	209 100.00 71.09
Current non-member	27 31.76 11.69	39 45.88 90.70	10 11.76 90.91	9 10.59 100.00	85 100.00 28.91
Total	231 78.57 100.00	43 14.63 100.00	11 3.74 100.00	9 3.06 100.00	294 100.00 100.00

However, together with age and student status (with the two obviously being highly correlated), another relevant distinction between members and non-members may arise from privacy attitudes and privacy concerns.

Before we asked questions about FB, our survey ascertained the privacy attitudes of participants with a battery of questions modelled after the Alan Westin's studies [15], with a number of modifications. In particular, in order not to prime the subjects, questions about privacy attitudes were interspersed with questions about attitudes towards economic policy and the state of the economy, social issues such as same-sex marriage, or security questions related to the fear of terrorism. In addition, while all instruments asked the respondent to rank agreement, concern, worries, or importance on a 7-point Likert scale, the questions ranged from general ones (e.g., "How important do you consider the following issues in the public debate?"), to more and more specific (e.g., "How do you personally value the importance of the following issues for your own life on a day-to-day basis?"), and personal ones (e.g., "Specifically, how worried would you be if" [a certain scenario took place]).

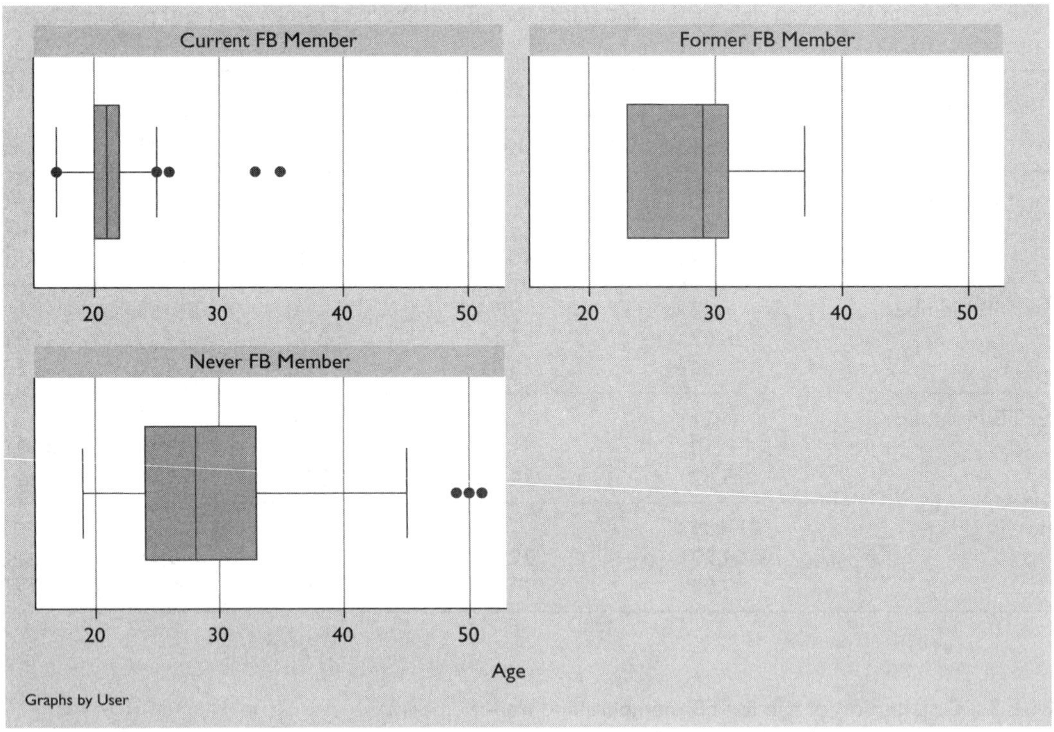

FIGURE 4 Box-plots of age distribution for different membership status.

"Privacy policy" was on average considered a highly important issue in the public debate by our respondents (mean on the 7-point Likert scale: 5.411, where 1 is "Not important at all" and 7 is "very important"; *sd*: 1.393795). In fact, it was regarded a more important issue in the public debate than the threat of terrorism ($t = 2.4534$, $Pr>t = 0.0074$; the statistical significance of the perceived superiority was confirmed by a Wilcoxon signed-rank test: $z = 2.184$ $Pr>|z| = 0.0290$) and same sex marriage ($t = 10.5089$, $Pr>t = 0.0000$; Wilcoxon signed-rank test: $z = 9.103$ $Pr>|z| = 0.0000$); but less important than education policy (mean: 5.93; *sd*: 1.16) or economic policy (mean: 5.79; *sd*: 1.21). The slightly larger mean valuation of the importance of privacy policy over environmental policy was not significant. (These results are comparable to those found in previous studies, such as [16].)

The same ranking of values (and comparably statistically significant differences) was found when asking for "How do you personally value the importance of the following issues for your own life on a day-to-day basis?" The mean value for the importance of privacy policy was 5.09. For all categories, subjects assigned slightly (but statistically significantly) more importance to the issue in the public debate than in their own life on a day-to-day basis (in the privacy policy case, a Wilcoxon signed-rank test returns $z = 3.62$ $Pr>|z| = 0.0003$ when checking the higher valuation of the issue in the public debate).

Similar results were also found when asking for the respondents' concern with a number of issues directly relevant to them: the state of the economy where they live, threats to their personal privacy, the threat of terrorism, the risks of climate change and global warming. Respondents were more concerned (with statistically significant differences) about threats to their personal privacy than about terrorism or global warming, but less concerned than about the state of the economy.

Finally, we asked how worried respondents would be if a number of specific events took place in their lives. The highest level of concern was registered for "A stranger knew where you live and the location and schedule of the classes you take" (mean of 5.78, with 45.58% of respondents choosing the 7th point in the Likert scale, "very worried," and more than 81% selecting Likert points above 4). This was followed by "Five years from now, complete strangers would be able to find out easily your sexual orientation, the name of your current partner, and your current political views" (mean of 5.55, with 36.39%—the relative majority—choosing the 7th point in the Likert scale, and more than 78% with points above 4), followed, in order, by the 'global warming' scenario ("The United States rejected all new initiatives to control climate change and reduce global warming"), the security scenario ("It was very easy for foreign nationals to cross the borders undetected"), the 'contacts' scenario ("A friend of a friend that you do not even know knew your name, your email, your home phone number, and your instant messaging nickname"), and the 'same-sex' scenario ("Two people of the same sex were allowed to marry in your State").

Privacy Attitudes and Membership Status

Privacy concerns are not equally distributed across FB members and non-members populations: a two-sided t test that the mean Likert value for the "importance" of privacy policy is higher for non-members (5.67 in the non-members group, 5.30 in the members group) is significant ($t = -2.0431$, $Pr<t = 0.0210$). Similar statistically significant differences arise when checking for the level of concern for privacy threats and for worries associated with the privacy scenarios described above. The test becomes slightly less significant when checking for member/non-member differences in the assigned importance of privacy policy on a day-to-day basis.

Importantly, in general no comparable statistically significant differences between the groups can be found in other categories. For example, worries about the global warming scenario gain a mean Likert valuation of 5.36 in the members sample and 5.4 in the non-members sample. (A statistically significant difference can be found however for the general threat of terrorism and for the personal worry over marriage between two people of same sex: higher values in the non-members group may be explained by their higher mean age.)

We also used two-sample Wilcoxon rank-sum (Mann-Whitney) tests to study the distributions of the sensitivities to the various scenarios. We found additional evidence that the sensitivity towards privacy is stronger among non-members than members. In the "A stranger knew where you live and the location and schedule of the classes you take" scenario, concerns are higher in the non-member population—the Mann-Whitney test that the two distributions are the same returns $z = -3.086$ $Pr>|z| = 0.0020$. Similar results are found for the "Five years from now, complete strangers would be able to find out easily your sexual orientation, the name of your current partner, and your current political views" scenario ($z = -2.502$ $Pr>|z| = 0.0124$), and the "A friend of a friend that you do not even know knew your name, your email, your home phone number, and your instant messaging nickname" scenario. Importantly, no such differences were found to be significant for the same sex marriage scenario, the illegal aliens scenario, and the US rejecting initiatives to control climate change scenario.

Overall, the distributions of reported intensity of privacy concerns tend to be more skewed towards higher values, and less normally distributed for non-members. For the most invasive scenarios, however, both members and non-members' distributions are not normal, with the distribution for non-members more skewed towards the higher values on the right (see Figure 5). These results do not change after accounting for people who do not know about FB—the t tests simply become more significant.

Disentangling Age, Student Status, and Privacy Concerns

An obvious hypothesis about FB membership is that individual privacy concerns will be inversely correlated with the probability of joining FB. However, while FB non-members seem to have higher average privacy concerns than members (over the scenarios we tested), we cannot directly conclude that the higher one's general privacy concerns, the less likely he will be a FB member. Figure 6, for instance, shows the distribution of levels of concern for privacy threats for both FB members and non-members. A measure of correlation provided by Pearson χ^2 (6) is not significant (χ^2 (6) = 8.0467, $Pr = 0.235$). (Pearson χ^2 is significant when studying the "stranger knows where you live scenario": χ^2 (6) = 16.5665, $Pr = 0.011$; likelihood-ratio χ^2 (6) = 17.4785, $Pr = 0.008$.)

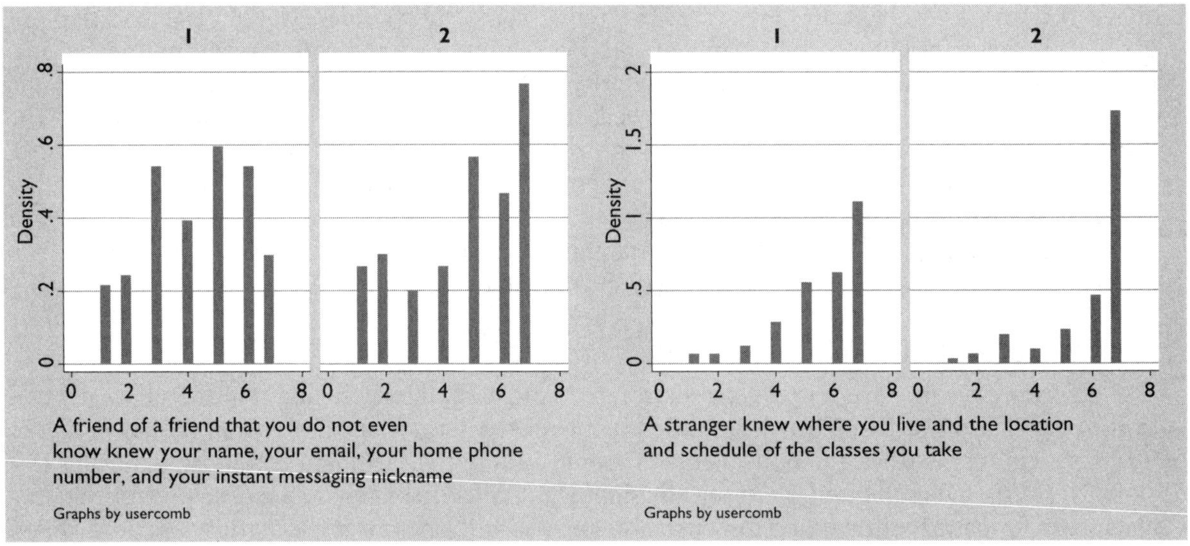

FIGURE 5 Distribution of privacy attitudes for FB members (columns marked "1") and non-members (columns marked "2"; this set includes both people that never had a profile and those who had a profile but deactivated it) for an exemplary scenario.

FIGURE 6 Distribution of levels of concern for threats to personal privacy for FB members and non-members.

	Threats to your personal privacy						
usercomb	**1**	**2**	**3**	**4**	**5**	**6**	**7**
Current member	6 2.87 100.00	15 7.18 71.43	20 9.57 80.00	28 13.40 68.29	46 22.01 71.88	52 24.88 76.47	42 20.10 60.87
Current non-member	0 0.00 0.00	6 7.06 28.57	5 5.88 20.00	13 15.29 31.71	18 21.18 28.13	16 18.82 23.53	27 31.76 39.13
Total	6 2.04 100.00	21 7.14 100.00	25 8.50 100.00	41 13.95 100.00	64 21.77 100.00	68 23.13 100.00	69 23.47 100.00

In addition, privacy concerns may also be correlated with gender,[4] and status (undergraduate, graduate, faculty, staff).[5] This makes it difficult to understand the actual impact of privacy attitudes and concerns and various other personal characteristics on FB membership.

For instance, when we focus on the undergraduate respondents in our sample, we find that even the undergraduates who expressed the highest level of concern for threats to their personal privacy are still in vast majority joining the Facebook: 89.74% of them. We also find that the mean level of concern is not statistically different between undergraduates who are members and those who are not. (Among undergraduate students, 2 were former members who were no longer members at the time of the survey; their expressed level of concern for threats was 5 and 7; one user was still a member but claimed to never login—his concern level is 6). On the other hand, among respondents who are not undergraduates, the mean concern level of non-members (controlling for those who have heard about the FB) is 5.41; the mean for members is 4.81. A two-side Student t test shows that the difference is mildly significant: *Ha : diff <t* = –1.5346 and *Pr<t* = 0.0646). In fact, the ratio of members to non-members decreases with the intensity of concern.

In order to disentangle these complex relations between age, respondent type, and privacy concerns—that we hypothesize are all factors affecting FB membership—we employed multivariate regression analysis.

In a first approach, we used k-means multivariate clustering techniques [17] to cluster respondents according to their privacy attitudes: we used all the 7-Likert scale responses relevant to privacy (from importance assigned to privacy policy, to worries about specific scenarios) and created a new categorical variable called *_Iprivacy_*. We employed that variable in logistical regressions (logit and probit) over the dependent variable *user_logit,* a dichotomous variable representing membership to the FB network (*user_logit*=1; or lack thereof, *user_logit*=0). We also used age (*age*), a dummy variable representing gender (male if *gender*=1), and a dummy variable representing student status (undergraduate if *undergrad*=1) as independent variables. We restricted the analysis to respondents who had heard about FB. The results of the regression are reported in Figure 7. The model has a good fit, explaining more than half of the variance between members and non-members of FB. As expected, age and undergraduate status are significant while gender is not. The signs of the regression are as expected: being an undergraduate increases the probability of being a member, and age decreases it. Interestingly, at least one of the categorical clusters for privacy attitudes (represented by the variables *_Iprivacy_~2,3,4,5,6,7,* measured against the base cluster *_Iprivacy_~1*—the one with the highest level of concerns) is significant, with a large positive impact on the probability of being a member.

In a second approach, we took the means of all the 7-Likert scale responses relevant to privacy and constructed a new categorical variable (*privacy_at~n*), that we used in the second regression reported in Figure 8.

The results are comparable to those from the previous regression. Both regressions show that even when controlling for age, status, and gender, one's privacy concerns have *some* impact on the decision to join the network, and the student status has some impact independent from age.[6] However, and importantly, this impact really only exists for the non-undergraduate population: when restricting the analysis to the undergraduate population, neither the privacy cluster nor the privacy mean variables are significant. They are, however, significant (*Pr>z* : 0.024) when focusing on the non-undergraduate population. In other words: privacy concerns may drive older and senior college members away from FB. Even high privacy concerns, however, are *not* driving undergraduate students away

FIGURE 7 Results of logit regression on FB membership using demographical characteristics and k-means clustered privacy attitudes (unstandardized effect coefficients).

Logit estimates	**Number of obs** =	**260**
	LR χ^2 (9) =	**130.27**
	Prob > χ^2 =	**0.0000**
Log likelihood = –63.57221	**Pseudo R^2** =	**0.5061**

user_logit	**Coef.**	**Std. Err.**	**z**	***P* > |z|**	**[95% Conf. Interval]**	
age	–.5632942	.1385778	–4.06	0.000	–.8349017	–.2916866
undergrad	.9950484	.5923704	1.68	0.093	–.1659763	2.156073
Iprivacy~2	4.906447	2.119911	2.31	0.021	.7514989	9.061395
Iprivacy~3	.2758868	.5935938	0.46	0.642	–.8875356	1.439309
Iprivacy~4	1.043322	.8747547	1.19	0.233	–.6711654	2.75781
Iprivacy~5	1.598968	1.113793	1.44	0.151	–.5840251	3.781961
Iprivacy~6	.6435271	.6934733	0.93	0.353	–.7156556	2.00211
Iprivacy~7	1.451756	1.173762	1.24	0.216	–.8487741	3.752287
gender	.1879386	.4760592	0.39	0.693	–.7451203	1.120997
_cons	12.81519	3.278655	3.91	0.000	6.389147	19.24124

FIGURE 8 Results of logit regression on FB membership using demographical characteristics and mean privacy attitudes (unstandardized effect coefficients).

Logit estimates				**Number of obs**	**=**	**260**
				LR χ^2 (4)	**=**	**121.80**
				Prob > χ^2	**=**	**0.0000**
Log likelihood = –67.804697				**Pseudo R²**	**=**	**0.4732**

user_logit	**Coef.**	**Std. Err.**	**z**	***P* > |z|**	**[95% Conf.**	**Interval]**
age	–.4953475	.1236441	–4.01	0.000	–.7376856	–.2530095
undergrad	1.025618	.5752887	1.78	0.075	–.1019268	2.153164
privacy_at~n	–.5152155	.2212539	–2.33	0.020	–.9488651	–.0815658
gender	.1798292	.4576217	0.39	0.694	–.7170929	1.076751
_cons	14.69864	3.322535	4.42	0.000	8.186591	21.21069

from it. Non-members have higher generic privacy concerns than FB members. These results suggest that FB membership among undergraduates is *not* just a matter of their not being concerned, in general, about their privacy—other reasons must be explored.

4.3 Reported Facebook Usage

In order to understand what motivates even privacy concerned individual to share personal information on the Facebook, we need to study what the network itself is used for. Asking participants this question directly is likely to generate responses biased by self-selection and fear of stigma. Sure enough, by far, FB members deny FB is useful to them for dating or self-promotion. Instead, members claim that the FB is very useful to them for learning about and finding classmates (4.93 mean on a 7-point Likert scale) and for making it more convenient for people to get in touch with them (4.92), but deny any usefulness for other activities. Other possible applications of FB—such as dating, finding people who share one's interests, getting more people to become one's friends, showing information about oneself/advertising oneself—are ranked very low. In fact, for those applications, the relative majority of participants chooses the minimal Likert point to describe their usefulness (coded as "not at all" useful). Still, while their mean Likert value remains low, male participants find FB slightly more useful for dating than female.

And yet, when asking participants to rate how often, on average, their peers use FB for the same activities, the results change dramatically: learning about classmates and the convenience factor of staying in contact are still ranked very highly, but now "Showing information about themselves/advertising themselves," "Making them more popular," or "Finding dates" suddenly become very popular. See how the distributions almost invert in Figure 9.

INFORMATION PROVIDED

What information do FB members provide, and of what quality? Many members are quite selective in the type of information they provide—for instance, most publish their birthdays but hide their cell phone numbers. However, interestingly, our survey participants' answers imply that if a certain type of information is provided at all, it is likely to be of good quality: complete and accurate (see Figure 10).[7]

When controlling for participants who have abandoned the Facebook, we find out that as users they were less likely than continuing members to provide information such as their birthday (85.71% do not provide this information, while 86.12% of current members claim they *do* provide it—Pearson χ^2 (2) = 33.9440 *Pr* = 0.000), AIM (Pearson χ^2 (2) = 14.2265 *Pr* = 0.001), cell phone number, home phone number, personal address, political orientation, sexual orientation, and partner's name (the differences between non-members and members across the last six categories however are not statistical significant).

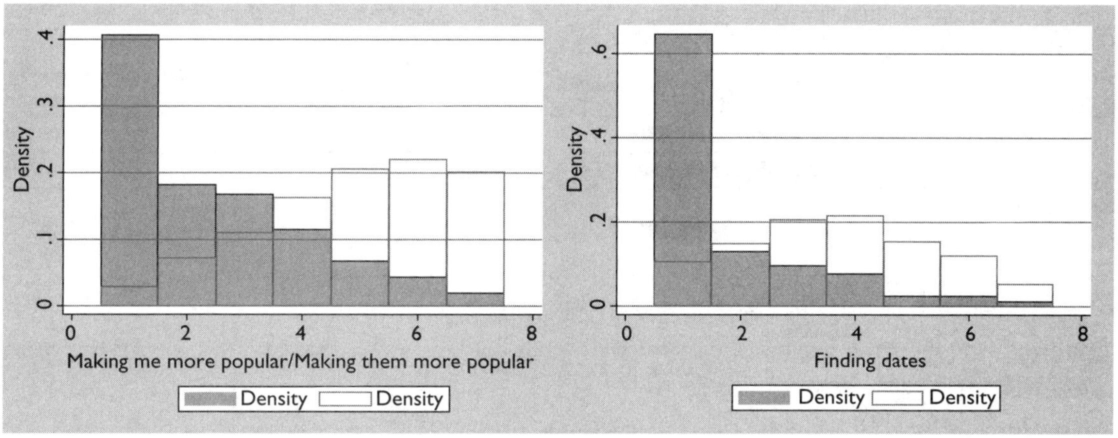

FIGURE 9 Do as I preach, not as I do—How useful is FB for you (grey boxes) vs. how often do you believe other members use FB for (transparent boxes)?

FIGURE 10 Information provided by FB members.

What personal information do you provide on the FaceBook and how accurate is that information?

	I don't provide this information	I provide this information and it is complete and accurate	I provide this information but it is intentionally not complete or not accurate
Birthday	12% (29)	84% (195)	3% (8)
Cell phone number	59% (138)	39% (90)	2% (4)
Home phone number	89% (207)	10% (24)	0% (1)
Personal address	73% (169)	24% (55)	3% (8)
Schedule of classes	54% (126)	42% (97)	4% (9)
AIM	24% (56)	75% (173)	1% (3)
Political views	42% (97)	53% (122)	6% (13)
Sexual orientation	38% (88)	59% (138)	3% (6)
Partner's name	71% (164)	28% (65)	1% (3)

Female members are not more or less likely than male members to provide accurate and complete information about their birthday, schedule of classes, partner's name, AIM, or political views. However, they are much *less* likely to provide their sexual orientation (Pearson χ^2 (2) = 11.3201 Pr = 0.003), personal address (Pearson χ^2 (2) = 10.5484 Pr = 0.005), and cell phone number (Pearson χ^2 (2) = 10.9174 Pr = 0.004). This confirms the results reported in [4], where less than 29% of females were found providing cell phone information, compared to 50% of male.

Self-Selection Bias?

Often, survey participants are less privacy conscious than non-participants. For obvious reasons, this self-selection bias is particularly problematic for survey studies that focus on privacy. Are our respondents a biased sample of the Institution's FB population—biased in the sense that they provide more information than the average FB members?

We did not find strong evidence of that. Since we mined the network before the survey was administered, we were able to compare information revelation by survey participants and survey non-participants. It is true that, on average, our survey takers provide slightly more information

On average, how often do you login to FaceBook?			
		Response Percent	Response Total
More than once a day		21.6%	50
Once a day		**25.4%**	**59**
More than once a week, but in general less than once a day		22.4%	52
Once a week		8.6%	20
Less than once a week		6%	14
More than once a month, but in general less than once a week		6%	14
Once a month		6%	14
I still have a profile, but I never login/use it.		3.9%	9

On average, how often do you update your profile on FaceBook?			
		Response Percent	Response Total
More than once a week		2.6%	6
Once a week		6%	14
More than once a month, but in general less than once a week		17.2%	40
Once a month		**50.4%**	**117**
Never		23.7%	55

FIGURE 11 Frequency of login and profile update.

than the average FB member. However, the differences in general do not pass a Fisher's exact test for significance, except for personal address and classes (where non-participants provide statistically significant less information) and political views (in which the difference is barely significant).

Attitudes vs. Behavior

We detected little or no relation between participants' reported privacy attitudes and their likelihood of providing certain information, even when controlling, separately, for male and female members. For instance, when comparing the propensity to provide birthday and the Likert values reported in the answers to the privacy threat question described at the beginning of Section 4.2, no statistically significant difference emerged: Pearson χ^2 (12) = 5.2712 Pr = 0.948. Comparable results were found when testing sexual orientation (Pearson χ^2 (12) = 10.7678 Pr = 0.549), partner's name (Pearson χ^2 (12) = 15.1178 Pr = 0.235), cell phone number (Pearson χ^2 (12) = 19.0821 Pr = 0.087), or personal address.

We obtained the same results when using the cluster variable that summarizes each respondent's privacy attitudes (see Section 4.2), both when using standard Pearson's χ^2 as well as when using Student's t test (the latter was used when comparing the mean privacy concern across respondents who provided or did not provide accurate information about various data types).

Combined with the results discussed in 4.2, the above evidence may suggest that privacy attitudes have some effect on determining who joins the network, but after one has joined, there is very little marginal difference in information revelation across groups—which may be the result of perceived peer pressure or herding behavior.

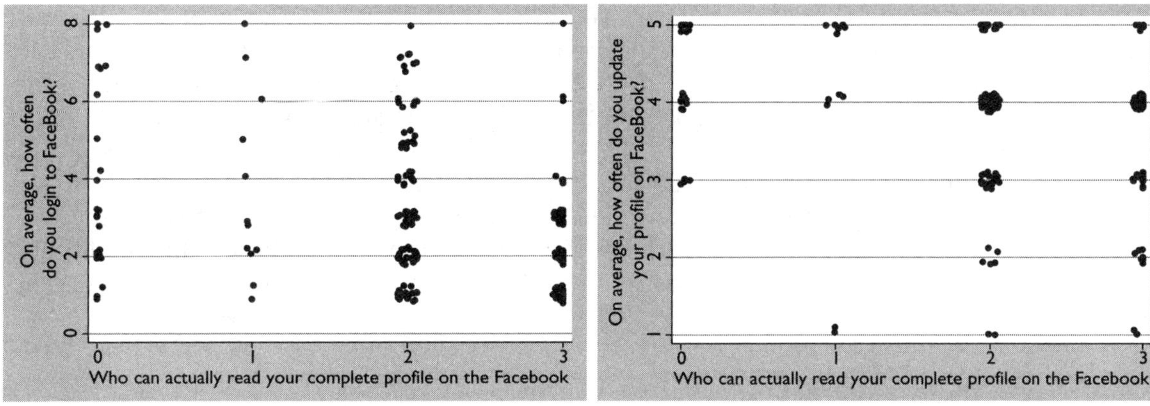

FIGURE 12 Self-awareness of ability to control who can see one's profile, by frequency of login (left) and frequency of update (right). On the x-axis, the value 0 means "Do not know" if there is any way to control; 1 means "No control"; 2 means "Some control" and 3 means "Complete control." On the y-axis, higher values mean *less* frequent login or update.

If anything, we found new confirmations of a privacy attitude/behavior dichotomy [18]. Almost 16% of respondents who expressed the highest concern (7 on the Likert scale) for the scenario in which a stranger knew their schedule of classes and where they lived, provide nevertheless both pieces of information (in fact, almost 22% provide at least their address, and almost 40% provide their schedule of classes).

Similarly, around 16% of respondents who expressed the highest concern for the scenario in which someone 5 years from now could know their current sexual orientation, partner's name, and political orientation, provide nevertheless all three types of information—although we can observe a descending share of members that provide that information as their reported concerns increase. Still, more than 48% of those with the highest concern for that scenario reveal at least their current sexual orientation; 21% provide at least their partner's name (although we did not control for the share of respondents who are currently in relationships); and almost 47% provide at least their political orientation.

4.4 Awareness of Facebook Rules and Profile Visibility

How knowledgeable is the average FB member about the network's features and their implications in terms of profile visibility?

By default, everyone on the Facebook appears in searches of everyone else, and every profile at a certain Institution can be read by every member of FB at that Institution. However, the FB provides an extensive privacy policy and offers very granular control to users to choose what information to reveal to whom. As mentioned above, relative to a FB member, other users can either be friends, friends of friends, non-friend users at the same institution, non-friend users at a different institution, and non-friend users at the same geographical location as the user but at a different university (for example, Harvard vs. MIT). Users can select their profile visibility (who can read their profiles) as well as their profile searchability (who can find a snapshot of their profiles through the search features) by type of users. More granular control is given on contact information, such as phone numbers.

And yet, among current members, 30% claim not to know whether FB grants any way to manage who can search for and find their profile, or think that they are given no such control. Eighteen percent do not know whether FB grants any way to manage who can actually read their profile, or think that they are given no such control. These numbers are not significantly altered by removing the 13 members who claim never to login to their account. In fact, even frequency of login does not explain the lack of information for some members. On the other hand, members who claim to login more than once a day are also more likely to believe that they have "complete" control on whom can search their profile.

Awareness of one's ability to control who can see one's profile is not affected by the frequency of login, but is affected by the frequency of update (a Pearson χ^2 (12) = 28.9182 *Pr* = 0.004 shows that the distribution is significant): see Figure 13. Note the difference between the two graphs and, specifically, the distribution by frequency of update for respondents who answered "Do not know" or "No control" (graph on the right).

Twenty-two percent of our sample do not know what the FB privacy settings are or do not remember if they have ever changed them. Around 25% do not know what the location settings are.

To summarize, the majority of FB members claim to know about ways to control visibility and searchability of their profiles, but a significant minority of members are unaware of those tools and options.

Self-Reported Visibility

More specifically, we asked FB members to discuss how visible and searchable their own profiles were. We focused on those participants who had claimed never to have changed their privacy settings (that by default make their profile searchable by everybody on FB and visible to anybody at the same Institution), or who did not know what those settings were.

Almost every such respondent realizes that anybody at their Institution can search their profile. However, 24% incorrectly do not believe that anybody on FB can in fact search their profile. Misunderstandings about visibility can also go in the opposite direction: for instance, 16% of current members believe, incorrectly, that *anybody* on FB can read their profile.

In fact, when asked to guess how many people could search for their profile on FB (respondents could answer by selecting the following possible answers from a drop-box: a few hundred, a few thousands, tens of thousands, hundreds of thousands, millions), the relative majority of members who did not alter their default settings answered, correctly, "Millions." However, *more than half* actually underestimated the number to tens of thousands or less.

In short, the majority of FB members seem to be aware of the true visibility of their profile—but a *significant minority* is vastly underestimating the reach and openness of their own profile. Does this matter at all? In other words, would these respondents be bothered if they realized that their profile is more visible than what they believe?

The answer is complex. First, when asked whether the current visibility and searchability of the profile is adequate for the user, or whether he or she would like to restrict it or expand it, the vast majority of members (77% in the case of searchability; 68% in the case of visibility) claim to be satisfied with what they have—most of them do not want more or less visibility or searchability for their profiles (although 13% want less searchability and 20% want less visibility) than what they (correctly or incorrectly) believe to have. Secondly, as we discuss further below in Section 4.5, FB members remain wary of whom can access their profiles, but claim to manage their privacy fears by controlling the information they reveal.

4.5 Attitudes Towards the Facebook

So far we have glanced at indirect evidence of a number of different reasons for the dichotomy between FB members' stated privacy concerns (high) and actual information hiding strategies (mixed, but often low also for members with high stated concerns). Those reasons include peer pressure and unawareness of the true visibility of their profiles.

Another possible reason is the level of trust FB members assign to the network itself. On average, FB members trust the system quite a bit (and in general trust its members *more* than members of comparable services, like Friendster or MySpace—see Figure 13).

This happens notwithstanding the fact that almost 77% of respondents claimed not to have read FB's privacy policy (the real number is probably higher); and that many of them mistakenly believe that FB does not collect information about them from other sources regardless of their use of the site (67%), that FB does not combine information about them collected from other sources (70%), or that FB does not share personal information with third parties (56%). (We note that hav-

FIGURE 13 How FB members assign trust.

	How much do you trust								
	Do not trust at all						Trust completely	N/A	Response Average
The FaceBook (the Company)	5% (10)	8% (17)	17% (38)	24% (53)	25% (56)	14% (32)	4% (8)	3% (7)	4.20
Your own friends on FaceBook	1% (3)	1% (3)	3% (6)	7% (15)	26% (57)	38% (84)	22% (49)	2% (4)	5.62
CMU Facebook users	3% (7)	6% (13)	14% (32)	26% (58)	32% (70)	15% (33)	3% (7)	0% (1)	4.35
Friends of your friends on FaceBook	6% (13)	8% (17)	15% (33)	27% (60)	24% (54)	14% (31)	4% (9)	2% (4)	4.17
On average, FaceBook users not connected to you	13% (28)	17% (38)	23% (51)	27% (60)	12% (27)	5% (10)	1% (3)	2% (4)	3.29
If you use or know about MySpace, MySpace users not connected to you	15% (33)	9% (19)	6% (14)	12% (26)	4% (9)	1% (2)	1% (3)	52% (115)	2.78
If you use or know about Friendster, Friendster users not connected to you	16% (35)	8% (17)	8% (17)	12% (27)	5% (12)	2% (4)	0% (1)	49% (108)	2.82

ing read, or claiming to have read, the privacy policy, does *not* make respondents more knowledgeable about FB's activities.)

While respondent are mildly concerned about who can access their personal information and how it can be used, they are not, in general, concerned about the information itself, mostly because they control that information and, with less emphasis, because believe to have some control on its access. Respondents are fully aware that a social network is based on information sharing: the strongest motivator they have in providing more information are reported, in fact, as "having fun" and "revealing enough information so that necessary/useful to me and other people to benefit from FaceBook."

However, psychological motivations can also explain why information revelation seems disconnected from the privacy concerns. When asked to express whether they considered the current public concern for privacy on social network sites such as the FaceBook or MySpace to be appropriate (using a 7-point Likert scale, from "Not appropriate at all" to "Very much appropriate"), the response average was rather high (4.55). In fact, the majority of respondents agree (from mildly to very much) with the idea that the information *other* FB members reveal may create privacy risks to *those* members (that is, the other members; average response on a 7-point Likert scale: 4.92)—even though they tend to be less concerned about their *own* privacy on FB (average response on a 7-point Likert scale: 3.60; Student's t test shows that this is significantly less than the concern for other members: $t = -10.1863$, $P<t = 0.0000$; also a Wilcoxon matched pair test provides a similar result: $z = -8.738$, $Pr <|z| = 0.0000$).

In fact, 33% of our respondents believe that it is either impossible or quite difficult for individuals not affiliated with an university to access FB network of that university. "Facebook is for the students" says a student interviewed in [19]. But considering the number of attacks described in [4] or any recent media report on the usage of FB by police, employers, and parents, it seems in fact that for a significant fraction of users the FB is only an *imagined* community.

5 Survey and Network Data

In order to justify conclusions informed by a survey, the validity of the answers provided by the subjects has to be addressed. For this study we were in the *unique* position to be able to directly compare the answers provided by the participants with visible FB profiles to the information they actually provide in the profile (downloaded and archived immediately before the survey was administered). This section compares the survey responses with profile data and examines survey impacts in the form of changes to FB profiles of survey participants.

5.1 Comparison between Reported Answers and Actual Data

In order to gauge the accuracy of the survey responses, we compared the answers given to a question about revealing certain types of information (specifically, birthday, cell phone, home phone, current address, schedule of classes, AIM screenname, political views, sexual orientation and the name of their partner) with the data from the actual (visible) profiles. We found that 77.84% of the answers were exactly accurate: if participants said that they revealed a certain type of information, that information was in fact present; if they wrote it was not present, in fact it was not. A little more than 8% revealed more than they said they do (i.e. they claim the information is not present when in fact it is). A little more than 11% revealed less then they claimed they do. In fact, 1.86% claimed that they provide false information on their profile (information is there that they claim is intentionally false or incomplete), and 0.71% have missing false information (they claimed the information they provide is false or incomplete, when in fact there was no information).

We could not locate the FB profiles for 13 self-reported members that participated in the survey. For the participants with CMU email address, 2 of them did mention in the survey that they had restricted visibility, searchability, or access to certain contact information, and 3 wrote that not all CMU users could see their profile.

5.2 Survey Impacts

For this analysis, we eliminated the survey responses for users whose profile we could not locate on the network, ending up with 196 profiles out of the 209 self-proclaimed FB member participants. We downloaded information from the network immediately before and after administering the survey, both for users who responded to it and those who did not, and then compared the profiles.

First, we found a statistically significant difference in the byte size of the resulting files. The mean byte size decreased in both the experiment and the control group, but the experiment group changed significantly more than the control group (paired t test $Pr<t = 0.0060$). See Figure 14 for histograms of the file size changes for both groups. However, no significant changes were found when evaluating *individual* data fields: 5 survey participants reduced the information they provided compared to 4 profiles in the control group that similarly removed specific information.

After further investigation, we found that what happened was the following: the 9 profiles with the highest byte change (all >10kb) were in fact the ones that completely changed the visibility of their profile. They represent slightly more than 5% of our sample of current FB members (whose profile before the survey was visible). Out of this group 6 were female and 3 male. In the control group only 2 profiles changed visibility. This difference is statistically significant (χ^2 $Pr<0.05$).

While the difference is significant and somewhat surprising, the magnitude in terms of number of members that changed their behavior is relatively small. One should note that this change happened even without us providing the survey participants with a real threat scenario. In addition, although privacy concerned individuals are on FB, only a fraction of them may have such high

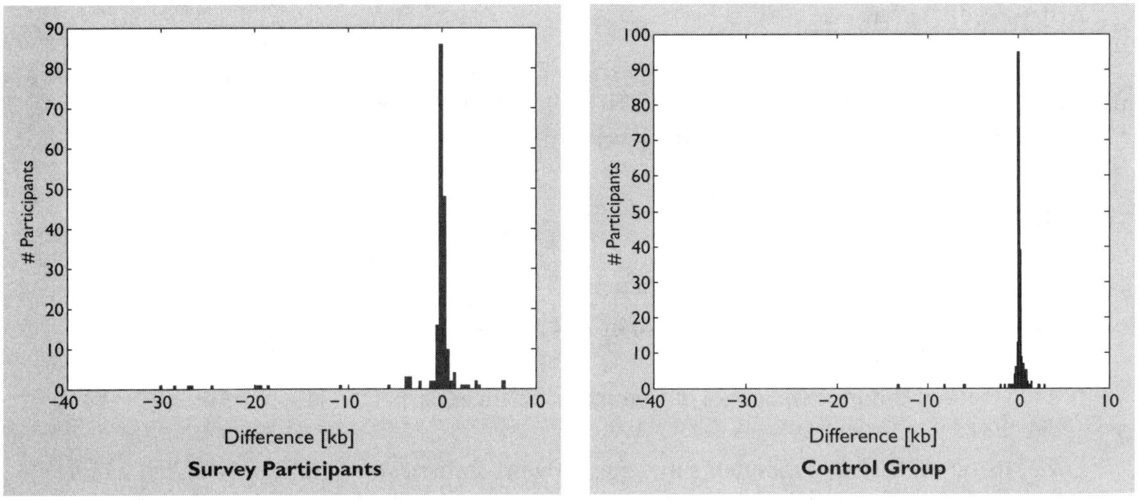

FIGURE 14 Changes in profile sizes for survey participants and a control group. The sizes for the survey participants changed significantly more.

concerns to be induced to abandon the network just by questions about its privacy implications. In fact, we found that this group of "switchers" have higher means in terms of average privacy attitudes, and their distributions of privacy attitudes are skewed towards the right (that is, towards higher concerns)—than non "switchers," although such differences are not statistically significant.

6 Discussion and Future Work

Online social networks offer exciting new opportunities for interaction and communication, but also raise new privacy concerns. Among them, the Facebook stands out for its vast membership, its unique and personally identifiable data, and the window it offers on the information revelation behavior of millions of young adults.

In our study we have combined survey instruments with data mined from a FB community at a North American college Institution. We looked for demographic or behavioral differences between the communities of the network's members and non-members, and searching for motivations driving the behavior of its members. Our analysis is going to be complemented by other experiments, but we can summarize here a number of initial results.

Age and student status obviously are the most significant factors in determining FB membership. However, we observe that privacy attitudes also play a role, but only for the non-undergraduate population. In fact, most of highly privacy concerned undergraduates still join the network. While a relative majority of FB members in our sample are aware of the visibility of their profiles, a significant minority is not. The 'aware' group seems to rely on their own ability to control the information they disseminate as the preferred means of managing and addressing their own privacy concerns. However, we documented significant dichotomies between specific privacy concerns and actual information revelation behavior. In addition, misunderstanding or ignorance of the Facebook (the Company)'s treatment of personal data are also very common.

It is interesting to note that a pilot study we ran in September 2005 provided similar results, but also small, yet significant differences in terms of members' awareness of their profile visibility and their ability to control it: respondents a few months ago appeared less aware of privacy risks and of means of managing their own profiles. This evidence may suggest that the widespread public attention on privacy risks of online social networks is affecting, albeit marginally, some of their users.

Acknowledgments

This research was supported by CMU Berkman Faculty Development Fund and CMU CyLab. We would like to thank Lorrie Cranor, Charis Kaskiris, Julia Gideon, Jens Grossklags, and Bradley Malin for helpful insights and suggestions in the development of the survey protocol.

Notes

1. Pre-proceedings version. Privacy Enhancing Technologies Workshop (PET), 2006
2. H. John Heinz III School of Public Policy and Management, Carnegie Mellon University, Pittsburgh, PA 15213
3. Data Privacy Laboratory, School of Computer Science, Carnegie Mellon University, Pittsburgh, PA 15213
4. For instance, female respondents in general report statistically significantly higher average concerns for privacy over the various scenarios and instruments we discussed above.
5. We did not find a significant correlation between age and a number of indicators of privacy concerns in our sample; however, our sample cannot be considered representative of the population of age over 25.
6. As noted above, in our sample age alone is not significantly correlated with privacy concerns.
7. Also such survey answers that elicit personal admissions about the quality of the data provided on FB may be, in turn, biased. However, since survey participants were not asked to disclose the actual information whose quality they were asked to evaluate, we have no reason to believe that their incentives to offer inaccurate answers were significant.

References

1. Parker, R.: Alcohol policy violated. Kansan.com February 28 (2006)
2. Youngwood, S.: Networking by the 'book'. The Times Argus February 26 (2006)
3. Kharif, O.: Big brother is reading your blog. BusinessWeek online February 28 (2006)
4. Gross, R., Acquisti, A.: Privacy and information revelation in online social networks. In: Proceedings of the ACM CCS Workshop on Privacy in the Electronic Society (WPES '05). (2005)
5. d. boyd: Reflections on friendster, trust and intimacy. In: Intimate (Ubiquitous) Computing Workshop—Ubicomp 2003, October 12–15, Seattle, Washington, USA. (2003)
6. d. boyd: Friendster and publicly articulated social networking. In: Conference on Human Factors and Computing Systems (CHI 2004), April 24–29, Vienna, Austria. (2004)
7. Donath, J., d. boyd: Public displays of connection. BT Technology Journal 22 (2004) 71–82
8. Liu, H., Maes, P.: Interestmap: Harvesting social network profiles for recommendations. In: Beyond Personalization—IUI 2005, January 9, San Diego, California, USA. (2005)
9. Jagatic, T., Johnson, N., Jakobsson, M., Menczer, F.: Social phishing. Communications of the ACM Forthcoming (2006)
10. Stutzman, F.: An evaluation of identity-sharing behavior in social network communities. In: Proceedings of the 2006 iDMAa and IMS Code Conference, Oxford, Ohio. (2006)
11. Sege, I.: Where everybody knows your name. Boston.com April 27 (2005)
12. Anderson, B.: Imagined Communities: Reflections on the Origin and Spread of Nationalism. Revised edn. Verso, London and New York (1991)
13. Wall, L., Christiansen, T., Orwant, J.: Programming Perl. 3rd edn. O'Reilly (2000)
14. Burke, S.: Perl & LWP. O'Reilly (2002)

15. Westin, A.F.: Harris-equifax consumer privacy survey (1991). Technical report, Equifax, Inc., Atlanta, GA (1991)
16. Acquisti, A., Grossklags, J.: Privacy and rationality in decision making. IEEE Security & Privacy January–February (2005) 24–30
17. Berry, M., Linoff, G.: Data Mining Techniques for Marketing, Sales and Customer Support. Wiley, New York (1997)
18. Acquisti, A.: Privacy in electronic commerce and the economics of immediate gratification. In: Proceedings of the ACM Conference on Electronic Commerce (EC '04). (2004) 21–29
19. Kornblum, J., Marklein, M.B.: What you say online could haunt you. USA Today March 8 (2006)

Discussion: MySpace and Deleting Online Predators Act (DOPA)

danah boyd and Henry Jenkins

The following interview with Henry Jenkins (co-director of the Comparative Media Studies Program at MIT) and danah boyd (PhD student at the School of Information, University of California-Berkeley) was conducted via email by Sarah Wright of the MIT News Office. An abbreviated version was published by the MIT News Office on 24 May 2006.

Q: What is MySpace? Why is it important? How big is it (and its cousins such as Facebook)?

danah: MySpace and Facebook are social network sites where individuals create profiles and link to others ("friends") within the system. The profile serves as an individual's digital representation (similar to homepages) of their tastes, fashion, and identity. In crafting this profile, individuals upload photos, indicate interests, list favorite musicians and describe themselves textually and through associated media. The social network feature allows participants to link themselves to others within the system, revealing their affiliations and peer group. These sites also allow friends to comment on each other's profiles. Structurally, social network sites are a cross between a yearbook and a community website.

These sites also provide numerous communication tools. Both have a messaging system similar to email; MySpace also has a bulletin board where people can post messages that all friends can read and a blogging service where people can post entries for either friends or the public at large. When youth login, their first task is typically to check messages in order to see who has written them. While email is still used to communicate with adults and authorities, MySpace is the primary asynchronous communication tool for teens. After checking personal messages, youth check friend additions, bulletin board posts, event announcements and new blog posts by friends. They visit their friends' pages to see new photos or check out each other's comments. The vast majority of social network site use amongst youth does not involve surfing to strangers' profiles, but engaging more locally with known friends and acquaintances.

MySpace has over 78 million registered accounts while Facebook has approximately 8 million. While over 85% of college students participate on Facebook if it exists on their campus, MySpace is a cultural requirement for American high school students. Or, as one teenager said, "If you're not on MySpace, you don't exist." Not all MySpace users are teenagers, but most American teenagers have accounts on MySpace.

These sites play a key role in youth culture because they give youth a space to hang out amongst friends and peers, share cultural artifacts (like links to funny websites, comments about TV shows) and work out an image of how they see themselves. They also serve as digital publics, substituting for the types of publics that most adults took for granted growing up, but are now inaccessible for many young people—neighborhood basketball courts, malls, parks, etc. Youth are trying to map out a public youth territory for themselves, removed from adult culture. They are doing so online because their mobility and control over physical space is heavily curtailed and monitored.

Q: What is the controversy over MySpace? Is it that site in particular or as a genre of web-based-social-networks?

danah: Like previous digital publics (blogs, discussion boards, chatrooms, newsgroups), MySpace is very open—anyone can join, participate and communicate with others. While MySpace allows 14 and 15 year old users to restrict who can see their page and contact them, most users opt to make their profiles public. The primary concern is that this openness puts youth at risk, making them particularly vulnerable to predators and pedophiles.

Henry: More broadly, there are concerns about what aspects of their lives teens reveal through their online profiles. Adults are confronting images of underage drinking or sex, discussions of drug use, and signs of bullying and other abusive behavior. In some cases, teens and adults have developed different notions of privacy: young people feel more comfortable sharing aspects of their lives (for example, their sexual identities) that previous generations would have kept secret. In some cases, teens do not fully understand the risks of making certain information public. In many cases, schools are being forced to respond to real world problems which only came to their attention because this information was so publicly accessible on the web. Schools are uncertain what level of responsibility they should have over what their students do online—some are worried about what they are doing on library computers and others seek to extend their supervision into what teens are doing on their own time and off school grounds. Much of the controversy has come not as a result of anything new that MySpace and the other social software sites contribute to teen culture but simply from the fact that adults can no longer hide their eyes to aspects of youth culture in America that have been there all along. All of this is coming to head with the proposal of new federal legislation which would require all schools and libraries which receive federal funds to restrict access to these digital tools and online communities.

Q: What is the direction of your current research on new media, and how does it relate to the controversy?

danah: For my doctoral dissertation, I am investigating why and how youth are engaging in digital publics like MySpace, how this affects identity development and how youth socialization has changed over the last century. This work is being funded by the MacArthur Foundation to help understand the nature of informal learning. Understanding why moral panics emerge when youth socialize is central to my research.

Henry: I am currently finishing up a white paper, commissioned by the MacArthur Foundation, which seeks to identify the core social skills and cultural competencies young people need in order to become full participants in the cultural, political, economic, and social life of the 21st century. In doing this research, we are reviewing the current state of educational research surrounding participatory culture and examining how teachers are currently deploying these technologies through schools. We want in the long term to develop new curricular materials which help parents and teachers build a more constructive relationship with new media. I also have a new book coming out this summer, *Convergence Culture: Where Old and New Media Collide,* which provides some frameworks for thinking about the new forms of participatory culture which are emerging in the digital era.

Q: What do 'social networking software programs' provide participants? What's their down side?

danah: By giving youth access to a public of their peers, MySpace provides a fertile ground for identity development and cultural integration. As youth transition from childhood, they seek out public environments to make sense of culture, social status and how they fit into the world. Interacting with strangers helps them understand who they are and communities of interest allow them to explore ideas and values. Although youth are able to socialize privately with one another in the homes of friends, most are not allowed to spend time hanging out in public, unaccompanied by parents or adults. They view MySpace as a place where they can be who they are, joke around with friends and make certain to stay in the loop about everything that is going on around them.

While integrating into cultural life is a critical process that takes place during these years, the actual process is not always smooth or pleasant. Bullying, sexual teasing, and other peer-to-peer harassment are rampant amongst teenagers, as these are frequently the tools through which youth learn to make meaning of popularity, social status, roles, and cultural norms. MySpace did not create teenage bullying but it has made it more visible to many adults, although it is not clear that the embarrassment online is any more damaging to the young victims than offline. Regardless of medium, the humiliation occurs when the entire school or social community knows of the attack; MySpace and other online mediums may help spread rumors faster, but they have always spread in the halls of schools pretty quickly. No one of any age enjoys being the target of public tormenting, but new media is not to blame for peer-to-peer harassment simply because it makes it more visible to outsiders. In fact, in many ways, this visibility provides a window through which teen mentors can help combat this issue.

Q: What skills do students/children learn in working in social networks? How does these contribute (or not) to their development?

Henry: As a society, we are at a moment of transition when the most important social relationships may no longer be restricted to those we conduct face to face with people in our own immediate surroundings but may also include a large number of relationships which are conducted over vast geographic distances. Over the past decade or so, we have been learning how to live in communities which are grassroots but not necessarily geographically local. We are learning how to interact across multiple communities and negotiate with diverse norms. These networking skills are increasingly important to all aspects of our lives. Social networking services are more and more being deployed as professional tools, extending the sets of contacts that people can tap in their work lives. It is thus not surprising that such tools are also part of the social lives of our teens. Just as youth in a hunting society play with bows and arrows, youth in an information society play with information and social networks. Our schools so far do a rather poor job of helping teens acquire the skills they need in order to participate within that information society. For starters, most adult jobs today involve a high degree of collaboration, yet we still focus our schools on training autonomous learners. Rather than shutting kids off from social network tools, we should be teaching them how to exploit their potentials and mitigate their risks.

Q: What educational use might/does MySpace or other social network software have?

Henry: Much of the current policy debate around MySpace assumes that the activities there are at best frivolous and at worst dangerous to the teens who participate. Yet, a growing number of teachers around the country are discovering that these technologies have real pedagogical value. Teachers are beginning to use blogs for knowledge sharing in schools; they use mailing lists to communicate expectations about homework with students and parents. They are discovering that students take their assignments more seriously and write better if they are producing work which will reach a larger public rather than simply sit on the teacher's desk. Teachers are linking together classrooms around the country and around the world, getting kids from different cultural backgrounds to share aspects of their everyday experience with each other and thus learn to communicate across differences. Social networking sites generate a great amount of statistical data about their users which can be used as raw materials for projects in social studies or math classes. Student-made videos are used for class projects, which get shared through free public sites like YouTube; students, in turn, get feedback on their work from broader audiences and begin to develop connections with other young artists which can push them to the next level. Classes are taping podcasts of lectures at other educational institutions, making the best speakers in the world available to students at the most remote locations. Students are doing their own research and contributing to sites like Wikipedia. Teachers are discovering that the design of web pages or personal profiles may function like autobiographical essays, encouraging adolescents to reflect more deeply on their own lives and identities and to exert more control over their self presentation. Many of these activities would be threatened by the proposed federal legislation which would restrict access to these sites via public schools or library terminals.

Q: What is the essence of the proposed legislation? If passed, how would it affect students? Teachers? Librarians? Parents?

danah: Recent federal legislation, Deleting Online Predators Act (DOPA) would require schools and libraries that receive federal aid "to protect minors from commercial social networking websites and chat rooms." The proposed law would extend current regulations that require all federally funded schools and libraries to deploy internet filters. The law is so broadly defined that it would limit access to any commercial site that allows users to create a profile and communicate with strangers. This legislation is targeting MySpace, but it would also block numerous other sites, including blogging tools, mailing lists, video and podcast sites, photo sharing sites, and educational sites like NeoPets.

Henry: In theory, the bill would allow schools to disable these filters for use in educationally specified contexts, yet in practice, most schools will simply lock down their computers and walk away. Teachers who wanted to exploit the educational benefits of these tools would face increased scrutiny and pressure to discontinue these practices. And students would lack the ability to explore these resources through independent research or social activities. Teens who lack access to the Internet at home would be cut off from their extended sphere of social contacts.

danah: Most major technology companies are moving in the direction of social software. They are using social features to help users find information, get recommendations, and share ideas. This would all be restricted. Even if its application were restricted solely to MySpace, this legislation assumes that nothing positive can be gained through the socialization that occurs there. For example, high school students currently contact college students through MySpace to learn about their schools and decide whether or not to apply.

Henry: Suppose, for the sake of argument, that MySpace critics are correct and that MySpace is, in fact, exposing large numbers of teens to high-risk situations, then shouldn't the role of educational institutions be to help those teens understand those risks and develop strategies for dealing with them? Wouldn't we be better off having teens engage with MySpace in the context of supervision from knowledgeable and informed adults? Historically, we taught children what to do when a stranger telephoned them when their parents are away; surely, we should be helping to teach them how to manage the presentation of their selves in digital spaces. The proposed federal legislation does nothing to help kids confront the challenges of interacting with online social communities; rather, it allows teachers and librarians to abdicate their responsibility to educate young people about what is becoming a significant aspect of their everyday lives. Our responsibilities as educators should be to bring reason to bear on situations which are wrought with ignorance and fear, not to hide our eyes from troubling aspects of teen culture.

danah: Police currently patrol MySpace, just as they patrol other areas where youth hang out. Many are thankful to know where youth go online because it helps them do their job. Too often, predators know youth haunts better than police and decentralized systems make it difficult for police to do their job. Blocking known sites will encourage teens to go further underground and seek out places to socialize that adults are unaware of. This puts youth at increased risk and means that neither educators nor law enforcement will be around to help.

Q: The proposed bill appears to offer protection to minors from online predators, by limiting their mutual access. Is predation a real danger with MySpace? Are there other issues people should be aware of in weighing this legislation?

danah: The media coverage of predators on MySpace implies that 1) all youth are at risk of being stalked and molested because of MySpace; 2) prohibiting youth from participating on MySpace will stop predators from attacking kids. Both are misleading; neither is true.

Unfortunately, predators lurk wherever youth hang out. Since youth are on MySpace, there are bound to be predators on MySpace. Yet, predators do not use online information to abduct children;

children face a much higher risk of abduction or molestation from people they already know—members of their own family or friends of the family. Statistically speaking, kids are more at risk at a church picnic or a boy scout outing than they are when they go on MySpace. Less than .01% of all youth abductions nationwide are stranger abductions and as far as we know, no stranger abduction has occurred because of social network services. The goal of a predator is to get a child to consent to sexual activities. Predators contact teens (online and offline) to start a conversation. Just as most teens know to say no to strange men who approach them on the street, most know to ignore strange men who approach them online. When teenagers receive solicitations from adults on MySpace, most report deleting them without question. Those who report responding often talk about looking for attention or seeking a risk. Of those who begin conversations, few report meeting these strangers.

The media often reference a *Crimes Against Children* report that states one in five children receive a sexual solicitation online. A careful reading of this report shows that 76% of the unwanted solicitations came from fellow children. This includes unwanted date requests and sexual taunts from fellow teens. Of the adult solicitations, 96% are from people 18–25; wanted and unwanted solicitations are both included. In other words, if an 18 year old asks out a 17 year old and both consent, this would still be seen as a sexual solicitation. Only 10% of the solicitations included a request for a physical encounter; most sexual solicitations are for cybersex. While the report shows that a large percentage of youth are faced with uncomfortable or offensive experiences online, there is no discussion of how many are faced with uncomfortable or offensive experiences at school, in the local shopping mall or through other mediated channels like telephone.

Although the media has covered the potential risk extensively, few actual cases have emerged. While youth are at minimal risk, predators are regularly being lured out by law enforcement patrolling the site. Most notably, a deputy in the Department of Homeland Security was arrested for seeking sex with a minor.

The fear of predators has regularly been touted as a reason to restrict youth from both physical and digital publics. Yet, as Barry Glassner notes in *The Culture of Fear,* predators help distract us from more statistically significant molesters. Youth are at far greater risk of abuse in their homes and in the homes of their friends than they ever are in digital or physical publics.

Q: You have written before on anti-electronic/anti-new media attitudes. Where does this proposed bill fit into those?

Henry: History shows us a recurring pattern surrounding the adaptation of any new communications technology. Young people are often early adopters: they are more open to new ideas and experiences; they are looking for ways to leave their mark on the world and they are seeking places where they can socially interact with minimal adult interference. Parents and teachers are often frightened by these new kinds of communication technologies which were not part of the world of their childhood: they don't really understand what their young people are doing with them and they don't know how to protect or supervise their children while they are engaged in these activities. The situation is thus ripe for moral panic.

A single high profile incident—some kind of tragedy or crime—can spark backlash. Political leaders, seeking headlines, and journalists, seeking readers, exploit those anxieties and feed those fears. Soon, there is a call to take action "even if it is wrong," a call to action which races well ahead of any serious research or thoughtful reflection on the matters at hand. The new legislation is being embraced by politicians in both parties eager to woo cultural conservatives and suburban voters as they enter what everyone knows is going to be a hotly contested election.

Over time, as these technologies become better integrated into everyday life, as the generation which grew up with these technologies takes on adult responsibilities, things calm down again. People develop a more balanced perspective which sees both the benefits and risks associated with these activities. Rather than restrict access, we educate our young people in the safe, ethical, and creative use of these technologies.

Right now, MySpace is at the most disruptive point in this cycle: people are reacting in ignorance and fear and in doing so, they increase the risks and discard the benefits of these emerging cultural practices.

Q: Most people know there is a digital divide—are the issues raised by this legislation related to the widening or narrowing of the gap or to other things?

Henry: Over the past decade, there has been intense public concern about the digital divide. Many of us have worked hard to insure that every kid has access to the Internet via schools and public libraries, if not through their own homes. And these efforts have been largely successful, outside of Indian reservations and some other rural pockets, in insuring at least minimal levels of access. Now, the problem shifts from concerns about technical access to concerns about participation in the key social and cultural experiences which are defining the emerging generation's relationship to these technologies. What a kid can do at home with unlimited access is very different from what a kid can do in a public library with ten or fifteen minutes of access at a time and with no capacity to store and upload information to the web. We further handicap these children by placing filters on the Internet which restrict their access to information which is readily available to their more affluent classmates. And now this legislation would restrict their ability to participate in social networks or to belong to online communities. The result will be to further isolate children from poorer economic backgrounds, to cut kids at risk from support systems which exist within their peer culture, and to limit the social and cultural experiences of kids who are already behind in acquiring important networking skills that will shape their professional futures. All of this will compound what we are now calling the participation gap. The early discussion of the digital divide assumed that the most important concern was insuring access to information as if the web were simply a data bank. Its power comes through participation within its social networks. The authors of the law are reading MySpace and other social software exclusively in terms of their risks; they are not focusing on the opportunities they offer for education and personal growth. In protecting children from those risks, they would cut them off from those educational benefits.

Q: You have said elsewhere (and several years ago) that virtual gaming experiences of today are analogous to the unfettered play in the backyards of the 1950s—very core & essential experiences. Have social networking like MySpace or games or other new media technology become core experiences now?

Henry: As I suggested above, most parents understand their children's experiences in the context of their memories of their own early years. For the baby boom generation, those defining experiences involved playing in backyards and vacant lots within suburban neighborhoods, socializing with their friends at the local teen hangout, and participating within a social realm which was constrained by the people who went to your local school. All of that is changing. Contemporary children and youth enjoy far less physical mobility, have less time outside of adult control, and have fewer physical places to hang out with their friends.

Much of this activity is being brought online. What teens are doing online is no better and no worse than what previous generations of teens did when their parents weren't looking. The difference is that as these activities are being digitized, they are also being brought into public view. Video games bring the fantasy lives of young boys into the family room and parents are shocked by what they are seeing. Social networks give adults a way to access their teens' social and romantic lives and they are startled by their desire to break free from restraints or act older than their age. Parents are experiencing this as a loss of control but in fact, adults have greater control over these aspects of their children's lives than ever before.

Indeed, one of the biggest risks of these digital technologies is not the ways that they allow teens to escape adult control but rather the permanent traces left behind of their transgressive conduct. Teens used to worry about what teachers or administrators might put in their permanent records since this would impact how they were treated in the future. Yet, we are increasingly discovering that everything we do online becomes part of our public and permanent record, easily recoverable by anyone who knows how to Google, and that there is no longer any statute of limitations on our youthful indiscretions.

danah: Because of mobile phones, current college students report greater ongoing communication with their parents than in previous generations. As Misa Matsuda has argued, networked technologies are allowing today's youth to maintain "full-time intimate communities." While the socialization that takes place in digital publics is equivalent to that which occurs in physical publics, new media is allowing youth to be more deeply connected to their peers and their family members, providing a powerful open channel for communication and sharing.

Q: The proposed bill is a political response to a social/technological development—could you offer a political framework for considering MySpace and laws to limit access to it?

Henry: Right now, MySpace and the other social network tools are being read as threats to the civic order, as encouraging anti-social behaviors. But we can easily turn this around and see them as the training ground for future citizens and political leaders. Young people are assuming public roles at earlier and earlier ages. They are interacting with larger communities of their peers and beginning to develop their own styles of leadership. Across a range of issues, young people are using social network software to identify and rally like-minded individualism, forming the basis for new forms of digital activism. Current research shows that teens who participate in massively multiplayer games develop a much stronger ability to work in teams, a greater understanding of how and when to take appropriate risks, an ability to rapidly process complex bodies of information, and so forth. At the same time, these teens are facing an array of ethical challenges which are badly understood by the adults around them. They have nowhere to turn for advice on how to confront some of the choices they make as participants within these communities. Part of the work we will be doing for the MacArthur Foundation involves the development of an ethics casebook which will help parents, teachers, and students work through some of these issues and make sensible decisions about how they conduct their online lives. We see this kind of pedagogical intervention as far more valuable than locking down all public computers and then sending kids out to deal with these issues on their own.

Q: What suggestions do you have for parents or other adults eager to learn more about MySpace and to understand what's going on with it, kids and political reactions?

Henry: Parents face serious challenges in helping their children negotiate through these new online environments. They receive very little advice about how to build a constructive relationship with media within their families or how to help their offspring make ethical choices as participants in these online worlds. As a culture, we have deeply conflicted assumptions about adolescence which functions as a period of transition: Most of us recognize that teens need to take on a greater degree of autonomy as they prepare for adult lives, even as they still need some degree of adult supervision to help them make sane and safe decisions; we simply disagree about the relative balance of freedom and autonomy that teens should receive. We respect the fact that the decisions families make about media reflect some of their most deeply held values; different families have different concerns and make different decisions. For that reason, we think decisions about youth access to digital technologies should be made in the context of individual families and not form the basis of one-size-fits-all federal legislation. Recognizing that different parents will approach these issues in different ways, we would still offer the following as our governing philosophy for dealing with MySpace and other social software.

1. Communication with your daughter or son is key. Build a trusting relationship through dialogue. It is important to talk with them about your concerns; it is even more important to listen to what they have to say about their online experiences and why these sites are such an important part of their interactions with their peers. You need to recognize that some unfamiliar experiences look scarier from the outside than they are. Take time to understand what you are seeing and what it means to participants.

2. Create an account to understand how the site works, but not to stalk your kids. They need room to explore, but if you are familiar with the media and technology that they consume, you can provide valuable guidance and suggestions. Surveillance, while possible, damages a trusting parent/child relationship.
3. Ask your kids how they choose to represent themselves and why. Use MySpace as a resource to start a conversation about contemporary fashion, ideals, and media images.
4. Talk about private/public issues with your kids. Help them to understand the consequences of making certain information publicly accessible. Get them to think through all of the possible audiences who might come into contact with their online information. Teens often imagine MySpace as a youth-only world. It isn't and they need to consider what the consequences would be if their grandparents, their teachers, admissions officers or a future employer read what they said about themselves. Helping your children learn how to negotiate such public environments is a great educational opportunity.
5. Talk through what kids should do if they receive unwanted attention online or if they find themselves the victims of cyberbullying. A growing number of sites provide useful information about how to confront such problems, including Net Family News, NetSmartz and SafeTeens. The "Safety Tips" section of MySpace also provides information for both parents and teens, including MySpace policies.

Henry and danah: We welcome further questions from parents. Our feeling is that there should be more public discussion of the opportunities and risks represented by MySpace and other social networks. Please send your questions to myspaceissues@mit.edu and we will do our best to respond.

About the Authors

Henry Jenkins is a co-director of the MIT Comparative Media Studies program. His current research, commissioned by the MacArthur Foundation, seeks to identify the core social skills and cultural competencies young people need in order to become full participants in the cultural, political, economic and social life of the 21st century. His new book, Convergence Culture: Where Old and New Media Collide, will be published this summer.

danah boyd is a Ph.D. student at the University of California at Berkeley in the School of Information. Her doctoral dissertation focuses on how youth negotiate identity formation through digital publics like MySpace. Her work is part of a broader project on Kids' Informal Learning with Digital Media, funded by the MacArthur Foundation. Prior to Berkeley, danah received an AB in computer science from Brown University and an MS in sociable media from the MIT Media Lab. She is also a researcher at Yahoo! and blogs regularly at Apophenia.

Index